HANS REICHENBACH: LOGICAL EMPIRICIST

VOLUME 132

HANS REICHENBACH: LOGICAL EMPIRICIST

Edited by

WESLEY C. SALMON
University of Arizona

D. REIDEL PUBLISHING COMPANY
DORDRECHT : HOLLAND / BOSTON : U.S.A.
LONDON : ENGLAND

Library of Congress Cataloging in Publication Data

Main entry under title:

Hans Reichenbach, logical empiricist.

(Synthese library; v. 132)
Bibliography: p.
Includes index.
1. Reichenbach, Hans, 1891–1953—Addresses, essays, lectures.
I. Reichenbach, Hans 1891–1953. II. Salmon, Wesley C.
B945.R42H36 1979 191 79–11032
ISBN 90–277–0958–0

Published by D. Reidel Publishing Company,
P.O. Box 17, Dordrecht, Holland

Sold and distributed in the U.S.A., Canada, and Mexico
by D. Reidel Publishing Company, Inc.
Lincoln Building, 160 Old Derby Street, Hingham,
Mass. 02043, U.S.A.

Printed in The Netherlands

This volume is dedicated to
Maria Reichenbach
For her unflagging efforts
to make Hans Reichenbach's philosophy
available to people throughout the world

HANS REICHENBACH

TABLE OF CONTENTS

PREFACE

Logical empiricism – not to be confused with logical positivism (see pp. 40–44) – is a movement which has left an indelible mark on twentieth-century philosophy; Hans Reichenbach (1891–1953) was one of its founders and one of its most productive advocates. His sudden and untimely death in 1953 halted his work when he was at the height of his intellectual powers; nevertheless, he bequeathed to us a handsome philosophical inheritance. At the present time, twenty-five years later, we can survey our heritage and see to what extent we have been enriched. The present collection of essays constitutes an effort to do just that – to exhibit the scope and unity of Reichenbach's philosophy, and its relevance to current philosophical issues.

There is no Nobel Prize in philosophy – the closest analogue is a volume in *The Library of Living Philosophers*, an honor which, like the Nobel Prize, cannot be awarded posthumously. Among 'scientific philosophers,' Rudolf Carnap, Albert Einstein, Karl Popper, and Bertrand Russell have been so honored. Had Reichenbach lived longer, he would have shared the honor with Carnap, for at the time of his death a volume on *Logical Empiricism*, treating the works of Carnap and Reichenbach, was in its early stages of preparation. In the volume which emerged, Carnap wrote, "In 1953, when Reichenbach's creative activity was suddenly ended by his premature death, our movement lost one of its most active leaders. But his published work and the fruit of his personal influence live on." We can, perhaps, regard the present volume as a member of a hypothetical series, *The Library of Living Philosophies*. Although there is no substitute for the philosopher's intellectual autobiography, or for his own replies to his critics and commentators, a collection of critical and expository essays such as these can bear witness to the historical and contemporary significance of his work.

The first essay in this collection attempts a fairly comprehensive survey of the full range of Reichenbach's philosophical work; the remaining essays deal with various specific areas. Some are expository, some are critical, some are historical, some show the immediate relevance of his work to current discussions, some attempt to extend the treatment of problems to which Reichenbach had addressed himself – indeed, most of the essays exhibit several of the foregoing approaches. Taken together, they provide a convincing display of the range and continuing significance of Reichenbach's monumental philosophical contributions.

BIBLIOGRAPHICAL NOTE

References to Reichenbach's principal philosophical works can be found in the Comprehensive Bibliography at the end of this volume; a more complete list of all of his published work up to 1959 appears in Reichenbach (1959). A nine volume German edition of Hans Reichenbach's *Gesammelte Werke*, under the joint editorship of Andreas Kamlah and Maria Reichenbach, is being published by Friedr. Vieweg & Sohn (Wiesbaden); the first two volumes have already appeared. My lead essay in this book, 'The Philosophy of Hans Reichenbach,' is a slightly altered version of my Introduction to the Vieweg edition, which, after translation into German by Maria Reichenbach, appears in the first volume. A two volume English edition, *Hans Reichenbach: Selected Writings, 1909–1953*, edited by Robert S. Cohen and Maria Reichenbach, is published in the *Vienna Circle Collection* by D. Reidel Publishing Company.

Tucson, Arizona, U.S.A. WESLEY C. SALMON
August, 1978

ACKNOWLEDGEMENTS

Some but not all of the essays in this book were previously published in *Hans Reichenbach, Logical Empiricist*, Parts I–V, in *Synthese*, volumes 34–35. They are reprinted here with only minor changes. New Postscripts with substantive content are appended to van Fraassen's 'Relative Frequencies' and to my 'Laws, Modalities, and Counterfactuals.' My new essay, 'Why Ask, "Why?"? – An Inquiry concerning Scientific Explanation,' originally appeared in *Proceedings and Addresses of the American Philosophical Association* **51**, No. 6 (August, 1978); it is reprinted here, with minor changes, by permission of the American Philosophical Association.

I should like to express my personal gratitude to all of the contributors to this collection. Their investments of time, thought, and effort have provided important added insight into the various facets of Reichenbach's work. I should also like to thank Jaakko Hintikka for his help, support, and encouragement in this undertaking.

W.C.S.

ACKNOWLEDGEMENTS

[illegible]

[illegible]

WESLEY C. SALMON

THE PHILOSOPHY OF HANS REICHENBACH

Among the greatest philosophers of science of all times one would surely have to include Aristotle, Descartes, Leibniz, Hume, and Kant. In an important sense Kant represents a culmination of this tradition on account of his strenuous attempts to provide an epistemological and metaphysical analysis appropriate to mature Newtonian science. There were, to be sure, significant developments in classical physics after Kant's death – e.g. Maxwell's electromagnetic theory – but these *seemed* more like completions of the Newtonian system than revolutionary subversions which would demand profound conceptual and philosophical revisions. It was only in 1905 that the first signs of fundamental downfall of classical physics began to be discernible; even though Planck's quantum hypothesis and the Michelson–Morley experiment had occurred earlier, their crucial importance was only later recognized.

It is especially appropriate that Hans Reichenbach, one of the most seminal philosphers of the twentieth century, should have been trained both in the Kantian philosophical tradition and in the physics and mathematics of the present century. Such preparation constituted an almost indispensable prerequisite to anyone who, like Reichenbach, was to make momentous contributions to the philosophy of twentieth century science.

Kant, it should be recalled, invoked the synthetic a priori to elevate philosophy from the 'coal pit' (as Thomas Reid picturesquely termed it) to which Hume's uncompromising empiricism had led. In ways that are too familiar to need recounting, the doctrine of the synthetic a priori fit with elegant smoothness into the analysis of space and time Newtonian physics seemed to require. In Kant's philosophy the ancient conflict between the two great traditions of rationalism and empiricism was articulated with unprecedented clarity. The issue, in its barest terms, is whether there are synthetic a priori propositions. As I understand Reichenbach's philosophical career, it is basically motivated, from beginning to end, by a thoroughgoing attempt to resolve this very question.

W. C. Salmon (ed.), Hans Reichenbach: Logical Empiricist, 1–84.

In the first book following his dissertation – *Relativitätstheorie und Erkenntnis Apriori* (1920) – Reichenbach retains the vestiges of a commitment to the synthetic a priori, but from that point on he remained the constant and diligent champion of empiricism. Throughout the remainder of his work runs the the deep theme of reconstructing knowledge in general – and scientific knowledge in particular – without recourse to the synthetic a priori in any form. This program is easy to state, but notoriously hard to implement. The difficulty inherent in maintaining a thoroughgoing empiricism is amply exhibited in the works of Bertrand Russell and Rudolf Carnap, two of the most influential and creative scientific philosophers of the twentieth century.

In *Human Knowledge, Its Scope and Limits*, his final and most comprehensive epistemological treatise, Russell – with great reluctance, I am sure – concedes the failure of a purely empirical theory of knowledge and invokes his infamous 'postulates of scientific inference'. The way the postulates are introduced is strongly reminiscent of a Kantian transcendental deduction.

Carnap devoted a major portion of his energies from the late forties until his death in 1970 to the development of a theory of degree of confirmation or inductive probability. Throughout all of these efforts he never succeeded in freeing the probability judgments which are used for prediction – as 'a guide of life', as Carnap often said, using the words of Butler's famous aphorism – from a dependency upon a priori distributions of prior probabilities. Such a priori probabilities can hardly be regarded as anything but synthetic a priori propositions. In the end, Carnap (1963, p. 978) maintained that the choice of these probability distributions had to be made a priori on the basis of 'inductive intuitions'. It is difficult to believe that the most creative and productive member of the Vienna Circle of logical positivists lightly abandoned empiricism or gladly embraced the synthetic a priori.

From the legacy of Hume and Kant, we are aware that the central issues giving rise to the problem of the synthetic a priori are space and time, causality and induction. These very concepts were Reichenbach's central concerns throughout his philosophic career. His chief fame rests upon his classic philosophical analyses of space and time in relativity theory, and his frequency theory of probability with its related 'pragmatic justification' of induction. Just as Hume was the great empiricist of the eighteenth century, so it may be, will Reichenbach be remembered as the great empiricist of the twentieth century.

To have any claim to cogency, an empiricist philosophy must come fully to grips with the most advanced scientific knowledge of its time. The deep involvement of Reichenbach's philosophy of science with the main developments of the physics of the first half of the twentieth century gives ample testimony to the commitment Reichenbach made to this fundamental principle of fruitful work in epistemology. The result, he seemed to be convinced, was an empiricist philosophy of science as true to twentieth century physics as was Kant's rationalist philosophy of science to the classical physics of the eighteenth and nineteenth centuries. Further, it must be added, Reichenbach was surely convinced that his philosophy was far more satisfactory, on strictly philosophical principles, than any version of rationalism. It seems to me that Reichenbach chose an issue he considered more basic than any other for his Presidential Address to the American Philosophical Association (Pacific Division, 1947): 'Rationalism and Empiricism: An Inquiry into the Roots of Philosophical Error'. As long as philosophy exists this may remain the most fundamental question of all. Its ramifications reach into every area of philosophy, and throughout his extraordinarily productive career, Reichenbach pursued this issue across almost the entire range of philosophic inquiry.

REICHENBACH'S LIFE[1]

Born in Hamburg on 26 September 1891, Hans Reichenbach was the third in a family of five children. His father, Bruno, was a prosperous wholesale merchant who had gone to work early in life, as a result of his father's premature death, in order to support his younger siblings. Bruno Reichenbach always regretted his lack of formal education, and the fact that he had not been able to enter a scientific career. Bruno's parents were Jewish, but Bruno, following an influential trend among Jews at the time, converted to the Reformed religion. He viewed this conversion not as a rejection of his Jewish heritage, but rather as a fulfilment of it.

Hans's mother, Selma (née Menzel), came from a family whose Protestant heritage dated back to the Reformation. Her family included many professionals – clergymen, pharmacists, doctors, engineers. Her father was employed as a construction engineer. She served for a time as a schoolmistress, having passed the teachers' examinations. She also had serious interests in music.

The Reichenbach household was comfortably situated financially while Hans was growing up. Educational expenses, summer trips, theatre and concert tickets posed no problems. Members of the family played various musical instruments, and there was considerable discussion of the then-contemporary literary works of such authors as Ibsen, Hauptmann, and Shaw. The family enjoyed a rich cultural life.

Hans's extraordinary intellectual gifts began to manifest themselves at an early age. At five he could defeat his elders at board games such as checkers; at about twelve he took up chess, for which he also displayed prodigious talent. He soon gave it up because, he realized, it would be likely to distract him from more important work. He excelled in school; from the first year until his university entrance examinations he stood at the top of his class. According to the report of his brother, Bernhard, he was driven by a tremendous thirst for knowledge, and although he enjoyed his academic successes, he also had an active social life both with his siblings and with his student friends. He participated enthusiastically in the adventures and pranks of his young comrades.

In 1910–11, Reichenbach attended the Technische Hochschule in Stuttgart, where he studied engineering. As a small child he had dreamed of becoming an engineer. At the age of six or seven, when asked what he wanted to become, he answered, "engineer"; when asked what he was then, he answered, "one-quarter engineer". When he actually began work in the subject, however, he did not find it intellectually satisfying, and so he turned to other fields, but it seems he retained an appreciation of technological achievements. On a drive in the hills near Pacific Palisades, California, where he spent the last years of his life, I recall his commenting with obvious admiration of the 'intelligence' of the then-new automatic transmission of his automobile, as it changed gears in response to changing steepness in the grades. He also pursued photography as a hobby for many years. The fact that 'engineering' was somehow a part of him may help to explain his success as a teacher and writer in giving concrete realization to highly abstract concepts. It also helped him to earn his livelihood before he found suitable academic employment.

During the next few years, Reichenbach studied mathematics, philosophy, and physics under such teachers as Born, Cassirer, Hilbert, Planck, and Sommerfeld at the Universities of Berlin, Göttingen, and Munich. He also served in the German Signal Corps at the Russian front, where he suffered a severe illness. For the rest of his life he felt a deep sense of the duty of all intellectuals to counteract the attitudes from which wars are spawned.

Throughout his student days, he was actively involved in the Free Student Movement; some of his earliest publications (1912, 1913) were eloquent tracts expounding its ideology. A staunch supporter of freedom of inquiry and education, he spoke out strongly in opposition to those ecclesiastical and political authorities which imposed intellectual constraints, and in support of the liberty of every individual to express any sincerely held conviction – even opinions that were opposed to the movement itself. At the same time, for very practical reasons, he resisted the extension of voting membership to those whose primary aim was to subvert the movement. He called attention to the economic injustices that put education beyond the reach of many, and proposed concrete remedies. Above all, he was a passionate advocate of individual self-determination of moral goals as well as intellectual deliberation. In these early writings we see clear indications of the non-cognitivist ethical theory he enunciated much later in *The Rise of Scientific Philosophy* and other writings, in which he analyzed ethical judgments as expressions of volitional decisions. His intellectual opposition to transcendental theological and metaphyscial doctrines was also unmistakeable in these early writings.

Reichenbach's doctoral dissertation (1916), which dealt with the applicability of the mathematical theory of probability to the physical world, was written largely without academic supervision, for he could not find any qualified person sufficiently interested in the subject. The dissertation consisted of two parts: a mathematical formulation of the probability calculus and an epistemological exposition. After a futile search at several universities for a sponsor competent in both areas, he finally found two individuals at Erlangen, a philosopher and a mathematician, each of whom was willing to sponsor a part. Jointly they accepted the whole, and the degree was conferred in 1915. Years later, in his elementary logic classes at the University of California, Reichenbach would cite this incident as an example of the fallacy of composition, but he would add, "I did not point that out to the good professors at that time!" In 1916 he took a degree in mathematics and physics at Göttingen by state examination.

For the next few years, until 1920, Reichenbach continued his academic work during the evenings, while holding a regular job in the radio industry. During this period Einstein offered his first seminar in the theory of relativity at the University of Berlin, and Reichenbach was one of the five who enrolled. It must have been a momentous occurrence in his life; he formed an enduring friendship and intellectual relationship with Einstein, and his interest in space–time–relativity constituted a major facet of his work for the rest of his life. In 1920 he moved back to the Technische Hochschule in Stuttgart, where he taught a variety of subjects ranging from relativity theory and philosophy of science to radio and surveying. In 1921 he married Elizabeth (née Lingener); they had two children, Hans Galama (1922) and Jutta (1924).

In 1926 Reichenbach returned to Berlin as Professor of the Philosophy of Physics. Einstein was largely instrumental in securing this position for him, since there was considerable opposition to the appointment on the part of influential members of the Berlin faculty. The chief obstacles seemed to lie in Reichenbach's publicly expressed disdain for traditional metaphysical systems and in his open espousal of radical ideals during his student days. Finally, however, the appointment was offered after Einstein had confronted the Berlin faculty with the question: "And what would you have done if the young Schiller had applied here for a position?" Reichenbach retained his post in Einstein's department until 1933.

Although Reichenbach was never formally a member of the Vienna Circle of logical positivists, he maintained close contact with several members of that group. In 1930 he and Rudolf Carnap founded the journal *Erkenntnis*, which they jointly edited. This important periodical was the major voice of the Vienna Circle of logical positivists and the Berlin School of logical empiricists until 1940, when publication was halted because of World War II.[2] In 1930 he also broadcast the series of radio lectures which were published in German in 1930 (*Atom und Kosmos*) and in English in 1932 (*Atom and Cosmos*).

Almost immediately after Hitler's rise to power in 1933, Reichenbach was dismissed from his university position, as well as his position with the state radio, but he was already on his way to Turkey before the notifications were delivered. He remained in Turkey until 1938, teaching philosophy at the University of Istanbul, and reorganizing the curriculum in philosophy. The government of Mustapha Kemal (Ataturk) made it possible for many of the ablest young people in the country to study in Istanbul under a number of outstanding German refugees. This sojourn provided Reichenbach with an opportunity to learn something of the Turkish language, a knowledge he put to good use in his later writings on the logic of conversational language.

In 1938 Reichenbach emigrated to the United States and took up his final position as Professor of Philosophy in the University of California at Los Angeles. In 1946, after the divorce from his first wife was final, he married Maria (née Moll). During his tenure at California, he frequently lectured at other universities and at professional meetings in the United States and Europe. For example, he presented his paper, 'The Verifiability Theory of Meaning', at the Conference of the Institute for the Unity of Science in Boston in 1950. The Institute for the Unity of Science was the American descendent of the Vienna Circle and the Berlin Group. In the summer of 1952 he presented a series of lectures entitled 'Les fondements logiques de la mécanique des quanta' at the Institut Henri Poincaré in France.

Reichenbach had suffered from a heart condition ever since a mountain-climbing holiday in the Alps in 1938. On 9 April 1953 he died as a result of a heart attack. At that time he was engaged in writing *The Direction of Time*, a work which was published posthumously without a final chapter he had intended to write.[3] His untimely death also prevented him from giving the William James Lectures, which he had been invited to deliver at Harvard University in the fall of 1953. Shortly before his death, a volume on logical empiricism, which would have been devoted to the works of Reichenbach and Carnap, was planned for Paul Schilpp's *Library of Living Philosophers*, but his death precluded completion of the project. It was, however, brought to partial fruition in *The Philosophy of Rudolf Carnap* in the same series.

In addition to being an extremely seminal philosopher, Reichenbach was also an unusually effective teacher. "His impact on his students was that of a blast of fresh, invigorating air," said Carl Hempel, who studied with him in Berlin; "he did all he could to bridge the wide gap of inaccessibility and superiority that often separated the German professor from his students."[4] In her biography of Reichenbach (1975) Cynthia Schuster, who studied with him at UCLA, reports, "His pedagogical technique consisted of deliberately over-simplifying each difficult topic, after warning students that the simple preliminary account would be inaccurate and would later be corrected." He possessed a unique talent for going to the heart of any issue, clearing away peripheral matters and cutting through irrelevancies, which made his teaching a model of clarity. It also endowed his creative philosophical work with genuine profundity.

On the basis of personal experience, I can testify to Reichenbach's qualities both as a teacher and as a man. I was a raw young graduate student with an M.A. in philosophy from the University of Chicago when first I went to UCLA in 1947 to work for a doctorate. At Chicago I had been totally immersed in Whitehead's philosophy; ironically, Carnap was at Chicago during those years, but I never took a course from him. My advisors barely acknowledged his existence, and certainly never recommended taking any of his classes.[5] Upon arrival at UCLA I was totally unfamiliar with Reichenbach or his works, but during my first semester I was stimulated and delighted by his course, 'Philosophy of Nature', based upon *Atom and Cosmos.* Simultaneously, I continued my intensive studies of Whitehead's *Process and Reality.* A severe intellectual tension emerged in my mind between Whitehead, the scientifically sophisticated metaphysician, and Reichenbach, the scientifically sophisticated anti-metaphysician.

To the best of my recollection, the tension grew to crisis proportions when I heard Reichenbach deliver his masterful Presidential Address, on rationalism and empiricism, to the Pacific Division of the APA at its meeting in Los Angeles in December of 1947. This lecture was precisely what I – as a naive graduate student – needed to make me face the crucial question: on what conceivable grounds could one make reasonable judgments concerning the truth or falsity of Whitehead's metaphysical claims? When I posed this question to myself, as well as to teachers and fellow graduate students sympathetic to Whitehead, I received nothing even approaching a satisfactory answer. By the end of that academic year I was a convinced – though still very naive – logical empiricist.

I continued to take every course Reichenbach had to offer, and the intellectual rewards were immense.

'Rationalism and Empiricism: An Inquiry into the Roots of Philosophical Error' may have appealed to a first-year graduate student such as I was at the time, but it did not receive universal approbation from the philosophical community, as one can well imagine even by considering nothing beyond the second part of the title. The themes of this lecture were shortly thereafter developed into the semi-popular book, *The Rise of Scientific Philosophy*. This book, which was quite widely read, aroused considerable antipathy among philosophers of other persuasions. Indeed, it would be false to the historical facts to omit mentioning that quite a few philosophers regarded this book as cavalier – especially in the treatment of important figures in the history of philosophy – and they considered its author narrow and doctrinaire. In his oral performances and in his writings, Reichenbach stated his position with an incisiveness which was sometimes taken for dogmatism.

Such criticisms were often leveled by philosophers who were unacquainted with the massive detailed background work Reichenbach had provided in his many technical books and papers. The truth is that he thought through the problems that concerned him with great care, and he stated his conclusions positively and forcefully. They were based upon carefully reasoned arguments, and even if these arguments were not always presented along with his conclusions, he was never reluctant to furnish and defend them. In many cases, an adequate defense might have consisted in a citation of another work in which he had analyzed the problem at issue in explicit detail.

When the time came for me to select a dissertation topic and advisor, I felt certain that Reichenbach would reject me because of my total lack of preparation in physics and mathematics. Instead, when I hesitatingly raised the question with him he reacted quite positively, expressing confidence that there were suitable topics not requiring much technical background of the sort I so miserably lacked. When finally I discovered John Venn's superb book, *The Logic of Chance* (1866), the first extended and thoroughgoing philosophical exposition of a frequency theory of probability, Reichenbach seemed delighted. We were both fascinated by Venn's clear insights regarding the probability concept combined with his almost total misunderstanding of the relations between probability theory and inductive logic. Reichenbach provided every kind of psychological and intellectual support I needed to carry out my research; I completed my dissertation on Venn in 1950. Without his generous sympathy and understanding I could never have done it.

PROBABILITY AND INDUCTION

At the time of his death, Reichenbach felt certain, I believe, that he had produced an essentially adequate theory of probability and induction.[6] Virtually all that remained to be done was to clarify the situation to those who did not recognize that fact. When he published *Theory of Probability* in 1949, it was explicitly designated "second edition" – not merely an English translation of his 1935 *Wahrscheinlichkeitslehre* – to indicate that he had made all revisions necessary to take into account any significant objections to the German original. The basic ingredients were three: (1) the elaboration of a limiting frequency concept of probability; (2) the doctrine that in all contexts in which probability concepts are used correctly, these concepts can be reduced to or explicated in terms of the limiting frequency definition; and (3) his pragmatic justification of induction. Needless to say, the philosophic community in general did not share Reichenbach's evaluation of his achievements in this area, and the issues he raised, as well as his approaches to these questions, still constitute material for important and extensive philosophical discussion.

(1) According to the limiting frequency interpretation, a probability is the limit with which an attribute occurs within an infinite reference class. In order for the concept of a limit to be employed, the reference class must be ordered; it therefore constitutes an infinite sequence. For example, we believe there is a probability of one-sixth that, among the class of tosses of a standard six-sided die, the side six will land uppermost. Under the limiting frequency interpretation of probability, this means (roughly) that the ratio of the number of tosses resulting in six to the total number of tosses becomes and remains arbitrarily close to one-sixth as the number of tosses increases without any bound.[7]

Reichenbach, whose doctoral dissertation had dealt with the problem of applications of probability in physical situations, was fully aware that coins are not tossed infinitely many times, and that gases do not contain infinite numbers of molecules. He repeatedly stressed the fact (as had John Venn) that the reference to infinite probability sequences was an idealization – a situation that almost always arises when mathematics is used in physics. Infinite aggregates are mathematically much simpler than large finite classes. Although, to the best of my knowledge, Reichenbach himself never attempted a rigorous finitization of his theory of probability, he regarded it as possible in principle. He was obviously pleased when Hilary Putnam, in a doctoral dissertation written under his supervision, carried out this project (1951).

It is trivial to show that the *elementary axioms* of the probability calculus – i.e., those axioms that do not refer to the internal order of probability sequences – are satisfied by finite class ratios (1949, §18). In Reichenbach's version of the frequency theory of probability (unlike that of von Mises) the application of probability is not restricted to random sequences. Instead, he constructs a more general theory, in which various types of random and nonrandom internal structure can be treated. In order to carry out this program, Reichenbach adds to the elementary axioms of the calculus of probability two additional *axioms of order* (1949, §28). Axioms of this latter sort are in general false if applied to finite sequences. In his dissertation, Putnam performed the non-trivial task of showing how they could be suitably modified so as to apply to finite sequences. Putnam further proved a result which came to Reichenbach as a pleasant surprise, namely, that for finite sequences one can demonstrate that the probability *must be* identified with the class ratio (which is, of course, the strict analogue of the limiting frequency in the infinite sequence), if the axioms for the finitized calculus are to be satisfied. For infinite sequences the probability need not be identified with the limiting frequency; in consequence of this fact Reichenbach introduced his Axiom of Interpretation α.[8] Putnam showed that no analogous axiom of interpretation is needed in the finitized version. Reichenbach clearly regarded Putnam's work as an important technical contribution to the perfection of his work on probability and induction. I do not know why it was never published.

In recent years there has been a strong movement, taking as its point of departure several papers by Karl Popper (1957, 1959, 1967), in support of a 'propensity interpretation' of probability. This new interpretation has many similarities to the frequency interpretation, and many of its advantages. The most basic advantage is that on either the frequency or the propensity interpretation, probabilities are objective features of the physical world. In this respect they contrast sharply with probabilities as construed by logical, subjective, or personalistic interpretations.

Popper first presented his propensity interpretation in connection with some extravagant, and to my mind entirely unjustified, claims about the possibility of resolving fundamental problems in quantum mechanics by introduction of the propensity interpretation. Leaving aside such issues in the philosophical foundations of quantum mechanics, we may note that Popper offers the following type of situation to show the inadequacy of the frequency interpretation – an interpretation which Popper had, until then, subscribed to. Suppose we have an (ideally) infinite sequence of tosses of a fair die, except for three tosses with a highly biased die which have been inserted somewhere in the sequence of tosses. Since the results of these three tosses with the biased die will have no effect upon the limit of the frequency with which a given side comes up, Popper claims that the frequency interpretation is unable to handle that kind of situation.

Reichenbach had recognized just such problems in his theory of probability, and had offered a way of dealing with them. His solution was the introduction of a two-dimensional 'probability lattice', which consists of an infinite sequence (arranged vertically) of probability sequences.[9] In these lattices one examines the limiting frequencies in the vertical columns as well as those in the horizontal rows. If, in the situation envisioned by Popper, we repeat the production of probability sequences, inserting tosses with a biased die at precisely the same positions in each such sequence, that fact will be revealed by a differing limiting probability for a given side in the three vertical columns produced by the biased die. Reichenbach is obviously not claiming that an infinite sequence of infinite sequences must actually exist; his claim is that the differing limiting frequency *would* appear *if* the production of the probability sequence were to be repeated. In many places, Popper has characterized propensities in terms of frequencies in *virtual* (as opposed to actual) sequences. It is difficult to discern a real, rather than merely verbal, difference between these two interpretations. Both, of course, involve counterfactual statements; I shall comment on Reichenbach's treatment of that topic in connection with his work on nomological statements.

Popper would emphatically reject the notion that the difference between the two interpretations is merely verbal. He insists that the propensity is a property of a physical set-up that produces such events as coin tosses or radioactive disintegrations rather than a property of a sequence of tosses or atomic histories. Statements of this sort have led some propensity theorists to distinguish between a 'virtual sequence' propensity theory and a 'single case' propensity theory.[10] In the latter interpretation, the probability – i.e., the propensity – is a physical property of the chance device that produces any *single* event. In this version there is a fundamental distinction between the propensity and frequency interpretations. Limiting frequencies are characteristics of infinite sequences of events; single case propensities are characteristics of a mechanism that produces a single event. The same mechanism may, of course, produce other similar events; in that case we may say that the same mechanism has the same property (propensity) on subsequent trials as it did on the first, much as an automobile that is blue today may have the same color tomorrow, without affecting the truth or falsity of the claim that it is blue today.

The problem of assigning probabilities to single events has been recognized by practically every frequentist as a crucial one. Some frequentists, following von Mises (1957), simply deny that there is any meaningful way to assign probability to the single case. Reichenbach (1949, §72) took a different tack, arguing that in an extended or 'fictitious' sense, probabilities can be assigned to single events. One assigns the single case in question to a reference class, and then 'transfers' the limiting frequency in that reference class to the occurrence in the single case. It is perhaps best, when discussing single cases, to use the term 'weight' when referring to the value of the probability that is transferred. I have argued (1967, pp. 90–96) that the ends of clarity might be served by invoking Carnap's extremely useful distinction between the theory of probability and rules of application (Carnap, 1950, §44–45), in the light of which we can say that dealing with probabilities as limiting frequencies is within the domain of probability theory or inductive logic properly speaking, and that the assignments of 'fictitious probabilities' or weights to single events falls within the purview of the rules of application.

The reason for this approach is rather obvious, for the problem we confront in attempting to assign a weight to a single case is the problem of selecting the *appropriate* reference class from which to borrow the value of the limiting frequency. Reichenbach's practical rule was to choose the *narrowest reference class for which reliable statistics are available*. While it seems to me that the intent of this maxim is clear enough, Reichenbach's formulation has led to needless criticism. I have therefore suggested that the rule for selecting the reference class be reformulated as follows: choose the *broadest homogeneous reference class* (1967, pp. 91–92). A homogeneous reference class is one in which no statistically relevant subdivision can be effected. The general idea is simply to include the single case in question within the class of cases which are relevantly similar, but to exclude from the reference class those sorts of cases that are relevantly different. In attempting to assign a probability (or better, a weight) to the survival of a particular person for an additional decade, it is important to take into account such factors as age, sex, state of health, and occupation, but to ignore such considerations as the color of his automobile and whether its license has an odd or an even number. He is thus placed in a reference class defined by characteristics of the former type, but not those of the latter sort, and the frequency of survival for ten more years among people in this class provides the weight in this individual case. I believe that this preserves Reichenbach's intent, and at the same time eliminates formulations that have tended to be confusing.

When we consider the single case version of the propensity interpretation of probability, it appears to me that precisely the same problem arises. One cannot say simply that the probability of an outcome is a property of some unspecified chance set-up; it is necessary, instead, to characterize the chance set-up in terms of its *relevant* characteristics. Single case propensity theorists readily admit that observed frequencies provide the evidence by means of which to ascertain the propensity of a given chance set-up to produce a result. In order to decide which chance set-ups to observe in order to collect the statistical data upon which to base a judgment about the value of a propensity, the propensity theorist must make precisely the same type of judgment as the frequentist concerning those features of the situation that are relevant to the occurrence of a given outcome, and those that are irrelevant. Thus, for example, in attempting to ascertain the propensity of *this* die to land with side six up on *this* toss, we must take account of the cubical shape of the die, the homogeneous distribution of mass within it, the fact that it is shaken sufficiently before throwing and is tossed with sufficient force. Contrariwise, we ignore the color of the die, the question of whether the throw is made during the day or the night, and whether the thrower is left or right handed. For the single case propensity theorist, as for the frequentist attempting to deal with the single case, the question of which factors are relevant and which irrelevant is equally vital. It thus appears that the single case propensity interpretation, no less than the virtual sequence propensity interpretation, provides no substantial improvement over Reichenbach's limiting frequency interpretation, but at most an interesting reformulation which *may* include some shift of focus. Reichenbach was fully aware of the problems which seem to have motivated the propensity interpretation, and he offered careful and detailed treatments of these very problems.

(2) In 1949, Reichenbach's *The Theory of Probability* was published, and in 1950, Carnap published *Logical Foundations of Probability*. These two works dominated philosophical discussion of probability and induction for a numbr of subsequent years, and to some extent they still do. Carnap, of course, continued to publish substantial work on this subject until his death in 1970. Nevertheless, Carnap's work subsequent to 1950 consisted in improvements, refinements, and extensions of the approach in *Logical Foundations*, rather than any radical change of viewpoint.[11]

Although he stated carefully and repeatedly that he believed in two concepts of probability – probability_1 or degree of confirmation, as well as probability_2 or relative frequency – most of his attention was devoted to explication and exploration of probability_1 or degree of confirmation. Referring to the reason for this disparity in amount of attention, Carnap's wife once remarked that they sometimes referred to probability_1 as 'little Rudi' and probability_2 as 'big Rudi'. It was little Rudi that needed support and defense, Carnap felt; big Rudi could take care of himself.

The important philosophical controversy that was engendered between Reichenbach and Carnap therefore hinged on the question of whether the frequency interpretation alone was adequate to all contexts in which probability concepts are correctly used, or whether the concept of degree of confirmation was also required. Unlike von Mises and other frequentists, Reichenbach did not take the view that it simply makes no sense to talk about probabilities of scientific hypotheses or theories, but rather, that this usage can be explicated in terms of the frequency concept. The key to this explication lies in the theorem of Bayes (1949, §21, 84, 85).

Most commentators on Reichenbach found his remarks on the probability of hypotheses at best confusing and obscure, at worst nonsensical. Indeed, even Ernest H. Hutten who, in collaboration with Maria Reichenbach, translated *Wahrscheinlichkeitslehre* into English (and who, I might add, was very sympathetic to Reichenbach's views on many other issues), sided with Carnap on this particular point (1956, chap. 6). As a devoted pupil, I thought Reichenbach was right, and that it was only a matter of stating his point more clearly and convincingly. During the first couple of years after finishing my dissertation, I endeavored to carry out the task, but without much success. The problem I faced was quite simple; I could look at Bayes's formula, but I couldn't see how to interpret the letters in the formula that stood for classes – in particular, if the formula is written

$$P(A \cdot C, B) = \frac{P(A, B) \times P(A \cdot B, C)}{P(A, B) \times P(A \cdot B, C) + P(A, \bar{B}) \times P(A \cdot \bar{B}, C)}$$

it was the letter '*B*' that had me particularly puzzled.[12]

In the winter of 1951–52, Reichenbach consented to visit Pullman, Washington, where the State College was located and where I was then teaching, to give a public lecture and an informal seminar. He and his wife Maria flew into Spokane, where I met them to take them to Pullman. The weather was bad and the roads were hazardous, so I decided, for safety's sake, to use the Greyhound bus rather than my own auto. I remember vividly the bus trip from Spokane to Pullman; it was after dark and the light inside the bus was dim. At one point I told Reichenbach of my problems about interpretation of Bayes's theorem, and particularly the letter '*B*'. He did not respond at once; instead, we sat in what seemed to me an almost interminable silence – so long that I began to wonder if I had offended him. Finally, he said simply, "The letter '*B*' must, I think, stand for the class of true hypotheses."

The visit was a complete success. Reichenbach lectured to an audience that overflowed the largest auditorium on campus; he was an extremely effective public lecturer, and received a most enthusiastic response. Later, at a seminar attended mostly by faculty members from various departments, Reichenbach made a favorable remark about the work of Bertrand Russell. One of the audience asked, "What is Russell? The mathematicians say he is a mathematician, and the philosophers say he is a philosopher. What do you say to that?" Without an instant's hesitation Reichenbach replied, with a characteristic twinkle of the eye, "Well, I think it's much better that way than the other way around."

In the following months I wrote a fairly lengthy paper on the probability of hypotheses in the frequency theory of probability, using the vital suggestion he had offered on our bus trip. In order to carry out this analysis, two fundamental points have to be recognized. First, any particular hypothesis is either true or false, and therefore constitutes a single case, quite analogously to a single coin toss which must result in either a head or a tail. When we try to assess the probability of a scientific hypothesis (or the degree to which it is confirmed by available evidence), we must assign it to an appropriate reference class of hypotheses, and transfer the relative frequency with which hypotheses of this class are true to the single hypothesis as a 'weight'. Second, assuming we have found such a class of hypotheses, we immediately face the question of how to arrive at any reasonable decision as to which hypotheses in the class are true and which false. It looks as if we must use Bayes's theorem over again in *each* such case, leading us into vicious circularity or infinite regress. Reichenbach's way of avoiding this difficulty lies in his important distinction between *primitive knowledge* and *advanced knowledge.*[13] In a state of primitive knowledge, logically speaking, when we have no hypothesis regarding which we can make judgments as to truth or falsity, we must resort to his basic *rule of induction* – the method of induction by enumeration.[14] This use of primitive induction can provide a set of general hypotheses that we can accept (in the sense of *posits*),[15] and from there we can move on to the use of Bayes's theorem as a method appropriate to advanced knowledge. Thus, Reichenbach claims, inductive logic consists in its entirety of the rule of induction and the mathematical calculus of probabilities, where probabilities are construed as limiting frequencies.

I sent my paper off to him, and although he made many suggestions and comments, he endorsed the basic ideas. I read versions of this paper at various meetings and colloquia, and it was rejected by some of the best journals in America. It eventually became Chapter VII of my *Foundations of Scientific Inference*, and has since been taken a bit more seriously. Although Reichenbach never got around to writing up this material himself, the ideas are entirely his. I only hope to have succeeded in stating them in a way that is easier to understand than are his somewhat cryptic remarks in *The Theory of Probability*.

The fate of Bayes's theorem was profoundly affected by L. J. Savage's *Foundations of Statistics* (1954),[16] one of the most provocative books in statistics in the present century. In it, Savage developed a modern subjective interpretation of probability which he preferred to call 'personal probability'. Bayes's theorem figures so prominently in this approach that its followers are often called 'bayesians'. Basic issues in the foundations of statistics are the subject of current and lively debate, among statisticians and philosophers alike. Much of the controversy focuses upon the viability of the bayesian approach, and many of the misgivings of anti-bayesians arise from the use of subjective or personal probabilities. What Reichenbach sought to offer (and what I have tried to support and elaborate in his views) is an *objective* bayesian approach. Bayes's theorem figures just as prominently in his view as it does in Savage's; the difference hinges on the question of whether the prior probabilities are objective or subjective entities. This issue, I believe, is one of vital current importance.

If Reichenbach's 'objective bayesian' approach can be carried through, then we can meaningfully refer to the probability or confirmation of hypotheses without invoking a second type of probability. Big Rudi can do the whole job.

(3) In the early stages of his thought, Reichenbach regarded Hume's problem of the justification of induction as a pseudo-problem, to be dismissed along with many other pseudo-problems which had been handed down to us through the history of philosophy.[17] He soon realized, however, that the key to the problem lay, not in the 'proof' of a metaphysical principle of uniformity of nature (e.g., Kant's synthetic a priori principle of universal causation), but rather in a pragmatic justification of a *rule*. Rules, as he and Herbert Feigl forcefully pointed out, are not statements whose truth values must be ascertained, but rather tools whose adoption rests upon their practical efficacy to do some specified task. For Reichenbach, with his frequency conception of probability, the task could be clearly designated, namely, the ascertainment (within some desired degree of approximation) of the limit of a relative frequency

within a given probability sequence. Reichenbach's well-known pragmatic justification, or 'vindication' to use Feigl's (1950) helpful terminology, consists in showing that his 'rule of induction' – i.e., the rule which directs us to infer (posit) that the observed frequency matches (approximately) the long-run frequency – will succeed in ascertaining limits if these limits exist and if the rule is used persistently (1949, §91). This rule is often called 'induction by simple enumeration'. It is the basic rule for making inductive posits.

Reichenbach's pragmatic justification takes Hume's critique of induction seriously in that it admits from the outset that we have no way of knowing in advance whether a particular sequence has a limiting frequency or not. This corresponds to Hume's realization that we can have no a priori assurance that nature is uniform. He argues, nevertheless, that we have everything to gain and nothing to lose by using induction, for if the sequence has a limit, persistent use of induction will enable us to ascertain it; whereas, if the sequence has no limit, no method will find it. This assertion, as Reichenbach repeatedly observes, is a rather trivial analytic consequence of the definition of the limit and the definition of his 'rule of induction'. He is, therefore, free from the need to postulate or prove any synthetic assumption.

Needless to say, there have been numerous objections. It has sometimes been said, for example, that induction may be a good way to try to ascertain limits of relative frequencies, but who cares about limits of frequencies? If, however, Reichenbach is right about the application of probabilities to single cases and about the possibility of handling the confirmation of hypotheses via the frequency interpretation, then all of our common sense knowledge, our scientific knowledge, and our practical decisions rest upon ascertainment of limiting frequencies.

The objection has been raised that a sequence may have a limit, but that convergence to that limit may occur so far along in the sequence that it will be of no use to us to try to find it. True. This means that, for practical purposes, we are dealing with non-convergent sequences, and that is just our bad luck. No method can succeed systematically in the face of that kind of misfortune. As Hume showed so forcefully, we cannot prove in advance that nature is uniform, or uniform enough to be of any use to us.

Reichenbach's pragmatic justification is designed to deal with just this sort of objection – either nature is tractable or she isn't; if she is, induction will work, if she isn't, nothing will. Reichenbach makes it quite clear that it is the existence of 'practical limits' that counts.

It has also been argued that we never observe more than a finite initial portion of a sequence, and *any* frequency in *any* finite initial section is compatible with *any* limiting frequency. This statement is also true, but it shows only (though quite dramatically) that we are dealing with an inductive, not a deductive, problem.

The most serious challenge to Reichenbach's justification is one that he himself pointed out. One can define a class of 'asymptotic rules' which have precisely the same convergence properties as Reichenbach's rule of induction. One can thus argue equally validly for *any* member of this class of rules that, if the sequence has a limit, *that* rule will ascertain the limit to any desired degree of accuracy within a finite number of applications. It is impossible to prove, moreover, that any particular asymptotic rule will enable us to ascertain the limit of any particular sequence more rapidly than some other asymptotic rule. Reichenbach was aware of these facts. He resolved the situation by saying that, since they are all equivalent in the long run, we should choose the rule of induction on grounds of *descriptive* simplicity (1949, §87). Here, it seems to me, Reichenbach took an easy, and unwarranted, way out of a stubbornly recalcitrant problem. Although there is a sense in which the asymptotic rules are equivalent in the long run, that doesn't help, for as Lord Keynes remarked, "In the long run we will all be dead."

I have found this aspect of Reichenbach's treatment of induction disquieting. Although the asymptotic rules are convergent, they are not uniformly convergent. This means that you may arbitrarily select *any* sample size and *any* logically possible value for the observed frequency in that sample; moreover, you may select arbitrarily *any* value between zero and one (endpoints included). Within the class of asymptotic rules there is at least one directing you to infer (posit) *that* arbitrarily selected limiting frequency on the basis of the foregoing arbitrarily specified sample size and frequency. Since we do not, in Carnap's charming phrase, live to the ripe old age of denumerable infinity, the class of asymptotic rules is, for all human purposes, a set of rules as radically non-equivalent as any set can be.[18]

I have sought (1967) to strengthen Reichenbach's pragmatic justification (in ways which I have no way of knowing whether he would have endorsed) by introducing certain additional requirements. The 'normalizing conditions', for example, prohibit adoption of any inductive rules that would allow us to infer limiting relative frequencies smaller than zero or greater than one. As a consequence, these conditions require our inductive rules to yield a set of inferred values for limiting frequencies of mutually exclusive and exhaustive outcomes that add up to one. In addition, the 'criterion of linguistic invariance' requires that our inferences concerning limiting frequencies *not* be functions of the language in which the observed sample or the limiting frequency is described – merely verbal reformulations in equivalent terms should *not* affect the value of the relative frequency we infer on the basis of our observation of a sample. This criterion is analogous to the principle that scientific results should be equally valid whether described in the metric or the English system of units. Though these additional requirements seem to provide good pragmatic grounds for ruling out large subclasses of the class of asymptotic rules, they do not (as I thought for a while they did) rule out all but Reichenbach's rule of induction. In 1965, at a conference at Bedford College, London, Ian Hacking (1968) showed precisely what requirements are necessary and sufficient to vindicate uniquely Reichenbach's rule. At present I do not know how to justify these requirements, or whether they are justifiable. Nor do I know that, among the asymptotic rules that have not been eliminated, the Reichenbach rule of induction is the one that is justifiable if any is. Perhaps a different asymptotic rule is the correct one, and is amenable to pragmatic justification. However, none of the alternatives that have been formulated in the course of these discussions (usually as a counterexample to some claim or other) is the least bit tempting.

Where does that leave us vis à vis Reichenbach's program of providing a pragmatic justification of induction? Several main alternative approaches have received considerable attention in the years since 1950. For some time a large and influential group of 'ordinary language' philosophers – the best known is Peter Strawson (1952, chap. 9) – have maintained that the problem is not genuine; I have often tried to show (1957a; 1967, pp. 48–52) that their arguments are defective. Two prominent philosophers, Max Black and R. B. Braithwaite, have offered inductive justifications of induction; I have, I believe, successfully shown (1957a; 1967, pp. 12–17) that their arguments are viciously circular (as

Hume had clearly anticipated). I have analyzed, and found wanting, the above-mentioned approaches of Carnap and Russell that led them to the abandonment of empiricism (see Salmon, 1967, pp. 43–48, 68–78; 1967a, 1974).

Finally, there is Karl Popper's claim that in 1927 he solved Hume's problem of induction by showing that induction has no place in science. It is my view, elaborated in a number of places, that Popper's 'deductivism' is actually another form of the very sort of thing Reichenbach set out to do, namely, to provide a pragmatic justification or vindication of certain forms of scientific inference.[19] Popper, I need hardly add, vehemently rejects this interpretation of his approach.

The problem of the justification of induction is, I believe, still very much with us. In my view, Reichenbach's general program of providing a pragmatic justification of induction, though it is as yet incomplete, is the most promising contender as a way of continuing to attack the problem. And attack it we must, I think, for to do otherwise is to relegate science to the same status as various forms of superstition that have plagued and continue to plague mankind.

SPACE, TIME, AND RELATIVITY

It is a tragedy of major proportions that a volume on Reichenbach never appeared in the *Library of Living Philosophers*. Such volumes offer a dramatic exhibition of the scope of the philosopher's work, and excellent opportunity for clarification of issues and resolution of misunderstandings raised by critics. But even more importantly, in Reichenbach's case, unlike Russell, Einstein, Carnap, and Popper, we do not have an intellectual autobiography. Thus, we have no first hand account of his philosophical beginnings – of his first acquaintance with the works of Helmholtz, Riemann, and Boltzmann; of the early influence of the neo-Kantian Cassirer (or whether he even knew of Russell's neo–Kantian *Foundations of Geometry* (1897), which sought to reconcile Kant's theory of space with the existence of non-Euclidean geometries); of his early reactions to the teachings of Hilbert; nor, indeed, an intimate account of the initial impact of Einstein. These are matters of great historical interest. We do, of course, have Maria Reichenbach's excellent intellectual biography in her 'Introduction to the English Edition' of *The Theory of Relativity and A Priori Knowledge* (1965), but it is not an autobiography, and she was not acquainted with Reichenbach during his earlier years in Germany.

We do know, of course, that Reichenbach was one of the five who attended Einstein's first seminar on relativity theory, that he and Einstein were friends and colleagues in Germany before Hitler's rise to power, and that they had numerous conversations in later years when Einstein was at the Princeton Institute for Advanced Study. And we have three major books and numerous articles on the philosophical foundations of space, time, and relativity that Reichenbach published in the decade 1920–30.

In the first, *Relativitätstheorie und Erkenntnis Apriori* (1920), though not fully divested of every kind of commitment to the synthetic a priori, he does argue convincingly that one cannot, in the light of relativity theory, retain the synthetic a priori character of all of the principles for which Kant claimed that status – at any rate, not unless one is willing to abandon normal inductive methods of science, a move to which Reichenbach was unwilling to resort (as would Kant himself surely have been, given his concern to understand how scientific knowledge is possible).

The second major book, *Axiomatik der Relativistischen Raum–Zeit–Lehre* (1924), actually carries out a relatively formal axiomatic reconstruction of Einstein's special and general theories of relativity, based on axioms that express the elementary physical facts upon which those theories rest. The axiomatization is not mathematically elegant, nor is it intended to be. It is intended rather as a logical analysis that would enable one to locate precisely the distinctions among physical facts, mathematical truths, and conventional definitions which are thoroughly intermingled in the usual presentation of the theory. Here, in concrete detail, we see an example of the kind of epistemological analysis advocated by the logical positivists in Vienna and their close associates in the Berlin group of logical empiricists. There is a close analogy between what Hilbert (1902) had done regarding the foundations of geometry, what Russell–Whitehead (1910–13) had done in *Principia Mathematica*, and what Reichenbach was doing for the *physical* theory of relativity. Certainly the motivations in all of these cases are closely allied.

The third major book, *Philosophie der Raum–Zeit–Lehre* (1928), is the crowning achievement of this period of Reichenbach's career. Indeed, in my opinion, it heads the list of candidates for *the single greatest work* in philosophy of the natural sciences of the twentieth century, and it seems gradually to be achieving its well-deserved status as a genuine classic. If not by itself, then surely in conjunction with the other two books, it supersedes such other works as Poincaré's famous *Science and Hypothesis* (1952), Carnap's dissertation *Der Raum* (1922) and Schlick's *Space and Time in Contemporary Physics* (1963). Indeed, in summarizing his evaluation of this book, Carnap says, "Even more outstanding than the contributions of detail in this book is the spirit in which it was written. The constant and careful attention to scientifically established facts and to the content of the scientific hypotheses to be analyzed and logically reconstructed, the exact formulation of the philosophical results, and the clear and cogent presentation of the arguments supporting them, make this work a model of scientific thinking in philosophy" (1958a, p. vii).

Einstein also had deep respect for Reichenbach's work on space, time, and relativity. In *Albert Einstein: Philosopher–Scientist* Reichenbach (1949a) reiterates some of the central theses of the 1928 book. Einstein replies, "Now I come to the theme of the relation of the theory of relativity to philosophy. Here it is Reichenbach's piece of work which, by the precision of deductions and the sharpness of his assertions, irresistibly invites brief commentary . . .". After a charming dialogue, composed by Einstein, between Reichenbach and Poincaré (or some unidentified non-positivist), he remarks, "I can hardly think of anything more stimulating as the basis for discussion in an epistemological seminar than this brief essay by Reichenbach . . ." (1949, pp. 676–679). I can attest from personal experience that this work of Reichenbach on space and time is indeed a source of intense intellectual stimulation to serious students both in philosophy and in the sciences.

It is difficult to say what is most significant about a work of the order of magnitude of *Philosophie der Raum–Zeit–Lehre*. In the first chapter Reichenbach clarifies the relationship between mathematical systems of geometry and physical space. Showing the evident oversimplification of the view that the very existence of non-Euclidean geometries refutes Kant's views on the nature of space, Reichenbach offers a sophisticated analysis of the crucial epistemological role of the concept of visualization. He develops his profoundly important 'theory of equivalent descriptions', showing the indispensable role of a physical coordinative definition of congruence. He shows how the geometric structure of physical space can, in principle, be ascertained empirically, *if* a coordinative definition of congruence has been given. At the same time, he shows how suitable changes or adjustments in the definition of congruence can enable us to preserve any particular geometry – presumably Euclidean – as the correct description of physical space. This leads to the consequence, which he explains with admirable clarity, that one may choose *either* a definition of congruence *or* an abstract geometry as a matter of convention. Once one of these choices has been made, however, determination of the other requires ascertainment of physical facts regarding physical space. He shows that there are equivalent descriptions which are equally adequate from a physical standpoint, but that nonetheless there are other descriptions which are not equivalent and, consequently, empirically unacceptable. The chapter provides a thorough and penetrating analysis of the apparently simple question, "What is the geometry of physical space?" This analysis is far deeper and more adequate than any simple version of conventionalism or operationism.

There is, perhaps, one unfortunate aspect to Reichenbach's discussion of this question, for in order to expound his main point in vivid terms he introduces the concept of a 'universal force'. Poincaré, incidentally, used the same heuristic device. What Reichenbach failed to make sufficiently clear is that such a manner of speaking is metaphorical. He has thus been criticized for invoking 'ad hoc hypotheses' in his discussion of equivalent descriptions (Nagel, 1961, pp. 264–265). In his foreword to the English edition, Carnap cites the use of the concept of universal forces as a singular virtue of Reichenbach's book, but Carnap, in a later book (1966, p. 169), suggests that they might better be regarded as 'universal effects'. A still better alternative would probably be to forego all such metaphorical uses of terms like 'force' and 'effect', in order to keep clearly in mind that we are talking about nothing more nor less than coordinative definitions of congruence.[20]

In the second chapter, Reichenbach turns to time. Among the most important themes to emerge in this chapter is the problem of ascertaining simultaneity relations in a single inertial frame – before we become involved in the relativity that arises when we look at reference frames in motion with respect to one another. The status of simultaneity in a single reference frame, which receives very brief and passing mention in Einstein's famous 1905 paper on special relativity, is elaborated with care by

Reichenbach and clearly related to the maximality of the velocity of light. Here again, we find equivalent descriptions playing a fundamental role; in this case it is evident that their very possibility depends, not merely upon some logical or semantical consideration, but upon the *physical* fact that light is, in Reichenbach's terms, a 'first signal', that is, a signal whose speed of propagation cannot be exceeded by a signal of any other type, even on a one-way trip.

In the third and final chapter, Reichenbach enters upon the philosophical discussion of relativity theory proper, i.e., the problems that arise when one combines the consideration of space and of time. This chapter is more technical, and rests more heavily upon the results of his *Axiomatik* (1924). The chapter is divided into three parts: Part A deals with the special theory of relativity, Part B takes up general relativity, and part C discusses the most general and abstract characteristics of the space–time manifold.

It is tempting, in discussions of special relativity, to follow the lead of Einstein's classic 1905 paper and take the concept of inertial reference frames more or less for granted, remarking perhaps that they are the physical systems in which Newton's first law holds. The situation is not that simple, however, because Newton's first law refers to uniform motion, and constant velocity makes sense only in a context in which the geometrical character of the space has been determined and clocks are synchronized along the paths of such motion. In the second chapter, Reichenbach had already provided a detailed analysis of simultaneity and clock synchrony, showing that it involves certain non-trivial conventions. Similar results concerning geometry had emerged from the first chapter. Relying upon the results of the *Axiomatik*, Reichenbach offers, in Part A of the third chapter, a profound analysis of the fundamental concept of an inertial system; indeed, he provides a detailed treatment of the entire problem of construction of the space–time metric of special relativity. He discusses the significance of Minkowski's indefinite metric, bringing out clearly the precise relationships between space and time in the special theory. He explains the relation between the physical geometry based upon light rays and that based upon solid rods, showing that the identity of these two geometrical constructions is a fundamental principle of the special theory – this is, in fact, the basic import of the Michelson–Morley experiment. It is in this context that the principle of the constancy of the average round-trip speed of light over a closed path is to be understood. This empirical fact about light is then combined with the physical fact that light is a first signal. Together, these *facts* are distinguished from the *convention* (discussed in detail in chapter two) that the speed of light is the same on both one-way segments of a round trip. In this exposition Reichenbach shows how the well-known phenomena of length contraction, time dilation, and composition of velocities arise directly out of the relativity of simultaneity, given the standard convention regarding the one-way speed of light. We must not conclude, however, that length contraction and time dilation are entirely matters of convention, for there are physical length contraction and time dilation phenomena which are objectively observable, quite apart from any simultaneity conventions we adopt.[21]

In part B of this chapter, Reichenbach removes the restriction to inertial frames of reference and moves on to consider physical phenomena in reference frames in arbitrary states of motion. These are the types of systems with which the general theory of relativity deals. Taking as his point of departure the problem of rotational motion, as it arises in Newton's famous bucket experiment, he analyses Mach's and Einstein's solutions.[22]

He shows that a *coordinative definition* of rest is required as a basis for dealing with motion at all. He then discusses the principle of equivalence and Einstein's treatment of gravitation, showing with admirable clarity how the *covariant* character of the components of the tensor representing the gravitational field is related to the *invariant* nature of the tensor itself. This leads to another important instance of Reichenbach's theory of equivalent descriptions.

> The coordinate systems themselves are not equivalent, but every coordinate system with *its* corresponding gravitational field is equivalent to any other coordinate system together with *its* corresponding gravitation field. Each of these covariant descriptions is an admissible representation of the *invariant* state of the world. (1958, p. 237)

With characteristic clarity, Reichenbach analyses the geometry appropriate to general relativity – i.e., Riemannian geometry – just as he had done in Part A with the Minkowski geometry of special relativity. He points to the fact that the metric in Riemannian geometry is given by a quadratic form, and observes that this represents an important fact of nature, namely, that measuring instruments behave according to

Euclidean geometry in infinitesimal regions. This general circumstance certainly cannot be taken as a priori; it is entirely conceivable that nature would demand a metric of a different order – e.g., a quartic metric – in which case Riemannian geometry would be inapplicable. The fact that physical space is Euclidean in the infinitesimal is, of course, closely related to the basic claim of general relativity that special relativity holds to a high degree of approximation in systems that are sufficiently small. General relativists are today deeply concerned with two of the most basic questions to which Reichenbach addresses himself in Part B. One issue concerns the fundamental justification for adopting a Riemannian geometry; the other concerns the physical means for ascertaining the geometrical structure of physical space–time.[23]

In concluding his discussion of general relativity, Reichenbach considers various types of gravitational field, progressing from rather special cases to more general cases, always taking care to state explicitly what restrictions are being relaxed at each step of the process of generalization. This part of the chapter culminates with an explicit and forceful statement of his commitment to a *causal theory of space and time*. "The system of causal ordering relations, independent of any metric, presents therefore the most general type of physical geometry . . .", he says; "The causal chain is the real process that constitutes the immediate physical correlate of the purely ordering geometry of Riemannian spaces" (1958, p. 268). The causal relation is, according to Reichenbach, the fundâmental invariant of the physical world. Reichenbach argues powerfully for the causal theory of space–time; this thesis is one of the perennial issues in philosophy of space and time.

Part C of chapter three, building upon foregoing technical results, takes up such problems as the ultimate nature of the distinction between space and time and the reason for attributing three dimensions to physical space. Reichenbach distinguishes the metrical from the topological considerations, showing how the causal structure of the world underlies the topological structure of space and of space–time. The dimensionality of space, a topological characteristic, arises from the decision to make causal relations satisfy the *principle of action by contact*. This principle plays a role in determining the topology of space analogous to the role played by the *principle of elimination of universal forces* in determination of the spatial metric. He argues that the principle of action-by-contact uniquely determines the three-dimensionality of space, at least as far as the macrocosm is concerned. Whether and how that principle can be satisfied for the quantum domain is a question Reichenbach raises; it remains, I believe, a fascinating mystery. He concludes,

> The fact that an ordering of all events is possible within the three dimensions of space and the one dimension of time is the most fundamental aspect of the physical theory of space and time. In comparison, the possibility of a metric seems to be of subordinate significance. (1958, p. 285)

Near the end of his book Reichenbach gives a concise statement to what might be considered the overall purpose of his analyses of space and time in both the *Axiomatik* and *Philosophie der Raum-Zeit-Lehre*: "... physics makes statements about reality, and it has been our aim to free the objective core of these assertions from the subjective additions introduced through the arbitrariness in the choice of the description" (1958, p. 287). In this remark he encapsulates, not only the importance of making the distinction between fact and convention in physical theory, but also the indispensability of his theory of equivalent descriptions to the execution of this task.

During the years between 1930 and 1950, roughly, little of significance seems to have been achieved in philosophy of space and time. Right around 1950, however, Adolf Grünbaum, who was never a student of Reichenbach, completed his doctoral dissertation on Zeno's paradoxes, and published several important papers on various aspects of that topic.[24] At the time of his death in 1953, Reichenbach was hard at work on *The Direction of Time*, which I shall discuss below. Grünbaum's work on Zeno has, as I shall show, important relevance to Reichenbach's last work on time. But shortly after 1960, considerable interest seems to have been regenerated in the problems that had concerned Reichenbach in the period 1920–30, much of which was occasioned by several of Grünbaum's works. The first was a monographic article in *Minnesota Studies in the Philosophy of Science* (1962) and then a major book, *Philosophical Problems of Space and Time* (1963), both of which explicitly take Reichenbach's work as a point of departure and re-examine many of the issues he raised. Grünbaum's work is similar in spirit to that of Reichenbach, though he differs with Reichenbach on many particulars, and extends Reichenbach's analyses on many points on which they are in substantial agreement. A second edition of Grünbaum's 1963 book has been published (1973) which is more than twice the size of the first. This fact alone gives some indication of the renaissance of interest in this area of philosophy of science in the last decade.[25]

One reaction to Grünbaum's work, relating closely to Reichenbach's was a spirited attack by Hilary Putnam entitled 'An Examination of Grünbaum's Philosophy of Geometry' (1963), which elicited a carefully documented monumental reply by Grünbaum (1968, chap. III). Another important exchange involved Reichenbach's views on the nature of simultaneity. In 1962, P. W. Bridgman's posthumous book, *A Sophisticate's Primer of Relativity*, including an epilogue by Grünbaum, was published; it raised certain questions about the use of transported clocks as a basis for establishing synchrony relations. In 1967, Ellis and Bowman exploited a somewhat similar idea in an attack on Reichenbach's position regarding conventional elements in the relation of simultaneity. A panel, consisting of Grünbaum, myself, Bas van Fraassen, and Allen Janis, responded in 1969, defending esssentially the original position of Reichenbach. In addition, John Winnie (1970) in an ingenious response along another line, developed a 'convention-free' formulation of special relativity. Reichenbach had maintained that, while the average round trip speed of light over a closed path is a matter of empirical fact, the one way speed involves the conventional choice of a number ε $(0 < \varepsilon < 1)$. If a light signal, sent from A to B at t_1, is reflected back to A, and arrives back at A at t_3, then its time of arrival at B can be given as $t_2 = t_1 + \varepsilon(t_3 - t_1)$. In order to show that the determination of the numerical value of ε is, indeed, a matter of convention, as maintained by Einstein as well as Reichenbach, Winnie carried through a formulation of special relativity in which ε remains an unspecified variable throughout, and yet all of the observational consequences expected in special relativity are derivable. This is a powerful answer to those who, like Ellis–Bowman, maintained that the Einstein–Reichenbach choice of the value of one-half (which is tantamount to the statement that on a round trip, light travels at the same speed in each direction) is not a matter of convention but of empirical fact.[26]

Although Grünbaum has held much the same view as Reichenbach about the definitional status of the relation of congruence, he offers a different basis for it. Reichenbach invoked his version of the verifiability theory of meaning to maintain that there can be no factual content in the claim that a measuring rod maintains the same length (or does not) when it is transported from one location to another. Grünbaum rests his case, instead, on what he calls the 'metrical amorphousness' of space. This concept, too, has aroused considerable objection, and a lengthy explication and defense has been given by Grünbaum in chapter 16 of the second and enlarged edition of *Philosophical Problems of Space and Time.* But whether one prefers to base his argument upon Reichenbach's verifiability principle, or upon Grünbaum's metrical amorphousness (which, essentially, he derives from Riemann), the result is that the geometrical structure of physical space is not an intrinsic property of it, but rather is a property of the space in relation to the behavior of such extrinsic measuring devices as measuring rods, light rays, and clocks.

It appears, historically, that after the dramatic success of general relativity in predicting apparent displacement of stars during solar eclipse in 1919, interest on the part of physicists in general relativity sunk to a rather low level for some time. In recent years, however, there has been renewed interest in general relativity, involving gravitational collapse, black holes, attempts to detect gravitational waves, etc. One of the most active figures in this area has been the noted Princeton physicist John Archibald Wheeler, who has enjoyed a considerable following among physicists and philosophers, and who has been (until recently) elaborating and defending a view of general relativity which he has called 'geometrodynamics'. The research of Wheeler and associates has recently issued in a heavy (2.5 kg) book entitled *Gravitation* (1973).

As Wheeler himself has formulated the geometrodynamic position he held, it is this:

> Is space–time only an arena within which fields and particles move about as 'physical' and 'foreign' entities? Or is the four-dimensional continuum all there is? Is curved empty geometry a kind of magic building material out of which everything in the world is made: (1) slow curvature in one region of space describes a gravitational field; (2) a rippled geometry with a different type of curvature somewhere else describes an electromagnetic field; (3) a knotted-up region of high curvature describes a concentration of charge and mass–energy that moves like a particle? Are fields and particles foreign entities immersed *in* geometry, or are they nothing *but* geometry?
>
> It would be difficult to name any issue more central to the plan of physics than this: whether space–time is only an arena, or whether it is everything. (1962, p. 361)

Where does this place Reichenbach's work of 1920–30, especially his *Philosophie der Raum–Zeit–Lehre*? Squarely in the center of what is presently one of the liveliest areas of discussion in physics and philosophy. If Reichenbach–Grünbaum (and one should add Poincaré) are right, space itself, in and of itself, has no geometric structure – no intrinsic curvature – to serve as a 'kind of magic building material'. But whatever one's point of view on this issue, it seems incontrovertible that the philosophy of space–time–relativity has once again become an active field of interest, discussion, and controversy. Reichenbach's work, especially his 1928 book, is the indispensable starting-point for anyone interested in participating in this fascinating field. And no one who had seriously studied even the first chapter of Reichenbach's 1928 book could use such terms as 'space' and 'geometry' virtually interchangeably, almost as if they were synonyms – a usage that is conspicuous in the foregoing quotation from Wheeler and in the writings of other geometrodynamicists.

THE DIRECTION OF TIME

Although around 1930 Reichenbach's writings began to turn to other areas, he never lost his fundamental concern with space and time. At the time of his death in 1953 he had completed five of the six projected chapters of his new book, *The Direction of Time*. Many of the fundamental ideas in the work had been anticipated in earlier papers, especially 'Die Kausalstruktur der Welt und der Unterschied von Vergangenheit und Zukunft' (1925a), but their systematic development, analysis, and refinement were reserved for *The Direction of Time*. The five chapters were a bit rough, and would surely have been improved had Reichenbach lived to polish them; nevertheless, with minor editing, they were published posthumuously. This book has received relatively little attention, I regret to say; it is, nevertheless, a veritable gold-mine of philosophically valuable material.

Beginning with a brief introductory chapter on the emotive significance of time, Reichenbach moves on in the second chapter to a discussion of time direction in the laws of classical particle mechanics. To no one's surprise, these laws are time-symmetric, and provide no basis for a physical time direction. He does argue, however, that classical mechanics provides an order – that is, a linear time structure embodying betweenness relations – though this order lacks any fundamental asymmetry.

Under the influence of Eddington, many philosophers seemed to believe that the second law of thermodynamics provides the temporal asymmetry of the physical world – 'time's arrow'. In its phenomenological form, this law strongly suggests temporal asymmetry, namely, entropy tends to increase (with increasing time). The third chapter contains a very careful and detailed analysis of the second law, as it is understood in classical statistical mechanics. Reichenbach shows that, in the light of considerations such as the Loschmidt reversibility objection, this law does not actually possess a time-asymmetric character. The temporal direction of the world arises from certain physical initial conditions in conjunction with the second law of thermodynamics. Thus, the asymmetry of time is not a consequence of the laws of nature alone – it is not a nomic necessity – but depends upon nomically contingent initial conditions as well. The direction of time arises, it turns out, not from the application of the second law of thermodynamics to an ideally closed physical system, but from its application to ensembles of physical systems that are quasi-closed, and which possess as an essential feature a finite lifetime. Reichenbach details the nature of these systems and their relations to their larger environment under the heading of 'the hypothesis of the branch structure'.

These considerations lead Reichenbach to some interesting speculations about 'the sectional nature of time direction', suggesting that the universe as a whole may have temporal epochs in which the temporal directions are opposite to one another. Perhaps our present environment in the universe is a result of a highly improbable chance fluctuation in the universal entropy curve which happens to provide us with an epoch in which there is a fair amount of available energy. But when one thinks quantitatively about the probabilities of fluctuations of such universal magnitude, such ideas seem to me to vanish. For instance, in a box containing forty nitrogen molecules and forty oxygen molecules, there is a probability much, much less than one-half that they will spontaneously separate themselves, with all of the nitrogen molecules in one half of the box and all of the oxygen molecules in the other half, in a time equal to our best present estimates of the entire history of the universe. When we contemplate the same situation except that the box contains Avogadro's number of molecules, the time span required to have any reasonable chance of any such extreme fluctuation is utterly staggering. To extend such considerations to the entire universe absolutely boggles even a cosmological imagination![27]

Given that the laws of thermodynamics, by themselves, do not provide a temporal asymmetry – that the asymmetry appears only in the conjunction of the law with suitable initial conditions – one might naturally raise the question (which to the best of my knowledge Reichenbach did not raise) whether the laws of classical mechanics *in conjunction with suitable initial conditions* might also provide a basis for temporal asymmetry. Anyone who has played the game of pool knows that the answer to this question is affirmative. To see the regular array of objects balls 'broken' upon the impact of the cue ball is commonplace. To see the regular triangular array emerge from the 'random' motions of the fifteen balls about the table is something no pool player ever hopes to see. Yet we may say, with only reasonable idealization, that the motion of the pool balls is governed strictly by the laws of classical mechanics. Grünbaum (1974) and Popper have both discussed this point.

Among the topics treated in this third chapter, Reichenbach's careful analysis of the 'lattice of mixture' repays close study. It provides a detailed mathematical analysis of the mixing process as governed by probabilistic considerations – such as the mixing of nitrogen and oxygen molecules in a container with two compartments, one containing only nitrogen and the other containing only oxygen, after a window between the two compartments has been opened. His discussion of the relation between the 'space ensemble', the configuration of all molecules at any given moment, and the 'time ensemble', the life history of any given molecule, helps enormously to clarify the relation between physical fact and mathematical theorem in this general region of the probabilistic interpretation of the second law of thermodynamics.

One task Reichenbach undertakes in his study of the direction of time is clarification of the physical status of such temporal concepts as *now*, *the present*, and *becoming*. He makes the very strong claim that careful scrutiny of the way in which time enters the equations and theories of physics resolves such philosophical problems.

There is no other way to solve the problem of time than the way through physics. More than any other science, physics has been concerned with the nature of time. If time is objective the physicist must have discovered that fact, if there is Becoming the physicist must know it; but if time is merely subjective and Being is timeless, the physicist must have been able to ignore time in his construction of reality and describe the world without the help of time. . . . If there is a solution to the philosophical problem of time, it is written down in the equations of mathematical physics. (1956, pp. 16–17)

He then makes what may seem a small modification of this statement, but it is a correction of major significance.

Perhaps it would be more accurate to say that the solution is to be read between the lines of the physicist's writings. Physical equations formulate specific laws, general as they may be; but philosophical analysis is concerned with statements *about* the equations rather than with the content of the equations themselves. (p. 17)

Grünbaum, who shares Reichenbach's general view about the function of philosophy of physics, has taken very strong exception to one conclusion of Reichenbach's analysis of physical time. While agreeing that time itself, with its relations of earlier and later, is fully an objective feature of the physical world, he nevertheless takes Reichenbach severely to task for maintaining that becoming or nowness are likewise fully objective. In the first chapter of *Modern Science and Zeno's Paradoxes* (1967) Grünbaum argues with force, and I think cogency, that becoming is not a totally objective feature of the inanimate world, but is, in senses he attempts to make quite precise, a mind-dependent quality. A more recent and somewhat revised account is given in Grünbaum (1971).

It is worth mentioning, however, that the 'Appendix' to *The Direction of Time* contains a brief quotation from Reichenbach's lectures on quantum mechanics at the Poincaré Institute (1953), which gives some hints at what the projected sixth chapter of *The Direction of Time* might have contained had he lived to complete it. In this passage, in rather sharp contrast to remarks that occur earlier in the book, the apparent differences between him and Grünbaum on the nature of 'now' and 'becoming' seem considerably less chasm-like. Indeed, Reichenbach approaches a mind-dependent view when he says, "What we call the time direction, the direction of becoming, is a relation between a registering instrument and its environment; and the statistical isotropy of the universe guarantees that this relation is the same for all such instruments, including the human memory" (p. 270). Clearly there is still an important difference between Reichenbach and Grünbaum; Grünbaum demands an act of conceptualizing conscious awareness, while Reichenbach requires only an act of recording by some device or other which certainly need not be conscious.

But perhaps this is a bridgeable gulf. After mentioning the token-reflexive character of the word 'now' Reichenbach elaborates further.

> An act of thought is an event and, therefore, defines a position in time. If our experiences always take place within the frame of *now* that means that every act of thought defines a point of reference. We cannot escape the *now* because the attempt to escape constitutes an act of thought and, therefore, defines a *now*. There is no thought without a point of reference because the thought itself defines it. (p. 270)

The most original and creative part of *The Direction of Time*, in my opinion, lies in its fourth chapter, 'The Time Direction of Macrostatistics'. In order to explain why I hold this view, it will be necessary to embark on an apparent digression.

Anyone having any acquaintance with the main themes of philosophy of science over the past quarter century is well aware of the voluminous literature on *scientific explanation*. The natural point of departure in this development is the classic 1948 article, 'Studies in the Logic of Explanation', by Hempel and Oppenheim. To give some idea of the weightiness of this literature, I might simply cite three outstanding examples: Braithwaite's *Scientific Explanation* (1953), Nagel's *The Structure of Science: Problems in the Logic of Scientific Explanation* (1961), and Hempel's *Aspects of Scientific Explanation* (1965) – especially the monographic article that bears the same title as the book. Or one could look at the comprehensive bibliography in Rescher's *Scientific Explanation* (1970).

In the 1948 article, Hempel–Oppenheim set forth what has become widely known as the D-N or *deductive–nomological* model of explanation. From the outset, they acknowledged the need for another model – a model of statistical or inductive explanation – but little attention was paid by them or anyone else to this topic until Hempel's famous 'Deductive–Nomological vs. Statistical Explanation' (1962), the main results of which were refined, modified, and extended in the 'Aspects' article of 1965. In that article he details what he calls the I-S or *inductive–statistical* model of explanation. If he achieves nothing else, Hempel certainly demonstrates that the attempt to construct a model of inductive explanation analogous to his well-known model of deductive explanation is no simple or routine exercise, but is fraught with difficulties of the most profound sorts.

In the years 1963–66, I became seriously puzzled about the problems that arise in attempting to account for statistical explanations, and decided that the way to attack the problem, at least as far as the explanation of individual events goes, was to base the treatment squarely upon Reichenbach's treatment of the probability of the single case. As noted above, the crucial issue in the single case problem is the selection of the appropriate reference class. The appropriateness of the reference class is determined by matters of statistical relevance – the reference class should be partitioned relevantly and not partitioned in terms of irrelevant factors. This led to my elaboration of the so-called 'statistical–relevance' or 'S-R model' (Salmon et al., 1971). Thus, the basic tools for the construction of this new model of scientific explanation were borrowed directly from Reichenbach's treatment of the single case.

Moreover, and here I return to the discussion of the fourth chapter of *The Direction of Time*, in order to deal with certain obvious difficulties connected with an elaboration of the model, a concept introduced by Reichenbach in that context proved extremely useful, namely, the *screening-off relation* (1956, §22). It is, basically, a relation of making irrelevant – a factor that is, by itself, statistically relevant to the occurrence of a certain outcome, may *in the presence of another factor* be rendered totally irrelevant. The factor made irrelevant in this way is screened off by the new factor which, so to speak, absorbs the relevance of the screened-off factor. The screening-off relation does much to enable one to distinguish pseudo-explanations in terms of symptomatic relations from genuine explanations in terms of causally relevant factors.

Unlike many contemporary philosophers of science, Reichenbach never wrote a book or article on scientific explanation, nor does any section or chapter of his works bear such a title. But section 18 of *The Direction of Time*, 'Cause and Effect: Producing and Recording', may very well contain the germ of a theory of scientific explanation which is a vast improvement over the received Braithwaite–Hempel–Nagel view. Reichenbach writes:

> Explanation in terms of causes is required when we meet with an isolated system displaying a state of order which in the history of the system is very improbable. We then assume that the system was not isolated at earlier times: explanation presents order in the present as the consequence of interaction in the past. (1956, p. 151)

Such remarks, taken in the light of the full context of this chapter, strongly suggest the possibilities of constructing a genuine theory of causal explanation.

Among the most serious difficulties I find with the D-N model of explanation as elaborated in Hempel (1965), two are directly relevant to the present discussion. In the first place, Hempel's model demands that the explanans contain at least one law-statement, but it does not stipulate that the law which is invoked must be a causal law. Hempel is clearly aware that some universal laws (e.g., the ideal gas law) are not causal laws (1965, pp. 252–253). In the second place, the D-N model does not include any requirement to assure the temporal asymmetry of the relation of explanation. According to the D-N conception, the explanation of an eclipse by astronomic configurations subsequent to the event seems just as good as explanation in terms of prior configurations. Surely something is deeply wrong here.

Reichenbach's brief remarks, quoted above, explicitly introduce both causality and temporal asymmetry into the context of explanation. These two features are, for Reichenbach, closely connected, since the asymmetry of the cause–effect relation is precisely the asymmetry of the before–after relation. This connection is made even more apparent in terms of Reichenbach's *principle of the common cause*, which is also enunciated in this chapter (§19). According to this principle, apparently improbable coincidences are to be explained by common causes (which are, of course, temporally prior) rather than by common effects. Indeed, the principle of the common cause can be used to provide a macro-statistical basis for the asymmetrical structure of physical time itself. Reichenbach provides a careful analysis of the close relations between the principle that orderly states of a system are to be explained by earlier interactions and the principle of the common cause (1956, pp. 163–167). I have found the principle of the common cause especially helpful in my attempts to develop a causal account of scientific explanation which is, I believe, fully in keeping with the spirit of Reichenbach's approach.[28]

In this chapter, Reichenbach offers an explication of causality (about whose adequacy I have some serious doubts) strictly in terms of statistical relevance relations. Whether successful or not, it constitutes a serious and noteworthy attempt to provide an analysis of causation which is, in the most important sense, Humean. While not explicating the causal relations in terms of constant conjunction, contiguity in space and time, and temporal precedence, it does attempt to reconstruct the causal relation in terms of more or less elaborate patterns of statistical relevance, without invoking any of the metaphysical notions of 'power' or 'necessary connection' to which Hume's strictures apply. And if Reichenbach's analysis is not completely successful it certainly constitutes substantial progress on the road to the achievement of an adequate explication of the causal relation. According to Reichenbach, unlike Hume, temporal asymmetry rests upon causal asymmetry, not vice-versa. A conception of causality which bears strong resemblances to Reichenbach's but which is more 'Humean' in taking temporal precedence as a relation given prior to the analysis of causation, is offered by Patrick Suppes (1970).

It is also worth remarking, in passing, that chapter 4 of *The Direction of Time* contains one of the earliest clear expositions of the relationship between thermodynamic entropy and communication-theoretic information. This relationship has been frequently cited in much subsequent literature.

In his earlier works on space and time, especially the *Axiomatik* and *Philosophie der Raum–Zeit–Lehre*, very little attention was paid to the question of time direction. Reichenbach apparently felt that this problem could be handled simply and adequately by his 'mark method' (1958, §21; 1969, §5, 6, 26). Subsequent reflection and criticism convinced him that this was not the case; *The Direction of Time* consequently provides a much more complicated, detailed, and elaborate treatment of the problem of temporal direction. It should also be mentioned, I believe, that anyone tempted to suppose that the difference between the future and the past lies in the fact that we can make experiments which affect the future, while there is no way in which we can 'intervene' in the past, should give careful attention to section 6, 'Intervention', in which Reichenbach to my mind conclusively demonstrates the question-begging character of that very simple appeal.

I shall defer comment upon the final chapter of *The Direction of Time* until I take up his views on quantum mechanics, for that is the topic with which that chapter deals.

Interim Remark

As far as I know, Adolf Grünbaum and I have been the two people who have devoted the largest amount of effort to carrying on the work of Reichenbach – Grünbaum mainly in the area of space–time–relativity, while most of my efforts have been in the areas of probability–induction–explanation. As indicated above, I was Reichenbach's student; Grünbaum, aside from a few brief meetings and personal correspondence, knew Reichenbach only through his written works. Neither of us has been a slavish follower – certainly not Grünbaum, at any rate – we have, instead, been inspired by Reichenbach's work and have attempted to carry it forward. An intellectual giant, Reichenbach has provided each of us with a mighty shoulder on which to stand. I trust that the foregoing discussion of topics in which Grünbaum and I have been actively engaged is sufficient to demonstrate that Reichenbach's ideas and influence are central to large regions of philosophy of science which are scenes of enormous current philosophical activity. But these areas by no means exhaust Reichenbach's important contributions to problems that are matters of deep philosophic concern. It is to some of these additional topics that I shall now turn.

GENERAL EPISTEMOLOGY

Reichenbach has often been called a 'logical positivist', but as a matter of historical fact, he was never a member of the Vienna Circle. He was, instead, the leader of the Berlin group which preferred other designations – 'logical empiricism' became the favorite term. During the twenties and early thirties there was a spirit of close cooperation and intellectual kinship between the two groups, shown most clearly perhaps in the fact that Carnap, who was a member of the Vienna Circle, and Reichenbach jointly edited the journal *Erkenntnis*. There seemed to be a clear realization on the part of both groups that their philosophical differences – though surely significant – were small in comparison with the very fundamental disagreements separating both from the other philosophic movements they opposed. Both groups shared a deep admiration for the kind of work exemplified by Whitehead and Russell in *Principia Mathematica*, and both looked to the rapid new developments in empirical science, especially physics, for considerable philosophical inspiration (as well as raw material for logical analysis).

The fact that Reichenbach is not to be considered a logical positivist is not, however, a geographical technicality, for he maintained that there were fundamental philosophical issues on which he differed from them.

These were exhibited in detail in *Experience and Prediction* (1938) which he considered, if not a refutation, then certainly a correction of standard positivist doctrines.[29] The chief aim of the book was to show the indispensable role of the probability concept in the theory of knowledge; this development then had serious implications for the phenomenalism–realism issue.

Reichenbach, along with the logical positivists, was a staunch supporter of a 'verifiability theory of meaning'. This was a cornerstone of his empiricism.[30] He realized quite early – a lesson most of the positivists learned sooner or later – that the criterion of cognitive meaningfulness cannot be stated in terms of strict deductive verifiability, but must invoke instead some sort of probabilistic verifiability (or confirmability) to some degree. For Reichenbach, then, a statement is to be held cognitively meaningful only if it is possible in principle to obtain evidence which will lend the statement some degree of probability or weight. Obviously, such probabilistic verification may be either supportive or undermining, for the resulting weight may be high, middling, or low. Thus, probabilistic confirmability or refutability in principle becomes the criterion of cognitive or factual meaningfulness. In addition, he also enunciated a criterion of sameness of cognitive meaning: two statements have the same cognitive meaning if and only if every possible item of empirical evidence will lend exactly the same probability or weight to one as to the other. This criterion of sameness of cognitive meaning constitutes the very basis of Reichenbach's 'theory of equivalent descriptions' which played such a central role in his philosophy of space and time, his analysis of quantum mechanics, and indeed, his entire epistemology.

It should be emphasized that Reichenbach *never* maintained that there are no sorts of meaning other than cognitive meaning; he was fully aware of such types as emotive and metaphoric meaning. It is a sheer mistake to suppose that he ever held the view that whatever is not empirically verifiable is utter nonsense. This shows, incidentally, that Popper's strident insistence upon the complete disparity of his attempt to construct a 'criterion of demarcation' from the attempts of such philosophers as Reichenbach to formulate a theory or criterion of cognitive meaningfulness is quite without merit.[31] There are, to be sure, differences in the ways in which Popper attempts to carry out the task, but to claim that it is an utterly different task is a grotesque distortion.

As Reichenbach points out explicitly in the Preface to *Experience and Prediction*, it is not possible to provide a full elaboration of an epistemology such as his without having a well-developed theory of probability and induction. He felt that he had presented just such a background theory in *Wahrscheinlichkeitslehre*. It is interesting to note, by contrast, how quickly A. J. Ayer's famous attempts to provide a criterion based upon 'weak verifiability' came to grief, not once, but twice, when he attempted casually to encapsulate that notion in a paragraph. Realizing that scientific laws could never hope to qualify as cognitively meaningful if cognitive meaningfulness were to be explicated in terms of 'strong verifiability' – i.e., deducibility from some finite set of observation statements – Ayer attempted, in *Language, Truth and Logic*, to formulate a weaker criterion which would do justice to scientific generalization.[32]

In practice, it appears, scientific hypotheses are tested by examining their observational consequences; however, generalizations do not, *by themselves*, have any observational consequences. In his original formulation, therefore, Ayer allowed a statement to qualify as 'weakly verifiable' if, in conjunction with some other premise, it entails an observation statement. The complete inadequacy of this attempt to capture the crux of scientific confirmation was realized immediately. Given *any* statement or pseudo-statement *S* (e.g., 'The Absolute is lazy') and any observation statement *O* (e.g., 'This pencil is yellow'), one need only use 'If *S* then *O*' as an auxiliary premise, and the observable consequence follows immediately from *S* and that additional statement (1946, p. 11).

It was obvious that some reasonable restrictions had to be placed upon the additional premises used in the deductions of observational consequences, and Ayer made a brief attempt to spell them out in the Introduction to the second edition (1946, p. 13). The result was a somewhat more complex explication of 'weak verifiability' – and, it must be admitted – one which sounded a good deal more plausible than the first. Nevertheless, by an ingenious and elegant argument, Alonzo Church (1949) showed that the second formulation was just as vacuous as the first, for, given any statement or pseudo-statement *S*, either it or its negation would qualify as weakly verifiable, even under the new explication.

In attempting to state a criterion of weak verifiability, Ayer does not seem to notice that inductive inference and the concept of probability are at all relevant to what he is trying to do. In contrast to Ayer, Carnap, whose 'Testability and Meaning' (1936–37) is a classic contribution to the confirmability approach, spent the last decades of his life attempting to elaborate a satisfactory conception of confirmability. It must be emphatically noted, moreover, that Church's devastating critique shows only the inadequacy of Ayer's analysis of 'weak verifiability'; it has *no bearing whatever* on the fundamental issue of whether cognitive meaningfulness is to be explicated in terms of empirical verifiability. It thus seems to me an error of major proportions to conclude, primarily because of the failure of Ayer's superficial attempts, that no satisfactory verifiability criterion can be formulated.[33]

Reichenbach never retreated from his conviction that a verifiability criterion of cognitive meaning is an indispensable part of an empiricist epistemology. Furthermore, as we have seen, he backed this conviction with a well-developed theory of probability and induction in terms of which the concept of empirical verifiability could be explicated, and whose inescapably inductive character he clearly recognized. Though it may mark me as a hopeless philosophical reactionary, I cannot resist expressing the view that, mainly under the influence of Hempel, the verifiability criterion was given a premature burial.[34]

One of the major themes of *Experience and Prediction* is a rejection of any form of phenomenalism, a doctrine to which a number of early positivists had been committed.[35] Reichenbach's resistance to phenomenalism seems to have had its roots in his early Kantianism. In *Relativitätstheorie und Erkenntnis Apriori* he had essentially rejected the synthetic a priori status of Euclidean geometry and causality, but the synthetic a priori retained a role in the constitution of objects. This means, as I understand it, that Reichenbach rejected the notion that ordinary physical objects can be constructed solely out of sense data – some constitutive principle is also required. He was thus, at that time, unable to embrace the sensationalistic empiricism of Hume or of Mach. When, however, he saw clearly the distinction between the *context of discovery* and the *context of justification*, he realized that the question of the psychological origin of concepts – e.g., of physical objects – is irrelevant; what matters is the possibility of verifying statements about them. Since, for him, empirical verification is probabilistic or inductive, the way became clear to adopt a thoroughgoing empiricism without becoming involved in any form of phenomenalism. Reichenbach's *probabilistic* theory of cognitive meaningfulness thus freed him from any need to retreat into phenomenalism, as many early positivists had done, in order to maintain a consistent empiricist epistemology.

Within this epistemological framework there was no particular temptation to regard ordinary everyday physical objects as constructions out of sense data, or to insist that statements about such ordinary objects be logically equivalent to a finite list of phenomenal report (protocol) statements. Again, for Reichenbach, the relation between impressions (sense data) on the one hand and middle-size physical objects on the other is strictly probabilistic. While Reichenbach acknowledged that one has a choice as to whether to make sense data statements, physical object statements, or even microphysical statements the most fundamental in one's epistemology, it is natural to choose ordinary physical object statements as basic evidential statements (1951a, pp. 50–51), and to regard pure impressions (sense data) as theoretical entities of psychology, just as electrons are in physics.[36]

At this point an objection is likely to arise concerning our lack of certainty regarding ordinary physical object statements, based, for example, on the traditional argument from illusion. C. I. Lewis, who gave one of the most detailed phenomenalist epistemologies, maintained that nothing can be built on uncertain foundations – "If anything is to be probable, then something must be certain" – was his succinct dictum (1946, p. 186). In a famous symposium with Lewis and Nelson Goodman, Reichenbach (1952) maintained that even phenomenal reports (Lewis's certainties) are not certain. He obviously never supposed that physical object statements are, or can ever be, certain.

Reichenbach's concept of the *posit* plays a key role in his epistemology, and I think it is pertinent in this context. A posit, briefly, is a statement that is (at least temporarily) treated as true even though we do not know that it is true (though, of course, it would not be posited if it were known to be false). Further evidence may occasion revision or rejection of a posit. Reichenbach's posit is strikingly similar to Popper's conjecture (Salmon, 1974, §7). In the theory of probability and induction, Reichenbach distinguished two types of posits, blind (anticipative) and appraised (1949, §90; 1938, §34, 39). Suppose that we examine a sample from a population – an intial section of a probability sequence – and note that a particular attribute occurs with a given frequency in the sample. If we have no other information (say in the form of higher level probabilities) relevant to the probability of that attribute in that sequence, we *posit* (using his rule of induction, which is a rule for making posits) that the limit of the relative frequency is approximately equal to the observed frequency in the sample. Such a posit is a blind posit, for we do not know how good it is. That does not mean, as some of Reichenbach's detractors have claimed, that it is no good at all (Russell, 1948, pp. 412–413). The posit is a statement based upon evidence: it is a 'blind' posit because we do not know how strong the evidence is. If we then proceed to construct a 'probability lattice', we can appraise the former posit by invoking higher level probabilities. The higher level probability statement, however, now represents a blind posit. This process can be continued indefinitely; *any* blind posit can be appraised by invoking still higher level probabilities, but each time this process occurs a *new* blind posit at the higher level emerges.

Posits are used in other ways. When one predicts a single occurrence on the basis of the transfer of probability from the whole sequence to the single case, thus assigning a high weight to the single occurrence, the prediction represents a posit.[37] This is an appraised posit, for our knowledge of the probability in the overall sequence (which is, of course, another posit) tells us how good our prediction is. A similar type of posit may be involved when we predict that a long run probability will manifest itself as a short run frequency to some reasonable degree of approximation.[38] This type of short run posit would be exemplified if we predict, on the basis of our knowledge of the half-life of a radioactive isotope, that about half of a particular sample of that substance will have decayed in the specified period.

There is no reason, as far as I can see, why perceptual judgments about ordinary physical objects cannot also be regarded as posits, and the judgments of memory as well for that matter.[39] If, for example, I toss a penny a hundred times to see how often it lands with the head up, I need these data to make an inductive inference – a posit – about the limiting frequency with which heads will show in the long run if this particular penny is tossed in this particular way. But observing the result of each toss involves a perceptual posit, and to regard the accumulated results of the hundred tosses as evidence (whether the results are written down or merely held in memory) involves memory posits. There certainly are circumstances in which further evidence might lead us to revise, correct, or reject any perceptual or memory posit. This, I take it, is what Reichenbach explains quite informally in his discussion of the 'perfect diary' in *The Rise of Scientific Philosophy* (1951, pp. 260–267). If a person were to write down all of his perceptual judgments, he would, upon reflection, classify some of them as non-veridical – as resulting from dreams, illusions, or hallucinations – because of their failure to fit appropriately with the remaining parts of our body of knowledge. Regarding all of empirical knowledge as a system of posits, subject to revisions on the basis of considerations such as coherence with well-established laws, makes it possible for Reichenbach to advance an empirical theory of knowledge which has no need for certainty *at any level* within it.

Before leaving the concept of the posit, which seems to me to play an absolutely central role in Reichenbach's epistemology, a remark is in order on the subject of 'logical evidence'. In *The Elements of Symbolic Logic* (1947), when discussing the fundamental justification for accepting logical truths, he makes appeal to an intuitive sort of judgment of logical evidence. This constitutes another use of the concept of the posit – though the posits invoked in this context differ in at least one fundamental way from all of the other types of posits I have mentioned. With very little exercise of imagination, one can easily see how a perceptual posit, a memory posit, an inductive posit, a single case posit, or a short run posit can fail to be true – even though, in a certain situation, we were fully justified in treating it as true. The posits involved in logical evidence, however, are tenable only as long as our most careful and ingenious reflection reveals no possible way in which these posits could turn out to be false. This is the reason, I believe, for the vast differences (denied by some philosophers) between the problems of justification of induction and justification of deduction. Logical posits may, however, turn out to be false; even logical evidence is no ironclad guarantee of truth.

If we accept the legitimacy of talking about ordinary physical objects that we observe in everyday life (at least in the form of collections of posits), the question can then be raised regarding the status of the same types of objects when they are not actually being observed. In the introductory chapter of *Philosophic Foundations of Quantum Mechanics* (1944), Reichenbach makes some puzzling statements about such objects as trees and houses. He maintains that we can, without contradicting any of our empirical evidence, claim that an ordinary tree splits into two trees when no one is looking, or that it ceases to exist altogether, popping back into existence whenever anyone looks in the appropriate direction. The startling feature of this view is that it sounds as if he is retreating into the very positivistic phenomenalism he has unequivocally rejected. Reichenbach was fully aware of this apparent conflict; indeed, he had already described it as astonishing in *Experience and Prediction*, where he raised the question of an outright contradiction in this combination of views (1938, p. 138). At that point he offered a detailed argument to show that the problem of the existence of external objects (i.e., the relation of observed physical objects to impressions) is quite distinct from the problem of the existence of unobserved physical objects (i.e., the relation of observed physical objects to unobserved physical objects). He argues that the former case involves a 'projective' relation; this means that no equivalence can exist between a statement about an external physical object and any set of statements about impressions. The latter case, in contrast, involves a 'reductive' relation; this means that equivalent descriptions can exist which differ from one another on the status of unobserved objects.

It is obvious that adoption of the description in which the ordinary tree, for example, splits in two when it is unobserved involves certain inconveniences. One has to say that, while one observed tree casts one shadow on a normally sunny day, the laws of optics apply to unobserved trees in a manner that requires two *unobserved* trees to cast one shadow under the same conditions. Similar adjustments would have to be made in our account of the automatic photographing of an unobserved tree when no one was present.[40] Reichenbach distinguishes two types of descriptions of the world: (1) In a *normal system* the laws of nature which apply to unobserved objects are the same as the laws which apply to observed objects; and (2) In a *non-normal system* the laws of nature differ for unobserved objects. In our daily dealings with macroscopic objects we use the normal description. The fanciful example given above, involving the splitting of the unobserved trees into two trees, is obviously a non-normal description. But, Reichenbach insists, such descriptions are equivalent to one another, according to the criterion of sameness of cognitive meaning he has enunciated and adopted.

How, we might then ask, can Reichenbach maintain the legitimacy of talking about the existence and nature of unobserved physical objects. His answer is that we must adopt certain 'extension rules' of our language. In this particular case, when we are referring to ordinary macroscopic physical objects which are unobserved, we adopt the extension rule stipulating that the normal system be employed. Once that rule has been accepted, we can proceed in the usual way to deal with unobserved macroscopic objects (1951a, pp. 55–58).

The crucial feature of this situation is the fact that, for ordinary macroscopic objects, a normal system exists. The existence of the normal system is emphatically *not* a matter of logical necessity, Reichenbach maintains, for in the quantum domain there is *no* description of the microscopic entities that fulfills the conditions of a normal system. If we ask whether unobserved entities of a certain type exist, without providing any extension rule for the language, the question is ill-formed and incomplete. The question becomes meaningful only if extension rules are provided. One cannot automatically stipulate that the normal system is to be adopted; whether a normal system exists is a factual question. But if we have good reason to believe that there is a normal system, then it is natural to impose the extension rule that selects the normal system as the preferred description. Ordinary questions about unobserved chairs and houses and trees then make sense. The important nonconventional and factual aspect of such questions lies in the existence or non-existence of the normal system. One might therefore say that the question of the existence and nature of unobserved objects is a question about the entire class of equivalent descriptions.[41] In this case the question is whether the class contains a certain member, namely, the normal system.

It is important, I think, to emphasize the analogy between this situation and those we discussed in connection with Reichenbach's treatments of space and time. It will be recalled that the question as to the geometry of physical space is incomplete unless a rule (a definition of congruence) is given. Many equivalent descriptions of physical space can be constructed, depending upon which coordinative definition of congruence is adopted. The preferred definition of congruence, according to Reichenbach, is the one which 'eliminates universal forces'.[42] Once this rule has been laid down, the question about the geometrical structure of physical space becomes a determinate empirical matter. The feasibility of adopting the preferred definition of congruence, we must recall, depends upon certain basic facts about the coincidence behavior of solid bodies (measuring rods).

An even closer parallel is found in Reichenbach's discussion of the topological structure of physical space–time. He points out that various equivalent descriptions can be given, some of which involve causal discontinuities. Again, the factual question must be raised regarding the entire class of equivalent descriptions – does it contain a member which preserves spatio-temporal continuity for causal processes? As long as we confine our attention to the macrocosm, the answer seems to be affirmative. This description, which eliminates causal anomalies, is then selected as the preferred member of the class of equivalent descriptions.[43] In the quantum domain, as we shall see below, the situation is much less straightforward.

Just as Reichenbach's probabilistic orientation enabled him to avoid any involvement with phenomenalism, so also did it steer him away from the instrumentalist view of scientific theories, another pitfall of early positivism. Having escaped the temptation to regard physical object statements as meaningless unless they could be shown logically equivalent to a set of sense-data reports, he could likewise avoid the thesis that scientific theories, containing references to types of entities that are unobservable in principle (which he called 'illata'), are without meaning unless the theoretical terms can be *explicitly defined* in terms of observables. It is sufficient if theories can be probabilistically confirmed or disconfirmed. In this context I must remark, however, that I have found it extremely difficult to understand how one can make even a probabilistic transition from empirical generalizations (regarding observable entities) to genuine theories (which deal with such unobservables – 'illata' – as atoms or molecules) within Reichenbach's own theory of probability and induction. A crucial part of his attempt to answer this question involves his analogy, presented in *Experience and Prediction*, of the 'cubical world' – an analogy I have found profoundly puzzling for many years (1938, §14). I think I may now have some clue as to what he was driving at (1975).

Reichenbach invites us to consider a fictitious observer confined within the walls of an imaginary cubical world. The observer cannot penetrate these walls, but they are made of a translucent material, and he can observe birdlike shadows on the ceiling and on one of the walls. He notices a close correspondence between the patterns on the ceiling and those on the wall. When, for example, a bird-shadow on the ceiling pecks at another bird-shadow on the ceiling, a corresponding bird-shadow on the wall pecks at another corresponding bird-shadow on the wall. Our fictitious observer is justified, Reichenbach maintains, in inferring with some probability that shadows of the same pair of birds are somehow being projected on both the ceiling and the wall. But on what basis, we wonder, is this inference justified? It might seem that, given many similar observed correlations between ceiling and wall shadows, our observer would simply be justified in inferring inductively that such correlations will persist. Why should he postulate an unobservable mechanism to explain the observed correlations? By what kind of argument can he infer that the existence of such a mechanism is probable?

The answer to these questions may lie, it seems to me, in Reichenbach's discussion of the *principle of the common cause*, which he enuciated much later in *The Direction of Time* (1956, §19). This principle figures importantly, I believe, in the theory of causal explanation implicit in Reichenbach's work. When events occur together with a probability exceeding the product of their separate probabilities, the principle of the common cause asserts, such coincidences are to be explained by a common causal antecedent. Thus, for example, pecking events of the type mentioned above occur with a certain small frequency on the ceiling, and they occur with the same frequency on the wall. The frequency with which a pecking-event occurs on the ceiling simultaneously with a similar pecking-event on the wall is much larger than the product of the two separate frequencies. In this case, according to the principle of the common cause, we must postulate a common causal antecedent. To say that no common cause is needed would be to say that the pecking events on the ceiling are actually independent of those on the wall, and the apparent correspondence is only a spurious chance coincidence – a highly unlikely state of affairs. The far more likely alternative, according to Reichenbach, is that these correlated events are neither statistically nor physically independent.

It seems evident to me that Reichenbach intends to claim that the common cause is related to each of the separate, but correlated, effects by a spatio-temporally continuous causal process. The fact that all or part of the connecting process is unobservable does not affect the argument. Thus, Reichenbach's realism concerning the status of unobservable theoretical entities seems to rest upon considerations closely related to his principle that the extension rules of the language should pick out the normal system if such a system exists. He has asserted repeatedly that there is no objection to employment of the normal system even if it makes reference to entities or events which are unobserved or unobservable. Indeed, he explicitly remarks (1951a, p. 57) that Heisenberg's indeterminacy principle, if it were solely a limitation upon possibility of measuring precise values of physical parameters, would pose no obstacle to the use of a normal system on the quantum level. In fact, however, quantum mechanics asserts much more than mere limitation on possibility of measurement; according to Reichenbach's 'principle of anomaly', quantum theory excludes, in principle, the possibility of a normal system. This point will be further elaborated below. Whatever the merits of Reichenbach's analysis of quantum mechanics, however, his profound elaboration and defense of the thesis of scientific realism have considerable bearing upon current discussions of this very issue.

Reichenbach's epistemological concerns were not confined to scientific subject matter; from his earliest student days he was concerned about the epistemic status of moral propositions. In offering a statement of the ideology appropriate to the free student movement (1913) he was careful to refrain from any attempt to prove the truth of any statement of aims of ends. His intent was, rather, to provide an appealing set of principles to which one could make a personal commitment. The volitional character of the commitment was recognized from the beginning.

Writing many years later, for example, in *The Rise of Scientific Philosophy* (1951, chap. 4), he emphasized the influence of the doctrine of 'ethico-cognitive parallelism' – the thesis that fundamental ethical principles can be established as true. This doctrine may take the form of maintaining that such moral principles can be factually confirmed in a manner somewhat analogous to that in which scientific theories are established, or even more perniciously, that they can be demonstrated as absolutely true by a super-scientific or transcendental philosophic method. One of the most important applications of the verifiability theory of cognitive meaning is to show that such 'propositions' have no legitimate claim to cognitive status.

In times of stress and conflict, we all feel a strong – perhaps compelling – desire to prove the truth of *our* basic moral principles. Comforting as it might seem to be able to do so – and here lies the deep emotional appeal of the ethico-cognitive parallelism – Reichenbach consistently maintained that we must recognize the fundamentally non-cognitive character of ethical utterances. This view in no way detracts from the human importance of value judgments – a point which has, I believe, been emphatically advanced in very different terms by existentialist philosophers.[44] "Reason is, and ought only to be, a slave of the passions", Hume affirmed, "and can pretend to no other office but to serve and obey them" (1888, p. 415). Reichenbach staunchly maintained that there is no use in pretending otherwise. Healthy and humane attitudes are what matter – not the illusion that we have discovered the ultimate moral truth.[45]

In view of the far-reaching role of the verifiability criterion of cognitive meaning, the question naturally arises as to its status. Reichenbach had a clear answer. On the one hand, he held that this criterion is actually used in science, at least in its rationally reconstructed form. On the other hand, the adoption of this criterion in its normative form is a matter of volitional decision.[46] One can choose to entertain as possible knowledge claims only those kinds of propositions that can be supported or undermined, in principle and with some degree of probability, on the basis of empirical evidence, or one can choose to admit, for example, statements of theology or transcendental metaphysics whose decidability on evidential grounds is outside the realm of possibility. Recognizing the verifiability theory as a matter of decision evades the question which some philosophers have found embarrassing: if you apply the verifiability criterion to itself, does it turn out to be meaningful? Reichenbach's answer is direct; it does not have the status of a cognitive statement – a proposition that is either true or false. One might say that it is a proposition in a more basic sense of the term – a proposal to be accepted or rejected.

If we continue by asking the perfectly appropriate question of what grounds there are for accepting this proposal, the answer seems to me to come in two parts. First, we want to be intellectually responsible in our beliefs or knowledge claims; we do not want to admit as candidates for belief or knowledge any 'propositions' for which there can be no conceivable positive or negative evidence. But one has to *choose* to be responsible in this sense; philosophical argument does not grab you by the throat and force it upon you. Second, there is the claim of empiricism itself: given that we wish to base our beliefs and knowledge claims upon evidence, the only kind of evidence there can be for factual statements is empirical evidence. This second part of the justification for the verifiability theory of cognitive meaning thus amounts to the denial that there are synthetic a priori propositions – an issue which, as I remarked at the outset, lay at the very heart of Reichenbach's philosophical concerns.

There is a further feature of Reichenbach's version of the verifiability criterion which deserves explicit note. When one asks in what sense the term 'possible' is to be construed when we say that a statement is to be considered cognitively meaningful only if there is *possible* evidence for or against it, Reichenbach unequivocally answers that he means *physical* possibility (1951a, pp. 53–54). Evidence is physically possible only if the existence or occurrence of such evidence does not contradict a law of nature. Thus, for example, certain statements about synchrony of widely separated clocks are unverifiable, not because of any logical contradiction involved in making such simultaneity judgments, but because light, which is a first signal, travels at a finite velocity, and this is a law of nature – a physical law. Adopting a verifiability criterion of cognitive meaning in which possibility is construed in this way seems to me to have profound philosophical consequences. In some sense, it seems to me (1966), this means that the laws of physics determine certain fundamental features of our logic. I shall return briefly to this point when I discuss Reichenbach's philosophy of quantum mechanics.

LAWS, MODALITIES, AND COUNTERFACTUALS

In the last few decades philosophers have become increasingly aware, I think, that science cannot confine its attention solely 'to the facts'. Laws of nature form an essential part of science, and these laws apply to all possible situations, not merely to those that actually occur. Thus, statements which purport to represent laws of nature must be distinguished from general statements that just happen to be true in all cases.[47] Such *lawlike generalizations* as 'No signals travel faster than light' differ fundamentally from such *accidental generalizations* as 'No atomic nucleus has a mass number greater than 280.'[48] The laws of nature delineate the realm of the physically possible, much as the laws of logic define logical possibility. Many physicists believe it possible to create atomic nuclei with much larger mass numbers (though, of course, they may be wrong about this), even though, we suppose, no such nuclei actually exist.[49] In contrast, we believe it is physically impossible to create a signal that travels faster than light. Indeed, physical laws are often stated in terms of impossibility. The first and second laws of thermodynamics assert the impossiblity of perpetual motion machines of the first and second kinds, respectively. Science cannot escape involvement with the physical *modalities* of necessity, possibility, and impossibility.

Science must also be prepared to deal with hypothetical situations – to tell us what *would* happen *if* certain conditions were to be realized, even though we may know that these very conditions will never obtain (or we may be unsure whether they will or not). We can confidently assert that the electric current *would* flow in the circuit again *if* the blown fuse were to be replaced with a bar of solid copper, even though we have no intention of doing anything so fool-hardy. The philosopher of science must, therefore, attempt to deal with the so-called *counterfactual* or *subjunctive conditional* statement, a problem that has turned out to be amazingly recalcitrant.

The subjunctive conditional is, in turn, closely related to the concept of a *dispositional property* – e.g., solubility, fragility, or flammability – a property which will manifest itself if certain conditions are imposed upon the object possessing the property. Salt, for example, is soluble; it dissolves *if placed in water*. But – and this is the connection with counterfactuals – a particular bit of salt is soluble whether it is placed in water or not. Even operationists, who consider theirs the most hard-headed philosophy of science, typically frame their operational definitions in dispositional terms[50] – e.g., to say that a table has a length of one meter means its ends will coincide with the ends of a meter stick *if* a meter stick is placed along its edge. The table surely retains this property when it is not actually being measured.[51]

To emphasize the fundamental importance of the closely related concepts of laws, modalities, counterfactual (or subjunctive) conditionals, and dispositional properties, let us see how they figure in several major areas of Reichenbach's work. The first four examples relate directly to topics we have already discussed; the fifth pertains to quantum mechanics, which is the subject of the following section. We shall then consider, very briefly, certain aspects of his efforts to provide a precise explication of these concepts.

(1) In the theory of relativity, the Minkowski light cone plays a crucial role – one which has come up repeatedly in Reichenbach's philosophical works on space and time from the *Axiomatik* to *The Direction of Time*. Events which are causally connectable with one another by any sort of causal process are said to have either a light-like or a time-like separation. Those which are connectable *only* by light rays (or some form of electromagnetic radiation), but not by other types of causal processes (such as sound waves or motions of material particles), have a light-like separation. Events which are not causally connectable with one another are said to have space-like separations. This is perhaps *the most basic* feature of the structure of physical space–time. Clearly, not all events that are causally connectable are causally connected; hence, the concept of a *possible* connection enters at a very fundamental level. Reichenbach was fully aware of this fact in his discussions in the *Axiomatik*, where he speaks of the structure based upon all possible connections as a 'limiting configuration'. He is obviously aware that some sort of idealization is involved, and with characteristic frankness, he points it out explicitly.[52]

(2) In his theory of probability, as I had occasion to point out in comparing his frequency theory with latter-day propensity interpretations, Reichenbach allowed himself to make free use of *counterfactual* formulations, and to talk about the limiting frequency that *would* occur *if* the penny were to be tossed an infinite number of times. He would not have been deterred from such talk even if the penny were melted down and made into copper wire before it ever was tossed. Such counterfactual statements are indispensable to his theory of probability, and he was obviously aware of that fact. Indeed, in *Nomological Statements and Admissible Operations* (1954, appendix) he explicitly discusses the circumstances under which it is legitimate to assign probabilities with

respect to empty reference classes. Propensity theorists, it might be added, often speak of a propensity as a disposition.

(3) In his adoption of the verifiability theory of cognitive meaning, Reichenbach makes free and explicit use of the notion of *physical possibility* of verification. The concept of a *law of nature* is, as we have seen, central to the very formulation of this criterion.

(4) All of the interesting models of scientific explanation discussed above have the requirement that the explanans contain essentially at least one lawlike statement. This is true of Hempel's deductive-nomological and inductive-statistical models, as well as the statistical–relevance and causal models of explanation I have been trying to extract from Reichenbach's work. Again, the concept of a *law of nature* – or a lawlike generalization – is crucial.

(5) A striking instance of the use of counterfactual conditionals occurs in quantum mechanics, in connection with the fundamental problem posed by Einstein, Podolsky, and Rosen (1935). Consider two physical systems, e.g., two atoms, which interact for a while and then become physically isolated from one another. In such so-called "correlated systems" it may be possible to ascertain the value of a parameter of one member of the pair by means of a measurement made upon the other member of the pair. It may be known, for instance, that the total angular momentum for the combined system is zero; then the measurement of a component of the angular momentum of one of the correlated systems makes it possible to deduce the value of that component for the other. *If*, however, we had chosen to measure a different component of the angular momentum of the first system, we *would have been able* to deduce the value of *that* component for the other correlated system. These assertions are not in dispute. It is agreed, moreover, that one can choose to measure either of these parameters but not both, and that the actual choice can be made after all interaction between the two systems ceases.

The problem is this. Since the two systems are physically isolated from one another, a measurement performed on one cannot affect the state of the other. Since measurement of either parameter on one system determines the value of that parameter for the other system, the second system (on which no measurement is made) must actually possess a value for each of two complementary parameters. Inasmuch as quantum mechanics does not provide for such simultaneous values of complementary variables, Einstein–Podolsky–Rosen argued that quantum mechanics is an incomplete physical theory. Niels Bohr, in an equally famous reply (1935), attempted to show how this conclusion can be avoided.

The use of counterfactual conditionals in this context deserves explicit mention because the Einstein–Podolsky–Rosen 'paradox' is a classic instance of a fundamental philosophical problem in the interpretation of quantum mechanics. Reichenbach uses this example, in the penultimate section of *Philosophic Foundations of Quantum Mechanics*, to illustrate his use of a three-valued logic, and this same problem is still being discussed by serious contemporary authors who are concerned with the foundations of physics.[53]

Anyone unacquainted with the literature on this subject might well ask why there is any particular difficulty involved in all of these concepts. In conversational language, as Reichenbach noted (1954, pp. 1–2), we skillfully use terms which involve physical modalities or subjunctive moods. The problem is that it is extremely difficult to make explicit the rules we unconsciously use so ably. This difficulty stems, in turn, from the fact that the 'material implication' defined in elementary logic is patently unsuited for the counterfactual or subjunctive usage. This results from the fact that a *material* conditional statement is true whenever it has a false antecedent. Thus, interpreted as material conditionals, all of the following statements are true:

(i) If the blown fuse is replaced with a copper bar, the current will flow in the circuit.

(ii) If the blown fuse is replaced with a copper bar, the current will not flow in the circuit.

(iii) If the blown fuse is replaced with a copper bar, the moon is made of green cheese.

Clearly, this is scientifically unsatisfactory.

Reichenbach recognized that similar problems arise concerning truth-functional connectives other than implication. Suppose, for example, that a man steps from a dark alley and, pointing a gun at your head, announces, "You will give me your money or I will blow your brains out." Terrified, you hand him your wallet, but immediately thereafter you notice that it was a toy gun in his hand. You chide him for issuing a false threat. Making passing reference to the truth-table for the disjunction, he points out that you did hand over your money, and so the disjunction was true. In consequence, Reichenbach generalized the problem of *subjunctive conditionals* to the problem of *admissible* (reasonable) *operations*. Early in his treatment of symbolic logic (1947, §7) he introduced two different ways of reading truth tables, one of which yields the 'adjunctive' (standard truth-functional) interpretation, while the other yields a 'connective' interpretation (which corresponds with what we ordinarily consider a 'reasonable' logical operation).

The problem of subjunctive conditionals is, as we have noted, closely related to the problem of lawlike generalizations. We should not, therefore, expect to be able to handle the problem of admissible operations without dealing with the problem of laws. If we think of laws as simple universal generalizations of the form, 'All *F* are *G*', we immediately encounter a problem analogous to that of the counterfactual conditional: if *F* is an empty class, the generalization is vacuously true. Using a slight modification of Reichenbach's example, we see that both statements

(iv) All men who have walked on Mars have worn red neckties.
(v) All men who have walked on Mars have not worn red neckties.

are true.[54] Once more, Reichenbach recognized that natural laws may assume much more complicated forms, and generalizations may be rendered non-lawlike by a variety of different defects. Consequently, he attempts to set forth criteria that will discriminate between the lawlike and the non-lawlike among all statements that can be formulated in first-order logic. In this way he endeavored to provide a general explication of the concept of a law of nature.

In the final chapter of *Elements of Symbolic Logic* (1947) Reichenbach attempted to characterize the kinds of statements that can express laws – nomological statements, as he called them – and in *Nomological Statements and Admissible Operations* (1954) he offered an extended and improved treatment of the same topic. It is extremely technical and complex. In this introductory essay I shall not attempt to explain these complexities,[55] but one can get some idea of their extent by noting that *twenty-six* definitions, some quite abstruse, are required before Reichenbach can lay down the *nine* conditions that characterize *original nomological statements*. The original nomological statements constitute a basic set of laws from which other laws – *derivative nomological statements* – can be deduced.

Part of the complexity of Reichenbach's treatment of original nomological statements is due to the fact that this class of statements contains both laws of nature (synthetic statements) and laws of logic (analytic statements). It seems evident that some simplification could have been achieved by distinguishing the two types of laws and offering separate characterizations of them. Reichenbach chose, quite deliberately, not to do so, because he saw a deep analogy between the problems associated with laws of the two types. To take one basic example, he noted that there is something 'unreasonable' about the 'law of logic' which says that any statement whatever is entailed by a contradiction. This is parallel to the unreasonableness of the vacuous truth of a material conditional with a false antecedent. It is this very problem of 'unreasonable entailments' that has given rise to much discussion of the nature of entailment, and to the systematic development of formal logics of 'relevant entailment'.[56] In spite of this analogy, pedagogical purposes might better have been served by developing separate explications of original nomological statements of the two types.

When Reichenbach moves on to the treatment of derivative nomological statements, he finds an added complexity arising out of the fact that a bifurcation needs to be introduced between two types. Any statement that can be deduced from any set of original nomological statements qualifies as a *nomological statement in the wider sense* (or simply *nomological statement*), while the imposition of certain restrictions defines a *nomological statement in the narrower sense* (or *admissible statement*). This is not complexity for complexity's sake. Nomological statements in the wider sense provide the basis for the analysis of modalities – necessity, possibility, and impossibility – while nomological statements in the narrower sense provide the basis for the definition of admissible operations.

Nomological Statements and Admissible Operations has, unfortunately, received very little attention. The chief criticism of the book seems to be its complexity; I am not aware of any demonstration that the treatment of these problems is inadequate. In the years since its publication there has been an enormous literature on laws and counterfactual conditionals, a full-blown logic of relevant entailment has been constructed through the efforts of many logicians, and modal logic has been the object of extensive development and research. Yet it is fair to say, I believe, that there is no treatment of lawlike statements, physical modalities, or subjunctive conditionals that is widely accepted as adequate. In view of this fact, I wonder whether we have reached a point at which we should entertain the possibility that the topic is genuinely complicated, and that no simple treatment will do the trick.

I do not know of any other theory of lawlike statements which enables us to say precisely why 'No signal travels faster than light' is a law of nature, while 'No gold cube is larger than one cubic mile', even if true, is not.[57] Perhaps the time has come to make the necessary effort and see whether Reichenbach did, after all, have a reasonable answer to this and the related fundamental philosophical questions of dispositions and counterfactuals. I am inclined to think that Reichenbach's work in this area does repay close study, for his treatment seems deeper in many respects than other well-known and standard treatments.

It was in 1952, as I recall, that I sent Reichenbach a copy of a paper I had written on the concept of avoidability in the context of moral responsibility – a paper which, mercifully, never was published. My approach was similar to that of Charles Stevenson (1944, chap. XIV), but I attempted to carry out the analysis without the use of counterfactuals – e.g., 'I could have done otherwise if I had desired', or 'He would have done otherwise if circumstances had been different'. In this paper I remarked, 'The conditional contrary to fact is shrouded in mystery.' Reichenbach commented that it is no more, since his (then unpublished) book cleared it all up. Perhaps he was right. At the very least, he made a profound and sustained attack on problems that are central to epistemology, philosophy of science, philosophy of language, philosophy of logic, and moral philosophy.

QUANTUM MECHANICS

It was inevitable that Reichenbach, with his longstanding interest in probability and causality, as well as his careful attention to twentieth century physics, should become involved in philosophical problems relating to quantum mechanics. In addition to *Philosophic Foundations of Quantum Mechanics* (1944), the fifth chapter of *The Direction of Time* was devoted to problems in the same area. In 1952, moreover, Reichenbach presented four lectures on this subject at the Institut Henri Poincaré; these were published in 1953 under the title, 'Les fondements logiques de la mécanique des quanta'.

Reichenbach's work in this area is remembered chiefly for his proposal to use a three-valued logic to resolve the problems of quantum theory. He was not the first to propose the use of a non-standard propositional logic in quantum mechanics; in a classic article, 'The Logic of Quantum Mechanics' (1936), Birkhoff and von Neumann proposed a different non-standard logic for dealing with quantum mechanical statements. Reichenbach's logic, though not the traditional two-valued logic, is still a truth functional logic – i.e., the truth value of a compound proposition is fully determined by the truth values of its components. The Birkhoff–von Neumann logic, in contrast, is a modular lattice logic; it is not truth functional, though it is two-valued. A fundamental formal difference between these two systems of non-standard logic is that the distributive laws (for disjunction and conjunction) of standard logic also hold in Reichenbach's three-valued logic (1944, p. 156) but they are not valid in the Birkhoff–von Neumann lattice logic.

After quite an extended period when quantum theorists seemed to pay very little attention to the idea of a non-standard logic, there has been a renewed interest in this topic in the last few years. Quite a number of recent works have dealt with this idea in one way or another; Jauch's *Foundations of Quantum Mechanics* (1968) is perhaps the best known exposition.[58] It is apparent that Reichenbach's three-valued logic has played very little role in this renewed concern with non-standard logics – much more attention being paid to the Birkhoff–von Neumann type. There are, I believe, sound reasons for this choice. First, the arguments offered by these authors for rejecting the distributive laws seem compelling (Birkhoff et al., 1936, 830–831).

Second, given the well-known difficulties in the interpretation of quantum mechanical phenomena, it is clear that certain kinds of problems are bound to arise in connection with such compound propositions as conjunctions or disjunctions.[59] In the Birkhoff–von Neumann treatment, as well as Reichenbach's, it is the disjunctions that exhibit pecularities.[60] In any case, it seems evident that the situation with compound propositions will depend crucially upon whether or not the statements which constitute the compound refer to conjugate parameters. Given two pairs of statements, each having the same pair of truth values, one pair being about conjugate parameters while the other pair is not, it is doubtful – to say the least – that we should expect the same truth value for the same compound in both cases. It therefore seems implausible to suppose that any kind of truth-functional logic will do the job that the non-standard logic is called upon to do.

Although Reichenbach's work on non-standard logic may not be likely to have great lasting impact, there are other aspects of his treatment of quantum phenomena which certainly deserve serious consideration. In the fifth chapter of *The Direction of Time* he employs his theory of equivalent descriptions in illuminating discussions of such topics as the Bose–Einstein and Fermi–Dirac statistics, and in the Feynman interpretation of the positron as an electron moving backward in time. He offers a conjecture (1956, p. 156) which is, I think, to be taken very seriously indeed; namely, that time is essentially a macroscopic phenomenon arising statistically from the aggregate behavior of extremely large numbers of micro-entities. A similar conjecture, it seems to me, ought seriously to be entertained regarding the structure of physical space.

These ideas may be closely connected – it is really hard to say with any confidence at all – to some highly visionary remarks by Wheeler, et al., toward the end of their enormous book, *Gravitation* (1973, chap. 44, §4–5) in which they acknowledge something more primitive than space which they refer to as 'pre-geometry'. Such an acknowledgment is, of course, an explicit rejection of Wheeler's earlier geometrodynamics. It is fascinating – but because of the sketchy character of the remarks it is hard to say more – to learn that pre-geometry is identified with the calculus of propositions (but not necessarily the old two-valued Boolean one, I presume)! This type of assertion exemplifies once more the tendency, noted above, of geometrodynamicists to use interchangeably terms which denote physical reality ('space') and terms that denote mathematical systems used to describe that reality ('geometry'). Certainly Wheeler, et al., do not mean to assert that the physical world is literally constructed out of an abstract logical system; rather the claim is that it is constructed out of something appropriately *described by* the calculus of propositions.

This intent is somewhat clarified by such additional remarks as these:

> A larger unity must exist which includes both the quantum principle and general relativity. Surely the Lord did not on Day One create geometry [read 'space'] and on Day Two proceed to 'quantize' it. The quantum principle, rather, came on Day One and out of it something was built on Day Two which on first inspection looks like geometry [sic!] but which on closer examination is at the same time simpler and more sophisticated. (Wheeler, 1971, p. viii)

The fact is that the microstructures of both space and time are currently deeply perplexing issues.

For many years after the advent of the Heisenberg–Schrödinger quantum mechanics, there was considerable discussion of the way in which interference by observation of a physical system is related to Heisenberg's indeterminacy relation.[61] Largely in response to Heisenberg's (1930) famous gamma-ray microscope thought experiment, many people maintained that the interference gives rise to the indeterminacy. Reichenbach (1929) was one of the first to realize that the situation is precisely the reverse; it is because of the indeterminacy relation that there must be an indivisible coupling of the physical microsystem with the apparatus of observation, leading to an *uncorrectable* (or *unpredictable*) interference with the microsystem as a result of observation. Reichenbach pointed out that interference with a physical system by observation or measurement is an integral part of classical physics. In classical physics, however, one can, in principle, make corrections for such interference, and one can, in principle, reduce the degree of interference without limit (Reichenbach, 1944, p. 16; 1956, §11). Quantum mechanics differs from classical mechanics in placing a limit in principle, given by the Heisenberg indeterminacy relation, upon the degree to which the interference can be reduced by physical refinement or compensated by calculation.[62]

Reichenbach's main contribution to the philosophy of quantum mechanics is, in my opinion, his 'principle of anomaly'. From the beginnings of quantum mechanics there has been extensive discussion of the determinism–indeterminism issue, and it continues vigorously in the context of 'hidden variable' theories. In *The Direction of Time* Reichenbach casts doubt upon the deterministic status even of classical physics. One comes out with the feeling that it does not make a really profound difference whether classical statistical thermodynamics is based upon an underlying deterministic theory or not (1956, p. 269). This is a philosophically interesting issue, but it should not be too difficult to accept the idea that the world, even of classical physics, is an indeterministic one involving some sort of irreducibly statistical element. It should, consequently, not be too difficult to accept the irreducibly indeterministic aspects of quantum mechanics, recognizing that our physical theory can, at best, furnish only probabilities for the occurrence of various types of events. In short, it is not such a strain to admit that Lucretius may have been right after all in saying that the atoms (in the Stern–Gerlach apparatus, for example) may do a little swerving for which no strict causal account can be made.

To use the modern vernacular, it is the causal anomalies – always involving some form of action at a distance – that really 'blow the mind'. In the two-slit experiment with electrons,[63] how is it that the electron, if it is a particle going through one slit, 'knows' whether the other slit is open? Or if we think of the electron as a wave, spread out in space, how is it that the wave (which can have astronomic dimensions) deposits all of its energy in a highly localized detector instantaneously? Examples of this sort can be multiplied indefinitely – the famous Einstein–Podolsky–Rosen experiment on correlated systems may be an excellent case in point (Reichenbach, 1944, §36). Reichenbach finds similar anomalies in Bose–Einstein and Fermi–Dirac statistics, and in the Feynman interpretation of the positron (1956, §26, 29, 30).

In *Philosophic Foundations of Quantum Mechanics* (1944, pp. 19, 23–34), Reichenbach defines a *normal system* as a description within which unobserved objects obey the same laws as observed objects. In a system which is not normal, causal anomalies occur. Reichenbach points out that it is easy to construct strange descriptions of ordinary objects which involve causal anomalies – e.g., trees splitting in two when no one is watching – but these descriptions are harmless (and unappealing) because, as we have noted above, they are simply equivalent to normal descriptions. In classical physics it was assumed that among the class of equivalent descriptions there was always at least one which would qualify as a normal system. It might have been tempting for classical physicists to assume as an a priori principle that a normal system is always possible in principle – perhaps this is a way of reformulating at least part of Kant's principle of universal causation.

In order to articulate the fundamental difference between quantum mechanics and classical physics, Reichenbach makes a distinction between *phenomena* and *interphenomena* (1944, pp. 20–21). Phenomena are events which are actually observable, or are readily inferrable from those which are observable; a particle interacting with a Geiger counter is an example of a phenomenon. Interphenomena are events which are less closely associated with observables; an electron passing through a narrow slit with no detecting device in the immediate vicinity is an example of an interphenomenon. Any adequate physical theory must deal with the phenoma; whether it must provide an account of the interphenomena as well is more controversial. Reichenbach distinguishes between *exhaustive* interpretations, which do offer descriptions of interphenomena, and *restrictive* interpretations, which do not (1944, §8). The principle of anomaly states that, *among the exhaustive interpretations*, there is none that is free from causal anomalies (1944, p. 44; 1956, pp. 216–218). This principle formulates the basic difference between quantum mechanics and classical physics (as well as relativity theory).

Causal anomalies can be 'suppressed' by adopting a restrictive interpretation. One such restrictive interpretation, due primarily to Heisenberg and Bohr (now widely known as 'the Copenhagen interpretation'), considers all statements about interphenomena to be meaningless. Reichenbach's interpretation in terms of the three-valued logic is another restrictive interpretation; it assigns the truth-value *indeterminate* to such statements (1944, §33).

The difference between these two restrictive interpretations is not a trivial matter of terminology. Certain compound statements in the three-valued logic are either true or false, even though some of their constituents have the truth-value indeterminate. Combinations which contain both statements and meaningless pseudo-statements, in contrast, must be meaningless.[64] Reichenbach's restrictive interpretation is, therefore, substantively stronger than the 'Copenhagen interpretation'; it allows us to assert a good deal more about the microcosm. Reichenbach's principle of anomaly does, however, come close to expressing some of the things Bohr said about complementarity, for example, that it is never possible to provide in a single consistent formulation both a space–time description and a causal description of a quantum mechanical system (1944, pp. 22, 157–159).

Ever since its inception, quantum mechanics has engendered serious and sustained controversy over the question raised in the title of the famous paper by Einstein, Podolsky, and Rosen: can quantum mechanical description of reality be considered complete? There are, obviously, at least two possible schools of thought on this issue, namely those who answer in the affirmative and those who answer in the negative. An affirmative answer involves adoption of what Reichenbach calls *the synoptic principle*, which asserts that all possible physical information about a quantum mechanical system is contained in a psi-function (1956, p. 214). In many textbooks (e.g., von Neumann, 1955) this principle is laid down as a fundamental axiom of the theory. Taken in conjunction with other well-established properties of the psi-function, this principle entails the conclusion that it is impossible ever to verify as true any exhaustive interpretation, such as the wave or particle interpretation. The synoptic principle therefore guarantees the possibility of 'suppressing' causal anomalies. Statements about interphenomena, asserting for example that a particular entity is wave-like or corpuscular, can legitimately be assigned the truth-value indeterminate (or, à la Copenhagen, be classed as meaningless) precisely because of the limitations imposed by the synoptic principle (1956, pp. 218–219).

Those who agree with Einstein in giving a negative answer to the question about the completeness of quantum mechanics deny the synoptic principle. They affirm the existence of 'hidden variables' which make it possible in principle (though we may not yet know how in practice) to provide information about physical systems beyond that contained in the psi-function.[65] Knowledge of the values of such hidden variables would enable us to circumvent the restrictions formulated in the Heisenberg indeterminacy relation.

Reichenbach explicitly recognizes the truism that the synoptic principle is no more than a well-established physical theory which could, in the light of future evidence, be overturned. He then proceeds to examine its consequences. In the last chapter (§26) of *The Direction of Time*, and more elegantly and concisely in his (1953) lectures to the Poincaré Institute, he argues that abandonment of the synoptic principle leads directly to the introduction of causal anomalies into physical theories. In briefest outline, the argument runs as follows. As long as bosons (e.g., alpha particles or He^4 atoms) are considered genuinely indistinguishable, Bose–Einstein statistics – which account, for example, for superfluidity of liquid helium at very low temperatures – involve no causal anomalies.

Genuine indistinguishability implies, however, that when two alpha particles collide and recoil it is impossible in principle to ascertain which particle recoiled in which direction. This means that no definite trajectory can be associated with each particle – indeed, such particles simply lack the property of genidentity. If, however, it *were* possible to associate a definite trajectory with each boson, establishing genidentity and rendering them distinguishable in principle, then Bose–Einstein statistics would involve statistical dependencies which would be tantamount to action at a distance. The reward for rejecting the synoptic principle would be the re-establishment of determinism; the price would be very high, however, for it would be a form of determinism shot through with causal anomalies. Is a strict determinism infused with causal anomalies better, Reichenbach asks, than the indeterminacy which results from the synoptic principle? Should we not be glad that the synoptic principle, which appears to be very strongly established, is available to screen us from the causal anomalies and render them relatively innocuous? We must not – obviously – decide such issues as the truth or falsity of the synoptic principle in terms of our desires or emotions. It is, nevertheless, salutary to consider the consequences of the alternatives, and to examine the nature of the determinism we may find ourselves hoping to re-establish in the microcosm (1956, p. 220).

It is important to understand fully the reason that the causal anomalies are harmless in Reichenbach's interpretation of quantum mechanics. Once more, his theory of equivalent descriptions plays a crucial role. Reichenbach says,

Given the world of phenomena we can introduce the world of interphenomena in different ways; we then shall obtain a class of equivalent descriptions of interphenomena, each of which is equally true, and all of which belong to the same world of phenomena . . . . In the class of equivalent descriptions of the world the interphenomena vary with the descriptions, whereas the phenomena constitute the invariants of the class . . . interpolation of unobserved values can be given *only* by a *class* of equivalent descriptions. (1944, p. 23, my emphasis)

Further analysis of the nature of exhaustive interpretations leads to *the principle of eliminability of causal anomalies*: "We do not have one normal system for *all* interphenomena, but we do have a normal system for *every* interphenomenon" (ibid., my emphasis). This principle explains, incidentally, the facility with which physicists switch back and forth between the corpuscular and wave interpretations, depending upon the particular problem at hand.

The principle of anomaly is much harder to accept than is the idea of indeterminism. Reichenbach believed it to be an intrinsic part of quantum mechanics – and it had the appeal of further undermining the synthetic a priori. It gave him another occasion to say in effect (if I may be permitted a paraphrase of a much overworked line from Shakespeare), "There are more things in modern physics, Immanuel, than are dreamt of in your philosophy!"

LOGIC AND PHILOSOPHY OF MATHEMATICS

'Logical empiricists' coined that name, as I understand it, partly to distinguish themselves from earlier empiricists – chiefly John Stuart Mill – whose philosophies of logic and mathematics they found inadequate, and partly to emphasize the importance they attached to the use of symbolic logic or logistic techniques in dealing with philosophical problems. As we noted above, for example, Reichenbach's interpretation of quantum mechanics rested squarely upon the use of a three-valued logic. It should also be noted that he developed a 'probability logic' (1949, chap. 10) – a logic with a continuous scale of truth values – in connection with his probability theory.[66] Moreover, although he was not primarily a logician, he wrote a textbook in logic – *The Elements of Symbolic Logic* (1947). By contemporary standards, the formal aspects of this book are not very rigorous, and it is not easy pedagogically, but it does have considerable importance for other reasons.

As Reichenbach explains in his Preface, a large part of the motivation for the book is the *application* of the techniques of symbolic logic to a wide range of problems. Much that is of current interest lies in the last two chapters. The last chapter, VIII, 'Connective Operations and Modalities', contains his earliest attempt to deal with laws and counterfactuals; it is, essentially, superseded by *Nomological Statements and Admissible Operations* (1954), which provides, as we saw above, a more detailed and satisfactory approach to the same problems. This chapter of *Elements of Symbolic Logic* is, nevertheless, virtually a prerequisite for the understanding of the subsequent book.

Chapter VII, 'Analysis of Conversational Language', provides some excellent and remarkable examples of what might be called 'analysis of ordinary language', carried out in the days before the school that came to bear that name had any wide currency among Anglo-American philosophers. This chapter contains (§50–51) a still-classic analysis of the tenses of verbs, as well as a treatment of such token-reflexive terms as 'I', 'here', 'this', and 'now'. Tenses and token-reflexivity are obviously relevant to the construction of temporal logics, and to issues surrounding such concepts as nowness and temporal becoming. These are topics of considerable current interest.[67] Another important part of this chapter (§48) attempts to develop a formal logic in which the individual variables range over events or facts, rather than things; in the course of this analysis the formal relations between the event-language and the thing-language are exhibited (§49). Reichenbach offers, in addition, formal treatments of a variety of types of existence: fictional existence (e.g., Hamlet), immediate existence (e.g., objects in dreams), and intentional existence (e.g., objects of belief or desire). In the discussion of these various sorts of existence, he also advances formal analyses of statements about perceiving, believing, knowing, and desiring. All of these topics are of perennial interest to metaphysicians, philosophers of science, epistemologists, and philosophers of language.

Taken together, the two final chapters of *Elements of Symbolic Logic* exhibit the broad scope Reichenbach envisioned for the application of formal methods. It is fair to say, I believe, that this range embraces the whole of rational discourse, including interrogative, exclamatory, and imperative – as well as declarative – sentences.[68] It encompasses subjunctive and modal statements, as we have already seen. The entire assortment of intentional contexts is, as just noted, also included. To assert that all of these analyses are entirely satisfactory would be rash, indeed, but it can reasonably be claimed that Reichenbach achieved a surprising degree of success in his ambitious venture. He surely laid solid foundations for further investigations. The widespread rejection of formal methods by proponents of 'ordinary language analysis' has, in my opinion, been an unfortunate road-block, during much of the time since Reichenbach's death, to progress along the lines that he (and many other like-minded philosophers) were so fruitfully pursuing.

Following Russell, Reichenbach was well aware of the fact that the superficial grammar of a statement may differ radically from its fundamental logical form – in current terms, there is an enormous discrepancy between the deep structure and the superficial structure – and in agreement with Russell, Reichenbach held that the traditional grammatical analysis of the forms of statements could be positively misleading (§45, 49). One of the charming aspects of this book, written originally in English by a German fluent in French, who gained some familiarity with Turkish as a refugee professor in that country, is the way it is generously sprinkled with examples and comparisons taken from many natural languages.[69] At the beginning of his discussion of conversational language, Reichenbach held out the "hope that the results of symbolic logic will some day, in the form of a modernized grammar, find their way into elementary schools" (1947, p. 255). The realization of this hope seems to have made substantial beginnings, chiefly due to the influence of Noam Chomsky, in the work of modern linguists who are well schooled in the fundamentals of contemporary formal logic.[70]

Reichenbach was a profound thinker in the literal sense; whenever he dealt with a topic he worked deeply into its foundations. His work on logic and mathematics is no exception. In the *Preface* to *Elements of Symbolic Logic* he expresses his great intellectual debts to David Hilbert, who was one of his teachers, and to Bertrand Russell, whose works he read and admired, and who was a colleague in Los Angeles for a year. In 1946, in the lead article, 'Bertrand Russell's Logic', in *The Philosophy of Bertrand Russell*, he goes into greater detail in this expression of admiration, but in the course of the discussion, it seems to me, he voices a view of his own that was closer to Hilbert's formalism than Russell's logicism.

Reichenbach acknowledges that, in a given formal language, a particular formula may be a tautology (a term Reichenbach uses to refer to any kind of logical truth), but its status as a tautology is relative to a given logical language, and the statement that it is a tautology in that language is, itself, an empirical statement.[71] Moreover, the choice of a language is a matter of convention – convention dictated, it must be emphasized, by practical considerations in the light of physical facts. *Tertium non datur* is a tautology in ordinary two-valued logic, and this logic works well for most everyday affairs. The fact that two-valued logic works well in the macrocosm is not, however, a logical necessity; if the physical world had a radically differential structure, a different logic might be preferable.

Indeed, in the quantum domain, Reichenbach urged, a three-valued logic constitutes a better choice, and in such a logic *tertium non datur* is not valid.[72] For certain other purposes, it will be recalled, Reichenbach advocated use of an infinite-valued probability logic (1949, chap. 10).

Moreover, Reichenbach's strict adherence to a version of the verifiability theory of meaning that incorporates *physical* possibility of verification may have basic consequences regarding the choice of a logic. It is not immediately evident in all cases that the possibility in principle of verifying each of two statements implies the possibility in principle of verifying disjunctions and conjunctions into which they enter, nor is it obvious that the verification of a compound statement entails the verifiability of its components (Salmon, 1966). Quantum mechanics suggests itself immediately as a domain in which such problems might arise. Thus, Reichenbach does not seem to commit himself to the existence of logical truth, per se, but only to the validity of some formulas in certain formal systems. Whether he held that there are some statements in some interpretations of some logical systems that express objective and independent logical truths is difficult to discern from his logical writings. I am inclined to think, however, that his answer would have been in the negative had he been pressed to give one.[73]

In his philosophy of mathematics and in his philosophy of logic, Reichenbach gave considerable attention (as did Hilbert, of course) to the problem of proving the consistency of formal systems. Where logic itself is concerned, this led ultimately to the concept of *logical evidence*; statements whose truth is asserted on the basis of logical evidence were explicitly regarded as posits (1947, §34). Although none of his published writings, to the best of my knowledge, explicitly deals with the consequences of Gödel's theorems for the Hilbert program of formalism, he did deal with this question in the course in advanced logic I had the privilege of taking from him. In this course he distributed a mimeographed manuscript, which was intended as a supplement to the logic book. After presenting some routine background material on topics in foundations in mathematics, he gave an informal exposition of Gödel's theorems on incompleteness and the impossibility of proving consistency of a consistent system of arithmetic within that system. In all of this material there is nothing very novel that is not widely available today.

Reichenbach then proceeds to discuss the implications of these results concerning consistency proofs both with respect to pure, but more especially, with respect to applied, mathematics. His treatment of the problem of consistency for systems of applied mathematics is, to the best of my knowledge, original and unique.[74] It is a topic that has been largely neglected by philosophers of mathematics – whose chief interests seem usually to lie in the area of pure mathematics and logic – but Reichenbach's fundamental orientation toward physics and the natural sciences brings him to consider the problem carefully. He produces an ingeniously simple argument to the effect that for applied mathematics, *strictly speaking*, no formal consistency proof is required. A system of applied mathematics is, after all, an axiom system that is interpreted; the aim is to produce a system of axioms that *truly* describes some aspect of the world.[75] Since no inconsistent system can have a true interpretation, the risk of error arising from the use of inconsistent systems in applied mathematics is encompassed within the risk of failure to provide an admissible interpretation of the axiom system, provided that the rules of derivation within the system are truth-preserving. The truth preserving character of the rules, which are ultimately the rules of deductive logic, is certified on the basis of *logical evidence*. If the axioms have an admissible interpretation and if the rules by which theorems are derived are truth preserving no threat of error due to inconsistency exists.

The situation is somewhat complicated, however, by the fact that the applied mathematician often – if not usually – employs mathematical techniques or systems that are more powerful than those minimally necessary for the application at hand. In this respect, then, proofs of consistency for systems of pure mathematics can be of vital interest to the applied mathematician. Given the elementary fact that anything can be deduced from a contradiction, it is useless to try to prove the consistency of a system within that system even if it is powerful enough to express the statement of its own consistency. To have any point at all, the proof must be given in a metalanguage. This gives rise immediately to the problem of whether the metalanguage is consistent, and the threat of an infinite regress seems to raise its ugly head. Reichenbach's ingenious way out of this unpleasant situation lies in his view that the metalinguistic assertion of consistency – which states that it is impossible to derive both a formula and its negation within the object language – is itself an empirical statement. This claim is closely related to his view that metalinguistic assertions about the tautological character of object language formulas are empirical statements. It is, therefore, possible for him to argue that the question of the consistency of a given object language – even if the object language is a system of *pure* mathematics – is a problem of *applied* mathematics in the metalanguage. But the problem of consistency of systems of applied mathematics has already been dispatched, as explained above. Hence, no infinite regress need arise.

EPILOGUE

At the time of his death, Reichenbach was planning to write a sixth chapter of *The Direction of Time* on the human brain as a physical recording instrument. It is our grave misfortune that he never had the opportunity to do so. In the first chapter of that book, however, he had discussed the deep interest with which thoughtful people regard time. As he saw it, this profound concern stems from the inevitable prospect of death. He explained the appeal of various metaphysical treatments of time in terms of their success in dealing with the emotions engendered by the thought of death. Reichenbach held no theology – no transcendental metaphysic – no belief in an afterlife. Unlike David Hume, he was fortunate not to experience a lengthy terminal illness, but like Hume, he died calmly and without fear. A decade later, Carnap wrote, "In 1953, when Reichenbach's creative activity was suddenly ended by his premature death, our movement lost one of its most active leaders. But his published work and the fruit of his personal influence live on" (1963a, p. 39). This influence is even more significant today than it was more than ten years ago when Carnap's words were published, for it has important bearing upon the work of a new generation of philosophers who have worked with Grünbaum, Hempel, myself, and others who knew Reichenbach during his lifetime.[76]

University of Arizona

NOTES

[1] Biographies of Reichenbach can be found in Achinstein (1967), Schuster (1975), and Salmon (forthcoming). Maria Reichenbach's Introduction to Reichenbach (1965) contains an intellectual biography. I am grateful to Dr Cynthia Schuster for showing me an advance copy of her article and to Dr Charles Gillispie for providing me with proofs of her article prior to publication. I am also grateful to Maria Reichenbach for making available to me a letter about Hans Reichenbach's family and childhood written by his brother, Bernhard.

[2] Publication of *Erkenntnis* was recently resumed under the editorship of Carl G. Hempel, Wolfgang Stegmüller, and Wilhelm K. Essler (D. Reidel Publishing Co., Dordrecht and Boston; Felix Meiner Verlag, Hamburg), vol. IX, 1975.

[3] A comprehensive bibliography of Reichenbach's works is contained in Reichenbach (1959).

[4] Quoted in Schuster (1975).

[5] In 1954 Carnap moved to UCLA to fill the chair vacated by Reichenbach's death.

[6] According to Martin Strauss (1972), Reichenbach regarded this as his most significant achievement. See pp. 273–285 for his memoir, 'Reichenbach and the Berlin School'.

[7] For a precise formulation see his (1949, §16).

[8] This axiom reads, "If an event C is to be expected in a sequence with a probability converging toward 1, it will occur at least once in the sequence." As Reichenbach notes, "This is a rather modest assumption" (1949, §65).

[9] See (1949, §34) for a detailed description of the probability lattice along with a discussion of the motives for introducing it.

[10] I find Popper's treatment of propensities ambiguous with regard to these two versions. Mellor (1971, chap. 4) makes the distinction clearly and adopts the 'single case' version.

[11] The second edition (1962) of Carnap (1950) contains an important new preface not found in the first edition. Carnap (1952) is an extension and generalization of (1950). Carnap (1962, 1963, 1971) deal with his later systems of inductive logic. A subsequent part of the system in (1971) remains to be published.

[12] It may have been a psychological idiosyncracy on my part to have been more troubled about the interpretation of 'B' than about the interpretation of 'A' and 'C'. All three are highly problematic: 'A' must stand for the reference class of hypotheses relevantly similar to the hypothesis being evaluated, and 'C' must represent a class of hypotheses which have been tested in a relevantly similar way. Traditionally, however, the antecedent or prior probability, $P(A, B)$, has been considered the most problematic part of Bayes's theorem, and my efforts were concentrated upon an attempt to provide a frequency interpretation of that particular probability.

[13] This indispensable distinction is explained in (1949, §70, 86–87). N.B., the treatment of this distinction in the English edition is one of the significant differences between it and the German edition. See 'Changes and Additions in the English Edition'.

[14] This rule will be discussed under item (3) below.

[15] The concept of the posit will be discussed in some detail below in the section 'General Epistemology'.

[16] This book, perhaps more than any other, helped to convince those interested in inductive logic and statistics that Bayes's theorem might have some application to the evaluation of scientific hypotheses.

[17] See the concluding paragraphs of Reichenbach (1930a). This passage bears striking resemblance to the views of Peter Strawson, to be mentioned below.

[18] See Salmon (1957). It is ironic that Reichenbach, whose characterization of the distinction between descriptive and inductive simplicity in his (1938, §42) is classic, should have misused the concept of descriptive simplicity in this manner.
[19] See Salmon (1967, pp. 21–27, 114–115; 1968; 1974) for my discussion of Popper's approach. Watkins (1968) attempts a defense of Popper's deductivism against my charges. His answer seems to me patently inadequate.
[20] This theme has been carefully elaborated in Grünbaum (1963, pp. 81–97).
[21] This point has been forcefully articulated in Winnie (1970) and (1972).
[22] Reichenbach discussed this problem from a historical standpoint in his (1924a). H. Stein (1967) provides an illuminating historical treatment of this subject, criticizing Reichenbach's analysis.
[23] These problems are discussed with careful attention to contemporary work, and with references to the relevant literature in Grünbaum (1973, pp. 730–750).
[24] This work later developed into a book, Grünbaum (1967).
[25] In the spring of 1974, two conferences, one at the Minnesota Center for Philosophy of Science and one at the Institute for Relativity Studies at Boston University – involving physicists, mathematicians, and philosophers – were devoted in large part to issues arising from the new edition of Grünbaum's book. The proceedings of these two conferences are to be published jointly in a forthcoming volume of *Minnesota Studies in the Philosophy of Science*.
[26] Reichenbach (1925) discussed the issue of simultaneity by slow clock transport in connection with Römer's determination of the speed of light on the basis of observations of the moons of Jupiter. The problem of distant simultaneity – or, equivalently, the one-way speed of light – continues to generate considerable discussion. For example, the physicist Eugene Feenberg (1974) endorses the Ellis–Bowman rejection of the conventional status of distant simultaneity. I am currently working on an article, 'The Philosophical Significance of the One-Way Speed of Light', for inclusion in a forthcoming issue of *Nous* which will be devoted to philosophical problems concerning space and time. In this article I shall survey a variety of proposals for empirical ascertainment of the one-way speed of light, and I shall attempt a general assessment of the current status of the problem.
[27] See Grünbaum (1963, pp. 261–264) for a further critique of Reichenbach's views on the 'hypothesis of branch structure' and the 'sectional nature of time direction'.
[28] My recent attempts to extend the account of scientific explanation beyond Salmon et al. (1971) into a full-blown account of causal explanation can be found in Salmon (1975) and (1976).
[29] It is an astonishing historical fact that this work, which clearly was Reichenbach's major epistemological treatise, has not been translated into any language other than the original English in which it was written.
[30] For a comprehensive statement of his views regarding this criterion, written more than a decade after *Experience and Prediction*, see 'The Verifiability Theory of Meaning' (1951). This article begins with some illuminating historical remarks on Reichenbach's relations with members of the Vienna Circle of logical positivists. It constitutes his final detailed treatment of this topic.
[31] See, for example, Popper, *The Logic of Scientific Discovery* (1959a), Appendix *1, pp. 311–17. This book is an English translation of *Logik der Forschung* (1934), but the appendix was added to the English translation and does not appear in the original German edition.
[32] This book, first published in 1936, was the most widely read early popular presentation of logical positivism to the English-speaking world. As such it was extremely influential.

[33] I have discussed this matter in some detail in Salmon (1966).
[34] See, for example, Hempel (1951). A somewhat revised and updated combination of this article and another famous article by Hempel on the same topic can be found in 'Empiricist Criteria of Cognitive Significance: Problems and Changes' in Hempel (1965).
[35] An early unequivocal statement of this anti-Machian view was given by Reichenbach in *Handbuch der Physik*, Vol. 4, Berlin, 1929, pp. 16–24. Though not published until 1929, the article was written in 1923.
[36] See Reichenbach (1938, §19, 26). Reichenbach used the term 'illata' to designate what are now usually called 'theoretical entities'. By contrast, he used the term 'abstracta' to refer to entities which, though not directly observable, can be fully and explicitly defined in terms of observables.
[37] Rudolf Carnap has offered a profound discussion of the problems involved in applying probability to single events – the sort of situation in which Reichenbach uses the *appraised* posit to assign a weight. Carnap's treatment of this particular problem seems to me to provide a distinct improvement over Reichenbach's treatment, but it can be regarded as an extension of or a supplement to Reichenbach's approach; it need not be taken as any sort of refutation. See Carnap (1950, §50–51). Carnap did not treat the problem of *blind* posits because, according to him, probability (degree of confirmation) statements are either analytic or self-contradictory.
[38] The problem of the short run has for many years impressed me as one of the most fundamental, and perhaps most intractable, of the epistemological problems in the general area of probability and induction. This problem was dramatically formulated by C. S. Peirce in (1931, Vol. II, §2.652). In my (1955), I attempted a pragmatic resolution of this problem, along lines quite parallel to Reichenbach's pragmatic justification of induction, but in (1957) I show that extremely fundamental difficulties beset both Reichenbach's justification of induction and my attempted resolution of the problem of the short run.
[39] I have elaborated this theme in my (1974a).
[40] My example.
[41] The question obviously must not be construed as asking which of the equivalent descriptions is *the true one*. Either they are all true or all false. However, the question as to what consideration makes the normal system the preferred description is legitimate. I presume Reichenbach would have offered descriptive simplicity as the operative principle.
[42] The desire for descriptive simplicity certainly motivates this choice.
[43] The answer to the question as to the justification of this choice seems to me less clear. Since, as we shall see below, the question of causal anomalies and action at a distance seems to play a central role in Reichenbach's views on scientific realism, this problem deserves careful consideration. I have attempted to offer some hints toward an answer in my (1975).
[44] See Meyerhoff (1951).
[45] See Reichenbach (1951, chap. 17); see also 'On the Explication of Ethical Utterances' in Reichenbach (1959), 193–198.
[46] (1938, chap. I, esp. §8) In this discussion Reichenbach places great importance upon the nature of the *entailed decisions* that follow upon any particular volitional decision.
[47] Strictly speaking, I regard a law of nature as a regularity which actually holds in the world; hence, laws of nature are *not* linguistic entities. We use statements – linguistic entities – to express or represent these laws. For stylistic reasons, however, I shall sometimes speak as if the statements themselves were the laws, when no confusion is likely to result from this manner of speaking.

[48] According to my *Handbook of Physics and Chemistry* (1972–73) the isotope with the largest mass number known to date is $_{104}Ku^{260}$.
[49] There are theoretical reasons to believe that stable nuclei of much greater mass number are possible, with perhaps 120 protons and 180 to 200 neutrons, and that they may be created by colliding heavy nuclei with one another. Machines that may someday be able to perform such experiments are currently under construction.
[50] Even hard-headedness is a dispositional property: one who is hard-headed will reject nonsense *if* confronted with it.
[51] For an enlightening discussion of operational definitions see 'Logical Appraisal of Operationalism' in Hempel (1965, pp. 123–133).
[52] Carnap, in (1958), is *not* careful to distinguish causal connectedness from causal connectability, and this constitutes a serious flaw in his attempt to formalize relativity theory.
[53] A recent encyclopedic discussion of the Einstein–Podolsky–Rosen problem is given by Clifford A. Hooker in his monographic and extensively documented article (1972). For an explicit discussion of the introduction of modalities into 'quantum logic' see Bas C. van Fraassen (1973).
[54] At the time of Reichenbach's death, the class of men who had walked on the moon was empty.
[55] In my Foreword to Reichenbach (1976), I attempt to explain this material in terms designed to make it more accessible to contemporary readers. This Foreword is reprinted below under the title 'Laws, Modalities, and Counterfactuals'.
[56] Anderson, et al. (1975) is the *magnum opus* on this topic.
[57] These are Reichenbach's examples (1954, pp. 11–12). In my essay 'Laws, Modalities, and Counterfactuals' below I try to explicate Reichenbach's somewhat puzzling treatment of them.
[58] Other illuminating discussions of alternative logics for quantum mechanics are van Fraassen (1974), the articles by van Fraassen and by Greechie and Gudder in Hooker (1973), and the articles by Finkelstein and Fine in Colodny (1972). Hooker (1975) is a large anthology devoted to this topic.
[59] Reasons for this claim, based upon the verifiability theory of meaning are given in Salmon (1966).
[60] Although the truth table for disjunction is just what one would expect in generalizing from two to three truth values, Reichenbach (1944, pp. 161–162) discusses the peculiarities that accompany this generalization.
[61] Reichenbach evidently preferred the term 'indeterminacy' to the more common term 'uncertainty' because of possible subjective connotations of the latter term.
[62] In his charming little dialog, *Are Quanta Real*?, J. Jauch overlooks this fundamental point (1973, pp. 35–48).
[63] Reichenbach (1944, §7). Because of the short wave-length of the electron, this is a 'thought experiment', but it is nevertheless a sound illustration of the principle under discussion.
[64] See Reichenbach (1944, p. 142). As I try to show in my (1966), this statement about meaninglessness of compounds must be handled with some care; however these niceties do not affect this comparison between the Reichenbach and the Bohr–Heisenberg interpretations.
[65] David Bohm has been one of the chief proponents of this view; for a recent statement see his (1971). Much of the recent work on hidden variable theories has focused on the issue of

'locality'. See, for example, Bell (1971). This work is discussed in Bub (1973). Roger Jones's essay 'Causal Anomalies and the Completeness of Quantum Theory', published below, provides a survey of these considerations.

[66] In discussing Reichenbach's use of three-valued logic in quantum mechanics, I maintained that it is unlikely that a truth functional logic is suitable in that context. Unlike the three-valued logic, the probability logic is not truth functional. The value assigned to compound propositions depends not only upon the truth values of the components, but also upon a certain kind of linkage between them; see (1949, Table 9, p. 400). It is just this sort of linkage that seems called for in quantum logic. I do not know of any explicit discussion by Reichenbach of his reasons for rejecting his probability logic (or some similar logic) for quantum mechanics. Perhaps such logics are unsuitable for obvious reasons, but I do not know of any such considerations. Although the probability logic is not truth-functional, it is 'extensional' in an important sense; see (1949, p. 403).

[67] Bertrand Russell treated token-reflexive terms under the heading of 'egocentric particulars' in (1940, chap. VII) and (1948, Part II, chap. IV). For recent work on temporal (or tense) logic see Rescher et al. (1971), which the authors describe in the Preface as "an introduction to temporal logic, a now flourishing branch of philosophic logic. . ." For critical discussions of 'now', 'becoming', and related terms see Grünbaum (1971).

[68] Important recent work has been done on the logic of questions and commands; see, for example, Rescher (1966), which contains an extensive bibliography. One important monograph on the logic of questions is Harrah (1963). Further bibliographical information can be found in Hamblin (1967).

[69] In addition to examples from English, German, French, and Turkish, there are others from Latin, Greek, Old French, and Old English in chap. VII.

[70] Some of Reichenbach's work has recently been taken seriously by professional linguists; see, for example, Bierwisch (1971).

[71] This statement must be construed with care. To say that a statement that fits a given description is a tautology may be a tautology in the metalanguage, but to say that a particular inscription is tautological is always an empirical statement in the metalanguage. See (1947, sec. 34, esp. pp. 186–88).

[72] See (1944, sec. 32) Reichenbach did *not* maintain that interpretations embodying other logical systems are impermissible; see sec. 37, the concluding section of the book. See also note 73 below.

[73] Support for this view can be drawn, it seems to me, from the concluding paragraph of Reichenbach (1953, chap. II) where he writes: "Does this result show that the true quantum logic is three-valued? I do not believe that we are able to speak of the truth of a logic. A system of logic is empty, that is to say, it has no empirical content. Logic expresses the form of a language, but does not formulate any physical laws. Nevertheless, we can consider the resulting consequences of the choice of the logic for the language; and, in so far as these consequences also depend upon physical laws, they reflect properties of the physical world. It is thus the combination of logic and physics which reveals the structure of reality that concerns the physicist. A quantum physics formulated by means of a two-valued logic exists; but it includes causal anomalies, whereas a quantum physics which employs a three-valued logic does not have any such anomalies. This conclusion, which is the result of all the experiments encompassed by wave mechanics, expresses the strange structure of the microcosm . . ." (p. 136, my translation).

[74] Reichenbach's treatment of this problem is discussed more fully in M. Salmon, 'Consistency Proofs for Applied Mathematics', published below.

[75] Providing a true physical interpretation of an axiom system is, of course, a 'proof' of consistency, but it is an empirical 'proof' and not a formal proof.

[76] I should like to express my deepest gratitude to Professors Adolf Grünbaum, Carl G. Hempel, Andreas Kamlah, Maria Reichenbach, Cynthia Schuster, and Merrilee Salmon for reading a preliminary version of this article and making numerous valuable criticisms and suggestions. Further *special* thanks go to Professors Grünbaum and Hempel for devoting their thought and attention to this matter at times of considerable pressure from other obligations. I should like to add a word of apology to the many philosophers whose important works on issues discussed in this article have not been cited among the References. I am painfully aware of many such omissions, but considerations of space precluded fuller bibliography. It is my hope that references found in the works cited will lead the interested reader into the relevant literature.

REFERENCES

Achinstein, P., 1967, 'Hans Reichenbach', in P. Edwards, ed., *The Encyclopedia of Philosophy*, Vol. 8, Macmillan Publishing Co., New York, 1967, 115–118.

Anderson, A., et al., 1975, *Entailment*, Princeton University Press, Princeton.

Ayer, A., 1946, *Language, Truth and Logic*, 2nd ed., Dover Publications, New York.

Bell, J., 1971, 'Introduction to the Hidden Variable Question', in B. d'Espagnet, ed., *Foundations of Quantum Mechanics*, Academic Press, New York, 1971.

Bierwisch, M., 1971, 'On Classifying Semantic Features', in D. Steinberg et al., eds., *Semantics*, The University Press, Cambridge, 1971.

Birkhoff, G., et al., 1936, 'The Logic Of Quantum Mechanics', *Annals of Mathematics*, **37**, 823–843.

Bohm, D., 1971, 'On the Role of Hidden Variables in the Fundamental Structure of Physics', in T. Bastin, ed., *Quantum Theory and Beyond*, The University Press, Cambridge, 1971, 95–116.

Bohr, N., 1935, 'Can Quantum Mechanical Description of Physical Reality be Considered Complete?', *Physical Review* **48**, 696–702.

Braithwaite, R., 1953, *Scientific Explanation*, The University Press, Cambridge.

Bridgman, P., 1962, *A Sophisticate's Primer of Relativity*, Wesleyan University Press, Middletown, Conn.

Bub, J., 1973, 'On the Completeness of Quantum Mechanics', in C. Hooker, ed., (1973).

Carnap, R., 1922, *Der Raum*, Reuter und Reichard, Berlin.

Carnap, R., 1936–37, 'Testability and Meaning', *Philosophy of Science* **3**, 420–471; **4**, 1–40.

Carnap, R., 1950, *Logical Foundations of Probability*, University of Chicago Press, Chicago.

Carnap, R., 1952, *The Continuum of Inductive Methods*, University of Chicago Press, Chicago.

Carnap, R., 1958, *An Introduction to Symbolic Logic and its Applications*, Dover Publications, New York.

Carnap, R., 1958a, 'Introductory Remarks to the English Edition' of Reichenbach (1958).

Carnap, R., 1962, 'The Aim of Inductive Logic', in E. Nagel, et al., eds., *Logic, Methodology and Philosophy of Science*, Stanford University Press, Stanford, 1962, pp. 303–318.

Carnap, R., 1963, 'Replies and Systematic Expositions', in P. Schilpp, ed., *The Philosophy of Rudolf Çarnap*, Open Court, La Salle, Ill., 1963.

Carnap, R., 1963a, 'Intellectual Autobiography', in P. Schilpp, ed., *The Philosophy of Rudolf Carnap*, Open Court Publishing Co., La Salle, Ill., 1963.

Carnap, R., 1966, *Philosophical Foundations of Physics*, Basic Books, New York. Reissued as *An Introduction to the Philosophy of Science*, Harper Torchbooks.

Carnap, R., 1971, 'A Basic System of Inductive Logic', in R. Carnap, et al., eds., *Studies in Inductive Logic and Probability*, Vol. I, University of California Press, Berkeley and Los Angeles, 1971.

Church, A., 1949, Review of *Language, Truth and Logic*, *Journal of Symbolic Logic* **14**, 52–53.

Colodny, R., ed., 1972, *Paradigms and Paradoxes*, University of Pittsburgh Press, Pittsburgh.

Einstein, A., 1905, 'On the Electrodynamics of Moving Bodies', in Einstein et al., *The Principle of Relativity*, Dover Publications, New York, n.d.

Einstein, A., et al., 1935, 'Can Quantum Mechanical Description of Physical Reality be Considered Complete?', *Physical Review* **47**, 777–780.

Einstein, A., 1949, 'Remarks Concerning the Essays Brought Together in this Cooperative Volume', in P. Schilpp, ed., *Albert Einstein: Philosopher–Scientist*, Tudor Publishing Co., New York, 1949, 665–688.

Ellis, B., et al., 1967, 'Conventionality in Distant Simultaneity', *Philosophy of Science* **34**, 116–136.

Feenberg, E., 1974, 'Conventionality in Distant Simultaneity', *Foundations of Physics* **4**, 121–126.

Feigl, H., 1950, 'De Principiis Non Disputandum . . . ?', in M. Black, ed., *Philosophical Analysis*, Cornell University Press, Ithaca, 119–156.

Grünbaum, A., 1962, 'Geometry, Chronometry, and Empiricism', in H. Feigl et al., eds., *Minnesota Studies in the Philosophy of Science*, Vol. III.

Grünbaum, A., 1963, *Philosophical Problems of Space and Time*, Alfred A. Knopf, New York. 2nd ed., 1973, D. Reidel Publishing Co., Dordrecht and Boston.

Grünbaum, A., 1967, *Modern Science and Zeno's Paradoxes*, Wesleyan University Press, Middletown, Conn. 2nd ed., slightly revised, 1968, George Allen and Unwin, London.

Grünbaum, A., 1968, *Geometry and Chronometry in Philosophical Perspective*, University of Minnesota Press, Minneapolis.

Grünbaum, A., et al., 1969, 'A Panel Discussion of Simultaneity by Slow Clock Transport in the Special and General Theories of Relativity', *Philosophy of Science* **36**, 1–81.

Grünbaum, A., 1971, 'The Meaning of Time', in E. Freeman, et al., eds., *Basic Issues in the Philosophy of Time*, Open Court Publishing Co., La Salle, Ill., 1971, 195–228.

Grünbaum, A., 1974, 'Popper's Views on the Arrow of Time', in P. Schilpp, ed., *The Philosophy of Karl Popper*, Open Court Publishing Co., La Salle, Ill., 1974, 775–797.

Hacking, I., 1968, 'One Problem About Induction', in I. Lakatos, ed., *The Problem of Inductive Logic*, North-Holland Publishing Co., Amsterdam, 1968, 44–59.

Hamblin, C., 1967, 'Questions', in P. Edwards, ed., *The Encyclopedia of Philosophy*, Vol. 7, The Macmillan Co., New York, 1967, 49–53.

Harrah, D., 1963, *Communication: A Logical Model*, The MIT Press, Cambridge, Mass.

Heisenberg, W., 1930, *The Physical Principles of the Quantum Theory*, University of Chicago Press, Chicago. Reprinted in paperback by Dover Publications, New York.

Hempel, C., et al., 1948, 'Studies in the Logic of Explanation', *Philosophy of Science* **15**, 135–175. Reprinted in Hempel (1965).

Hempel, C., 1951, 'The Concept of Cognitive Significance: A Reconsideration', *Proceedings of the American Academy of Arts and Sciences* **80**, 61–77.

Hempel, C., 1962, 'Deductive–Nomological vs. Statistical Explanation', in H. Feigl et al., eds., *Minnesota Studies in the Philosophy of Science*, Vol. III, University of Minnesota Press, Minneapolis, 1962, 98–169.

Hempel, C., 1965, *Aspects of Scientific Explanation*, The Free Press, New York.

Hilbert, D., 1902, *Foundations of Geometry*, Open Court Publishing Co., La Salle, Ill.

Hooker, C., 1972, 'The Nature of Quantum Mechanical Reality', in Colodny (1972), 67–302.

Hooker, C., ed., 1973, *Contemporary Research in the Foundations and Philosophy of Quantum Mechanics*, D. Reidel Publishing Co., Dordrecht & Boston.

Hooker, C., ed., 1975, *The Logico-Algebraic Approach to Quantum Mechanics*, vol. I, D. Reidel Publishing Co., Dordrecht & Boston.

Hume, D., 1888, *A Treatise of Human Nature*, L. Selby-Bigge, ed., Clarendon Press, Oxford.

Hutten, E., 1956, *The Language of Modern Physics*, The Macmillan Co., New York.

Jauch, J., 1968, *Foundations of Quantum Mechanics*, Addison-Wesley Publishing Co., Reading, Mass.

Jauch, J., 1973, *Are Quanta Real?*, Indiana University Press, Bloomington.

Jeffrey, R., 1965, *The Logic of Decision*, McGraw Hill Book Co., New York.

Lewis, C., 1946, *An Analysis of Knowledge and Valuation*, Open Court Publishing Co., La Salle, Ill.

Mellor, D., 1971, *The Matter of Chance*, The University Press, Cambridge.

Meyerhoff, H., 1951, 'Emotive and Existentialist Theories of Ethics', *Journal of Philosophy* **48**, 769–783.

Nagel, E., 1961, *The Structure of Science*, Harcourt, Brace & World, New York & Burlingame.

Peirce, C., 1931, *The Collected Papers of Charles Sanders Peirce*, eds. C. Hartshorne et al., Harvard University Press, Cambridge, Mass.

Poincaré, H., 1952, *Science and Hypothesis*, Dover Publications, New York.

Popper, K., 1957, 'The Propensity Interpretation of the Calculus of Probability, and the Quantum Theory', in S. Körner, ed., *Observation and Interpretation*, Butterworths Scientific Publications, London, 1957, 65–70.

Popper, K., 1959, 'The Propensity Interpretation of Probability', *British Journal for the Philosophy of Science* **10**, 25–42.

Popper, K., 1959a, *The Logic of Scientific Discovery*, Basic Books, New York.

Popper, K., 1967, 'Quantum Mechanics without "The Observer"', in M. Bunge, ed., *Quantum Theory and Reality*, Springer-Verlag, New York, 1967, 7–44.

Putnam, H., 1963, 'An Examination of Grünbaum's Philosophy of Geometry', in B. Baumrin, ed., *Philosophy of Science: The Delaware Seminar*, John Wiley & Sons, New York, 1963, 205–255.

Reichenbach, H., 1912, 'Studentenschaft und Katholizismus', in W. Ostwald, ed., *Das Monistische Jahrhundert*, No. 16, zweites Novemberheft 1912, 533–538. English translation to be published in R. Cohen et al. (forthcoming).

Reichenbach, H., 1913, 'Die freistudentische Idee. Ihr Inhalt als Einheit', in *Freistudententum. Versuch einer Synthese der freistudentischen Ideen*. In verbindung mit Karl Landauer, herausgegeben von Hermann Kranold, Max Steinebach, München, 1913, pp. 23–40. English translation to be published in R. Cohen et al. (forthcoming).

Reichenbach, H., 1916, 'Der Begriff der Wahrscheinlichkeit für die mathematische Darstellung der Wirklichkeit', *Zeitschrift für Philosophie und philosophische Kritik* **161**, 210–239; **162**, 98–112, 223–253.

Reichenbach, H., 1920, *Relativitätstheorie und Erkenntnis Apriori*, Springer, Berlin. English translation: H. Reichenbach (1965).

Reichenbach, H., 1924, *Axiomatik der Relativistischen Raum–Zeit–Lehre*, Vieweg, Braunschweig. English translation, Reichenbach (1969).

Reichenbach, H., 1924a, 'Die Bewegungslehre bei Newton, Leibniz und Huyghens', *Kantstudien* **29**, 416–438. An English Translation, 'The Theory of Motion According to Newton, Leibniz and Huyghens', appears in Reichenbach (1959).

Reichenbach, H., 1925, 'Planetenuhr und Einsteinsche Gleichzeitigkeit', *Zeitschrift für Physik* **33**, 628–634.

Reichenbach, H., 1925a, 'Die Kausalstruktur der Welt und der Unterschied von Vergangenheit und Zukunft', *Bayerische Akademie der Wissenschaften, Sitzungsberichte*, Nov., 1925, 133–175.

Reichenbach, H., 1928, *Philosophie der Raum–Zeit–Lehre*, Walter de Gruyter, Berlin & Leipzig. English translation, Reichenbach (1958).

Reichenbach, H., 1929, 'Ziele und Wege der physikalischen Erkenntnis', *Handbuch der Physik* **4**, 1–80.

Reichenbach, H., 1930, *Atom und Kosmos*, Deutsche Buch-Gemeinschaft, Berlin.

Reichenbach, H., 1930a, 'Kausalität und Wahrscheinlichkeit' *Erkenntnis* **1**, 158–188. An English translation, 'Causality and Probability', appears in Reichenbach (1959).

Reichenbach, H., 1932, *Atom and Cosmos*, George Allen & Unwin, London.

Reichenbach, H., 1935, *Wahrscheinlichkeitslehre*, A. W. Sijthoff's Uitgeversmaatschappij N. V., Leiden.

Reichenbach, H., 1938, *Experience and Prediction*, University of Chicago Press, Chicago.

Reichenbach, H., 1944, *Philosophic Foundations of Quantum Mechanics*, University of California Press, Berkeley & Los Angeles.

Reichenbach, H., 1946, 'Bertrand Russell's Logic', in P. Schilpp, ed., *The Philosophy of Bertrand Russell* (The Library of Living Philosophers, Inc.), Evanston, Ill., 1946, 23–54.

Reichenbach, H., 1947, *Elements of Symbolic Logic*, The Macmillan Co., New York.

Reichenbach, H., 1948, 'Rationalism and Empiricism: An Inquiry into the Roots of Philosophical Error', *Philosophical Review* **57**, 330–346. Reprinted in H. Reichenbach (1959).

Reichenbach, H., 1949a, 'The Philosophical Significance of the Theory of Relativity', in P. Schilpp, ed., *Albert Einstein: Philosopher-Scientist*, Tudor Publishing Co., New York, 1949, 287–311.

Reichenbach, H., 1949, *The Theory of Probability*, 2nd ed., University of California Press, Berkeley and Los Angeles.

Reichenbach, H., 1951, *The Rise of Scientific Philosophy*, University of California Press, Berkeley & Los Angeles.

Reichenbach, H., 1951a, 'The Verifiability Theory of Meaning', *Proceedings of the American Academy of Arts and Sciences* **80**, 46–60.

Reichenbach, H., 1952, 'Are Phenomenal Reports Absolutely Certain?', *The Philosophical Review* **61**, 147–159.

Reichenbach, H., 1953, 'Les fondements logiques de la mécanique des quanta', *Extraits des Annales de l'Institut Henri Poincaré* **13**, 109–158.

Reichenbach, H., 1954, *Nomological Statements and Admissible Operations*, North-Holland Publishing Co., Amsterdam.

Reichenbach, H., 1956, *The Direction of Time*, University of California Press, Berkeley & Los Angeles.

Reichenbach, H., 1958, *The Philosophy of Space and Time*, Dover Publications, New York. English Translation of Reichenbach (1928).

Reichenbach, H., 1959, *Modern Philosophy of Science*, Routledge & Kegan Paul, London.

Reichenbach, H., 1965, *The Theory of Relativity and A Priori Knowledge*, University of California Press, Berkeley & Los Angeles. English translation of Reichenbach (1920).

Reichenbach, H., 1969, *Axiomatization of the Theory of Relativity*, University of California Press, Berkeley & Los Angeles. Translation of Reichenbach (1924).

Reichenbach, H., 1976, *Laws, Modalities, and Counterfactuals*, University of California Press, Berkeley & Los Angeles. Reprint of Reichenbach (1954).

Rescher, N., 1966, *The Logic of Commands*, Dover Publications, New York.

Rescher, N., 1970, *Scientific Explanation*, The Free Press, New York.

Rescher, N. et al., 1971, *Temporal Logic*, Springer-Verlag, New York.

Russell, B., 1897, *An Essay on the Foundations of Geometry*, The University Press, Cambridge. Reprinted, 1956, Dover Publications, New York.

Russell, B., 1940, *An Inquiry into Meaning and Truth*, W. W. Norton & Co., New York.

Russell, B., 1948, *Human Knowledge, Its Scope and Limits*, Simon & Schuster, New York.

Salmon, W., forthcoming, 'Hans Reichenbach', in *Dictionary of American Biography*, Charles Scribner's Sons, New York.

Salmon, W., 1955, 'The Short Run', *Philosophy of Science* **22**, 214–221.

Salmon, W., 1957, 'The Predictive Inference', *Philosophy of Science* **24**, 180–190.

Salmon, W., 1957a, 'Should We Attempt to Justify Induction?', *Philosophical Studies* **8**, 33–48.

Salmon, W., 1966, 'Verifiability and Logic', in P. Feyerabend et al., eds., *Mind, Matter, and Method*, University of Minnesota Press, Minneapolis, 1966, 354–376.

Salmon, W., 1967, *The Foundations of Scientific Inference*, University of Pittsburgh Press, Pittsburgh.

Salmon, W., 1967a, 'Carnap's Inductive Logic', *Journal of Philosophy* **64**, 725–740.

Salmon, W., 1968, 'The Justification of Inductive Rules of Inference', in I. Lakatos, ed., *The Problem of Inductive Logic*, North-Holland Publishing Co., Amsterdam, 1968, 24–43.

Salmon, W., et al., 1971, *Statistical Explanation and Statistical Relevance*, University of Pittsburgh Press, Pittsburgh.

Salmon, W., 1974, 'Russell on Scientific Inference *or* Will the Real Deductivist Please Stand Up?', in G. Nakhnikian, ed., *Bertrand Russell's Philosophy*, Gerald Duckworth & Co., London, 1974, 183–208.

Salmon, W., 1974a, 'Memory and Perception in *Human Knowledge*', in G. Nakhnikian, ed., *Bertrand Russell's Philosophy*, Gerald Duckworth & Co., London, 1974, 139–167.

Salmon, W., 1975, 'Theoretical Explanation', in S. Körner, ed., *Explanation*, Basil Blackwell, Oxford, 1975, 118–145.

Salmon, W., 1976, 'A Third Dogma of Empiricism', in R. Butts et al., eds., *Logic, Methodology and Philosophy of Science*, D. Reidel Publishing Co., Dordrecht & Boston, 1976.

Savage, L., 1954, *Foundations of Statistics*, John Wiley & Sons, New York.

Schlick, M., 1963, *Space and Time in Contemporary Physics*, Dover Publications, New York.

Schuster, C., 1975, 'Hans Reichenbach', in C. Gillispie, ed., *Dictionary of Scientific Biography*, Vol. 11, Charles Scribner's Sons, New York, 1975, 355–359.

Stein, H., 1967, 'Newtonian Space-Time', *Texas Quarterly* **10**, 174–200.

Stevenson, C., 1944, *Ethics and Language*, Yale University Press, New Haven.

Strauss, M., 1972, *Modern Physics and Its Philosophy*, D. Reidel Publishing Co., Dordrecht and Boston.

Strawson, P., 1952, *Introduction to Logical Theory*, Methuen & Co., London.

Suppes, P., 1970, *A Probabilistic Theory of Causality*, North-Holland Publishing Co., Amsterdam.

van Fraassen, B., 1973, 'Semantic Analysis of Quantum Logic', in Hooker (1973), 80–113.

van Fraassen, B., 1974, 'The Labyrinth of Quantum Logics', in R. Cohen et al., eds., *Logical and Epistemological Studies in Contemporary Physics*, D. Reidel Publishing Co., Dordrecht & Boston, 1974, 224–254. Reprinted in Hooker (1975).

Venn, J., 1866, *The Logic of Chance*, Macmillan & Co., London and Cambridge.

von Mises, R., 1957, *Probability, Statistics and Truth*, The Macmillan Co., New York.

von Neumann, J., 1955, *Mathematical Foundations of Quantum Mechanics*, Princeton University Press, Princeton.

Watkins, J., 1968, 'Non-Inductive Corroboration', in I. Lakatos, ed., *The Problem of Inductive Logic*, North-Holland Publishing Co., Amsterdam, 1968, 61–66.

Wheeler, J., 1962, 'Curved Empty Space-Time as the Building Material of the Physical World', in E. Nagel, et al., eds., *Logic, Methodology and Philosophy of Science*, Stanford University Press, Stanford, 1962, 361–374.

Wheeler, J., 1971, Foreword to *The Conceptual Foundations of Contemporary Relativity Theory* by J. Graves, The MIT Press, Cambridge, Mass., 1971.

Wheeler, J., et al., 1973, *Gravitation*, W. H. Freeman & Co., San Francisco.

Whitehead, A. et al., 1910–13, *Principia Mathematica*, The University Press, Cambridge.

Whitehead, A., 1929, *Process and Reality*, The Macmillan Co., New York.

Winnie, J., 1970, 'Special Relativity without One-way Velocity Assumptions', *Philosophy of Science* **37**, 81–99, 223–238.

Winnie, J., 1972, 'The Twin-Rod Thought Experiment', *American Journal of Physics*, **40**, 1091–1094.

F. JOHN CLENDINNEN

INFERENCE, PRACTICE AND THEORY

ABSTRACT. Reichenbach held that all scientific inference reduces, via probability calculus, to induction, and he held that induction can be justified. He sees scientific knowledge in a practical context and insists that any rational assessment of actions requires a justification of induction. Gaps remain in his justifying argument; for we can not hope to prove that induction will succeed if success is possible. However, there are good prospects for completing a justification of essentially the kind he sought by showing that while induction may succeed, no alternative is a rational way of trying.

Reichenbach's claim that probability calculus, especially via Bayes' Theorem, can help to exhibit the structure of inference to theories is a valuable insight. However, his thesis that the 'weighting' of all hypotheses rests only on frequency data is too restrictive, especially given his scientific realism. Other empirical factors are relevant. Any satisfactory account of scientific inference must be deeply indebted to Reichenbach's foundation work.

1. THE PRACTICAL ROLE OF INDUCTION

To adopt empiricism is to acknowledge that our beliefs must be based on the data gained in experience. But this leaves open the question of how our beliefs about what has not been directly experienced are to be based on what has been.

A radical empiricism may seem to avoid this question by insisting that the only beliefs we may claim as justified are those derived directly from experience. However, taking this view seriously leads to solipsism of the present moment. This doctrine is unacceptable to some philosophers because it conflicts with what we normally mean by the word 'knowledge'. To others there is a more important issue: namely the grounding of all practical activity. Our actions are based on our expectations; we act in the belief that we can bring about or avoid certain occurrences. The issue is whether there are legitimate critical standards which may be employed in assessing the beliefs which guide our actions.

A radical empiricism which denies that any belief going beyond what is given in experience is justified, thereby denies any basis for rational discrimination among the different actions which are open to us in any situation.

W. C. Salmon (ed.), Hans Reichenbach: Logical Empiricist, 85–128. *All Rights Reserved.*

An action can only be intentional if it is done in the belief that certain consequences will follow it. The bases of all deliberate actions are conditional beliefs which go beyond experience, even if they may be based on it. If empiricism is to allow that thought and reflection have any place in human affairs, it must allow that there are rational criteria for assessing beliefs about the consequences of present happenings. It may well embrace a moderate scepticism which warns us that no such beliefs are beyond doubt, but it must reject a radical scepticism which tells us that any such belief is as good as any other since neither is legitimate.

In developing a philosophical position, one cannot, in conscience, simply adopt or reject certain positions in the light of common-sensical plausibility. One must go where the argument leads. The very notion of philosophical enquiry precludes our rejecting certain positions at the outset of our enquiry. Thus radical empiricism, and the scepticism it entails, can not be ruled out without a hearing. What is important, however, is to be aware of what is involved. If the seriousness of a problem is not recognized it may remain unsolved, even though a solution is possible. What is at issue here is the very idea that there is some point to thought and reflection. If nothing can be said as to why certain ways of forming expectations are superior to other ways, then nothing can be said in favour of thinking or urging others to think – there can be no basis for distinguishing between the fool who rushes in, and the angel who stops to think. To accept this would be to accept something of profound significance for the life-style common to us all. Those philosophers who say there is no question of justifying any kind of ampliative inference, and that all that can be hoped for is an analysis of how people do form beliefs, should be aware of the full import of what they say.

Hans Reichenbach never lost sight of the relationship between philosophy and practical activity. This marks him off from those philosophers of science who emphasize the theoretical and explanatory role of science, and forget the way it guides decisions in medicine, engineering and various other branches of technology. Karl Popper's thesis that inductive inference is not involved in science gains what plausibility it has by viewing science from this one-sided perspective. He argues with some force that the properties of a theory which make it valuable for scientific explanation can be measured by its 'degree of corroboration'. He defines this concept so that its measure at any time is determined by properties and relations which the hypothesis possesses at that time. Thus a claim that a hypothesis is well corroborated is in no way a claim about the future. And in choosing the best corroborated hypothesis the scientist chooses the hypothesis because of properties present to it at the time of choice. Thus no problem of induction is involved. So goes Popper's argument (see e.g. 1972, pp. 17–21). However, even Popper is forced to acknowledge that theories are sometimes of concern to the 'man of action', as well as to the 'pure theoretician'. This leads to the question: why choose the best corroborated theory as a guide for practical action? Popper simply brushes this question aside and says that it is obviously rational to "*prefer* as a basis for action the best-tested theory" (1972, pp. 21–22).

As Wesley Salmon has pointed out with care on a number of occasions (1966, pp. 155–161, 1968a, pp. 25–29, 1968b, pp. 95–97), this does not amount to avoiding inductivism. Popper is, in effect, putting forward a theory of practical decision about actions which are concerned with the future; and this is precisely the problem of induction as Reichenbach saw it. He, no more than Popper, thought that we can acquire *knowledge* of the future. He did see that there was a problem in saying why we ought to be guided in practice by theories which have succeeded in explaining known facts. Popper's preoccupation with the explanatory and theoretical aspect of science leads him to overlook the weight of this problem.

Reichenbach acknowledged a descriptive task in epistemology, but insisted that there was also a *critical* task (1938, p. 7). That is, epistemology is concerned with the justification of various thought processes and not merely with the fact that they are typically accepted in everyday or scientific thinking. "We can not share Hume's quietism. We shall not deny that induction is a habit; of course it is. But we wish to know whether it is a good habit or a bad one" (1951, p. 89). The use of the word 'quietism' is significant. It expresses Reichenbach's deep conviction that philosophy is not divorced from human concerns and problems. The understanding that science gives is essentially linked to its predictive power; and hence to the increased control it gives us over nature. And philosophy has the task of critically assessing the patterns of inference employed in science. For if scientific thinking is to be preferred to superstition, an account of why this is so is part of philosophy.

Reichenbach had little respect for sceptics who avoided the implications of their own scepticism (1938, pp. 345–348). If we have no rational basis for induction we have no basis for turning the steering wheel to the right, rather than the left, when we wish our car to turn right. But we all agree that such scepticism should be set aside in driving and "a philosopher who is to put aside his principles any time he steers a motor car is a bad philosopher" (ibid., p. 347). "Inductive inference cannot be dispensed with because we need it for the purpose of action" (ibid., p. 346). Consequently, Reichenbach saw the justification of induction as a central issue. Philosophers should seek diligently for a justification and should acknowledge the implications of holding that no such justification is possible.

There is, as Hume had pointed out, no hope for proof that inductively based predictions must be true. However, Reichenbach was not concerned with a quest for certainty. What needs to be justified are actions rather than beliefs (1949, p. 481). These remarks must be borne in mind if we are to avoid being misled by the way Reichenbach speaks in other places. He uses the term 'inductive inference' and speaks of its justification. From the passage referred to it is clear that what is to be justified are the actions based on the inductive method, and it seems clear that what he calls inference is the acceptance of a statement as a *guide for action*, not as *true*.

The rationality of an action does not depend on knowing that it will succeed. He pointed out that we often act rationally when we know that our action may fail to achieve its goal. "The physician who operates on a patient because he knows that the operation will be the only chance to save the patient will be justified, though he can not guarantee success" (1949, p. 481). It was a justification of this kind that Reichenbach sought. He was convinced that such a justification must be possible, for it seemed clear that actions based on scientific theories are better than those based on superstition. What he utterly rejected was the pseudo-scepticism which holds that there is no justification, but unconcernedly recommends that we keep on behaving as if there were – presumably on the grounds that to do otherwise would be rather bad form.

2. REICHENBACH'S PRACTICAL JUSTIFICATION

For Reichenbach the justification of induction is the key to justifying scientific method as a whole. He developed a method which he believed covered all the non-deductive, or ampliative, inferences which are involved in scientific reasoning. This method is based on a frequency interpretation of probability. On this interpretation probability statements are laws of nature, universal laws being a special case where the frequency is unity. Also the theorems of the calculus of probability become arithmetical truths. In this way certain probability statements may be derived, by essentially deductive inference, from other probability statements. Ampliative inference comes in at two points; firstly in induction, that is in adopting probability statements from a finite sets of observational data, and secondly, in applying probability statements, which are about infinite sets of events, to particular cases. Both these steps are called 'posits' by Reichenbach, for in both cases we treat as true a statement which we do not know to be so. It is not possible to prove that predictive posits made according to this method will always turn out true, but we may hope to show that such predictions are the best which can be made. Thus Reichenbach claims that all ampliative components of inferences to generalizations and theories reduce to enumerative induction and that a justification for employing induction as a guide to action is available. I want to examine both these theses, starting with the latter.

A practical justification, or vindication as Herbert Feigl calls it, must be relative to some end pursued. If induction is to be shown to be the best method, we must first decide what it is intended to achieve. Reichenbach's close argument and analysis is directed to establishing that induction stands in a certain relation to the aim of discovering probability laws. However he acknowledges quite explicitly that our ultimate aim in using induction is correctly predicting the future (1949, p. 480; 1938, p. 349). While Reichenbach does discuss the reasons for using probability laws in making specific predictions, I will argue that what he says does not constitute a satisfactory argument to the conclusion that the only reasonable way of predicting is via general laws. Hence I hold that he would have failed to justify induction relative to its ultimate aim, even if he had a satisfactory argument that induction is the best way of attempting to discover laws. However, Reichenbach's argument on this point was not satisfactory and the sustained efforts of Wesley Salmon have failed to close all the gaps which he left.

The core of Reichenbach's justification was the simple proof that the inductive method is convergent; that is that the repeated use of induction will eventually detect with any required degree of accuracy probability laws where they exist. Induction consists in using the frequency with which an attribute is observed in a sample to predict that the long run frequency in the whole set will be approximately the same value. Without examining in detail Reichenbach's explication of 'long run frequency' it is easy to see that this method must eventually give a good estimate of any frequency which does settle down to a stable value as larger and larger subsets of the whole set are examined. That induction is convergent is analytic, and according to Reichenbach this shows that if, for a certain kind of occurrence, there is a probability law, "the use of the rule of induction is a sufficient instrument to find it" (1949, p. 474). However this statement is a somewhat misleading summing up of what he had spelled out more carefully; he should have said that the repeated use of induction will eventually find the probability law. This is the analytic property of induction which is to provide its justification.

A difficulty was immediately recognized by Reichenbach; there are infinitely many alternatives to induction, all of which share the property of convergence. To justify induction we must give some reason for preferring induction to any of these alternatives. If we are concerned to determine the frequency with which A's are B and we have a sample of nA's of which r are B, the rule of induction directs us to predict that in the long run $\frac{r}{n}$ of entities which are A are B. If C_n is any function of n which converges to 0 as n increases, then the rule which instructs us to predict that the long run frequency is $\frac{r}{n} + C_n$ is also convergent. We can say of any such rule, as well as of induction, that its repeated use will eventually detect laws where they exist.

On what basis are we to choose induction; that is put C_n identically equal to zero? In 1938 Reichenbach mistakenly thought that induction could be shown to involve the smallest risk in estimating long run frequencies (1938; p. 355). However it is clear that on any finite set of data the inductive estimate may be completely wrong and in such a case there will exist some other convergent rule which will give the correct estimate. In the face of this it is not possible to give any account of 'smallest risk' which allows the above claim to stand. Reichenbach recognized his error and in 1948 rejected this claim and held that the grounds on which induction is to be chosen from the set of convergent rules is the criterion of descriptive simplicity (1949, pp. 447n).

Salmon has recognized that this is unsatisfactory. Reichenbach had said that "since we are considering a choice among methods all of which will lead to the aim, we let considerations of a technical nature determine our choice"; and these technical considerations were that "the rule of induction has the advantage of being easier to handle, owing to its descriptive simplicity" (1949, pp. 475–476). This seems to suggest that induction gives exactly the same result as any other convergent method; which is, of course, not the case. At this point Reichenbach forgot to say that the different methods only lead to the same result in the *long run*, and that may be very far in the future. At any given time the law arrived at by any non-inductive convergent rule will differ from that arrived at by induction; many will differ so much that they will lead to quite opposite expectations. Thus if simplicity is used to select induction it is not descriptive simplicity, which on Reichenbach's own account is merely a criterion for selecting between logically equivalent formulations of a theory (1938, p. 374). That is, the choice is not merely a matter of convenience and if simplicity is appealed to, the use of this criterion must be justified.

Salmon set out to formulate requirements which he believed could be legitimately imposed on a predicting method and which he hoped would rule out all the convergent methods other than induction. As he points out (1968a, p. 37), he has not been able to bring this to a satisfactory completion. The requirements which he formulates are not sufficient to select induction and he cannot see how to justify those additional requirements which would achieve this.

3. SALMON'S DEVELOPMENT OF THE VINDICATION

The question we must consider is this: if a justification of induction, or some other rule of ampliative inference, is to be sought, is Reichenbach's criterion of convergence the appropriate starting point? It may well be possible to provide a pragmatic justification which consists in establishing some analytic property of the method, that is, a justification of the general kind sought by Reichenbach, but which does not start with a proof of the convergence of the method. Salmon has not explored this possibility.

And, on the other hand, the unsupported assumption that the completion of Reichenbach's argument is the only possible way to vindicate induction has sometimes been employed to argue that no justification is possible (e.g. Katz, 1962). I will argue that the general conception of a practical justification is correct but that the emphasis on convergence has been a mistake.

Salmon points out (1966, p. 222) that it is always possible to formulate some convergent rule of inference which will give us a law asserting any frequency we care to name from data about the occurrence of an attribute in some reference class. (C_n can be specified to have values of any nominated value in the immediate or middle range future and yet to have the limit zero.) Thus by requiring only that we use a convergent rule we impose no limitation on the conclusion which may be inferred from data. This underlines that a selection of induction on the basis of convenience from among convergent rules, as proposed by Reichenbach, is no justification. It also shows how large a burden must be borne by subsidiary requirements, such as are proposed by Salmon. The justification of induction rests on the justification of these requirements.

The introduction of a multiplicity of desiderata into the justification of induction complicates the issue significantly as compared to the original notion of a vindicating argument. The situation is fairly straightforward if we are to establish some relationship between the method and the one end we seek in employing it. But if some other property is appealed to in choosing between methods, all of which stand in the same relation to the goal, then we must ask why this property is important. If we could show how the possession of this property contributed to the achievement of the goal, then there would be no new desideratum. The situation would be that the relationship established between method and goal would be different in one case to that in the alternatives. The effectiveness of such a justification would depend on the difference in relationships which would show that one method was the better way of pursuing the goal. If, however, a genuinely new desideratum is introduced to select among a number of possible methods, all standing in the same relationship to the goal aimed at in employing the method, then the achievement of this desideratum must be seen as another goal which we also seek. And the justification of the one method as against any of the alternatives is relative to this second goal as well as the original one.

Salmon imposes a normalizing condition and a criterion of linguistic invariance on putative inference rules. He says that both of these "may be regarded as consistency requirements" (1968a, p. 37). Now if the only desideratum expressed by these requirements were that the conclusions of inferences should be non-contradictory statements, there would be no problem. Consistency, in this sense, is no new aim, for a self-contradiction makes no assertion and is not a prediction. However the issue is not so straightforward. At least one of the criteria mentioned is not a consistency requirement in this direct sense and, in any case, Salmon imposes a criterion in addition to these which is clearly not a consistency requirement.

The normalizing condition is the closest to a straight consistency requirement. This condition requires that all the probability statements derivable by a method from a set of data could be true. It rules out methods which give, either directly or in conjunction with logic and the probability calculus, probabilities outside the range 0 to 1. If it is allowed that the methods under consideration provide inferences to statements, then this condition does come down to a straightforward consistency requirement. However this has been challenged by Hacking (1968), who holds that enumerative induction should be characterized as an estimating rule, and that the normalizing conditions do not apply to it as a matter of logic.

The criterion of linguistic invariance has less claim to being a consistency requirement. It requires that the prediction determined by a method should be independent of the language employed. Salmon supports this criterion with the claim that its rejection would result in contradictions. Suppose there is a method which is quite general with respect to the kinds of phenomena it can be applied to, which is quite explicit and consistent in the predictions it makes in any given language but such that its predictions do depend on certain features of the language employed. It is clear that such a method would lead to contradictions if predictions about the same phenomena were made in different languages and, via translations, all these predictions were expressed in one language and conjoined. On the other hand, if the method were employed by only one person who spoke but one language, no contradictions would result. This would also be so if it were used only by a unilingual group of people who were willing to trust each other but eschewed all alien contacts.

Even when different languages are employed in an intellectual community it is possible, as Brian Skyrms has pointed out (1965, p. 263), to modify the above method so as to ensure that no contradictions will result. This is achieved by the simple expedient of writing into the inference rule the requirement that the data be translated into one specified language in which the inference is to be carried out. Salmon, indeed, makes essentially the same point when he allows that it is possible "to maintain that there exists one particular privileged language of science" (1963, p. 31). He goes on to claim that the status of such a language has to be established *a priori* and so rejects this procedure as metaphysical. He makes a related point in his 'Reply to Black' appended to the same article (pp. 52–53). He notes that some contributors suggested that it was quite proper that the use of different languages should lead to different inductive conclusions on the grounds that different languages embody different conceptual schemes. It was suggested that conceptual schemes should play a role in our inferences. Salmon replies by insisting that a conceptual scheme is a very general theory and that while theories include components which may be considered linguistic they clearly have empirical content. Salmon allows that many inductions are influenced by theoretical considerations. The principle of linguistic invariance is only aimed at excluding those differences between languages which depend only on the language employed and do not reflect empirically-supported generalizations.

We may agree with Salmon in both his rejection of one privileged language of science and his distinction between purely linguistic and theoretical aspects of a system, yet insist that the criterion of linguistic invariance is stronger than the principle of non-contradiction which can be ensured without invoking the former. The more positive content of this criterion is that it excludes a certain kind of arbitrary consideration from inductive reasoning. It seems that there are excellent grounds for this exclusion; however, our concern at the moment is with Salmon's argument. What has emerged is that linguistic invariance is a desideratum which goes beyond the requirement of consistency and therefore must itself be justified before it can play a role in the vindication of induction.

Other things equal, it would certainly be nice if we had a predicting method which gave the same result when applied in any language. But this is a minor concern compared with our aim of making correct predictions. To suggest that the reason we prefer induction to some of the odd 'methods' which are convergent but not linguistically invariant has nothing to do with the aim of correct prediction, but depends on a less central aim that we have, is surely incorrect. We believe that such an arbitrary method is thereby an unreliable way of aiming for true predictions.

Thus in introducing an additional desideratum which is not itself justified with respect to a central aim of correct predicting, Salmon seems to be weakening the justification in an unacceptable way. This goes even further in the way he handles Goodman's paradox. The possibility of queer predicates such as 'grue' show that Reichenbach's explication of induction is unsatisfactory; it can give counter intuitive results and, what is worse, it can lead to different and incompatible predictions depending on which predicates are employed in applying the rule of induction. This latter characteristic showed that the classical specification of induction failed to live up to Salmon's criterion of linguistic invariance. The way that Salmon met this problem was to write in the requirement that enumerative induction should employ only ostensive predicates.

Salmon's argument is that enumerative induction is a primitive rule of inference and that it is therefore reasonable to limit its application to basic, that is, purely ostensive predicates (1963, p. 40). He has defined 'basic predicates' as those which are purely ostensive on the grounds of their special suitability in giving meaning to the empirical terms of science (1963, p. 38). It is clear that any empirical language must be ultimately related to non-linguistic entities by some technique which is not simply linguistic. Salmon's account of ostensive definitions seems to fill this bill and these predicates are consequently basic in a quite clear sense. Salmon is not, of course, committed to holding that the primitive terms of any empirical system are purely ostensive. He is only committed to the claim that an empirical formal system must be directly or ultimately interpreted by a language in which the undefined terms are basic in his sense. There is no need to challenge Reichenbach's distinction between primitive and advanced knowledge, but we must examine Salmon's assumption. It is clear that the appeal here is not to consistency. Inductions using Goodman-type predicates are to be ruled out because they contravene the principle that primitive inferences should employ only premises stated in basic predicates. The basis of this principle is not obvious and it is not clear how it relates to a practical justification. I would argue that this principle is not valid and that it is perfectly reasonable to contravene it in certain cases; for instance if we discover that certain dispositional predicates, which are not ostensive, apply to all observed entities of a certain kind. However, it is enough for my purposes to point to the problem which arises when such a requirement is introduced into what set out to be a practical justification of the kind described by Reichenbach.

4. HOW OFTEN WILL INDUCTION SUCCEED?

Let us now turn to what I believe is the central weakness of the whole Reichenbach–Salmon approach to the vindication of induction. This is the gap between relating induction to the discovery of laws and to its ultimate aim of correctly predicting. Let us examine how it might be closed.

Reichenbach is inclined to move from the aim of correct predicting to the discovery of laws as if the connection is obvious. For instance:

> ... we must begin with a determination of the aim of induction. It is usually said that we perform inductions with the aim of foreseeing the future; This determination is vague; let us replace it by a formulation more precise in character:
> *The aim of induction is to find series of events whose frequency of occurrence converges towards a limit.* (1938, pp. 349–350)

He goes on to briefly give his reasons for choosing this formulation: "We found that we need probabilities and that a probability is to be defined as the limit of a frequency." Elsewhere Reichenbach claims that a satisfactory definition of 'predict' "will turn out to entail the postulate of the existence of certain series having a limit of the frequency" (1938, p. 360). He presumably has in mind what he has said about the policy of making predictions according to probability. "The preference of the more probable event is justified on the frequency interpretation by the *argument in terms of behaviour most favourable on the whole*: if we decide to assume the happening of the most probable event, we shall have in the long run the greatest number of successes" (1938, pp. 309–310; cf. 1949, p. 373). His argument seems to be that since we can have no direct knowledge of the truth of a prediction at the time it is made, it is rational to employ a method which we know analytically will pay off in the long run.

Can we know that probability-based predictions will succeed more often than not? A full examination of this issue would call for an examination of Reichenbach's treatment of long run frequency, which is complicated. From the point of view of his mathematical probability theory, he assumes that probabilities are frequencies in infinite reference classes. This means that they cannot be treated as simple ratios. He assumes that the elements of the reference class are ordered. This order will determine a sequence of subsets of this class as more and more elements, in order, are added. The probability of an attribute in the infinite reference class is then defined as the limit of the frequencies in this sequence of subsets. Brian Skyrms shows that on this definition the frequency of an attribute in an infinite reference class may depend on the ordering of elements (1965, pp. 255–256). He further points out that the frequency with which predictions based on a probability law are correct will tend to a limit equal to the probability only given the quite implausible assumption that predictions are made in precisely the order in which the elements of the reference class are arranged (p. 258).

It might be possible to defend Reichenbach from this criticism by pointing out that the probability which occurs in his mathematical theory is an idealization. The sets of objects which are and will be observed by humans are finite – although very large. The probability of physical events (as distinct from the mathematical concept) will be what he calls a 'practical limit' (1949, p. 347). Although he still speaks of sequences, he makes it clear that the sets involved are finite, so presumably practical limits could be identified as ratios of large but finite size. There would be some difficulties in this line of defence, since Reichenbach says very little about practical limits and gives the impression that we can think of probabilities in mathematical terms in justifying induction. However, even if we treat probabilities as ratios in finite sets an argument very similar to Skyrms' will hold. We could only claim that predictions based on probability will be correct more often than not if predictions are made about a substantial percentage of the elements of the reference class, and this assumption is blatantly false in almost every case where we use a law to predict. In practice it is normal to use a law to make only a very small fraction of the predictions that come under it in principle (for the most industrious predicter must be limited in the number of initial conditions he can observe). Hence it is possible that the ratio of success to failure may differ radically from the probability which guides the predictions. Suppose we have a finite population of A's and of these 90% are *B*'s. Suppose we make predictions about 10% of this population. In each case we will make a highly weighted prediction that the *A* in question is a *B*. However, it is possible to be wrong in every one of these predictions. (Some may be tempted to respond by saying that this, while possible, is a very unlikely eventuality. However this can only stand if a justification for assigning degrees of expectation has been given. Reichenbach is here proposing a justification for assigning 'weight' to predictions based on probabilities, so he cannot appeal to the very principle he is seeking to justify.)

So far we have spoken about predictions made in accordance with actual probabilities. Even on this assumption we see that little can be proved about the sequence of predictions. However, this assumption does not hold in actual cases. In making predictions about particular cases we can only employ probability values which we believe are true; we can never be sure in advance what the probability laws are. Given that we want to arrive at laws of a certain kind there may be an excellent justification for a certain inference procedure. Given that laws of the kind under discussion are known with certainty there may be an excellent justification for using them in a certain way in predicting. These two justifications, even if established beyond doubt, would not constitute a justification for the composite method of making a singular prediction. This is a two stage process, but the total inference is from a set of data to hand to an event which is logically independent of this data. Such an inference is fallible; and it must be possible that every such inference may turn out false.

Suppose we have a finite population of *A*'s with *p* members and initially have access to a sample of *n* members. On our knowledge of the frequency of *B*'s in the sample we use the straight inductive rule to predict the frequency in the population at large and use this to predict whether the next *A* examined will be *B*. Each prediction is checked before the next is made and the result of the check is incorporated into the data. This procedure is convergent with respect to estimating the frequency of the population. Clearly there is a frequency to determine and induction will eventually do it. However, this convergence of the method of estimating, or predicting, frequency has no bearing on the success ratio of predictions about individuals. As we have seen from general considerations, this ratio could be zero. For this to happen a rather strange pattern would have to occur in the sequence of individuals about which predictions were made. For the purpose of this exercise let us tighten up Reichenbach's rule for single case predicting from estimated probabilities. The rule is indeterminate where the probability of A's being B is estimated as 0.5. We will arbitrarily extend the rule thus: in this case predict that the next A is B. Now suppose in the initial sample of n individuals there were more B's than non-B's. We would start off predicting that A's were B. For every prediction to be false all the A's which were predicted about would have to not-B until of the total examined A's, the number of B's was one less than the number of non-B's. The next A would have to be B and from then on the A's would have to alternate between being B and not-B.

It might be suggested that such a sequence of error could not persist through any significantly large population, for an intelligent person would detect the pattern of B's and not-B's and would use this pattern to predict. Now the use of induction to detect such oscillations in a sequence is certainly allowed by Reichenbach (1938, p. 358). However, the person who changes his predicting method in this way cannot hope to thereby guarantee success. The oscillating pattern may cease at the very moment that he starts to count on it, and then recommence when he reverts to direct induction. At this stage he would doubtless conclude that in some way the predictions he made were causally operative on their outcome. He may try making a prediction and then quickly changing it while, as it were, no one was looking. However, a few more failures and he would doubtless recognize the omniscience of the Cartesian Demon. Persistent failure remains a logical possibility in a method that uses past data to make predictions of the future.

We can see the hollowness of establishing deductively that a method of predicting frequencies is convergent. Repeatedly modified estimates of the frequency in a finite set must eventually succeed just because eventually we work our way through almost all, and then all, of the set. The point is that as the examined sample becomes a larger and larger proportion of the population, the genuinely predictive content of a predicted frequency becomes less and less. (The situation with an infinite sequence is not so straightforward, for we can never know that we have reached a point where convergence has occurred if it is going to. So the best that can be established analytically is that at some time or other we will get a correct prediction of the limiting frequency if there is one and if we continue to examine elements of the sequence in their order.) If, however we are concerned only with predictions which are exclusively about as-yet-unexamined entities, it is clear that we could have no deductive proof about the possible ratio of success to failure, even when dealing with a finite population.

If we specify as the goal of a predicting method that it should make more correct predictions than false about individuals, then it is impossible to show for any method that it will attain this goal. Since the goal is clearly attainable, it is impossible to establish between any method and this goal the relationship which Reichenbach cites as the justifying relationship between induction and its aim; namely that induction must succeed if success is possible.

5. DOES PREDICTABILITY ENTAIL THE SUCCESS OF INDUCTION?

Reichenbach does argue at one point that "the applicability of the inductive procedure is a necessary condition of predictability", where predictability is taken to mean no more than that some method of prediction, however strange it may be, succeeds more often than not (1938, pp. 357–9). If this could stand, he would have established a relationship between the use of induction and the ultimate aim of correct predicting, and not merely with the secondary aim of discovering laws where they exist. His point is that if any method of predicting is, in the timeless sense, successful, then there is thereby a regularity which could be discovered by induction.

Reichenbach considers the possibility of a world in which "there is no series having a limit" and yet there is a "clairvoyant who knows every event of a series individually". He concludes that this is not possible since on the given descriptions there will be a series of predictions by the clairvoyant in which the frequency of correct predictions approaches a limit significantly higher than 0.5 (p. 358).

Skyrms holds that this account is a simplification and that strictly what would be established by induction would be a generalization about a certain predicting rule; that is a proposition in the next language level. This he holds is different to showing that induction works at the same level as the successful predicting rule (1975, pp. 44–46). However I think that this criticism can be avoided. It can be argued that the job done by a non-logical predicting rule can always be done by the combination of a generalization in the object language together with the ordinary logical rules of inference (provided metaphysical constraints are not placed on what may be said in the object language). Thus for any predicting rule which is known to be successful there will be an object language generalization with inductive support to a corresponding degree. It seems clear that any way of predicting which could be successful could be stated as a non-logical rule in the necessary sense; for it must be possible to specify what is an instance of that kind of prediction. In Reichenbach's example the object language generalization 'Whenever the clairvoyant says that *x*, then *x*' will gain inductive support to precisely the degree that the predicting rule 'Trust the clairvoyant' succeeds. Thus I believe that Skyrms' point is no real objection to the argument; there may, however, be other objections which do stand.

There is one modification to Reichenbach's argument which must be made immediately. It is not possible for him to hold that the inductivist will have grounds for precisely the set of predictions which the clairvoyant makes. There will be no inductive grounds for believing the first few predictions which he makes. All that can be argued is that there will eventually be inductive grounds for believing the predictions of a clairvoyant who continues to succeed when science (in the ordinary sense) continues to fail. However we must consider whether this justifies the conclusion that the applicability of induction is a necessary condition of predictability. The success of a predicting method is not an all-or-nothing matter. Two methods may both be successful more often than not, but one may have a better success ratio than the other. Now we can easily show that the inductively based method is not necessarily the method with the highest success ratio. Consider a modification of Reichenbach's case. In a field in which science has made no progress clairvoyant *C* starts to predict and from t_1 to t_2 is always successful. After t_2 he continues to predict but is usually wrong. However at t_2 clairvoyant *D* comes on the scene and starts to predict and always succeeds. In retrospect it is easy to pick a method which would have been more successful than any application of induction: namely, from t_1 to t_2 trust *C* and thereafter trust *D*. (Indeed, someone who espoused this method might claim inductive support for it after *C* had some measure of success in just the way that someone may claim inductive support for the hypothesis that emeralds are grue. But if Goodman's 'New Riddle of Induction' has a solution, and I argue elsewhere that it has, then both these will be excluded as inductive inferences.)

If we consider a clairvoyant competing with science, it is clear that the clairvoyant can have a marginally higher success rate while using a method which does not reduce to induction. For instance, he normally accepts scientific predictions, but on two or three occasions makes different and conflicting predictions which turn out correct. If we consider a world with a very large number of clairvoyants whose predictions are mostly wrong, it is clear that there will be a method which succeeds more often than induction. This method will consist of normally trusting scientific predictions but on certain particular occasions trusting certain specified clairvoyants. In retrospect it would be easy enough to formulate such a method which would have succeeded. But prior to the events induction could provide no guide. For any specification of such a method which covered events up to a given present, there would be a vast number of possible projections into the future and hence a vast number of total methods all of which would share the same success in the initial part; and so this could not be used to distinguish between them. However one or other of these methods is destined to do better than science (unless science is always correct whenever a clairvoyant makes a counter-scientific prediction). We can certainly not establish that induction succeeding more often than any competing method is a necessary condition of predictability.

We make inductively based predictions on various particular occasions. On any such occasion there are alternatives open and it should be possible, in each case, to provide a justification for employing that method. A general vindication of induction should be general in the sense that it may be applied to any particular situation to justify the use of induction on that occasion. Suppose someone believes that a certain method will certainly succeed more often than any possible alternative. Does it follow that on a given occasion this person should employ this method? Granted that it is not known that the method will always succeed, that is, that the use of the method is always a sufficient condition for correct prediction, it is always possible that the prediction derived from the method is destined to fail. Given this possibility why should reason exclude his making the prediction in some other way on this occasion? Why not guess *this* time? A justification of induction requires an answer to this question.

Reichenbach's statement at the end of *Theory of Probability* (1949, p. 480) – namely, "we can devise a method that will lead to correct predictions if correct predictions can be made" – must be rejected as simply false. The requirement of convergence, even when supplemented by the criteria of normalizing conditions, linguistic invariance and the use of only purely ostensive predicates, fails to select induction as the rational method of inferring to laws. And now we see that even if this justification were completed another task remains before induction is vindicated in terms of its basic aim.

5.1. *Is Rationality Enough*?

We have seen that considerable difficulties stand in the way of developing a satisfactory justification which starts with the requirement of convergence. Before considering whether some other approach to the vindication of induction is possible, I want to deal with some difficulties which stand in the way.

Salmon, in his rejoinder in a symposium, deals with Kyburg's claim that an appeal to 'inductive intuition' must be employed in justifying induction (1965). He admits reluctantly that in the last analysis this may be necessary, but urges that this step be taken only as a last resort. The use of intuition in this way would be a serious inroad into the empiricism which is basic to Salmon's position. However it is what he says about the kind of intuition which could be acceptable that is of particular interest. I believe that it expresses some misconceptions which could limit the prospects for a vindication, whether or not intuition is to play a part therein.

Kyburg held that the role of intuition was to distinguish rational from irrational expectations. Salmon disagrees:

> Furthermore, if inductive intuition is required it will, I believe, have to be an intuition concerning a relation between induction and frequency of truth-preservation, not between induction and rationality as Kyburg would have it. I know what it means to intuit that a particular inductive method is frequently truth-preserving. Though such an intuition might be utterly unreliable, I could recognize it if I had it. If, however, the rationality of an inductive method had nothing whatever to do with frequency of truth-preservation, I really do not know what it would be like to intuit its rationality. (1965, p. 280)

Salmon seems to be insisting that some more direct relationship than 'the most reasonable way of trying' must be established between induction and its goal if it is to be justified. If so it is hard to see how he can avoid requiring the impossible – that the method be justified by showing that it will succeed. From this point of view intuition certainly seems to be the only solution. And Salmon's brand of intuition is more objectionable than Kyburg's. An intuition about what is rational is normative and thus not directly at the mercy of future experience in the way that Salmon's intuitions about matters of fact are.

Salmon is claiming that the one and only desideratum in adopting a method of prediction is the frequency of truth-preservation that it will achieve. He concludes that rationality has no place as an alternative, or even subsidiary, desideratum. This rests on a false dichotomy. As Reichenbach has pointed out, we often justify a means with respect to an end when we do not know that it will succeed, but can show that it is a sensible way of trying. Again, a means may not be the only way of seeking a certain end, but it may be the most sensible way. Thus we often justify means by showing them to be the best way of trying to seek a certain end. That is, we justify a means by showing it to be the most reasonable way of seeking the end. Thus it is clear that to allow rationality as a consideration in a vindication is not to erect a new and distinct end. Making true predictions can be accepted as our sole aim in induction but since we know that it cannot be shown that any method will maximize this goal, our concern must be to establish one method as the most reasonable way of attempting to achieve it. If this can be done without any appeal to intuition, so much the better. If intuition is to be employed it must be shown to be legitimate.

5.2. *Black's Objections*

In two papers Max Black has criticized the idea of a practical justification of induction. Some of his arguments deal directly with actual arguments put forward by authors such as Reichenbach and Feigl, but others are directed against the very possibility of any vindication.

Black points out that there can be no way of proving anything about the proportion of predictions made by a method which will be successful (1962, p. 200). Thus he concludes that no justifying relation can be established between induction and the primary aim of making correct predictions; and that the way that practical justifications have proceeded, and indeed must, is by dealing with a modified goal which can not be justified relative to the basic goal. Black's main argument against the possibility of a practical justification (1954) goes as follows. In the typical practical argument which justifies an action or policy with respect to some end sought, we employ as premises generalizations which tell us the probable outcomes of various actions. But such generalizations are established by induction, and by the standards of those who seek a practical justification, are not available in the argument which provides a justification for using this method of inference (1954, pp. 184–187). The justifying argument must be deductive and the premises of the argument must be analytic; for observation statements could be of no use here and any empirical generalization would presuppose inductive inference (cf. Reichenbach 1954, p. 479). (An appeal to synthetic *a priori* truths would, of course, cut across the whole empiricist point of a pragmatic justification.) Black remarks that this removes the very conditions "we found to be *essential* to the possibility of rational choice" (p. 184).

Black's point is that unless experience can provide a guide as to the outcome of different actions, no rational choice is possible. For how can we have any idea about what action is appropriate in seeking a certain goal if we have no idea at all about what will be the consequence of any action? And how can we have any idea about consequences if we must argue deductively from analytic premises? This argument seems very plausible. And he substantiates his claim that justifying arguments which have been offered depend on replacing the basic aim of induction by another aim which is not justified by appeal to the basic aim.

However, the question of the possibility of any vindication is not yet settled. Firstly Black has assumed that a particular relationship must be established to justify a means with respect to an end; namely that the means is sufficient to achieve the end if it is achievable. It is certainly difficult to see how this relationship could be established between a method and the goal of successful prediction without appeal to empirical premises. However, it may be that some other justifying relationship will suffice. Secondly, Black admits that "deductive arguments using necessary statements about problematic or recondite notions may have great value in philosophical analysis" (pp. 185–6). However, this depends on the arguments being non-trivial. The fault with the justifications he criticizes is held to be their triviality – that they reduce to establishing that "these inductive policies are the only ones that could achieve the aims distinguishing these inductive policies from all other policies" (ibid.) Thus he is rightly not ruling out the possibility that deductive elaboration of what is implicit in concepts such as 'induction' and 'rational' may bring to light non-obvious relationships which are significant for practical activities. And it remains possible that there is some justifying relationship between induction as a means and the end of correct predicting which can be established by such deductive elaboration.

5.3. *The Logic of Practical Arguments*

There is a development from Black's line of argument which may seem to settle the matter. An argument justifying induction must lead to the conclusion that induction is the rational way of predicting. This assertion is surely synthetic for it carries normative implications (cf. Urmson, 1956). In saying that it is rational to do something we recommend an action and a recommendation can not be an analytic proposition. Now given that any argument justifying induction must be deductive and have analytic premises, it follows that there can only be such an argument if its conclusion is analytic. We now see that this is not possible; so it seems that no justification is possible.

To assess this conclusion we must distinguish two steps in any practical argument which justifies some action as a means with respect to a given end. In the first place it will be necessary to establish some relationship between the proposed means and the end. Secondly it is necessary to conclude from the existence of this relationship that it is reasonable to employ this means in seeking this end. We must acknowledge that in vindicating induction the first stage of this argument is deductive, however we must proceed cautiously before pronouncing on the argument as a whole. The grounds for holding that the whole argument must be deductive is that it cannot include any inductive step. But does the dichotomy between deduction and induction embrace practical arguments?

The non-inductive nature of the second stage of the argument is in no way unique to the vindication of induction but is true of any practical argument. Any argument which moves from an assertion of a relationship to a recommendation is clearly neither a simple nor a concatenated induction. However there seem to be good reasons for also holding that such an argument is not deductive. If the conclusion is normative, as it seems to be, it cannot be implicit in a premise which is descriptive or analytic. If this is correct we must conclude that practical arguments have a quite distinctive logical status.

We must carefully consider whether the arguments in question are indeed non-deductive, for this is not beyond debate. In the first place we must remember that the conclusion in the arguments we are concerned with is not a simple recommendation but a conditional one. (We are finally concerned with establishing that induction is the rational procedure when attempting to predict. Indeed, to simply say 'induction is rational' fails to make a univocal statement, for there is no suggestion that a person's entire time should be taken up with induction, yet something more is recommended than that we should all try induction once.) Secondly, when we choose certain examples it seems that nothing is added in the recommendation of the conclusion to what is asserted in the premise. Consider this example:

(I) *M* is necessary and sufficient for *E*
So it is rational to employ *M* as a
means in seeking *E*.

The premise implies that *M must* be brought about in order to achieve *E* and the conclusion says something less than this. Hence, it may be argued, the inference is deductive.

There is a rebuttal to the above argument. Firstly, the form of argument is misrepresented in (I); secondly, even when correctly stated, (I) is far from typical of practical arguments and finally the gap between a non-normative premise and normative conclusion cannot be ignored. On the first point, consider an agent *X* who does not know and could not be expected to know that *M* is necessary and sufficient for *E*. In such cases the conclusion of (I) would be false even although the premises were true. (E.g. *X* must make an appointment on the other side of town. He knows that the underground is normally faster than a taxi but he does not know that a train has been derailed. In fact the only way he could make the appointment would be by taking a taxi. However in waiting at the station he acted rationally.) This consideration shows that the rationality of an action is determined by the reasonable beliefs of the agent rather than by the objective facts. This in turn shows that the conclusion of (I), in referring to what is rational rather than what is rational for some agent, makes an abstraction from the epistemic context which may well be misleading. More directly, this shows that the argument form considered is misleading in its first premise.

Let us replace (I) by:

(II) X reasonably believes that M is necessary and sufficient for E.
So it is rational for X to employ M as a means in seeking E.

In this argument form it is easier to see that a distinct recommendation which is not present in the premises is introduced in the conclusion. The move to this recommendation rests on the conviction that we ought to be guided in our actions by our beliefs. Perhaps this principle, more carefully stated, should be seen as the basic principle of rationality.

On the second point, there are beliefs about relationships other than those employed in (II) which establish the rationality of a certain action; for example consider the following argument:

(III) X reasonably believes that M is necessary to achieve E and knows of no other action which makes the occurrence of E more probable.
So it is reasonable for X to employ M as a means in seeking E.

Such arguments do not seem to be deductive in the way that (I) does and the recommendation to base action on belief and its limitations is more obvious than in (II).

We are also reminded that in asserting the rationality of a means to an end we are not asserting any single relationship. Consequently it is not possible to hold that saying a certain action is the rational means to a certain end is no more than to assert that a particular relation (e.g. necessity and sufficiency) holds between action and end. It may be held that what is asserted is a disjunction of relations; however this is less than plausible. We are all confident we know what we mean in an assertion that something is rational, but it is not a simple matter to list with confidence the relationships covered by this claim. Rather, we can take various relationships and consider for each whether it establishes a means as rational for a certain end. It seems that in each case we can recognize how knowledge of the relationship provides a basis for recommending the relationship to someone who would pursue the end.

I conclude that in a practical argument we move from non-normative premises to a conclusion which is normative, albeit conditional-normative. The argument is neither deductive nor inductive; which is not surprising if we see a practical argument as a move from a statement to a recommendation. What I have said merely touches on the logic of practical arguments – I believe a good deal of work remains to be done here. However, it is clear that in general we see no difficulty in moving from belief in a relationship between *M* and *E* to an acceptance of *M* as the proper way of seeking *E*. And this move presents no special problem for the vindication of induction.

The final word on the vindication of induction will not be said until a thoroughgoing analysis and vindication of practical arguments is available. However we will have proceeded a long way if we can provide a practical argument of a form which is generally accepted without question as valid. This was the task as Reichenbach saw it.

6. THE PROSPECTS FOR AN ALTERNATIVE VINDICATION

What are the prospects for proving by deductive argument from analytic premises some relationship between induction and the goal of making true predictions, the knowledge of which in the case of a normal practical argument would be taken to establish the means as the rational way of pursuing the end? This is the question we must now consider.

It is clear that we cannot hope to prove by a non-inductive argument that the use of induction is either necessary or sufficient to make a correct prediction. This rules out not only these two relationships but also any other more complex relationship which includes one or other of these. (For example, sufficient when no other method is known to be.) Does this make the task impossible? I will argue that it does not.

Let us consider an example which Reichenbach employed to show that sufficiency for success is not required to justify an action. He cites the case of a doctor who operates knowing that the operation may fail but that it is the only chance of saving the patient (1949, p. 481). He suggests that in this case the operation is necessary for success. As he describes the case, the surgeon does believe this; however in the typical case this would be a misleading account of the situation. Rarely does a doctor have positive grounds for believing that one line of treatment is necessary if the patient is to survive. Rather he knows that the operation may succeed, in some cases it initiates processes which lead to recovery; he also knows that neither he nor other medicos know of any alternative treatment of which this is true. Reichenbach has looked for an objective relationship between means and end the knowledge of which constitutes the justification for an agent's action. He has not seen clearly that it is an agent's knowledge *and the limits of his knowledge* which establish the rationality of an action.

Reichenbach's own example demonstrates this point: a means may be neither necessary nor sufficient for its end. If we know that M may bring about E and we have no reason to believe that any other action will do so, then it is reasonable to employ M when we wish to achieve E. Let us consider whether there is any hope of establishing such a relationship between induction and the making of true predictions. Firstly it is clear that we cannot hope to establish that induction may succeed in the way that the surgeon establishes that the operation may succeed. The surgeon appeals to beliefs about probable consequences of operating and other forms of treatment. These beliefs are based on induction. Thus, if circularity is to be avoided, we cannot hope for a justification which exactly parallels the surgeon's case. However, we certainly know that induction may succeed, and if we can establish that there is something positively unreasonable about any possible alternative to induction, then we will have established a justification for induction which parallels in certain broad respects the justification which the surgeon has for operating. In both cases we show that the proposed means may succeed and that it is unreasonable to employ any other means. And we recognize this as a legitimate and familiar relationship in justifying a means with respect to an end.

There may be a temptation to insist that a proper analysis of the surgeon's practical argument must be in terms of probability; and hence reject the parallel I propose between the two cases. The concept of probability has often been employed to explicate vague and imprecise arguments, and if it can be so employed here, it would seem that it should be. While this is a reasonable proposal, we must insist that any explication does justice to the original argument. In many cases a surgeon will have data which facilitate an estimate of the probable outcome of an operation. However, sometimes a particular kind of operation is done for the first time. In some such cases the only helpful knowledge is that if the cause of the condition is such and such then the operation will probably cure it. In such cases it would be misleading to speak of the surgeon's knowledge of the probabilities. We can say that he knows that the operation may work, but it would be wrong to say that his knowledge amounts to anything more precise than this. And there are many other cases in practical situations where this is so. Consequently I conclude that in some case where the knowledge which justifies an action is expressed by 'it is known that *M* may succeed while no alternative is a reasonable means to the end sought', the case may be more precisely described in terms of knowledge of probabilities; but in other cases this is not so. There are cases where there is no more precise knowledge.

In the case of induction, we employ the regularities present in known facts as a guide in our expectations of the as-yet-unknown. It is clear that this method may succeed on any particular occasion and it may succeed in the great majority of cases; for it is clearly possible that known regularities may persist. Consequently, if we can establish by conceptual analysis that there is something rationally defective in employing any possible alternative to induction we will have provided a justification for induction of a quite familiar kind. We will have established that this method may achieve its goal and that there is no reasonable alternative way of trying to achieve this. What would be known here about the relationship between method and end would be precisely the kind of knowledge which often serves as a justification for practical activity. What marks this justification off from the typical case is that our knowledge is derived from conceptual analysis.

Such a justification would be very much in the spirit of what Reichenbach aimed for but would differ in detail. It seems that the most promising approach to the problem of induction is to seek a justification of this kind.

7. NON-INDUCTION AS ARBITRARY–AN OUTLINE

It has been my main concern to argue that, in spite of the problems that stand in the way of bringing to completion the justification which starts where Reichenbach started, the basis of his approach was correct.

Induction needs a justification for if there is no justification here, there can be no distinction between reasonable and unreasonable behaviour. This has been argued forcefully and eloquently by Salmon. And I have argued that the possibility of a practical justification of induction remains open and have suggested what its form might be. To make the point more forcefully, let me indicate in outline the way I believe the vindication may be completed.

I believe that it can be argued that there is only one way in which we can be guided by known facts in forming our expectations of the future; and this is induction. Other methods may indeed employ facts, but they will involve arbitrary decisions about how the facts are used. (I have argued this elsewhere. See Clendinnen 1966. However, the reason I gave in that article for holding that the non-arbitrary method is superior I now consider unsatisfactory.) Now a conclusion reached by procedures involving an arbitrary step is on the same footing with any alternative conclusion which would have been reached if the arbitrary step had been taken in some different way. We can, of course, simply choose to take the step one way rather than another; but once we consider what we have done and acknowledge that we have not been guided in making that step, we must recognize that our conclusion is no better than any other conclusion. We may now argue that it is unreasonable to place any confidence in a statement if there is an extremely large set of alternative statements, all of which are on the same footing; that is where there is no reason for preferring any one statement to any other.

Thus, I would argue, we may reject any arbitrarily based prediction as unreasonable, since it is no better than many other competing arbitrarily based predictions. However, the inductively based prediction is not arbitrary; in accepting it we are guided by the facts. Of course we may go wrong in using induction, for the regularities we know may cease, or the actual regularities may be different to the way they seem on our limited knowledge. But we cannot allow for such possibilities; there are an unlimited number of ways that a pattern may change and in recognizing the mere possibility of such a change we have no basis for selecting among this set of possibilities.

In the second last paragraph of *Theory of Probability* Reichenbach sums up his justification of induction in the following terms.

> We do not know whether tomorrow the order of the world will not come to any end; tomorrow all known physical laws may be invalidated, the sun may no longer shine, and food no longer nourish us – or at least our own world may come to an end, because we may close our eyes forever. Tomorrow is unknown to us, but this fact need not make any difference in considerations determining our actions. We adjust our actions to the case of a predictable world – if the world is not predictable, very well, then we have acted in vain. (1949, p. 482)

I believe that this passage can stand as a justification of induction. At this level of generality Reichenbach is correct. But it must be stressed that what is entertained when we say that the world may not be predictable is that the world may be such that no *reasonable* way of predicting will succeed. Yet we must allow that however the world may be, it is always logically possible that any guess may turn out to be true. There is no way that the world could be such that statements about the future are all, or mostly, necessarily false. Thus the term 'unpredictable' is open to misinterpretation. It is not possible to say that if induction fails every non-inductive prediction must fail and we cannot establish, as Reichenbach claims we can on the page before the passage cited above, that induction is a "method which will lead to successful trials if success is attainable at all". Where he went wrong was to assume without justification that we must predict via laws. What must be shown is that predictions not based on asserted regularities, or those based on regularities which are not derived inductively, are unreasonable; for we cannot show that they must fail.

My proposal as to how induction may be vindicated is given only in barest outline above. The way in which any non-inductively derived inference is arbitrary compared to an inductively based prediction and the irrationality of predictions which are arbitrary in this sense both need to be developed with attention to possible objections. However, there is another issue of a rather different kind which bears on the sequel. If induction is to be justified we must be clear about what it is. Various people, especially Goodman, have brought out some of the problems in specifying precisely what this method is. We can certainly not specify induction as: 'adopt any generalization with which all past experience is consistent'. And it does not help to require that the generalization has, or could have, facilitated true predictions, for this would still allow an inductive argument to the conclusion that all emeralds are 'grue'. I believe that Goodman's paradox can be overcome by the requirement of inductive simplicity. A problem remains, however, as to how we are to determine the breadth of the generalization we aim to formulate. The answer is that since there can be no formula which is generally applicable, we must view the system of laws as a whole. Thus the general principle of induction directs us to seek the simplest overall systematization of knowledge. The breadth of the laws and theories we formulate will be determined by seeking to maximize this goal.

I would further argue that induction explicated in this form can be vindicated, for it is only by adopting the simplest generalization compatible with known facts that we are guided by the facts themselves. To allow the adoption of a generalization which is more complex than is necessary to cover known facts is to allow components in the law which are purely arbitrary; and there are infinitely many ways in which a law may be modified so as to make it more complex (Clendinnen, 1966, pp. 225–226, also Schlesinger 1974, ch. 11, esp. p. 35).

8. INDUCTION AND THEORY

According to Reichenbach, all inference to scientific theory is essentially induction. This will typically be disguised by the complexity of the data used in an inference. He emphasizes that simple enumerative inductions are rare in scientific thinking. This would only occur in the primitive stages of a science when no theory had been accumulated. In the more usual case, accepted theory will serve as a guide to the conclusions which are drawn from experimental or observational data. However, he holds that in such cases the inference, when analyzed, can be seen to be a 'concatenation of inductions' – that is the derivation, via the probability calculus, of new probability statements from ones already accepted (1949, § 84, esp. p. 432). All the basic probability statements will be derived by enumerative induction and subsequent inference will be via the probability calculus, the axioms of which are arithmetical truths on Reichenbach's frequency interpretation.

The claim that theory normally guides the interpretation of, and inference from, data in science is certainly borne out by actual practice. It is not so clear that this happens in the way that can be reasonably reconstructed as conforming to Reichenbach's account. There are two major reasons for doubt. Firstly, the very certainty of inferences according to a frequency interpreted probability calculus indicates the limitations of such inferences. If the probability calculus certifies an inference from two probability statements, which have been established by enumerative induction, to a third probability statement, then it is hard to imagine circumstances in which the data for the enumerative inductions would not include data which could have been used for a direct enumerative induction to the third probability statement. For example, suppose we calculate the probability that something is both A and B given that it is C by the multiplication theorem; namely:

$$P(C, A \cap B) = P(C, A) \times P(C \cap A, B) .$$

(I use Reichenbach's method of expressing probabilities in which the reference class is named first and attribute class is named after the comma.) If we establish the first probability on the right hand side by enumerative induction, we must have a sub-set of C (which we have no reason to believe to be atypical) and determine the proportion which are also in the set A, and hence in the set $C \cap A$. To establish the second probability we need a sample sub-set of $C \cap A$, in which we determine the proportion which are members of $C \cap A \cap B$. Now the second determination may be made by using the sub-set of $C \cap A$ which was used in determining the first probability; for the latter sub-set is randomly chosen. (A random selection of set C was made and from this a selection was made of elements included in A.) In determining the first probability we are obliged to select a sample of the set C and then count the number which are A. Then we may further count how many are in $A \cap B$. Having done so we have the data for inductive estimates of the two probabilities on the right hand side of the equation; but we also have the data for a direct inductive estimate of the probability on the left hand side.

On Reichenbach's account all the laws of science, that is probability statements, must fall into two categories; those which are established by simple enumerative induction and those which are derived from the former set via the probability calculus. But the above argument indicates that for any law in the second category which is derived directly from laws in the first category there will be inductive data which could have been used so the law would have been located in the first category. Following the same line of argument through, we can conclude that none of the laws in the second category are necessarily located there. In other words, all laws of science could have been derived by simple enumerative induction.

Where this is not so, it is presumably merely because of convenience in collecting samples. For instance, in the example considered above, if the sample of C investigated to establish the probability $P(C, A)$ was dispersed before the need to determine $P(C \cap A, B)$ was recognized, then it may have been more convenient to select a new sample of $C \cap A$. (However since both probabilities were established by direct induction, it would have been possible to collect a new and larger sample of C from which both probabilities could be estimated – then $P(C, A \cap B)$ could also have been established in the same way.)

This account minimizes the role that theory plays in guiding scientific inference – according to it a theory-guided inference cannot go beyond what could have been inferred by enumerative induction.

The second general consideration indicating the inadequacy of Reichenbach's account of scientific inference is as follows. According to this account all theories should consist merely of frequency laws. However, many theories assert the existence of various entities which are not open to direct observation, so that the grounds for accepting these existential claims involve inferences. Micro-theories assert the existence of certain entities which are held to constitute macroscopic objects. Certainly these claims will be in the form of generalizations to the effect that all macroscopic entities of a certain kind are composed of such and such micro-entities. However, such a generalization could only be the conclusion of an inductive argument if there were a reasonably large number of cases in which we could establish directly that a particular macro-entity of the kind in question had the appropriate micro-structure. But this micro-structure is not open to direct observation and we must consider what kind of inference may be employed in concluding that any particular entity has a certain micro-structure. This inference cannot be inductive. And it is at least equally clear that assertions about the remote past, such as we find in the theories of geology or biological evolution, are not inductive conclusions. Yet inferences to microstructures and the remote past are an essential part of scientific reasoning.

Some philosophers deny that science is concerned with inferences to unobservable entities or events. According to thorough-going instrumentalists, any statement which seems to refer to micro-entities or the remote past is not to be taken literally. Such sentences are components of theories; and the whole meaning of a theory consists of the material implications between direct observation sentences which are entailed by the theory. On this view the only inference involved in accepting a theory is that the regular patterns of observable facts which are entailed by the theory will continue. And this inference is straightforward induction. The only difference between a theory and an empirical generalization is that in the former the regularity is expressed in a very complex way. Thus the above objections to the thesis that all scientific inference is reducible to induction would not hold if instrumentalism is correct in holding that theories are formal systems which are partially interpreted in the observation language.

In the last decade there has been a growing volume of argument against instrumentalism and for scientific realism; however, it would take us too far from the main theme to review these arguments. It is sufficient to note that Reichenbach did not adopt an instrumentalist position. His general position on ontology was realism. He rejected positivism and held that the meaning of an observation-based material object statement is not reducible to the meaning of statements about impressions (1938, ch. II, esp. pp. 111 & 146).

This realism concerning the objects of the everyday world he extended to many theoretical entities. Scientific statements about some putative entities are reducible, without residue, to statements about 'concreta'; that is, the objects of everyday experience. Such putative entities he calls 'abstracta', citing as examples the political state, the spirit of the nation, the soul and the character of a person (1938, § 11). However, science speaks of other entities which are claimed to exist in the same way that the concreta exist but which are not open to direct observation. He calls such entities *illata* or 'inferred things' and says: "The existence of these things is not reducible to the existence of concreta because they are inferred by probability inferences from concreta" (1938, p. 212). The probabilistic character of the inference shows that it is ampliative; the meaning of statements about atoms, for example, is not contained in the meaning of any set of statements about observable things. This puts Reichenbach in the camp of scientific realism; and it is clearly not possible for him to hold that adopting a theory which deals with *illata* is no more than an induction to the conclusion that regularities observed in the past will continue.

Where Reichenbach speaks directly about inferences to *illata*, he gives little in detail about the structure of these inferences, which he calls 'projections' (1938, § 25). However, he does explicitly commit himself to the claim that all inferences to scientific general statements rest on enumerative inductions together with the concatenation of these via the probability calculus (1938, §§ 32 & 41; 1949, §§ 84, 85 esp. p. 432). A very simple instance of a concatenation is where an induction over one class is corrected, or strengthened, by an induction over classes of the same kind. Thus a generalization about the degree of constancy of plumage colour within species of birds may guide our inductions about the plumage colour of a particular species. (Reichenbach gives a biologically incorrect example about the colour of swans 1949, p. 430.) He goes on to talk about inferences to hypotheses and theories and presumably this is meant to cover cases where *illata* are involved. Such inferences he calls explanatory inductions and involve causal explanations. They are crudely and incorrectly represented by those who speak of confirmation by the hypothetico-deductive method. A deeper analysis, he holds, shows that such inferences are concatenated inductions employing Bayes' Theorem (1949, pp. 431–432).

If I am correct in the general considerations I advanced above, Reichenbach must be wrong in holding inferences to the existence of unobservable theoretical entities are concatenated inductions. I have argued that any concatenated induction must be reducible, at least in principle, to a simple enumerative induction, and, secondly, that inferences to *illata* cannot be simple enumerative inductions. Let us examine in more detail the way that Reichenbach's thesis that inferences to scientific hypotheses may be reduced to an application of Bayes' Theorem.

9. THE ROLE OF BAYES' THEOREM

Reichenbach's claim that inferences to theoretical hypotheses have the form of Bayes' Theorem seems plausible in the light of the following considerations. Firstly, if an hypothesis explains certain facts, then the probability of these facts given the hypothesis is much higher than their initial probability. Secondly, Bayes' Theorem relates the probability of an hypothesis given certain facts to the probability which is conferred on those facts when we assume the hypothesis. Thus it seems that when evidence is explained by an hypothesis, Bayes' Theorem should show how the probability of the hypothesis is increased by this evidence. And it does seem clear that scientific hypotheses, including those which involve *illata*, do explain those facts which count as evidence for the theory.

Bayes' Theorem is a direct consequence of the Multiplication Axiom of Probability Theory. Using conventional probability notation (so as to leave open for the moment the question of interpretation), Bayes' Theorem can be expressed as follows:

$$\text{Prob.}\left(\frac{h}{e\ \&\ b}\right) = \text{Prob.}\left(\frac{h}{b}\right) \times \text{Prob.}\left(\frac{e}{h\ \&\ b}\right) \Big/ \text{Prob.}\left(\frac{e}{b}\right)$$

where h is some hypothesis, b is our general background knowledge and e is a set of facts which serve as evidence for h. The probability of h (or e) given only b may be called the 'antecedent probability'. And those who are sceptical about this theorem are so because they are dubious about the possibility of establishing the antecedent probabilities. No-one denies that the theorem is a valid consequence of the probability theory. I believe that there are good grounds for allowing that Bayes' Theorem does reflect the structure of inference to scientific hypotheses, but only if an exclusively frequency interpretation of probability is abandoned.

Salmon has discussed in some detail the way that inferences to scientific hypotheses can be analyzed as conforming to Bayes' Theorem (1966, pp. 242–266). He sees the main task as giving a frequency interpretation of the antecedent probability of an hypothesis H. His solution is to take this as the frequency with which hypotheses which are similar to H in relevant respects are true. There are a number of difficulties with this interpretation which are not, I believe, resolved by Salmon. In the first place he sees no problem with the attribute class: that is the class of true hypotheses. However if Bayes' Theorem is the typical method by which hypotheses are assessed, and hence accepted as true, there would seem to be a danger of an infinite regress. Any use of Bayes' Theorem needs a substantial number of hypotheses to have been established as true or false, and in each case this must have been done via Bayes' Theorem.

Salmon holds that the reference class will be identified by the methodological rule which must always be used when probability is to be applied to a singular case. His version of this rule is that we are to take the broadest homogeneous reference class which includes the individual in question (1966, p. 225). (This rule is a modification of and improvement on Reichenbach's.) Thus we are to take all hypotheses which are similar to H in characteristics which affect the frequency of truth. Hypotheses which differ from H by a characteristic C will be included if the frequency with which hypotheses are true remains unchanged when C is used to select a sub-set. This is offered as a straightforward empirical procedure.

This account seems plausible enough if we can assume the existence of a large universe of hypotheses of various kinds which can be examined. However, a problem arises in locating this universe of hypotheses. It would not, of course, be possible for an empiricist to take this as the set of all logically possible hypotheses; it must be a set of entities which can be counted. The obvious solution would be to include just those hypotheses which have been formulated. The first difficulty with this proposal is that the only hypotheses we can normally find out about are those which have been reported in scientific literature and these will be ones which have been considered plausible enough to be worthy of discussion. This would mean that the sample of the reference class to which we had access would be biased. More significantly there seems to be circularity involved, for considerations of plausibility seem to be involved in determining the data from which prior probability, and hence plausibility, are to be determined. This point would remain even if we could gain access to all hypotheses which are formulated, for plausibility considerations of some kind surely influence a scientist in the very formulation of hypotheses. Another consequence of this proposal is that determining plausibility becomes an exercise in the history of science – what has happened up to now. This might result in very low prior probabilities for comparatively novel hypotheses introduced to deal with data from a new area; yet in such a case the novel hypothesis may be quite plausible because nothing more familiar even looks like accounting for the data. The strict implementation of the frequency method may result in an unreasonable barrier to the formulation of new hypotheses, merely because they are novel.

Salmon considers the other probabilities in the formula and in each case suggests a frequency interpretation. What he does not do is give an overall analysis of the theorem showing how the attribute and reference classes in one probability can be identified with the corresponding classes wherever the same term re-appears in the formula.

Speaking of Bayes' Theorem expressed in Reichenbach Terminology, viz.:

$$P(A \cap C, B) = P(A, B) \times P(A \cap B, C)/P(A, C)$$

Salmon says:

From an hypothesis H and statements of initial conditions I, an observational prediction O is deducible. For purposes of this discussion we assume I to be true and unproblematic. Under this assumption H implies O. We can provide a loose and preliminary interpretation of Bayes' Theorem, even though many difficult problems of interpretation remain to be discussed. Let 'A' refer to hypotheses like H; let 'B' refer to the property of truth; and let 'C' refer to the observed result with respect to the prediction O. If positive confirmation occurs 'C' means that O obtains; in the negative case 'C' designates the falsity of O. This interpretation makes the expression on the left-hand side of Bayes' Theorem refer to precisely the sort of probability that interests us; $P(A \cap C, B)$ designates the probability that a hypothesis of the sort in question, for which we have found the given observational result, is true. This is the probability we are looking for when we deal with the confirmation of scientific hypotheses. (1966, p. 251)

In this account it is clear enough that A and B are sets. However, this is not so clear in the case of C. What is said explicitly about this term makes it difficult to see what set it names. If we consider what Salmon says about the interpretation of the left hand side of the theorem, the most plausible interpretation for C would seem to be:

> "The set of hypotheses which are related to established facts in the way that H is related to O given I."

Perhaps this could be tightened up so that only the number of predictions derivable from the hypothesis and the ratio of these which are confirmed would count in determining the set. This would give the following interpretation for the posterior probability; that is the left-hand side of the formula:

> "The frequency with which elements of the intersection of the set of hypotheses similar to H with the set of hypotheses related to established facts in the way that H is, are true."

This is not too bad as a frequency interpretation of this expression. However, this interpretation of C results in $P(A, C)$ meaning:

> "The frequency with which hypotheses similar to H are hypotheses related to established fact in the way H is."

This cannot be understood as the antecedent probability of the evidence. Indeed it is difficult to see how this can possibly be achieved given the interpretation which Salmon unequivocally gives to A; namely the set of hypotheses similar to H.

If the frequency interpretation is to work, A, B, and C must be sets and must all intersect. Salmon's interpretations for A and B are satisfactory in this respect. So would be my proposal for C, but as we see, it is unsatisfactory on other counts. It may seem that a solution could be achieved by taking all the sets A, B, and C as composed of elements which are ordered pairs, one of which being an hypothesis and the other a set of evidence. B would simply be the set of all such ordered pairs in which the hypothesis was true, and A those in which it was of the same kind as H. C would be the set of all ordered pairs in which the evidence stood in some specified relation to the hypothesis. This might work for $P(A \cap C, B)$ and $P(A, B)$, however difficulties would remain for $P(A, C)$ and also for $P(A \cap B, C)$, if they are to retain the meanings assigned to them in the original account of how Bayes' Theorem functions.

These problems bear out the more general arguments against the possibility of reducing to enumerative induction all inferences to hypotheses.

10. NON-FREQUENCY-BASED WEIGHTS

Both Reichenbach and Salmon hold that the only acceptable way of assigning weight to an hypothesis is on the basis of frequency data. Anything else would be a contravention of empiricism (see for example Salmon 1966, p. 262). This is the basis of their interpretation of probability. It is not suggested that an epistemic, or credibility, interpretation of probability is impossible or defective; merely that such an interpretation is not necessary. Indeed, in his *Theory of Probability*, Reichenbach takes the basic notion of probability as a relation between two assertions, and this relation is akin to implication (1948, p. 45). Later in the book he shows how a logical interpretation of the probability calculus is possible. Indeed his notion of 'weight' is essentially what some epistemic theorists call 'probability'. The core of his thesis might be put by saying that epistemic probabilities must be based on frequency probabilities. This means that a calculus of epistemic probability is not necessary. Any manipulations and inferences among probability statements may be carried out in terms of frequency probability until we have that frequency probability which is appropriate for the weighting of the posit in which we are interested.

Some probability theorists have disagreed with Reichenbach by claiming that there are purely logical considerations which must play a role in weighting hypotheses. Carnap is the obvious example (1950 & 1952). In the latter work he takes the confirmation of an hypothesis as the weighted mean of an empirical factor and a logical factor. The value of the logical factor is determined entirely by the structure of the language and of the hypothesis. In Carnap's system it is possible to derive the purely logical probability of an hypothesis by taking the confirmation of the hypothesis given only a tautology. This will be identical to the logical factor. The important role that this factor plays in Carnap's system is shown in the first work where the confirmation of a hypothesis given certain evidence is analyzed as the quotient of two logical probabilities (that of the conjunction of h and e divided by that of e). Since many aspects of a language are open to arbitrary choice, it is not surprising that philosophers who have a strong commitment to empiricism should find this unsatisfactory.

Salmon (1961) argues against Carnap's confirmation function on the grounds that it conflicts with his criterion of linguistic invariance. If there should be any empirical grounds in favour of choosing one set of descriptive predicates, and hence one language in preference to another, and if this choice properly influences our assessment of an hypothesis, then it should be possible to express the confirmation formula so as to take account directly of this empirical data. However, in Carnap's system, the unguided choice of a language will influence the confirmation of an hypothesis on a given set of evidence. This Salmon finds unacceptable.

The choice is not between a system such as Carnap's and the Reichenbach–Salmon frequency theory; for it remains possible that there are empirical considerations which should bear on the rational assessment of a theory, yet cannot be summarized in terms of the frequency of one set in another. This possibility is not allowed for in either system, but must be taken seriously. Let me suggest in very crude outline a consideration of just this kind. In assessing an hypothesis, scientists will take into account how well that hypothesis fits in with the body of laws and theories already accepted. Since the accepted body of scientific beliefs is empirically based, an appeal to these laws is in turn empirical.

To require that new hypotheses 'fit in' as well as possible with extant theory is to require that scientific knowledge remains as systematized and unified as possible while it grows. This will be the practical application of an even more basic principle which may be expressed as follows:

> Always seek the system of laws and theory which is the simplest generalization of regularities exhibited in known facts.

Where we are confident of a body of laws and theories this will require us to fit an explanation of new facts as neatly as possible into the system. However, it is always possible that someone may come up with a new theoretical perspective, and when this happens, the proposal will be judged by considering whether the new system is simpler than the old. I believe that this is essentially what is done when a new theory is accepted as better than an old one on the grounds of providing a better explanation of known facts.

I have argued earlier that an explication of induction which will avoid paradoxes such as result from predicates like 'grue' must include a principle of simplicity. Further I have suggested that there is no criterion for determining the scope of a generalization except for the way this contributes to the development of knowledge as a whole. Hence the general principle of induction requires us to seek the simplest overall system of laws, and it is induction thus explicated for which I suggested a vindication.

The basic principle of induction now emerges as the guide for the more sophisticated inferences required to assess theories. However, we must recognize that these latter inferences involve steps which are not the generalizing of a pattern detected in observed facts. The point is that we have inductive grounds for believing that there are many facts of which we do not know directly. We have similar grounds for the general structure of these facts. In many cases we are guided by simplicity considerations in the beliefs we form about particular unobserved facts. The general spatio-temporal structure of the physical world and the quasi-permanence of material objects are essentially concepts which are epistemologically very basic. Their acceptance is, in a broad sense, inductive; for they rest on our ability to so order the experiences we have had. This provides a basic structure in which we can locate unobserved facts, and which we fill out and develop with our inductively based laws and theories. Given this structure we can sometimes deduce the details of unobserved facts from those observed, but this is often not possible.

General inductions assure Sherlock Holmes that events of some general kind occurred which caused the death of the victim. But when he claims to 'deduce' the details of these events, he uses the word in a way foreign to logic; and it would be just as misleading to say that he 'induced' his conclusion. He is not, in the typical instance, able to cite many cases in which there were clues of the same kind as those present and in which these clues related to the guilty party in the way present clues relate to one of the suspects. The way he proceeds is to seek the simplest explanation of the known facts. And this is the way we all proceed in many more prosaic situations; when, for instance, one concludes from an unlocked door and packages on the table that one's spouse has already been home.

Much of scientific theorizing consists in postulating facts for known gaps in the accepted basic conceptual framework. Such are the inferences to what Reichenbach called *Illata*. Hypotheses of this kind are characterized by Feigl as 'existential', for they assert the existence of entities, or facts, which are not open to direct observation. Thus such hypotheses are not merely inductive generalizations. It does not follow that the principle of induction plays no role in guiding such hypotheses. On the contrary, micro-theories and hypotheses about occurrences of the remote past are formulated with an eye to the subsumption of empirical generalization about medium sized objects under much more basic and more general laws; that is with achieving the simplest overall system of laws. The only other principle involved is that the complexity of postulated entities should be kept to a minimum; and this seems a very natural extension of the principle of inductive simplicity.

I would suggest that these ideas can be used to give a much more satisfactory account than Salmon's of how inferences to theories conform to Bayes' Theorem. Consider antecedent probability. Here, I would hold, the main concern is how a proposed hypothesis fits in with extant theory. This allows for the open texture of scientific theories. In judging that a new hypothesis fits in with extant theory, we are often judging what kinds of elaborations or modification may be required to achieve a deductive systematization.

The open texture of theories also creates difficulties for Salmon's account of the way an hypothesis makes probable a body of evidence. There is often neither a deductive nor a frequency–probabilistic relationship involved. However, on the proposed account we may allow that a theory which in its present form does not imply certain facts nevertheless makes them probable if we judge that this will be achieved by a simple set of existential hypotheses or by minor modifications to accepted laws.

The account I propose employs a notion of probability which in many cases could not be measured. All that is possible is to make judgements about inequalities of probability. To many this will seem to be a defect. However, the proposal competes with accounts of scientific inference in which it is merely theoretically possible to measure the probabilities of hypotheses; it is only in a fairly special category of statistical hypotheses that there is any generally accepted way of making such measures. In the light of this, not too much should be made of the inability of a theory of inference to provide a measure of probability in all cases. The crucial consideration will be how adequate an explication of actual practice is provided by the theory and whether it is possible to vindicate the method based on it. I have tried to indicate in outline the case on both of these issues.

11. CONCLUSION

I have criticized Reichenbach on a number of points. There are two main theses against which I have argued. Firstly his claim that the convergence of the inductive method must be the starting point for a vindication of induction; secondly the thesis that frequency data is the only basis for assigning weights, or degrees of confidence, to hypotheses.

My central concern, however, has been to show how far Reichenbach has gone in laying a foundation for a satisfactory theory of scientific inference. His monumental work in developing a rigorous theory for a frequency interpretation of probability is generally recognized, and it has clarified a host of issues for those who disagree as well as for those who agree with his interpretation. His conviction that the principle of induction is basic to other more complex kinds of scientific inference seems to me sound. This is also true for his realist position concerning theoretical entities. His views on this matter have been largely overlooked in recent criticisms of instrumentalism. The realist view of theories raises special problems for a theory of scientific inference. Although I think that his solution was unsatisfactory, it is clear that Reichenbach was well aware of these problems and sought to solve them.

I have argued that there are a number of aspects of Reichenbach's treatment of induction which must be incorporated into any satisfactory solution. First and foremost is Reichenbach's uncomprised empiricism. This is complemented by his recognition of the practical issues that lie behind problems in epistemology. It is by no means a simple matter of how the word 'knowledge' is used. Rather we are concerned with how decisions for action are to be made. We may or may not conclude that it is proper to say that we have rational grounds for certain beliefs about the future; this is a matter of linguistic usage. But to say that there are no rational criteria for preferring one action to another as a means to an end, is to challenge the basis of all thought that goes beyond reverie; that is, of everything that is distinctively human about us.

From this perspective the justification of induction becomes a matter of vital importance to any philosophy which does not turn its back on all practical activity. It is a necessary starting point for any rational assessment of any decision. This perspective also led Reichenbach to recognize the kind of justification which is needed. It is impossible to prove that inductive conclusions are true; what is needed is an argument to show that it is rational to use them as a basis for action. He saw clearly the general outline of this argument. We know that past uniformities may cease or that there may be complexities in the overall pattern which do not show in what has been observed; but there is no way of allowing for these possibilities. Hence the best we can do is make decisions in a way which we know will lead to success in a sufficiently uniform world; this is why we should be guided by induction.

University of Melbourne

BIBLIOGRAPHY

Barker, S., 1961, 'On Simplicity in Empirical Hypotheses', *Philosophy of Science* **28**, 162–71.

Black, M., 1954, *Problems of Analysis*, Cornell University Press, Ithaca N.Y., Ch. X.

Black, M., 1962, *Models and Metaphors*, Cornell University Press, Ithaca N.Y., Ch. XI.

Carnap, R., 1950, *The Logical Foundations of Probability*, University of Chicago Press, Chicago.
Carnap, R., 1952, *The Continuum of Inductive Methods*, University of Chicago Press, Chicago.
Clendinnen, F. J., 1966, 'Induction and Objectivity', *Philosophy of Science* **33**, 215–229.
Clendinnen, F. J., 1970, 'A Response to Jackson', *Philosophy of Science* **37**, 444–448.
Feigl, H., 1950, 'De Principiis Non Disputandum . . .', *Philosophical Analysis*, Black (ed.), Cornell University Press, Ithaca N.Y., pp. 113–147.
Feigl, H. and Maxwell, G., 1961, *Current Issues in the Philosophy of Science*, Holt Rinehart Winston, N.Y.
Goodman, N., 1965, *Fact Fiction and Forecast*, 2nd ed., Bobbs–Merrill.
Hacking, I., 1968, 'One Problem about Induction', in Lakatos (1968), pp. 44–59.
Katz, J. J., 1962, *The Problem of Induction and Its Solution*, University of Chicago Press.
Kyburg, H. E. Jr. and Nagel, E. (eds.), 1963, *Induction, Some Current Issues*, Wesleyan University Press.
Lakatos, I. (ed.), 1968, *Inductive Logic*, North Holland Publ. Co., Amsterdam.
Popper, K. R., 1972, *Objective Knowledge*, Clarendon Press, Oxford.
Reichenbach, H., 1938, *Experience and Prediction*, University of Chicago Press.
Reichenbach, H., 1949, *The Theory of Probability*, University of California Press.
Reichenbach, H., 1951, *The Rise of Scientific Philosophy*, University of California Press.
Salmon, W. C., 1961, 'Vindication of Induction', in Feigl and Maxwell (eds.), 1961, pp. 245–256.
Salmon, W. C., 1963, 'On Vindicating Induction', in Kyburg and Nagel (eds.), 1963, pp. 27–41.
Salmon, W. C., 1965, 'The Concept of Inductive Evidence', in 'Symposium on Inductive Evidence', *American Philosophical Quarterly*, Vol. 2, pp. 1–16.
Salmon, W. C., 1966, 'The Foundations of Scientific Inference', in Colodny (ed.), *Mind and Cosmos*, University of Pittsburgh Press, pp. 135–275.
Salmon, W. C., 1968a, 'The Justification of Inductive Rules of Inference', in Lakatos (ed.), 1968, pp. 24–43.
Salmon, W. C., 1968b, 'Reply' (to discussion on 1968a), in Lakatos (ed.), 1968, pp. 74–93.
Schlesinger, G., 1963, *Methods in the Physical Sciences*, R & K P London, Chap. 1.
Schlesinger, G., 1974, *Confirmation and Confirmability*, Clarendon Press, Oxford.
Skyrms, B., 1965, 'On Failing to Vindicate Induction', *Philosophy of Science*, Vol. 32, pp. 253–68.
Skyrms, B., 1975, *Choice and Chance*, 2nd ed., Dickenson.
Urmson, J. O., 1956, 'Some Questions concerning Validity', in A. Flew (ed.), *Essays in Conceptual Analysis*, pp. 120–133.

BAS C. VAN FRAASSEN

RELATIVE FREQUENCIES*

The probability of an event is the limit of its relative frequency in the long run. This was the concept or interpretation developed and advocated in Reichenbach's *The Theory of Probability*. It cannot be true.

After so many years and so much discussion, it is a bit *vieux jeu* to attack this position. But the arguments tell equally against such related views as those of Kolmogorov and Cramer, which surround the standard mathematical theory. Even so, I would not attempt this, if it were not that I also want to defend Reichenbach. For the heart of his position is that the statistical theories of physics intend to make assertions about actual frequencies, and about nothing else.

In this, I feel, Reichenbach must be right.

I. ABSOLUTE PROBABILITIES

1. *The Inadequacy of Relative Frequency*

What is now the standard theory of probability is due to Kolmogorov, and it is very simple. A *probability measure* is a map of a Borel field of sets into the real number interval [0, 1], with value 1 for the largest set, and this map countably additive.[1] Somewhat less standard is the notion of a *probability function*, which is similar, but defined on a field of sets and finitely additive. Relative frequency lacks all these features except that it ranges from zero to one.

This is not news; I need simply marshall old evidence (though I must say that it was a surprise to me, and I painfully reconstructed much of this along the way as I studied Reichenbach).

Let the actual long run be counted in days: 1 (today), 2 (tomorrow), and so on. Let $A(n)$ be an event that happens only on the nth day. Then the limit of the relative frequency of the occurrence of $A(n)$ in the first $n+q$ days, as q goes to infinity, equals zero. The sum of all these zeroes, for $n = 1, 2, 3, \ldots$ equals zero again. But the union of the events $A(n)$ – that

W. C. Salmon (ed.), Hans Reichenbach: Logical Empiricist, 129–167.

is, $A(1)$-or-$A(2)$-or-$A(3)$-or . . .; symbolically $\cup\{A(n): n \in N\}$ – has relative frequency 1. It is an event that happens every day.

So relative frequency is not countably additive.[2] Indeed, its domain of definition is not closed under countable unions, and so is not a Borel field. For let B be an event whose relative frequency does not tend to a limit at all. Let the events $B(n)$ be as follows: $B(n) = A(n)$ if B happens on the nth day, while $B(n) = \Lambda$, the 'empty' event, if B does not occur on the nth day. The limit of the relative frequency of $B(n)$ exists, and equals zero, for each number n. But B is the union of the events $B(n)$, and the limit of its relative frequency does not exist.

A somewhat more complicated argument, due to Rubin and reported by Suppes, establishes that the domain of relative frequency in the long run is not a field either.[3] Let us divide the long run of days into segments: $X(n)$ is the segment which stretches from segment $X(n-1)$ to day 2^n inclusive ($X(1)$ is just the first and second day). We note that $X(n)$ is as long as the sum of all preceding segments. Call $X(n)$ an odd segment if n is odd; an even segment otherwise. Let A be an event that happens every day in every odd segment; but on no other days. In that case, the relative frequency of A has no limit in the long run. A is not in the domain of definition of relative frequency.

We let B and C be two events that overlap inside A, in a regular way: Let B happen on all the even dates on which A happens, and on all the odd dates when A does not happen. And let C also happen on all the even dates on which A happens, and in addition, on all even dates when A does not happen. About B and C it is true to say that in each segment, each of them happens every other day. So each has relative frequency $\frac{1}{2}$.

But the intersection $B \cap C$ is curious. Up to the end 2^n of segment $X(n)$, there have been exactly half as many $(B \cap C)$ days as there were A days. So if $B \cap C$ had a relative frequency in the long run, so would A. And A does not.

Let me sum up the findings so far. The domain of relative frequency is not closed under countable unions, nor under finite intersections. But still, countable additivity fares worse than finite additivity. For when the relative frequency of a countable union of disjoint events exists, it need not be the sum of the relative frequencies of those components. But if the relative frequencies of B, C, and $B \cup C$ all exist, while B is disjoint from C, then the relative frequency of $B \cup C$ is the sum of those of B and C.

We cannot say therefore that relative frequencies are probabilities. But we have not ruled out yet that all probabilities are relative frequencies (of specially selected families of events). For this question it is necessary to look at 'large' probability spaces; specifically, at geometric probabilities.

It is sometimes said that a finite or countable sample space is generally just a crude partition of reality – reality is continuous. Of course you cannot have countable additivity in a sensible function on a countable space! But the problem infects continuous sample spaces as well. In the case of a mechanical system, we would like to correlate the probability of a state with the proportion of time the system will spend in that state in the limit, that is, in the fullness of time. But take a particle travelling forever in a straight line. There will be a unique region $R(n)$ filled by the trajectory of the particle on day (n). The proportion of time spent in $R(n)$ tends to zero in the long run; but the particle always stays in the union of all the regions $R(n)$.

Geometric probabilities such as these have further problems. Take Reichenbach's favorite machine gun example, shooting at a circular target. The probability that a given part of the target is hit in a given interval, is proportional to its area, we say. But idealize a bit: let the bullets be point-particles. One region with area zero is hit every time: the set of points *actually* hit in the long run. Its complement, though of area equal to the whole target, is hit with relative frequency zero.

This example is very idealized; but it establishes that in certain cases, it is not *consistent* to claim that all probabilities are relative frequencies.

2. *Weaker Criteria of Adequacy*

When an interpretation is offered for a theory such as that of probability, what criteria should be satisfied? In logic, we tend to set rather high standards perhaps. We expect *soundness* (all theorems true in the models offered) and *completeness* (each non-theorem false in some of those models). In addition we hope for something less language-bound and more informative: the relevant features of all models of the theory must be reflected in the models offered by the interpretation. This is imprecisely phrased, but an example may help. Take the truth-functional interpretation of propositional logic. Any model of that logic is a Boolean algebra. Every Boolean algebra can be mapped homomorphically into the two-element $\{T, F\}$ Boolean algebra pictured in the truth-tables. And if $a \neq 1$ in the original, the map can be such that the image of a is not T.

Translating this to the case of probability theory, we would ask of the relative frequency interpretation first of all that the structures it describes (long runs, with assignments of real numbers to subsets thereof through the concept of limit) be models of the theory. But that would require relative frequency to be countably additive which it is not. Secondly we would ask that if we have any model of probability theory, we could map its measurable sets into the subsets of a long run, in such a way that if the originals have distinct probabilities, then their images have distinct relative frequencies. The machine gun example shows this is not possible either.

In his book, Reichenbach certainly considered many questions concerning the adequacy of the relative frequency concept of probability. His exact formulation I shall examine in Part III below. But his book appeared in 1934, too early to be addressed to Kolmogorov's mathematical theory (which appeared in 1933), and much too early to guess that the latter would become the standard basis for all work in the subject. So Borel fields and countable additivity were not considered by Reichenbach. In addition, he construed probability theory on the paradigm of a logical calculus. This introduces the extra feature that probabilities are assigned to only denumerably many entities, for these are all denotations of terms in the language. So Reichenbach's discussion is concerned with weaker criteria of adequacy than I have listed.

Turning to representability, remember that Reichenbach could or would not consider more than denumerably many distinct probabilities at once. To show that he was right about this, in a way, I will have to give some precise definitions. A *probability space* is a triple $S = \langle K, F, P \rangle$, where K is a non-empty set, F a Borel field on K, P a map of F into $[0, 1]$ such that $P(K) = 1$ and P is countably additive. The infinite product $S^* = \langle K^*, F^*, P^* \rangle$ is formed as follows: K^* is the set of all maps σ of N (the natural numbers) into K; F^* is the least Borel field on K^* which includes all the sets

$$\{\sigma \in K^* : \sigma(1) \in A_1, \ldots, \sigma(n) \in A_n\} = A_1 \times \cdots \times A_n \times K^*$$

for sets $A_1, \ldots, A_n$ in F; and P^* is the probability measure on F^* such that

$$P^*(A_1 \times \cdots \times A_n \times K^*) = P(A_1) \cdots P(A_n)$$

which exists by a theorem of Kolmogorov. A *long run* is simply a member σ of K^*.

The relative frequency of A in σ – where A is a member of F – call it relf (A, σ), is defined by

$$\text{relf}\,(A, \sigma) = \underset{n \to \infty}{\text{limit}} \frac{1}{n} \sum_{i=1}^{n} a(\sigma(i))$$

where a is the characteristic function of A; that is, $a(x)$ equals 1 if x is in A and equals zero otherwise. The Strong Law of Large Numbers[4] has the consequence

$$P^*(\{\sigma \in K^* : \text{relf}\,(A, \sigma) = P(A)\}) = 1$$

from which we can infer, as a minor consequence, that there will be at least one long run σ such that the probability $P(A)$ is the relative frequency of A therein. I shall attach no other importance to the function P^*, using it *only* to establish the existence of certain long runs.

The intersection of countably many sets of probability measure 1, must have measure 1 again. Thus we generalize that if $A_1, \ldots, A_k, \ldots$ are all members of F, then

$$P^*(\{\sigma \in K^* : \text{relf}\,(A_i, \sigma) = P(A_i) \text{ for } i = 1, 2, \ldots\}) = 1$$

also, and therefore

> If X is a denumerable subfamily of F, there exists a 'long run' σ in K^* such that, for each set A in X, $P(A) = \text{relf}\,(A, \sigma)$.

In other words, any denumerable family of probabilities can consistently be held to be reflected in the relative frequencies in the actual long run.

3. *Square Bullets*

Although relative frequencies are not countably additive, and so are not probabilities, we have just seen something very encouraging. Any countable family of probability assertions can consistently be interpreted as a family of assertions of relative frequencies. This creates the temptation to do a bit of Procrustean surgery. Perhaps large probability spaces can be approximated by small ones, and a *small* probability space can be identified with a suitably chosen *part* of a relative frequency model. (On the subject of choosing a suitable part of a relative frequency model, Reichenbach explicitly rejected von Mises' restriction to 'random sets'. I feel that Reichenbach had good reasons for this; and also, I see little gain in the partial representation theorems the randomness tradition provides, so I shall say no more about this.)

Let me give a simple example to show the possibilities in this line, and also its limitations. Let the machine gun still fire point-bullets, but let the target be a line segment, conveniently coordinatized by distances as the interval $[0, 1]$. All the bullets fall on it, and the probability that interval $[a, b]$ is hit equals its length $b-a$. More generally, the probability a Borel set $E \subseteq [0, 1]$ is hit equals its Lebesgue measure. We would have the same problems as before trying to interpret this situation exhaustively in terms of relative frequencies. But why try, when perception thresholds are not infinitely fine? Given the essential human myopia, appearances must be a fragmentary part of reality. (Anyway, bullets are not point-sharp; they may be round or square.)

Any real interval $[a, b]$ is of course suitably approximated by the rational intervals $[a', b']$ such that $a' < a < b < b'$ and a' and b' are rational fractions. There are only countably many such rational intervals. Let us take the field F they generate (still countable), and find a long run in which the relative frequency of hits in a set in that field equals the Lebesgue measure of that set. Let that long run be σ, and let us begin our reconstruction.

(1) $\quad m(A) = \text{relf}\,(A, \sigma)$ if A is in F.

Now in addition, if $B_1 \supseteq \cdots \supseteq B_k \supseteq \cdots$ is a series of sets in F, and B their intersection, let us say

(2) $\quad m(B) = \lim_{n\to\infty} \text{relf}\,(B_n, \sigma)$

when that exists.

This function m is a pretty good shot at the probability. (Perhaps even identical with it; let us not stop to inquire.) Since it assigns zero to every countable point-set, it is not identifiable with any relative frequency. But it is *approximated* by the function relf (X, σ) *restricted to* field F.

Let us suppose for a moment that all and only functions thus approximated by relative frequencies are probabilities. This will remind us of very typical mathematical moves. If a structure is not topologically closed – if the limits of converging series in the structure do not always belong to the structure – then they are 'put in'. That is, the mathematician's attention switches to a larger structure, which is closed under such operations. The original topic of concern is looked upon as an arbitrarily hacked out fragment of the important structure. (Consider the definitions of Hilbert space and of tensor products of Hilbert spaces, as example.) So did probability theorists simply widen the original topic of concern – relative frequencies – in the traditional, smooth paving, mathematical fashion?

But no, that cannot be. The family of probability measures does not include relative frequencies in long runs, except on finite sample spaces. The limit points, if that is what they are, are not *added* onto the original family, but the originals are thrown out. Moreover, those limit functions are not extensions or extrapolations of relative frequencies. In the above example, $m(X)$ is not an extension of relf (X, σ), for the two disagree on countable sets; $m(X)$ is only an extension of the result of restricting relf (X, σ) to field F.

The construal of probabilities I have just described in a simple case is explained by Reichenbach for more interesting examples in his Chapter Six. Let the target be a region A on the Cartesian plane. A geometric probability will be a measure m on the plane that gives 1 to A, and for which

$$(3) \qquad m(B) = \iint_B \phi(x, y)\, dx\, dy$$

for each subregion of A, for a function ϕ – the *probability density* – which is characteristic of m.

We can proceed as follows: Cut the region A into n subregions of equal area, call them $E_1, \ldots, E_n$. The probability space which has m restricted to the Borel field generated by $E_1, \ldots, E_n$ is a finite probability space. Therefore the probability assertions about it can be interpreted through relative frequencies. Call this space S_1.

Construct space S_2 by refining S_1: each region E_k is subdivided into n regions of equal area, and the probabilities assigned are again the restriction of m to the (Borel) field generated by this (finite) set of small regions. And so forth.

Given now a series of sets $B_1 \supseteq \cdots \supseteq B_k \supseteq \cdots$ where B_k is a measurable set in S_k, it is clear that m will still be defined for their intersection, and the value there assigned by m is the limit of the values $m(B_1), \ldots, m(B_k), \ldots$. From this Reichenbach infers that there must be a parallel 'empirical construction'. The probability of B_k, he says, is 'empirically determined' by estimating the long run relative frequency on the basis of finite samples. If the limit of these probabilities exists, we then assign that limit to the intersection $\cap\{B_k : k = 1, 2, 3, \ldots\}$. With the probabilities so determined, we form the probability space S which has as measurable sets the least Borel field containing all the measurable sets in each space S_k.

The idea is clearly that S is the mathematical limit of the series of spaces $S_1, \ldots, S_k, \ldots$ and that the representation of S as such a limit of finite spaces constitutes a representation in terms of relative frequencies. For each space S_k has only finitely many measurable sets, so there will be a long run σ of points such that $m(B) = \text{relf}(B, \sigma)$ for each measurable set B of S_k.

But in fact we have nothing like a representation in terms of relative frequencies here. Let S_∞ be the space that has as measurable sets all the ones from the spaces $S_1, \ldots, S_k, \ldots$; and let the probability of B in S_∞ be exactly what it is in those spaces S_k in which B is measurable. Then S is the least upper bound, in a natural sense, of the series $S_1, \ldots, S_k, \ldots$. In addition, the measurable sets of S_∞ form a field – but not a Borel field. However, there are only countably many of them so there is a long run of points σ' such that the probability of B in S_∞ equals relf (B, σ'). Surely, from the frequency point of view, S_∞ is the reasonable extrapolation from the finite spaces for which the probabilities are empirically determined.

The space S contains many more measurable sets than S_∞ does, and there is no single long run in which the relative frequencies reflect the probabilities in S. Thus it is not reasonable to present the relation between S and the finite spaces S_k as showing that the probabilities in S are a mere extrapolation from, or representable in terms of, relative frequencies.

4. *Representation of Probabilities*

We already know that we cannot have a representation of probabilities as relative frequencies. But an approximation theorem, if strong enough, is a sort of representation theorem. If all probability measures are suitably approximated by (series of) relative frequency functions, then we can say that probabilities are at least representable *in terms of* (if not *as*) relative frequencies. And the preceding section holds out this hope.

But this would not be a *good* representation in this case. Indeed, I suspect that Reichenbach was misled in this. Let us try to picture the practical context in which we take samples and estimate probabilities. This work is done in what we might call a 'practical language'. In this we state experimental and sampling results – always proportions in finite classes – and also extrapolate these results (by induction, if that exists; by guessing, if that is any different) into hypotheses about probabilities.

In any experiment or observation I can *explicitly check* only finitely many samples, and these of finite size. Moreover, my language has only countably many expressions in it, so I can *explicitly extrapolate* only to countably many sets. This must be what suggested to Reichenbach that he need only concern himself directly with finite sample spaces, extrapolate to countable series of these, and worry only about finite unions and intersections. After all, the language cannot have explicit designations for all the countable unions of classes designated in it.

But if this was the train of thought, then he was misled. For our practical language does certainly contain expressions like '*limit*', $\sum^{\infty}$, and $\bigcup^{\infty}$. It has numerals and number variables. And even though it *is* countable, it has a systematic way of designating countable unions and limits of series. Our extrapolations, always *in* this language, are *to* countable unions and infinite sums. And the method of extrapolation, however haphazard in its inductive leaps, rigidly follows countable addition. These assertions I make about the practical activity of going from sampling data to the framing of probabilistic hypotheses, models, and theories; and what I have to say further will be baseless if in fact scientists are happily violating countable additivity when they propose their hypotheses – but I do not believe so.

If all this is true, the picture drawn in the preceding section (and following Reichenbach's Chapter Six) is not realistic. For there the practical language was assumed to contain itself only descriptions of the finite sample spaces $S_1, S_2, \ldots$; the 'nearest' limit was a frequency space S_∞ which was ignored; and the extrapolation was to be outside the practical language to a function m defined on a large family of sets not designatable in the language itself. The only criterion of adequacy considered is that m agrees with all the *finitary* extrapolations from sample data (by ordinary conjunction, disjunction, and negation), because that is all that is assumed to rear its head in the practical context.

I shall now use the Law of Large Numbers – or rather the corollaries to it in Section 2 above – to give an explicit representation of all probabilities *in terms of* (not *as*) relative frequencies. This will be different from the representation by approximations so far discussed, in that it will assume that what we extrapolate to is relative frequencies on countable families of sets, and that the extrapolation to countable unions is indeed by countable addition.

A *probability space* is a triple $S = \langle K, B, P\rangle$ where K is a non-empty set, B a Borel field on K, P a countably additive map of B into $[0, 1]$ such that $P(K) = 1$. A *frequency space* is a couple $M = \langle\sigma, F\rangle$ for which there is a set K such that

(a) σ is a countable sequence of members of K
(b) F is a family of subsets of K
(c) relf (A, σ) exists for each A in F.

I shall call $M = \langle\sigma, F\rangle$ a *special frequency space* exactly if relf (X, σ) is countably additive on F in so far as F is closed under countable unions; and F is a field:

(d) F is a field on K
(e) if $\{A_i\}$, $i = 1, 2, \ldots$ is a countable family of disjoint sets, all in F, and their union A is also in F, then relf $(A, \sigma) = \sum_{i=1}^{\infty}$ relf (A_i, σ).

Let us shorten 'special frequency space' to 'sfs'. We have already seen in Section 2 that for each probability space $S = \langle K, B, P\rangle$ and countable subfamily G of B there is a sequence σ such that $P(A) =$ relf (A, σ) for all A in G. Without loss of generality we can say: countable subfield G of B, for if G is countable, so is the field it generates. Moreover, because P is countably additive on B, so is the function relf $(-, \sigma)$ on G. So $\langle\sigma, G\rangle$ is an sfs.

A *good family* will be a family Z of sfs, such that (a) the union of all the fields G of members $\langle\sigma, G\rangle$ of Z is a Borel field, and (b) if $\langle\sigma, G\rangle$ and $\langle\sigma', G'\rangle$ are both in Z, and A in both G and G', then relf $(A, \sigma) =$ relf (A, σ'). The *space associated* with a family Z of frequency spaces is a triple $S(Z) = \langle K, F, P\rangle$ such that F is the union of all the second members of elements of Z, K is the union of F, and P the function: $P(A) = q$ iff relf $(A, \sigma) = q$ for all $\langle\sigma, G\rangle$ in Z such that A is in G.

It will be clear that the space $S(Z)$ associated with a good family Z of sfs is a probability space. Moreover, the weak representation result of the section before last shows at once that there is for each probability space a good family of sfs with which it is associated.

II. CONDITIONAL RELATIVE FREQUENCIES

1. *The Inadequacy of Defined Conditional Probabilities*

The standard (Kolmogorov) theory defines the conditional probability by

(1) $\quad P(B/A) = P(B \cap A)/P(A)$ provided $P(A) \neq 0$.

Thus the probability that it will rain on a given day if the sky is overcast, is well-defined if and only if the probability that the sky is overcast is not itself zero.

There are other theories, specifically those of Renyi [8] and Popper [6] in which $P(B/A)$ is taken as well-defined also in (some) cases in which $P(A) = 0$. The relevant features of those theories here are:

(2) $\quad P(B \cap C/A) = P(B/A)P(C/B \cap A)$ if all these terms are defined,

(3) $\quad$ If $P(-/A)$ and $P(-/B)$ are both well-defined functions, they have the same domain as $P(-/K)$,

when the sets B, C, A are measurable sets in a relevant sort of conditional probability space with 'universe' K – i.e., when $P(A/K)$, $P(B/K)$, $P(C/K)$ are defined.

Clearly Reichenbach's theory does not belong to the first sort. The consideration of relative frequencies is a main motive for dissatisfaction with the definition 1. But as I shall also show, conditional relative frequencies do not quite fit Renyi's or Popper's theory.

When a submarine dives, filling its ballast tanks to some extent with water, and then the engine is switched off, it remains stationary.[5] Its center of gravity is at a certain exact depth; that is, lies in one of the planes cutting the water horizontally. The probability of its coming to lie in that exact plane, is zero. What is the probability that it lies in that plane and north of the forty-first parallel, given that it lies indeed in that plane?

I imagine the practical mind can always find some way around such questions. Recall, however, the enormous distance of idealization we have seen between actual repetitive phenomena, *even* extrapolated to the fullness of time, and probability spaces. I do not think the way around should be introduced with the statement such 'unreal' or 'impractical' problems have no place in the business-like mind of the probability theorist.

But my point is not to ask for improvements in probability theory here. I only want to point out a further difference between probabilities and relative frequencies. Even when A has relative frequency zero, the relative frequency of B's among the A's may just as well make sense. Suppose we toss a die forever, and even numbers come up only ten times. Unlikely, but possible. In such a world, the probability that a given toss yields an even number is, on the frequency view, zero. But surely, even there, the probability that a particular toss yields six, if this toss is one of those that yield even, must be well-defined. It is exactly the proportion of sixes in those ten tosses.

More abstractly, let σ be just the series of natural numbers. If A is the set of integral powers of 93, then relf (A, σ) is zero. But those powers themselves form a subseries σ'. If B is the set of *even* powers of 93, then relf (B, σ') is not only well-defined, but is already $\frac{1}{2}$. And should this not exactly be the relative frequency of the even powers among the integral powers?

The subject cannot be simple, however. If A is a set of members of σ, let $\sigma(A)$ be the subseries of σ consisting of members of A. I shall worry about exact definitions later. Let the *conditional* (*relative*) *frequency* relf $(B/A, \sigma)$ be construed as relf $(B, \sigma(A))$. In that case we would like to know when that conditional frequency is defined.

As I pointed out, some theories of probability (like Renyi's or Popper's) take conditional probability as basic. But these have feature 3 above; so that if $P(B/K)$, $P(C/K)$, and $P(B/A)$ are all defined, then so is $P(C/A)$. We can get nothing so nice for conditional frequencies.

For let A and B be as in the natural number example of the second paragraph before last. In addition, let C be a subset of A whose relative frequency in $\sigma(A)$ does *not* tend to a limit. In that case, relf $(A, \sigma) = 0$, and since B and C are subsets of A, relf $(B, \sigma) = 0 =$ relf (C, σ). Moreover, A is a perfectly good 'antecedent condition' because relf $(B/A, \sigma)$ is defined and equals $\frac{1}{2}$. But relf $(C/A, \sigma)$ is not defined.

We cannot localize the problem by noting that C has measure zero from the point of view of σ. First, so did B and relf $(B/A, \sigma)$ is defined. Second, there are no powers of 93 which are even numbers, but if E is the set of even numbers, then relf $(E \cup C, \sigma) = \frac{1}{2}$ while relf $(E \cup C/A, \sigma)$ is not defined.

The subject of conditional frequencies is therefore not correctly treated by *either* Kolmogorov, *or* Popper, *or* Renyi. In the next two sections I shall indicate ways in which one might go about providing a general theory. Reichenbach's attempts I shall consider in Part III.

2. *The Natural Frequency Space*

In Part I, I defined a *frequency space* as a couple $\langle \sigma, F \rangle$ for which there exists a set K such that

(a) σ is a countable sequence of members of K
(b) F is a family of subsets of K
(c) relf (A, σ) exists for each A in F.

We can easily map any such structure into what I shall call the natural frequency space $\langle \omega, F(\omega) \rangle$ where

(a) ω is the natural number series $(1, 2, 3, \ldots)$
(b) $F(\omega)$ is the family of sets A of natural numbers such that relf (A, ω) exists.

What exactly is in the family $F(\omega)$ is very mysterious. The map $A \rightarrow \{i : \sigma(i) \in A\}$ will relate $\langle \sigma, F \rangle$ to $\langle \omega, F(\omega) \rangle$ in a natural way, preserving relative frequencies. It also preserves the set operations on the members of F which are sets of members of σ.

If K has in it members foreign to σ, these will of course play no role in the determination of relative frequencies, and are ignored in this map.

Before turning to conditionalization, I wish to make relative frequency and conditional relative frequency precise. If A is in $F(\omega)$, let a be its characteristic function: $a(i)=1$ if i is in A, and $a(i)=0$ otherwise. In that case:

$$+a(n)=\sum\{a(m):m\leqslant n\}$$

$$\text{rel}\,(a,n)=\frac{+a(n)}{n}$$

$$\text{relf}\,(A)=\lim_{n\to\infty}\text{rel}\,(a,n)$$

$$\text{relf}\,(B/A)=\lim_{n\to\infty}\frac{\text{rel}\,(ba,n)}{\text{rel}\,(a,n)}$$

$$=\lim_{n\to\infty}\frac{+ba(n)}{+a(n)}$$

where ba is the characteristic function of AB, namely $ba(n)=b(n)a(n)$.

What I would like to show now is that the natural frequency space is *already* closed under conditionalization (insofar as it can be). That is, there is for each pair of sets B and A in $F(\omega)$ another set $A\rightharpoonup B$ such that

> relf $(A\rightharpoonup B)$ exists if and only if relf (B/A) exists; and if they exist, they are equal.

We find this set by constructing its characteristic function $(a\rightharpoonup b)$. As a first approximation, I propose

$$(a\rightharpoonup b)(i)=\begin{cases}1 \text{ if the } i\text{th member of } A \text{ is in } B\\ 0 \text{ otherwise}.\end{cases}$$

The members of A have of course a natural order: if A is $\{1, 3, 5\}$ then 3 is its *second* and 5 its *third* member. If A is infinite, the above will do very well, for we shall have

$$\text{rel}\,(a \rightharpoonup b, n) = \frac{+(a \rightharpoonup b)(n)}{n}$$

= (The number of B's among the first n A's)$/n$

$$= \frac{+ab(m)}{+a(m)}$$

where m is the number at which the nth A occurs. The variables n and m go to infinity together, and the lefthand limit equals the righthand limit.

If A is finite, we need a slight emendation. For example, the relative frequency of the even numbers among the first 10, should be $\frac{1}{2}$. If we kept the above definition, it would be zero. So we call k the *index* of A if its characteristic function a takes the value 1 exactly k times. Each number i is of course a multiple of k plus a remainder r (ranging from one to k; this slightly unordinary usage of 'remainder' makes the sums simpler). We let $a \rightharpoonup b$ take value 1 at i exactly if the (r)th A is a B. In that way,

$$\text{rel}\,(a \rightharpoonup b)(k)$$

$$= \frac{\text{the number of } B\text{'s among the first } k\ A\text{'s}}{k}$$

$$\text{rel}\,(a \rightharpoonup b)(km)$$

$$= \frac{m \times \text{the number of } B\text{'s among the first } k\ A\text{'s}}{mk}$$

$$= \frac{+(a \rightharpoonup b)(k)}{k}$$

which, if there are only k A's, is just correct. If we call that number x/k, we also see that

$$\frac{+(a \rightharpoonup b)(km+r)}{km+r} \leqslant \frac{mx+(k-1)}{km+(k-1)}$$

which series converges to x/k as m goes to infinity. So this is indeed the correct characteristic function.

As a minor exercise, I shall now make this precise. Let 0 and 1 stand also for those characteristic functions which belong to Λ and N respectively, that is, take constant values 0 and 1.

(i) index $(a) = k$ iff $\sum \{a(n) : n \in N\} = k$

[let k here be any natural number, or ∞]

(ii) $i \dot{-} k = r$ iff *either* $k = \infty$ and $i = r$, *or*

$(\exists m)(mk + r = i)$ where $0 < r \leqslant k$.

(*iii*) $\# a(r) = i$ iff $a(i) = 1$ *and* $+a(i) = r \dot{-} \text{index}(a)$

(iv) $(a \rightharpoonup b) = \begin{cases} b \# a & \text{if } a \neq 0 \\ 1 & \text{otherwise} \end{cases}$

Here (iii) must be read as defining 'the *r*th (modulo the index of A!) A occurs at i'. Thus if A is $\{1, \ldots, 10\}$, then the 5th A occurs at 5; also the 25th (modulo the index 10) occurs at 5. So $\# a(5) = \# a(25) = 5$ in this example. If this "25th" A occurs at 5, then $a \rightharpoonup b$ should take the value 1 at 25 exactly if b takes the value 1 at 5. Thus $(a \rightharpoonup b)(25) = b(\# a(25)) = b(5)$. This is what (iv) says. In the case where a never takes the value 1, I have arbitrarily given $a \rightharpoonup b$ the value 1 everywhere. This is only a trick to keep the object well-defined.

The class $A \rightharpoonup B$ is the one which has characteristic function $a \rightharpoonup b$. While the operation $\rightharpoonup$ is very 'linear', it is not really interesting. It is certainly not a conditional in the sense of a logical implication, because even the analogue to *modus ponens* would not hold. It serves its purpose of showing that the natural frequency space already contains all the *conditional relative frequencies*; but in its own right, it is no more than a mathematical *objet trouvé*.

3. *A Partial Algebra of Questions*

I shall now continue the general theory of relative frequencies, taking the conditionality for granted. That is, we know at this point that relf $(B/A, \sigma)$ can be reduced to an assertion of 'absolute' form relf (X, σ). Henceforth, the conditional concept relf $(B/A, \sigma)$ will therefore be used without comment.

Let us view suitable relative frequencies in the long run as answers to questions of the form 'What is the chance that an A is a B?' The terms A and B I take to stand for subsets of a large set K; the set of possible situations or states or events. The question I shall reify as the couple $\langle B, A\rangle$. Since the answer would be exactly the same if we replaced 'a B' in the question by 'an A which is a B', I shall simplify the matter by requiring that $B \subseteq A$. This couple is a *question on* K, and I shall call B the *Yes-set* and A the *Domain* of the question. The *answer* relative to long run σ (a countable sequence of members of K) will be relf $(B/A, \sigma)$, if indeed relf $(B/A, \sigma)$ exists; otherwise the question is mistaken relative to σ.

The word 'chance' was proposed by Hacking as a neutral term. I use it here, but mean of course exactly what Reichenbach thought we should mean with 'probability'. The occurrence of *actual* long run σ in the determination of the answer makes the question empirical.

(a) *Questions.* A *question on* set K is a couple $q = \langle qY, qD\rangle$ with $qY \subseteq qD \subseteq K$. Call questions q and q' *comparable* exactly if $qD = q'D$. The set $[q]$ of questions comparable to q is clearly ordered through the relations on its first members (*Yes-sets*), and we could define

$$q \cap q' = \langle qY \cap q'Y, qD\rangle - q = \langle qD - qY, dD\rangle$$

and so on, to show that $[q]$ is a family isomorphic by a natural mapping to the powerset of qD.

The operation of *conditionalization* which I shall now define, takes us outside $[q]$:

$$q \to q' = \langle qY \cap q'Y, qY\rangle$$

unless of course $qY = qD$. This leads us to the next topic.

(b) *Unit questions.* If q is a question and $qY = qD$, I shall call it a *unit question.* Its answer must always be 1; and in $[q]$ it plays the role of unit, that is, supremum of the natural partial ordering.

Henceforth let Q be the set of questions on K, and U the set of unit questions thereon; let $u, v, \ldots$ always stand for unit questions.

The unit questions are not comparable to each other. But they are easily related nevertheless; they form a structure isomorphic to $P(K)$

under the natural map: A to $\langle A, A\rangle$. Let us use the symbols $\neg, \wedge, \vee, \leqslant$ in this context, to maintain a distinction between the natural ordering of the unit questions and that of the questions comparable to a given one:

$$\neg u = \langle K - uY, K - uY\rangle$$

$$u \wedge v = \langle uY \cap uY, uY \cap vY\rangle$$

$$u \leqslant v \text{ iff } uY \subseteq vY$$

where of course $u = \langle uY, uD\rangle$ and so on. Every question is a conditionalization of unit questions:

$$q = \langle qY, qD\rangle = \langle qD, qD\rangle \to \langle qY, qY\rangle\,.$$

Therefore define

$$uq = \langle qD, qD\rangle$$

$$vq = \langle qY, qY\rangle$$

$$q = (uq \to vq)\,.$$

We note that if q and q' are *comparable*, the operations on them are definable in terms of those on unit questions:

$$-q = uq \to \neg vq$$

$$(q \wedge q') = uq \to (vq \wedge vq')$$

$$(q \cup q') = uq \to (vq \vee vq')$$

$$q \subseteq q' \text{ iff } vq \leqslant vq'\,.$$

It would not make much sense to generalize these except for $-$ which is of course defined for all questions. However, I want to add one operation on all questions

$$q \cdot q' = uq \to (vq \wedge vq')$$

which reduces to $\cap$ when q and q' are comparable, but is otherwise not commutative. However, it gives the comparable question to q that comes closest to being its conjunction with q'.

(c) *The logic of questions.* Since I have not given a partial ordering of all questions, it may seem difficult to speak of a logic at all. Should we say that q *implies* q' if the Yes-set of q is part of the Yes-set of q'? Or should we require in addition that their Domains are equal? Or that the No-set (Domain minus Yes-set) of q' be part of the No-set of q? The minimal relation is certainly that the Yes-set of q be part of the Yes-set of q'. We may think of a Yes-No question q as related to a proposition which is *true* at x in K if $x \in qY$, *false* at x if $x \in (qD - qY)$, and *neither true nor false* in the other cases. In that context the relation of semantic entailment is just that minimal relation (corresponding to valid arguments) of 'if true, then true'. So let us define

$$q \Vdash q' \text{ iff } qY \subseteq q'Y$$

$$\text{iff } q \to q' \text{ is a unit question}.$$

Then we note that the analogue to *modus ponens* holds, but only because something stronger does

$$q \cdot (q \to q') \Vdash q \to q' \Vdash q$$

which should not be surprising, because these conditionals are just like Belnap conditionals (a Belnap conditional says something only if its antecedent is true; in that case it says that its consequent is true – see [2], [11]).

(i) $q \to q' = vq \to vq'$

(ii) $q_0 \to (q \to q') = q_0 \to (vq \to vq')$

$$= q_0 \to \langle vqY \cap vq'Y, vqD \rangle$$

$$= q_0 \to (vq \wedge vq')$$

$$= q_0 \to (q \cdot q')$$

Corollary: $u \to (v \to v') = u \to (v \wedge v')$

(iii) $(q_0 \to q) \to q' = \langle q_0 Y \cap qY, q_0 Y \rangle \to q'$

$$= \langle q_0 Y \cap qY \cap q'Y, q_0 Y \cap qY \rangle$$

$$= (vq_0 \wedge vq) \to vq'$$

$$= (q_0 \cdot q) \to q'$$

Corollary: $(u \to v) \to v' = (u \wedge v) \to v'$

(iv) if $u_0 \leqslant u \leqslant u'$ then $u \to u_0 = (u' \to u) \to (u' \to u_0)$.

The second and third show that iteration is trivial; the fourth is a trivial corollary which I mention because of the way it recalls the 'multiplication axiom'. Which brings us to the next topic.

(d) *The multiplication axiom.* Reichenbach's fourth axiom was the 'theorem of multiplication'. In our symbolism, it states

(AM) If relf $(B/A, \sigma)=p$ and relf $(C/A \cap B, \sigma)=r$ exist, then relf $(B \cap C/A, \sigma)$ also exists and equals $p \cdot r$.

The answer to the question q, relative to σ is

$$\text{relf}\,(qY/qD, \sigma)=m(q).$$

Just for the moment, let A, B, C stand equally for the unit questions, $\langle A, A\rangle$, $\langle B, B\rangle$, etc. Then the axiom clearly says:

If $m(A \to B)=p$ and $m(A \cap B. \to C)=r$ exist, then $m(A \to B \cap C)$ also exists and equals $p.r$.

Because $X \to Y = X \to X \cap Y$, there are only three sets really operative here: $A \cap B \cap C$, $A \cap B$, A. So we can phrase this also as follows:

If $u_0 \leqslant u \leqslant u'$, and $m(u' \to u)=p$ and $m(u \to u_0)=r$ exist, then so does $m(u' \to u_0)$ and equals $p.r$.

Thus in the favorable case of $p \neq 0$, which here means only that a certain conditional probability is not zero,

$$\text{(AM*)} \quad m(u \to u_0)=\frac{m(u' \to u_0)}{m(u' \to u)} \text{ if } u_0 \leqslant u \leqslant u'.$$

But at the end of subsection (c) we just saw that with $u_0 \leqslant u \leqslant u'$ given, $u \to u_0=(u' \to u) \to (u' \to u_0)$. So we have

$$m[(u' \to u) \to (u' \to u_0)]=\frac{m(u' \to u_0)}{m(u' \to u)}.$$

Let us now generalize this to conditionals for which the antecedent and consequent are not specially related:

$$\begin{aligned} m(q \to q') &= m(vq \to (vq \wedge vq')) \\ &= \frac{m(uq \to (vq \wedge uq'))}{m(uq \to vq)} \\ &= \frac{m(q \cdot q')}{m(q)} \qquad \text{provided } m(q) \neq 0 \end{aligned}$$

where I went from the first line to the second by the reflection that $uq \leqslant vq \leqslant vq \wedge vq'$ always, and applying (AM*).

We have seen that there is in the general theory of questions "What is the chance that an A is a B?" there is a logically reasonable conditionalizing operation; relative frequency of the conditional object looks quite familiar in the relevant special cases.

III. REICHENBACH'S THEORY

Reichenbach had two aims in writing *The Theory of Probability*. He wanted to present a good axiomatization of probability theory, and also to defend his frequency interpretation. The two aims interfered somewhat with each other. But the book is a monumental and instructive work, with many fascinating features that a narrower or less committed enterprise would certainly have lacked.

1. *The Formulation*

Logic had a bad influence. The language is essentially that of quantificational logic, with the quantifier ranging over the positive integers. There are special function terms x, y, . . . such that x_i is an individual (event) if i is an integer, and class terms A, B, ... The operators are Boolean (on the classes); the single predicate is the binary $\in$ of membership; and the connectives are those of propositional logic plus a special variable connective

$$\underset{p}{\ni\!\!-}$$

where p is any numerical expression denoting a real number (if you like, between zero and one inclusive). This special connective is eventually allowed to combine any sentences, but to begin is considered only in the context

(1) $(i)(x_i \in A \underset{p}{\ni} y_i \in B)$.

To understand this, reason as follows: If this coin be tossed, the probability of its showing heads equals $\frac{1}{2}$. This asserts a correlation between two sequences of events: the event x_n is the nth toss of the coin, and the event y^n the nth landing of the coin.

There is no gain, formally speaking, in this consideration of more than one sequence of events. If $\{x_i, y_i\}, \ldots$ are the sequences of events designated, we could construct a single sequence $\{\omega_i\}$ such that for each n, ω_n is the sequence $\langle x_n, y_n, \ldots \rangle$. In that case the class terms $A, B, \ldots$ would have to be reinterpreted accordingly; to the old formula "$y_i \in B$" would correspond a new formula "$\omega_i \in B'$" with the same truth-conditions. (That is, B' would designate the class of all sequences $\langle a, b, \ldots \rangle$ such that b is in the class designated by B.) In that way, all basic probability assertions would take the form

(2) $(i)(\omega_i \in A \underset{p}{\ni} \omega_i \in B)$.

The event ω_n is the complex totality of all the events happening on the nth day, say, and the sequence of elements $\omega = \{\omega_n\}$ plays the role of the long run σ. Thus 2 would in my symbolism so far be

(3) $P(B/A, \omega) = p$

for the truth-conditions of 2 and 3 (as explained respectively by Reichenbach and in part II above) are the same.

2. *The Multiplication Axiom*

In a paper published in 1932, preceding the book, Reichenbach used special 'reversal axioms' to govern inverse calculations. An example would be the inference (using Reichenbach's own abbreviations):

(1) If A and B are disjoint, then $P(A \cup B) = P(A) + P(B)$. Therefore, in that case, if $P(A \cup B) = p$ and $P(A) = q$, then $P(B) = p - q$.

This inference contains a hidden assumption, namely that all the relevant terms are well-defined (all the probabilities exist). Strictly speaking, the premise should be:

(2) If $P(A)$ and $P(B)$ exist, and A and B are disjoint, then $P(A \cup B)$ exists and equals $P(A)+P(B)$.

And a *second* premise needed in the inference is then

(3) If $P(A \cup B)$ and $P(A)$ exist, and A and B are disjoint, then $P(B)$ exists.

The special reversal axioms were apparently too weak (see page 61 of *The Theory of Probability*), and in the book their role was played by a single 'Rule of Existence' (page 53). This rule seems to be correct, and Reichenbach is very careful in its use.

But extreme care does seem to be needed. On page 62 Reichenbach gives as his fourth axiom the 'Theorem of Multiplication'.

(4) $(A \underset{p}{\ni} B).(A.B \underset{p}{\ni} C) \supset (\exists w)(A \underset{w}{\ni} B.C) \cdot (w = p.u)$

where $(A \underset{p}{\ni} B)$ is short for "$(i)(x_i \in A \underset{p}{\ni} y_i \in B)$" and so forth. He concludes via the Rule of Existence that

(5) $P(B \cap C/A) = P(B/A)P(C/A.B)$

"can be solved according to the rules for mathematical equations for each of the individual probabilities occurring". This is certainly true, but only a very careful reader will not be tempted to infer if $P(B \cap C/A)$ and $P(B/A)$ both exist and equal zero, then $P(C/A \cap B)$ exists though its numerical value cannot be determined. This inference would be invalid on the frequency interpretation. (This can be seen from the example at the beginning of Part II: replace 'A' by 'K' and 'B' by 'A' in equation 5, to get $P(C/K) = P(A/K)P(C/A)$ given that $A \cap C = A$ and all are subsets of K.) Reichenbach does not make this inference, but neither does he point out its invalidity, which is a very special feature of his theory, because it is involved with the fact that $P(B/A)$ may exist while $P(C/A)$ does not, although $P(C/K)$ does – a feature not shared by the theories of Kolmogorov, Popper, and Renyi.

I should mention the very nice deduction Reichenbach gives of the multiplication axiom from the weaker axiom that $P(B \cap C/A) = f[P(B/a), P(C/A \cap B)]$ for some function f. His proof assumes that f is differentiable, but he says that this requirement can be dropped.

3. *The Axiom of Interpretation*

Reichenbach's second axiom (Normalization) has the corollary

(1) $(i)(x_i \notin A) \supset (A \underset{p}{\not\Rightarrow} B)$ for *all* p.

In other words, if an event A never occurs, then the probability of A does not exist. Much later on (Section 65) Reichenbach states also the Axiom of Interpretation

(2) If an event C is to be expected in a sequence with a probability converging toward 1, it will occur at least once in the sequence.

It is not clear whether this axiom is meant as part of the formal theory, but Reichenbach does say that "whoever admits [a contrary] possibility must abandon every attempt at a frequency interpretation" (page 345).

Either of these axioms is sufficient to rule out geometric probability altogether. For any sequence $\{x_i\}$ will be countable, and so receive probability zero in a geometric example. The complement of the set of points in the sequence thus receives 1, but this is an event which never occurs in the sequence.

4. *Geometric Probability*

In Chapter Six, Reichenbach tries to deal with geometric probability. He says that this provides an interpretation for his probability calculus. Without qualification, this is not true; but it would be true if we delete the Axiom of Normalization (and of course, never add the Axiom of Interpretation). It would also be true if we allow the class terms A to stand only for finite or empty regions.

On page 207, Reichenbach actually says that the geometric and frequency theories are isomorphic. There is no reasonable qualification which makes this true. He discusses the special feature of geometric probability that

(1) if $B_1 \supseteq \cdots \supseteq B_k \cdots$ is a series of measurable sets converging to measurable set B, then $P(B) = \lim_{k \to \infty} P(B_k)$.

His comments on this are curiously phrased. They suggest that it is a special feature, which was not mentioned before, because he wished to deal with finite families of events (as well). But of course, 1 does not fail in the finite case; it holds trivially there. It fails only in the mathematically curious case in which the space is a countable sequence, and P the limit of relative frequencies. If 1 were added to the calculus, that would rule out the relative frequency interpretation; again, unless the designation of the class variables $A, B, \ldots$ were restricted so as to eliminate violations.

It would of course be contrary to Reichenbach's basic intentions to solve any problem by restricting the range of the class variables. For he wishes to say that the probability of A exists if the limit of the relative frequency of A in the long run exists – as a general assertion explicating probability.

5. *Higher-level Probabilities*

In Chapter 8, Reichenbach considers probabilities of probabilities, assertions such as we might like to symbolize

(1) $P(P(C/B) = p/A) = q$.

He writes these in the forms

(2) $(k)(x_k \in A \underset{q}{\ni} [(i)(y_{ki} \in B \underset{p}{\ni} z_{ki} \in C)])$.

That would be a *second-level* assertion. It is abbreviated

(3) $(A^k \underset{q}{\ni} (B^{ki} \underset{p}{\ni} C^{ki})^i)^k$.

A *third-level* assertion would take the form

(4) $(A^k \underset{r}{\ni} (B^{mk} \underset{q}{\ni} (C^{mki} \underset{p}{\ni} D^{mki})^i)^m)^k$.

The way to understand it, I think, is to think of long runs of experiments done in the Central Laboratory at the rate of one per day – these are designated by x_i, y_i, ... and so on. On every day, however, long runs of experiments are done in the Auxiliary Laboratory; the ones done on the kth day are designated by x_{ki}, y_{ki}, ...

In that way, we certainly get non-trivial iterations of the probability implication. I doubt that it is the right explication of what we mean. For although $(B \underset{p}{\ni} C)$ looks like it is a constituent of formula 3, it really is not. The truth conditions of 3 and those of

(5) $$B \underset{p}{\ni} C$$

have nothing to do with each other, because in 5 we correlate sequences of experiments in the Central Laboratory, with respect to B and C, while in 3 we correlate sequences in the Auxiliary Laboratory with respect to B and C.

Let me suggest something that might be a slight improvement. First, I want to do everything in terms of a single function symbol t rather than the diverse ones $x, y, z, \ldots$. For any finite sequence $ijk \ldots n$ of positive integers, let $t_{ijk\ldots n}$ be a countable sequence of events. So one event would be $t_{21}(3)$ for example. Now, in some of these long runs of events, such as t_{21}, we find they are all A. This is a long run selection from class A. And we might ask: in those selections from A, how probable is it that half the B's are C's?

(6) $$\text{relf}\,(\{k : \text{relf}\,(C/B, t_k) = q\}/\{k : (i)(t_{ki} \in A)\}, \sigma) = q$$

where σ is the natural number series, is then the explication of assertion 1.

A slightly different construction puts the probability implication in the antecedent. Thus

(7) $$P(A/P(C/B) = p) = q$$

would on the Reichenbachian explication indicate how often an event in class A occurs in the Central Laboratory, on days when the Auxiliary Laboratory reports a relative frequency p of C's among B's. On my version it would be explicated as

(8) $$\text{relf}\,(\{k : (i)(t_{ki} \in A)\}/\{k : \text{relf}\,(C/B, t_k) = p\}, \sigma) = q\,.$$

That is, it reports how many long runs in which there is a proportion p of C's among B's, are selections from the class A.

My version is semantically just as odd as Reichenbach's, though it has the advantage that in both sides of the slash mark, the same class $\{t_k : k$ a positive integer$\}$ is being talked about. To finish let me just indicate what a third level sentence would look like.

(9) $$P(P(P(D/C) = p/B) = q/A) = r$$

is explicated by Reichenbach with formula 4, and here with

(1) $$\text{relf}(\{k : \text{relf}(\{i : \text{relf}(D/C, t_{ki}) = p\}/\{i : (j)(t_{ki}(j) \in B)\}, \sigma) = q\}/\{k : (i)(t_k(i) \in A)\}, \sigma) = r.$$

IV. THE TRANSCENDENCE OF PROBABILITIES

There is something factual about relative frequencies, and something counterfactual about probabilities. What is probable is a gradation of the possible. And what is likely to happen is what would happen most often if we could realize the same circumstances many times over.

This is the logic of the concept. It implies, to my mind, nothing about the ontology we must accept. But if a physical theory contains probability assertions, then it contains modal assertions, for that is what they are.[6] It is important to see first what such a theory literally says, before we ask what accepting the theory involves.

The view I shall develop here uses a modal interpretation of statistical theories.[7] But it is also a version of the frequency interpretation. Literally construed, probabilistic theories may posit irreducible modal facts – propensities – in the world. But a philosophical retrenchment is possible: we can accept such a theory without believing more than statements about actual frequencies.

1. *Propensities Construed*

Probabilities cannot (all, simultaneously) be identified with the relative frequencies in the actual long run. If we wish to identify them with something real in the world, we must postulate a physical counterpart. In the case of modalities and counterfactuals, realist philosophers postulate dispositions. Probabilities are graded modalities; graded dispositions are propensities. And propensities are just right, because they are postulated to be so.

What exactly is the structure of propensities? Some account is needed, if they are not to be merely an *ad hoc* 'posit'. ('Posit' is a verb which Reichenbach used only in reasonable ways, to indicate empirical hypotheses. In realist metaphysics, however, the operation seems to be conceived on the model of laying eggs, and equally productive.) Kyburg attempted to provide such an account, which represents propensities by means of relative frequencies in alternative possible worlds [5]. In this way he tried to show that there is no real difference between 'hypothetical frequency' and 'propensity' views of probability. This attempt did not quite work because the failure of countable additivity in each possible world infected the function that represented the propensity.

However, the representation I gave in Part I, Section 4, shows how this can be done. In each possible world we select a *privileged* family of sets, with respect to which relative frequency is countably additive in so far as that family contains countable unions. In other words, a *model structure* for the language of probability, will be a set of possible worlds, which together form a *good family of special frequency spaces*. The representation theorem then shows how to interpret probability. So Kyburg was right in the main.

Thus propensities do have structure, which can be described in terms of relative frequencies in different possible worlds. It also allows the following to happen; a certain coin has a propensity of $\frac{1}{2}$ to land heads up, but in fact never does. (The Rosencrantz and Guildenstern Problem, or, if you like, the Tom Stoppard Paradox.) It allows this in the sense of implying that it is possible, occurs in some possible world; which is not to say that our theories would be perfectly all right if it happened.

Of course, if we had a real coin never landing heads up, and since we do not have an ontological telescope which discloses propensities, we would more likely conjecture that this coin has a propensity *zero* to land heads up. This is a typical feature of realist metaphysics: the appearances do not uniquely determine the reality behind them. But it is more than that. It is a typical feature of scientific theories that actual occurrences do not uniquely determine one and only one model (even up to isomorphism) of the theory. Just consider the old problem of mass in classical mechanics: an unaccelerated body may have any mass at all (compatible with the empirical facts), but in any model of mechanics it has a unique mass. There are many such examples. There are some subtle distinctions to be drawn about possibility here, to which I shall return in the last section.

2. *Nominalist Retrenchment*

The bone of contention between medieval realists and nominalists was not so much properties *überhaupt* as causal properties. Fire burns by virtue of its heat, a real property whose presence explains the regularity in fire-involving phenomena. The phenomena do not uniquely determine the world of real properties behind them. As Ockham pointed out, God could have created the world lawless with respect to any connection between fire and burning, but simply have decreed in addition an actual regularity (He directly causing the burning, e.g., the wood turning to charcoal, on all and only those occasions when fire is sufficiently proximate). So we cannot *infer* the real causal properties, dispositions and hidden powers. But we can postulate them, and then reap the benefit of their explanatory power. At this point, the nominalist can reject only the why-questions the realist wants answered ("No, there is no underlying reason for the regularities in nature – they are actual but not necessary") and maintain that science has the air of postulating hidden powers only because it wishes to systematize the description of actual regularities.

To give a systematic description of the actual regularities is to exhibit them as an arbitrary fragment of a larger unified whole. Since actuality is all there is, there is only one picture we can form of that larger whole: it is the system of all possible worlds. Thus Kant in his *Inaugural Dissertation*:

> . . . the bond constituting the *essential* form of a world is regarded as the principle of *possible interactions* of the substances constituting the world. For actual interactions do not belong to essence but to state. ([9], page 40.)

To a nominalist, this picture of the actual as but one of a family of possibles can be no more than a picture; but it can be granted to be the picture that governs our thinking and reasoning.

A distinction must be drawn between what a theory *says*, and what we believe when we accept that theory. Science is shot through and through with modal locutions. The picture the scientist paints holds irreducible necessities and probabilities, dispositions and propensities. Translation without loss into a 'nominalist idiom' is impossible – *but it is also a mistaken ideal.* The nominalist should focus on the *use* to which the picture is put, and argue that this use does not involve automatic commitment to the reality of all elements of the picture.

3. *Epistemic Attitudes*

To explain how a nominalist retrenchment is possible in the specific case of probabilistic theories, I shall try to answer two questions: what is it to accept a (statistical) theory? and, what special role is being played by the 'privileged' sets in my representation?

Elsewhere I have given a general account of acceptance versus belief.[8] A scientific theory specifies a family of models for empirical phenomena. Moreover, it specifies for each of these models a division into the observable parts ('empirical substructures') and the rest. The theory is true if at least one of its models is a faithful replica of the world, correct in all details. If not true, the theory can still be *empirically adequate*: this means that the actual phenomena are faithfully represented by the empirical substructures of one of its models. To believe a theory is to believe that it is true; but accepting a theory involves only belief in its empirical adequacy.

This is a distinction between two epistemic attitudes, belief and acceptance. I am not arguing here, only presenting my view, in a quick and summary manner. A similar distinction must be drawn in the attitudes struck to a particular model, when we say that the model 'fits' the world. This can mean that it is a faithful replica, or merely that the actual phenomena are faithfully mirrored by empirical substructures of this model.

When the model is of a family of possible worlds, then the latter belief, which is about the actual phenomena, only requires that at least one of these possible worlds exhibits the requisite fit. This is to the point when the theory is a probabilistic theory, for in that case it will never narrow down the possibilities to one. It will always say merely: "The actual world is among the members of this family of possible worlds, so-and-so constituted and related."

Now it is possible to see how a statistical theory in general must be constructed. Each model presents a family of possible worlds. In each world, certain substructures are identified as observable. Empirical adequacy consists in this: the observable structures in our world correspond faithfully to observable structures in one of these possible worlds in one of these models. But all this fits in well with my representation of a probability space (which is what a model of a statistical theory must be). For I represent it as a family of possible worlds, in each of which a particular family of events is distinguished from the other events (as 'privileged'). These privileged events can be the candidates for images of observable events.

To see whether this holds, we must look at the testing situation. It is very important to be careful about the modality in 'observable'. Recall that we are distinguishing two epistemic attitudes to theories, in order to see how much ontological commitment is forced upon *us* when *we* accept a theory. Since we are trying to answer an anthropocentric question, our distinctions must be anthropocentric too. It will not do to say that DNA is observable because there could be creatures who have electron microscopes for eyes, nor that absolute simultaneity is an observ*able* relationship, because it could be observed if there were signals faster than light. What is observable is determined by the very science we are trying to interpret, when it discusses our place in the universe. In the testing situation for a statistical theory, we use a language in which we report actual proportions in finite sets, and estimate (hypothesize, postulate, extrapolate) relative frequencies in many other sets. The explicitly checkable sets are those which can be described in this language – and these are the sets which we try to match against the privileged sets in some possible world in the model.

I have already indicated in Part I how I see this. We consider the totality of all sets on whose relative frequency we shall explicitly check in the long run – *whatever they be*. We estimate their probabilities on the basis of actual finite proportions in samples. Since in our language we have the resources to denote limits, by means of expressions like $\text{limit}_{n\to\infty}$, $\sum^{\infty}$, $\bigcup^{\infty}$, it follows that we shall sometimes estimate relative frequencies of limits of sets we have treated already; and we make this estimation by a limiting construction. But that means that we do so by countable addition. For in the presence of finite additivity, the postulates

$$(1) \qquad m(B) = \underset{n\to\infty}{\text{limit}}\, m(B_n) \text{ if } B_1 \supseteq \cdots \supseteq B_k \supseteq \cdots$$

converges to B

(2) $m(B) = \sum_{n=1}^{\infty} m(B_n)$ if $\{B_n\}$, $n = 1, 2, \ldots$ is a countable disjoint family

are equivalent.

Now the sets *explicitly* checked and subject to such estimating hypotheses, are all explicitly named in our language. This language is not static, it can be enriched day by day. But even so, it will be a language with only countably many expressions in it, even in the long run. So the family of privileged sets – the ones encountered explicitly in our dialogue with nature – *whichever they may be*, shall be countable.

The probabilistic theory says that our world is a member of a certain good family of special frequency spaces. Let us consider the Rosencrantz and Guildenstern problem again. Suppose that the theory implies that coins of that construction land heads with probability one-half.[9] Imagine also that this coin is tossed every day from now on, and lands heads every time. In that case the theory is false, and that because it is not empirically adequate. For we have here a class of events described in our language, whose actual relative frequency is not equal to the probability the theory attributes to it. If this particular coin's behaviour is an isolated anomaly, the theory may still be a useful one, in some attenuated way. But of course, what I am describing here is not a situation of epistemic interest. What I say is an extrapolation from this: we would know only that the first x tosses had landed heads, and we would count this as *prima facie* evidence against the theory; but as long as we accept the theory we shall assert that in the long run, this coin too will land heads half the time. It seems mistaken to me to interpret statistical theories in such a way that they can be true if the relative frequencies do not bear out the probabilities. For if that is done (*and various propensity views suggest it*) then it is possible to hold this theory of our present example, and yet not assert that in the long run Rosencrantz' coin will follow suit – or indeed, *any* coin, *all* coins. The empirical content disappears.

What is important about Rosencrantz' coin is that the model should have in it some possible world (which we emphatically deny to be the actual one) in which an event described by us here shows aberrant relative frequencies. What we must explain is the assertion that this is always possible, but never true. (To be logical: as long as we are looking at a model of *this* theory, then in each possible world therein it is *true* that if X is a described event, the probability of X equals its relative frequency, and it is *also true* in each world that *possibly*, the relative frequency of X is not its probability. The language in which X is described, is here given a semantics in the fashion of 'two-dimensional modal logic', so that a sentence can be true in every world where that sentence expresses a proposition, but still it does not express a necessary proposition. Consider 'I am here' which is true in every world in which there are contextual factors (a speaker, a place) fixing the referents of the terms, but which expresses in our world the contingent proposition that van Fraassen is in Toronto.)

To summarize, suppose that a physical theory provides me with probabilistic models. Each such model, as I interpret science, is really a model of a family of possible worlds. The empirical content of the theory lies in the assertion that some member of that family, for some model of the theory, fits our world – insofar as all empirically ascertained phenomena in the long run are concerned, whatever they may be.

To accept a theory, in my view, is only to believe it to be empirically adequate – which is, to believe that empirical content which I just indicated. So to accept a probabilistic theory involves belief only in an assertion about relative frequencies in the long run.

In this, I feel, Reichenbach was right.

V. POSTSCRIPT

After writing this paper I remained dissatisfied with the details of the modal frequency interpretation proposed in Part I, Section 4 and Part IV, Sections 1 and 3. Happily I had the opportunity to develop this interpretation further as part of my lectures on the foundations of probability theory at the 1977 Summer School on the Foundations of Physics of the Enrico Fermi Institute (proceedings published in *Il Nuovo Cimento*, 1978). Using a theorem due to Polya and reported by von Mises, the privileged frequencies in a given possible world can, it turns out, be taken to be those of the observable (rather than the explicitly observed) events.

In my paper 'Relative Frequencies' the privileged events in a world are those which are explicitly observed during that world's history. The observation may be mechanical, of course, but it must be actual.

However, in my general 'constructive empiricist' account of theories, to accept a theory is to believe that it is correct with respect to all observable phenomena. This means: all the events which are humanly discriminable; not merely those which, due in part to the experimenter's whim, are actually observed. Richmond Thomason argued strongly that a corresponding course should be followed in the interpretation of probabilistic or statistical theories; but at first, I did not see how this could be done, because if the basic (atomic) set of discriminable events forms a countable partition of the sample space, the sigma-field of events it generates is uncountable. But von Mises' *Mathematical Theory of Probability and Statistics* (New York: 1964; pp. 18–20) reports the following theorem due to Polya:

Let $\{p_i\}$, $i = 1, 2, \ldots$ be a set of real numbers such that

(a) $0 \leqslant p_i \leqslant 1$ and $\sum_{i=1}^{\infty} p_i = 1$

(b) $p_i = \text{limit}_{m \to \infty} \; m_i/m$

Then $\sum \{p_i : i \in I\}$ exists and equals $\text{limit}_{m \to \infty} \sum \{m_i/m : i \in I\}$ for any set $I \subseteq \{1, 2, \ldots\}$.

The remainder of this postscript is a summary of Section 4.3 'A Modal Frequency Interpretation' of my lectures to the Fermi Institute (to be published in the proceedings, see above). The focus of the interpretation has shifted from the 'privileged' events to the 'significant' ones, the latter notion providing a precise version of the notion of 'observable phenomenon' in a statistical theory.

In an actual experiment we generally find numbers attaching to large but finite ensembles (whether of similar coexistent systems, or of the same system brought repeatedly into the same state) which we compare with theoretical expectation values.

But this comparison is based on the idea that the number found depends essentially on a (finite) frequency in just the way that the theoretical expectation value depends on a probability. Moreover, we demand closer agreement between the number found and the expectation value as the size of the ensemble is increased.

Hence it is reasonable to take as an *ideal* (*repeated*) *experiment* an experiment performed infinitely many times under identical conditions or on systems in identical states. The relation between the ideal and the actual should then be this: the actual experiment is thought about in terms of its possible extensions to ideal repeated experiments (its ideal extensions). If this is correct, we compare an actual experiment with a conceptual model, consisting in a family of ideal repeated experiments. In that model there should be an intimate relation between frequencies and probability so that the model can be directly compared with the theory under consideration. Secondly, the theory of statistical testing should be capable of being regarded as specifying the extent to which the actual 'fits' or 'approximates' that model of the experimental situation.

In any actual experiment we can only make finitely many discriminations. We can, for instance, determine whether a given spot appears with an x-coordinate, in a given frame of reference, of $y \pm 1$ cm; with $-10^5 \leqslant y \leqslant 10^5$. The possible values of the x-coordinate are all real numbers, but we focus attention on a finite partition of this outcome space. The *first idealization* is to allow that partition to be countable, that is, finite or countably infinite. Secondly, in the actual experiment we note down successively a finite number of spots; the *second idealization* is to allow this recorded sequence of outcomes also to be countable.

Note that theory directs the experimental report at least to the extent that the outcome space (or sample space) is given; in different experiments of the same sort we focus attention on different partitions of that space. So we begin with a space $\langle K, F\rangle$, F a Borel field, and describe one ideal experiment by means of a countable partition $\{A_n\} \subseteq F$ and a countable sequence σ of members of K.

We also note that in the idealization, we could have proceeded in different ways; for instance, we could have merely lifted the lower and upper limits to y, or we could also have made the discrimination accurate to within 1 mm as opposed to 1 cm. So our model will incorporate many ideal experiments with partitions $\{A_n^\alpha\}$, all of which are finer than the partition in the actual experiment. These represent different experimental set-ups, and we cannot expect the series of outcomes to be the same, partly because the change in the set-up may affect the outcomes, and partly because we must allow for chance in the outcomes: if many ideal experiments *were* performed, just as in the case of actual ones, we would expect a spread in the series of outcomes.

But we are building a model here of what would happen in ideal experiments, and this model building is guided either by a theory we assume, or by the expectations we have formed after learning the results of actual experiments. Hence we expect not only a spread in the outcomes, but also a certain agreement, coherence. In actuality, such agreement would again be approximate, but here comes the *third idealization*: we assume that the agreement will be exact.

To make this precise, let us call *significant events* in an ideal experiment with partition $\{A_n\} = G \subseteq F$ exactly the members of the Borel field $BG \subseteq F$ generated by G. We now *stipulate first* that, if that ideal experiment has outcome sequence σ, then relf(A_n, σ) is well-defined for each A_n in G and that $\sum_{n=1}^{\infty}$ relf$(A_n, \sigma) = 1$. By Polya's result it follows then that refl(—, σ) is well-defined and countably additive on BG. Only the relative frequencies of those significant events generated by the partition will be considered in the use or appraisal of any ideal experiment. We *stipulate secondly* that if A is a significant event in several ideal experiments or the countable union of significant events in other such experiments then the frequencies agree as required. To answer the question how many ideal experiments the model should contain, we *stipulate thirdly* that the significant events together form a Borel field on the set K of possible outcomes. This must include, for instance, that if we consider experiments with partitions $\{A_n^m\}$, $m = 1, 2, \ldots$, and $A_1^m \downarrow A$, then there must be some experiment in the model in which A is a significant event.

Together, these three stipulations give the exact content of the third idealization. And the three idealizations together yield the notion of a model of an experimental situation by means of a family of ideal experiments ('*a good family*'), which I shall now state precisely.

A good family (*of ideal experiments*) is a couple $Q = \langle K, E \rangle$ in which K is a non-empty set (the 'possible outcomes' or 'worlds') and E is a set of couples $\alpha = \langle G_\alpha, \sigma_\alpha \rangle$ (the 'possible experiments') such that

(i) G_α is a countable partition of K and σ_α a countable sequence of members of K (the 'outcome sequence' of the experiment α);

(ii) if $A_1, A_2, \ldots$ are in $BG_{\alpha_1}, BG_{\alpha_2}, \ldots$, the Borel fields generated by $G_{\alpha_1}, G_{\alpha_2}, \ldots$ (the Borel fields of 'significant events' of $\alpha_1, \alpha_2, \ldots$)

then there is an experiment β in E such that $B = \bigcup^{\infty} A_i$ is in β and $\text{relf}(B, \beta) = \sum \{\text{relf}(A_i, \alpha_i)\colon i = 1, 2, 3, \ldots\}$ if the A_i are disjoint.

(iii) $\text{relf}(A, \sigma_\alpha)$ is well-defined for each A in G_α.

(iv) $\sum \{\text{relf}(A, \sigma_\alpha)\colon A \in G_\alpha\} = 1$;

(v) If $A \in BG_\alpha \cap BG_\beta$ then $\text{relf}(A, \sigma_\alpha) = \text{relf}(A, \sigma_\beta)$.

As noted, in the experiment $\langle G_\alpha, \sigma_\alpha \rangle$ we call σ_α the *outcome sequence* and the members of BG_α the *significant events.* Polya's result guarantees, from (iii) and (iv), that $\text{relf}(A, \sigma_\alpha)$ is defined for each significant event A. Finally, we call $\langle K, F \rangle$, where $F = \{G_\alpha : \alpha \in E\}$ the *sample space of* good family Q. It will be noted from condition (ii) that F is itself a Borel field, for, if $A_1, A_2, \ldots$ are significant events of $\alpha_1, \alpha_2, \ldots$, their union is in BG_β and hence in F.

If $Q = \langle K, E \rangle$ is a good family with sample space $\langle K, F \rangle$, *we define*: $PQ(A) = r$ if and only if $\text{relf}(A, \sigma_\alpha) = r$ for all α in E such that A is a significant event in α.

RESULT I. If $Q = \langle K, E \rangle$ is a good family with sample space $\langle K, F \rangle$, then $\langle K, F, PQ \rangle$ is a probability space.

RESULT II. If $\langle K, F, P \rangle$ is a probability space, then there is a good family $Q = \langle K, E \rangle$ with sample space $\langle K, F \rangle$ such that $P = PQ$.

We now have the desired representation result: probability spaces bear a natural one-to-one correspondence to good families of ideal experiments. It is possible therefore to say:

the probability of event A equals the relative frequency with which it would occur, were a suitably designed experiment performed often enough under suitable conditions.

This is the modal frequency interpretation of probability.

University of Toronto

NOTES

* The author wishes to thank Professors R. Giere (Indiana University), H. E. Kyburg Jr. (University of Rochester), and T. Seidenfeld (University of Pittsburgh) for their help, and the Canada Council for supporting this research through grant S74-0590.

[1] I use Kolmogorov's term 'Borel field' rather than the more common 'σ-field', because I shall use the symbol sigma for other purposes.

[2] R. von Mises claimed countable additivity in [12], Chapter I; but the failure of this property for relative frequencies has long been known. See [4], pages 46 and 53 and [3], pages 67–68; I owe these two references to Professors Seidenfeld and Giere respectively.

[3] This argument is reported in [10], pages 3–59 and 3–60; I owe this reference to Professor Giere.

[4] Professor Kyburg pointed out to me that the conclusion does not follow from the Bernoulli theorem. Needed for the deduction is also the Borel–Cantelli lemma. For details see Ash [1], Chapter 7; especially the theorem called Strong Law of Large Numbers for Identically Distributed Variables.

[5] This subject, and Kolmogorov's discussion of related problems, will be treated in a forthcoming paper 'Representation of Conditional Probabilities'.

[6] This point of view about modality in scientific theories was argued strongly in a symposium at the Philosophy of Science Association 1972 Biannual Meeting, by Suppes, Bressan, and myself; the emphasis on probability assertions as modal statements was Suppes'.

[7] In other publications I have developed a modal interpretation of the mixed states of quantum mechanics which can be regarded as a special case of the present view.

[8] In my 'To Save the Phenomena'; a version was presented at the Canadian Philosophical Association, Western Division, University of Calgary, October 1975; a new version will be presented at the American Philosophical Association, Eastern Division, 1976.

[9] I am keeping the example simple and schematic. Our actual beliefs, it seems to me, imply only that in the set of all tosses of all coins, which is most likely finite but very large, half show heads. The same reasoning applies. Our belief is false if the proportion of relative frequency is not one-half. This is quite possible – we also believe *that.* But would be a mistake to infer that we believe that perhaps the proportion is actually not one-half. We believe that our beliefs could be false, not that they are false – what we believe are true but contingent propositions. (This problem can also not be handled by saying that a theory which asserts that the probability is heads equals one-half implies only that the actual proportion is most likely one-half; for then the same problem obviously arises again.)

REFERENCES

[1] Ash, R. B., *Real Analysis and Probability*, Academic Press, New York, 1972.

[2] Belnap, Jr., N. D., 'Conditional Assertion and Restricted Quantification', *Nous* **4** (1970), pp. 1–12.

[3] Fine, T., *Theories of Probability*, Academic Press, New York, 1973.

[4] Kac, M., *Statistical Independence in Probability, Analysis and Number Theory*. American Mathematical Association, Carus Mathematical Monograph #12, 1959.

[5] Kyburg, Jr., H. E., 'Propensities and Probabilities', *British Journal for the Philosophy of Science* **25** (1974), pp. 358–375.

[6] Popper, K. R., *The Logic of Scientific Discovery*, Appendices *iv and *v. revised ed., Hutchinson, London, 1968.

[7] Reichenbach, H., *The Theory of Probability*, University of California Press, Berkeley, 1944.

[8] Renyi, A., 'On a New Axiomatic Theory of Probability', *Acta Mathematica Hungarica* **6** (1955), pp. 285–333.
[9] Smith, N. K. (ed.), *Kant's Inaugural Dissertation and Early Writings on Space*, tr. J. Handiside, Open Court, Lasalle, Ill., 1929.
[10] Suppes, P., *Set-Theoretical Structures in Science*, Mimeo'd, Stanford University, 1967.
[11] van Fraassen, B. C., 'Incomplete Assertion and Belnap Connectives', pp. 43–70 in D. Hockney *et al.* (ed.) *Contemporary Research in Philosophical Logic and Linguistic Semantics*, Reidel, Dordrecht, 1975.
[12] von Mises, R., *Mathematical Theory of Probability and Statistics*, Academic Press, New York, 1964.

BEN ROGERS

THE PROBABILITIES OF THEORIES AS FREQUENCIES*

From the beginning of his career, Reichenbach studied the role that probability played both in modern physical theory and in epistemology.[1] He was, with Richard von Mises, one of the foremost proponents of the frequency theory of probability and axiomatized a very general form of it. He further claimed that all reasonable uses of the concept of probability, both in physical theory and as an evaluative concept in epistemology, are to be explicated in terms of the frequency theory.

Central to Reichenbach's epistemological position is the very strong claim that all inductive inference is reducible, in principle, to inference by simple enumeration.[2] Briefly, one can infer probability statements about classes of events by means of the rule of simple enumeration; and, once such probability statements are available, other probability statements can be inferred from these by means of the probability calculus. Since these later inferences are analytic, all inductive inferences are reducible to the application of the enumeration rule which supplies the probabilities in the first place. In particular, he accepts the idea that the relation between a theory or hypothesis and the evidence for or against it can be construed in terms of probabilities.[3] Via such probabilities of hypotheses, construed as frequencies, all inductive inferences about scientific theories then in principle can be reduced to inference by simple enumeration.

Most recent work concerning the probability of theories has approached the problem either by attempting to develop the concept as a logical relationship as in the work of Carnap or via a personalist Bayesian interpretation as in the work of de Finetti and Savage. Except for the extensions of his work made by Salmon, Reichenbach's attempt to interpret the probabilities of hypotheses in terms of the frequency theory has been little discussed in the literature, Nagel's comments in *Principles* and in his review of the English edition of *Theory of Probability* constituting almost the sole discussion.[4] On the other hand, Reichenbach's frequency interpretation and his justification of induction have been

W. C. Salmon (ed.), Hans Reichenbach: Logical Empiricist, 169–185.

widely discussed. Given the dearth of published commentary, one can only speculate as to why his attempt has received so little attention. Perhaps, the general difficulties with the frequency interpretation have made it seem otiose to show the complexities and defects in his attempt to handle the probability of hypotheses in this interpretation, or it could have been the intrinsic appeal of the logical interpretation.

However, I conjecture that one reason for the lack of discussion is that there is the following very evident prima facie objection to the interpretation as exposited by both Reichenbach and Salmon. Given the form of their exposition, there *seems* to be no way to assign probabilities to statistical hypotheses except in extremely limited circumstances. This objection has long kept me from giving this interpretation other than the most casual study and has been mentioned by almost everyone I have discussed the interpretation with. After explaining how Reichenbach intends the interpretation to be carried through, I will exhibit the objection in detail and then offer a partial resolution of the difficulty. I hope that the removal of this prima facie objection and the subsequent clarification of the nature of the difficulties which still face the interpretation will lead to a more definitive reappraisal of Reichenbach's work on this important topic in inductive inference.

Reichenbach analyzes inductive inference in terms of two fundamental items: a rule of induction by simple enumeration which licenses the assertion of probability statements, and the probability calculus which establishes deductive relationships between probability statements, some of which must be taken as already established. The rule of inference allows one to infer limit statements of the relative frequency of properties of things taken in ordered sequences on the basis of frequencies in finite ordered sequences. The connection between the rule of inference and the probability calculus is established by showing that the identification of the limit so inferred with probability results in an admissible interpretation of that calculus.

His central thesis is that given a suitable choice of statements taken as true[5] the enumeration rule will license inferences to probabilities of other statements; and hence, actual inductive inference can be understood in terms of this interpretation. The fundamental difficulty in understanding his interpretation is in getting clear on the manner in which statements of different levels of generality are related to each other by the interpretation so that true theories, if there are any among those under consideration, will get assigned a probability near one in the long run.

As is usual in studies of epistemic probability, the interpretation is chosen so that Bayes' Theorem may be used to calculate the increase or decrease in the probability of a hypothesis as new evidence is considered. In a simple form the theorem states:

$$P(A \cap C, B) = \frac{P(A, B) \times P(A \cap B, C)}{P(A, B) \times P(A \cap B, C) + P(A, \bar{B}) \times P(A \cap \bar{B}, C)}$$

For the purposes of the present interpretation, let A represent theories of a certain kind, B represent true theories, and C represent evidence of a certain kind.[6] The $P(A \cap C, B)$ represents the probability of truth, given theories of kind A and evidence of kind C; that is, it represents the posterior probability of a theory of kind A on evidence C. Similarly, $P(A, B)$ represents the probability of the truth of A prior to the evidence provided by C, or the prior probability of A. An analogous interpretation is given to $P(A, \bar{B})$ and $P(A \cap \bar{B}, C)$, where $\bar{B}$ represents the class of false theories.

A particular theory T is not assigned a probability on Reichenbach's account because it is an individual theory and individual instances of a class (e.g., events) cannot have a probability on the relative frequency account; probability is a property of classes. However, *weight* is assigned to individual instances by assigning the individual instance to any appropriate reference class,[7] and these weights form the basis for practical action or evaluation. For example, weights can function as betting quotients in games of chance where bets are placed on forthcoming individual events. So, the weight of an individual theory T on evidence E is posited by assigning T to an appropriate class of theories A and an appropriate class C of evidence statements like E, and it is such posits about weight which are used for the evidential comparison of individual theories.

In order to use Bayes' Theorem to calculate the posterior probability of theories of kind A, it is necessary to have the probabilities on the right side of the equation. Following Reichenbach, Salmon suggests that these probabilities may be obtained from a certain probability lattice. Each horizontal row in the lattice represents a particular theory, T_i.

> To incorporate a theory into the lattice as a row we consider the separate observations which serve as tests of that theory. Some of these observations will be positively confirmatory, others will tend to disconfirm the theory. For a confirmatory instance we will put 'E' and for a disconfirmatory instance '$\bar{E}$'. This will yield a sequence such as this:
>
> $\bar{E}EEEE\bar{E}E\bar{E}EEEE\bar{E}EEE$[...]...
>
> ... A is the class of horizontal rows in the lattice and derivatively the class of theories which contribute rows.... B is the class of true theories – represented by rows in which the limit of the relative frequency of E is very close to 1. C is the class made up of rows all of which have similar initial sections.[8]

The probabilities $P(A, B)$, $P(A \cap B, C)$, and $P(A \cap \bar{B}, C)$ are inferred directly from the lattice by applying the enumeration rule to the appropriate classes. For example, $P(A, B)$ is obtained by examining, in the vertical direction in the lattice, those theories T_i which are of a kind A and then obtaining for this reference class the relative frequency of true theories, B, which are represented by horizontal rows in which the limit of the relative frequency of E's is near 1. Since $P(A, \bar{B})$ equals $(1 - P(A, B))$, on this interpretation we have all the probabilities required to calculate $P(A \cap C, B)$ by means of Bayes' Theorem.

Reichenbach distinguishes between the use of the induction by enumeration in contexts of primitive knowledge, where no probability values are known, and in situations of advanced knowledge where some probability values are already known.[9] The probabilities $P(A, B)$, $P(A \cap B, C)$, and $P(A \cap \bar{B}, C)$ in the interpretation above are clearly cases of the application of the enumeration rule in the context of advanced knowledge because they depend on prior inferences concerning the limit of the E's in each horizontal row. These latter limits are inferred from the initial sequences of E's in each row by the enumeration rule. The inference about the limit of E's in each individual lattice row appears to be an inductive inference in a state of primitive knowledge because no probability knowledge appears to be presupposed in the inference. These row probabilities form the basis for 'the concatenation of inductions' via either the probability calculus (principally the use of Bayes' Theorem) or enumerative inductions in advanced knowledge (as in the inference in the vertical direction on the lattice above to obtain $P(A, B)$). It is these concatenated inductions, and not enumerative induction in primitive knowledge, which characterize most of our actual inductive inference.

For Reichenbach, all inductive inference is based on and is reducible to enumerative induction from observations. But as has just been described, he argues that most practical inductive inference occurs in a context of advanced knowledge, where the probabilities of some statements are known; and such inference is best construed as applications of Bayes' theorem which yield the posterior probability of the truth of the statements under consideration on the available evidence. But ultimately the probabilities used in inference in advanced knowledge are based on the lattice row probabilities which are associated with individual theories, hypotheses, or other statements about physical events. Hence, it is important to see in detail what is involved in the characterization of these row probabilities and to see with what epistemic characteristics they are endowed. To accomplish this we must examine in more detail the role played by the enumerative rule in assigning probabilities to theories.

First, let us consider induction by simple enumeration as characterized by Reichenbach in his pragmatic justification of induction. He considers a sequence of events of some specified kind, A, which constitute the reference class for the inference. With each event A is associated another event, and it is to be determined whether the associated event has an attribute of interest, B. For example, the A's might be tosses of a given die and B might be the property of the ace being turned up on the toss. Let $F^n(A, B)$ be the relative frequency of B's among the first n members of A. The rule of induction by simple enumeration allows the inference of a probability on the basis of an observed frequency.

> The rule of induction by simple enumeration: Given $F^n(A, B) = m/n$ one may infer that the $\lim_{n \to \infty} F^n(A, B) = m/n \pm d$.

That is, given the frequency interpretation of probability, one has inferred that $P(A, B) = m/n \pm d$, since $P(A, B)$ is identified with the limit (as $n \to \infty$) of $F^n(A, B)$.

Reichenbach claims that we are justified in using the rule of induction by simple enumeration because continued application of the rule will lead to the assertion of the actual limit in a finite time if the limit exists.[10] The probability which is asserted on the basis of the rule establishes a relation between a *class* of events of kind A and a *class* of events of kind B. Our practical interest is, however, with the prediction and explanation of individual events of these kinds; and, strictly speaking, the probabilities so defined do not apply to the individual events but to the classes only. But he argues that if the reference class, A, is chosen appropriately, then the probability, $P(A, B)$, can be transferred to individual A's as a weight or posit which may serve as a betting quotient for betting on the occurrence of individual B's.

In the fundamental interpretation of probability given by Reichenbach, the things which are related are events classified by their physical properties, and the probability assertions which are licensed by the rule of enumerative induction are assertions about the long run relative frequencies of these events. Thus probability statements are claims about general *physical facts*; they are claims which, if true, are physical laws.[11] By analogy, it would seem that if one infers from events which are confirming instances of a theory to the probability of the theory by means of enumerative induction then the probability of the theory must also be construed as a physical fact. But to construe the evidential relationship between a theory and the evidence for or against it, which is what the probability of a theory presumably expresses, as a physical fact must seem, for many philosophers, simply a category mistake. Surely, they might say, the evidential relationship is a logical relationship, not a relationship expressible as a physical fact or law. The prejudice against Reichenbach's interpretation is reinforced, I suspect, by the prima facie objection to his interpretation which I alluded to in the opening paragraphs of this essay. It is to the development of the details leading to this prima facie objection that we now turn.

As explained above, the interpretation of the probabilities of theories is given in terms of a probability lattice. Each row in the lattice represents a different theory. A particular theory, T, is represented in its row by E's and $\bar{E}$'s which stand for respectively *confirmatory instances* of T and *disconfirmatory instances* of T.[12]

The most basic question here is, 'What is a confirmatory instance of a theory?' In characterizing what counts as a confirming instance of a theory, one must start with an account of the kinds of things about which assertions are made in the theory. For the sake of simplicity, I shall assume that the theory T contains predicates which classify *individual* physical events into three kinds: D, F, and G. Also, let I represent a set of initial conditions which, relative to T, will yield the set of events E, which are confirming instances of T. Let us further suppose that T entails assertions of probabilistic hypotheses, H, of the form $P(D, F) = r, 0 \leqslant r \leqslant 1$, which state relations between classes of events of the kinds D and F.

If the theory T entails universal statements H of the form $(x)(Dx \supset Fx)$, then the occurrence of an individual event D would constitute an instance of an initial condition I and if F occurred, then a confirming instance E of the theory is established; and if F fails to occur, a disconfirming instance $\bar{E}$ of the theory is established. From the resulting E's and $\bar{E}$'s a lattice row is constructed which is associated with H and thereby with T. The limit of this sequence of E's and $\bar{E}$'s is inferred by the enumeration rule. The property of truth is ascribed to H if the limit of $F^n(I, E)$ is near 1.

When T contains statements of other than strictly universal form, it is not so clear what is to count as a confirmaion instance of the theory. Consider the case where T entails statements H which ascribe low probability to *individual* events of a certain kind, e.g., $P(D, F) = \frac{1}{6}$. On Reichenbach's account such a statement may be *asserted* on the basis of the enumeration rule from the observed frequency, $F^n(D, F)$, of the individual events of the kinds D and F. But it is not at all clear how one can come to assert that the probability of the statement $P(D, F) = \frac{1}{6}$ is near 1, because it is not at all clear on Reichenbach's theory what is to count as a confirmatory instance of a statement like this. The difficulty is that one cannot simply count the *individual* occurrences of F's as confirming it because if $P(D, F) = \frac{1}{6}$ is true one would expect $F^n(D, F)$ to be near $\frac{1}{6}$ and thus would not approach a limit near 1, which is expected of confirmatory instances when the statement tested is true. But if the individual events F are not confirming instances of the statement $P(D, F) = \frac{1}{6}$, what are?

Reichenbach, in expositing his interpretation of the probability of theories, considered only universal statements and probability statements which attributed very high probabilities to individual events.[13] For such statements there seems to be little question as to what counts as confirming instances of them. But surely if one is to have an adequate account of the probability of hypotheses and theories, one must provide for the confirmation and disconfirmation of statistical statements. That Reichenbach did not seem to do so constitutes a powerful prima facie objection to the attempt to construe the probability of hypotheses in terms of the relative frequencies of the events the hypotheses are about. For example, if it is claimed that the probability of aces on the toss of a die is $\frac{1}{6}$, then the occurrence of an ace, in and of itself, does not constitute evidence for the claim. This prima facie objection reinforces the intuition that the relative frequency interpretation of the probability of hypotheses rests on a category mistake, that the relation between a hypotheses and the evidence relevant to it is a logical relation and *not* an empirical one.

I will now consider the possibility of characterizing confirmation instances in keeping with Reichenbach's interpretation. To do so I shall take orthodox statistical practice as a guide, but certainly one to be followed warily. The idea is to construct a confirmation class, E, such that if certain initial conditions, I, are fulfilled then for a particular observation 0, the $P(I, 0 \in E)$ is near 1 if the probabilistic hypothesis H: $P(D, F) = r$ is true. In general, one chooses a class of D's which constitute the set of initial conditions, I. For example, I, might be sets of n consecutive D's. Then one calculates the confirmation class E such that $0 \in E$ iff the observed $F^n(D, F)$ is in the range $r_1 \leqslant F(D, F) \leqslant r_2$ and $P_H(I, O \in E)$ is near 1.[14]

If one can calculate a class E with these properties and if when $0 \in E$ one enters E in the lattice row which represents the theory T and enters $\bar{E}$ when $0 \notin E$, then that row has the property that if H is true the limit of the relative frequency of E's in the row will be near 1. Thus by following the guide of orthodox statistical theory we can characterize the concept of the confirmation instance of a probabilistic hypothesis which meets Reichenbach's condition that the limit of confirming instances is near 1 if the hypothesis is true.

What is needed in order to select an I and its correlated E which has the desired property that $P(I, 0 \in E)$ is near 1? First, one must have H stated. Secondly, the structure of the sequence of ordered pairs $\langle D, F \rangle$, must be characterized. Proceeding with our earlier example where H is $P(D, F) = \frac{1}{6}$, we need to choose a set of n D's which will be the initial condition for the confirmation class E, so that $P(I, E)$ can be established. If the F's are distributed randomly in the sequence of $\langle D, F \rangle$'s, then we might choose $n = n_1$. But suppose we knew that F's tend to form runs whose average length is longer than would be expected if the sequence is random. Then we would choose $n = n_2$ so that $n_1 \ll n_2$, because we would need a much longer sequence of D's in order to get a fair sample of $F^n(D, F)$ than if the sequence of F's is random. Thus, in general, a necessary condition for the calculation of an E which meets the criterion that $P(I, 0 \in E)$ is near 1 is that the structure of the sequence of F's be characterized, as well as H stated.

Once it is recognized that one cannot define a confirmation instance for the probabilistic hypothesis H unless one can also characterize the structure of the sequence of ordered pairs of events $\langle D, F \rangle$, which H is about, one can see that there is a certain difficulty in carrying through Reichenbach's schema for defining the probability of hypotheses in terms of the relative frequency of their confirmation instances. The difficulty concerns the epistemic status of the characterization of the structure of the sequence of $\langle D, F \rangle$'s.

One option is to consider the class of E's, not as confirmation instances of the probabilistic hypothesis H alone, but as confirmation instances of the conjunction of H and S, where S is a hypothesis about the structure of the sequence of $\langle D, F \rangle$'s. If this option is exercised, then one cannot assign a probability to any arbitrary empirical hypothesis via Reichenbach's schema, but only to certain *joint* hypotheses, like H and S, which both attribute a probability to events of a certain kind in a sequence and which also concern the probabilities of events in the subsequences of the principal sequence of $\langle D, F \rangle$'s. I shall call such an interpretation a *joint interpretation* because the confirmation instance E confirms H and S taken conjointly but is not taken to confirm H (or S) taken alone. The row probability of the conjoint hypothesis H and S is defined in terms of the confirming instances; and since the characterization of a confirming instance does not depend on knowing some other empirical generalization, the row probability is inferred in the context of primitive knowledge. In attaining a joint interpretation one has paid the price of being unable to assign a row probability by enumeration over confirming instances to any arbitrary probabilistic hypothesis, but only to those which are conjoined with a hypothesis about the structure of the sequence of events referred to by the first hypothesis.

There are serious objections to the joint interpretation as defined above. First, Reichenbach's overall program seems to require that there be no significant restriction on what kind of hypothesis can be assigned probabilities. Especially, straightforward probabilistic hypotheses surely must not be excluded, for after all these are exactly the kind of statements which are assertable on the basis of the enumeration rule and, as such, have a fundamental place in his treatment of induction. A second consideration against the joint interpretation is that it goes against scientific practice to the extent that, by whatever poorly understood means, we do seem to judge the probability of individual statements within a theoretical context, the typical Duhemian argument notwithstanding. Third, as I shall argue subsequent to my discussion of the second option below, there remains an essential arbitrariness to the preceding definition of confirmation instance which is intuitively unacceptable and to which the joint interpretation is particularly subject.

In contrast to the joint interpretation characterized above, let us say that an interpretation is an *individual interpretation* when a confirmation instance E confirms an individual hypothesis H. In the preceding discussion, I showed that in the individual interpretation it is in general impossible to define a confirmation instance E for H in the context of primitive knowledge, because the definition depends on the characterization by S of the structure of the sequence of $\langle D, F \rangle$'s. For the joint interpretation, there remains open the option of defining a confirmation instance so that the row probability of the conjoint hypothesis H and S is inferred in the context of primitive knowledge. A second option must be taken in the case of the individual interpretation, which is to assert S by means of induction by enumeration. If this is done, then the confirmation class E for a probabilistic hypothesis H is calculated on the basis of the structure S, which itself has been asserted by an application of the rule of enumerative induction. In this way, Reichenbach'a claim that the probability of hypotheses is based solely on induction by enumeration and inference via the probability calculus would be fulfilled. But if this second procedure is followed, in the resulting individual interpretation the characterization of a confirming instance of H depends on the structure as characterized by S, is not defined therefore purely with respect to H, and is justified only if the assertion about the character of S is true. In the individual interpretation, the enumeration rule must be used to make assertions over two entirely different domains in the process of ascertaining the row probability of H. First, it is used over the domain of individual events like D and E, which domain forms the inductive basis for the assertion of the structure statement *S*. Then it is used a second time over the domain of confirmation instances to obtain the row probability for *h*. Thus, the row probability of *H*, on the individual interpretation, is a probability asserted in the context of advanced knowledge.

Similarly, on the individual interpretation, if one wants to calculate the probability of a structural hypothesis like *S*, then it will be necessary, in order to calculate a confirmation class for *S*, to assert some limit statements about the events which determine the primary sequence of $\langle D, F \rangle$'s, i.e., statements like *H*, on the basis of induction by enumeration.

Reichenbach himself analyzed carefully the nature of orthodox statistical theory. In so doing he clearly recognized that most of the inferences licensed by this theory are made in the context of advanced knowledge. In fact, he described in detail how one might proceed to determine the nature of the sequence structure by repeated use of the enumeration rule.[15] What is missing in his writing, and which I am attempting to supply, is the extension of these ideas to the problem of the probability of hypotheses, so that it is treated in its full generality. The aim is to overcome a prima facie inadequacy of the interpretation and to lay bare the more fundamental difficulties of this position.

With respect to his interpretation of the probability of hypotheses, the conclusion is that the lattice row probability of an arbitrary individual hypothesis can be asserted only on the basis of a prior inference by induction to another generalization. This certainly goes against the impression Reichenbach sometimes gives that the row probability of any hypothesis can be given in the context of primitive knowledge;[16] that is, that all of the uses of the enumeration rule in determining a row probability of *H* are of the same kind, namely over confirming instances alone. Does this situation undercut the self-corrective nature of repeated uses of the enumerative rule? The corrective nature seems to be retained only if the characterization of a confirmation instance relative to a particular *H* may change as one moves out the lattice row, which is somewhat awkward and certainly not something Reichenbach tells us about his method. More will be said about this problem toward the end of the paper.

In the preceding pages, I have been exploring the difficulties which the Reichenbach analysis is heir to because the confirmation instance of a hypothesis can only be defined if the structure of the primary sequence is characterized. There is another, and I think, more serious difficulty facing his account on the score of characterizing this concept, which has to do with a certain arbitrariness in the choice of the confirmation class E. Earlier the confirmation class E was characterized as a class defined on a class I of D's such that the probability of a particular observation 0 having an observed $F^n(D, F)$ falling in the interval $r_1 \leq F^n(D, F) \leq r_2$ is near 1. But the class E in general is *not* unique. There are typically a large number of classes E (the set is infinite for a continuous probability distribution) so that $P_H(I, 0 \in E)$ is near 1. These alternative classes are constructed by taking different ranges into which $F^n(D, F)$ must fall, all the ranges being subject to the condition that $P_H(I, 0 \in E)$ is near 1. Some of these classes are highly counterintuitive as confirmation classes. For example, suppose $P(D, F) = 0.2$ and the sequence of $\langle D, F \rangle$'s is random in F. If one takes an observation of 1000 D's then the probability of getting exactly 200 F's is small; but still 200 F's is a possible observation, no other single observation has a higher expectation, and most other single values have a much lower expectation. Yet one can construct a confirmation class E so that $P(I, 0 \in E)$ is near 1 and 200 F's is not in the confirmation class at all.

When faced with this difficulty, one seeks some criterion which provides a means of reducing the number of classes which will count as confirmation classes and which are also justifiable or in some manner non-arbitrary. Orthodox statistical theory at this point suggests choosing a confirmation class which not only makes $P_H(I, 0 \in E)$ near 1 but which also maximizes the probability of showing that H is false if H is false. But in order to do this one must consider the class of hypotheses H' which may be true if H is false. If H' contains all the other hypotheses which are logically possible, then the class is too diverse to allow one to maximize the probability that is required.[17] However, in many important cases if H' is properly characterized, a confirmation class can actually be constructed so that $P_H(I, 0 \in E)$ is near 1 and so that E maximizes the probability of H being discovered as false when one of the members of H' is true. The confirmation class chosen will have the property of maximizing the probability of discovering that H is false if it is false only if the true hypothesis is in the class $H \cup H'$. But one can make the inference that the true hypothesis is in $H \cup H'$, when $H \cup H'$ is not logically exhaustive, only on the basis of positive empirical evidence. Hence, from

Reichenbach's point of view, the choice of a confirmation class using the criteria applied in orthodox statistical testing can only take place in the context of advanced knowledge.[18]

It is impossible both to follow the orthodox hint in characterizing confirmation instances and to have a joint interpretation where the row probabilities are inferred in the context of primitive knowledge. As shown previously, in order to have such a joint interpretation one must define a confirmation instance E for the joint hypothesis S and H. But the considerations just adduced show that there is a large class of E's which have the property that the sequence of E's so defined has a limit near 1 if S and H is true. In order to narrow the class of E's by following the orthodox hint, one would have to consider a set of hypotheses S and $(H \cup H')$. If the latter set contains all the hypotheses logically possible relative to S, then it will be impossible to define an E with the desired properties; and if this set is a proper subset of all the hypotheses logically possible relative to S, then the true hypothesis must be in this set for the E so defined to have the characteristics claimed for it. In the latter case, the claim that the true hypothesis is in the restricted set is an empirical claim, and on Reichenbach's general empiricist position, can only be asserted on the basis of positive empirical evidence, which contradicts the supposition that the interpretation allows the inference to the row probabilities of H in the context of primitive knowledge. Hence, if we follow the hint given by orthodox statistical theory as to the proper way to characterize a confirmation instance, such an interpretation is impossible. I know of no other way of restricting the class of E's which is at all in harmony with Reichenbach's general empirical orientation. Hence, I conclude that if we are to make sense of the probability of hypotheses on Reichenbach's account the interpretation must be an individual interpretation or a joint interpretation where the row probabilities are inferred in the context of advanced knowledge.[19]

Now consider the individual interpretation. Suppose that the structure of the sequence is inferred by induction by enumeration. Does the overall process – induction plus testing for confirmation instances – retain the characteristic of giving a high probability to the true hypothesis if there is one?

Before we attempt to answer that question let us look a little further into the dynamics of the individual interpretation. Suppose that we wish to test a theory T. A statistical hypothesis H is derivable from T which concerns a set-up of a particular kind. We run a sequence of n individual trials on a set up of this kind. On the basis of those trials we use induction by enumeration and infer that the trials are independent. Then we define a confirmation class E for H which will be about frequencies in selected subsets of the primary sequence. Then using the same primary sequence of n individual trials, we set up the sequence of E's and $\bar{E}$'s from which the row probability for H is inferred. Now suppose we continue the individual trials until we have $n+k$ individual trials. We must continue making inductions by enumeration on the structure of the extended primary sequence, otherwise the self-corrective structure of the inductive inference won't come into play. Nevertheless, Reichenbach's interpretation of the probability of hypotheses can be carried through in this manner. The interpretation has the required characteristic that hypotheses which are true, if any true ones are among those considered, will attain a probability near one in the long run. Thus, Reichenbach's claim that in principle inductive inference about theories can be reduced, via the probabilities of hypotheses, to inference by inductive enumeration is correct. Hence, his justification of induction carries through to these inductive inferences in advanced knowledge. Even though his interpretation can be carried through, his general program for justifying the enumerative rule is not without its difficulties, some of which cause particular problems in the present context.

We shall now review some of the general difficulties with the position. Reichenbach's justification of induction applies not only to the rule of simple enumeration but to *any* rule which will converge to the limit, if such a limit exists. Call the class of rules which have this convergence property the class of asympototic rules. Salmon has shown that the class of such rules is clearly too broad a class to constitute a proper basis of inductive inference and has made substantial progress in eliminating subsets of these rules which generate pathological inferences.[20] Nonetheless, the task of justifying the rule of simple enumeration is not complete in that it cannot as yet be accorded status as uniquely justified.

Let us now look at a related problem with respect to the topic of this essay. We noticed that, on the basis of continuing applications of the enumerative rule to the primary sequence, differing conclusions as to the structure of the sequence might be inferred at different stages of inquiry.

Thus the use of the probability lattice to calculate the probabilities for use in Bayes' Theorem must in reality be seen as the use of a sequence of such lattices. At any particular stage in inquiry, one uses the last lattice in the sequence. But in terms of anything like practical inference, one must not move from lattice to lattice too quickly; the situation requires a modicum of stability. The rule states that

$$\lim_{n \to \infty} F^n = m/n \pm d\,.$$

The stability of the system of inferences is related to the role of *d* in applications of the rule. If you do not have a *d*, every finite primary sequence which is not uniform will have at least one subsequence in which the inferred probability will differ from the probability inferred for the primary sequence. So without *d* as part of the rule, one can never conclude that the sequence is random, unless the sequence is uniform. Also, typically the structure attributed to the sequence will change with each new bit of evidence. So, to get any practical stability one must have a *d*. But the choice of a particular value for *d* affects the rate at which truly different sequence structures are discriminated. One intuition about the selection of *d* is that as the primary sequence gets longer *d* ought to decrease in size and so enable one to discriminate among more and more similar structures. But the particular function chosen to relate the length of the primary sequence to the value of *d* will tend to favor inferences to one characterization over other possible ones.

Salmon, in discussing the role of *d* in Reichenbach's statement of the rule, suggests that the choice of *d* is a pragmatic question and that it should be chosen in relation to the practical inference at hand.[21] When the rule is being used where practical inferences are being made this suggestion is a reasonable one. But where the question of the choice of *d* is raised in the theoretical context of establishing the probabilities of a whole set of hypotheses, the choice takes on a more systematic character. In effect the choice of a particular value for *d* partially determines a metric in a whole system of inductive inferences, much in the way that the choice of a *c*-function determines an inductive logic in early Carnapian systems.[22] Unless there is some means of determining a value for *d*, the interpretation is subject to a very large degree of arbitrariness. This and the related arbitrariness involved in choosing a rule from the class of asymptotic rules remains one of the outstanding problems of Reichenbach's inductive logic.

Wichita State University

NOTES

* Part of the research for this paper was done during study supported by the National Science Foundation. I benefited from comments by James A. Fulton, Deborah H. Soles, and James W. Nickel.

[1] Wesley C. Salmon, 'The Philosophy of Hans Reichenbach', *Synthese* **34** (1977), 5–88.

[2] Hans Reichenbach, *The Theory of Probability* (Berkeley and Los Angeles: University of California Press, 1949), pp. 432–433. Also see his *Experience and Prediction* (Chicago: The University of Chicago Press, 1938), p. 304.

[3] *Theory of Probability*, pp. 434–442. See especially pp. 438–441.

[4] Salmon's work on the probabilities of hypotheses is summarized in his *The Foundations of Scientific Inference* (Pittsburgh: The University of Pittsburgh Press, 1967), Section VIII. Nagel's comments are in: Ernest Nagel, 'Principles of the Theory of Probability', in Otto Neurath, Rudolf Carnap, and Charles Morris (eds.), *Foundations of the Unity of Science*, Vol. 1, Chicago, 1955 (originally published separately in 1939), pp. 404–408; and 'Review of Reichenbach, *The Theory of Probability*', *The Journal of Philosophy* **47**, 551–555.

[5] In *Experience and Prediction*, Reichenbach proceeded from the general empiricist position that no statement could be known as true; and so only probability weights, and not truth, are attributed to individual observation statements. In *Theory of Probability*, he took the admittedly simplified position of treating such statements as true. I follow the latter course in this paper.

[6] This follows Salmon's exposition in *Foundations of Scientific Inference*, p. 117.

[7] See Reichenbach, *Theory of Probability*, Section 72, and Salmon, *Foundations of Scientific Inference*, pp. 90–96. One of Nagel's criticisms of Reichenbach was that there are not enough different theories to constitute a proper reference class. Modern analytic techniques, however, allow scientists to formulate and compare many different but similar theories. For example, see Clifford M. Will, 'Gravitation Theory', *Scientific American* (1974), 24–33, where a large class of alternatives in Einstein's general theory of relativity is examined.

[8] Wesley C. Salmon, 'The Frequency Interpretation and Antecedent Probabilities', *Philosophical Studies* **4** (1953), 44–48. [*Guest Editor's note*: I now consider this article hopelessly naive.]

[9] *Theory of Probability*, p. 364.

[10] *Ibid.*, Section 91; *Experience and Prediction*, Section 39.

[11] The exact characterization of the nature of a physical law is a complex matter in Reichenbach's later work. See his *Nomological Statements and Admissable Operations* (Amsterdam: North-Holland, 1954). The probability statements I am considering would probably be considered derivative laws in the language of this book.

[12] 'Confirmatory instance' is the term used by Salmon and not by Reichenbach; its use seems to be consistent with Reichenbach's intention.

[13] *Theory of Probability*, pp. 434–439. Salmon seems to make the same assumption in 'The Frequency Interpretation'.

[14] '$P_H(I, 0 \in E)$' means '$P(I, 0 \in E)$ calculated on the basis of H'.

[15] *Theory of Probability*, pp. 463–465.

[16] *Ibid.*, p. 439. "We now establish the degree of probability for each row [a first level probability] by means of a posit based on the inductive rule. Assuming the posits to be true, we count in the vertical direction and thus construct the probability of the second level holding for the statement that the probability of a row is $= p$." Because of the possibility of very long, but finite runs, no hypothesis but the hypothesis that the sequence is uniform can really be treated so that the row probability is inferred in the context of primitive knowledge.

[17] For a discussion of this point and of the general theory of orthodox statistical inference, see my 'Material Conditions on Tests of Statistical Hypotheses', in Roger C. Buck and Robert S. Cohen (eds.), *PSA 1970* (Dordrecht-Holland: Reidel, 1971), pp. 403–412. For a more complete discussion, see Ronald N. Giere, 'The Epistemological Roots of Scientific Knowledge', in Grover Maxwell and R. M. Anderson, Jr. (eds.), *Induction, Probability, and Confirmation*: *Minnesota Studies in the Philosophy of Science*, Vol. V (Minneapolis: University of Minnesota Press, 1975). pp. 212–261. See especially pp. 238–240.

[18] There might be theoretical reasons to expect the true hypothesis to belong to a restricted set of hypotheses, or one might proceed by inferring a set $H \cup H'$ by using the enumeration rule with a fairly large d, thus in effect giving a set of hypotheses. See the discussion about the role of d at the end of this paper.

[19] From here on when I talk about the individual interpretation, I shall assume that to give a proper individual interpretation one must be in the context of advanced knowledge. Although one can carry through a joint interpretation in the context of advanced knowledge, as described here, no further discussion of this possibility is given because there are no epistemic issues raised by this approach that do not arise for the individual interpretation.

[20] *The Foundations of Scientific Inference*, Section VI, and the references cited therein.

[21] *Ibid.*, p. 138, n. 111.

[22] Rudolf Carnap, *Logical Foundations of Probability* (Chicago: University of Chicago Press, 2nd ed., 1962), Section 110.

JAMES H. FETZER

REICHENBACH, REFERENCE CLASSES, AND SINGLE CASE 'PROBABILITIES'*

Perhaps the most difficult problem confronted by Reichenbach's explication of physical probabilities as limiting frequencies is that of providing decision procedures for assigning singular occurrences to appropriate reference classes, i.e., *the problem of the single case.*[1] Presuming the symmetry of explanations and predictions is not taken for granted, this difficulty would appear to have two (possibly non-distinct) dimensions, namely: the problem of selecting appropriate reference classes for *predicting* singular occurrences, i.e., the problem of (single case) prediction, and the problem of selecting appropriate reference classes for *explaining* singular occurrences, i.e., the problem of (single case) explanation. If the symmetry thesis is theoretically sound, then these aspects of the problem of the single case are actually non-distinct, since any singular occurrence should be assigned to one and the same reference class for purposes of either kind; but if it is not the case that singular occurrences should be assigned to one and the same reference class for purposes of either kind, then these aspects are distinct and the symmetry thesis is not sound.[2]

The decision procedure that Reichenbach advanced to contend with the problem of the single case, i.e., the policy of assigning singular occurrences to '*the narrowest reference class for which reliable statistics can be compiled*', moreover, strongly suggests that one and the same reference class should serve for purposes of either kind. Reichenbach himself primarily focused attention on the problem of (single case) prediction, without exploring the ramifications of his resolution of the problem of the single case for the problem of (single case) explanation.[3] The theories of explanation subsequently proposed by Carl G. Hempel and by Wesley C. Salmon, however, may both be viewed as developments with considerable affinities to Reichenbach's position, which nevertheless afford distinct alternative solutions to the problem of (single case) explanation.[4] In spite of their differences, moreover, when Hempel's and Salmon's formulations are understood as incorporating the frequency criterion of statistical relevance, they appear to be saddled with theoretical difficulties whose

W. C. Salmon (ed.), Hans Reichenbach: Logical Empiricist, 187–219.

resolution, in principle, requires the adoption of an alternative construction.

The purpose of this paper is to provide a systematic appraisal of Reichenbach's analysis of single case 'probabilities' *with particular concern for the frequency conceptions of statistical relevance and of statistical explanation*, especially as they may be related to the theories of explanation advanced by Hempel and Salmon. Among the conclusions supported by this investigation are the following:

(a) that the frequency criterion of relevance is theoretically inadequate in failing to distinguish between two distinct kinds of 'statistical relevance';

(b) that reliance upon this defective criterion of relevance suggests that, on frequency principles, there are no irreducibly statistical explanations; and,

(c) that these difficulties are only resolvable within the frequency framework by invoking epistemic contextual considerations.

As a result, taken together, these reflections strongly support the contention that the problem of (single case) prediction and the problem of (single case) explanation require distinct (if analogous) resolution, i.e., that the symmetry thesis is unsound; and indirectly confirm the view that the meaning of *single case* probabilities should be regarded as fundamental, where a clear distinction may be drawn between *causal* relevance and *inductive* relevance on the basis of a statistical disposition conception, i.e., that Reichenbach's limiting frequency construction should be displaced by Popper's propensity interpretation for the explication of probability as a physical magnitude.

1. REICHENBACH'S ANALYSIS OF SINGLE CASE 'PROBABILITIES'

The theoretical foundation for Reichenbach's analysis of the single case, of course, is provided by the definition of 'probability' itself: "In order to develop the frequency interpretation, we define probability as the *limit of a frequency* within an infinite sequence".[5] In other words, the probability r of the occurrence of a certain outcome attribute A within an infinite sequence of trials S is the limiting frequency with which A occurs in S, i.e.,

(I) $P(S, A) = r =_{df}$ the limit of the frequency for outcome attribute A within the infinite trial sequence S equals r.

Reichenbach himself assumes no properties other than the limiting frequency of A within S as necessary conditions for the existence of probabilities and thereby obtains an interpretation of the broadest possible generality.[6] It is important to note, however, that Reichenbach envisions *finite* sequences as also possessing 'limits' in the following sense:

Notice that a limit exists even when only a finite number of elements x_i belong to S; the value of the frequency for the last element is then regarded as the limit. This trivial case is included in the interpretation and does not create any difficulty.[7]

One justification for the inclusion of such 'limits', moreover, is that when the sequence S contains only a finite number of members, those members may be counted repetitiously an endless number of times to generate trivial limiting frequencies.[8]

As the basis for a theoretical reconciliation of these concepts, therefore, let us assume that a sequence S is *infinite* if and only if S contains at least one member and the description of its reference class does not impose any upper bound to the number of members of that class on syntactical or semantical grounds alone. Although 'limits' may be properties of finite sequences under this interpretation, they are not supposed to be properties of single individual trials *per se* and may only be predicated of single individual trials *as a manner of speaking*:

I regard the statement about the probability of the single case, not as having a meaning of its own, but as representing an elliptical mode of speech. In order to acquire meaning, the statement must be translated into a statement about a frequency in a sequence of repeated occurrences. The statement concerning the probability of the single case thus is given a *fictitious meaning*, constructed by a *transfer of meaning from the general to the particular case.*[9]

Strictly speaking, therefore, probabilities are only properties of single trials as members of reference classes collectively; but Reichenbach nevertheless countenances referring to 'probabilities' as properties of such trials distributively "not for cognitive reasons, but because it serves the purpose of action to deal with such statements as meaningful".[10] Indeed, in order to distinguish the meaning of 'probability' with respect to the occurrence of singular trials and of infinite trial sequences, Reichenbach introduces a different term, i.e., 'weight', for application to the single case.

In spite of this difference in meaning, the numerical value of the weight to be assigned to attribute A as the outcome of a singular trial T, in principle, is determined by the limiting frequency with which A occurs within a trial sequence of kind K, where $T \in K$. The existence of a single case weight for the occurrence of attribute A thus requires (i) the existence of an infinite sequence of trials of kind K (ii) with a limiting frequency for A equal to r, where (iii) it is not the case that there exists some other infinite sequence of kind K such that the limiting frequency for A is not equal to r; hence, '$P(K, A) = r$' is true if and only if there exists an infinite sequence of trials of kind K such that the limiting frequency for A is equal to r and, for all trial sequences S, if S is an infinite sequence of kind K, then the limiting frequency for outcomes of kind A is equal to r.[11] In effect, therefore, every single individual trial T that happens to be a trial of kind K with respect to the occurrence of attribute A must be regarded as a member of a unique trial sequence K consisting of all and only those single trials that are trials of kind K with respect to the occurrence of outcome attribute A; otherwise, violation of the uniqueness condition would generate explicit contradictions of the form, '$P(K, A) \neq P(K, A)$'.[12]

Any single individual trial, however, may be exhaustively described if and only if *every* property of that trial is explicitly specifiable including, therefore, the spatial and the temporal relations of that instantiation of properties relative to every other. Let us assume that any single individual trial T is a property of some object or collection of objects x such that, for each such individual trial, 'Tx' is true if and only if '$F^1x \cdot F^2x \cdot \ldots$', is true, where '$F^1$', '$F^2$', . . . , and so on are predicates designating distinct properties of that single individual trial. Then the single individual trial T is subject to exhaustive description, in principle, if and only if there exists some number m such that F^1 through F^m exhausts every property of that single individual trial; otherwise, the single individual trial T is not exhaustively describable, even in principle, since there is no end to the number of properties that would have to be specified in order to provide an exhaustive description of that trial *T*. The last time I turned the ignition to start my car, for example, was a single individual trial involving a 1970 Audi 100LS 4-door red sedan, with a Kentucky license plate, a recorded mileage of 88 358.4 miles, that had been purchased in 1973 for $2600.00 and driven to California during the Summer of the same year, and so on, which was parked in the right-most section of a three-car wooden garage to the rear of a two-story building at 159 Woodland Avenue in Lexington, on a somewhat misty morning at approximately 10:30 A.M. on 12 March 1976, one half-hour after I had drunk a cup of coffee, and so forth. As it happened, the car would not start.

The significance of examples such as this, I surmise, has two distinctive aspects. On the one hand, of course, it indicates the enormous difficulty, in principle, of providing an exhaustive description of any such individual trial; indeed, it strongly suggests *a density principle for single trial descriptions*, i.e., that for any description '$F^1x \cdot \ldots \cdot F^mx$' of any singular trial *T* that occurs during the course of the world's history, there exists some further description '$F^1x \cdot \ldots \cdot F^nx$' such that the set of predicates {'F^1', ..., 'F^m'} is a proper subset of the set of predicates {'F^1', ..., 'F^n'}.[13] On the other hand, it is at least equally important to notice that, among all of the properties that have thus far been specified, only some *but not all* would be viewed as contributing factors, i.e., as relevant variables, with respect to the outcome attribute of starting or not, as the case happened to be; in other words, even if this single individual trial is not exhaustively describable, it does not follow that the *causally relevant* properties of this trial arrangement *T* are themselves not exhaustively describable; for the properties whose presence or absence contributed to bringing about my car's failure to start would include, perhaps, accumulated moisture in the distributor, but would not include, presumably, my wearing a flannel overshirt at the time. It is at least logically possible, therefore, that among the infinity of properties that happened to attend this single individual trial, only a finite proper subset would exhaust all those exerting any influence upon that outcome attribute on that particular occasion.

The theoretical problem in general, therefore, may be described as follows, namely: for any single individual trial *T*, (i) to determine the kind of trial *K* that *T* happens to be with respect to the occurrence of outcome attribute *A*; and, (ii) to ascertain the limiting frequency *r* with which attribute A occurs within the infinite sequence of trials of kind K, if such a sequence and such a limit both happen to exist. The theoretical 'problem of the single case', therefore, is precisely that of determining the kind of trial K that any single individual trial T happens to be with respect to the occurrence of an outcome attribute of kind A, i.e., the problem of the selection of a unique reference sequence K for the assignment of a unique individual trial T. Reichenbach's solution to this problem is therefore enormously important to his analysis of the single case:

We then proceed by considering *the narrowest class for which reliable statistics can be compiled.* If we are confronted by two overlapping classes, we shall choose their common class. Thus, if a man is 21 years old and has tuberculosis, we shall regard the class of persons of 21 who have tuberculosis [with respect to his life expectancy]. Classes that are known to be irrelevant for the statistical result may be disregarded. A class C is irrelevant with respect to the reference class K and the attribute class A if the transition to the common class $K \cdot C$ does not change the probability, that is, if $P(K \cdot C, A) = P(K, A)$. For instance, the class of persons having the same initials is irrelevant for the life expectation of a person.[14]

A property C is *statistically relevant* to the occurrence of an attribute A with respect to a reference class K, let us assume, if and only if:

$$\text{(II)} \qquad P(K \cdot C, A) \neq P(K, A);$$

that is, the limiting frequency for the outcome attribute A within the infinite sequence of $K \cdot C$ trials differs from the limiting frequency for that same attribute within the infinite sequence of K trials itself.[15]

Since a unique individual trial T is supposed to be assigned to the narrowest reference class K 'for which reliable statistics can be compiled', it seems clear that for Reichenbach, at least, a decision of this kind depends upon the state of knowledge $\mathcal{K}$ of an individual or collection of individuals z at a specific time t, i.e., the set of statements $\{\mathcal{K}\}$ accepted or believed by z at t, no matter whether true or not, as follows:

Whereas the probability of a single case is thus made dependent on our state of knowledge, this consequence does not hold for a probability referred to classes .... The probability of death for men 21 years old concerns a frequency that holds for events of nature and has nothing to do with our knowledge about them, nor is it changed by the fact that the death probability is higher in the narrower class of tuberculous men of the same age. The dependence of a single-case probability on our state of knowledge originates from the impossibility of giving this concept an independent interpretation; there exist only substitutes for it, given by class probabilities, and the choice of the substitute depends on our state of knowledge.[16]

Reichenbach allows that statistical knowledge concerning a reference class K may be fragmentary and incomplete, where the problem is one of "balancing the importance of the prediction against the reliability available". Nevertheless, as a general policy, Reichenbach proposes treating an individual trial T as a member of successively narrower and narrower reference classes $K^1, K^2, \ldots$ and so on, where each class is specified by taking into account successively more and more statistically relevant properties F^1, F^2, and so forth, where $K^1 \supset K^2$, $K^2 \supset K^3$, and so on and $P(K^1, A) \neq P(K^2, A)$, $P(K^2, A) \neq P(K^3, A)$, and so forth.

Reichenbach observes that, strictly speaking, the choice of a reference *class* is not identical with the choice of a reference *sequence*, since the members of that class may be ordered in different ways, which may sometimes differ in probability.[17] The intriguing question, however, is precisely how narrow the appropriate class in principle should be if our knowledge of the world were complete:

According to general experience, the probability will approach a limit when the single case is enclosed in narrower and narrower classes, to the effect that, from a certain step on, further narrowing will no longer result in noticeable improvement. It is not necessary for the justification of this method that the limit of the probability, respectively, is $=1$ or is $=0$, as the hypothesis of causality [i.e., the hypothesis of determinism] assumes. Neither is this necessary *a priori*; modern quantum mechanics asserts the contrary. It is obvious that for the limit 1 or 0 the probability still refers to a class, not to an individual event, and that the probability 1 cannot exclude the possibility that in the particular case considered the prediction is false. Even in the limit the substitute for the probability of a single case will thus be a class probability, ...[18]

The evident reply on Reichenbach's analysis, therefore, is that the appropriate reference class K relative to a knowledge context $\mathcal{K}zt$ containing every sentence that is true of the world and no sentence that is false would be some reference class K^i where $T \in K^i$ and, for every class K^j such that $T \in K^j$ and $K^i \supset K^j$, it is not the case that $P(K^i, A) \neq P(K^j, A)$; i.e., with respect to attribute A for trial T, K should be an *ontically homogeneous reference class* in the sense that (i) $T \in K$; (ii) $P(K, A) = r$; and, (iii) for all subclasses K^j of K to which T belongs, $P(K^j, A) = r = P(K, A)$.

Notice that the class K itself is not necessarily unique with respect to the set of properties $\{F^i\}$ specified by the reference class description of K; for any class K^j such that $K \supseteq K^j$ and $K^j \supseteq K$ (where K is an ontically homogeneous reference class relative to attribute A for trial T) will likewise qualify as an ontically homogeneous reference class relative to attribute A for trial T even though the set of properties $\{F^j\}$ specified by the reference class description of K^j differs from that for K, i.e., $\{F^i\} \neq \{F^j\}$. Consequently, although any classes K and K^j which happen to be such that $K \supseteq K^j$ and $K^j \supseteq K$ will of course possess all and only the same trial members and will therefore yield the same frequencies for various specific attributes with respect to those same trial members, their reference class descriptions, nevertheless, will not invariably coincide. From this point of view, therefore, the resolution of the reference class problem by assigning a single case to an ontically homogeneous reference class does not provide a unique *description* solution but rather a unique *value* solution. Nevertheless, on Reichenbachian principles, it may be argued further that there *is* a unique description solution as well as a unique value solution, namely: that any single individual trial T should be assigned to the *narrowest* class, i.e., *the class whose description includes specification of the largest set of properties of that individual trial*, "for which reliable statistics can be compiled".

The largest set of properties of an individual trial T that might be useful for this purpose, of course, is not logically equivalent to the set of all of the properties of that individual trial; for any individual trial T may be described by means of predicates that violate the provision that reference classes may not be logically limited to a finite number of members on syntactical or semantical grounds alone. Let us therefore assume that *a predicate expression is logically impermissible for the specification of a reference class description* if (a) that predicate expression is necessarily satisfied by every object or (b) that predicate expression is necessarily satisfied by no more than a specific number N of objects during the course of the world's history.[19] Predicates that happen to be satisfied by only one individual object during the course of the world's history, therefore, are permissible predicates for reference class descriptions so long as their extensions are not finite on logical grounds alone, as, e.g., might be any predicate expression essentially requiring proper names for its definition or such that the satisfaction of that expression by some proper name would yield a logical truth.[20] Let us further assume that no predicate expression logically entailing the attribute predicate or its negation is permissible. Then the largest set of permissible predicates describing a single trial T is not logically equivalent to the set of all of the properties of that individual trial; but nevertheless it will remain the case that, in general, such *narrowest* reference class descriptions are satisfied by only one individual event during the course of the world's history *as a matter of logical necessity.*[21]

These considerations suggest an insuperable objection to the applicability, in principle, of the frequency interpretation of probability; for if it is true that each individual trial T is describable, in principle, by some set of predicates such that (i) every member of that set is a permissible predicate for the purpose of a reference class description and (ii) that reference class description itself is satisfied by that individual trial alone, then the indispensable criterion of statistical relevance is systematically inapplicable for the role it is intended to fulfill. For under these circumstances, every individual trial T is the solitary member of a reference class K^* described by a reference class description consisting of the conjunction of a set of permissible predicates $F^1, F^2, \ldots, F^n$, i.e., $\{F^n\}$, where, moreover, lacking any information concerning the statistical relevance or irrelevance of any property of any singular trial – other than that the attribute A did occur (or did not occur) on that particular trial – it is systematically impossible to specify which of the properties of the trial T are statistically relevant and which are not; for the occurrence of that outcome, whether A or not-A, in principle, *must be attributed to the totality of properties present at that individual trial.*[22] Reliable statistics, after all, are only as reliable as the individual statistics upon which they are based; so *if the only statistical data that may be ascertained, in principle, concern the occurrence of outcome attributes on singular trials where each singular trial is the solitary member of a reference class, there is no basis for accumulating the 'reliable statistics' necessary for the applicability of the frequency criterion.* Of each individual trial $T^1, T^2, \ldots$, it is possible in principle to specify a homogeneous reference class description $K^{*1}, K^{*2}, \ldots$; but since each trial is the solitary member of its particular reference class, it is impossible to employ the frequency criterion to ascertain which properties, if any, among all of those present on each such singular trial, are statistically irrelevant to its outcome.

The theoretical problem of the single case, let us recall, requires, for any single individual trial T, (i) to determine the kind of trial K that T happens to be with respect to the occurrence of outcome attribute A; and, (ii) to ascertain the limiting frequency r with which attribute A

occurs in the infinite sequence of trials of kind K, when such a sequence and such a limit both happen to exist. With respect to any single individual trial T that actually occurs in the course of the world's history, however, these conditions are, in effect, automatically satisfied; for (i) any such trial T may be described by some set of permissible predicates $\{F^n\}$ specifying a kind of trial K^* of which T is the solitary member, where, nevertheless, (ii) the reference class K^* is not logically limited to any finite number of members and Reichenbach's limit concept for infinite sequences of this kind is trivially satisfied by 0 or 1. Any other such singular trial T^i, after all, may likewise be described by some set of permissible predicates, i.e., $\{F^i\}$, where $\{F^i\} \neq \{F^n\}$, a condition of individuation that distinct events surely have to satisfy. With respect to each such singular trial T, of course, outcome attribute A either occurs or does not occur. Assume that T belongs to reference class K for which the probability of A is $\neq 0$ and $\neq 1$; i.e., (a) $T \in K$; and, (b) $P(K, A) = r$, where $0 \neq r \neq 1$. Then there necessarily exists a subclass of that class K^* such that $P(K^*, A) = 0$ or $= 1$, namely: the class specified by any set of permissible predicates $\{F^n\}$ of which trial T is the solitary member. Hence, since $P(K^*, A) \neq P(K, A)$ and $K \supset K^*$, the properties differentiating between K^* and K are statistically relevant with respect to the occurrence of attribute A, necessarily, by the frequency criterion; moreover, for any such reference class K such that (a) $T \in K$ and (b) $P(K, A) = r$, where $0 \neq r \neq 1$, it is theoretically impossible that K is a homogeneous reference class for trial T with respect to attribute A in Reichenbach's sense. Indeed, on the basis of the frequency criterion of statistical relevance, such an assumption is always invariably false.

The point of the preceding criticism, therefore, may be stated as follows: The only data available for ascertaining the reliable statistics necessary for arriving at determinations of the statistical relevance or irrelevance of any property F^i with respect to any outcome attribute A is that the frequency with which outcome A accompanies trials of kind $K \cdot F^i$ differs from that with which A accompanies trials of the kind K. Every single individual trial T^i occurring during the course of the world's history may be described as a member of some K^* reference class specified by at least one conjunction of permissible predicates K^{*i} that is satisfied by that individual trial. As a necessary logical truth, every such trial happens to belong to one and only one *narrowest* class of this kind;

however, since the K^* reference classes are not likewise limited to any finite number of members, the occurrence of the attribute A may still be assigned a probability, which, in this case, will be $=1$ or $=0$, depending upon whether that outcome occurred or failed to occur on that individual trial. The frequency criterion of statistical relevance, presumably, should permit distinguishing between the statistically relevant and irrelevant properties of each such trial. However, *since each trial T^i happens to be different in kind*, the only conclusion supported by that criterion of relevance is that every single property distinguishing those individual trials with respect to the occurrence of a specific attribute A is statistically relevant to the occurrence of that outcome; for those properties certainly 'made a difference', insofar as in one such case the attribute A occurred, while in another such case A did not occur. Thus, under circumstances of this kind, it is systematically impossible to obtain the reliable statistics necessary to support determinations of statistical relevance; for in the absence of that statistical evidence, such 'conclusions' merely take for granted what that evidence alone is capable of demonstrating.

2. SALMON'S 'RELEVANCE' ACCOUNT OF STATISTICAL EXPLANATION

It is interesting to observe that Reichenbach himself tended to focus upon the problems involved in the *prediction* of singular events rather than those involved in their *explanation*. The differences that distinguish explanations from predictions, however, may figure in significant ways within the present context; for if it happens to be the case that the purpose of a prediction is to establish grounds for *believing that* a certain statement (describing an event) is true, and the purpose of an explanation is to establish grounds for *explaining why* an event (described by a certain statement) occurs, i.e., if predictions are appropriately interpreted as *reason-seeking* why-questions, while explanations are appropriately interpreted as *explanation-seeking* why-questions, it might well turn out that at least part of the problem with the frequency criterion of statistical relevance is that it is based upon an insufficient differentiation between explanation and prediction contexts.[23] Consider the following:

(a) Reason-seeking why-questions are relative to a particular epistemic context, i.e., a knowledge context $\mathcal{K}zt$ as previously specified, where, with respect to an individual hypothesis H whose truth is not known, *a requirement of total evidence* may appropriately be employed according to which, for any statement S belonging to $\mathcal{K}zt$, *S is inductively relevant to the truth of H* if and only if the inductive (or epistemic) probability EP of H relative to $\mathcal{K}zt \cdot S$ differs from the inductive probability of H relative to $\mathcal{K}zt \cdot -S$, i.e.,

$$\text{(III)} \qquad EP(\mathcal{K}zt \cdot S, H) \neq EP(\mathcal{K}zt \cdot -S, H);$$

where, in effect, any statement within $\mathcal{K}zt$ whose truth or falsity makes a difference to the inductive probability of an hypothesis H must be taken into consideration in determining the inductive probability of that hypothesis.[24]

(b) Explanation-seeking why-questions are relative to a specific ontic context, i.e., the nomological regularities and particular facts of the physical world W, where, with respect to the explanandum-event described by an explanandum statement E whose truth is presumably known, *a requirement of causal relevance* should appropriately be employed according to which, for any property F belonging to W, *F is explanatorily relevant to the occurrence of E* (relative to reference class K) if and only if the physical (or ontic) probability PP of E relative to $K \cdot F$ is not the same as the physical probability of E relative to $K \cdot -F$, i.e.,

$$\text{(IV)} \qquad PP(K \cdot F, E) \neq PP(K \cdot -F, E);$$

where, in effect, any property whose presence or absence relative to reference class K makes a difference to the physical probability of an explanandum-event E must be taken into consideration in constructing an adequate explanation for that event.[25]

If predictions belong to the epistemic reason-seeking context, then it is entirely plausible to take into account any property whose presence or absence changes the inductive probability of an hypothesis H with respect to a knowledge context $\mathcal{K}zt$; indeed, a requirement of total evidence would require that any such property G of any individual object x belonging to the knowledge context $\mathcal{K}zt$ has to be taken into consideration in calculating the inductive probability of any H where the statement G^* attributing G to x is such that

$$\text{(V)} \qquad EP(\mathcal{K}zt \cdot G^*, H) \neq EP(\mathcal{K}zt \cdot -G^*, H);$$

where, as it might be expressed, the statistical relationship between property G and the attribute property A described by the hypothesis H may be merely one of statistical correlation rather than one of causal connection, i.e., it is not necessary that G be a property whose presence or absence contributes to bringing about the occurrence of an outcome of kind A relative to some reference class K.

It is not at all obvious, however, that the frequency criterion provides any theoretical latitude for differentiating between statistical relations of these quite distinctive kinds; on the contrary, it appears as though any factor at all with respect to which frequencies differ is on that account alone 'statistically relevant' to the probability of an hypothesis H or an explanandum E, respectively, without reflecting any theoretical difference between the explanation-seeking and the reason-seeking situations themselves. If every member of the class of twenty-one year old men K were also a member of the class of tuberculous persons T, for example, then the property of being tuberculous would be statistically irrelevant to the outcome attribute of death D. But although this property might reasonably be ignored for the purpose of prediction (as a matter of statistical correlation), it would not be reasonable to ignore this property for the purpose of explanation (when it makes a causal contribution to such an individual's death). If there is a significant difference between these kinds of situations, therefore, then those principles appropriate for establishing relevance relations within one of these contexts may be theoretically inappropriate for establishing relevance relations within the other. Perhaps the crucial test of the utility of Reichenbach's criterion of statistical relevance is found within the context of explanation rather than the context of prediction; for an examination of the statistical relevance account of explanation advanced by Salmon may provide an opportunity to evaluate the criticisms of that principle thus far considered.

Salmon departs from Reichenbach's formulations, not in deviating from his criterion of statistical relevance, but rather in assigning a single case not to the *narrowest* relevant class but to the *broadest* relevant class instead:

If every property that determines a place selection is statistically irrelevant to A in K, I shall say that K is a homogeneous reference class for A. A reference class is homogeneous if there is no way, even in principle, to effect a statistically relevant partition without already knowing which elements have the attribute in question and which do not .... The aim in selecting a reference class to which to assign a single case is not to select the narrowest, but the widest, available class .... I would reformulate Reichenbach's method of selection of a reference class as follows: choose the broadest homogeneous reference class to which the single event belongs.[26]

Precisely because Salmon preserves the frequency criterion of relevance, his own formulations encounter difficulties analogous to those previously specified; but the introduction of the concept of a partition and of a 'screening-off' rule may be viewed as significant contributions to the frequency theory of explanation as follows:

(i) On Salmon's analysis, *a partition of a reference class K* is established by a division of that class into a set of mutually exclusive and jointly exhaustive subsets by means of a set of properties $F^1, F^2, \ldots, F^m$ and their complements where each ultimate subset of that class $K \cdot C^1, K \cdot C^2, \ldots, K \cdot C^n$ is homogeneous with respect to the outcome attribute A.[27] This procedure may be regarded as effecting a refinement in the application of Reichenbach's criterion, since it thus assumes that a property C is *statistically relevant* to the occurrence of attribute A with respect to a reference class K if and only if:

$$\text{(II*)} \qquad P(K \cdot C, A) \neq P(K \cdot -C, A);$$

that is, the limiting frequency for the outcome attribute A within the infinite sequence of $K \cdot C$ trials differs from the limiting frequency for that same outcome within the infinite sequence of $K \cdot -C$ trials.

(ii) Furthermore, a property *F screens off a property G* with respect to an outcome attribute A within the reference class K if and only if:

$$\text{(VI)} \qquad P(K \cdot F \cdot G, A) = P(K \cdot F \cdot -G, A) \neq P(K \cdot -F \cdot G, A);$$

where the equality between the limiting frequency for A within the classes $K \cdot F \cdot G$ and $K \cdot F \cdot -G$, on the one hand, establishes that the property G is not statistically relevant with respect to attribute A within the reference class $K \cdot F$; and the inequality between the limiting frequency for A within the class $K \cdot -F \cdot G$ and classes $K \cdot F \cdot G$ and

$K \cdot F \cdot -G$, on the other hand, establishes that the property F is statistically relevant with respect to the attribute A within the reference class $K \cdot G$.[28]

Salmon consolidates these ingredients as the foundation for his explication of explanation on the basis of the principle that screening-off properties should take precedence over properties which they screen-off within an explanation situation.[29] According to Salmon, an explanation of the fact that x, a member of the reference class K, is a member of the attribute class A as well, may be provided by fulfilling the following set of conditions, i.e.,

(1) $K \cdot C^1, K \cdot C^2, \ldots, K \cdot C^n$ is a homogeneous partition of K relative to A;

(2) $P(K \cdot C^1, A) = r^1, P(K \cdot C^2, A) = r^2, \ldots, P(K \cdot C^n, A) = r^n$;

(3) $r^i = r^j$ only if $i = j$; and,

(4) $x \in K \cdot C^m$ (where $m \in \{1, 2, \ldots, n\}$).[30]

Consequently, the appropriate reference class to specify in order to explain an outcome A for the trial T is the ontically homogeneous reference class $K \cdot C^m$ such that (a) $T \in K \cdot C^m$; (b) $P(K \cdot C^m, A) = r^m$; and, (c) for all homogeneous reference classes $K \cdot C^1, K \cdot C^2, \ldots, K \cdot C^n$ relative to outcome A, $r^i = r^j$ only if $i = j$, which, presumably, is intended to insure that there is one and only one reference class to which T may appropriately be assigned, namely: the *broadest* one of that kind. From this point of view, therefore, Salmon provides a unique description solution as well as a unique value solution to the single case problem.

Salmon's conditions of adequacy, let us note, are sufficient for the purpose of assigning individual trials to broadest homogeneous reference classes *only if* broadest homogeneous reference classes may be described by means of what may be referred to as *disjunctive properties*, i.e., predicate expressions of the form, '$K \cdot F^1 \vee F^2 \vee \ldots \vee F^n$', where a statement of the form, '$P(K \cdot F^1 \vee F^2 \vee \ldots \vee F^n, A) = r$', is true if and only if

'$P(K \cdot F^1, A) = P(K \cdot F^2, A) = \ldots = P(K \cdot F^n, A) = r$' is true, as Salmon himself has explicitly pointed out.[31] Otherwise, Salmon's conditions would be theoretically objectionable, insofar as it might actually be the case that there is an infinite set of reference classes, $\{K \cdot C^i\}$, whose members satisfy conditions (1) and (2) only if they do not satisfy condition (3), and conversely. The difficulty with this maneuver as a method of preserving Salmon's conditions (1)–(4) as sufficient conditions of explanatory adequacy, however, is that it entails the adoption of *a degenerating explanation paradigm*; for the occurrence of attribute A on trial T may be explained by referring that trial to a successively more and more *causally heterogeneous* reference class under the guise of the principle of reference class homogeneity. For if condition (3) is retained, then if, for example, the division of the class of twenty-one year old men K on the basis of the properties of having tuberculosis F^1 or heart disease F^2 or a brain tumor F^3 establishes a subclass such that, with respect to the attribute of death D, $P(K \cdot F^1 \vee F^2 \vee F^3, D) = r$ is an ultimate subset of the homogeneous partition of that original class, i.e., $\{K \cdot F^1 \vee F^2 \vee F^3\} = \{K \cdot C^m\}$, then the explanation for the death of an individual member i of class K resulting from a brain tumor, perhaps, is only explainable by identifying i as a member of class $K \cdot C^m$, i.e., as a member of the class of twenty-one year old men who either have tuberculosis or heart disease or a brain tumor.

The significance of this criticism appears to depend upon how seriously one takes what may be referred to as *the naive concept of scientific explanation for singular events*, namely: that the occurrence of an outcome A on a single trial T is to be explained by citing *all and only* those properties of that specific trial which contributed to bringing about that specific outcome, i.e., a property F is *explanatorily relevant* to attribute A on trial T if and only if F is a *causally relevant* property of trial T relative to attribute A.[32] From this perspective, Salmon's explication of explanation is theoretically objectionable for at least two distinct reasons:

First, *statistically relevant* properties are not necessarily *causally relevant* properties, and conversely.[33] If it happens that the limiting frequency r for the attribute of death D among twenty-one year old men K differs among those whose initials are the same F when that class is subject to a homogeneous partition, then that property is explanatorily relevant, necessarily, on the basis of the frequency criterion of statistical relevance;

and it might even happen that such a property screens-off another property G, such as having tuberculosis, in spite of the fact that property G is causally relevant to outcome D and property F is not. Indeed, Salmon himself suggests that "relations of statistical relevance must be explained on the basis of relations of causal relevance",[34] where relations of statistical relevance appear to fulfill the role of evidential indicators of relations of causal relevance.[35]

Second, the admission of disjunctive properties for the specification of an ultimate subset of a homogeneous partition of a reference class does not satisfy the desideratum of explaining the occurrence of an outcome A on a single trial T by citing *all and only* properties of that specific trial, whether causally relevant, statistically relevant, or otherwise. This difficulty, however, appears to be less serious, in principle, since a modification of Salmon's conditions serves for its resolution, namely: Let us assume that a reference class description 'K' is *stronger than* another reference class description '$K+$' if and only if 'K' logically entails '$K+$', but not conversely.[36] Then let us further assume as new condition (3*) Salmon's old condition (4), with the addition of a new condition (4*) in lieu of Salmon's old condition (3) as follows:

(4*) $K \cdot C^m$ is a strongest homogeneous reference class;

that is, the reference class description '$K \cdot C^m$' is stronger than any other reference class description '$K \cdot C^j$' (where $j \in \{1, 2, \ldots, n\}$) such that $x \in K \cdot C^j$.[37] The explanation of the death of an individual member i of class K resulting from a brain tumor, therefore, is thus only explainable by identifying i as a member of a *strongest* class $K \cdot C^m$, i.e., as a member of the class of twenty-one year old men who have brain tumors, which is a broadest homogeneous reference class of the explanatorily relevant kind, under this modification of the frequency conception.

It is important to observe, however, that none of these considerations mitigates the force of the preceding criticism directed toward the frequency criterion of statistical relevance itself. For it remains the case that any single trial T which belongs to a reference class K for which the probability of the occurrence of attribute A is $\neq 0$ and $\neq 1$ will likewise belong to innumerable subclasses $K \cdot C^j$ of K and, indeed, T itself will necessarily belong to some subclass K^* of K such that $P(K^*, A) = 0$ or $= 1$, which, on Salmon's own criteria, requires that trial T be assigned to

K^*, so long as K^* is a broadest (or a strongest) ultimate subclass of a homogeneous partition of the original class K. In effect, therefore, those properties that differentiate trial T as a member of class K from all other trial members of that class establish the basis for effecting a homogeneous partition of that class, a partition for which the probability for attribute A within every one of its ultimate subclasses (whether strongest or not) is $=0$ or $=1$, necessarily. And this result itself may be viewed as reflecting a failure to distinguish those principles suitable for employment within the context of explanation from those principles suitable for employment within the context of prediction, promoted by (what appears to be) the mistaken identification of statistical relevance with explanatory relevance.

It should not be overlooked that the properties *taken to be* statistically relevant to the occurrence of attribute A on trial T relative to the knowledge context $\mathscr{K}zt$ are not necessarily those that actually *are* statistically relevant to that attribute within the physical world. For this reason, Salmon's analysis emphasizes the significance of the concepts of *epistemic* and of *practical* homogeneity, which, however, on Salmon's explication, actually turn out to be two different kinds of *inhomogeneous* reference classes, where, for reasons of ignorance or of impracticality, respectively, it is not possible to establish ontically homogeneous partitions for appropriate attributes and trials.[38] Salmon's analysis thus does not provide an explicit characterization of the conception of reference classes that are believed to be homogeneous whether or not they actually are; nevertheless, there is no difficulty in supplementing his conceptions as follows: Let us assume that a reference class K is an *epistemically homogeneous reference class* with respect to attribute A for trial T within the knowledge context $\mathscr{K}zt$ if and only if the set of statements $\{\mathscr{K}\}$ accepted or believed by z at t, considerations of truth all aside, logically implies some set of sentences, $\{S\}$, asserting (a) that $T \in K$; (b) that $P(K, A) = r$; (c) that for all subclasses K^j of K to which T belongs, $P(K^j, A) = r = P(K, A)$; and (d) that it is not the case that there exists some class $K^i \supset K$ such that $P(K^i, A) = r = P(K, A)$. The satisfaction of conditions (a), (b), and (c), therefore, is sufficient to fulfill the epistemic version of the concept of an *ontically homogeneous* reference class K for trial T with respect to attribute A, while satisfaction of (d) as well is sufficient to fulfill the epistemic version of the *broadest homogeneous* reference class for T relative to A, within the knowledge context $\mathcal{K}zt$. In order to differentiate this concept from Salmon's original, however, let us refer to this definition as the *revised* conception of epistemic homogeneity, while acknowledging the theoretical utility of both.

3. HEMPEL'S REVISED REQUIREMENT OF MAXIMAL SPECIFICITY

The conclusions that emerge from the preceding investigation of Salmon's own analysis of statistical explanation support the contention that, on the frequency criterion of statistical relevance, statistical explanations are only *statistical* relative to a knowledge context $\mathcal{K}zt$, i.e., as the matter might be expressed, "God would be unable to construct a statistical relevance explanation of any physical event, not as a limitation of His power but as a reflection of His omniscience".[39] It is therefore rather intriguing that Salmon himself has strongly endorsed this conclusion as a criticism of *Hempel's account* of statistical explanation, for as the following considerations are intended to display, the fundamental difference between them is that Hempel's explicit relativization of statistical explanations to a knowledge context $\mathcal{K}zt$, as it were, affirms *a priori* what, on Salmon's view, is *a posteriori* true, namely: that for probabilities r such that $0 \neq r \neq 1$, the only homogeneous reference classes are epistemically homogeneous. What Salmon's criticism fails to make clear, however, is that this difficulty is a necessary consequence of adoption of the frequency criterion of statistical relevance for any non-epistemic explication of explanation, including, of course, Salmon's own ontic explication. Additionally, there are at least two further important issues with respect to which Hempel's analysis and Salmon's analysis are distinctive, in spite of their initial appearance of marked similarity.

In order to establish the soundness of these claims, therefore, let us consider the three principal ingredients of Hempel's epistemic explication. First, Hempel introduces the concept of an *i-predicate* in $\mathcal{K}zt$ which, in effect, is any predicate 'F^m' such that a sentence '$F^m i$' asserting the satisfaction of 'F^m' by the individual i belongs to $\mathcal{K}zt$, regardless of the kind of property that may be thereby designated.[40] He then proceeds to define *statistical relevance* as follows:

'F^m' will be said to be *statistically relevant* to 'Ai' in $\mathcal{K}zt$ if (1) 'F^m' is an i-predicate that entails neither 'A' nor '$-A$' and (2) $\mathcal{K}zt$ contains a lawlike sentence '$P(F^m, A) = r$' specifying the probability of 'A' in the reference class characterized by 'F^m'.[41]

Insofar as sentences of the form, '$P(F^m, A) = r$', are supposed to be *lawlike*, it is clear that, on Hempel's analysis, (a) their reference class descriptions must be specified by means of permissible predicates and (b) these sentences are to be understood as supporting subjunctive (and counterfactual) conditionals.[42] Salmon likewise assumes that the limiting frequency statements that may serve as a basis for statistical explanations are lawlike, although the theoretical justification for attributing counterfactual (and subjunctive) force to these sentences should not be taken for granted, since it may entail the loss of their extensionality.[43]

The condition that the knowledge context $\mathcal{K}zt$ contain a set of probability statements, of course, might be subject to criticism on the grounds that it requires $\mathcal{K}zt$ to contain an enormous number of lawlike sentences.[44] This objection lacks forcefulness, however, when the following are considered, namely:

(i) the sentences belonging to $\mathcal{K}zt$ are accepted or believed by z at t, i.e., they represent what z takes to be the case;

(ii) this set of sentences, therefore, may be believed to be exhaustive with respect to attribute A and trial T, whether that is actually true or not; and,

(iii) presumably, this explication is intended to specify the conditions that must be fulfilled in order to provide an adequate explanation relative to a specified knowledge context without presuming that these conditions are always (or even generally) satisfied.

Second, Hempel defines the concept of *a maximally specific predicate* 'M' *related to* 'Ai' *in* $\mathcal{K}zt$ where 'M' is such a predicate if and only if (a) 'M' is logically equivalent to a conjunction of predicates that are statistically relevant to 'Ai' in $\mathcal{K}zt$; (b) 'M' entails neither 'A' nor '$-A$'; and, (c) no predicate expression stronger than 'M' satisfies (a) and (b), i.e., if 'M' is conjoined with a predicate that is statistically relevant to 'Ai' in $\mathcal{K}zt$, then the resulting expression entails 'A' or '$-A$' or else it is logically equivalent to 'M'.[45] Thus, for outcome A on trial T, 'M' is intended to provide a description of that trial as a member of a reference class K determined by the conjunction of every statistically relevant predicate 'F^m' of trial T with respect to outcome A, i.e., 'M' is the conjunction of all and only the statistically relevant *i*-predicates (or the negations of such *i*-predicates) satisfied by trial *T* in $\mathcal{K}zt$.

Third, Hempel formulates the *revised requirement of maximal specificity* as follows: An argument

$$\text{(VII)} \qquad \frac{\begin{array}{c} P(F, A) = r \\ Fi \end{array}}{Ai} [r],$$

where *r* is close to 1 and all constituent statements are contained in $\mathcal{K}zt$, qualifies as an explanation of the explanandum-phenomenon described by the explanandum-sentence '*Ai*' (or of the fact that *i* is a member of the attribute class *A*), within the knowledge context $\mathcal{K}zt$, only if:

(*RMS**) For any predicate, say '*M*', which either (a) is a maximally specific predicate related to '*Ai*' in $\mathcal{K}zt$ or (b) is stronger than '*F*' and statistically relevant to '*Ai*' in $\mathcal{K}zt$, the class $\mathcal{K}$ contains a corresponding probability statement, '$P(M, A) = r$', where, as in (VII), $r = P(F, A)$.[46]

Since a predicate expression 'F^j' is stronger than a predicate expression 'F^i' if and only if 'F^j' entails but is not entailed by 'F^i', moreover, a predicate which is logically equivalent to the conjunction of '*F*' with *any other permissible i-predicate* 'F^k' such that $\mathcal{K}zt$ contains the corresponding probability statement, '$P(F \cdot F^k, A) = r$', will satisfy condition (b) of this requirement, provided that, as in (VII), $r = P(F, A)$.

The motivation of Hempel's introduction of a requirement of this kind, let us recall, was the discovery that *statistical explanations* suffer from a species of explanatory ambiguity arising from the possibility that an individual trial *T* might belong to innumerable different reference classes $K^1, K^2, \ldots$ for which, with respect to a specific attribute *A*, the probabilities for the occurrence of that outcome may vary widely, i.e., the reference class problem for single case explanations. In particular, Hempel has displayed concern with the possibility of the existence of alternative explanations consisting of alternative explanans which confer *high probabilities* upon both the occurrence of an attribute *A* and its non-occurrence $-A$, relative to the physical world itself or a knowledge context $\mathcal{K}zt$, a phenomenon which cannot arise in the case of explanations involving universal rather than statistical lawlike statements.[47] However, it is important to note that there are *two* distinct varieties of explanatory ambiguity, at least one of which is not resolved by Hempel's maximal specificity requirement. For although Hempel provides a unique *value* solution to the single case problem (which entails a resolution to the difficulty of conflicting explanations which confer high probabilities upon their explanandum sentences), Hempel's approach does not provide a unique *description* solution to the reference class problem, a difficulty which continues to afflict his conditions of adequacy for explanations invoking universal *or* statistical lawlike statements within an epistemic *or* an ontic context. Hempel's explication, therefore, appears to afford only a restricted resolution of one species of statistical ambiguity, which, however, does not provide a solution to the general problem of explanatory ambiguity for explanations of either kind. For explanations involving universal laws as well as statistical laws continue to suffer from the difficulties that arise from a failure to contend with the problem of providing a unique *description* solution for single case explanations.

Hempel resolves the problem of statistical ambiguity, in effect, by requiring that, within a knowledge context $\mathcal{K}zt$ as previously specified, the occurrence of outcome attribute A on trial T is adequately explained by identifying a reference class K such that (a) $T \in K$; (b) $P(K, A) = r$; and, (c) for all subclasses K^j of K to which T belongs, $P(K^j, A) = r = P(K, A)$; but he does not require as well that (d) it is not the case that there exists some class $K^i \supset K$ such that $P(K^i, A) = r = P(K, A)$. For on the basis of condition (b) of (*RMS**), if, for example, the probability for the outcome death D within the reference class of twenty-one year old men who have tuberculosis K is equal to r and the probabilities for that same outcome within the reference classes $K \cdot F^1, K \cdot F^1 \cdot F^2, \ldots$ of twenty-one year old men who have tuberculosis K and who have high blood pressure F^1, or who have high blood pressure F^1 and have blue eyes $F^2, \ldots$ are likewise equal to r, then the explanation for the occurrence of death for an individual who belongs not only to class K but to class $K \cdot F^1$ and to class $K \cdot F^1 \cdot F^2$ and so on is adequate, *regardless of which of these reference classes is specified by the explanans in that single case.*[48] Consequently, on Hempel's explication, there is not only no unique explanation for the occurrence of such an explanandum outcome but, if the density principle for single trial descriptions is sound, *there may be an infinite* *number of adequate explanations for any one such explanandum*, on Hempel's explication. And this surprising result applies alike for explanations invoking laws of universal form, since if salt K dissolves in water A as a matter of physical law (within a knowledge context or without), then table salt $K \cdot F^1$ dissolves in water, hexed table salt $K \cdot F^1 \cdot F^2$ dissolves in water, ... as a matter of physical law as well; so if a single trial involves a sample of hexed table salt an adequate explanation for its dissolution in water may refer to any one of these reference classes or to any others of which it may happen to belong, providing only that, for all such properties F^i, it remains the case that all members of $K \cdot F^i$ possesses the attribute A as a matter of physical law.

Comparison with Salmon's explication suggests at least two respects in which Hempel's explication provides theoretically objectionable conditions of adequacy. The first, let us note, is that Hempel's requirement of maximal specificity (*RMS**) does not incorporate any appropriate relevance criteria that would differentiate statistically relevant from statistically irrelevant properties in the sense of principle (II*). For Hempel has defined the concept of statistical relevance so generally that a property F^i such that there exists some probability r with respect to the attribute A where $P(F^i, A) = r$ necessarily qualifies as 'statistically relevant' independently of any consideration for whether or not there may exist some class K such that $K \supset F^i$ and $P(K, A) = r = P(F^i, A)$; in other words, *Hempel's concept of statistical relevance is not a relevance requirement of the appropriate kind.*[49] In order to contend with this difficulty, therefore, major revision of Hempel's definition is required along the following lines:

> 'F^m' will be said to be *statistically relevant* to 'Ai' relative to 'K' in $\mathcal{K}zt$ if and only if (1) 'F^m' and 'K' are *i*-predicates that entail neither 'A' nor '$-A$'; (2) $\mathcal{K}zt$ contains the lawlike sentences, '$P(K \cdot F^m, A) = r^i$' and '$P(K \cdot -F^m, A) = r^j$'; and, (3) the sentence, '$r^i \neq r^j$', also belongs to $\mathcal{K}zt$.

The second is that Hempel's explication of explanation is incapable of yielding a unique description solution to the single case problem because it is logically equivalent to the *revised concept of epistemic homogeneity*, i.e., the epistemic version of the concept of an *ontically homogeneous* reference class, rather than the epistemic version of the concept of a *broadest homogeneous* reference class. There appears to be no reason, in principle, that precludes the reformulation of Hempel's requirement so as to incorporate the conditions necessary for fulfilling the desideratum of providing a unique description solution (for arguments having the form (VII) previously specified) as follows:

(*RMS***) For any predicate, say '*M*,' (a) if '*M*' is weaker than '*F*' and is statistically relevant to '*Ai*' in $\mathcal{K}zt$, then the class $\mathcal{K}$ contains a corresponding probability statement asserting that $P(M, A) \neq r$; and, (b) if '*M*' is stronger than '*F*' and is statistically relevant to '*Ai*' in $\mathcal{K}zt$, then the class $\mathcal{K}$ contains a corresponding probability statement asserting that $P(M, A) = r$; where, as in (VII)), $r = P(F, A)$.

Condition (*RMS***), therefore, not only requires that any property of trial *T* that is statistically relevant to '*A*' but nevertheless not explanatorily relevant must be excluded from an adequate explanation of that outcome on that trial in $\mathcal{K}zt$, but also requires that trial *T* be assigned to the *broadest* homogeneous reference class of which it is a member, in the sense that '*F*' is the *weakest* maximally specific predicate related to '*Ai*' in $\mathcal{K}zt$. Moreover, on the reasonable presumption that the definition of a maximally specific predicate precludes the specification of reference classes by *non-trivial* disjunctive properties, i.e., disjunctive properties that are not logically equivalent to some non-disjunctive property, these conditions actually require that trial *T* be assigned to the *strongest* homogeneous reference class of which it is a member. From this point of view, therefore, the reformulation of Hempel's requirement appears to provide a (strongest) unique description solution as well as a unique value solution to the single case problem within the spirit of Hempel's explication.

The revised formulation of Hempel's requirements (incorporating appropriate relevance conditions and (*RMS***) as well) and the revised formulation of Salmon's requirements (incorporating his original conditions (1), (2), and (3*), together with new condition (4*) as well), of course, both accommodate the naive concept of scientific explanation to the extent to which they satisfy the desideratum of explaining the occurrence of an outcome *A* on a single trial *T* by citing *all and only relevant properties of that specific trial.* Neither explication, however, fulfills the expectation that a property F is *explanatorily* relevant to outcome A on trial T if and only if F is *causally* relevant to outcome A on trial T, so long as they remain wedded to the frequency criterion of statistical relevance. Nevertheless, precisely because Hempel's explication is *epistemic*, i.e., related to a knowledge context $\mathcal{K}zt$ that may contain sentences satisfying the conditions specified, it is not subject to the criticism that the only adequate explanations are *non*-statistical, i.e., those for which probability $r = 0$ or $r = 1$. On the other hand, it *is* subject to the criticism that there are no *non*-epistemic adequate explanations for which probability $r \neq 0$ and $r \neq 1$, i.e., there are no ontic (or true) *statistical* explanations. As it happens, this specific criticism has been advanced by Alberto Coffa, who, while arguing that Hempel's epistemic explication is necessitated by (1) implicit reliance upon the frequency interpretation of physical probability, in conjunction with (2) implied acceptance of a certain 'not unlikely' reference class density principle, unfortunately neglects to emphasize that Salmon's *ontic* explication is similarly afflicted, precisely because the fatal flaw is not to be found in the *epistemic*-ness of Hempel's explication but rather in *reliance on the frequency criterion of statistical relevance for any ontic explication.*[50]

If Coffa's argument happens to be sound with respect to Hempel's rationale, then it is important to observe that, provided (1) is satisfied and (2) is true, there are, *in principle*, no non-epistemic adequate explanations for which probability $r \neq 0$ and $r \neq 1$; that is, an epistemic explication is not only *the only theoretically adequate kind of an explication*, but an explication remarkably similar to Hempel's specific explication appears to be *the only theoretically adequate explication.* So if it is not the case that an epistemic explication of the Hempel kind is the only theoretically adequate construction, then either (1) is avoidable or (2) is not true. Intriguingly, the density principle Coffa endorses, i.e., the assumption that, for any specific reference class K and outcome A such that trial $T \in K$, there exists a subclass K^i of K such that (i) $T \in K^i$, and (ii) $P(K^i, A) \neq P(K, A)$, is demonstrably false, since for all subclasses K^i such that $K \supseteq K^i$ and $K^i \supseteq K$, this principle does not hold, i.e., it does not apply to *any* homogeneous reference class K, whether or not T is the only member of K.[51] Insofar as every distinct trial is describable, in principle, by a set of permissible predicates, $\{F^n\}$, such that that trial is the solitary member of the kind K^* thereby defined, however, evidently Coffa's density principle is not satisfied by even *one single trial* during the course of the world's history. Coffa's principle is plausible, therefore, only so long as there appear to be *no* homogeneous reference classes; once the existence of reference classes of kind K^* is theoretically identified, it is obvious that this density principle is false. Nevertheless, another density principle in lieu of Coffa's principle does generate the same conclusion, namely: the density principle for single trial descriptions previously introduced. Coffa has therefore advanced an unsound argument for a true conditional conclusion, where, as it happens, by retaining one of his premisses and replacing the other, that conclusion does indeed follow, albeit on different grounds.

In his endorsement of Coffa's contentions, Salmon himself explicitly agrees that, on Hempel's explication, there are no *non-epistemic* adequate explanations for which probability $r \neq 0$ and $\neq 1$; thus he observes, "There are no homogeneous reference classes except in those cases in which *either* every member of the reference class has the attribute in question *or else* no member of the reference class has the attribute in question".[52] With respect to his own explication, by contrast, Salmon remarks, "The interesting question, however, is whether under any other circumstances K can be homogeneous with respect of A – e.g., if one-half of all K are A".[53] It is Salmon's view, in other words, that Hempel's position entails an *a priori* commitment to determinism, while his does not. But the considerations adduced above demonstrate that determinism is a consequence attending the adoption of the frequency criterion of statistical relevance alone, i.e., determinism is as much a logical implication of Salmon's own position as it is of Hempel's. This result, moreover, underlines the necessity to draw a clear distinction between *statistical* relevance and *causal* relevance; for, although it is surely true that two distinct events are describable, in principle, by different sets of permissible predicates, it does not follow that they are necessarily not both members of a *causally homogeneous* reference class K for which, relative to attribute A, $P(K, A) = r$ where $0 \neq r \neq 1$, even though, as we have ascertained, they may *not* both be members of some *statistically homogeneous* reference class K^* for which, with respect to attribute A, $P(K^*, A) = r$, where $0 \neq r \neq 1$. Although it is not logically necessary *a priori* that the world is deterministic, therefore, adoption of the frequency criterion of statistical relevance is nevertheless sufficient to demonstrate determinism's truth.

When the reformulated versions of Salmon's and Hempel's explications which entail assigning each singular trial T to the 'strongest' homogeneous reference class K of which it is a member (with respect to attribute A) are compared, the revised Hempel explication provides, in effect, a meta-language formulation of the revised Salmon object-language explication, with the notable exception that the Hempel-style explication envisions explanations as arguments for which high probability requirements are appropriate, and the Salmon-style explication does not. The issue of whether or not explanations should be construed as arguments is somewhat elusive insofar as there is no problem, in principle, in separating *explanation-seeking* and *reason-seeking varieties of inductive arguments*, i.e., as sets of statements divided into premises and conclusions, where the premises provide the appropriate kind of grounds or evidence for their conclusions.[54] But however suitable a high probability requirement may be relative to the reason-seeking variety of inductive argument, there appear to be no suitable grounds for preserving such a requirement relative to the explanation-seeking variety of inductive argument in the face of the following consideration, namely: that *the imposition of a high probability requirement between the explanans and the explanandum of an adequate explanation renders the adequate explanation of attributes that occur only with low probability logically impossible, in principle.*[55] Indeed, it appears altogether reasonable to contend that no single consideration militates more strongly on behalf of conclusive differentiation between 'inductive arguments' of these two distinct varieties than this specific consideration.

The theoretical resolution of these significant problems, therefore, appears to lie in the direction of a more careful differentiation between principles suitable for employment within the explanation context specifically and those suitable for employment within the induction context generally. The problem of single case explanation, for example, receives an elegant resolution through the adoption of the *propensity* interpretation of physical probability; for on that explication,

(VIII) $P^*(E, A) = r =_{df}$ the strength of the dispositional tendency for any experimental arrangement of kind E to bring about an outcome of kind A on any single trial equals r;

where, on this statistical disposition construction, a clear distinction should be drawn between *probabilities* and *frequencies*, insofar as frequencies display but do not define propensities.[56] The *propensity criterion of causal relevance*, moreover, provides a basis for differentiating between causal and inductive relevance relations; for, on the propensity analysis, a property F is *causally relevant* to the occurrence of outcome A with respect to arrangements of kind E if and only if:

$$\text{(IX)} \qquad P^*(E \cdot F, A) \neq P^*(E \cdot -F, A);$$

that is, the strength of the dispositional tendency for an arrangement of kind $E \cdot F$ to bring about an outcome of kind A differs from the strength of the dispositional tendency for that same outcome with an arrangement of kind $E \cdot -F$ on any single trial. By virtue of a probabilistic, rather than deductive, connection between probabilities and frequencies on this interpretation, it is not logically necessary that probabilities vary if and only if the corresponding frequencies vary; but that

$$\text{(X)} \qquad P(E \cdot F, A) \neq P(E \cdot -F, A);$$

i.e., that long (and short) run frequencies for the attribute A vary over sets of trials with experimental arrangements of kind $E \cdot F$ and $E \cdot -F$, nevertheless, characteristically will provide information which, although neither necessary nor sufficient for the truth of the corresponding probability statement, may certainly qualify as inductively relevant to the truth of these propensity hypotheses.[57]

On the propensity conception, it is not the case that every distinct trial T must be classified as a member of a *causally homogeneous* reference class $\{K^*\}$ of which it happens to be the solitary member merely because T happens to be describable by a set of permissible predicates $\{F^n\}$ such that T is the only member of the corresponding *statistically homogeneous* reference class. Consequently, a single individual trial T may possess a statistical disposition of strength r to bring about the outcome A (a) whether that trial is the only one of its kind and (b) whether that outcome actually occurs on that trial or not. On this analysis, the question of determinism requires an *a posteriori* resolution, since it is not the case that, on the propensity criterion of causal relevance, any two distinct trials are therefore necessarily trials of two different causally relevant kinds. And, on the propensity criterion of causal relevance, it is not the case that the only adequate explanations for which the probability $r \neq 0$ and $\neq 1$ are invariably epistemic; for a set of statements satisfying the revised Salmon conditions (1)–(4*) or the corresponding revised Hempel conditions (in their ontic formulation) will explain the fact that x, a member of K, is also a member of A, provided, of course, those sentences are true.[58] For both explications, thus understood, may be envisioned as fulfilling the theoretical expectations of the naive concept of scientific explanation, where the only issue that remains is whether or not, and, if so, in what sense, statistical explanations should be supposed to be inductive arguments of a certain special kind.

Perhaps most important of all, therefore, is that Reichenbach's frequency interpretation of physical probability, which was intended to resolve the problem of providing an objective conception of physical probability, has indeed contributed toward that philosophical desideratum, not through any demonstration of its own adequacy for that role, but rather through a clarification of the conditions that an adequate explication must satisfy. For the arguments presented above suggest:

(a) that *the concept of explanatory relevance* should be explicated relative to a requirement of causal relevance;

(b) that *the requirement of causal relevance* should be explicated relative to a concept of physical probability; and,

(c) that *the concept of physical probability* should be explicated by means of the single case propensity construction;[59] and, moreover,

(d) that *the concept of inductive relevance* should be explicated relative to a requirement of total evidence;

(e) that *the requirement of total evidence* should be explicated relative to a concept of epistemic probability; and,

(f) that *the concept of epistemic probability* should be explicated, at least in part, by means of the long run frequency construction.[60]

If these considerations are sound, therefore, then it is altogether reasonable to suppose that the recognition of the inadequacy of the frequency conception as an explication of *physical* probability may ultimately contribute toward the development of an adequate explication of *epistemic* probability, where the frequency criterion of statistical relevance is likely to fulfill its most important theoretical role.

University of Cincinnati

NOTES

* The author is grateful to Wesley C. Salmon for his valuable criticism of an earlier version of this paper.

[1] Hans Reichenbach, *The Theory of Probability*, Berkeley, University of California Press, 1949, esp. pp. 366–378.

[2] For the same attribute A on the same trial T in the same world W or knowledge context $\mathcal{K}zt$. Recent discussion of the symmetry thesis is provided by Adolf Grünbaum, *Philosophical Problems of Space and Time*, New York, Alfred A, Knopf, 1963, 2nd ed. 1973, Ch. 9; and in James H. Fetzer, 'Grünbaum's "Defense" of the Symmetry Thesis', *Philosophical Studies* (April 1974), 173–187.

[3] Reichenbach apparently never explicitly considered the question of the logical structure of statistical explanations; cf. Carl G. Hempel, 'Lawlikeness and Maximal Specificity in Probabilistic Explanation', *Philosophy of Science* (June 1968), 122.

[4] Especially as set forth in Hempel, 'Maximal Specificity', pp. 116–133; and in Wesley C. Salmon, *Statistical Explanation and Statistical Relevance*, Pittsburgh, University of Pittsburgh Press, 1971.

[5] Reichenbach, *Probability*, p. 68.

[6] Reichenbach, *Probability*, p. 69.

[7] Reichenbach, *Probability*, p. 72.

[8] Cf. Hilary Putnam, *The Meaning of the Concept of Probability in Application to Finite Sequences*, University of California at Los Angeles, unpublished dissertation, 1951.

[9] Reichenbach, *Probability*, pp. 376–77.

[10] Reichenbach, *Probability*, p. 377.

[11] This condition may be viewed as circumventing the problems posed by the requirement of randomness that might otherwise be encountered; but issues of randomness (or normality) will not figure significantly in the following discussion.

[12] As will be explained subsequently, this condition imposes a unique *value* solution but does not enforce a unique *description* solution to the problem of the single case.

[13] Assuming, of course, an unlimited supply of predicates in the object language. A different *density principle for reference classes* is discussed in Section 3.

[14] Reichenbach, *Probability*, p. 374. Variables are exchanged for convenience.

[15] It is not assumed here that '$P(K, A)$' stands for '$P(K \cdot -C, A)$'; see Section 2.

[16] Reichenbach, *Probability*, p. 375.

[17] Reichenbach, *Probability*, p. 376.

[18] Reichenbach, *Probability*, pp. 375–376.

[19] More precisely, perhaps, the term 'object' may be replaced by the term 'event' (or 'thing') to preserve generality; cf. Hempel, 'Maximal Specificity', p. 124.

[20] An extended discussion of this issue is provided by Hempel, 'Maximal Specificity', pp. 123–129.

[21] If a 'narrowest' reference class description is satisfied by more than one distinct event, then it is not a *narrowest* class description, since there is some predicate 'F^i' such that one such event satisfies 'F^i' and any other does not; otherwise, they would not be distinct events. This conclusion follows from the principle of identity for events and may therefore be characterized as a matter of logical necessity; cf. James H. Fetzer, 'A World of Dispositions', *Synthese* (forthcoming). A reference class description may be satisfied by no more than one distinct event, however, without being a narrowest reference class description. Thus, an

ontically homogeneous reference class may have only one member but is not therefore logically limited to a finite number of such members.

[22] There are therefore two theoretical alternatives in specifying a homogeneous reference class for the single individual trial T with respect to outcome A, namely: (a) a reference class description 'K^*' incorporating only *one* predicate 'F^i' (or a finite set of predicates {'F^n'}) satisfied by that individual trial alone; or, (b) a reference class description 'K^*' incorporating *every* predicate 'F^i', 'F^j',... satisfied by that individual trial alone. Since the set of predicates {'F^i', 'F^j', ...} satisfied by that individual trial alone may be an infinite set (and in any case will be a narrowest reference class description), alternative (a) shall be assumed unless otherwise stated. Note that $P(K^*, A) = P(F^i, A) = P(F^j, A) = \ldots = P(F^i \cdot F^j, A) = \ldots$, but nevertheless each of these predicates turns out to be statistically relevant.

[23] Cf. Carl G. Hempel, in *Aspects of Scientific Explanation*, New York, The Free Press, 1965, pp. 334–335; and Fetzer, 'Grünbaum's "Defense"', pp. 184–186.

[24] This requirement is discussed, for example, in Hempel, *Aspects*, pp. 63–67.

[25] The relationship between the requirement of *total evidence* and a requirement of *explanatory relevance* (whether causal or not), moreover, is a fundamental issue. See, in particular, Hempel, *Aspects*, pp. 394–403; Salmon, *Statistical Explanation*, pp. 47–51 and pp. 77–78; and esp. Hempel, 'Maximal Specificity', pp. 120–123.

[26] Salmon, *Statistical Explanation*, p. 43. Variables are exchanged once again.

[27] Salmon, *Statistical Explanation*, pp. 42–45 and pp. 58–62.

[28] Salmon, *Statistical Explanation*, p. 55.

[29] Salmon, *Statistical Explanation*, p. 55.

[30] Salmon, *Statistical Explanation*, pp. 76–77.

[31] Wesley C. Salmon, 'Discussion: Reply to Lehman', *Philosophy of Science* (September 1975), 398.

[32] Where the role of laws is to certify the relevance of *causes* with respect to their *effects*; cf. James H. Fetzer, 'On the Historical Explanation of Unique Events', *Theory and Decision* (February 1975), esp. pp. 89–91.

[33] Cf. J. Alberto Coffa, 'Hempel's Ambiguity', *Synthese* (October 1974), 161–162.

[34] Salmon, 'Reply to Lehman', p. 400.

[35] Wesley C. Salmon, 'Theoretical Explanation', in *Explanation*, ed. by Stephan Körner, London, Basil Blackwell, 1975, esp. pp. 121–129 and pp. 141–145.

[36] Cf. Hempel, 'Maximal Specificity'; p. 130.

[37] This result may likewise be obtained by prohibiting the specification of any homogeneous reference class by *non-trivial* disjunctive properties, which is apparently implied by Hempel's definition of a maximally specific predicate, as discussed in Section 3.

[38] Salmon, *Statistical Explanation*, p. 44.

[39] Cf. Wesley C. Salmon, 'Comments on "Hempel's Ambiguity" by J. Alberto Coffa', *Synthese* (October 1974), 165.

[40] Hempel, 'Maximal Specificity', p. 131.

[41] Hempel, 'Maximal Specificity', p. 131. Variables are again exchanged.

[42] Cf. Hempel, *Aspects*, pp. 338–343; and Hempel, 'Maximal Specificity', pp. 123–129.

[43] Salmon, *Statistical Explanation*, p. 81. Reichenbach's epistemological program, including (a) the verifiability criterion of meaning, (b) the rule of induction by enumeration, and, (c) the pragmatic justification of induction, suggest a profound commitment to establishing wholly extensional truth conditions for probability statements, an effort that may be regarded as culminating in his introduction of the 'practical limit' construct. For discussion of certain difficulties attending the reconciliation of these desiderata, see James H. Fetzer,

'Statistical Probabilities: Single Case Propensities vs Long Run Frequencies', in *Developments in the Methodology of Social Science*, ed. by W. Leinfellner and E. Köhler, Dordrecht, Holland, D. Reidel Publishing Co., 1974, esp. pp. 394–396.
[44] Cf. Henry E. Kyburg, Jr., 'Discussion: More on Maximal Specificity', *Philosophy of Science* (June 1970), 295–300.
[45] Hempel, 'Maximal Specificity', p. 131.
[46] Hempel, 'Maximal Specificity', p. 131.
[47] Hempel, *Aspects*, pp. 394–396; and Hempel, 'Maximal Specificity', p. 118.
[48] As his own examples illustrate; cf. Hempel, 'Maximal Specificity', pp. 131–132.
[49] This contention may be taken to be Salmon's basic criticism of Hempel's view; cf. Salmon, *Statistical Explanation*, pp. 7–12 and esp. p. 35.
[50] Coffa, 'Hempel's Ambiguity', esp. pp. 147–148 and p. 154.
[51] Coffa, 'Hempel's Ambiguity', p. 154. Niiniluoto interprets Coffa's density principle analogously, concluding that it is true *provided* all the subclasses K^i of K required to generate these reference classes happen to exist. Ilkka Niiniluoto, 'Inductive Explanation, Propensity, and Action', in *Essays on Explanation and Understanding*, ed. J. Manninen and R. Tuomela, Dordrecht, Holland, D. Reidel Publishing Co., 1976, p. 346 and pp. 348–349.
[52] Salmon, 'Comments on Coffa', p. 167.
[53] Salmon, 'Comments on Coffa', p. 167.
[54] Cf. James H. Fetzer, 'Statistical Explanations', in *Boston Studies in the Philosophy of Science*, Vol. XX, ed. by K. Schaffner and R. Cohen, Dordrecht, Holland, D. Reidel Publishing Co., 1974, esp. pp. 343–344.
[55] Cf. Salmon, *Statistical Explanation*, pp. 9–10; Fetzer, 'Statistical Explanations', pp. 342–343; and esp. Richard C. Jeffrey, 'Statistical Explanation vs Statistical Inference', in Salmon, *Statistical Explanation*, pp. 19–28.
[56] Among those adhering to one version or another are Peirce, Popper, Hacking, Mellor, Giere, and Gillies. See, for example, Karl R. Popper, 'The Propensity Interpretation of Probability', *British Journal for the Philosophy of Science* **10** (1959), 25–42; James H. Fetzer, 'Dispositional Probabilities', in *Boston Studies in the Philosophy of Science*, Vol. VIII, ed. by R. Buck and R. Cohen, Dordrecht, Holland, D. Reidel Publishing Co., 1971, pp. 473–482; R. N. Giere, 'Objective Single-Case Probabilities and the Foundations of Statistics', in *Logic, Methodology, and Philosophy of Science*, ed. by P. Suppes *et al.*, Amsterdam, North Holland Publishing Co., 1973, pp. 467–483; and the review article by Henry E. Kyburg, 'Propensities and Probabilities', *British Journal for the Philosophy of Science* (December 1974), 358–375.
[57] See, for example, Ian Hacking, *Logic of Statistical Inference*, Cambridge, Cambridge University Press, 1965; Ronald N. Giere, 'The Epistemological Roots of Scientific Knowledge', in *Minnesota Studies in the Philosophy of Science*, Vol. VI, ed. by G. Maxwell and R. Anderson, Jr., Minneapolis, University of Minnesota Press, 1975, pp. 212–261; and James H. Fetzer, 'Elements of Induction', in *Local Induction*, ed. by R. Bogdan, Dordrecht, Holland, D. Reidel Publishing Co., 1976, pp. 145–170.
[58] Hempel's specific explication is not the only adequate explication (or the only adequate kind of an explication), therefore, precisely because condition (1) – reliance upon the frequency interpretation of physical probability – is not the only theoretical option. While Hempel's and Salmon's explications are both logically compatible with either the frequency or the propensity concepts, the choice between them is not a matter of philosophical preference but rather one of theoretical necessity (although Salmon suggests otherwise; cf. Salmon, *Statistical Explanation*, p. 82). For related efforts in this direction, see James H.

Fetzer, 'A Single Case Propensity Theory of Explanation', *Synthese* (October 1974), 171–198; and James H. Fetzer, 'The Likeness of Lawlikeness', *Boston Studies in the Philosophy of Science*, Vol. XXXII, ed. by A. Michalos and R. Cohen, Dordrecht, Holland, D. Reidel Publishing Co., 1976. Also, note that even when Hempel's conception is provided an *ontic* formulation (by deleting the conditions that render statistical explanations inductive arguments for which a high probability requirement within a knowledge context $\mathcal{K}zt$ is appropriate), it is still not the case that Hempel's explication and Salmon's explication are then logically equivalent. For Salmon's (original) conditions may be criticized as requiring the explanans of (at least some) explanations to be 'too broad', while Hempel's conditions even then may be criticized as permitting the explanans of (at least some) explanations to be 'too narrow'. Consequently, the relationship between Hempel's and Salmon's requirements is more complex than I previously supposed; cf. Fetzer, 'A Propensity Theory of Explanation', pp. 197–198, note 25, and Fetzer, 'Statistical Explanations', p. 342. For convenience of reference, finally, we may refer to the model of explanation elaborated here as the *causal-relevance* (or C–R) explication, by contrast with Hempel's (original) *inductive-statistical* (or I–S) model and with Salmon's (original) *statistical-relevance* (or S–R) model examined above.

[59] It is significant to note, therefore, that Hempel himself has endorsed the propensity interpretation of physical probability for statistical 'lawlike' sentences; Hempel, *Aspects*, pp. 376–380, esp. p. 378, Note 1. Hempel's view on this issue is critically examined by Isaac Levi, 'Are Statistical Hypotheses Covering Laws?', *Synthese* **20** (1969), 297–307; and reviewed further in Fetzer, 'A Propensity Theory of Explanation', esp. pp. 171–179. Salmon, in particular, does not believe that the choice between the propensity and the frequency conceptions of physical probability is crucial to the adequacy of his explication of explanation, contrary to the conclusions drawn above; Salmon, *Statistical Explanation*, p. 82.

[60] It is also significant to note that Hempel has recently expressed the view that Reichenbach's policy of assigning singular occurrences to 'the narrowest reference class' should be understood as Reichenbach's version of the requirement of total evidence *and that the requirement of total evidence is not an explanatory relevance condition*; 'Maximal Specificity', pp. 121–122. From the present perspective, moreover, Hempel's own revised requirement of maximal specificity (*RMS**), when employed in conjunction with the frequency interpretation of statistical probability, may itself be envisioned as applying within the *prediction context* generally rather than the *explanation context* specifically; indeed, for this purpose, Hempel's original definition of statistical relevance would appear to be theoretically unobjectionable, provided, of course, that these probability statements are no longer required to be *lawlike*. See also Fetzer, 'Elements of Induction', esp. pp. 150–160.

CLARK GLYMOUR

REICHENBACH'S ENTANGLEMENTS*

Looking back, Reichenbach's views about knowledge and meaning seem in most respects entirely typical of those of empiricists of the day. His epistemological differences with Carnap, C. I. Lewis and others seem far less significant than the agreements they shared. If Reichenbach is distinctive, it is because his epistemological concerns seem to have had their source in careful analyses of innovations in physical theory, and because his developing views were constantly buttressed with physical examples. It is this interplay of epistemological doctrine and scientific theory that makes Reichenbach's work especially appealing, vivid and forceful. Reichenbach's epistemological views are, by now, unpopular enough that they scarcely need criticism, and even though I will criticize them that is not my chief purpose. My purpose is to provide a perspective on the interaction of Reichenbach's epistemological doctrines and his analyses of scientific inference, a perspective which I hope will reveal something both new and true about the problems that motivated Reichenbach's work. It is exactly because Reichenbach's views are in many ways typical of empiricism present and past that the enterprise is worthwhile; the points I shall make about Reichenbach could equally be made about Carnap, Lewis, or any number of contemporary writers.

In so far as I have a thesis, it is threefold. First, that the categories in terms of which Reichenbach thought it necessary to analyze scientific theories drew their plausibility from mistakes about scientific inference which Reichenbach made early in his career. Second, that the account of scientific inference which generated the problems Reichenbach saw as most central to understanding knowledge was an entirely different account from the one which he developed to meet those problems. Finally, that his later account of scientific inference and of meaning relations did not resolve the problems that motivated it. I know of no better way to make my points than by surveying the early development of Reichenbach's views about knowledge and meaning.

Reichenbach believed that the advent of non-Euclidean geometries and of the theories of relativity had made Kant's analysis of knowledge

W. C. Salmon (ed.), Hans Reichenbach: Logical Empiricist, 221–237.

untenable. His first important book-length work (1920) was devoted to arguing that contention, and to sketching, in what can only be described as a very unsatisfactory way, the views which he thought ought to suceed Kant's. The details of Reichenbach's exegesis of Kant do not concern us, but his positive proposals do. According to Reichenbach, there are two distinct conceptions of the *a priori*: on the one hand, what is *a priori* is what is necessarily true and not disconfirmable by any possible experience; on the other hand, what is *a priori* is what "constitutes the concept of the object". The claims of Kantians to have provided a set of *a priori* truths of the first kind is refuted by modern physics but there are, nonetheless, *a priori* truths of the second kind. Reichenbach gives a new and unclear significance to what it is to "constitute the concept of the object"; the principles that are *a priori* in this second sense are those general claims which contain 'general rules' according to which connections between physical things and their states may take place. These *a priori* principles include, for example, the 'principle of normal induction', the principle of covariance, various causality principles, and so on. What they do, allegedly, is to 'coordinate' something in language with something in sensation – I am still unsure what Reichenbach took the respective somethings to be. In any case, such *a priori* principles are 'principles of coordination' which somehow establish what it is that scientific laws are to relate; principles of coordination contrast with principles of connection, which are just ordinary scientific laws relating one kind of state of affairs to other kinds of states of affairs. Unlike Kant's, Reichenbach's *a priori* principles are empirical and contingent; they can change if what constitutes knowledge changes. We can only discover them empirically by analyzing how it is our theories are formed and constituted, and while individual principles of coordination, in isolation, may be untestable, in combination they can have testable consequences and so can be refuted. All that differentiates Reichenbach's *a priori* principles from ordinary, everyday scientific laws is that the former determine the objects of scientific knowledge. They determine the relata that humdrum laws relate.

There is another feature of the account of knowledge given in *The Theory of Relativity and A Priori Knowledge* which is of first importance, and that is Reichenbach's analysis of how scientific theories are tested and confirmed. As we will see shortly, this analysis played an important role in

Reichenbach's emerging conventionalism. A set of coordinative principles and connecting laws is tested by obtaining two or more independent values of for a quantity or quantities of the theory. If the values so obtained are always the same then, says Reichenbach, the coordination is unique and the theory is true. The values for the quantities are obtained by applying theoretical principles to the results of experience, or, in some cases, from theory alone. What is important is that the values be obtained in different ways, at least one of which begins with some experience. Reichenbach's account is crude perhaps, but it represents a common theme in the epistemological analyses of the day, and one which has some claim to accuracy. The theme was stated better by Reichenbach himself some years later. Discussing the testing of a quantitative hypothesis, Reichenbach wrote:

We write the function to be tested in the form

$$F(p_1 \ldots p_r) \quad (1)$$

where $p_1 \ldots p_r$ are parameters that characterize the process ...

Now we wish to compare the function with the observational data. For this purpose, we take observed sets of correlated values $p_1^* \ldots p_r^*$, which we shall call observed points p^* and see whether these points p^* lie on the surface F_r, i.e., whether the sets of values ... satisfy condition (1) ...

Let us now imagine that the test has been performed. We cannot expect that all points p^* will lie exactly on the surface F_r, but if the deviation is too great, we shall modify our theory. Let us assume that the deviations are considerable. Then we can avail ourselves of two methods of changing our hypothesis:

(1) We assume that our data are false. The magnitudes $p_1 \ldots p_r$, however, are not directly observed, but are usually inferred by means of theoretical considerations from the indications of the measuring instruments, i.e., they are derived with the help of functions

$$p_i = P_i(q_1 \ldots q_m)$$

from other data $q_1 \ldots q_m$. The mistake therefore could lie in the functions P_i. Yet, if one has made use of functions that are highly confirmed through frequent previous observations, and the test has been performed according to the schema we described, we shall regard the values $p_1^* \ldots p_r^*$ as correct.

(2) We assume that the observed phenomenon is not adequately described by expression (1) and replace it by

$$F'_{r+s}(p_1 \ldots p_r, p_{r+1} \ldots p_{r+s}) = 0\,. \quad (2)$$

We call the magnitudes $p_{r+1} \ldots p_{r+s}$ additional parameters ... They are variables, and are determined for every set of values $p_1^* \ldots p_r^*$ in such a way that (2) is satisfied with the desired exactness. (1959, pp. 114–115)

Reichenbach says that the new relation (2) should meet several conditions, the first (a) of which is that the observed points should lie sufficiently close to the surface determined by the equation. Further, (b) "the function F'_{r+s} should be the *simplest* among all those functions satisfying (a) . . . a function F'_{r+s} needing a smaller number of parameters is simpler than another one". (1959, p. 116) And finally condition (c), the function should be *justified*:

it should be possible to incorporate the additional parameters into other relationships. There should exist functions

$$P_{r+i}(q_1 \dots q_k) = p_{r+i}$$

and the special values p^*_{r+i} which the parameters p_{r+i} assume relative to the $p^*_i \dots p^*_r$ should be related by means of the functions P_{r+i} to observational data $q^*_1 \dots q^*_k$ which have been found experimentally to belong to the respective set of values $p^*_1 \dots p^*_r$. The functions P_{r+i} should be confirmed empirically in the manner described here for the function F'_{r+s}. (1959, p. 116)

If one gets lost among the p's, the q's and the asterisks, still the basic idea is clear enough: Hypotheses are tested and confirmed by producing instances of them; to produce instances of theoretical hypotheses one must use other theoretical relations to determine values for theoretical quantities; these other theoretical relations are tested, in turn, in the same way. Ideally, one might hope for bodies of evidence that permit each hypothesis to be tested independently, and so on. It is this strategy, I think, that Reichenbach had in mind when, in *the Theory of Relativity and A Priori Knowledge*, he required that our theories establish a 'unique coordination'. Although Reichenbach's formulation is perhaps the most explicit and detailed I have come upon, essentially the same view of how theories are tested was held by many of the foremost scientists and philosophers of the day. Weyl (1963) certainly held it; Carnap's 'Testability and Meaning' (1936–37) is devoted to a formal confirmation theory embodying similar ideas; Bridgeman's insistence (1961) that every physical quantity be 'operationally definable' at least by 'paper and pencil' operations is probably most generously understood along the lines of Reichenbach's requirement (c) above; and, even earlier, Poincaré (1952, esp. pp. 99–100) seems to have taken the same line and drawn radical epistemological conclusions from it. Given the general prevalence of this qualitative view of confirmation forty years ago, one must wonder how it came to be almost completely replaced by less intricate hypothetico-deductive views, so that today it is a novelty among accounts of scientific method.

In four years following the publication of *The Theory of Relativity and A Priori Knowledge*, Reichenbach's epistemological views appear to have changed significantly and permanently. The radical empiricism of his earlier piece was replaced by a muted conventionalism; 'coordinative principles' were replaced by 'coordinative *definitions*' which were in no way empirical. The task of epistemology came to be seen as the location of conventional elements in our knowledge. I have no idea what brought about these changes, but the arguments that Reichenbach produced for them depend on the view of theory testing just described, indeed, on a particular elaboration of that view.

Reichenbach's new view were developed in his second major work, *Axiomatization of the Theory of Relativity*. The chief aim of the book was to exhibit the epistemic structure of the special and general theories by showing what was required in order to derive the postulates of the theories from the evidence for them. 'The evidence' was not, as with Poincaré, some hypothetical phenomenal basis but rather ordinary physical statements established without resort to relativistic principles:

> this investigation starts with elementary facts as axioms; all facts whose interpretation can be derived by means of simple theoretical principles . . . all axioms of our presentation have been chosen in such a way that they can be derived from experiments by means of pre-relativistic optics and from experiments by means of pre-relativistic optics and mechanics. (1969, pp. 4–5)

It remained characteristic of Reichenbach's analyses that he did not try to found theories on any sort of phenomenal basis, but rather on the truths or approximate truths of preceding theory and ordinary belief. Now the essential epistemological task of Reichenbach's reconstruction of relativity was to sort out the 'empirical statements' from the 'coordinative definitions', that is, to separate the synthetic from the analytic. The principle used to effect the sorting seems to have been roughly as follows: Suppose that in order to obtain a value or values for a quantity or quantities in a certain hypothesis, *H*, we have to use another hypothesis, *G*. And suppose further that in order to produce values for quantities in *G* we have to use *H*. Then either *H* or *G* must really be a coordinative definition, a matter of convention, for otherwise the arguments for the theory would be viciously circular. The same holds if the circle is larger, consisting of more hypotheses than H and G, but none of the hypotheses can be tested without using some of the others.

No point is more important to understanding Reichenbach, for throughout the rest of his career the business of separating conventions from empirical claims was a point, perhaps the main point, of his epistemology, and the argument of the paragraph above is really the *only* self-sufficient argument for the distinction. More, it provided Reichenbach's criterion for separating conventional from non-conventional parts of a theory. Essentially the same kind of argument was given earlier by Poincaré for the conventionality of geometrical and mechanical laws, and his arguments were familiar to Reichenbach. But if the line was not entirely original with Reichenbach, still it most definitely was Reichenbach's line:

> Since the necessity of a coordinative definition for the application of the concept of simultaneity has often been overlooked, I should like to add a few remarks. We are speaking of simultaneity at different space points, not at the same place. It is impossible to *know* whether two events at different space points are simultaneous. Let these space points be P_1 and P_2, and let the events be E_1 and E_2. In order to compare the times of these events, a causal chain, a signal, would have to travel from E_2 at P_2 to P_1 which would arrive at P_1 later than E_1 at the time t_1. By measuring the time different $t_1 - E_1$ at P_1 and keeping the velocity of the signal in mind, one could infer at what time the signal had left P_2: only in this way is a time comparison of E_1 and E_2 possible. The measurement of simultaneity presupposed the knowledge of a velocity. On the other hand, the measurement of a velocity presupposed the knowledge of the simultaneity, because time measurements at two different places are required. At one space point only the sum of the time intervals for a round trip, that is, an average velocity, can be measured – from which nothing can be inferred concerning the velocity in *one* direction. The attempt to decide about simultaneity by *measurements* leads to a circular argument. Simultaneity of distant events therefore can only be *defined*, and all measurements will yield only that simultaneity which has previously been determined by definition. Even if one tries to circumvent the use of a signal for the determination of simultaneity and determine it by the transport of clocks . . . one does not change the definitional character of such a procedure. All measurements that are made relative to such a synchronization reflect the synchronization in their numerical values and can never be a test of its truth. From these values one can infer only that those events are simultaneous by the rule. A different rule would therefore not lead to contradictions . . . (1969, pp. 10–11)

Thus, according to Reichenbach, the impossibility of testing certain hypotheses in isolation from others makes it impossible to know, or to have good reasons to believe, any of the entangled hypotheses. Wherever such entanglements occur there are alternative hypotheses which will save the phenomena equally well. Not only is the choice between such alternative hypotheses conventional in the sense that no one choice has any rational or empirical warrant that others lack, it is conventional in the sense that there is no fact of the matter as to which choice is correct:

> The preference for a specific definition of simultaneity in the special theory is due to considerations of *simplicity* . . . This simplicity has nothing to do with the truth of the theory. The truth of the [empirical, not theoretical] axioms decides the empirical truth, and every theory compatible with them which does not add new empirical assumptions is equally true. (1969, p. 11)

Thus Reichenbach avoided scepticism by adopting a radical verifiability theory of meaning: Two theories which have the same empirical consequences are materially (and, it would seem, even logically) equivalent. At least at this point in his career, Reichenbach's verifiability views seem to have been unanalyzed and undefended; but given the lack of emphasis Reichenbach put on a characterizable phenomenal basis for knowledge, and thus the vagueness of 'empirical consequences', the emerging theory of knowledge already had inharmonious parts.

Undoubtedly, Reichenbach's best read and most influential book is *The Philosophy of Space and Time* (1958). Yet it contains scarcely any epistemological advance on his earlier books. What is new is the extension of the earlier ideas to new scientific domains and the vivacity and ingenuity with which these applications are presented. Just because of its influence, it is worth recalling the chief philosophical arguments of the first chapter of the book. The chapter is devoted to space. After introducing the now familiar doctrines of equivalent descriptions and coordinative definitions, Reichenbach argued that the self-congruence of any object under transport is a matter of coordinative definition. The argument is that to test such any hypothesis, to ascertain that a particular rod, say, is or is not self-congruent under transport, we must *use* some assumption or other which is not supported directly and independently by observation. Thus we cannot determine that the rod is self-congruent by transporting another rod of the same length along with it unless we assume that the second rod is self-congruent under transport. The coextension of such rods under transport is consistent with the alternative assumption that there is universally a change in length of transported bodies. Further, "An optical comparison . . . cannot help either. The experiment makes use of light rays and the interpretation of the measurement of the lengths depends on assumptions about the propagation of light." (1958, p. 16)

Later in the chapter this same argument, or a closely related one, is given a more formal guise. Suppose, in roughly the context of classical physics, that one is given a physical theory P specifying the forces acting on bodies of various kinds, the sources of fields and their laws, etc. Suppose, further, that the theory postulates geometrical laws G; if the theory is a field theory in covariant form, we can suppose that the laws of the geometry in effect specify a metric tensor field g on space, and that the spatial projections of the trajectories of particles moving subject to zero net force are geodesics of g. Now consider a contrasting theory which requires the metric tensor to be g' and which postulates a 'universal force' given by a tensor field u such that $g = g' + u$. Further, in the new theory bodies subject to no forces according to the original theory P will be subject only to the universal force, u. The two theories thus make exactly the same predictions about the motions of bodies. With such a contrasting theory in hand, according to Reichenbach, we see that experiments that might purport to establish G must tacitly use the assumption that universal forces are zero, and we further see that this assumption is capable of no independent confirmation.

The views that Reichenbach endorsed in the late twenties and early thirties are subject to obvious difficulties, and it is not unreasonable to suppose that Reichenbach became aware of them. First, there seems to be no objective feature which would enable us to single out one from among a set of entangled hypotheses and determine that that one is conventional, whereas the other hypotheses with which it is entangled are empirical. If A cannot be tested without supposing B, and B cannot be tested without supposing A, what makes A conventional rather than B, or vice-versa? Second, even given a clear distinction between what is observable and what is not, Reichenbach's crude verifiability criterion leads to obvious difficulties: if an hypothesis were true if and only if all of its observational consequences were true, then the most elementary rules of inference, *modus ponens* for example, would not be truth preserving.[1] Whether because he realized these difficulties, or for other reasons, Reichenbach modified his views in *Experience and Prediction*, published in the late thirties (1938). In this book we are still gaffed by distinctions – observational/non-observational, empirical/conventional, and so on. But single hypotheses in a tangled set are no longer simply conventional while others in that same set are empirical; rather, Reichenbach says, it is itself a matter of convention that some one hypothesis in a tangled set counts as conventional while others in that same set count as empirical. Such distinctions are made solely for convenience in rational reconstruction. In reality, no one hypothesis of such a set is any more or less conventional; each of them has a conventional element. Reichenbach's conventionalism had become holistic, not to say etherial. All the members of a set of entangled hypotheses could not be simply conventional, for the members of such a set, together, typically have clearly testable consequences (just as sets of Carnap's reduction sentences may have such consequences), and so the conventionality, not to be denied, was distributed among them. What had happened is that a distinction had come unraveled. It is worth nothing that the same distinction had come to grief for exactly the same reason in Carnap's 'Testability and Meaning' published only one year before *Experience and Prediction*.

In *Experience and Prediction* Reichenbach replaced his earlier theory of equivalent descriptions with a new account similar in spirit. According to the 'probability theory of meaning' a sentence has empirical meaning if and only if it can be confirmed or disconfirmed by some physically possible observation, that is, by an observation whose possibility is consistent with the laws of nature. Two sentences have the same empirical meaning if they receive the same confirmation by every physically possible observation. A sentence is confirmed by an observation if, given the observed event, the posterior probability of the state of affairs described by the sentence is greater than the prior probability of that state of affairs. Posterior probabilities are obtained from prior probabilities by Bayes' Rule, that is, by forming the conditional probability on the new evidence. Since Reichenbach's interpretation of probability was strictly frequentist, the prior probability of a theory had to be the limit of a relative frequency of an appropriate attribute in an appropriate reference class. The attribute class for the prior probability of an hypothesis, Reichenbach seems to have thought, is the class of *true* hypotheses of a certain kind; the reference class is the set of hypotheses of that kind. What the appropriate kinds are is left obscure.[2] Pursuing Reichenbach's idea, Wesley Salmon (1971) has lately suggested that the reference class should be the broadest, statistically homogeneous class, and that various characteristics of hypotheses – their formal relations with other hypotheses, the sources from which they derive, their simplicity, etc. – may be statistically relevant to truth and thus may enter into the determination of the appropriate reference class.

The great novelty of *Experience and Prediction* is the 'probability theory of meaning'; it is used throughout the book to analyze epistemological issues. And yet it is clearly a very strange theory. Synonymy is analyzed in terms of the probability theory of meaning, which in turn is analyzed in terms of truth. It seems that this order of analysis corresponds to a real order in our understanding of hypotheses; we may come to believe that a certain array of experimental results cannot discriminate between two hypotheses only because we have some understanding of what would make the hypotheses true and of what other truths such states of affairs would imply; we may come to believe that certain experimental results have no relevance at all to an hypothesis only because, again, we have some understanding of the hypothesis. It is only because we can understand something of hypotheses and make rough judgements of their implications that we can test them. Not the other way round. That granted, it is hard to see how the 'probability theory of meaning' can be a theory of meaning: it is certainly not *because* we see or don't see how to test an hypothesis that we understand it or not. But Reichenbach's analysis could be correct in the sense that there is a correspondence between our understandings and inferences and those given by his probability theory of meaning. Whether or not that theory gives us, for example, the criterion for synonymy, it may still give us an indicator of synonymy. We should, however, expect some argument for such a correspondence; perhaps an argument to show that synonymy as analyzed in the probability theory generates ordinary rough criteria for synonymy, syntactic intertranslatability for example. Reichenbach gave us no arguments for his theory of meaning, not of this kind or any other.

That is as far as I shall take the development of Reichenbach's views, but the salient features of that development merit reemphasis. The analysis of scientific inference as a procedure in which some hypotheses are used to obtain, from observation, instances of others, led Reichenbach to the conclusion that the principles of most scientific theories are underdetermined by the evidence, for any purported test of such entangled principles must be viciously circular. To avoid scepticism, Reichenbach advocated conventionalism, and to make 'convention' come to something more than 'unwarranted assumption' he adopted, first a simple verifiability theory of meaning and later his probability theory of meaning. In the course of this development, the analysis of scientific inference from which the problems derived was itself left behind, to be replaced in *Experience and Prediction* and *The Theory of Probability* by a probabalistic, Bayesian account of scientific inference. The latter account, however, provided no basis for the problems and examples that motivated and illustrated the conventionalist thesis: Nothing in Reichenbach's earlier books tends to show that alternative hypotheses about simultaneity or about congruence are equally confirmed by all physically possible observations. To make such a showing one would need to demonstrate that all of these alternatives have the same prior probability,[3] something that Reichenbach was never prepared to do, and something which Salmon's sketch of how Reichenbach's account of confirmation might be extended suggests is not even true.

If a system is a coherent, harmonious and mutually supporting set of views, then Reichenbach had no system. What he had was a view of knowledge as a structured set of beliefs with a foundation, and from that foundation a series of independently established theoretical propositions. What he had was a set of problems, one generic problem really, that arose from his tacit foundationalism and from a genuine analysis of scientific inference which he applied to a number of theories. His epistemological writings contain a familiar, and I think implausible, appeal to verifiability to resolve these problems, and a philosophical theory of scientific inference and of meaning designed to bolster the appeal to verifiability. The philosophical theory of inference, that is, the probabalistic, Bayesian theory, was largely unconnected with the understanding of the structure of scientific inference that led to Reichenbach's problems in the first place.

Reichenbach's problems were genuine enough. They have sometimes been dismissed on the grounds that they presuppose an observation language whereas there is no such thing. To dismiss them on that ground would miss a good deal of what is at stake. Regardless of whether or not there is a language of observation, it is undeniably the case that the evidence that we muster for theories is often stated in a vocabulary that does not exhaust the vocabulary of the theory, and that to produce instances of hypotheses of our theory from our evidence we often must use other hypotheses of that same theory in such a way that the hypotheses form the kind of entanglements that Reichenbach judged to be viciously circular. The problems were genuine enough, but I think they took their dreadful form chiefly because Reichenbach himself was not careful enough about his own early account of theory testing. Conventionalism was necessary to avoid scepticism, and scepticism threatened because theoretical principles were apparently radically underdetermined by observational evidence and rational inference. Any why underdetermination? Perhaps because Reichenbach subscribed to the following principle, or to something close to it.

A. If two consistent theories both logically entail a given class of observation statements, then the truth of any statements in that class provides no grounds for preferring one theory to the other.

Principle A, or something like it, has been supposed by a great many people, but on Reichenbach's own account of testing it is false. Suppose that the axioms of one theory can be *instantiated* and tested by starting with the observation sentences and *using* parts of the theory itself to deduce theoretical states of affairs or values for theoretical quantities. Suppose further that the axioms of the other theory cannot be so instantiated and tested. Then it seems we have straightforward grounds for preferring one theory to the other. It is neither logically nor historically difficult to see how this sort of situation can arise. If some theoretical quantity of a theory is replaced, wherever it occurs in the theory, by an algebraic combination of several distinct new theoretical quantities, then the theory that results from this odd substitution will not be testable in the way described. Historically, there certainly are cases in which two theories entailed a common class of phenomena but, in retrospect, were not equally tested by the phenomena. Both Ptolemaic and Copernican astronomy, with suitable values for their parameters, could generate the observed positions of the planets on the celestial sphere (to within commonly recognized observational error), and both theories contained a common hypothesis about the order of the planetary distances from the sun. Yet that common hypothesis was tested by the observations within the Copernican scheme but not within the Ptolemaic scheme.[4]

There are other ways in which Reichenbach's own account of testing, his early account, suggests that Principle A may fail. Suppose that the observation statements of the given class are, in one theory, all explained by a common set of theoretical principles so that these very principles are tested by each of the observation statements – and thus the theoretical principles in one theory are tested repeatedly. Suppose, by contrast, that in the other theory each observation must be explained by a distinct set of theoretical principles, and tests only the principles in that set, so that in this theory each theoretical principle is tested at most once. Then we would certainly think the first theory better warranted by the evidence. It is trivial to construct pairs of theories and a hypothetical body of evidence in which such a difference applies. Historically, once more Copernican and Ptolemaic theories provide examples, albeit examples a little too complicated for me to describe here.[5]

There are still other reasons why A should be false. The hypotheses of one theory that are tested by the observations might, for example, be central to that theory in some important and specifiable way, whereas the hypotheses of the other theory that are tested might be peripheral. These sorts of differences are possible exactly because the account of theory testing which Reichenbach's early discussion suggests is not radically holistic and can parcel out praise and blame to parts of theories.

Besides Principle A, there are other reasons why Reichenbach may have thought our theories to be underdetermined unless some of their hypotheses are disguised conventions. I certainly believe that Reichenbach held the following principle, or something close to it:

B. Let *P* and *Q* be empirical claims. An argument for *Q* which uses *P* and an argument for *P* which uses *Q* are, together, viciously circular. If, based on a given body of evidence, every test of *Q* uses *P* and every test of *P* uses *Q*, then that body of evidence provides no grounds either for *P* or for *Q*.

Principle B is certainly false. To see that it is, let *P* and *Q* be the same hypothesis; then B says that observations provide no evidence for *P* if, in order to produce instances of *P* from the observations, we must use *P* itself. But suppose *P* is an hypothesis that contains an empirical constant; for example, suppose *P* is the ideal gas law. Then to determine instances of *P* that test *P* it is perfectly ordinary and legitimate to use *some* of the data *and P itself* to determine a value of the constant occurring in *P*, and then to use this value and the remainder of the data to provide confirming instances of *P*.

Principle B is false even if *P* and *Q* are logically distinct. There need be nothing vicious about the circularity involved in using *P* to test *Q* and using *Q* to test *P*. The procedure would be circular if it somehow *guaranteed* that positive instances of *Q* or positive instances of *P* would result. But it need do neither. In order to determine whether planets moved in accordance with Kepler's second law – his rule about the area swept out by the vector from the sun to a planet – seventeenth century astronomers had to make use of an hypothesis about the orbits of the planets. Generally, they used the hypothesis of Kepler's first law: planets move in ellipitical orbits. Conversely, in order to test Kepler's first law they had to use his second. But it is evident that the procedure of determining the elements of an orbit by using the hypothesis that the orbit is elliptical does not of itself guarantee that the resulting orbit will satisfy Kepler's second law. Only if the observed positions of the planets exhibit very special patterns will Kepler be vindicated.[6]

Why then do arguments of the sort that Principle B mistakenly prohibits *seem* circular? Perhaps because we imagine them to be arguments of roughly the following form:

Premise 1: The observations are such and such.
Premise 2: It is the case that *P*.
Hence: The observations confirm *Q*.
Hence: Inductively, it is the case that *Q*.
Hence: By premise 1 and the conclusion just drawn, the observations confirm *P*.
Hence: Inductively, it is the case that *P*.

Conclusion: It is the case that *P* and *Q*.

Circular indeed, but why must we suppose that the inductive reasoning has this form? The argument seems to presuppose that confirmation is just a binary relation between a piece of evidence and an hypothesis – but suppose instead that it is a ternary relation holding among a piece of evidence, an hypothesis, and whatever other claims are used in relating the evidence to the hypothesis by Reichenbach's procedure. To use *P* to obtain an instance of Q from the data need not involve us in asserting P; to establish that the data provide us with an instance of Q *relative to* P or, if you prefer, *on the supposition that* P, does not require that we claim to know that P is the case.

If our confirming instances are really instances of a ternary relation of the kind suggested, how do we choose from among competing theories? I think we choose if we can that theory which is best confirmed with respect to itself by the observations available to us. What goes into 'best confirmed' is surely a complex matter, but it must include at least those features of confirmation discussed in objection to Principle A. Principle B really supposes that before we can use an hypothesis in confirming others we must independently establish the first hypothesis. The process of confirmation is thought of rather like the construction of a conventional building – first the foundation, then .... Confirmation doesn't have to be that way, and generally it isn't. Teepees make better analogies.

There is a third principle about confirmation which I think it is clear that Reichenbach held:

C. If some of the hypotheses of a theory form an entangled collection with respect to a given set of observation statements, so that some of the hypotheses in this collection must be used to test others from the observation statements, and vice-versa, then there is another theory, inconsistent with the first, such that the given class of observation statements provides no grounds for preferring one of the theories to the other.

I have no idea whether or not Principle C is true. It is easy enough to show that even with plausible restrictions about simplicity there can be two theories that logically entail a common class of observation statements – even a non-finitely-axiomatizable class of observation statements – but are contradictory and do not even satisfy syntactic criteria for intertranslatability (Glymour, 1971). An analogous proof is not so easy when it is required that the two theories be equally well confirmed, each with respect to itself, by the observation statements. For, in the first place, a mathematically clear account of a ternary relation of confirmation is bound to be rather complex; and, in the second place, we lack a definitive set of criteria framed in terms of such a ternary confirmation relation for comparing theories. The matter of the truth of Principle C is interesting and important and merits philosophical attention.

I have argued that Reichenbach's later probabalistic account of confirmation has little bearing on the problems which occasioned his conventionalism and does not vindicate Reichenbach's claims about putative examples of equivalent descriptions. The problems arose in the framework of an earlier and more accurate account of theory testing whose implications Reichenbach misunderstood. Had Reichenbach been more careful of the thrust of his early account, the general arguments he produced for the conclusion that the entanglement of hypotheses is a symptom of conventions would have seemed jejune and unfounded. The question remains whether the sorts of reasons why we should doubt the correctness of Principles A and B are also reasons why we should doubt that some of Reichenbach's ingenious examples of conventional alternatives in physics are really that at all. Perhaps these cases, some of them anyway, are really cases in which one of a family of competing theories is best supported by the imagined evidence. At least in the case of Reichenbach's arguments for the conventionality of geometry in the context of classical physics, I believe this to be so. Insistence on explicit, covariant field theoretic formulations of theories with and without universal forces would show, I believe, that the latter are better tested by the sorts of evidence usually supposed for such theories. The argument is much too technical to consider here, but if I am right then one of Reichenbach's most interesting and most discussed examples does not illustrate the role of conventions in our knowledge; it illustrates, instead, the power of Reichenbach's early conception of confirmation to discriminate among empirically equivalent theories. Still, an example is only that. The general questions which Reichenbach raised, contributed to answering, and passed on to us remain as profoundly disturbing as ever: what is empirical equivalence; how do our methods of confirmation descriminate even between empirically equivalent theories; are our theories radically underdetermined by what we recognize as evidence for them and by our inductive strategies; what rational basis, if any, have we for choosing one from among many equally tested and confirmed theories?

University of Oklahoma

NOTES

* Research supported by the National Science Foundation.
[1] For a development of this argument see Glymour (1971).
[2] Here I am relying not only on Reichenbach (1938) but also on Reichenbach (1949).
[3] At least so it would seem on any Bayesian account of 'confirms' that I know of.
[4] For a discussion of this case see T. Kuhn (1957).
[5] Details are given in Glymour (forthcoming).
[6] For historical details see C. Wilson (1969).

REFERENCES

Bridgman, P., 1961, *The Logic of Modern Physics*, Macmillan, New York.

Carnap, R., 1936–37, 'Testability and Meaning', *Philosophy of Science* **3**, 420–471; **4**, 1–40.

Glymour, C., 1971, 'Theoretical Realism and Theoretical Equivalence', in R. Buck *et al.* (eds.), *PSA 1970*, D. Reidel Publishing Company, Boston and Dordrecht.

Glymour, C., Forthcoming, 'Physics and Evidence', in the *University of Pittsburgh Series in the Philosophy of Science*, University of Pittsburgh Press, Pittsburgh.

Kuhn, T., 1957, *The Copernican Revolution*, Harvard University Press, Cambridge.

Poincaré, H., 1952, *Science and Hypothesis*, Dover Publications, New York.

Reichenbach, H.: 1920, *Relativitätstheorie und Erkenntnis Apriori*, Springer, Berlin. English translation, *The Theory of Relativity and A Priori Knowledge*, University of California Press, Berkely and Los Angeles, 1965.

Reichenbach, H., 1938, *Experience and Prediction*, University of Chicago Press, Chicago.

Reichenbach, H., 1949, *The Theory of Probability* (2nd ed.), University of California Press, Berkeley and Los Angeles.

Reichenbach, H., 1958, *The Philosophy of Space and Time*, Dover Publications, New York. Translation of *Philosophie der Raum-Zeit-Lehre*, Walter de Gruyter, Berlin und Leipzig, 1928.

Reichenbach, H., 1959, 'The Principle of Causality and the Possibility of Its Empirical Confirmation', in H. Reichenbach, *Modern Philosophy of Science*, Routledge & Kegan Paul, London, 1959. Translation of 'Die Kausalbehauptung und die Moglichkeit ihrer empirischen Nachprufung', *Erkenntnis* **3** (1932), 32–64.

Reichenbach, H., 1969, *Axiomatization of the Theory of Relativity*, University of California Press, Berkeley & Los Angeles. Translation of *Axiomatik der relativistischen Raum-Zeit-Lehre*, Friedrich Vieweg & Sohn, Braunschweig, 1924.

Salmon, W., 1971, 'Statistical Explanation', in W. Salmon *et al.*, *Statistical Explanation and Statistical Relevance*, University of Pittsburgh Press, Pittsburgh.

Weyl, H., 1963, *Philosophy of Mathematics and Natural Science*, Atheneum, New York.

Wilson, C., 1969, 'From Kepler's Laws, So-called, to Universal Gravitation: Empirical Factors', *Archive for the History of Exact Sciences* **6**.

KEITH LEHRER

REICHENBACH ON CONVENTION

Reichenbach articulated and defended a distinction between conventional and factual components within a scientific theory. For example, he argued that the axioms of geometry are factual while the coordinative definitions for geometry are conventional.[1] The scientific spirit with which Reichenbach approached philosophical discussion is illustrated by the manner in which he clarified his doctrine when confronted with criticism. His reply to Einstein represents, I shall argue, a fundamental discovery that refutes both the conventionalism of Poincaré and Quine's claim that no categorical distinction can be drawn between those statements that are true as a matter of fact and those that are true by convention. I shall conclude with a comparison between Reichenbach's doctrine and a similar one proposed by Carnap.

I. REPLY TO EINSTEIN

According to Reichenbach, the statements of geometry are empirical provided that coordinative definitions are laid down as a convention. These definitions are statements within a scientific system articulating relations of congruence. In a *normal* system, they imply that there are no universal forces. The choice of a normal system with that implication is a matter of convention. A different set of conventions, one postulating universal forces, might, when combined with a different geometry, yield the same empirical consequences. We obtain factually equivalent descriptions by combining different coordinative definitions with different geometries, but the choice of coordinative definitions remains conventional.

Einstein (1949, pp. 677–679) replied to Reichenbach on behalf of Poincaré who maintained that geometry itself was conventional. His reply was in the form of a dialogue and is subject to various interpretations. Perhaps the most subtle has been offered by Grünbaum (1960, pp. 79–84). A principal objection to Reichenbach's doctrine contained in

W. C. Salmon (ed.), Hans Reichenbach: Logical Empiricist, 239–250.

Einstein's remarks, and in the writings of Quine, is by now familiar. As Reichenbach (1951, pp. 133–134) notes, a geometrical system alone does not give us a factual description of the structure of the physical world. Only a *complete* description including both coordinative definitions and a geometry gives us a factual description of the physical world. If various combinations of different coordinative definitions and different geometries give us factually equivalent descriptions, then why could we not regard the choice of a geometry as being conventional? If we first adopt coordinative definitions by convention, then the geometry we combine with it will appear to be the factual component in the system. But if we first adopt a geometry by convention, the coordinative definitions we combine with it will, name aside, appear to be the factual component in the system. Thus, it would seem that Poincaré was as well justified in averring that geometry is conventional as Reichenbach was in affirming that coordinative definitions are conventional. Moreover, the proper conclusion appears to be that of Quine, that no sharp distinction can be drawn between those statements that are conventional and those that are factual.[2]

Reichenbach's reply is pellucid, and has been curiously neglected in recent discussion. I quote from Reichenbach (1951, pp. 136–137):

> Assume that empirical observations are compatible with the following two descriptions:
>
> CLASS I
>
> (a) The geometry is Euclidean, but there are universal forces distorting light rays and measuring rods.
>
> (b) The geometry is non-Euclidean, and there are no universal forces.
>
> Poincaré is right when he argues that each of these descriptions can be assumed as true, and that it would be erroneous to discriminate between them. They are merely different languages describing the same state of affairs.
>
> Now assume that in a different world, or in a different part of our world, empirical observations were made which are compatible with the following two descriptions:
>
> CLASS II
>
> (a) The geometry is Euclidean, and there are no universal forces.
>
> (b) The geometry is non-Euclidean, but there are universal forces distorting light rays and measuring rods.
>
> Once more Poincaré is right when he argues that these two descriptions are both true; they are equivalent descriptions.
>
> But Poincaré would be mistaken if he were to argue that the two worlds I and II were the same. They are objectively different. Although for each world there is a class of equivalent descriptions, the different *classes* are not of equal truth value. Only one class can be true for a given kind of world; which class it is, only empirical observation can tell. Conventionalism

sees only the equivalence of the descriptions within one class, but stops short of recognizing the differences between the classes. The theory of equivalent descriptions, however, enables us to describe the world objectively by assigning empirical truth to only one class of descriptions, although within each class all descriptions are of equal truth value.

Instead of using classes of descriptions, it is convenient to single out, in each class, one description as the *normal system* and use it as a representative of the whole class. In this sense, we can select the description for which universal forces vanish as the normal system, calling it *natural geometry*.

II. SEMANTIC INTERPRETATION

The reply, though clear, leaves us with a question. Which statements are true by convention, and which are factual? The objection said that no distinction could be drawn between conventional and factual statements. Reichenbach replies that whether we choose (a) or (b) within a class as a description is a matter of convention but which class is true for a world is a matter of empirical observation. The problem is that the reply does not explicitly answer the question as to which *statements* are true by convention and which are true as a matter of fact. It is not at all difficult, however, to extract an answer.

To formulate the answer we require a semantic conception of truth instead of the empirical conception of truth Reichenbach employs when he says that systems within a given class are *equally true*. As Reichenbach is using the expression 'true', a system of statements is true, or, more precisely, *empirically* true if and only if all the empirically testable consequences of the system are true, and the latter are true if and only if relevant observational test conditions are satisfied. Thus, if system (a) in Class I is empirically true, then so is system (b). This conception of empirical truth must be supplemented with a semantic conception of truth in order to obtain a semantic analysis of logical relations, for example, logical incompatibility. Though (a) and (b) in Class I are equally true *empirically*, they are logically incompatible. For (a) in Class I states that there are universal forces, and (b) states that there are no universal forces. Thus, if system (b) is true, then, in one sense of 'true', system (a) is not true. Since both systems may be empirically true, we require another sense of 'true' to express the incompatibility semantically. The semantic conception of truth introduced by Tarski (1944) will suffice. A system or sentence is true under the semantic conception if and only if the satisfaction conditions for it are fulfilled. Satisfaction conditions for a sentence are not the same as empirical test conditions. The latter conditions are restricted to the domain of the observable and the former are not. To distinguish those statements that are true by convention from those that are true as a matter fact, it will be useful to depart from Reichenbach's use of 'true' to refer to empirical truth and to use the word to refer the semantic conception. The question of selecting a normal system within each class to represent the class arises because systems within a class are logically incompatible. It is logically perspicuous, therefore, to use the term 'true' semantically rather than empirically and be able to say that if one system within a class is true, perhaps because of a convention to select a normal system, then the other logically incompatible systems within the same class are not true. I shall, henceforth, use the word 'true' according to the semantic conception and rearticulate Reichenbach's argument accordingly.

Let C_{I} be a set containing as one member the statement (a) in Class I, and as another member the statement (b) in that class, and let C_{II} be analogously constructed from the members in Class II. Now consider the following two statements:

A. Some member of C_{I} is true.
B. Some member of C_{II} is true.

From *A* together with the semantic theory of truth, we can derive the testable consequences of the systems in Class I, and from *B* we can derive the testable consequences of the systems in Class II. Thus, statements *A* and *B* are factual. Whether these statements are true depends on whether the empirically testable consequences of the systems in each class are true.

Moreover, since it is a matter of convention which system within each class we adopt as our description of the world, a statement affirming that one particular system within a class is true if any system in the class is, will be true by convention. Let C_{I}(a) be the statement referred to by (a) in Class I, and so forth for C_{I}(b), C_{II}(a), and C_{II}(b). Then consider the following four statements:

A_1. If *A*, then C_{I}(a) is true.
A_2. If *A*, then C_{I}(b) is true.
B_1. If *B*, then C_{II}(a) is true.
B_2. If *B*, then C_{II}(b) is true.

If we single out one description in each class as representative of the class, this is equivalent to adopting the convention to select one statement from A_1, A_2 as true by convention and one statement from B_1, B_2 as true by convention. Following Reichenbach's proposal, therefore, A_2 and B_1 are true by convention because system (b) in Class I and system (a) in Class II are the normal systems. Statements A_1, A_2, B_1, and B_2 are ones, which, if they are true, are true by convention.[3] The consequents of the first pair are factually equivalent as are the consequents of the latter pair. Hence, the selection of one statement from the pair A_1, A_2, and one from the pair B_1, B_2, is a matter of convention.

Thus, contrary to Poincaré and Quine, Reichenbach's argument provides us with a means for distinguishing between statements which, if true, are factually so, and those which, if true, are conventionally so. On the other hand, the distinction between factual and conventional statements occurs in a semantic metalanguage for the scientific systems within each class. Moreover, in a refined articulation of the argument, this feature would be essential. The number of factually equivalent systems within each class is infinite, for, as Grünbaum (1963, pp. 99-105) has shown, there is an infinity of different coordinative definitions that may be combined with geometries to yield the same empirical consequences. So an attempt to recast statements A_1, A_2, and so forth as conditionals in the object language with a disjunction of all the systems in each class as the antecedent and a single system in the consequent would fail (within standard logic) because the set of systems in each class is infinite.

What this shows is that Reichenbach had discovered a new way of drawing the distinction between fact and convention that rendered previous formulations otiose. The position of his adversaries rested on the assumption that the distinction between factual and conventional statements must be articulated in terms of the statements *within* the scientific system, in this case geometries combined with coordinative definitions. It must be conceded that either set of statements within a system, either the coordinative definitions or the geometry, could be chosen as true by convention. We are not bound to select from A_1, A_2 and B_1, B_2, in such a way as to make the coordinative definitions true by convention. If, for instance, we select A_1 and B_1 as true by convention, that would be equivalent to selecting Euclidean geometry as true by convention, which is equivalent to construing the axioms as implicit definitions, as opposed to Reichenbach's selection of A_2 and B_1 as true by convention, which is equivalent to excluding universal forces by convention.

However, and this is Reichenbach's insight, no matter what you single out as true by convention, the distinction between statements that are true by convention and those that are true as a matter of fact remains. Statement A is, if true, true as a matter fact, as is statement B. By contrast, whatever convention we adopt, statements A_1 or A_2 are, if A is true, true by convention, as are statements B_1 or B_2 if B is true.

We are now in a position to adjudicate the dispute between Reichenbach and his critics. His critics contend that there is no *feature* of statements *within* a scientific system that enables us to distinguish some of those statements as true by convention and others as true by empirical fact. Reichenbach's argument under our analysis shows that even if there is no feature of statements within a scientific system that enables us to explicate this distinction, the distinction may be formulated in a metalanguage of the scientific system. Moreover, once the distinction is thus articulated, we may readily explain why no feature of the statements within a scientific system enables us to explicate the distinction between fact and convention. How statements within a scientific system are to be interpreted is not determined by any feature within the system, but rather by a decision that is made *about* the system. Which statements within a scientific system are true by convention is a consequence of a decision to adopt a convention concerning that system. Such conventions, like conventions assigning meaning to words in a language, are formulated in the metalanguage of that system.[4] Since it is a result of a convention articulated in the metalanguage that certain statements within a scientific system are true by convention, there is no way by which one can distinguish between those statements within a scientific system that are true by convention and those that are true as a matter of fact by considering only the statements within the system itself. The convention lies outside the system and is formulated in a metalanguage of that system.

By way of illustrating the foregoing, let us suppose we adopt Reichenbach's proposed convention to select the normal systems in each class. As we noted, that is equivalent to the decision to accept statements A_2, B_1. Given that we adopt this convention, the question of whether space is

Euclidean becomes a factual question of whether A and B is true. If A is true, then, as a matter fact, space is non-Euclidean, and if B is true, then, as a matter fact, space is Euclidean. Thus, given a convention in the metalanguage to the effect that A_2 and B_1 are true by convention, the statement that there are no universal forces is true by convention. If we consider only the statements within these systems, we have no way of ascertaining that this is so. But that is a simple consequence of looking for something in the wrong language.

Finally, it should be noted that, though it might be pragmatically perverse, a finer articulation of the membership of factually equivalent classes could provide us with means for adopting a convention that would select metalinguisitic statements as true by convention that did not correspond to any statements within the scientific system itself. Suppose that Class I consists of three descriptions instead of two, where description (a) in Class I is subdivided into (ai) and (aii), as follows:

(ai) The geometry is Euclidean, but there are universal forces F_i distorting light rays and measuring rods

and

(aii) The geometry is Euclidean, but there are universal forces F_{ii} distorting light rays and measuring rods

where F_i and F_{ii}, though described differently, yield the same empirical consequences when combined with Euclidean geometry. Similarly, suppose that (b) in Class II is subdivided into (bi) and (bii) by referring to forces G_i and G_{ii} respectively which, though described differently, yield the same empirical consequences when combined with Euclidean geometry. (Whether F_i is the same as G_i and F_{ii} the same as G_{ii} will not matter for our argument.) Now suppose we adopt the convention that the following two conditionals are true:

If A, the C_I(ai) is true

and

If B, then C_{II}(bii) is true.

Simple inspection reveals that there is no statement within the scientific systems that corresponds to this convention because (ai) tells us that geometry is Euclidean and the distorting force is F_i while (bii) tells us that geometry is non-Euclidean and the distorting force is G_{ii}. So the choice of a convention selecting statements in the metalanguage as true by convention that correspond to a specific statement in the language of the scientific system is not logically inevitable. The choice of a convention that selects as true by convention statements in the metalanguage that correspond to statements within the scientific system is a pragmatic decision. Neither the empirical facts nor any feature of the statements within the scientific system compel such a choice.

In summary, Reichenbach's argument demonstrates that there is a distinction between the conventional and factual status of statements within a scientific system. Which statements are conventional and which are factual depends on a convention made about that system in a semantic metalanguage. Such a convention should be grounded in sound pragmatic considerations, and ordinarily such a convention in a metalanguage of the scientific system will correspond to a convention to select some statements within the system as true by convention. In any event, a convention selects some statements in the metalanguage as true by convention, and others in the metalanguage, if true, will be so as a matter of empirical fact.

III. COMPARISON WITH CARNAP

In conclusion, it will be useful to compare Reichenbach's argument as interpreted above with a similar proposal made by Carnap. Carnap was concerned with the problem of distinguishing the factual component of a scientific system from the semantic component which provides a meaning postulate. To keep the cases parrallel, let the system be a geometry together with coordinative definitions. Thus let S_I(a) be the axiom system and the coordinative definitions corresponding to (a) in Class I. Now let us represent the Ramsey sentence for this sytem, in which we existentially quantify over theoretical terms in the system, as RS_I(a).

Carnap (1963, pp. 964–966) proposed that the factual component of a system be represented by the Ramsey sentence. The reason, noted by Ramsey (1931, pp. 212–215, 231), is that the Ramsey sentence for the system has the same empirical consequences as the system itself. Consider, then, the four Ramsey sentences corresponding to the original four systems in Class I and Class II.

RA_1. $RS_{I}(a)$
RA_2. $RS_{II}(b)$
RB_1. $RS_{II}(a)$
RB_2. $RS_{II}(b)$.

To express the semantic component of $S_I(a)$, the meaning postulate, we form a conditional with the Ramsey sentence as an antecedent and the system as a consequent. Thus we form the meaning postulates for the four systems as follows:

MA_1. $RS_I(a) \supset S_I(a)$
MA_2. $RS_I(b) \supset S_I(b)$
MA_2. $RS_{II}(b) \supset S_I(b)$
MB_1. $RS_{II}(a) \supset S_{II}(a)$
MB_2. $RS_{II}(b) \supset S_{II}(b)$.

The latter four statements, though not truths of formal logic, are treated as meaning postulates.

There are similarities and differences between Carnap's proposal and our analysis of Reichenbach's argument. The most striking similarity is the distinction between a factual component and a conventional component. What meaning postulates we adopt is a convention based on pragmatic considerations according to Carnap. So on both proposals we have conditional statements true by convention which have a factual component as an antecedent implying a scientific system. One difference is that the conventional statement in Carnap's analysis is not formulated in the metalanguage, but, allowing for quantification over predicates, in the language of the scientific system itself.

There is, however, a second difference of greater importance. The Ramsey statements RA_1 and RA_2 are empirically equivalent, that is, they have the same empirical consequences. The two Ramsey sentences have the same empirical consequences as the original two systems of Class I. Since the empirical consequences of the two Ramsey sentences are the same, the choice between these two Ramsey sentences must itself be a matter of convention. Moreover, by turning to the metalanguage, we can formulate the choice as one between the following two statements:

AR_1. If A, then RA_1
AR_2. If A, then RA_2.

Thus, the choice between factually equivalent Ramsey sentences is itself a matter of convention.

It should be noted that Carnap was primarily concerned with finding some means for distinguishing the analytic meaning postulates from the synthetic postulates which represent the factual content of the theory. Moreover, the various meaning postulates MA_1, MA_2, MB_1, and MB_2 may all be accepted as meaning postulates unlike the statements A_1, A_2 and B_1 and B_2 which are such that only one member from each pair can be adopted by convention. The reason is that the antecedents of the meaning postulates are incompatible in pairs, while the antecedents of A_1 and A_2 are the same and the consequents are incompatible in pairs. The distinction that Carnap formulated, though it does enable us to formulate a meaning postulate for a system, falls short of fully articulating the distinction between the conventional and factual component of the theory.

One final comparison between the two philosophers may be useful. Carnap (1963, pp. 871–873) distinguished between what he called internal and external questions. An internal question was formulated and answered within a linguistic framework. By contrast, an external question concerns whether it was reasonable to adopt a linguistic framework. External questions, though important, are answered on the basis of pragmatic considerations. Now such considerations, as we have noted, might lead us to adopt conventions that would, in combination with empirical observation, lead us to adopt a linguistic framework corresponding to system S_I(a). Having made such a decision, the existential statements within that system obtained by existential quantification over theoretical terms, for example, a Ramsey sentence, become internal sentences. If considerations of fact lead us to the conclusion that some system in Class I is true, and we then adopt the convention that system S_I(a) is true, we thereby adopt a linguistic framework in which certain existential statements, among them Ramsey sentences, are also true. In the semantics for such a system, we must allow for a domain of entities that are values for the variables existentially generalized. For Carnap (1963, p. 871), this means that if we adopt such a framework, the postulation of the domain of entities, unobserved theoretical entities for example, becomes a meaning postulate and trivial.

Reichenbach (1951), p. 185) was, I believe, rather less sanguine about such postulation. Causal anomalies in the quantum domain deprive us of a normal system and lead him to remark that it is "impossible to speak of unobserved objects of the microcosm in the same sense as of the macrocosm." As a result, Reichenbach insisted that factually equivalent descriptions are equally true. Reichenbach (1951, p. 186) thus remained sceptical about the postulation of a domain of unobserved entities that behave unreasonably. If convention leads us to single out one description of the world rather than another for pragmatic considerations, then even if, as Carnap insisted, theoretical reflection enters into pragmatic decision, that is not sufficient reason to commit oneself to a domain of irrational entities. Existentially quantified statements, such as Ramsey sentences, may be a logical consequence of a convention we adopt but such a convention selects but one from among a multitude of equivalent descriptions of reality. We would, therefore, be unreasonable to suppose that the syntactical form of the conventionally chosen system was a reliable guide to what exists in the physical world.

In conclusion, let us note that Reichenbach's argument does not depend on our being able to draw a sharp distinction between the terms of theory and those of observation, even if he did, in fact, hold to such a distinction. The distinction between empirically testable consequences of a theory and those components of theory which we cannot test need not be drawn in terms of diverse vocabularies. However one chooses to draw the former distinction, there will be scientific systems that are equivalent in terms of the testable consequences we can draw from them. From this equivalence, Reichenbach revealed a distinction between what is true by convention and what is true as a matter of fact. The distinction holds for theories in general, for those of translation and belief as well as for geometry and physical forces. That there is no feature of statements within such systems that mark some as true by convention and others as factually true is unimportant. The distinction is a consequence of conventions we adopt concerning such systems rather than any internal feature of the systems themselves. Once the convention is adopted, there is a precise division, in the metalanguage, and, ordinarily, within the scientific system itself, between those statements that are true by convention, and those others, which, if true at all, are so as a matter of fact.[5]

University of Arizona

NOTES

[1] The first comprehensive presentation by Reichenbach is [8], 14–27, and the most recent discussion is in [7], 131–140.

[2] Of a system of science Quine remarks, "It is a pale grey lore, black with fact and white with convention. But I have found no substantial reasons for concluding that there are any quite black threads in it, or any white ones." In [5], 406.

[3] If the conditionals A_1 and A_2 are interpreted as material conditionals then, of course, A_1 and A_2 might, as a matter of fact, both turn out to be trivially true if A were to be false. However, such factual considerations would leave some questions concerning the truth of some such conditionals undecided, for whichever class of systems has empirically true consequences will yield a set of conditionals none of which are trivially true. Moreover, the conditionals are best interpreted as strong conditionals warranting subjunctive inferences such as, if A were true, then $C_I(b)$ would be true. The truth of such conditionals would not follow from the falsity of their antecedents.

[4] Cf. Reichenbach [7], 132.

[5] I am greatly indebted to W. C. Salmon for clarification of Reichenbach's doctrine. I am indebted to W. F. Sellars and K. Fine for suggestions concerning the relation between Carnap and Reichenbach.

BIBLIOGRAPHY

[1] Carnap, R., 1963, Replies and Systematic Explositions', in [9].

[2] Einstein, A., 1949, 'Reply to Criticisms', in [10].

[3] Grünbaum, A., 1960, 'The Duhemian Argument', *Philosophy of Science* **27**, No. 1.

[4] Grünbaum, A., 1963, *Philosophical Problems of Space and Time*, A. Knopf, New York.

[5] Quine, W. V., 1963, 'Carnap and Logical Truth', in [9].

[6] Ramsey, F. P., 1931, *The Foundations of Mathematics and other Logical Essays*, Routledge and Kegan Paul, London.

[7] Reichenbach, H., 1951, *The Rise of Scientific Philosophy*, University of California, Berkeley and Los Angeles.

[8] Reichenbach, H., 1958, *The Philosophy of Space and Time*, Dover, New York, translated from *Philosophie der Raum-Zeit-Lehre*, 1926, M. Reichenbach and J. Freund (transl.).

[9] Schilpp, P. A., 1963, *The Philosophy of Rudolf Carnap*, Open Court, La Salle.

[10] Schilpp, P. A., 1949, *Albert Einstein: Philosopher-Scientist*, Tudor, Evanston.

[11] Tarski, A., 1944, 'The Semantic Conception of Truth and the Foundations of Semantics', *Philosophy and Phenomenological Research* **4**, 341–375.

ANDREAS KAMLAH

HANS REICHENBACH'S RELATIVITY OF GEOMETRY

ABSTRACT. Hans Reichenbach's 1928 thesis of the relativity of geometry has been misunderstood as the statement that the geometrical structure of space can be described in different languages. In this interpretation the thesis becomes an instance of 'trivial semantical conventionalism', as Grünbaum calls it. To understand Reichenbach correctly, we have to interpret it in the light of the linguistic turn, the transition from 'thought oriented philosophy' to 'language oriented philosophy', which mainly took place in the first decades of our century. Reichenbach – as Poincaré before him – is undermining the prejudice of thought oriented philosophy, that two propositions have different factual content, if they are associated with different ideas in our mind. Thus Reichenbach prepared the change to language oriented philosophy, which he also accepted later.

1. THE DIFFICULTY OF UNDERSTANDING REICHENBACH'S 'RELATIVITY OF GEOMETRY'

It was about half a century ago that Hans Reichenbach published his highly celebrated *Philosophie der Raum-Zeit-Lehre*[1] (*Philosophy of Space and Time*, abbreviated as PST), according to Carnap "still the best book in the field" (Carnap, 1958, p. vi). In this book Reichenbach's 'theory of the relativity of geometry' plays an important role. He repeatedly states essentially the same thesis under different names and applied to different subjects. He calls it 'Mach's principle in the wider sense' in application to accelerated inertial systems (PST, p. 216), and 'the philosophical theory of relativity' if applied to space–time (PST, p. 177). This shows that his relativity thesis seemed to him to be a most fundamental insight and achievement of modern philosophy of science.

The geometrical version of the thesis says that a geometry can only be true or false relative to the coordinative definitions, which determine how geometrical quantities like length are to be measured and how geometrical relations like congruence are to be ascertained (PST, §8). We may say for example, that physical space is Euclidean according to one set of coordinative definitions and hyperbolic with regard to another. A geometry is only empirically meaningful if coordinative definitions are

W. C. Salmon (ed.), Hans Reichenbach: Logical Empiricist, 251–265. *All Rights Reserved.*

added to it. If these are lacking, it is only a set of implicit definitions without any empirical content.

We may already suspect that a historical interpretation of Reichenbach's book may be helpful after such a long time, when we look at the later discussion of Reichenbach's thesis of the conventionality of geometry. To some of Reichenbach's readers his conventionality thesis seemed to be trivial at first sight. They wondered what could be the nontrivial meaning of this thesis, since Reichenbach of course could not have laid so much emphasis on a claim to which everybody would readily agree without any hesitation. This difficulty arises in the following way: It is clear that the uncommitted sound 'The distance between the points x and y is r', mathematically $r = d(x, y)$, has to be associated with a meaning in some way. These symbols have no meaning by themselves. It is also clear that this can be done in different ways, and therefore that the meaning of the function $d(x, y)$ is conventional, as is the meaning of any other term or expression of any language. So, for example, we cannot know beforehand the meaning of the word 'gift'. In German it means 'poison', which is a gift of quite a special sort not to be presented to one's own friends. This word has different meanings in the different semantical systems of different languages: the meaning is not automatically given by the sound. Thus, the conventionality of the signs of language is quite a trivial truism. We all are of course adherents of 'trivial semantical conventionalism' as Grünbaum calls it (Grünbaum, 1963, p. 24ff; 1968, p. 96ff.).

Reichenbach gives reason for this puzzling interpretation, when he writes in his later book *The Rise of Scientific Philosophy* (1951) on page 133:

> This consideration shows that there is not just one geometrical description of the physical world, but that there exists a class of *equivalent descriptions*; each of these descriptions is true, and apparent differences between them concern, not their content, but only the languages in which they are formulated.

Is he really trying to convince us, that the same state of affairs is subject to different descriptions in different languages, since the relations between symbols and their meanings differ from language to language? Can we believe, that he wants to bring coals to Newcastle in that way?

Adolf Grünbaum has argued that Reichenbach is surely not a proponent of trivial semantical conventionalism, and has defended his well known 'geochronometrical conventionalism' which he believes to be partly Reichenbach's position. The authors of some recent articles even refer to the 'Reichenbach–Grünbaum thesis' (for example, Whitbeck, 1969, p. 329). Grünbaum's thesis is, that space and space–time have no intrinsic metric, and that the metrical properties are relational in the same way as the unclehood of somebody, which depends on something external to him (in this case on the fact, that there exists a person, who is his nephew) – (Grünbaum, 1976, sec. 4, i). This thesis states something about the ontological status of the metric. There are indeed passages in Reichenbach's writings, which can be read in the sense of Grünbaum (PST, p. 57). But if one reads these paragraphs carefully in the context in which they appear, one is aware of the fact that Reichenbach had no intention of making an ontological statement about space, and that he is talking about other topics at this point. Reichenbach possibly would have agreed with Grünbaum, that space is metrically amorphous. But he was concerned to express what he held to be the philosophical essence of Einstein's theory of relativity, in which he saw a revolutionary achievement of mankind. Grünbaum's thesis however (which is certainly of interest for philosophy of science) can hardly be understood as a claim having the same fundamental importance for our scientific world view which Reichenbach attributed to his own 'philosophical theory of relativity'. Therefore we cannot explain, by the geochronometrical interpretation alone, the emphasis with which Reichenbach expressed the claim that geometry is relative, even if it may still be a correct interpretation of one aspect of Reichenbach's thesis.

Hilary Putnam sees the important content of Reichenbach's assertion in another aspect of his philosophical theory:

> He asserts *both* that *before* we can discuss the truth or falsity of any physical law all the relevant theoretical terms must have been *defined* by means of 'coordinating definitions' *and* that the definitions must be *unique*, i.e., must uniquely determine the extensions of the theoretical terms. (Putnam, 1963, p. 243).

It is not difficult to find the quotations in Reichenbach's works which support this interpretation. But we easily agree with Hilary Putnam, that this statement 'is a quite special epistemological thesis'. Thus also Putnam can not tell us what seemed so important for Reichenbach.

2. THOUGHT ORIENTED AND LANGUAGE ORIENTED PHILOSOPHY

If we are now still puzzled by the great emphasis laid by Reichenbach on the possibility of alternative metrizability of space, we should look for reasons for our perplexity. Possibly we have difficulties to understand a work which arose from a different historical situation. And therefore we should investigate Reichenbach's philosophical background before we try to understand his relativity thesis.

The German philosophy of that time was largely governed by Neokantianism and by Husserl's school of transcendental phenomenology. In Vienna Carnap, Schlick and their friends were reading Wittgenstein's *Tractatus* (see Carnap, 1963, p. 24), and some of Bertrand Russell's books were already known in their German translation (for example, Russell, 1926). One year after the publication of Reichenbach's *Philosophie der Raum-Zeit-Lehre* the Vienna Circle presented itself for the first time as a philosophical school under its name by publishing a manifesto with the title *Der Wiener Kreis* (Schleichert, 1975). This was done at the first of a series of conferences, which made the logical empiricists conscious of the fact that they formed a worldwide community. The first issue of Erkenntnis appeared two years after Reichenbach's book. Thus logical empiricism was not yet constituted as a movement at that time. The new ideas of the German and Austrian logical empiricists were discussed by very few persons in Vienna, Berlin (Reichenbach, Dubislav, Grelling) and Münster (Scholz; see Jørgensen, 1951). We should therefore not expect to find logical empiricism in its later form already in Reichenbach's book from 1928. We should rather look for the affinities of Reichenbach's early philosophy to the schools then prevailing.

This leads us to the study of the main difference between the schools of traditional philosophy and the modern approaches to philosophical reasoning, which arose from the Vienna Circle, the Berlin group, and related schools. The old philosophy was concerned with the *thinking* of man. Truth was frequently characterized as 'adaequatio rei et *intellectus*', one was interested in *conceiving* the world correctly. The proposition was the connection of two *ideas*. Thus some philosophers tried to analyse the human mind in order to work out a theory of knowledge. Titles like *An*

Inquiry Concerning Human Understanding and *Kritik der reinen Vernunft* (*of pure reason*) are typical for this endeavour. Concepts and propositions were held to be mental entities. I call this old philosophy 'thought oriented philosophy'. Similar expressions, which are already in use, like 'conceptualistic', 'psychologistic' and others are different in their meaning. Therefore we had better coin a new term. We contrast the old approach with 'language oriented philosophy', which is largely an achievement of our century. One could also call it 'analytic philosophy' in the sense of Arthur Pap (Pap, 1949), and subdivide it into the philosophy of ideal language (represented by Carnap, Quine and others) and ordinary language philosophy. But even here we get into trouble with common usage, since for many people analytic philosophy *is* the analysis of natural language. Thus a neutral term not yet bound to a meaning is preferable here too. It is of course unnecessary to describe this new species of philosophy to the reader of this journal, since in most cases it will constitute the uncontested background of his philosophical education.

But for just this reason many contemporary philosophers have difficulties with seeing, in which respect the thought oriented philosophy is fundamentally different from their own. And I think that the difficulties in understanding Reichenbach's early philosophy are partly due to the inability to imagine oneself in the position of a thought oriented philosopher. For Reichenbach is just criticizing the thought oriented way of reasoning. Therefore we should remind ourselves of what is peculiar to our own language oriented approach. Frequently the importance given to the analysis of meaning of the words used in philosophical, scientific, and ordinary discourse has been considered as the new achievement. This characterization is surely correct. But it should be emphasized that this attitude contains an important presupposition. It takes for granted that we want to have our sentences clear and not our thoughts. Since language is intersubjective, we can investigate it, while thoughts are only indirectly accessible through the medium of language.

For the old approach, words are only signs standing for ideas or concepts. To grasp the meaning of a word is to imagine the idea associated with it. Thus the old thought oriented and the new language oriented philosophy differ in their identity criteria of meaning. Two different propositions are different in their meaning for the old approach, if they involve different ideas or mental images. Thus it makes a difference, if we imagine that the world, including all objects, animals, people and measuring instruments, doubles its size overnight, or if we imagine that it remains unchanged. In the one case we see the world grow with our mental eyes, in the other case we do not. For the language oriented standpoint the meaning of a statement is independent from the mental images associated with it. It does not matter what people are thinking; the process of visualizing a state of affairs is only of psychological interest. In the case of the expanding world we can show, that with respect to the relevant criteria of empirical meaning the expanding world is identical with a world of constant size, as was already shown by Poincaré (Poincaré, 1914, chap. II, 1), who was one of the first philosophers to undermine the thought oriented view. The old approach is in some way the more natural, the new approach the less naïve. Each of us has been a proponent of thought orientation in his prephilosophical state as a schoolboy or undergraduate. We usually consider thoughts of the people to be more important than their utterances; when we ask somebody what he means with a sentence, we want to know what he thinks or imagines. Therefore it should be easy after all to remind oneself of the difficulties one would have had with Poincaré's example of the expanding world in ones own philosophical childhood.

We now come back to the historical situation of 1928 when Reichenbach's *Philosophie der Raum-Zeit-Lehre* was published. It is clear that the change from thought oriented to language oriented philosophy, sometimes called the 'linguistic turn' (Rorty, 1967), was no sudden change. It was indeed a Gestalt switch in philosophical method, but not one without any transitional state. The transition took place in different philosophical traditions. There were the pragmatists in America, Frege in Germany, and Poincaré in France; all of them broke with the old criteria for the identity and difference of meaning. For Reichenbach Poincaré's conventionalism was of crucial importance. Poincaré's writings, though they are sparkling with ideas, offer no systematic exposition of a philosophy of knowledge or science. Such a treatment however was given by Moritz Schlick, especially in his famous book *Allgemeine Erkenntnislehre* of 1918 (*General Theory of Knowledge*). We are referring here to Schlick I, who was not yet under the influence of Wittgenstein, not to the Wittgensteinian Schlick II. Here we find a theory of concepts, which is very similar to Reichenbach's in his article in the *Handbuch der Physik*: 'Ziele und Wege der physikalischen Erkenntnis' (1929). For Schlick a concept is something like the idea of a sign. He argues, as Berkeley did already before him, that general ideas are inconceivable:

> This much is certain: it is completely impossible that I can form an intuitive image of a dog that is neither a St. Bernard, nor a Newfoundland nor a dachshund nor some other particular sort of dog; that is neither brown nor white, neither large nor small: briefly, an animal that would be nothing more than precisely a dog in general.[2]

What we have in our mind, is not the general idea of the dog, but a sign or symbol for this species. Thus Schlick says:

> The nature of concepts is exhausted by the fact that they are signs which we coordinate in thought to the objects about which we think.[3]

We see, how Schlick's theory is placed on the borderline between the thought oriented and the language oriented view. Concepts are still in the mind, but they are symbols or signs as terms are in language oriented philosophy. Therefore they have no meanings by themselves like general ideas; they are rather established with a meaning by *coordination* [Zuordnung].

Reichenbach's theory of concepts is largely under the influence of Schlick, but Reichenbach's formulations are much clearer. We find an example in the following quotation:

> Thereby we arrive at a theory of concepts: concepts are configurations of perceptions .... Concepts exist like all complex mental experiences, and are therefore things; they exist as long as the person exists who thinks them, independent of the temporal position of the things coordinated to them. They receive their meaning as signs through [a] coordination ..., which establishes a mediation between the totality of all things and these special things. One might refer for comparison to the image of a map, which is placed on the soil *within the represented land*: it coordinates all points of a large spatial field to a small selection among them.[4]

We see here too, that concepts have to get their meaning by *coordination*. For this purpose Reichenbach introduces his well known coordinative definitions. These are not found in Schlick's *Allgemeiner Erkenntnislehre*, though Reichenbach says himself, that the idea of coordination is Schlick's. This causes a certain ambiguity of the term 'coordinative definition'. On the one side it seems at first glance to signify a definition in the proper sense, which connects words with words: all examples given by Reichenbach are of that sort. On the other side the function attributed to these definitions is the coordination of concepts to objects of the real world. We wonder if pointing gestures should also be considered as coordinative definitions as for instance: 'Look here, this is the standard meter'. It is clear that the first words of our language can only be taught by such 'ostensive definitions' as Schlick calls them later (Schlick II; see Schlick, 1949, p. 148). Possibly Reichenbach would have included these pointing gestures in the class of coordinative definitions, though they are very different from what one would usually call a definition. So for example he says that the size of the unit length "can only be established by reference [Hinweis] to a physically given length such as the standard meter in Paris." (PST, p. 15). The german word *Hinweis* could also have been translated by 'ostension' or 'pointing gesture'.

The other important feature of Reichenbach's coordinative definitions is, that they define concepts, not words (PST, p. 14). This might be merely a terminological peculiarity of Reichenbach. But if we read the above quotation about the notion of 'concept' carefully, we see that he really wants to define what he calls a concept, a 'configuration of perceptions' which has the function of a symbol. This will become important, when we see, that for Reichenbach the images of geometrical visualization also have to be provided with a meaning by coordinative definitions. If concepts are mental entities, then geometrical images are not much different from them, and it seems to be very natural to apply coordinative definitions to the elements of visualization as well.

Reichenbach's theory of concepts is already very near to a language oriented theory. What is yet lacking here is the insight that it is completely sufficient to talk about signs in all semantic considerations. Carnap sees this already very clearly at the same time:

Since meaningful talk of concepts always concerns concepts that are designated, or at least are in principle designable, through signs; basically then, one always speaks of these signs and of the laws of their use.[5]

Some years later, 'concept' is for Carnap simply a synonym for 'predicate'. Here the Gestalt switch to language orientation has been accomplished. (See Carnap 1928, § 28.)

3. AN INVESTIGATION OF REICHENBACH'S PRINCIPLE

We are now prepared for our interpretation. Therefore let us go back to Reichenbach's exhibition of the 'relativity of geometry'. Reichenbach formulates his principle in the following way:

> Theorem θ. Given a geometry G' to which the measuring instruments conform, we can imagine a universal force F which affects the instruments in such a way that the actual geometry is an arbitrary geometry G, while the observed deviation from G is due to a universal deformation of the measuring instruments. (PST, p. 33).

He is thus saying that something can be *imagined* in a different way. Everybody who has read the preceding section of this article, will already see how this quotation differs from the following from 1951:

> Statements about the geometry of the physical world, therefore, have a meaning only after a coordinative definition of congruence is set up. If we change the coordinative definition of congruence, a different geometry will result. This fact is called the *relativity of geometry*. (Reichenbach, 1951, p. 132; 1965, p. 153).

While the latter formulation can indeed be ascribed easily to trivial semantical conventionalism, the former cannot, since it is not a statement about the meaning of sentences and words; it is rather stating something about our possible imaginations, and how these can be provided with a meaning. That concepts, ideas and images have no meaning by themselves is – as we have already seen – the new message here.

But before we come to any definite conclusion, we wish to investigate Reichenbach's early theory in further detail, in order to supply more evidence for our interpretation. The context, in which Reichenbach's thesis appears, is the refutation of Kant's theory of pure intuition. The whole first chapter of his 1928 book is directed against Kant and especially against the Neokantians. Therefore it might be considered as a shortcoming of this article that little is said in it about the distinction '*a priori – a posteriori*'. The distinction 'thought orientation – language orientation' however seems to me to be more fundamental, since the thought oriented view is the philosophical background of Kant's reasoning. The refutation of Kant is therefore best done by destroying the basis on which his philosophy rests. Therefore it might be justified after all, if in this paper the refutation of apriorism is not discussed extensively. The author is more explicit about this issue in another publication (Kamlah, 1976).

Kant's theory of pure intuition is refuted by Reichenbach in two steps. In the first step Reichenbach shows that pure intuition, if it exists, does not give us the scientifically relevant information about space, since the imagination of *any* geometry is compatible with *any* empirical evidence obtained by measuring rods. In the second step Reichenbach analyzes the capacity for visualizing geometrical facts by itself. We are here only concerned with the first step, in which Reichenbach's 'theory of relativity of geometry' appears.

Reichenbach asks here:

> Let us first assume, that it is correct to say, that a special ability of visualization exists, and that the Euclidean geometry is distinguished from all other geometries by the fact, that it can easily be visualized [my literal translation: that it is the one, which is distinguished by its visualizability]. The question arises: what consequences does this assumption have for physical space? (PST, p. 32).

Reichenbach says to the Kantian:

> I will not discuss here your opinion, that Euclidean geometry is intuitively evident. Suppose, you are right. But what does this mean for the empirical world?

After this comes Reichenbach's theorem θ. It claims that pure intuition is absolutely useless for empirical science, since it is compatible with any empirical evidence. It excludes nothing. Thus the Kantian is not yet defeated, but he has lost one of his most important weapons, the power of making synthetic apriori statements about physical space.

After all that has been said until now, it will be clear, that only the original version of Reichenbach's principle of relativity – which is not yet formulated in the language oriented spirit – is suitable for his purpose. The theory, that Euclidean geometry is a true intuition about the world, can only be refuted by a principle which is also about mental images, ideas or other mental entities. And indeed, Reichenbach uses the word 'imagine' [denken] in his theorem θ. It has to be understood directly as 'imagine by geometrical visualization'. We are thus led to Reichenbach's theory of geometrical images, which fills almost the entire second half of chapter I, and consists strictly speaking of two theories on two different types of visualization, which are not clearly separated in the text. One type of geometrical intuition is described by a quotation from Helmholtz: to visualize is "to imagine the series of perceptions, we would have, if something like it occured in an individual case" (PST, p. 63; Helmholtz 1962, p. 227). Reichenbach calls this type of intuition 'physical' or 'empirical visualization' (PST, pp. 81–83, 103, 104). It is about the empirical world, and anticipates possible sensory experiences. It is evident that this is not what Reichenbach has in his mind when he formulates his theorem θ. For Helmholtz, visualization anticipates the behavior of real physical objects. We predict by this sort of visualization, that a wooden stick will behave in a certain way if carried around, while in §8 ('The Relativity of Geometry') Reichenbach claims, that Euclidean intuition *does not* imply any physical predictions. The image of the other type of visualization is not yet interpreted. It is no realistic picture of a wooden stick, it is rather a figure consisting of mathematical lines which are not specified in any way. These lines may be identified with light rays, or with edges of rigid bodies, or with something else. The second type of visualization makes no assumption about the physical interpretation of the geometrical figures. Like concepts, the geometrical images or ideas are symbols in our mind, and they have to be interpreted by coordinative definitions. Thus an important feature of Reichenbach's theory of concepts is extended to his theory of the second type of visualization, which he calls 'mathematical' or 'pure visualization' (PST, pp. 81-83), the elements of which represent physical objects in the same way as symbols do. Probably this even means, that according to Reichenbach's theory of concepts we can consider these images as concepts of a special sort; they differ from Reichenbach's concepts in no important respect. After all they are 'configurations of perceptions' with all the features mentioned in the quotation in Section 2 of this article.

Our interpretation of Reichenbach's mathematical visualization can be improved by further evidence obtained from §15 ('What is a Graphical Representation'). There he writes: "The so-called visual geometry is thus already a graphical representation, a mapping of the relational structure *A* upon the system *a* of real objects" (PST p. 104). One should also read the foregoing page: (PST, p. 103!) "Now we understand the meaning of graphical representations. They signify nothing but a coordination of the system *a* to the systems *b*, *c*, . . . of other physical objects, which is possible only because all these systems are realizations of the same conceptual system *A*" (PST, p. 106). This, I suppose, underlines very clearly the symbolic character of geometrical images. Reichenbach tries to show, that mathematical visualization is relative and has no meaning by itself.

(For further details on Reichenbach's theory of visualization see Kamlah, 1976.)

4. RESULTS AND CONCLUDING REMARKS

Let us now summarize the result of our interpretation. Hans Reichenbach wants to refute Kant's theory of pure intuition. In order to do this, he destroys one of the presuppositions, on which Kant's theory rests: thought oriented semantics. According to this semantics, which also conforms to common sense, the meaning of a word is what somebody is thinking when he expresses the word, the mental image, idea, or something like this (see Kretzmann, 1967, for example on Locke at p. 380). Reichenbach shows, that empirical meaning is something still different, which has to be coordinated to the image or idea by a coordinative definition. With this thesis Reichenbach follows Poincaré, who however had been less clear about the necessity of coordination. Kurt Grelling sees in Reichenbach's coordinative definitions the substantial progress of Reichenbach compared with Poincaré (Grelling, 1930, pp. 111–118).

Before we accept the definite result of this interpretation, we should first ask the author himself, where he sees the substantial significance of his relativity thesis. Reichenbach believed himself to be only reproducing the philosophical lesson of Einstein:

> The philosophical significance of the theory of relativity consists in the fact that it has demonstrated the necessity for metrical coordinative definitions in several places where empirical relations [orig. text: *Erkenntnisse* = cognitions, knowledge] had previously been assumed (PST, p. 15).[6]

Once again we see here, that we must not read this quotation in a language oriented interpretation. Scientists before Einstein could not have believed that scientific words are meaningful simply as bare sounds. Of course even in the pre-Einsteinian era scientists and philosophers have been adherents of trivial semantical conventionalism. To understand better what is meant by Reichenbach's statement, let us look at his treatment of simultaneity, which is quite analogous to the treatment of geometry. There he proves, "that simultaneity is not a matter of knowledge, but of coordinative definition" (PST, p. 127). How could it be a matter of knowledge? Suppose, I imagine, that my friend in a town one thousand miles away from here is just now lighting a cigarette. I do this by having just now in my mind the image of this action of my friend. If I believe that he has lit his cigarette one minute earlier, I have the same image in my mind and think moreover, that this has been a true image one minute ago. Thus it seems to be intuitively clear what is meant by 'just now one thousand miles away'. According to thought oriented semantics, it means that my mental image, which I have just now, is a true picture of what is happening one thousand miles away. If I want to know if this picture is indeed true, I have just to go and look or ask somebody afterwards who has seen the represented event. Thus we can ascertain if the lighting of a cigarette happens just now, without asking for the meaning of our mental images. Reichenbach now claims this meaning to be a matter of convention. When I see my friend lighting a cigarette with my mental eyes, I still do not know the procedure required to ascertain if this picture is true. Thus there is something to be defined before we can have any knowledge of a state of affairs. And this is the meaning of Reichenbach's statement quoted above. We may conclude, that Reichenbach's mentioned remark offers no difficulties for our interpretation.

We have seen that Reichenbach expressed a very important insight at a time when many people were unable to accept Einstein's theory of relativity, since for them simultaneity or absolute rest were intuitively given concepts. Even today the undergraduate student, who tries to understand the elements of relativity, has to free himself from the strong influence of his own visualization. But at the moment when we want to talk about these problems in terms of the language oriented approach and shift the discussion of ideas, images, and concepts to the linguistic level, it becomes completely trivial: since we have to transform Reichenbach's claim into one about symbols and words. The relativity of visualization and imagination is transformed into the relativity of descriptions. In the fully developed language oriented philosophical discourse the original problem is so completely eliminated that we cannot even talk about it. We may say with Wittgenstein: "The solution of the problem . . . is seen in the vanishing of the problem." (*Tractatus*, 6.521). Wittgensteins ladder has been thrown away by the language oriented philosopher.[7]

Universität Osnabrück

NOTES

[1] In all quotations of PST I refer to the English translation. The new German edition (Reichenbach, 1976, vol. 2) contains a synopsis of the numbers of pages for the German and the English editions, which allows the reader to find the quoted passage in the German text too.

[2] The German original in Schlick (1925) p. 17 is:

Soviel steht jedenfalls fest: ganz unmöglich kann ich mir eine anschauliche Vorstellung bilden von einem Hunde, der weder ein Bernardiner, noch ein Neufundländer, noch ein Dackel, noch sonst irgendein bestimmter Hund ist, der weder braun noch weiß, weder groß noch klein, kurz, ein Tier, das weiter nichts wäre als eben ein Hund im allgemeinen.

An English translation is Schlick (1918). The quotation will there be found at p. 18.

[3] German in Schlick (1925) at p. 37:

Das Wesen der Begriffe war darin erschöpft, daß sie Zeichen sind, die wir im Denken den Gegenständen zuordnen, über die wir denken.

English translation in Schlick (1975) at p. 40.

[4] Original German text in Reichenbach 1929 at p. 23:

Damit gelangen wir zu einer Theorie der Begriffe: Begriffe sind Gebilde aus Wahrnehmungen, ... Begriffe haben Existenz wie alle psychischen Erlebnisse komplexer Art, sind also Dinge; sie existieren, solange der Mensch existiert, der sie denkt, unabhängig von der zeitlichen Position des ihnen zugeordneten Dinges. Ihre Bedeutung als Zeichen erhalten sie durch ... Zuordnung ..., die eine Vermittlung zwischen der Gesamtheit aller Dinge und diesen speziellen Dingen herstellt. Man darf vielleicht zur Verdeutlichung auf das Bild einer Landkarte hinweisen, die *innerhalb des dargestellten Landes* auf den Boden hingelegt wird: sie ordnet alle Punkte eines großen Raumgebiets einer kleinen Auswahl unter ihnen zu.

[5] Original German text in Carnap (1926) at p. 4:

denn wenn von Begriffen sinnvoll die Rede ist, so handelt es sich stets um durch Zeichen bezeichnete oder doch grundsätzlich bezeichenbare Begriffe; und im Grunde ist dann stets die Rede von diesen Zeichen und Ihren Verwendungsgesetzen.

[6] The same remark is found frequently in Reichenbach's writings. See PST, p. 176, 177; Reichenbach, 1951, p. 169; 1955, p. 192; 1929, p. 34.

[7] The author thanks Dr. George Berger for checking the English style of this paper and for translating the German quotations.

BIBLIOGRAPHY

Carnap, R., 1926, *Physikalische Begriffsbildung*, Braun, Karlsruhe; reprint: Wissenschaftliche Buchges., Darmstadt, 1966.

Carnap, R., 1928, *Der logische Aufbau der Welt*, Berlin-Schlachtensee; reprint: Meiner, Hamburg, 1961.

Carnap, R., 1958, 'Introductory Remarks to the English Edition' in *PST*, pp. v–vii; German translation in Reichenbach (1976).

Carnap, R., 1963, 'Intellectual Autobiography', in P. A. Schilpp (ed.), *The Philosophy of Rudolf Carnap*, Open Court, La Salle (Ill.).

Grelling, K., 1930, 'Die Philosophie der Raum-Zeit-Lehre', (extensive review), *Philosophischer Anzeiger* **4**, 101–128.

Grünbaum, A., 1963, *Philosophical Problems of Space and Time*, Knopf, New York; 2nd edition: 1973, Reidel, Dordrecht-Boston.

Grünbaum, A., 1968, *Geometry and Chronometry in Philosophical Perspective*, Univ. of Minnesota Press, Minneapolis.

Grünbaum, A., 1976, 'Absolute and Relational Theories of Space and Space–Time', in *Minnesota Studies in the Philosophy of Science*, Vol. 8, Univ. of Minnesota Press, Minneapolis.

Helmholtz, H. v., 1962, *Popular Scientific Lectures*, Dover, New York.

Jørgensen, J., 1951, 'The Development of Logical Empiricism', *Int. Encycl. of Unif. Science* **2**, No. 9, Univ. of Chicago Press, Chicago–London.

Kamlah, A., 1976, 'Erläuterungen, Bemerkungen und Verweise', in Reichenbach (1976), Vol. 2.

Kretzmann, N., 1967, 'Semantics, History of', in P. Edwards (ed.), *The Encyclopedia of Philosophy*, Macmillan, New York–London, Vol. 7, pp. 358–406.

Pap, A., 1949, *Elements of Analytic Philosophy*, Macmillan, New York.

Putnam, H., 1963, 'An Examination of Grünbaum's Philosophy of Geometry', in B. Baumrin (ed.), *Philosophy of Science, The Delaware Seminar*, Vol. 2, Interscience, New York.

Poincaré, H., 1914, *Science and Method*, London.

Reichenbach, H., 1928, *Philosophie der Raum-Zeit-Lehre*, de Gruyter, Berlin; 2. ed. Reichenbach (1976), vol. 2.

Reichenbach, H., 1929, 'Ziele und Wege der physikalischen Erkenntnis', in H. Geiger and K. Scheel (eds.), *Handbuch der Physik*, Springer, Berlin, Vol. 4, pp. 1–80.

Reichenbach, H., 1951, *The Rise of Scientific Philosophy*, Univ. of California Press, Berkeley-Los Angeles; German translation: Reichenbach (1965); 2. ed.: Reichenbach (1976), vol. 1.

Reichenbach, H., 1958 (PST) *Philosophy of Space and Time*, Dover, New York.

Reichenbach, H., 1965, *Der Aufstieg der wissenschaftlichen Philosophie* (translation of Reichenbach, 1951), Vieweg, Braunschweig.

Reichenbach, H., 1976, *Gesammelte Werke in 9 Bänden*, Vieweg, Braunschweig-Wiesbaden; Vol. 1: Reichenbach (1965); Vol. 2: Reichenbach (1928).

Rorty, R., (ed.), 1967, *The Linguistic Turn. Recent Essays in Philosophical Method*, Chicago–London.

Russell, B., 1926, *Unser Wissen von der Außenwelt* (German translation of *Our Knowledge of the External World*), Meiner, Leipzig.

Schleichert, H., (ed.), 1975, *Logischer Empirismus-Der Wiener Kreis* (Essays of the Vienna Circle) Fink, München.

Schlick, M., 1918, *Allgemeine Erkenntnislehre*, Springer, Berlin.

Schlick, M., 1925, 2. edition of Schlick (1918).

Schlick, M., 1949, 'Meaning and Verification', from *Philosophical Revue* **44** (1936); reprinted in H. Feigl and W. Sellars (eds.), *Readings in Philosophical Analysis*, Appleton Century Crofts, New York, pp. 146–170.

Schlick, M., 1974, *General Theory of Knowledge* (transl. of Schlick, 1918, 2nd edition) Springer, Wien–New York.

Whitbeck, C., 1969, 'Simultaneity and Distance', *Journal of Philosophy* **66**, 329–340.

J. ALBERTO COFFA

ELECTIVE AFFINITIES: WEYL AND REICHENBACH[1]

Ortega once defined a classic as someone who, like the Cid, can wage war and defeat us, even when he is dead. In disciplines such as ours, where change of topic threatens to become the *Ersatz* of progress, sheer self-interest should inspire us periodically to invoke the ghosts of our classics, and challenge them to battle. Having suffered the halberd once too often, however, my more prudent goal will be to summon two of our classic heroes to quarrel with each other.

My heroes are Reichenbach and Weyl; their quarrel concerns the theory of relativity. Few people have thought more highly of that theory and disagreed more deeply on its content. The disagreement cannot be consigned to ignorance or confusion for, by any standards, they both understood Einstein's theory thoroughly. Rather, the divergence manifests a difference in philosophical perspective. Perhaps I need now add that this makes the difference all the more interesting.

My aim in this paper is not to decide who was right and who was wrong, but only to understand the nature of the disagreement. Even when so limited, the task is far too complex to be developed within the confines of one paper. Many essential areas of conflict will have to be ignored (prominent among them, their disagreement on the empirical determination of the metric). Furthermore, since Weyl's opinions are, by far, the least known among philosophers, we have devoted more space to them than to Reichenbach's. We have had to assume that the reader is familiar with the main components of Reichenbach's epistemology and philosophy of science (for an excellent survey, see Salmon's *Introduction* to this volume). All in all, what follows is no more than an invitation to pursue a topic, and some reasons that show this is worth doing.

Our goals will be two: (i) to explain Weyl's conception of relativity, tracing its connections with his views on geometry; (ii) to illustrate the nature of the disagreement between Reichenbach and Weyl through their divergent responses to two of the basic conceptual problems which the theory of relativity faced around 1920. Sections 1–3 are devoted to the former task, section 4 to the latter.

W.C. Salmon (ed.), Hans Reichenbach: Logical Empiricist, 267–304.

1. CONNECTIONS

One of the basic premises of natural philosophy, Weyl thought, is that nature should be understood through an analysis of its behavior in the infinitely small. In the age-old struggle between the metaphysics of action-at-a-distance and action-by-contact, Weyl stood decidedly on the side of the latter, extending its validity beyond the domain of physics to that of pure geometry.[2]

In the first edition of *Raum-Zeit-Materie* (1918a) Weyl had hailed Riemann's work as the fulfillment of the infinitesimal principle in mathematics:

> To the transition from Euclid's distant-geometry to Riemann's contact-geometry corresponds the transition from action-at-a-distance physics to action-by-contact physics . . . The principle that the world should be understood through its behavior in the infinitely small is the leading epistemological force behind both the physics of action by contact and Riemann's geometry, and it is, indeed, the dominating motive in the rest of Riemann's grandiose life-work . . . (1918a, p. 82)

Only a few months later, however, Weyl had changed his mind, and he was now complaining that "Riemann's geometry only goes half-way towards the ideal of a purely infinitesimal geometry" (1922a, p. 102). What had happened in the meantime is that Weyl had achieved a new understanding of the nature of geometry and had built on its basis a generalization of the idea of a Riemannian manifold. Indeed, he had good reason to place upon this new theory his highest hopes; for, as he developed its varied themes, he came to regard it as a key to the solution of major problems in physics, mathematics and philosophy.

The task of this section and the two subsequent ones is to examine the geometric, philosophical and physical implications of Weyl's infinitesimal geometry. We begin with the mathematical component of this trilogy, for at the source of the causal chain which ends with Weyl's unified field theory stands a purely mathematical event, indeed, a seemingly harmless instance of normal (mathematical) science: Levi-Civita's definition of parallel displacement in Riemannian manifolds.

1a. *The Italian Connection*

In (1917) Levi-Civita had introduced the idea of parallel displacement for Riemannian manifolds. Using the old 'coordinate language' – which was the standard one at the time and for decades to come – Weyl, following Levi-Civita, explained the basic point along the following lines: The task is to define what it means to say that a vector **v** in a Riemannian manifold is displaced in parallel fashion from one point to another. Assume that **v** is given byi ts coordinates (ξ^i) in a reference frame and that it is located at some point $P = (x^i)$ of the manifold. Our Euclidean (action-at-a-distance) prejudices may tempt us to produce a definition of parallel transport aiming to identify the coordinates of **v** once parallel-transported to a finitely distant point Q. But, clearly, there can be no such definition since a vector parallel-transported on, say, a sphere from P to Q will coincide with different vectors at Q depending on the path it has followed. Whereas others had drawn the conclusion that parallel displacement makes no sense in Riemannian manifolds, Levi-Civita wondered whether we shouldn't rather conclude that the notion is not definable *for finite domains.*

The redefined aim is to characterize the parallel transplant of (ξ^i) from P to the *infinitesimally near* point $P' = (x^i + dx^i)$. After the displacement of **v** to P' we say that its coordinates will be $(\xi^i + d\xi^i)$; the problem now becomes that of finding an expression for $d\xi_i$. Assuming that the Riemannian manifold is embedded in a higher-dimensional Euclidean manifold, Levi-Civita had calculated that the expression for $d\xi^i$ would be

$$(1) \qquad d\xi^i = - \Gamma^i_{jk} \xi^j dx^k,$$

where the Γ^i_{jk}'s are Christoffel's symbols.

The geometric significance of the idea of parallel displacement derives from the fact, noted by Levi-Civita, that both curvature and straightness are definable on its basis. In particular, Levi-Civita observed that if we think of straight lines the way Euclid seems to have, that is, as lines such that their tangents are parallel-displaced along themselves, then, the idea of parallel displacement allows us to characterize the straightest lines in the manifold quite independently of metric concepts.

Weyl immediately saw the pointlessness of Levi-Civita's embedding space and formulated an intrinsic version of parallel displacement (in 1918b). Moreover, he endeavored to capture the fully general concept of the affine structure by liberating the components of the affine connection from the need to agree with a metric. In general, Weyl explained, there is no need to conceive of parallel transport in such a way that under displacements of that sort vectors should preserve their length; indeed, there need not even be a metric (hence, a notion of length) in the manifold with which the affine structure might agree or conflict. Thus, the gamma-coefficients in (1) need not be construed as derived from a metric. Weyl proposed as the most general reasonable criterion to define this broader notion of parallel displacement the condition that a vector field in a neighborhood of P defines a legitimate parallel displacement at P if there is a coordinate system in the neighborhood such that the components of the vectors in the field (relative to the coordinate frame field) do not change in an infinitesimal neighborhood of P. (In the less inspiring language sanctioned by set-theory, we would say that there must be a coordinate system around P in which the gamma-coefficients in (1) vanish at P). Weyl showed that the necessary and sufficient condition for this to obtain (apart from smoothness assumptions) is that the components of the affine connection be symmetric in their subindexes. Therefore, given any coordinate system in a region and any choice of (appropriately smooth) functions Γ^i_{jk} such that $\Gamma^i_{jk} = \Gamma^i_{kj}$ there will be a legitimate concept of parallel displacement and, therefore, a legitimate affine structure in the manifold (for different coordinate systems the Γ^i_{jk} transform in agreement with well known laws).[3] In the presence of a metric, of course, the straightest and the shortest lines need no longer coincide.

In order to underscore the analogy between classical (linear) affine spaces and the metric-free structures obtained by adding to a differentiable manifold any of the newly defined affine structures Weyl coined the expressions 'affine connection'[4] and 'affine manifold' to refer to his generalization of Levi-Civita's notion and to the manifolds defined through it. It was Weyl more than anyone else who taught mathematicians and physicists how to detach from Riemannian manifolds their purely affine structure and – as we shall soon see – how to put that structure to good use in the domain of physics.

Levi-Civita's idea was also influential in determining the next major step in Weyl's geometric speculations. But before we proceed to examine it, a *caveat* is in order.

If my reader is a young philosopher with a good training in logic, the infinitesimal language employed up to this point must have proved offensive to his set-theoretic sensitivities[5]. Indeed, we know how set-theory frowns upon language such as this:

> If P' is a point infinitely close to P, we say that P' is affinely connected to P when, for every vector in P it is determined which vector in P' corresponds to it as a result of parallel displacement (Weyl, 1918b, p. 6)
>
> In Riemannian geometry vectors can be displaced only infinitesimally (Weyl, 1923c, p. 14)
>
> [A] stretch can be displaced in parallel fashion only to an infinitesimally nearby stretch (Weyl, 1923c, p. 15).

Of course, nothing could be furthest from Weyl's intuitionistic mind than to suggest that there is an actual infinitesimal neighborhood of P where parallel displacement is path-independent, or even an actual infinitesimal distance dx through which a vector can be displaced. To say that we define parallel displacement only in infinitesimal domains is like saying that the derivative is the quotient of two infinitesimal magnitudes: in both cases we are dealing with suggestive nonsense. But such nonsense is, I fear, the stuff of which creative mathematics is made. As we shall now see, it was the 'nonsensical' infinitesimal picture, in fact, the element in the preceding account of parallel displacement which gets lost under set-theoretic translation, that played a leading role in Weyl's revolutionary discovery.

1b. *The German Connection*

As soon as he heard about Levi-Civita's idea Weyl was struck not only by its mathematical and physical power, but also by the fact that the affine connection had set in relief the *only partial* success of Riemann's geometry in its attempt to abandon the Euclidean frame of ideas.

On the one hand, Levi-Civita's discovery had established that in Riemannian geometry the affine notion of parallel displacement is liberated from its Euclidean, action-at-a-distance assumptions. There is a path-free notion of parallelism, as Euclid thought, but only in infinitesimal domains. For finite domains one can only ask whether two distant vectors are parallel relative to a path, and not absolutely.

But once one noticed that there is a (path-independent) notion of parallel displacement in the infinitesimal and nowhere else, it became natural to ask, why does not the same situation obtain for the next major geometric structure, i.e., the metric? In Riemann, as in Euclid, there is a path-independent notion of congruence (congruence-at-a-distance). Thus, whereas the direction of a vector which has been parallel-transported is dependent upon the path it has followed, the length of a vector congruently transported does not depend upon the path *it* follows. This, Weyl thought, is a blemish, the last remainder of Euclideanism within Riemannian geometry. It was the recognition of this blemish which elicited the complaint quoted early in this section.

By analogy with the affine case, Weyl argued, the revolution started by Riemann can only be completed if we circumscribe the path-independent notion of congruence to the infinitely small. In order to do so we supplement the idea of an affine connection with that of a *metric connection*, whose role is to provide a path-independent extension of the point-wise metric only infinitesimally. Let us see how this is done.

In Weyl's infinitesimal geometry the metric is introduced in two stages: first one defines the metric (in the tangent space) at each point, then one connects the point-metrics within each infinitesimal neighborhood.

The metric at each point is defined in the usual way, by giving the metric at the corresponding tangent space; the only novelty is that the metric is defined only up to a scale (gauge, calibration) factor. We are thus given only the similarities, and not the isometries on each tangent space. The manifold is said to have a *measure determination* (Massbestimmung) when the similarities in *all* tangent spaces are given.

Each vector in the tangent space at P determines a stretch or distance (Strecke). Two vectors $\mathbf{u}$ and $\mathbf{v}$ in a tangent space determine the same stretch whenever $\mathbf{u}^2 = \mathbf{v}^2$ (where $\mathbf{x}^2$ is the metric quadratic form). Thus, the stretches associated with any given vector space form a 1-dimensional manifold. Once a unit of length (scale, calibration, gauge factor) is introduced in the tangent space at P, we say that the manifold is *calibrated* (geeicht) at P. A *gauge* or *calibration* of the manifold is simply a vector field G such that for each point P, $G(P)$ is a unit vector at the tangent space in P.

Once the measure determination is given, we obtain a Weyl-manifold by introducing the *metric connection.* A metric connection resembles an affine connection in that the former does for stretches and their congruence properties what the latter does for vectors and their orientation. A point P is metrically connected to the points in an infinitesimal neighborhood when for each stretch in P it is determined which stretch corresponds to it at each infinitely nearby point P' through a process of congruent transplantation. Just as vector components may change under parallel transport, length may change under congruent transport. And just as the affine connection was characterized through the demand that in a neighborhood N of P there should be a (geodesic) coordinate system in which infinitesimal parallel transport at P is represented by unchanging components ($d\xi = 0$), the metric connection is defined through the condition that there must be a gauge or calibration in N such that infinitesimal congruent transport at P is represented in it by vanishing length changes ($\dot{d}l = 0$).

Pursuing the analogy with the affine case, we should like to identify an expression which, in analogy with Levi-Civita's (1), determines the law according to which lengths alter under infinitesimal transport. Weyl's basic requirement on congruent transport from $P = (x^i)$ to $P' = (x^i + dx^i)$ is that the stretches at (the tangent space in) P must be related to those at P' by a *similarity* transformation. Thus, we have

$$dl = -l\,d\phi,$$

where $-d\phi$ functions, so far, as an undetermined (but stretch-independent) factor. In order to determine it we now appeal to the already mentioned constraint involving the existence of 'geodesic' calibrations. This constraint, Weyl argues, is satisfied precisely when $d\phi$ is a linear form $\phi_i dx^i$. The result, in striking analogy to (1), is that

$$(2) \qquad dl = -\phi_i l dx^i.$$

We conclude this preliminary survey of infinitesimal geometry by noting that in order to give the coordinate expression of a Weyl metric we must both coordinatize *and* calibrate the metrized region. Weyl used the expression "reference system" to designate the complex coordinatization-cum-calibration. Relative to a reference system the metric is fully characterized through *two* forms, and not just one as in Riemannian geometry: the quadratic form

$$g_{ik}dx^i dx^k$$

(the old Riemannian metric), and the new linear form

$$\phi_i dx^i$$

which occurs as a factor in Weyl's expression (2) for congruent transport. Both forms are invariant under coordinate transformations; but under gauge transformations they are not. A change of calibration at a point P by a factor λ will simply produce a change in the measure numbers assigned to each stretch at P by the given factor: lengths measured by l before recalibration will now be measured by λl. Thus the quadratic form under a gauge transformation will simply alter by a (point dependent, definite, continuous) factor λ. The one-form, on the other hand, will be diminished by a factor $d\lambda/\lambda$ (see, e.g., 1922a, §17). Let this suffice as a rough sketch of Weyl's idea of a metric manifold (we shall return to this subject in the following section).

It is worth noting that besides the mathematical motivation of Weyl's infinitesimal geometry, there is a philosophical idea which, Weyl thought, rendered his theory philosophically superior to Riemann's. The idea in question was embodied in what Weyl called 'the principle of the relativity of magnitude.' Regrettably, in spite of the frequency with which he appealed to it, Weyl never explained with any precision the content of this principle. Perhaps one way to understand it, and the role it plays within Weyl's philosophy of geometry, is by drawing a connection between the principle and a prominent aspect of contemporary geometric conventionalism.

I consider it likely that Weyl would have endorsed Grünbaum's contention that physical space has no intrinsic structure that determines whether the amount of space occupied by a measuring rod (or the amount of time measured by a clock) is equal to the amount of space (or time) occupied by the same rod (or clock) elsewhere. From this, Weyl concluded that there is no *a priori* reason to believe that measuring rods which occupy the same amount of space at a point will occupy the same amount of space at all points (likewise for clocks). Riemann's geometry is based on the opposite assumption (Riemann's concordance assumption). That rods which coincide at a place and time will coincide at all places and times regardless of how they have been transported was, for Weyl, not a geometric principle but a surprising physical fact very much in need of a physical explanation. To regard it as a geometric principle is to invite the confusion that no physical action is necessary in order to account for the concident behavior. In due course Weyl came to conjecture that the Universe, through its curvature properties, exerts an influence upon measuring instruments such as rods and clocks which forces them to adjust to the pattern described by Riemann's concordance assumption. But, whatever the explanation, he would insist that geometry ought not to prejudge physics, so that the integrability condition on congruent transport ought to be removed from Riemannian geometry.[6]

2. THE PROBLEM OF SPACE

In Weyl's eyes his infinitesimal geometry was the end result of a twice millenary process of phenomenological *Wesensanalyse*, in Husserl's sense (1922a, pp. 147–8). The philosophical power of his new theory was most clearly manifested by the fact that it opened a completely new perspective from which it now seemed possible to solve one of the major traditional philosophical puzzles concerning the nature of space. Specifically, Weyl thought – for a while, at any rate – that his view of geometry allowed one to solve the old Kantian problem concerning the *a priori* elements in space.

The expression 'the problem of space' was in currency during the late 19th and early 20th century as a designation for a number of only loosely related questions concerning the nature of space. From this multitude of topics, one basic concern stood out as a focus of attention for philosophically oriented geometers. The subject of this particular problem was Riemann's infinitesimal expression for the metric, his celebrated formula

(3) $ds^2 = g_{ik}dx^i dx^k$;

and the problem was, why ought we to accept it? Are the reasons empirical or *a priori*? And, in any case, what are they?

In his Inaugural Dissertation Riemann had examined this matter briefly, and had dismissed alternative forms for reasons that most geometers had found wanting. The first one to raise the problem as a major issue was Helmholtz, who also gave the sketch of a solution which would later be completed by Lie. Very roughly, the relevant facts are the following.

In Helmholtz's time (and until much later) Riemannian variable-curvature geometries were regarded as needless oddities that ought not to be taken too seriously. The homogeneity of space was considered to be inevitable, often because it was thought to be a condition of possibility of measurement, and this automatically ruled out spaces of variable curvature. Helmholtz conceived the program of deriving the necessity of (3) from homogeneity or free-mobility assumptions. Lie showed that the program could be accomplished.

But the Helmholtz-Lie solution to the problem of space could no longer be accepted after the theory of relativity had shown the legitimacy of variable-curvature spaces. What was now needed was a justification of (3) that did not restrict inadmissibly the class of possible metric spaces. It was at this point that Weyl decided to reopen the issue and examine the matter from the standpoint of his newly developed conception.

The first task Weyl undertook was that of defining in precise terms the problem of space. In order to do so, he introduced a distinction between what he called the 'nature' and the 'orientation' of the metric in the manifold. That the spatial metric has a *nature* means that the metric is the same in all tangent spaces; that the *orientation* of the metric may change means that the isometries at each tangent space need not coincide but must constitute mutually conjugate groups.

In order to illustrate the idea of orientation we may fix our attention on a Riemannian manifold (although, of course, the idea applies to a much wider domain of metric spaces), for example, a surface embedded in Euclidean 3-space. Let us first say that a base in a metric vector space has *the Pythagorean property* if, relative to that base, the length of any vector $\mathbf{v}$ is given by the Pythagorean formula

$$\mathbf{v}^2 = x_1^2 + x_2^2 + \cdots + x_n^2,$$

where the x_i's are the components of $\mathbf{v}$ relative to the given base. Imagine now a curve in our Riemannian manifold. Since the tangent space at each point must be Euclidean, at each point along the curve we can identify an orthonormal base which will have the Pythagorean property. Fix your attention now at a point P of the curve and a base b in the tangent space at P endowed with the Pythagorean property. If we push b along the curve without changing its orientation (now in an obvious intuitive sense) we may raise the question of whether at points other than P, b still has the Pythagorean property. If the manifold is Euclidean the answer will be affirmative. But in the general Riemannian case the moving frame will have to twist and turn along the curve in order to catch those components of the vectors at each tangent space that will determine their lengths in terms of the Pythagorean formula.

Having drawn the nature-orientation distinction Weyl was now in a position to analyze Kant's a priority claim into two sub-theses: (i) that the nature of the spatial metric is *a priori* (Euclidean), and (ii) that its orientation is also *a priori*. The latter claim is entirely defeated by Riemann's and Einstein's investigations which show orientation to be dependent upon the contingent matter-energy distribution. The former, however, is quite consistent with both Riemann's and Einstein's results. The problem of space as understood by Weyl, is to show that the former claim is true.

Weyl conjectured that there are three conditions which are *a priori* defensible and which entail (3). The first one seemed to him so obvious that he often did not bother to state it explicitly: space must have some nature or other; this is one of the several manifestations of the grip which the old principle of homogeneity still had on Weyl's work. The other two were, roughly, that the nature of space must impose no restrictions upon the possible metrical connections (this is where Weyl's new conception played a basic role) and that the metric must uniquely determine the affine connection. The content of these requirements was appealingly illustrated in his Spanish lectures devoted to the problem of space:

The essence of the Constitution of a State is in no way entirely settled if I require: the State Constitution must be binding for all members of the State; but within the limits imposed by the Constitution each citizen is endowed with absolute freedom. In order to set some bounds I must begin by introducing a positive requirement, perhaps the following: regardless of how the citizens may want to use their freedom, it lies in the essence of the Constitution that the good of the society must be guaranteed to a sufficient extent. At this point one can begin to speculate on how to come up with a Constitution that satisfies these requirements and, indeed, on whether there is no more than one that could possibly satisfy them.– The binding State law in the domain of space is the nature of the metric, the liberty of its citizens is the possibility of diverse mutual orientations of the metric at different points in space; and what is "the good of the society"? When one considers the structure of infinitesimal geometry as well as its applications to reality through the general theory of relativity, there is one fact that sets itself in clear relief with inescapable uniqueness as the decisive fact that makes the whole development possible: that the affine connection is determined by the metric field; it is here, it seems, that lies the good of the community within the domain of space and time. (1923c, p. 47)

Metaphors aside, the basic ideas are the following. We are given a differentiable manifold which is endowed with a metric at each tangent space. The metric must be the same at each tangent space (space has a nature) but, of course, it is not required at the outset that it be Euclidean. These metrics can be given by a group-field, assigning to each point of the manifold a group of linear transformations representing the isometries at each tangent space. The requirement that space should have a possibly variable orientation becomes the condition that the isometries in different tangent spaces should be mutually conjugate groups. The final condition to be formalized is that metric connections be maximally free. Appealing once more to an analogy with our expression (1) for the affine connection, Weyl derives the following law for the change in the components of a vector (ξ^i) under infinitesimal congruent transport:

$$(4) \qquad d\xi^i = \Lambda^i_{jk}\xi^j dx^k,$$

where the Λ^i_{jk}'s are the components of the metric connection. The postulate of liberty asserts that the range of admissible values of the Λ^i_{jk}'s must be maximized; that is, given any rotation group in an arbitrary point P, and given arbitrary values for the Λ^i_{jk}'s, (4) defines an admissible system of congruent displacements from P to any point in its infinitesimal neighborhood.

In the 4th edition of *Raum-Zeit-Materie* (1921a = 1922a) Weyl issued the conjecture that his three assumptions entailed (3), and he added that he could prove his conjecture for dimensions 2 and 3. A few months later he had established the general validity of his conjecture. The problem of space, he thought, had finally been solved.[7]

Of course, the derivation of (3) from Weyl's hypotheses does not really solve the problem of space unless there is reason to take the latter as *a priori* true. Weyl said next to nothing on this subject, but it is clear that he thought or hoped one could so take them. In (1927), for example, he writes

Only in the case of this particular group [i. e., the group of Euclidean rotations] does the contingent quantitative development of the metric field, however it be chosen within the framework of its *a priori* fixed nature, uniquely determine the infinitesimal parallel translation . . . It is possible that the postulate of the unique determination of 'straight progression' can be justified on the basis of the requirements posed by the phenomenological constitution of space. (1927, pp. 99–100; 1949a, p. 127)

As his commitment to phenomenology, and to a priorism in general, waned, Weyl may have come to wonder about the relevance of his theorem to the great Kantian question (see 1954, p. 634). Be that as it may, in the early twenties Weyl's remarkable theorem seemed to provide striking confirmation that he had managed to grasp the right concept of space.

Even so, the crowning achievement of his infinitesimal geometry did not lie either in the domain of mathematics or in that of philosophy. We are now ready to look at the way in which these ideas led Weyl to conceive the first project for a unified field theory.

3. Protogeometrodynamics

In order to understand how Weyl came to apply his infinitesimal geometry in the project for a unified field theory, we must begin by examining, if only briefly, his conception of the general theory of relativity.

There is a picture of relativity theory, the one from which it presumably derives its name, which appeals to considerations involving Mach's relationism, the relativity of motion and covariance. The picture derives from Einstein, and a superficial look at Weyl's writings might lead one to believe that he was an enthusiastic proponent of this quasi-Machian understanding of relativity.

But there is also another picture, much closer to Newton's than to Mach's ideas, in which covariance and the relativity of motion (as generally understood) play little or no role. This picture is also present in Einstein's work, but in a more hesitant and less defined way than in Weyl's. Weyl was able to draw its lines quite sharply thanks to the mathematical machinery which he had borrowed from others and which, to some extent, he himself had developed for the purpose; for the basic element of this picture is a peculiar understanding of the affine structure of the spacetime manifold.

3a. *Weyl's Picture of General Relativity*

Weyl used to present the theory of relativity as the outcome of two major revolutions: the first one concerned the inertial and causal aspects of nature, the second related to the physical reality of the inertial structure (see, e.g., (1934, Chapter IV)).

The special theory of relativity had uncovered the fact that the causal structure of the universe is not given by objective simultaneity cross-sections but by the richer light-cone structure. This insight was quite beyond the reach of physicists up to the 19th century, since its discovery depended essentially upon facts which only became known late in that century. Einstein's basic insight concerning inertia, however, was a master stroke of almost purely conceptual thinking, since the factual information involved in its discovery was known since Galileo and Newton.

The essence of Einstein's insight comes to light only through Levi-Civita's (post-relativistic) discovery of the affine connection. Roughly put, Weyl's idea is that it is a serious error to try to understand gravitation on the basis of the full metric structure of spacetime. For Einstein's explanation of gravitation is given purely on the basis of the affine structure which, even though physically dependent upon the metric, is conceptually quite independent of it. Levi-Civita's connection is the mathematical instrument which allows us to peel-off from each Riemannian manifold its metric crust, revealing underneath it the pure affine structure, which is the only one which should concern us when we are dealing with gravitation.

Recall that in Weyl's conception of the affine connection there is an infinity of possible affine structures with which one may endow any given differentiable manifold. Whereas some manifolds have no intrinsic affine connection – so that it is a merely conventional matter to decide which one to impose upon them – other manifolds come naturally endowed with an affine connection.

> If a manifold is of such a nature that at a point *P*, of all 'possible' concepts of parallel displacement only one is singled out as the *uniquely 'real'*, then we say that the manifold is endowed with an affine connection. (1923c, p. 12)

Galileo's principle of inertia offers a clear indication that physical space is endowed with an affine connection, since the principle tells us that given any point and any velocity, objects at that point and with that velocity, regardless of their constitution, exhibit a natural tendency to proceed along the (Euclidean) straight line characterized by those two elements. Thus, Galileo's principle associates with space what Weyl called a 'guiding field' (Führungsfeld), a field that leads or guides every object, regardless of its constitution, through a 'channel' which the object will not abandon unless affected by external forces. If there is to be *any* natural embodiment of the idea of a straightest path, Weyl thought, *this* must be it; hence the propriety of identifying the guiding field with the affine structure of physical space.

According to this conception, inertia defines the straightest natural paths of arbitrary objects. Forces, on the other hand, determine their deviation from such paths. Since Galileo and Newton we have learned to conceive of motion as the outcome of a struggle between inertia and forces. Relativity theory remains committed to this duality, but it invites us to relocate gravitation and, in consequence, to reinterpret inertia. This was, Weyl thought, Einstein's *basic* insight.

The classical conception of inertia included the seeds of its own demise. There was, on the one hand, the fact that the reference frames relative to which the 'natural' inertial paths turned out to be straight lines had no physical realization (that is, they could not be defined on the basis of physically realizable processes). Rather, such frames could only be defined implicitly, as the frames in which Newtonian inertia holds. More significantly, Galileo's law of free fall had underscored the surprising fact that, irrespective of their constitution, bodies freely falling in a gravitational field will move in exactly the same way, just as they do under the circumstances envisaged in the law of inertia. Thus, no physical distinction seems to correspond to the conceptual distinction between inertia and gravitation.

In the struggle between inertia and forces, the classical detachment of gravitation from inertia was displayed through the assignment of gravitation to the category of forces. The curved path of a planet and the parabolic trajectory of a projectile were explained as the deviation from their inertial paths due to the effect of gravitational *forces*. Einstein's reinterpretation of gravitation consists, quite simply, in the assignment of gravitation to the category of inertial actions. For him gravitation is not a force deviating a particle from its affine-inertial path but it is a part (often conveniently, always arbitrarily[8]) detached from the ontologically indivisible unit constituted by the inertial structure of spacetime.

Once we come to regard gravitation as a part of inertia, our picture of the universe alters dramatically (even though our beliefs about observations or 'spacetime coincidences' may remain unchanged). The natural paths of particles moving in the absence of forces are no longer given by Euclidean straight lines but by the geodesics of some affine manifold. Moreover, these straightest inertial lines are no longer rigid but they must be dependent upon the distribution of matter, since gravitation, known to be matter-dependent, is now somehow contained within the inertial structure. Cartan's (and Havas') Newtonian studies establish that this picture does not inevitably lead to a rejection of Newtonianism, or to an alteration of its observational content. But the picture readily suggests a metaphysics or a theoretical image of nature which leads far beyond Newtonian conceptions.

The metaphysical turn is most clearly apparent in the second stage of the relativistic revolution: the recognition of the physical reality of the affine structure, which entailed its capability of being acted upon, no less than its ability to act. For Newton space was a rigid, immutable God, capable of exerting overwhelming influences upon objects by imposing, through its rigid, Euclidean affine structure, their natural paths; but it was entirely incapable of being affected by them. Einstein, on the other hand, demotes space to the role of an immensely powerful giant still able to exert the most overpowering of actions, but now capable of being affected by these same objects it so severely affects. Notice, however, that spacetime has not lost an iota of its Newtonian reality, manifested now through the non-substantial ether filling spacetime and endowing it with a metric structure through its states. If anything, there is more *physical* reality in Einstein's space than in Newton's.

3b. *Weyl's Unified Field Theory*

In the late 10's and early 20's it was not wholly unreasonable to believe that physics had essentially two topics: gravitation and electromagnetism[9]. Weyl's analysis of Einstein's theory as developed in the preceding section had led him to the conclusion that the theory of relativity had succeeded in reducing gravitation to geometry: specifically, to the affine structure of spacetime. If one could manage to show that Maxwell's equations are also reducible to geometry, one would be in a position to conclude that "physics does not extend beyond geometry" (1922a, p. 129). Descartes' bold dreams of a purely geometrical physics would be thereby realized, no less than Clifford's ideal "to describe matter and motion in terms of extension only" (1922a, pp. 284, 129).

In late 1917 or early 1918 the dream suddenly seemed to become a reality. Weyl noticed a quite unexpected formal analogy between a certain feature of his infinitesimal geometry and Maxwell's equations, and he concluded that the geometrization of physics was now within grasp. In (1918b) and (1918c) Weyl proudly announced his discovery:

I have managed to construct a world-metric by liberating Riemannian geometry – which aims to be a pure 'contact geometry' – from an inconsequence which is currently present in it, the last ingredient of 'distant-geometry' which it preserved as a remnant of its Euclidean past. From this world-metric stem not only gravitation but also electromagnetic effects, so that one may reasonably assert that it offers an account of the totality of physical processes. According to this theory *everything real, everything occurring in the world, is a manifestation of the world-metric;* the physical concepts are none other than the geometric concepts. The only difference between physics and geometry lies in this, that geometry provides general foundations for what lies in the essence of metric concepts.

Physics, instead, looks for the law and its consequences through which the real world is distinguished from all the geometrically possible four dimensional metric spaces. (1918b, p. 2)

The link with electromagnetism had been established as follows. As Levi-Civita had noted in his (1917), a version of Riemannian curvature can be defined in a purely affine manner, roughly as a measure of the deviation from integrability of parallel transport at each point; the greater the curvature at a point, the greater the limit of the deviation in ever smaller parallel closed trips through it. By analogy, we can say that the non-integrability of congruent transport in Weyl's geometry is the symptom of another kind of curvature (known as the Weyl-curvature). Weyl observed that a tensor can be defined which measures at each point the extent of that non-integrability. Its components are given by

$$F_{ik} \equiv \frac{\partial \phi_k}{\partial x^i} - \frac{\partial \phi_i}{\partial x^k} \tag{5}$$

where the ϕ_j's are the functions introduced on p. 273. Thus, the Riemann curvature, affinely conceived, is a direction-curvature, whereas Weyl's is a stretch- or distance-curvature. In Euclidean manifolds both Riemann's and Weyl's curvatures vanish since both direction and length are unchanged under parallel and congruent transport respectively; in Riemannian manifolds Weyl's curvature vanishes; in Weyl's infinitesimal manifolds neither curvature need vanish. It was at this point that Weyl was struck by the obvious formal analogy between the distance curvature and the Faraday tensor. For if the factor ϕ is taken to be the electromagnetic potential our F_{ik}'s become the components of the Faraday (electromagnetic field) tensor. From (5) we can readily derive

$$\frac{\partial F_{ik}}{\partial x_j} + \frac{\partial F_{kj}}{\partial x_i} + \frac{\partial F_{ji}}{\partial x_k} = 0$$

which is Maxwell's first system of equations in its spacetime formulation. Furthermore, Weyl noted that the simplest possible integral invariant that could be constructed in a four-dimensional metric manifold is precisely a field action magnitude (1922a, p. 128; §§26, 35) from which the second system of Maxwell's equations derives.

All of this could hardly fail to suggest that there was an intimate connection between Weyl's geometry and electromagnetism. As he incorporated in his theory an action principle of the sort Mie had investigated in his recent speculations concerning a pure field theory of matter, Weyl could add to this set of striking analogies another one: according to his theory gauge invariance corresponds to the conservation of electric charges in the same way that coordinate invariance corresponds to energy-momentum conservation (1918c, pp. 37ff.) It was the loosening of this link, through the development of quantum theory, that seems to have started Weyl's estrangement from his unified field theory.

Weyl's theory contended that what had been so far regarded as manifestations of electromagnetic effects was to be seen henceforth (to some extent, at any rate) as a manifestation of the non-integrability of congruence in spacetime regions. It seemed reasonable to conclude that whereas in the absence of electromagnetic fields we need not fear a violation of Riemann's concordance assumption, in the presence of such fields we may expect the failure of integrability conditions, so that the period of atomic clocks and the length of moving rods will normally depend upon their past histories.

Einstein was the first to draw this consequence, and he found it entirely unacceptable. In an appendix to Weyl's (1918c) Einstein writes that if one allows the use of infinitely small rigid bodies and clocks it is possible to define the interval element up to a scale factor. He then adds:

> Such a definition . . . would be entirely illusory if the concept 'unit meter stick' and 'unit clock' rested on a fundamentally false assumption. This would be so if the length of a unit meter (or the period of a unit clock) depended upon their history. If this were the case in Nature there could not be chemical elements with spectral lines of determined frequencies; rather, the relative frequencies of two spatially neighboring atoms of the same kind should be, in general, different. Since this is not the case, I find I must reject the basic assumption of this theory, whose depth and boldness, however, are such as to fill every reader with admiration (1918c, p. 40)

Weyl's reply to Einstein's criticism (1918c, pp. 41–2) was to observe, first, that except for rather uncommon circumstances, there is very little that we know concerning the behavior of 'rigid' rods and clocks. (Indeed, in (1918a) Weyl had already manifested qualms concerning the notion of a rigid body in relativity theory and had consequently attempted to avoid their use in the empirical determination of the metric.) More to the point, he added that the mathematical process of vector displacement "has nothing to do with the real process of a clock's motion, whose course is determined through natural laws". The point is reiterated in (1919), and a methodological argument is introduced as further ammunition: If rods and clocks are to play the role of indicators of the metric field, it would be wrongheaded to try to use them also in order to *define* the metric field through them (p. 67).

The obvious difficulty with this answer is that it seems to rescue the theory from refutation at the price of depriving it of empirical content. If rods and clocks are not among the things with which a geometric theory is concerned, what *is* its proper domain[10]? Weyl must have felt the pressure of this problem, for in subsequent years he was led to modify his stand in a subtle but, I think, radical way. In (1920b) he introduced a fundamental distinction between two modes of physical action: persistence [Beharrung] and adjustment [Einstellung]. "A magnitude in Nature", he tells us, "can be determined either through persistence or through adjustment" and we ought to be very careful to distinguish between the two. A clear example of the pure effect of persistence is given by a rotating top, whose axis can receive any initial direction, but which will preserve the original direction under transport. A clear example of the pure effect of adjustment is offered by a compass needle moving in a magnetic field.

Weyl is now ready to reformulate the basic premise of his geometry as follows:

There is no *a priori* reason to believe that a transplantation *following the pure tendency of persistence* must be integrable, i.e., path-independent (1921b, p. 258, my italics);

and the reply to Einstein could now be reformulated in the following terms:

The length of a measuring rod is obviously determined by adjustment, for I could not give *this* measuring rod in *this* field-position any other length arbitrarily . . . in place of the length it now possesses, in the manner in which I can at will predetermine its direction. (1921c, pp. 261–2)

Thus, Weyl's theory doesn't tell us anything about what rods and clocks will do *categorically*, because the theory does not commit itself either to the presence or absence of adjustment factors. But the theory does talk about rods and clocks after all; for it does tell us, conditionally, that if adjustment factors do not operate, in all likelihood integrability will fail. Since, as Einstein noted, integrability doesn't fail, the conclusion we ought to draw is not that Weyl's theory is wrong but that there is (most likely) a natural force to which rods and clocks adjust as they move through space and time, and which explains the validity of Riemann's concordance assumption as well as its temporal and electromagnetic versions.

The ad hocish appearance of the preceding considerations to a large extent vanishes when one examines the way Weyl articulated his law of adjustment within the context of Einstein's modified relativity (with a cosmological constant). But this complex episode is no longer part of the story we set out to tell.[11]

We conclude this survey of Weyl's views *circa* 1920 with a paragraph from one of Weyl's letters to Reichenbach which conveys clearly and concisely the basic analogies which Weyl drew between his own theory and Einstein's:

Concerning my generalized theory of relativity, I cannot grant you that from an epistemological standpoint matters stand any differently in my theory than in Einstein's. To the relativity of motion corresponds the relativity of magnitude; to the unity of inertia (= force which conserves direction through infinitesimal transport) and gravitation corresponds the unity of the force which preserves length infinitesimally (and which had not yet been recognized as a force, but was located in geometry) and electromagnetism.[12]

We are now ready to take a look at Reichenbach's reaction.

4. WEYL MEETS REICHENBACH

Let me begin by presenting a record of the Reichenbach-Weyl exchanges, as I know them.

1. In (1920) Reichenbach devotes one page to Weyl's geometry and to his extension of relativity theory.

2. Reichenbach sends a copy of (1920) to Weyl who replies (on February 2, 1921) in a friendly but critical letter in which a few misunderstandings are cleared up.

3. Reichenbach briefly discusses Weyl's theory in (1921) (translated in 1959) with discussion of Weyl omitted). One of the charges raised in (1920) is now withdrawn. Weyl's views are praised for their mathematical sophistication, but Reichenbach fears that their physical aspect "comes dangerously close to a mathematical formalism which, for the sake of elegant mathematical principles, complicates physics needlessly" (1921, p. 367).

4. In a postcard from Spain (March 3, 1922) and a letter from Zürich (May 20, 1922), replying to an earlier letter from Reichenbach, Weyl reiterates his basic conception of relativity, explains why he regards alternative formulations of relativity theory as philosophically uninteresting, and gives a sketch of the mathematical treatment of the metric in the special theory of relativity via Möbius transformations, avoiding the use of rods and clocks.

5. Weyl reviews (Reichenbach, 1924) in (Weyl, 1924a). The review offers details of the spacetime metric construction in terms of light cones and inertial paths for special relativity (in the spirit of more recent constructions such as (Winnie, 1977)). Reichenbach has "succeeded in illuminating the presuppositions of the theory [of relativity] from all sides." But the fundamental question of the book, the axiomatization of relativity, "is not a philosophical question but a purely mathematical one, and it must con-

sequently be judged from a mathematical standpoint as well. From this perspective it is unsatisfactory, too complicated and too opaque. What is, properly speaking, of value, the axioms a), b) and c) and the transition from them to spacetime measurement, can be explained in a couple of pages appealing to coordinate spaces and Möbius geometry, thereby increasing the clarity and intelligibility of the whole matter" (column 2128). Weyl's opinions on this topic were later elaborated in his 1930 Göttingen lectures entitled *Axiomatik*.[13]

6. In (1925) Reichenbach introduces a distinction between deductive axiomatics and constructive axiomatics, the former being, presumably, Hilbert-style axiomatics, the latter stating as axioms "only such claims as are directly amenable to experiment." "In physics, 'axiom' means 'highest empirical proposition'" (p. 33). Referring to Weyl's review, Reichenbach urges the reader to compare his own axiomatization with Weyl's and to decide which of the two "is less 'complicated and opaque'. As for myself, I shall admit that I have always preferred the clarity of a step by step construction developed with the simplest logical operations to the opalescent mathematical fog with which some people prefer to surround their thoughts" (p. 37).

7. Finally, in (Reichenbach, 1928) Weyl's unified field theory is extensively discussed and dismissed.

Fireworks aside, both the published record and the surviving correspondence reveal a deep conflict of viewpoints concerning the correct interpretation of relativity[14]. A careful study of the nature and extent of this disagreement would be richly rewarding not only in historical insights but in philosophical clarification as well; it could not have been otherwise given the quality of the minds involved. Regrettably, this is not the opportunity for such a comparison which would demand, *inter alia*, a careful analysis of each of the documents alluded to above. However, I should like to offer at least a glimpse of what I take to be the basic point of contention between Reichenbach and Weyl: the conventionality of the affine structure. My strategy in the rest of this paper will be to let the different standpoints display themselves through their responses to two of the major conceptual problems faced by relativity around 1920: the dynamical inequivalence of motion and the geometrical interpretation of relativity.

4a. *The Dynamical Inequivalence of Motion*

Newton had argued that even though kinematics does not allow us to distinguish between different kinds of motions, dynamics does. Around 1920 it was common to describe this situation saying that whereas kinematics complies with the principle of the relativity of motion, dynamics does not. The delinquent phenomenon was often characterized as the dynamical inequivalence of motion.

Both Einstein and Weyl regarded this inequivalence as one of the basic concerns of relativity. But whereas Einstein had emphasized its dubious connection with covariance, Weyl stayed largely clear from that celebrated confusion and focused on a different aspect of the problem.

The difficulty to be solved had been clearly stated by P. Lenard, one of the most hostile critics of relativity, in his (1918). He illustrated the basic point as follows:

> Let our imaginary train undergo a markedly non-uniform motion. When, through the influence of inertial effects, everything in the train is wrecked whereas everything outside the train remains unaffected, anyone with a sound common sense will draw the consequence that it was the train and not its environment that has altered its motion with a sudden jolt. In its straightforward sense the generalized principle of relativity asserts that, even in this case, it is possible to say that it has been the environment that has experienced the change of motion, and that, therefore, the unfortunate occurrence in the train is a consequence of the jolt in the external world, mediated through a gravitational effect exerted by the external world upon the train. To the related question, why is it that the church right next to the train has not collapsed even though *it*, together with the rest of the environment, underwent a jolt – why the consequences of the jolt are *so one-sided* as to affect only the train, *even though* one is not allowed to draw a similarly one-sided conclusion concerning the location of the jolt – to this question, I say, evidently the principle can offer no satisfactory answer. (quoted in Einstein's 1918c, p. 700)

Weyl observed that there were essentially three possible solutions to this problem:

(i) Newton's (and Lenard's): Absolute space imposes its rigid Euclidean structure upon the natural course of phenomena. It prescribes the inertial, natural paths of every object, and generates inertial forces (thus creating dynamical inequivalences) whenever such paths are abandoned. It acts but cannot be acted upon.

(ii) Mach's: There is no space nor is there a space-structure that could be made responsible for dynamical inequivalences. In the domain of mechanics there is only relative motion. Thus, inertia will depend only upon the interaction of masses and is therefore determined by a form of action at a distance.

(iii) Weyl's: Spacetime is filled with a non-substantial ether (1918b, p. 15) whose states constitute the metric field. This field, in turn, imposes an intrinsic affine structure on spacetime which defines the inertial, natural paths of every object, and which generates inertial forces (thus creating dynamical inequivalences) whenever such paths are abandoned. The metric structure is not rigid but dynamic, so that it can be decisively influenced by physical objects and processes.

Einstein's stand on this trichotomy is not quite clear. Weyl says that he 'flirts' with Mach's solution (1927, p. 75; 1949a, p. 106), and there is reason to agree with him. Reichenbach, on the other hand, seems to have been a more committed supporter of the Machian line. The diversity of attitudes is revealingly illustrated in their diverse reactions to Lenard's challenge.

Einstein's reply to Lenard appeared in (1918c), where, after describing Lenard's objection, he proceeded to note that the two hypotheses contemplated by Lenard (motion of train vs. motion of environment) are not really two conflicting hypotheses but rather two forms of representation which depict the very same state of affairs, or the same 'spacetime coincidences'. "One can argue concerning which representation to choose only on grounds of simplicity, but not as a matter of principle" (p. 701). In a *footnote* to this remark Einstein adds:

> That the steeple does not collapse, according to the second form of representation, derives from the fact that it, together with the land and the whole earth, fall freely in a gravitational field (during the jolt), whereas the train is prevented from pursuing its free-fall path by external forces . . . (p. 701)

Whether Lenard's call for an explanation was justified or not, it seems clear that Einstein's reply in the text does not provide one; or, if it does, it is more in the spirit of a linguistic 'dissolution' than in that of a physical solution of the problem. As Lenard's last quoted sentence makes clear, his question was, why is it that we have a symmetry of viewpoints concerning the causal states of motion whereas we have no symmetry of dynamical effects? According to relativity, we can just as well say that the train jolts or that the rest of the world jolts; but we do not have an equal freedom of choice concerning the effects of the jolt. In language not available to Lenard, his question was, oughtn't the symmetry of causes be reflected in a symmetry of effects? Einstein's reply simply takes for granted that the asymmetry obtains (as, of course, it does) and does not acknowledge the need for any further explanation: under certain states of relative motion only one of the bodies exhibits non-inertial processes, and that is all there is to it. Einstein's footnote, on the other hand, offers the sketch of an explanation for the asymmetry, except for the odd fact that the remark is made to depend upon one mode of description or coordinatization.

To my knowledge, Reichenbach did not explicitly discuss Lenard's example; but it is not difficult to guess what form his solution would have taken.

In his study of the problem of motion in Leibniz and Huyghens, Reichenbach (1959) contends that "the decisive answer to Newton's argument concerning centrifugal force was given by Mach" through his interpretation of centrifugal force as "a dynamical effect of gravitation produced by the rotation of the fixed stars" (p. 50). Lacking Mach's argument "which alone can defend the relativity of motion," Leibniz was unable to support his relativistic viewpoint against Clarke's appeal to dynamical inequivalence. Leibniz was consequently forced to grant that 'true' motion is detectable dynamically (p. 57). Huyghens' analysis, on the other hand, unsatisfactory as it might be on purely physical grounds, "represents a consistent interpretation of dynamic relativity, of the idea that even from the occurrence of forces, only a relative motion of bodies can be inferred" (p. 65).

The Machian character of Reichenbach's solution becomes clearer when we consider Reichenbach's discussion of the relativity of motion in section 34 of (1928). At one point his discussion seems aimed at a Lenard-type objection:

> Relativity can be extended to dynamics if forces are reinterpreted relativistically. The same force that affects a body K_1 as the result of the rotation of K_1 according to one interpretation, affects it according to the other interpretation as the result of the rotation of K_2. We thus arrive at a complete reinterpretation of the concept of force . . . Forces are not absolute magnitudes but depend on the coordinate system. (1958, p. 214)

As applied to Lenard's case (with K_1 = the train and K_2 = the rest of the universe) what this tells us is essentially what Einstein had explained in the text of his reply. The same force that, under the first interpretation, affects Lenard's train as a result of *its* jolt, affects it, under the second interpretation, as a result of the jolt in the rest of the universe. Under the first interpretation we say that inertial forces affect the moving object; under the second one we say that they affect the object at rest. In consequence no absolute correlation can be drawn between inertial forces and the states of motion of their bearers, for the correlation is dependent upon coordinatizations.

Weyl's response to Lenard appears in (1920a) and takes a radically different form. Newton, he tells us, makes absolute space responsible for the dynamical inequivalence; the special theory of relativity offers a solution "which is not essentially different from Newton's" in that it also assigns the responsibility for the asymmetry to a *rigid* metric structure. The decisive improvement of general relativity over these two solutions lies in the fact that it recognizes the reality and concreteness of the metric and affine structures and their ability to be acted upon as much as they act.

> The antinomy disclosed by the case of the train-collision is to be solved precisely in the way in which common sense has always solved it: the train, but not the church, is deviated from the natural motion determined by the guiding field in virtue of the action of molecular forces arising from the collision. What is new [i.e., with respect to the Newtonian and special relativistic solutions] is that we no longer regard the guiding field as *a priori* given, but we search for the laws according to which it is engendered by matter. (1920a, p. 135)

Elsewhere he writes,

> The contradiction between the principle of the relativity of motion and the existence of inertial forces appears only if we consider the world as an unstructured manifold. But as soon as we realize that it is endowed with a guiding field, with an affine connection, the difficulty vanishes. (1923a, p. 221)[15]

Thus, the explanation of dynamical inequivalences lies in an asymmetry of causes. It is a fact, a coordinate-free fact, if one may say so, that the deviation from a natural inertial path involves the generation of forces such as the centrifugal force in Newton's bucket and the inertial forces in Lenard's train. The reason why the train suffers the consequences of the jolt and the church does not is *not* their state of motion relative to the fixed stars, but the fact that both train and church are moving (in spacetime) in a *highly inhomogeneous* medium, the ether which fills spacetime and whose states constitute the metric structure.

This explanation, Weyl thought, is the only real alternative to Newton's. For one can show that the Machian explanation is untenable even on its own grounds; in particular, one can show in agreement with Machian principles that there is no objective distinction between states of relative rest and relative motion. Weyl's argument for this claim appears in a variety of places (such as 1924d and 1949a. pp. 104–6) and proceeds, roughly, as follows. The objective features of the Universe coincide with those that are invariant under its automorphisms, that is, under those transformations that leave the 'ultimate furniture' of the Universe invariant. Now, the Machian assumes that as we list these ultimate elements all we find besides masses and charges is states of relative motion. If so, an automorphism will take the Universe from an arbitrary state of motion to a state of universal mutual rest. In order to illustrate the basic idea Weyl depicts spacetime as a mass of plasticine with the worldlines corresponding to its fibers. If the Universe had no more basic structure than the Machian allows, a continuous transformation carrying all plasticine fibers to a vertical position (state of relative rest) would be an automorphism. Consequently the idea of relative motion would not be objective. It follows that

> independently of the metric field, the relative motion of objects in the Universe is pure Nothing. [If the Machian assumption] were correct, processes in spacetime would only depend upon and be determined by the charge and the mass of elementary particles. Since this is obviously absurd . . . one must give up the [Machian] causal principle. (1924b, p. 479)

And elsewhere Weyl adds,

> No solution of the problem [of dynamical inequivalence] is possible as long as in adherence to the tendencies of Huyghens and Mach one disregards the structure of the world. But once the inertial structure of the world is accepted as the cause for the dynamical inequivalence of motions, we recognize clearly why the situation appeared so unsatisfactory. We were asked to believe that something producing such enormous effects as inertia . . . is a rigid geometrical property of the world, fixed once and for all. (1927, p. 74; 1949a, p. 105)

4b. *Pictures vs. Facts*

Reichenbach's (1928) concludes with an appendix entitled 'Weyl's Extension of Riemann's Concept of Space and the Geometrical Interpretation of Electromagnetism' (1928, pp. 331–373). This discussion of Weyl's unified field theory is of considerable intrinsic value; our main interest, however, relates to the light it sheds on Reichenbach's attitude towards affine structures, a subject glaringly absent from the body of his (1928). Since the English translation of (1928) does not include this appendix, I shall begin by presenting an outline of Reichenbach's considerations.

Reichenbach's criticism of Weyl's unified field theory is radically different from Einstein's. Einstein, you may recall, had contended that Weyl's theory is false in view of the validity of Riemann's concordance assumption. Weyl's response was that his theory has nothing (categorical) to say about measuring devices. At this point Reichenbach thought that Weyl had opened his flank to a philosophical objection. But instead of limiting himself to the purely negative task of showing the lack of additional content in Weyl's theory, Reichenbach chose to engage in the much more ambitious enterprise of undermining the whole Weylian conception of a unified field theory.

A unified field theory, à la Weyl, is a theory that incorporates gravitation and electromagnetism within a unified theoretical framework, and in which no appeal is made to concepts beyond the domain of pure geometry. Its basic aim is to perform an 'ideological' reduction of physics to geometry. The enterprise seems majestic, the goal unachievable. On the contrary, Reichenbach tells us, nothing is easier or less significant.

In order to explain how this goal can be achieved we must first turn to the domain of geometry. Weyl had recognized that a reduction of all physics to geometry demands an extension of the Riemannian framework of ideas. Reichenbach agrees and therefore starts by considering the different ways in which such extensions could be carried through.

The basic elements in a manifold are two: the metric and the affine connection. Engaging in a form of philosophical license, Reichenbach interprets parallel transport as inevitably involving the idea that lengths are preserved under such transport, even in the absence of a metric. If so, a metric and an affine connection may well conflict in their metric judgements. There is in principle nothing wrong with this, but Reichenbach prefers to exclude those cases in which discrepancies arise.

Under such circumstances there are essentially no more than two ways to construct a geometric manifold of the most general kind, depending on whether we let the metric or the affine connection play the dominating role. The first procedure starts with an arbitrary Riemannian metric and adjusts the (not necessarily torsion-free) affine connection to its dictates; the second one starts with an arbitrary affine connection and trims-off the metric into consistency with it. Manifolds of the first sort are called 'metric spaces', those of the second sort are 'displacement spaces.'

The construction of a metric space offers no difficulties, and we shall give an example of this sort in a moment. Displacement spaces, on the other hand, are somewhat more involved. It is best to describe their construction in three separate stages. In stage one we choose an arbitrary affine connection which we use in order to define length identity between parallel but distant vectors, relative to a given path; in stage two we introduce a method to compare lengths between both parallel and non-parallel vectors *at a point;* finally, in stage three, we determine how to compare the lengths of non-parallel vectors *at different points.* Let us examine each of these steps separately.

Stage one: Given a vector $\mathbf{u}$ at (the tangent space of) P and a path c from P to Q ($P \neq Q$) there is a uniquely determined vector $\mathbf{v}$ at Q that is parallel to $\mathbf{u}$ via c. We say that $\mathbf{v}$ has the same length as $\mathbf{u}$ via c and that all other vectors (except for $-\mathbf{v}$) on its straight line do not have the same length.

Stage two: The intended goal could be accomplished by simply defining a metric at each tangent space. Reichenbach prefers to endow the space with a standard Riemannian metric and to neutralize the excess information it contains by preserving only such information as remains invariant under multiplication of the metric by an arbitrary place-dependent factor $\lambda(x_1, \ldots, x_n)$. We are now in a position to compare lengths at each point since either characterization gives us the congruence classes at each point plus a linear ordering of such classes.

It is worth pausing for a moment to observe that if we had not added the qualification '$P \neq Q$' in our first stage we could now be involved in contradiction. Because when **u** (at P) is parallel-displaced around a closed path c, it may end up coinciding with a vector **v** (also at P) whose length may differ from that of **u** according to the metric chosen at P. Thus, strictly speaking, the affine connection is not entirely dominant in this construction but adjusts itself, if only slightly, to the pointwise metrics. To put it differently, it isn't really true that *every* case of parallel displacement is assumed to preserve length, but only parallel transport from one point to *another*.

This consideration sets in relief the need to relativize congruence judgements to paths. For, take a closed path c such as the one described in the preceding paragraph, and let Q be a point on c other than P. Call c_1 a portion of c from P to Q. and call c_2 the rest of c. Then there will be a vector **w** at Q, parallel to both **u** (via c_1) and **v** (via c_2) which must have the same length as **u** (via c_1) and the same length as **v** (via c_2). If we ignore the relativizing clauses we could now derive a contradiction.

Stage three: Let **u** at P coincide with **v** at Q via c and let **w** be a vector at Q in a direction different from that of **v**. Our problem now is to decide how to compare the lengths of **u** and **w** via c. We first choose an order between P and Q, say $\overline{PQ}$, then we transport **u** to Q via c and we finally rotate the tangent space at Q so that **v** goes to a multiple **t** of **w**. The length ratio between **u** and **w** is the same as that between **t** and **w**.[16]

Notice that in this third stage we have defined only the length ratio between **u** (at P) and **w** (at Q) via c and *in the direction from P to Q*. The italicized relativization is, once again, essential; for if **w** is transported to P via c, it may end up coinciding with a vector at P whose length ratio to **u** is different from that between **t** and **w**. This phenomenon is what Reichenbach calls the 'asymmetry' of length comparisons in general displacement spaces. We have now completed the characterization of a displacement space.

Weyl's spaces are obtained by imposing two restrictions on displacement spaces. The first one is that the affine connection must be symmetric or torsion-free. Weyl had assumed as a matter of course that this should be the case, since he had shown that this condition was necessary and sufficient for the existence of geodesic coordinate systems at each point (or, as Wheeler puts it, for the validity of the principle of equivalence). The second restriction is that length comparison should be symmetric in Reichenbach's sense, so that all we need give in order to fix the length-relation between two vectors unambiguously is the path which connects them.

Whereas Weyl's unified field theory is an application of Reichenbach's displacement spaces, Reichenbach's unified field theory is developed following the pattern of his metric spaces. Let us see how this is done.

To begin with, we borrow the metric from relativity and allow it to explain gravitation as in Einstein's theory. We now seek to introduce an affine structure which will provide an explanation of electromagnetism. In a nutshell, the idea is to construe geodesics not as the paths of freely falling objects, but as the paths of freely falling or charged unit-mass particles. Since these geodesics will not, in general, coincide with those defined by the relativistic metric, and since Reichenbach insists that the affine connection should be compatible with the metric, the intended goal cannot be achieved through a torsion-free connection. For Weyl's 'Fundamental Theorem' of infinitesimal geometry had shown that there is a unique symmetric affine connection compatible with the metric (for each infinitesimal manifold) and with this affine connection the affine-geodesics coincide with the metric-geodesics.

The strategy in agreement with which electromagnetism is stuffed into the affine connection is relatively straightforward. In spacetime notation the components of the Lorentz force acting on a particle of unit mass and charge e are given by

$$eF^k_j \frac{d\alpha_j}{dt}$$

where the F^k_j's are the components of the electromagnetic field tensor and the $d\alpha_j/dt$'s are the components of the four-velocity of the given particle. Since freely falling particles move along geodesics, their motion will satisfy the geodesic equation

$$\frac{d^2\alpha_k}{dt^2} + \Gamma^k_{ij} \frac{d\alpha_i}{dt} \frac{d\alpha_j}{dt} = 0$$

Thus, according to general relativity, the motion of the unit mass charged particle will be given by a function $\alpha(t)$ satisfying

$$\text{(6)} \qquad \frac{d^2\alpha_k}{dt^2} + \Gamma^k_{ij} \frac{d\alpha_i}{dt} \frac{d\alpha_j}{dt} + eF^k_m \frac{d\alpha_m}{dt} = 0$$

Reichenbach's idea is simply to look for an affine connection compatible with the metric, and which will turn (6) into a geodesic equation. *This* will be the affine connection which we impose upon the spacetime manifold.[17]

We are now ready to state Reichenbach's unified field theory. It says, in effect, that spacetime is endowed with both a metric and an affine structure. The metric is Einstein's,[18] and the affine structure is the one we have just described. The metric explains gravitation and the affine connection 'explains' electromagnetism. Parallel transport preserves length, but the straightest and the shortest lines need not coincide. In the absence of electromagnetic fields or charges they will, in fact, coincide; but in the presence of electromagnetic effects the affine geodesics will define the paths of charged unit masses which will differ from those of freely falling particles. It is to this extent that physics has been geometrized and that the fields have been unified.

Of course, nothing could be further from Reichenbach's mind than to offer a theory such as this for serious consideration as a physical theory. His main aim, as we said, was to exhibit the vacuity of Weyl's enterprise. Reichenbach's example is intended to make obvious what is hidden in Weyl's more complicated construction. For it is clear that in Reichenbach's theory the affine structure "is no more than the geometrical clothing for the law (6), an illustration but not a new physical thought" (1928, p. 365). That is, in fact, the basic problem with Weyl's theory: it is an illustration, a pretty picture which conveys no new physical thought. The fact that Weyl has failed to notice this is evidence of his inability to distinguish between the 'clothing' and the 'underlying body' of a physical theory, between the mathematical trappings and the empirical content[19]. And it is this very inability that vitiates Weyl's understanding of Einstein's theory.

Reichenbach must have thought it ironic that Weyl of all people, the champion of invariance as a criterion of objectivity, had failed to be sensitive to the collapse of his geometric picture of relativity in alternative but equivalent versions of that theory. The fact that there are alternative versions of relativity in which the metric and affine structures differ from those in the standard version should have led Weyl to recognize the irreality of those features not invariant under different versions. Or, to put it differently, it should have led him to recognize the conventionality of such factors. It may, indeed, be pleasing to the mind to produce poetic images such as Weyl's allusion to the struggle between forces and inertia and his account of how Einstein expelled gravitation from the domain of forces, but

> we can always insist in applying the force-picture which, before Einstein, we associated with point-masses; it is quite without meaning to say that this intuitive representation in terms of forces of attraction is *false*. (1928, p. 353)

When the conventional is sifted-out from the factual, we come to see that the heart of Einstein's geometric interpretation of gravitation resides in the mediating role that Einstein assigns to measuring instruments: as gravitational forces influence rods and clocks they, at the same time, display the metric properties of space and time:

> The geometric interpretation of gravitation is the expression of a real state of affairs: the real influence of gravitation on rods and clocks; here lies its physical value . . . All else is the product of phantasy, mere illustration. (1928, p. 352)[20]

Weyl never replied to Reichenbach's arguments. By the time these were published the development of quantum mechanics had led him to abandon his hopes for a unified field theory and to move into entirely new domains of physics. It is not difficult to see, however, to what extent he would have been unmoved by Reichenbach's criticism. In fact, in (1927) Weyl had already considered and dismissed an attempt to infer the conventionality of the metric from alternative metrizability. More than likely he would have acted accordingly with regard to the affine structure.[21]

If pressed to give his gut-reasons for dismissing the merits of equivalent formulations as an argument for conventionalism, my guess is that Weyl would have appealed to his view of science as the search for the transcendent. Spacetime coincidences are, indeed, all that is ever given to us in intuition (1931, p. 336), but we cannot seriously contend that such coincidences are all that science is concerned with.

> It has been said that physics is concerned only with the determination of coincidences . . . But, in all honesty, one must grant that our theoretical interests do not relate exclusively or even primarily to the 'real assertions' [i.e., those which can be realized in intuition] such as the observation that this needle coincides with this portion of the scale; rather, they relate to ideal postulations that *according to theory* display themselves in such coincidences, but whose meaning cannot be completely given in any given intuition (1928, p. 149)

If we believe (as Einstein did in 1918) that "laws of nature are assertions concerning spacetime coincidences" (Einstein, 1918a, p. 241) then Reichenbach's 'equivalent descriptions' will appear to be truly equivalent. But when one realizes that laws are also about a transcendent domain of things far removed from empirical intuition—such as the metric or the affine structure – the temptation to regard those descriptions as equivalent vanishes. In this conviction, Weyl would add, metaphysical speculation and scientific practice agree:

> The possibility must not be rejected that several different constructions might be suitable to explain our perceptions . . . [However] Einstein described the real epistemological situation with great justice as follows: 'The historical development has shown that among the imaginable theoretical constructions there is invariably one that proves to be unquestionably superior to all others. Nobody who really goes into the matter will deny that the world of perception determines the theoretical system in a virtually unambiguous manner' (1927, p. 113; 1949a, p. 153)

In Weyl's eyes, the superiority of his geometric picture over its 'equivalent' rivals was rendered obvious by its ability to solve a number of major physical and philosophical problems, such as that of the dynamical inequivalence of motion and the identity of inertial and gravitational mass. In contrast, Reichenbach's fictitious affine structures explain nothing, since even as we change the mass of a charged particle we are led to deviations from the alleged geodesic paths. Not behavior vis-à-vis spacetime coincidences but explanatory power is the mark that distinguishes those theoretical pictures which depict the features of the transcendent domain from their artificially constructed rivals.

5. CONCLUSION

Reichenbach was not only a metric conventionalist but also an affine conventionalist; and for essentialy the same reasons. *Regardless of the features of the manifold,* he thought, the question of the truth or falsehood of the attribution of an affine structure to a manifold makes no sense until after a coordinative definition is given. The purpose of that definition is to identify the processes that are to count as parallel displacements. The ascription of an affine structure to a manifold is therefore always a matter of convention, so that it is absurd to talk of a particular affine connection as if it were the only one adequate to a given manifold.

Weyl, on the other hand, thought that *depending on the features of the manifold,* we might not need to introduce a coordinative definition; that, so to speak, Mother Nature might sometimes do the choosing on our behalf. Of course, Weyl knew that every manifold has a multitude of possible affine connections; as we saw, he was largely responsible for the ideas that led to an understanding of this circumstance. His point was that in spite of the "alternative connectability"[22] of every manifold, some spaces, in virtue of their intrinsic features, select from the class of logically possible affine connections a single one. Spacetime is one such manifold, for one of the deepest and simplest of physical phenomena, the one revealed through Galileo's laws of free-fall and inertia, indicates that the intrinsic features of the world select from all possible sets of paths a single set which is physically realized by freely falling bodies, regardless of their constitution.

It is on this issue that the basic disagreement between Reichenbach and Weyl becomes clearest; not so much on the subject of the metric but on the affine structure. Weyl's views on the metric are uncommonly obscure[23]; some of the things he says about it indicate conventionalist leanings, others indicate the exact opposite. But his views on the affine structure are unequivocally anti-conventionalist. A physically real affine structure, in no way dependent on human convention, is at the heart of his version of the theory of relativity. With it, we can bring to completion centuries of speculation on the nature of space; without it, we are left with a Machian, incoherent picture of the Universe. For Reichenbach, on the other hand, Weyl's stand on this matter was what his friend Carnap would have called "a confusion due to the material mode of speech," a confusion between features of the formulation of certain facts and features of those facts. Once the conventionality of affinities is recognized, Reichenbach thought, both Weyl's picture of relativity and his unified field theory are seen for what they are: not new insights into the structure of the Universe, but misleading representations of well known facts.

It would be pleasant to be able to conclude the story of this conflict by announcing a winner. The debate is certainly still with us. Reichenbach's tradition is pursued in the works of Grünbaum, Salmon and their followers; Weyl's tradition is enjoying a most welcome revival under the influence of the 'invariant' approach to relativity, particularly (within philosophy of science) in the work of Winnie and others. But the lines can no longer be drawn as neatly as before. For the Weylians are more sensitive than Weyl ever was to issues of conventionalism, and the Reichenbachians no longer think—as Reichenbach seems to have thought—that under no conceivable circumstances could the spacetime metric or its affine structure be placed entirely outside the field of convention. Today few would dare to minimize, as Weyl did, the role of convention in theoretical knowledge or the relevance of equivalent descriptions to ontological issues. On the other hand, no one seems to dispute any longer the relevance of empirical and theoretical findings on the geodeticity of free fall to the issue of the intrinsicality of the affine structure. Who won? We did.

Indiana University

NOTES

[1]I should like to express my gratitude to Adolf Grünbaum, who awakened my interest in this subject and helped me understand some of Weyl's most obscure pronouncements, and to John Winnie, who convinced me of the philosophical relevance of Weyl's 'invariant' approach. Whether they like it or not, they are largely responsible for the standpoint from which this paper is written. My thanks also to Linda Wessels and her two daughters, who helped to improve the quality of this paper; to NSF for generous support; and to Mrs M. B. Weyl for permission to quote from Weyl's letters.

[2]In (1928) Russell formulated an idea similar to the one which inspired Weyl's researches in his Principle of Differential Laws: "Any connection which may exist between distant events is the result of integration from a law giving a rate of change at every point of some route from one to the other." (p. 101)

[3]For details see, e.g., Chapter II of (1923a) or (1922a). A note concerning references: Weyl's papers are reprinted in (Weyl, 1968) which is readily accessible; consequently, all page references to Weyl's *papers* are to this source rather than to the original publication. Thus (Weyl, 1918b, p. 6) refers to a remark on p. 6 of the reprint of (Weyl, 1918b) in volume II of (Weyl, 1968).

[4]For an illuminating account of the intuition behind this terminology see (Cartan, 1924, pp. 865–6)

[5]The new coordinate-free way of saying things is certainly clearer and avoids all conflict with 'static' set-theoretic intuitions. For example, parallel transport is treated as follows. An affine connection is a function ∇ which is symmetric, Leibnizian and additive. This function associates with any two vector fields **U** and **V** on the manifold another function $\nabla_{\mathbf{U}}\mathbf{V}$, such that for every point P, $\nabla_{\mathbf{U}}\mathbf{V}(P)$ is the derivative of **V** in the direction of $\mathbf{U}(P)$. As usual, a dynamic picture is now replaced by a static one: instead of talking about the displacement of a vector **v** from P to each of the points in a neighborhood N of P, we talk of a vector field **V** defined on N and such that $\mathbf{V}(P) = \mathbf{v}$ (where **V** may be conceived as the composite photograph of all the end-results of displacements of **v** within N). The task of distinguishing between those displacements that are parallel and those that are not becomes the task of distinguishing between certain vector fields and others. And the claim that displacements of **v** are parallel *infinitesimally*, or *at P*, becomes the claim that for every vector $\mathbf{u}_p$ at the tangent space of P (i.e., for every direction at P) the directional derivative (change) of the vector field **V** estimated at P is null (i.e., as **V** 'moves' away from P in every direction, the field remains initially parallel to **v**). Thus, to say that **V** is parallel at P simply means that for every $\mathbf{u}_p$ in the tangent space of P, $\nabla_{\mathbf{u}_p}\mathbf{V} \mid_P = 0$. It is a simple excercise to show that the coordinate expression for this invariant formula is Levi-Civita's (1), where the ξ^i's are now properly interpreted as follows: take any neighborhood N of P and coordinatize it; the ξ^i's are functions from N into the reals defining the components of **V** relative to the coordinate frame field.

[6]Einstein was not very impressed by this argument. In a letter to Weyl of June 1918, Einstein raised a 'slippery slope' objection: "Could one accuse the Lord of inconsequent behavior if he had declined to avail himself of the methods which you have discovered, in order to harmonize the physical world? I do not think so. If he had built the world according to you, Weyl II would have come to tell him reproachfully: 'Dear God, since you, in your wisdom, have chosen not to give an objective meaning to the congruence of infinitely small rigid bodies, so that when they are a finite distance appart from each other one cannot say whether they are congruent or not; why haven't you, Oh Incon-

ceivable!, chosen not to disdain also depriving angles and similarities of this property? If two infinitely small and initially congruent bodies K and K' are no longer congruent after K' has made a trip around space, why should the similarity between K and K' have been preserved through this trip? It seems more natural that the change of K' relative to K should not be affine but of some more general kind.' But the Lord has pointed out well before the development of theoretical physics that he will not be led by the opinions of this world, and that he will do as he pleases" (Seelig, 1952, p. 160). See also Weyl's recollections of his dialogues with Einstein on this matter in (Seelig, 1952, pp. 156–7).

[7]At the meeting of July 10, 1922 of the Academie des Sciences, Cartan reported that Weyl's conjecture (which he described as being of "great philosophical significance") was true for arbitrary *n*. His proof was published in (1923), a year after Weyl's (1922b). Both Cartan's and Weyl's papers are extremely involved. Cartan does not seem to have thought very highly of Weyl's proof; he dryly refers to Weyl's acknowledgement that, at points in his argument, he feels that he is dancing on the tight rope.

Laugwitz noted in (1958) that Weyl's problem, as stated in the 1919 edition of (1923b) differs from the group-theoretic formulation stated and solved in (1922b) and (1923c). Freudenthal (1960, 1964) acknowledges the difference but notes that the methods developed by Weyl in (1923c) to solve his group-theoretic problem provide also a solution to the original one. Scheibe (1957) includes another proof of Weyl's theorem (bridging gaps that he detects in both Weyl's and Cartan's arguments). For a reasonably uninvolved account of an aspect of Weyl's theorem see (Laugwitz, 1965, Chapter IV).

[8]See, e.g., (Weyl, 1920a, pp. 136–7).

[9]"Modern physics renders it probable that the only fundamental forces in nature are those which have their origin in gravitation and in electromagnetism" (Weyl, 1921c, p. 260).

[10]Light and free particles would, presumably, *define* the metric, so that the theory would not convey information about them either.

[11]See, e.g., (1923a, §40). By the late 20's Weyl had abandoned his hopes for a unified field theory, whereas Einstein had decided that the project was worth pursuing (Weyl, 1929c, 1929b). But Einstein would move in a direction quite opposite to Weyl's; rather than removing the integrability condition on congruent transport, Einstein assumed that parallel transport was path independent! For Weyl's shocked reaction see (1929b, p. 235; 1931, p. 343).

[12]Letter of February 2, 1921. I am grateful to Prof. Maria Reichenbach for allowing me to consult Weyl's letters to Reichenbach and for permission to quote from them.

[13]A copy of these notes is available at the Mathematics Library, Institute for Advanced Study, Princeton. I am grateful to Mrs. V. Radway for her help in connection with Weyl's lecture notes.

[14]It isn't clear who (if anyone!) had Einstein's *nihil obstat*. Early on, at any rate, he seems to have been pleased with Reichenbach's philosophy of relativity. See, for example, Einstein's supportive review of (Reichenbach, 1928) in (Einstein, 1928). Furthermore, Einstein never wavered in his unqualified rejection of Weyl's unified field theory. This reaction may be explained, in part, by Einstein's extreme hostility towards Weyl's Cartesian-Cliffordian vision. His only comment on Reichenbach's discussion of Weyl's theory in (Reichenbach, 1928) is to applaud Reichenbach's thesis "in my opinion, entirely justified, that it is untenable to hold that the theory of relativity is an attempt to reduce physics to geometry" (Einstein, 1928, column 20). For further (amusing) evidence of Einstein's irritation against geometricism, see (Ortega, 1969, vol. 5, p. 268).

[15]See also (1950, p. 426).

[16]Reichenbach's tacit appeal to free mobility in the tangent space is appropriate under the assumption that we are dealing with infinitesimally Euclidean spaces. But already in specetime this assumption collapses; it would have been better to say simply that the length ratio between **u** and **w** is the one which the metric at Q assigns to **v** and **w**.

[17]Reichenbach's affine connection is given by the components

$$\Gamma^k_{ij} = -\{^{\,k}_{ij}\} - g_{ir} F^k_j \frac{\partial F^{rs}}{\partial \alpha_s}.$$

Appealing to Maxwell's second system of equations and to the demand (obscurely stated by Reichenbach) that the affine connection should be compatible with the metric, Reichenbach shows that the geodesics of this affine connection satisfy (6).

[18]Actually, Reichenbach introduces a metric tensor which is the addition of Einstein's g_{ik}'s plus the components of the electromagnetic field tensor. Since this latter tensor is antisymmetric, its effects on the computation of the length element is nil.

[19]It was this very distinction between the clothing and the body of a theory that lay behind their first public disagreement concerning the axiomatization of relativity. See (Weyl, 1924a; Reichenbach, 1925) and Weyl's Göttingen lectures on Axiomatization at the Institute for Advanced Study, Princeton.

[20]At one point in the text of (1928) Reichenbach had addressed himself, if obliquely, to Weyl's picture of relativity. Although Weyl is never mentioned by name, it is not hard to see that he is the target of this remark: "According to Einstein's theory, we may consider the effects of gravitational fields on measuring instruments to be of the same type as known effects on forces. This conception stands in contrast to the view which interprets the geometrization of physics as an exclusion of forces from the explanation of planetary motion" (1958, p. 258).

[21]The argument (written, no doubt, with Reichenbach in mind) takes the following form. We are given two interpretations of Schwarzschild's solution, both designed to preserve the Euclidean character of space at the price of assigning to measuring instruments a highly non-standard behavior. Of these two distinct renderings of the facts Weyl tells us that "each introduces into the factual state of affairs an arbitrary element which has no perceptually confirmable consequences, which therefore must be eliminated . . . Each of the two theories can, if properly formulated, be split into two parts: the theory (E) of Einstein, and an addition (A) which is neither connected with (E) nor touching reality, and which must therefore be shed" (1927, p. 84; 1949a, pp. 118–9). At this point an issue of consistency arises within Weyl's philosophy; for if theories are the Quinean wholes that Weyl believes them to be, how are we to detach from the Euclideanized version of relativity its 'illegitimate' or content-free portion (A)? To put it differently, how do we make Quinean holism consistent with anti-Reichenbachian realism? (For evidence of Weyl's holistic views, see 1922a, pp. 60, 67, 93; 1928, p. 149; 1929a, p. 157; 1932, p. 78; 1934, pp. 44–6; 1946, p. 603; 1949a, pp. 132, 151–4; 1949b, pp. 309, 332; 1953, p. 534.)

[22]The expression is Salmon's. See his (1977), where he discusses the relevance of alternative connectability to the conventionality of the affine structure. In that paper Salmon wonders whether one could reasonably accept (1) that physical space has no intrinsic structure which determines whether two disjoint space-stretches are congruent or not, and (2) that physical space has an intrinsic structure that determines parallel transport. As we saw, Weyl certainly upheld the spacetime version of (2), and he probably appealed to (1) as a motivation for his metric geometry. Weyl's reasons for his acceptance of

(1) are unclear (as is his whole metric philosophy) but his arguments for (2) are, I think, straightforward. Provided that we are ready to allow for the *conceivability* of manifolds with intrinsic affine structures, we must be prepared to recognize criteria for the intrinsicality of such structures (otherwise affine conventionalism would become a logical thesis and, consequently, empty). Weyl's remark quoted on page 279 can be readily interpreted as a reasonable proposal for a criterion of the intended kind: the spacetime manifold has an intrinsic affine structure if given any point (event) and any direction (4-velocity) at that point there is a unique line through that point and in the given direction such that all objects *not subject to external forces* will have it as their world line. The italicized clause poses familiar problems, but we must recall how thin the line is that divides loose but acceptable qualifications from content-draining provisos. If Weyl's criterion is accepted, the state of knowledge in the 20's supported his conclusion that spacetime has an intrinsic structure. Whether our *current* state of knowledge would justify that conclusion is a very different matter (for a thorough analysis of the relevant facts, see Chapter 22 of Grünbaum, 1973).

[23]See, for example, Weyl's extremely puzzling views concerning the recovery of free mobility (homogeneity) for spacetime even in the presence of an inhomogeneous metric structure (1922a, p. 98; 1923c, p. 44; 1949a, p. 87), and his opinions on the relation between visual form, material content and congruence (1922a, pp. 99–101, wth changes in different editions). Compare also Einstein's concluding complaint in his otherwise remarkably enthusiastic review (Einstein, 1918).

BIBLIOGRAPHY

Cartan, Elie, 1923, 'Sur un théorème fondamental de M. H. Weyl dans la théorie de l'espace métrique,' *Journal des Mathématiques Pures et Appliqués* **9**, 167–192.

Cartan, Elie, 1924, 'Les recents généralisations de la notion d'espace,' *Bulletin des Sciences Mathématiques* **48**, 294–320.

Einstein, Albert, 1918a, 'Prinzipielles zur allgemeinen Relativitätstheorie,' *Annalen der Physik* **55**, 241–244.

Einstein, Albert, 1918b, Review of Weyl (1918), *Die Naturwissenschaften* **25**, p. 373.

Einstein, Albert, 1918c, 'Dialog über Einwände gegen die Relativitätstheorie,' *Die Naturwissenschaften* **48**, 697–702.

Einstein, Albert, 1928, Review of (Reichenbach, 1928), *Deutsche Literaturzeitung* **1**, columns 19–20.

Feudenthal, Hans, 1960, 'Zu den Weyl-Cartanschen Raumproblem,' *Archiv der Mathematik* **11**, 107–115.

Freudenthal, Hans, 1965, 'Lie Groups in the Foundations of Geometry,' *Advances in Mathematics* **1**, 145–190.

Grünbaum, Adolf, 1973, *Philosophical Problems of Space and Time*, 2nd edition, Reidel, Holland.

Lenard, P., 1918, *Ueber Relativitätsprinzip, Aether, Gravitation*, Leipzig.

Laugwitz, Detlef, 1958, 'Ueber eine Vermutung von Hermann Weyl zum Raumproblem,' *Archiv der Mathematik* **9** 128–133.

Laugwitz, Detlef, 1965, *Differential and Riemannian Geometry*, Academic Press, New York.

Levi-Civita, Tulio, 1917, 'Nozione di Paralelismo in una Varietà Cualunque e Conseguente Specificazione Geometrica della Curvatura Riemanniana,' *Rendiconti del Circolo de Palermo* **42**, 173–215.

Ortega y Gasset, José, 1965, *Obras Completas*, Revista de Occidente, Madrid.

Russell, Bertrand, 1928, *Analysis of Matter*, Dover, New York.

Reichenbach, Hans, 1920, *Relativitätstheorie und Erkenntnis A Piori*, Springer, Berlin.

Reichenbach, Hans, 1921, 'Der gegenwärtige Stand der Relativitätsdiskussion,' *Logos*, **10**, 316–378; translated (with discussion of Weyl's work omitted) in (Reichenbach, 1959).

Reichenbach, Hans, 1924, *Axiomatik der relativistischen Raum-Zeit-Lehre*, Vieweg, Berlin.

Reichenbach, Hans, 1925, 'Ueber die physikalischen konsequenzen der relativistischen Axiomatik,' *Zeitschrift für Physik*, **34**, 34–48.

Reichenbach, Hans, 1928, *Philosophie der Raum-Zeit-Lehre*, Vieweg, Berlin.

Reichenbach, Hans, 1958, *The Philosophy of Space and Time*, Dover, N.Y.

Reichenbach, Hans, 1959, *Modern Philosophy of Science*, Routledge, London.

Reichenbach, Hans, 1965, *The Theory of Relativity and A Priori Knowledge*, University of California Press, California.

Salmon, Wesley, 1977, 'The Curvature of Physical Space,'' in Earman, Glymour and Stachel (eds.), *Foundations of Space-Time Theories*, University of Minnesota Press, Minnesota, pp. 281–302.

Scheibe, E. 1957, 'Ueber das Weylsche Raumproblem,' *Journal für Mathematik*, **197**, 162–207.

Seelig, Carl, 1952, *Albert Einstein und die Schweiz*, Europa Verlag.

Weyl, Hermann, 1918a, *Raum-Zeit-Materie*, lst edition, Springer, Berlin.

Weyl, Hermann, 1918b, 'Reine Infinitesimalgeometrie,' *Mathematische Zeitschrift*, **2**, 384–411; also in Weyl, (1968), vol. II. pp. 1–28.

Weyl, Hermann, 1918c, 'Gravitation und Elektrizität,' *Sitzungsberichte der Königlich Preussichen Akademie der Wissenschaften zu Berlin*, pp. 465–480; also in Weyl (1968), vol. II, 29–42.

Weyl, Hermann, 1919, 'Eine Neue Erweiterung der Relativitätstheorie,' *Annalen der Physik* **59**, 101–133; also in Weyl (1968), vol. II, pp. 55–87.

Weyl, Hermann, 1920a, 'Die Einsteinsche Relativitätstheorie,' *Schweizerland*, 1920; also in Weyl, (1968), vol. II, pp. 123–140.

Weyl, Hermann, 1920b, 'Elektrizität und Gravitation,' *Physikalische Zeitschrift*, **21**, pp. 649–650; also in Weyl (1968), vol. II, pp. 141–142.

Weyl, Hermann, 1921a, *Raum-Zeit-Materie*, 4th edition, Springer, Berlin.

Weyl, Hermann, 1921b, 'Feld und Materie,' *Annalen der Physik*, **65**, 541–563; also in Weyl (1968), vol. II, pp. 237–259.

Weyi, Hermann, 1921c, 'Electricity and Gravitation,' *Nature*, **106**, 800–802; also in Weyl (1968), vol. II, pp. 260–262.

Weyl, Hermann, 1922a, *Space-Time-Matter*, Dover, New York (translation of Weyl (1921a)).

Weyl, Hermann, 1922b, 'Die Einzigartigkeit der Pythagoreischen Massbestimmung,' *Mathematische Zeitschrift* **12**, 114–146; also in Weyl (1968) vol. II, pp. 263–295.

Weyl, Hermann, 1923a, *Raum-Zeit-Materie*, 5th edition, Springer, Berlin.

Weyl, Hermann, 1923b, Commentary to Riemann's Inaugural Dissertation, in H. Weyl,

Das Kontinuum und Andere Monographien, Chelsea Publishing Company, New York (no date).
Weyl, Hermann, 1923c, *Mathematische Analyse des Raumproblems,* Springer, Berlin.
Weyl, Hermann, 1924a, Review of (Reichenbach, 1924), *Deutsche Literaturzeitung*, no. 30, columns 2122–2128.
Weyl, Hermann, 1924b, 'Massenträgheit und Kosmos. Ein Dialog,' *Die Naturwissenschaften* **12**, 197–204; also in Weyl (1968), vol. II, pp. 478–485.
Weyl, Hermann, 1927, *Philosophie der Mathematik und Naturwissenschaft,* Oldenbourgh München.
Weyl, Hermann, 1928, 'Diskussionsbemerkungen zu dem zweiten Hilbertschen Vortrag über die Grundlagen der Mathematik,' *Abhandlungen aus dem mathematischen Seminar der Hamburgischen Universität* **6**, 86–88; also in Weyl (1968), vol. III, pp. 147–149.
Weyl, Hermann, 1929a, 'Consistency in Mathematics,' *The Rice Institute Pamphlet* **16**, 245–265; also in Weyl (1968), vol. III, pp. 150–170.
Weyl, Hermann, 1929b, 'Gravitation and the Electron,' *The Rice Institute Pamphlet* **16** 280–295; also in Weyl (1968), vol III, pp. 229–244.
Weyl, Hermann, 1929c, 'Elektron und Gravitation,' *Zeitschrift für Physik* **56**, 330–352; in Weyl (1968), vol. III, pp. 245–267.
Weyl, Hermann, 1931, 'Geometrie und Physik,' *Die Naturwissenschaften* **19**, 49–58; also in Weyl (1968), vol. III, pp. 336–345.
Weyl, Hermann, 1932, *The Open World,* Yale University Press, New Haven.
Weyl, Hermann, 1934, *Mind and Nature*, University of Pennsylvania Press, Philadelphia.
Weyl, Hermann, 1946, 'Review: The Philosophy of Bertrand Russell,' *The American Mathematical Monthly* **53**, 8–14; also in Weyl (1968), vol. IV, pp. 599–605.
Weyl, Hermann, 1949a, *Philosophy of Mathematics and Natural Science,* Princeton University Press, Princeton, New Jersey.
Weyl, Hermann, 1949b, 'Wissenschaft als Symbolische Konstruktion des Menschen,' *Eranos-Jahrbuch,* 1948, pp. 375–431; also in Weyl (1968), vol. IV, pp. 289–345.
Weyl, Hermann, 1950, '50 Jahre Relativitätstheorie,' *Die Naturwissenschaften* **38**, 73–83; also in Weyl (1968), vol. IV, pp. 421–431.
Weyl, Hermann, 1953, 'Ueber den Symbolismus der Mathematik und mathematischen Physik,' *Studium Generale*, **6**, pp 219–228; also in Weyl (1968), vol. IV, pp. 527–536.
Weyl, Hermann, 1954, 'Erkenntnis und Besinnung, (Ein Lebensrükblick),' *Studia Philosophica, Jahrbuch der Schweizerischen Philosophischen Gesellschaft-Annuaire de la Société Suisse de Philosophie;* also in (Weyl, 1968), vol. IV, pp. 631–649.
Weyl, Hermann, 1968, *Gesammelte Abhandlungen,* 4 volumes, Springer, Berlin.
Winnie, John, 1977, 'The Causal Theory of Space-Time,' in Earman, Glymour and Stachel (eds.), *Foundations of Space-Time Theories,* University of Minnesota Press, Minneapolis, Minnesota.

LAURENT A. BEAUREGARD

REICHENBACH AND CONVENTIONALISM

1. INTRODUCTION

One major theme in the philosophy of Hans Reichenbach concerns the presence of elements of convention in human knowledge. The distinction between fact and convention is held to be essential for epistemology and for the philosophy of science (Reichenbach, 1938, Section 1). Indeed, much of Reichenbach's technical work in the philosophy of space and time is addressed precisely to the task of separating out explicitly the conventional component from the empirical core in physical geometry and in the theory of relativity. Thus in the concluding section of (1958) Reichenbach says. "... it has been our aim to free the objective core of [statements about reality made by physics] from the subjective additions introduced through the arbitrariness in the choice of description." And in the beginning section of his major work in the theory of knowledge (1938), Reichenbach refers to the task of separating fact from convention as constituting "an integral part of the critical task of epistemology." In this paper, I analyze Reichenbach's conventionalism. I shall be especially concerned with the conventionalism attendant upon certain space- and time-indeterminacies, and, throughout this paper, I shall persist in asking the same question over and over: What is the *source* of this indeterminacy? As we shall see, there is good reason to raise this question.

2. UNOBSERVED OBJECTS

In 1944, Reichenbach tackles the venerable problem about 'our knowledge of unobserved things' with a view toward illustrating the key features of his *theory of equivalent descriptions.*[1] He asks: "How do things look when we do not look at them?" The problem is that of the continuing existence of a tree even when nobody looks at it. He argues that we can have no inductive evidence about unobserved trees. The solution to which he finally arrives is this:

> ... there is more than one true description of unobserved objects ... there is a *class of equivalent descriptions*, and ... all these descriptions can be used equally well.

W. C. Salmon (ed.), Hans Reichenbach: Logical Empiricist, 305–320.

The particular description that we use is *a matter of convention*, though *the class itself* is objectively determined. The point, however, is that when we do *not* look at the tree, we have a choice as to what we might say about it without coming into conflict with observed facts (such as the fact of its shadow which we do see, for example):

D_1 the one tree exists and proceeds 'as usual';
D_2 the tree splits in two (when nobody looks);
D_3 the tree turns into a duck (when nobody looks);

and so on. Description D_1 would be, of course, the *normal choice* and it is certainly a matter of empirical fact that it is *possible* for us to make the normal choice. What Reichenbach denies is that the facts could ever make such a choice *necessary*. A suitable change in the laws of optics or ornithology could lead us to countenance such *perverse choices* as D_2 or D_3, but the 'perversity' here would pertain *only to matters of linguistic convenience.*

Again, in the second of his Poincaré Institute lectures, Reichenbach (1953) speaks of unobserved objects:

> The exigencies of practical life require us to add to our observations a theory of unobserved objects. This theory is quite simple: we suppose that unobserved objects are similar to those that we do observe. No one doubts that a house remains the same whether or not we observe it visually. The act of observation does not change things; this principle seems to be a truism. But a moment's thought shows that it is wholly impossible to verify it. A verification would require us to compare the unobserved object with the observed object, and hence to observe the unobserved object, which amounts to a contradiction.
>
> It follows that the principle ... is a matter of convention: one introduces the definition that the unobserved object is governed by the same laws as those which have been verified for observed objects. This definition constitutes a rule which allows us to extend the observation language to unobserved objects; we shall call it a *rule of extension of language*. Once this rule is established, one can find out whether the unobserved object is the same as the observed object or not; for example, we can conclude that the house stays at its place when we do not look at it, while the young girl does not remain in the box when the magician saws the box into two parts. Such statements would have no meaning if a rule of extension of the observation language had not been added. [p. 123, *my translation*]

It is no part of my purpose in this paper to criticize this view. It suffices to exhibit it and, in particular, to bring out its pertinence to Reichenbach's philosophy of space and time. What is the source of the need for a convention as regards unobserved objects? Reichenbach's answer is that it is logically impossible to observe an unobserved object.

It is worth noting at this point that Reichenbach's views about unobserved objects have not gone unnoticed by Adolf Grünbaum. Grünbaum claims to endorse Reichenbach's theory of equivalent descriptions *in the context of geochronometry*, but he explicitly disavows its application "to the theory of the states of macro-objects like trees" on the grounds that it seems to him "fundamentally incorrect because its ontology is that of Berkeley's *esse est percipi*." In Grünbaum's judgment,

> Reichenbach's application of the theory of equivalent descriptions to unobserved objects lacks clarity on precisely those points on which its relevance turns. (1968, p. 52)

So be it. The question now arises as to whether the situation is any different 'in the context of geochronometry', that is to say, in Reichenbach's philosophy of space and time.

3. COORDINATIVE DEFINITIONS

According to Reichenbach, physical knowledge coordinates concepts to real objects:

> In general this coordination is not arbitrary .... But certain preliminary coordinations must be determined before the method of coordination can be carried through any further: these first coordinations are therefore definitions which we shall call *coordinative definitions*. They are *arbitrary*, like all definitions; [yet] on their choice depends the conceptual system which develops with the progress of science. (1958, p. 14)

As an example of a coordinative definition, Reichenbach offers that of the choice of a unit of length. But what is the source of the need for such a coordinative definition? The answer is obvious. One wants to make a length measurement: therefore one has a *purely logical* need for a unit. In a perfectly trivial sense, it is logically impossible to measure a distance (or a time-duration) without having first decreed a length-unit (or a duration-unit). The logical situation here is exactly parallel to that with unobserved objects. Reichenbach holds that, in *any* human knowledge situation, certain elements of convention exist *by logical necessity*. The need for coordinative definitions is nothing less than pervasive.

4. CONGRUENCE

Consider, with Reichenbach, the problem of comparing the lengths of two rods *at different locations*. One must introduce a *third* rod to be transported from the one location to the other. But how do we know that this 'measuring rod' does not change during the transport? We are told that it is 'fundamentally impossible' for us to know: all we can *know* is that two rods are always equal in length *at the place where they are compared* (1958, p. 16). "Here we are . . . dealing with . . . a logical impossibility", says Reichenbach (1958, p. 29).

It is in this manner that Reichenbach defends the conventionality of spatial congruence. There is not the slightest hint of a Riemann-begotten Grünbaum vision of *intrinsic* metric *amorphousness.* And the situation with *temporal* congruence is precisely the same:

There is basically no means to compare two successive periods of a clock, just as there is no means to compare two measuring rods when one lies behind the other. We cannot carry back the later time interval and place it next to the earlier one (1958, p. 116).

Here again Reichenbach makes it unmistakably clear that what is at issue (as regards the fundamental *source* of the conventionality of temporal congruence) is a logical impossibility of verification. It is for *this* reason that one *must* "introduce the concept of a *coordinative definition* into the measure of time" and *this* is why

temporal congruence is not a matter of *knowledge* but a matter of *definition* (1958, p. 116).

As far as I know, there is not a shred of evidence that Reichenbach ever nods consent to a doctrine of the intrinsic metrical amorphousness of the time continuum. Nor does the empiricist Reichenbach ever speak of 'the domain of intrinsic facts or states of affairs'. Yet Grünbaum feels that the following historical claim is warranted:

Indeed, it is with respect to the domain of INTRINSIC facts or states of affairs that equivalence obtains among alternative descriptions that Reichenbach calls 'equivalent' in his theory of equivalent descriptions as applied to metric descriptions (1970, p. 577).

But this is manifestly false, unless we are to understand by 'intrinsic facts' facts of a purely logical kind.

5. LENGTHS OF MOVING RODS

Suppose one has two frames K and K', which are in a state of relative (uniform) motion. Let there be one rod in each frame. How can we compare the lengths of *these* two rods?

For the sake of definiteness, let us suppose that frame K is at rest, so that frame K' is 'the one that moves'. Then we can use rod-transport to determine the length of the K-rod. The above question then amounts to this: How can we find out the length of the moving rod?

The question is ambiguous. Are we asking for the length of a K'-rod as measured in K, or are we asking for the length of a K'-rod as measured in K'? Clearly it is a matter of pure logic to distinguish these two questions and thus to come to realize that there are two distinct concepts at issue here. And in each case, there is a purely logical need for a coordinative definition, a convention. And there is no physical hypothesis which served to guarantee the existence of either of these conventions. For we are here concerned *only* with what Reichenbach calls "the philosophical theory of relativity". It is a matter of coming to grips with the fact "that coordinative definitions are needed far more frequently than was believed in the classical theory of space and time" (1958, p. 177). This is a question of semantics, not of amorphousness.

6. REST

How can we tell whether something is resting or moving? Once more the need for a coordinative definition is imminent.

> Why can motion only be characterized as relative? This question leads again to the concept of coordinative definition; the unobservability of absolute motion indicates the lack of a coordinative definition . . . , [and] the fact [is] that purely logical considerations are involved. The question which system is in motion is not a defined question, and therefore no answer is possible. We are not dealing with a failure of knowledge but with a logical impossibility . . . (1958, p. 219).

One has here what Reichenbach would call an "epistemological relativity of motion", something to be sharply distinguished from a *physical* relativity of motion:

> . . . this [epistemological] relativity [of motion] which depends on the arbitrariness of a coordinative definition, applies even if the dynamic relativity of Mach cannot be carried through . . . . (*Ibid.*)

And the same constrast can be made within Einstein's special theory of relativity. Even if Einstein's *physical relativity* of uniform motion (or rest) could not 'be carried through', one's 'epistemological results' as regards the concept of motion (or rest) would remain unchanged. The reason is that, as Reichenbach says, "we are dealing [here] with a logical impossibility." At bottom, the conventionality of rest is *not* to be seen as receiving its support in the truth of some physical hypothesis.

In short, it is the doctrine of the pervasive need for coordinative definitions which lies at the heart of Reichenbachian conventionalism. It is what Reichenbach himself calls 'philosophical relativity', something not to be conflated with 'physical relativity'. Let us now look a little more closely at this Reichenbachian distinction.

7. RELATIVITY

Reichenbach speaks of the congruence conventionalities as constituting 'an epistemological foundation' for the relativistic theory of space and time. He says that this foundation

> is supplied by the discovery that coordinative definitions are needed far more frequently than was believed in the classical theory of space and time ... (1958, pp. 176–177).

He sees his own work as having achieved a sharp separation

> between the physical assertions of the relativistic theory of space and time and its epistemological foundation (*Ibid.*, p. 176).

In this context, he asserts that there are really *two* theories of relativity: a physical one, and a philosophical one. What is the philosophical theory of relativity? It is simply

> the discovery of the definitional character of the metric in all its details ... (*Ibid.*, p. 177).

And what is the status of this discovery? Reichenbach asserts that it

> holds independently of experience. Although it was developed in connection with physical experiments, it constitutes a philosophical result not subject to the criticism of the individual sciences (*Ibid.*, p. 177).

Thus Reichenbach holds that the existence of conventions obtains in an *a priori* manner. So far, all of this agrees perfectly with the interpretation of Reichenbach which we have urged.

In section 4 of (1958), there occurs the following passage:

> The philosophical significance of the theory of relativity consists in the fact that it has *demonstrated* the necessity for metrical coordinative definitions in several places where empirical relations had been previously assumed (p. 15, italics mine).

But why should it have taken an Einstein to have *demonstrated* this, if all that is involved is *logical analysis*? In keeping with my interpretation of Reichenbach, I take him to have understood this 'demonstration' to be of a pragmatic or heuristic character. Thus what Reichenbach is telling us here is that the theory of relativity offers us a gold mine of *instances* of the usefulness of the doctrine of the coordinative definition. In short, the physical theory of relativity affords an excellent *illustration* of the notion of epistemological relativity (conventionality). And the non-triviality of a convention, for Reichenbach (as I interpret him), would be a direct function of the extent to which *we were previously ignorant of the need for that convention.* I would add that if we apply *this* criterion for the non-triviality of conventions to such cases as the conventionality of the unit of length, the congruence conventionalities, the conventionality of rest, *and* the conventionality of simultaneity, we come out with results which are not implausible.

The point is an important one for those who wish to disentangle Reichenbachian conventionalism from *Grünbaumian* conventionalism. I believe that Wesley C. Salmon's criterion for the nontriviality of conventions, namely that the non-triviality of a convention is a direct function of the physical content of the hypotheses that guarantee the existence of that convention, captures perfectly the spirit of *Grünbaumian* conventionalism. The contrast between the two varieties of conventionalism here at issue is brought out very clearly in the following passage from the writings of Grünbaum:

> *With respect to certain particular physical theories,* I assert . . . the conventional status of congruence in space, time, and space–time, and of simultaneity. It has been mistakenly thought that these particular claims of mine can be construed as mere instances . . . of a quite *general* thesis which *some* philosophers have asserted alike for any and every theoretical term and for *every* physical theory. Yet . . . I would *deny* the conventionality . . . of congruence and simultaneity, *if* certain *other* physical theories were held to be true (1970, p. 471).

Clearly, Grünbaum's kind of conventionality is very different from Reichenbach's 'epistemological relativity' or 'philosophical relativity'. And then too, at the very end of 1958, Reichenbach goes out of his way to assert

> that the problem concerning space and time is not different from that of the description of any other physical state as expressed in physical laws (p. 288).

I have argued that what Grünbaum calls "a quite *general* thesis which *some* philosophers have asserted alike for any and every theoretical term and for *every* physical theory" fits *precisely* the position of Reichenbach.

8. SIMULTANEITY

The acid test for my interpretation of Reichenbach's conventionalism is surely going to be the analysis of what Reichenbach *means* when he speaks of "the relativity of simultaneity." Unfortunately, however, it appears that Reichenbach sometimes exhibits an unhappy way of speaking in connection with relativity and simultaneity. For example, he says:

The definitional character of simultaneity was recognized first by Einstein and has since become famous as the relativity of time (1958, pp. 123–124).

But just what *was* it that had become *famous* by 1928? The need for a coordinative definition? I cannot believe that. It was rather *the inter-frame relativity of simultaneity* (as in Einstein's example of the moving trains) and this is a far cry from 'the definitional character of simultaneity'.

There can be no doubt that Reichenbach takes 'the epistemological character of Einstein's discovery' to be

the result that the simultaneity of distant events is based on a coordinative definition. (*Ibid.*, p. 124.)

A close study of section 19 of 1958 reveals the following two salient items: (1) his treatment of simultaneity in that section stands in exact parallel to his treatment of spatial congruence in section 4 and to his treatment of temporal congruence in section 17; (2) in particular, the limiting character of the velocity of light plays no role whatever in the argument of that section. What, then, *is* the argument of that section?

Reichenbach is concerned to argue that *any* comparison of the time-instants of spatially separated events will have "the characteristic properties of a coordinative definition." For any such time-comparison presupposes *a causal chain* (a signal, say) connecting the one location with the other. One needs to know the velocity of the signal. But this is *logically impossible*:

. . . we are faced with a circular argument. To determine the simultaneity of distant events we need to know a velocity, and to measure a velocity we require knowledge of the simultaneity of distant events. *The occurrence of this circularity proves that simultaneity is not a matter of knowledge, but of a coordinative definition* (1958, pp 126–127, italics mine).

It is *the logical circle* that shows that a knowledge of simultaneity is 'impossible in principle', says Reichenbach. Such is the epistemological relativity of simultaneity. It holds *a priori*. The acid test has been passed.

But Reichenbach has much more to say about the relativity of simultaneity. For he is clearly aware of the existence of a *physical* relativity of simultaneity to which Einstein's *physical* theory of relativity is committed. It is in Reichenbach's *Axiomatization of the Theory of Relativity* that one finds his clearest statement on this. There he delineates the sense(s) in which *absolute simultaneity* would be physically meaningful if Einstein's theory were false:

Only [the following] interpretation of absolute simultaneity is meaningful: it is the assertion, "There is a rule which, if carried through in the same way in every coordinate system, will always determine the same events as being simultaneous."

Such a rule might employ an infinitely high velocity or the transport of.clocks.

It is an empirical question whether this assertion is true. In this sense, the determination of absolute time is an empirical matter (1969, p. 96).

There are two distinct physical hypotheses at issue here:

(1) THE LIGHT POSTULATE: Light is the fastest causal signal, and light travels at a finite speed. (Called *LP*.)

(2) THE CLOCK POSTULATE: If two clocks are synchronized locally at some place *A*, and are subsequently brought to another place *B*, either by different paths, or at different speeds, then these two clocks will be found to be out of synchrony when they come together at *B*. (Called *CP*.)

What Reichenbach is saying is that if *either* LP or CP were false, then there would be absolute simultaneity. But since *each* of the postulates LP and CP (of Einstein's *physical* theory) is *true*, one has the *physical relativity of simultaneity*.

But it now becomes clear that there are two distinct senses of this 'physical relativity of simultaneity', corresponding to two distinct senses of 'absolute simultaneity'. Let us say that we have *absolute signal time* if LP is false, and that we have *absolute transport time* if CP is false. And let us associate the *denial* of the existence of absolute *signal* time with the *intra-frame conventionality of simultaneity*, and the *denial* of the existence of absolute *transport* time with the *inter-frame relativity of simultaneity*. (We are linking the Clock Postulate with inter-frame relativity of simultaneity. At the end of this section, we assert that, in the presence of the Light Postulate, the Clock Postulate implies, and is implied by, the inter-frame relativity of simultaneity.)[2] Thus our analysis of what is meant by 'the relativity of simultaneity' requires that one distinguish carefully among three levels of such 'relativity':

(1) *The need for a coordinative definition for simultaneity*
(2) *The intra-frame conventionality of simultaneity*
(3) *The inter-frame relativity of simultaneity*

Pure logic supports (1); LP supports (2); and CP supports (3). And the corresponding three conceptions of ABSOLUTE TIME are:

(1′) *Epistemologically absolute time*
(2′) *Absolute signal time*
(3′) *Absolute transport time*

According to Reichenbach, conception (1′) is absurd or 'meaningless', while conceptions (2′) and (3′) are, at most, physically false. Each is, at any rate 'meaningful' and it is very easy to construct a world in which either or both of these latter two conceptions apply. This little scheme, incidentally, has implications as regards the analysis of the intellectual guilt of the Newtonians who believed in all three of these notions.

Reichenbach has still more to say about simultaneity – in particular about the status of absolute transport time and its connection (or rather, lack of connection) with what he calls the epistemological relativity of simultaneity. In the *Axiomatization*, he asserts that all "presentations in the defense of an absolute time"

overlook the fact that the new method achieves nothing but a definition of simultaneity, and that it does not disprove the *epistemological* relativity of simultaneity. It might disprove only the *physical* relativity of simultaneity and would then constitute an absolute time in the sense of . . . a uniform rule leading to *the same simultaneity independently of the state of motion of the spatial coordinate system* (1969, p. 123).

This means clearly that the truth of (3′) – or the falsity of (3) – cannot possibly affect the truth of (1) – that is, the falsity of (1′). A similar passage occurs in (1958):

[Even] if relativistic physics were wrong, and the transport of clocks could be shown to be independent of path and velocity, this type of time comparison could not change our epistemological results, since the transport of clocks can again offer nothing but a *definition* of simultaneity. Even if the two clocks correspond when they are again brought together, how can we know whether or not both have changed in the meantime? This question is as undecidable as the question of the comparison of length of rigid rods (p. 133).

If the argument of this paper is correct, then this passage should be interpreted to mean exactly the same: namely that (3′) does not imply (1′), or that the falsity of (3) does not imply the falsity of (1).

For the sake of our subsequent discussion, however, let us suppose that I am wrong about this. Let us suppose, in particular, that when Reichenbach speaks of "the epistemological relativity of simultaneity" he may be ambiguously referring to either or to both of levels (1) and (2) of the relativity of simultaneity. In that case, each of the foregoing passages could be taken to assert that (3′) does not imply (2′), or that the falsity of (3) does not imply the falsity of (2). Now if it is true – in addition – that the truth of (3) *does* imply the truth of (2), then we would have a solid basis for the following

PRIORITY CLAIM #1: The intra-frame conventionality of simultaneity is logically prior to the inter-frame relativity of simultaneity.

An additional priority claim would be

PRIORITY CLAIM #2: The Light Postulate is logically prior to the Clock Postulate.

I shall now demonstrate two things: First, that Reichenbach *is* committed to priority claim #1; and secondly, that the claim is perfectly correct.

In 1958, Reichenbach refers to a *mistake* which

> results from the derivation of the relativity of simultaneity from the different states of motion of various observers (p. 146).

Here Reichenbach is referring to level (3), the inter-frame relativity of simultaneity, and he is making the somewhat surprising claim that the relativity of simultaneity *in this sense* need not result from motional considerations. The passage continues as follows:

> It is true that one *can* define simultaneity differently for different moving systems, which incidentally is the reason for the simple measuring relations of the Lorentz Transformation, but such a definition is not *necessary*. We could arrange the definition of simultaneity of a system K in such a manner that it leads to the same results as that of another system K' which is in motion relative to K; in K, ε would not be equal to $\frac{1}{2}$ in the definition of simultaneity (section 19), but would have some other value. It is a serious mistake to believe that if the state of motion is taken into consideration, the relativity of simultaneity is necessary....
>
> The arbitrariness in the choice of simultaneity makes it possible to asign the value $\varepsilon = \frac{1}{2}$ to *every* uniformly moving system (*Ibid.*).

I believe that the first clause of the first sentence in this passage does not mean what it says (or does not say what it means). I take its real meaning to be precisely what is expressed by the *last* sentence of the passage just quoted. With that exegetical caveat, we may proceed. What Reichenbach is saying in this passage is simply that one can use the intraframe conventionality of simultaneity to wipe out the interframe relativity of simultaneity despite the presence of relative motion between frames K and K'. I take this to be tantamount to priority-claim #1, enunciated above. My next task is to vindicate it.

To this end, we follow Reichenbach in describing the intraframe conventionality of simultaneity in terms of a parameter called *epsilon*. To say that simultaneity is conventional in frame K is to say that one may choose ε to have any value between 0 and 1. Likewise, in frame K', one may choose ε' to have any value between 0 and 1. It is to be understood that the respective choices of ε and ε' are to be mutually independent: neither need constrain the other. Now let two spatially separated events occur such that they are observed in each of the two frames K and K', where K' moves at a velocity v relative to K. According to the usual interpretation of the relativity of simultaneity, if K' judges these events to occur simultaneously, then K will *not* judge the same events to occur simultaneously. The difference in judgment is generally supposed to have its source in the relative motion of K and K', as well as in the limiting character of light the invariant speed of which is denoted by c. From our present standpoint, however, the question arises as to whether the respective epsilon-choices in the two frames might not affect this "motional" relativity of simultaneity judgments. Reichenbach is asserting that it does.[3]

Results at which I first arrived in June 1974 indicate that Reichenbach is demonstrably correct. I have been able to derive the following general condition for the wiping out of the motional relativity of simultaneity with respect to a given pair (K, K') of frames: one need only choose ε and ε' such that

$$\varepsilon' = \frac{\varepsilon \cdot [1+(v/c)]}{1+(2\varepsilon-1)(v/c)} \tag{4}$$

henceforth to be known as the *RS-Wipe-Out-Condition*. Given this result, the detailed vindication of what Reichenbach says follows easily. One chooses $\varepsilon' = \frac{1}{2}$ for frame K', and the *RS-Wipe-Out-Condition* then reduces to

$$\varepsilon = \tfrac{1}{2}(1 - v/c). \tag{5}$$

In the passage in question, Reichenbach is claiming that, no matter which value of ε' in K' is chosen, subject to the condition $0 < \varepsilon' < 1$, there exists a value of ε in K for which $0 < \varepsilon < 1$ such that for any given pair of separated events, a simultaneity judgment made by K' of these events with respect to ε' *will agree with* a simultaneity judgment made by K of the same events with respect to ε. It is easy to see that our result (4) implies precisely this claim. For one has, for any two *distinct* frames K and K'.

$$0 < v/c < 1.$$

Now from (4), it follows that when v/c is close to zero, then ε is nearly equal to ε'; and that when v/c is close to unity then ε approaches zero and ε' approaches unity. And from the special case (5) of (4), we see that Reichenbach is correct in maintaining that a *standard* simultaneity metric in K' necessarily implies a *non-standard* simultaneity metric in K, *given that one is bent on wiping out inter-frame relativity of simultaneity*. For, clearly, for every value of v/c in the above range, it follows from (5) that $\varepsilon \neq \frac{1}{2}$. This concludes my support of Reichenbach's priority claim #1. One interesting by-product of the argument is that one sees very clearly just how and why the Newtonian chauvinist desideratum of wiping out relativity of simultaneity *can* be achieved within Einstein's universe, *but only at the Newtonianly unpalatable price of being forced to adopt a non-standard simultaneity metric in at least one frame*. Eliminate inter-frame relativity of simultaneity if you will, and then the intra-frame conventionality of simultaneity comes at you with a vengeance, I thereby vindicate not only Reichenbach but Anaximander as well.[4]

I turn now to that portion of the last quoted passage from Reichenbach which was deliberately omitted. It is this:

> Actually the relativity of simultaneity has nothing to do with the relativity of motion. It rests solely on the existence of a finite limiting velocity for causal propagation (1958, p. 146).

Does this follow from the foregoing priority considerations? What has been shown is that the inter-frame relativity of simultaneity *need not follow* from considerations regarding the state of motion of frames, taken together with the Light Principle. That is all. It happens that the inter-frame relativity of simultaneity *logically entails* (and in this sense, *presupposes*) a certain simultaneity-free Principle of Relativity of Uniform Motion (as well as the Light Postulate).[5] I take it that Reichenbach does not intend to assert the banality that the *intra-frame* conventionality of simultaneity has nothing to do with the relativity of motion. But clearly the *inter-frame* relativity of simultaneity has *plenty* to do with the relativity of motion: the former entails the latter.

What role does the Clock Postulate play in all this? One would guess, in the first place, that priority claim #2 is correct: the Light Postulate (which undergirds the intra-frame conventionality of simultaneity) is indeed logically prior to the Clock Postulate (which we take to undergird the inter-frame relativity of simultaneity). It would not, however, follow that the inter-frame relativity of simultaneity 'has nothing to do with' the Clock Postulate. In fact, given the Light Postulate, the two can be shown to be logically equivalent.[6]

I have tried to show that Reichenbach's insight that light-signalling considerations and intra-frame conventionality have a certain priority over clock-transport considerations and interframe relativity of simultaneity is well-founded. It leads us to a deeper understanding of the kinematics of the special theory of relativity. And it is clear that this insight into an important physical theory stands largely independent of any criticism of any 'outmoded' positivist doctrines that may be attributed (rightly or wrongly) to Reichenbach. If Reichenbach insists that *epistemologically* absolute time (1′) is logically impossible, then he also recognizes the meaningful status of *each* kind – absolute signal time (2′) and absolute transport time (3′) – of *physically* absolute time, and that the obtaining of either of these is, at most, physically impossible. It is in this sense, then, that Reichenbach can be said to have adumbrated what we have called Grünbaumian conventionalism.

9. CONCLUSION

There are two distinct components to Reichenbach's conventionalism: a 'positivistic component' which is based upon the notion of logical impossibility of verification, *and* what may appropriately be called a 'Grünbamian component' which is based upon the notion of physical impossibility of verification. The physical impossibilities are to be instantiated with the help of actual physical laws, the so-called physical principles of impotence such as the Limiting Light Postulate or, perhaps Heisenberg's Principle of Indeterminacy. And, although a physical impossibility of this kind will seem to many considerably more interesting and non-trivial than any number of logical impossibilities one can think up, I feel certain that Reichenbach would insist that each of these interesting physical impossibilities alike, if they are to be stated within a language, must needs presuppose certain coordinative definitions. At least this will be the case whenever one pretends to be talking about physical reality. It may be that Sections 1 – 7 of this paper constitute an unusual kind of preface to a study of Quine's *Word and Object.* Be it so. It is, however, very different with most of Section 8. Let it never be thought that the physical indeterminacy of simultaneity to which the theory of special relativity is committed has anything whatever to do with Quinian indeterminacy. For the space–time indeterminacies entailed by the theory of relativity are, and ought to be, *sui generis.* If there be ontological relativity, then let some well-established physical theory tell us this. And *that* is the difference between Grünbaum (or half of Reichenbach) and Quine.

5505 S.E. 48
Portland, Oregon 97206

NOTES

[1] The discussion of the unobserved tree occurs at the beginning of the book as a prelude to the discussion of the unobservable electron.
[2] See note 6.
[3] In this connection, see also Grünbaum (1973, pp. 359–368).
[4] From the one extant fragment: "... for they pay penalty and retribution to each other for their injustice according to the assessment of Time." A most remarkable anticipation! Kirk *et al.*, p. 117.
[5] This is Winnie's (1970) Passage-Time Principle.
[6] Elsewhere I show that the principle of Round-Trip Clock Retardation (a congruence-metric-determinate version of the Clock Postulate) is equivalent to a simultaneity-free formulation of the Principle of Relativity of uniform motion (namely, Winnie's Passage Time Principle). The equivalence holds, given the Light Postulate. However, given the latter postulate, it turns out that such a Principle of Relativity in its turn is equivalent to certain synchrony-free inter-frame relativity-of-simultaneity effects! It should be noted, in this connection, that the *RS-Wipe-Out-Condition* (4) above *cannot* in general be used to eliminate interframe relativity of simultaneity when we are dealing with *more than two* inertial frames. The Clock Postulate, or the absence of absolute transport time, can then be identified with inter-frame relativity as has been done above.

BIBLIOGRAPHY

Grünbaum, A., 1968, *Geometry and Chronometry in Philosophical Perspective*, University of Minnesota Press, Minneapolis.
Grünbaum, A., 1970, 'Space, Time, and Falsifiability', *Philosophy of Science* **37**, 469–588.

Grünbaum, A., 1973, *Philosophical Problems of Space and Time*, D. Reidel Publishing Co., Dordrecht and Boston.

Kirk, G. S. *et al.*, 1966, *The Presocratic Philosophers*, Cambridge University Press, Cambridge.

Reichenbach, H., 1938, *Experience and Prediction*, University of Chicago Press, Chicago.

Reichenbach, H., 1944, *Philosophic Foundations of Quantum Mechanics*, University of California Press, Berkeley and Los Angeles.

Reichenbach, H., 1953, 'Les Fondements Logiques de la Mécanique des Quanta', *Annales de l'Institut Henri Poincaré* **XIII**, fasc. II, pp. 108–158.

Reichenbach, H., 1958, *The Philosophy of Space and Time*, Dover Publications, New York.

Reichenbach, H., 1969, *Axiomatization of the Theory of Relativity*, University of California Press, Berkeley and Los Angeles.

Winnie, J., 1970, 'Special Relativity Without One-Way Velocity Assumptions', *Philosophy of Science* **37**, 81–99, 223–238.

ADOLF GRÜNBAUM AND ALLEN I. JANIS

THE GEOMETRY OF THE ROTATING DISK IN THE SPECIAL THEORY OF RELATIVITY*

1. INTRODUCTION

In his (1924; 1969), Hans Reichenbach discussed the status of a two-dimensional spatial geometry on a disk rotating with respect to an inertial frame in the Minkowski space-time of the special theory of relativity ('STR'). Since then, the literature on this topic has grown very considerably.[1]

In the present paper, our aim is to give a treatment which, to our knowledge, either deals with facets not considered by others or gives an analysis differing from theirs in some respects that we deem significant. Thus, the two-dimensional spatial metric $d\sigma^2$ on the disk is usually derived by using the four-dimensional space-time metric as a point of departure and considering a disk which is already rotating at some constant angular velocity (cf. Møller, 1972, pp. 272–273). But we shall give a deduction of $d\sigma^2$ – expressed in terms of polar coordinates on the disk – by considering the kinematics of the transition of a disk from a pre-rotational to a uniform rotational state, subject to specified assumed dynamical constraints. And this deduction will be carried out for the case in which the change in angular velocity from zero to some other constant value involves the twisting of the radial lines no less than for the alternative case of no such twisting.

Elsewhere, one of us (Grünbaum, 1973, pp. 542–545) has replied to doubts which have been raised as to the possibility of using the space-like hypersurfaces of Minkowski space-time to provide a *four*-dimensional space-time *legitimation* of the two-dimensional non-Euclidean (hyperbolic) spatial metric $d\sigma^2$ to be deduced below. In that reply, a suitable family $H(x^4)$ of non-intersecting space-like hypersurfaces is used to explain each of the following: (i) The non-Euclidean 2-metric $d\sigma^2$ to be derived below can be viewed as generated by using rigid (corrected) unit rods on the rotating disk to metrize the appropriate space-like 2-dimensional subsurface of each member of $H(x^4)$, and (ii) The non-Euclidean spatial metric $d\sigma^2$ on these latter space-like surfaces is, of

W. C. Salmon (ed.), Hans Reichenbach: Logical Empiricist, 321–339.

course, avowedly *not* the same as the metric $d\sigma_0^2$ that is *induced* on them by the 4-dimensional space-time metric of the embedding Minkowski space-time, since $d\sigma_0^2$ is demonstrably Euclidean! Thus, it is made apparent in the cited reply that $d\sigma^2$ and $d\sigma_0^2$ constitute *alternative* metrizations of one and the same family of space-like surfaces such that the resulting geometries are respectively curved and flat. The interest of the non-Euclidean $d\sigma^2$ derives from its being generated by rigid (corrected) rods on the rotating disk, unencumbered by its *not* being the 2-metric that is *induced* by the 4-metric of the embedding space-time.

2. CHANGE OF TWO-GEOMETRY ON THE DISK DURING THE TRANSITION FROM REST TO UNIFORM ROTATION

In the usual discussions of the 2-geometry on the rotating disk as initiated in Einstein's (1916, § 3), one considers a disk D which is already rotating uniformly with respect to an inertial frame I, and one compares measurements made on D with measurements made in I of various spatial projections from D into I. The latter comparison pertains to metrical findings *during* the uniform rotational state, and of course must be distinguished from the following: The specification of the *change* in the spatial measurements made by an observer stationed on a disk S first when S is initially at rest in I and then when S has been set into rotation so as to have attained a given subsequently constant angular velocity ω. Although angular velocity can be either positive or negative, hereafter 'ω' will denote its *absolute* value.

We shall now turn our attention to the change in the spatial metric on S as between its states of rest and of uniform rotational motion. And we shall then see that the spatial metric thus deduced for the rotational state agrees with the one often derived from a comparison of the already rotating D with I in regard to their respective metrical findings.

For the moment, let us treat only the *kinematical* aspects of S's transition from rest to rotation with constant angular velocity ω. Thereafter, we shall comment on the appropriate forces which – in the first approximation of Newtonian dynamics – would have to be exerted to assure the accelerations that are involved. Kinematically speaking, we treat the disk as a set of concentric infinitesimal circular strips. And we shall infer the results of the respective measurements made in S on both infinitesimal tangential and radial elements of each such strip by means of the following basic principles enunciated in Einstein's fundamental paper on the STR (1905, § 2, pp. 41–42):

Let there be given a stationary rigid rod; and let its length be l as measured by a measuring-rod which is also stationary. We now imagine the axis of the rod lying along the axis of x of the stationary system of co-ordinates, and that a uniform motion of parallel translation with velocity v along the axis of x in the direction of increasing x is then imparted to the rod. We now inquire as to the length of the moving rod, and imagine its length to be ascertained by the following two operations: –

(a) The observer moves together with the given measuring-rod and the rod to be measured, and measures the length of the rod directly by superposing the measuring-rod, in just the same way as if all three were at rest.

(b) By means of stationary clocks set up in the stationary system and synchronizing in accordance with § 1 (Einstein's light-signal synchrony), the observer ascertains at what points of the stationary system the two ends of the rod to be measured are located at a definite time. The distance between these two points, measured by the measuring-rod already employed, which in this case is at rest, is also a length which may be designated 'the length of the rod'.

In accordance with the principle of relativity the length to be discovered by the operation (a) – we will call it 'the length of the rod in the moving system' – must be equal to the length l of the stationary rod.

The length to be discovered by the operation (b) we will call 'the length of the (moving) rod in the stationary system'. This we shall determine on the basis of our two principles, and we shall find that it differs from l.

Hereafter, we shall refer to the two operations which Einstein here calls '(a)' and '(b)' respectively by using these same labels.

In its initial *rest* state, assume the disk S to be equipped with a network of polar coordinates ρ and θ such that the circle $\rho = \rho_0$ is the periphery. Let 'dl_t' denote the *pre*-rotational length in S of an arc element on the periphery (or on any other line of constant ρ), where the subscript 't' is intended to convey that the infinitesimal element in question is *tangential* as distinct from radial. And let 'dl_r' denote the *pre*-rotational length of an infinitesimal element on any radial line $\theta =$ constant. As long as S is still at rest in I, its pre-rotational geometry is assumedly Euclidean in the STR. Thus, for the peripheral circular strip, $dl_t = \rho_0 d\theta$ and $dl_r = d\rho$. And hence during the pre-rotational state, the length of the disk's periphery is $2\pi\rho_0$ in S no less than in I.

We can give a perspicuous comparison of the pre-rotational and rotational spatial metrics in S by expressing both metrics in terms of the original *pre*-rotational polar coordinates ρ and θ, in addition to expressing the rotational space-metric on S in terms of the *rotational* polar coordinates r and θ', which will turn out to be a network of polar coordinates *different* from the original network (ρ, θ). Hence assume that S *remains* equipped with its initial coordinates ρ, θ even when it is uniformly rotating. But this is assumed *without* any prejudice to either whether the coordinate values ρ retain their initial *metrical* significance as *distances* from the center, or whether the original family of radial geodesics $\theta = \text{constant}$ is identical with the family $\theta' = \text{constant}$.

We assume dynamical conditions such that pre-rotational circles $\rho = \text{constant}$ do remain circles after uniform rotation is achieved. But as we shall see, this is *not* tantamount to also making the *metrical* assumption that the *radius* of any given circle $\rho = \text{constant}$ will retain the value ρ. Moreover, once uniform rotation has begun, no external forces and no internal forces other than those present pre-rotationally are exerted in any *tangential* direction on the disk, although such forces do then indeed continue to be exerted in the radial directions. Hereafter it is to be understood that all the circles in S which concern us are centered on the origin of (ρ, θ).

Invoking the principle of relativity in the manner set forth above in the quotation from Einstein, we can now deduce the following infinitesimally: For any given tangential element on the periphery of the disk S, the performance of Einstein's operation (a) on that element, in an appropriate local inertial frame co-moving with S at the linear velocity v with respect to I, will show that the length in S of the given arc element *remains* $dl_t = \rho_0 \, d\theta$ once the state of uniform rotation is achieved. Hence, as measured in S, the length of the disk's periphery is $\int_0^{2\pi} dl_t = 2\pi\rho_0$ once S is rotating, no less than when S was still at rest.

Next we must determine whether there has been any change in the length of the disk's diameter when measured in S as between the non-rotating and rotating states. We are going to *allow* for the possibility of twisting during the onset of rotation in the sense that the rotational radial *geodesics* $\theta' = \text{constant}$ are not identical with the pre-rotational ones $\theta = \text{constant}$. Hence such identity is not necessarily assumed in our impending metrical comparison of the pre-rotational and rotational radii. By the same token, it is not necessarily assumed that the points on any pre-rotational radial element of a given circular strip remain a radial set rotationally. In order to effect the desired comparison of the pre-rotational and rotational radii *as measured in S* and also of the corresponding radial elements in S between two circles whose ρ-coordinates differ infinitesimally by $d\rho$, we shall proceed as follows: We shall make purely *heuristic* and *intermediate* use of the result obtained by Einstein's operation (b) when performed in I on the spatial projection into I of a peripheral arc element whose pre-rotational *and* rotational length in S we saw to be $dl_t = \rho_0\, d\theta$. In virtue of the definition of Einstein's operation (b), in the present context that operation is, of course, performed at a given instant in I when S is already rotating. The required heuristic preliminaries will occupy us until after we have derived Equations (1)–(4) below.

It now follows from the STR that the stated application of the operation (b) will yield the value $\rho_0\, d\theta\sqrt{1-v^2/c^2}$ (where c is the usual speed of light), although a *pre*-rotational measurement of the length of the given peripheral arc element *in the inertial frame* I would, of course, have yielded the greater value $\rho_0\, d\theta$. Since the simultaneity-projection of the circular periphery of S into I when S is rotating is a Euclidean circle in I, we integrate the result $\rho_0\, d\theta\sqrt{1-v^2/c^2}$ of operation (b) from $\theta = 0$ to $\theta = 2\pi$ and obtain $2\pi\rho_0\sqrt{1-v^2/c^2}$ as the value of the circumference of this circular projection in I. On the other hand, we recall that the corresponding *pre*-rotational projection into I yielded a circle of greater circumference $2\pi\rho_0$. Hence we see that, *as measured in I*, there has been a *shrinkage* of the circular spatial projections of S's periphery into I during S's transition from rest to uniform rotational motion. And, of course, the radius of the circular projection obtained in I once uniform rotation is achieved has the reduced value $r_0 \equiv \rho_0\sqrt{1-v^2/c^2}$ as compared to the corresponding radius ρ_0 of the *pre*-rotational projection. Since $v = r_0\omega$, we can rewrite this equation as

$$\rho_0 = \frac{r_0}{\sqrt{1-r_0^2\omega^2/c^2}}. \tag{1}$$

By parity of reasoning, the intermediate heuristic Equation (1) – which pertains to the projection of S's peripheral circle $\rho = \rho_0$ after the onset of uniform rotational motion – also holds, *mutatis mutandis*, for the projection into I of any concentric *inner* circle ρ = constant in S. The points of any such inner circle will have linear speeds $u < v$ in I. Hence we can generalize Equation (1) by writing

$$\rho = \frac{r}{\sqrt{1 - r^2\omega^2/c^2}}, \tag{2}$$

where $0 \leqslant \rho \leqslant \rho_0$ and $r\omega$ is the linear velocity in I of any point on the disk located at a distance r in I from the common origin. Of course, in the STR, the maximum linear velocity $r_0\omega$ of S's peripheral points must be less than c, so that $r_0 < c/\omega$, which is finite for $\omega \neq 0$. But for points on the disk, the values of r in I are restricted to the interval $0 \leq r \leq r_0$. Hence for any points on S, $r < c/\omega$.

Recall that *before* the onset of S's rotation, the ρ-coordinate ρ_0 of a point on the periphery also had the metrical significance of being the distance in S of any such point from the center, i.e., ρ_0 was the *pre*-rotational radius of the disk S. Now we can show by means of Equation (1) that whereas the unattainable limit c of the linear velocities $r_0\omega$ of peripheral points does indeed restrict r_0 to less than c/ω – as we already saw – the limiting role of c does *not* also restrict the pre-rotational size of the disk S, because it does *not* similarly restrict its *pre*-rotational radius ρ_0. To see that the 'light-barrier' does not impose a finite upper bound on the values of ρ_0, note that in Equation (1), $\rho_0 \to \infty$ as $r_0 \to c/\omega$. For when $r_0 \to c/\omega$, the numerator of Equation (1) approaches the latter finite value, while the denominator approaches zero. This means that the pre-rotational size ρ_0 of the radius of the disk S is *not* restricted by the fact that the linear velocities of the peripheral points of S may not exceed the value c.

In order to make heuristic use of Equation (2) in ascertaining the results of *radial* measurements in S after the onset of uniform rotation, observe that this equation can be rewritten by expressing r in terms of ρ as follows:

$$r = \frac{\rho}{\sqrt{1 + \rho^2\omega^2/c^2}} \tag{3}$$

Taking differentials in Equation (3) and simplifying, we obtain

$$dr = \frac{1}{(1 + \rho^2\omega^2/c^2)^{3/2}}\, d\rho \tag{4}$$

where it will be observed that the factor in front of $d\rho$ is *less* than 1 for $\rho > 0$.

As we shall soon see, the heuristic value of Equation (4) lies in the fact that it can be shown to govern the relation connecting the *pre*-rotational length $d\rho$ and the rotational length *dr in S* of the radial elements lying between two circles on S whose ρ-coordinates differ infinitesimally by $d\rho$. Hence Equation (4) will turn out to show that, as measured in S itself, during this transition from rest to uniform rotation the disk S has shrunk radially in the following way: For any given value of $\rho > 0$ and any positive values of $d\rho$, it must be true that

$$dr < d\rho,$$

but such that the loss in size $d\rho - dr$ incurred because of radial shrinkage is greater at points of greater ρ.

Some dynamical points must now be emphasized before we can make the promised heuristic use of our Equations (1)–(4):

(i) As already noted, in the case of our disk S, forces must continue to be applied in the *radial* directions when S is rotating uniformly in I. In the first approximation of Newtonian dynamics, we can say that in I centripetal forces must be applied to the disk in axially symmetric fashion so as to assure the accelerations in I requisite for the shrinkage in I which we have described kinematically. And insofar as these axially symmetric centripetal forces are not supplied by the disk's internal forces, they are to be applied to the disk externally by suitable agencies in I. The dynamical constraints which we have thus imposed as conditions governing our problem are not present, for example, in the rotational motion of pizza dough when the latter is twirled by a pizza maker.

(ii) Once there is uniform rotation, the aforestated *absence* of *tangential* forces that would act on tangential arc elements of the disk did allow us to invoke Einstein's operation (a) so as to conclude, via the principle of relativity, that in S the *pre*-rotational length $dl_t = \rho_0\, d\theta$ of a tangential arc element remains *unchanged* after uniform rotation is achieved. By contrast, the stated *presence* of *radial* forces during uniform rotation would make it quite unsound to adduce the principle of relativity as a basis for inferring that in the aforementioned co-moving local inertial frame – and hence in S – the length of a radial element whose coordinates differ by $d\rho$ in the uniformly rotating state is the same as the pre-rotational length of such an element!

Although the principle of relativity cannot thus be invoked, we *are* entitled to make infinitesimal use of the relevant Lorentz transformation equation in the state of uniform rotation to deduce the length in S of any radial element on S from the considerations that we shall now set forth. We label as lines $\theta' = \text{constant}$ those lines on S whose simultaneity-projections into I are radial lines in I, and we explicitly *assume* as an empirical fact that in the rotational metric $d\sigma^2$ on S, these lines $\theta' = \text{constant}$ will everywhere be *orthogonal* to the circles $\rho = \text{constant}$ on S. Now introduce on the disk what, for now, is merely a new *coordinate* r specified by the transformation of Equation (3). Then lines $\rho = \text{constant}$ will also be lines $r = \text{constant}$. (Furthermore, let us assume axial symmetry in the sense that none of the components of the metric tensor in the $d\sigma^2$-metric depend on θ'. Then the lines $\theta' = \text{constant}$ can be shown to be *geodesics*,[2] which we anticipated them to be when so characterizing them earlier.) Now be mindful of our explicitly stated assumption that lines $\rho = \text{constant}$ and hence lines $r = \text{constant}$ are orthogonal to lines $\theta' = \text{constant}$, and consider the rotating system S at some *given instant in I*. Note that at the given instant and at any point P on S, the direction in I of the tangent to the line of constant r at P is instantaneously parallel to the direction of motion in I of the co-moving local inertial frame at P. Then the appropriate Lorentz transformations enable us to say the following: Since the local co-moving inertial frame at P does not move with respect to I in the *radial* direction of the line $\theta' = \text{constant}$ through P, a radial element at P on that line will have the same rotational length *in S* as its likewise radial simultaneity-projection in I. But the length in I of the latter simultaneity projection is dr. Hence the rotational length of the given radial element in S must have the *same* value dr as the length of its simultaneity-projection into I. Thus we see that as measured in S, the lesser rotational length dr of a radial element is related by Equation (4) to the greater pre-rotational length $d\rho$ of such an element. And Equation (1) now also tells us that during the transition from rest to uniform rotation, there has been a shrinkage of the disk radius from ρ_0 to r_0 *even as measured within S itself*!

But we saw that during this transition there was no change at all in the length $2\pi\rho_0$ of the periphery. Therefore, if we refer to Equation (1), we can see that after the onset of uniform rotational motion, the ratio of the disk's circumference to its diameter is

$$\frac{2\pi\rho_0}{2r_0} = \frac{\pi}{\sqrt{1-v^2/c^2}},$$

which is *greater* than π, as in hyperbolic non-Euclidean geometry. Yet before the rotational motion, the corresponding ratio had the Euclidean value $\pi = 2\pi\rho_0/2\rho_0$. In this connection, Martin Strauss (1974, p. 107) reasons fallaciously from a correct premiss by writing:

Einstein's argument (for the spatial non-Euclidicity of the rotating S), in my view, is wrong: if the measuring rods laid along the circumference of the rotating disk are Lorentz contracted with respect to the inertial frame, so are the distances on the circumference they are supposed to measure; hence the two effects would cancel each other and the ratio C/D (circumference/diameter) would turn out to equal π as in the Euclidean plane. Thus, Einstein's thought experiment, far from establishing non-Euclidicity for the intrinsic geometry on the rotating disk, suggests in fact that this geometry be Euclidean.

As one of us has remarked elsewhere (Grünbaum, 1967, p. 139), the number of times which the unit tangential rod on the rotating disk fits into that disk's periphery must indeed be the *same* as the number of fits of the Lorentz-*contracted* simultaneity-projection in I from a *tangentially*-applied unit rod on S *into* the corresponding Lorentz-*contracted* simultaneity-projection from the disk's periphery. But, as is evident from the above analysis, Strauss's correct premiss as to the sameness of the number of fits takes no account of the *radial* measurements in both S and I. Nor of the fact that circumferences and diameters of circles are measured in I by *unit* rods in I, and *not* by such projections (shadows) into I from unit rods in S as are Lorentz-contracted in I. Thus, the correct premiss adduced by Strauss does not warrant at all his conclusion that "the ratio C/D (circumference/diameter) would turn out to equal π as in the Euclidean plane." It is to be understood, however, that this conclusion of Euclideanism is *not* the one which Strauss draws from his *own* analysis of the geometry of the rotating disk; instead, Strauss' contention is the following: If, with Einstein, the radius of the disk is considered invariant as between the disk and ground frames, then Einstein's thought experiment – upon being properly analyzed and amended – yields π for the ratio C/D and thereby *suggests* that the disk geometry is Euclidean after all. But Strauss deduces from his own treatment in his paper the *non*-Euclidean conclusion at which Einstein had arrived in a manner that Strauss rejects as fallacious. Moreover, Strauss promised there a generalized treatment in which the disk radius is no longer taken to be invariant. And he has informed us (private communication) that this generalization has now been completed.

It now emerges from our argument that in our Equation (3) – which is no longer merely heuristic but applies *within* the rotating S! – the value of r corresponding to any given value k of ρ also has the following status within the rotating S: This r-value is the *distance* between the center and any point on the particular circle $\rho = k$. The reason is that the value of r in question is the length $\int_{\rho=0}^{\rho=k} dr$ of the segment of any radial *geodesic* $\theta' =$ constant which links the center of the disk to a point on the circle $\rho = k$. Hence Equation (3) provides us with *one* of the ways in which we can introduce the values of r that satisfy it as *new radial coordinates* of a new system (r, θ') of *polar* coordinates on S whose origin is the center of S. And in virtue of the simultaneous *metrical* significance of the values of r, Equation (3) then also exhibits the implementation of our initial requirement that during uniform rotation, the pre-rotational circles $\rho =$ constant continue to qualify metrically as circles, although their size may change. For Equation (3) shows that lines of constant ρ are each also lines of constant r.

Let us express schematically the spatial metric prevailing on S *during uniform rotation* in the two coordinate systems (r, θ') and (ρ, θ) as follows:

$$d\sigma^2 = g'_{11}\, dr^2 + 2g'_{12}\, dr\, d\theta' + g'_{22}\, d\theta'^2 \tag{5}$$

and

$$d\sigma^2 = g_{11}\, d\rho^2 + 2g_{12}\, d\rho\, d\theta + g_{22}\, d\theta^2. \tag{6}$$

We are now ready to consider the deducibility of the respective sets of functions in (5) and (6) which represent the metric tensors in the two coordinate systems (r, θ') and (ρ, θ) during the states of uniform rotation. Being an invariant, the $d\sigma$ of the schematic Equation (5) is, of course, the same as the $d\sigma$ of Equation (6). Once the metric coefficients g_{ik} appropriate to Equation (6) have been made specific, let '$d\sigma(\rho, \theta)$' denote the specific representation of $d\sigma$ in the *original* coordinate system (ρ, θ), and let '$d\sigma(r, \theta')$' denote the corresponding representation in the new (r, θ')

coordinate system. As we shall see, the assignment of the new θ' coordinates depends on the particular twisting history of the disk's radial lines, whereas there is, of course, no such dependence in the assignment of either of the original coordinates ρ and θ. It will turn out that therefore the $d\sigma(\rho, \theta)$ representation will exhibit much more perspicuously than $d\sigma(r, \theta')$ the following important fact: Either representation of $d\sigma$ stands for an infinite class K of metrics on the surface S whose members generate different equivalence classes of intervals on S by being associated with a whole gamut of permissible twisting histories of the disk. Thus, for a given choice of unit there is one rotational metric in K which results if S's uniform rotational motion was not preceded by any twisting of the radial lines, while there will be infinitely many other members of K that correspond to different histories of non-vanishing twist.

A comparison of the class K of rotational metrics $d\sigma$ with the pre-rotational Euclidean metric $d\sigma_0^2 = d\rho^2 + \rho^2 \, d\theta^2$ will then show at a glance just how – after due correctional allowance for Reichenbachian 'differential' forces! – the coincidence behavior of a thus 'standard' rod, which physically realizes these metrics for $\omega \neq 0$ and $\omega = 0$, changes as between the pre-rotational and the uniformly rotating states of the disk. Finally, we shall consider critically an allegation of paradox against the conclusion that there has been such a change of coincidence behavior and/or of the metric tensor.

Before deriving the g_{ik} functions needed for Equation (5), some ancillary discussion of twisting is needed.

Recall the way in which we introduced the lines $\theta' = \text{constant}$ and that their geodesicity was then proven (cf. note 2). To say that the *original* radial lines $\theta = \text{constant}$ were twisted during the disk's transition from rest to uniform rotation is to say that once uniform rotation is achieved, the lines $\theta = \text{constant}$ no longer qualify as geodesics and hence are no longer 'radial lines'. Thus, when there is a non-zero twist, the radial geodesics $\theta' = \text{constant}$ are not identical with the lines $\theta = \text{constant}$. We have allowed for the possibility of twisting all along, because its occurrence would appear to be compatible with the requirements that define our problem.

Mathematically, the existence of non-zero twist rules out that, in the case of *arbitrary* point pairs, $d\theta' = d\theta$. For if the latter were quite generally true for arbitrary pairs of points, we could deduce that $\theta' = \theta + k$ (where k is a constant), a relation which is interdicted by non-zero twist because the relation asserts the *identity* of the two families of lines $\theta' =$ constant and $\theta =$ constant. But when there is a non-zero twist, it is *not* ruled out that *for* $\rho = constant$ (i.e., for $d\rho = 0$), $d\theta' = d\theta$: The latter at least *allows* that $\theta' = \theta + f(\rho)$, where f is some *non*-constant function, and thereby allows the twisting codified by that equation.

Indeed, the dynamical conditions which we postulate are such as to require that even when there is twisting, the following condition be satisfied: *If* $d\rho = 0$ for a given pair of points, then $d\theta' = d\theta$ holds for this pair. The latter condition expresses the requirement that a kind of axial or circular symmetry be preserved even under twisting. And it is of a piece with our earlier assumption that none of the metrical coefficients of the rotational metric may depend on θ'.

The newly stated assumption that for $d\rho = 0$, $d\theta' = d\theta$ enables us to determine more specifically the nature of the coordinate transformation $\theta' = \theta'(\rho, \theta)$ even when twisting takes place. Taking differentials of this transformation yields $d\theta' = (\partial\theta'/\partial\theta)\, d\theta + (\partial\theta'/\partial\rho)\, d\rho$. The imposition of our newly stated assumption then entails that $\partial\theta'/\partial\theta = 1$, so that

$$(7) \qquad \theta' = \theta + f(\rho)\,,$$

where f is an unspecified function. When there is no twist, $f(\rho) = k$, where k would naturally be taken to be zero. Any particular twisting history will issue in a particular outcome which is codified by a specific function f. Thus it appears from Equation (7) that for any given point P on S whose original (ρ, θ) coordinates are (ρ_1, θ_1),the θ'-coordinate assigned to P by Equation (7) will depend not only on θ_1 but also on the twisting function f. On the other hand, it is evident from Equation (3) that there is no dependence on f of the r-coordinate assigned to P: The value r_1 of P's r-coordinate is there seen to depend only on ρ_1.

Now, the new metrical coefficients g'_{ik} in Equation (5) are functions of the new coordinates (r, θ') *as assigned via Equations* (3) *and* (7). Hence the following state of affairs should not occasion any surprise: Although the very assignment of the new θ'-coordinates does depend on the twisting function f, $d\sigma(r, \theta')$ as such *conceals* the role of f in the initial assignments of the new θ' coordinates. Thus it will turn out that *once the new coordinates are univocally assigned* via Equations (3) and (7), the corresponding new metrical coefficients g'_{ik} and thereby the representation $d\sigma(r, \theta')$ *conceal* an important fact as follows: Qua metric defined on the points of the surface of *S*, *dσ does depend on the disk's twisting history*. On the other hand, since the assignment of the original ρ, θ coordinates does *not* depend on *f*, one can expect that the dependence of $d\sigma$ on the twisting history *will* be explicit in $d\sigma(\rho, \theta)$, and we shall show that this expectation is borne out via the first derivative of the twisting function *f*.

We are now ready to turn to the determination of the functions g'_{11}, g'_{12} and g'_{22} in Equation (5). Let $d\theta' = 0$. Then $d\sigma^2 = g'_{11}\, dr^2$. But as we saw, when $d\theta' = 0$, a radial element has length $d\sigma = dr$, so that $d\sigma^2 = g'_{11}\, dr^2$ becomes $dr^2 = g'_{11}\, dr^2$ which yields $g'_{11} = 1$. Since we showed that lines of constant ρ are also lines of constant *r*, our original assumption of orthogonality for the families $\theta' =$ constant and $\rho =$ constant is tantamount to the orthogonality of the families $r =$ constant and $\theta' =$ constant. Hence $g'_{12} = 0$.

Next let $dr = 0$ so that Equation (5) becomes $d\sigma^2 = g'_{22}\, d\theta'^2$. Recall via Equation (4) that $dr = 0$ iff $d\rho = 0$ and also our 'axial symmetry assumption' that even if there is twisting, if $d\rho = 0$, then $d\theta' = d\theta$. But since we recall from Einstein's operation (a) that the *rotational* length $d\sigma$ of a *tangential* arc element $d\theta$ is $\rho\, d\theta$, that length is also $\rho\, d\theta'$ and we can write for $dr = 0$,

$$d\sigma^2 = \rho^2\, d\theta'^2 = \frac{r^2}{1 - r^2\omega^2/c^2}\, d\theta'^2 = g'_{22}\, d\theta'^2$$

where we used Equation (2) to express ρ^2 as a function of *r*. It follows that $g'_{22} = r^2/(1 - r^2\omega^2/c^2)$. These functions g'_{11}, g'_{12} and g'_{22} of the polar coordinates on the rotating disk *S* are the same as those which are usually obtained by using the 4-dimensional space-time metric as a point of departure (cf. Equations (8.80) on p. 272 of Møller, 1972). Using these functions in Equation (5), we get

$$d\sigma^2 = dr^2 + \frac{r^2}{1 - r^2\omega^2/c^2}\, d\theta'^2 . \tag{8}$$

As expected, the $d\sigma^2(r, \theta')$ of Equation (8) conceals the dependence of its metric $d\sigma$ on the disk's twisting history.

To obtain the functions g_{ik} needed for the derivation of $d\sigma^2(\rho, \theta)$ from Equation (6), we use Equations (4) and (2) as well as Equation (7). Since

θ' is a function of θ and ρ in Equation (7), we have $d\theta' = (\partial\theta'/\partial\theta)\, d\theta + (\partial\theta'/\partial\rho)\, d\rho = d\theta + f'(\rho)\, d\rho$. Hence Equation (8) becomes

$$d\sigma^2 = \frac{1}{(1+\rho^2\omega^2/c^2)^3}\, d\rho^2 + \rho^2[d\theta + f'(\rho)\, d\rho]^2$$

or

$$(9)\qquad d\sigma^2 = \left[\frac{1}{(1+\rho^2\omega^2/c^2)^3} + \rho^2[f'(\rho)]^2\right] d\rho^2 + 2\rho^2 f'(\rho)\, d\rho\, d\theta + \rho^2\, d\theta^2 .$$

By 'non-zero twist' we mean precisely that $f(\rho) \neq$ constant and hence that $f'(\rho) \neq 0$, whereas zero twist means that $f(\rho)$ is constant and hence that $f'(\rho) = 0$. Therefore it is the first derivative of $f(\rho)$ rather than $f(\rho)$ itself which 'measures' the twist. It follows that for zero twist, Equation (9) becomes

$$(10)\qquad d\sigma^2 = \frac{1}{(1+\rho^2\omega^2/c^2)^3}\, d\rho^2 + \rho^2\, d\theta^2 .$$

Comparison of the rotational metrics of Equations (9) and (10) with the *pre*-rotational Euclidean metric $d\sigma_0^2 = d\rho^2 + \rho^2\, d\theta^2$ now yields the following important conclusions:

(1) Regardless of whether the twist is non-zero $[f'(\rho) \neq 0]$ or zero $[f'(\rho) = 0]$, each of our rotational metrics $d\sigma$ generates an equivalence class of intervals which is different from the one generated by the Euclidean metric $d\sigma_0$. But as we already know from Einstein's operation (a), for *purely tangential* intervals $d\theta (d\rho = 0)$, all of the $d\sigma$ metrics in Equation (9) *agree* with the Euclidean $d\sigma_0$ metric.

(2) Since the rotational metrics of Equation (10) and the pre-rotational metric $d\sigma_0$ are physically realized on the rotating S by the rigid unit rods of the co-moving local inertial frames, we can say the following: As measured by the rigid rods of the co-moving local inertial frames, an infinitesimal interval on S which is *not* purely tangential ($d\rho \neq 0$) typically changes its length in the course of the disk's transition from rest to uniform rotation, because the metrical coefficients g_{11} and g_{12} of the rotational metric $d\sigma(\rho, \theta)$ in Equation (9) generally have different values from the corresponding coefficients in the pre-rotational metric $d\sigma_0(\rho, \theta)$. Speaking *non*-infinitesimally, for non-tangential intervals, the rigid *unit*

rods of the *co-moving local inertial frames* typically do not coincide with the same intervals on S as did the rigid unit rods in S during the latter's *pre*-rotational state. This can be seen by noting that if the definite integral of $d\sigma_0$ yields 1 along the (geodesic) path of an (appropriate) interval i between the limits (ρ_1, θ_1) and (ρ_2, θ_2), then typically the corresponding definite integral of $d\sigma$ in Equation (9) will not yield 1. In the zero twist case of Equation (10), the metrical coefficient g_{11} is less than 1 (for $\rho \neq 0$), and calculation shows that in this special case, we can assert more strongly that the definite integral in question *never* yields 1 but always *less*.

The stated difference in coincidence behavior as between the pre-rotational and rotational metric findings can be formulated *elliptically* by saying that 'rigid rods on S change their coincidence behavior' during the transition in question. The latter statement is also an ellipsis for the claim that *after* being corrected for the effects of Reichenbachian 'differential' forces, which are associated with accelerations on the disk (e.g. centrifugal accelerations), solid rods that are *thus* metrical *standards* change their coincidence behavior as between the pre-rotational and rotational states of S. A solid rod, one of whose termini is *attached* to the rotating disk, will suffer the distortions associated with the accelerations sustained by it, whereas a unit rod in the *co-moving* local inertial frames escapes these distortions and can remain rigid (for pertinent details, see Møller, 1972, p. 254). It was this latter difference in rigidity between the measuring rod *attached* to S at one of its ends, on the one hand, and the co-moving measuring rod which is not thus attached, on the other, that prompted us to invoke the co-moving rods to give one of our descriptions of the change in coincidence behavior.

(3) The infinite class of functions $f'(\rho)$ has a subset such that this subset generates infinitely many metrics $d\sigma$ in Equation (9) which differ other than by a constant factor. And the latter infinitude of metrics, in turn, has a subset whose members generate infinitely many different equivalence classes of intervals on S.

3. A PURPORTED PARADOX

The preceding conclusions can be put into sharper focus by dealing with a putative allegation of paradox which may be as *prima facie* plausible as it is fallacious.

This allegation might take essentially the following form. For simplicity, consider the zero twist case of Equation (10). If the analysis in the preceding Section 2 were correct, it would also have to hold when we employ on S unit rods made of the same kind of solid material as the disk S itself. Now, if one such rod R did coincide *pre*-rotationally with a given radial segment PQ of the disk, then R's execution of a measurement of PQ after the onset of uniform rotation would require that R be kept at rest on S and in touch with PQ. But if this is to be accomplished, R must be subjected to appropriate forces no less than PQ qua part of S. Hence R will suffer the same accelerations as PQ itself, and – being made of the same kind and amount of material – PQ and R will sustain the same distortions relatively to I. Therefore R and PQ will preserve their pre-rotational coincidence even under uniform rotation. The objection might then continue: Let it be offered as a defense that once the rod is suitably corrected for its distortions by Reichenbachian 'differential forces' in the fashion detailed by Møller (1972, p. 254), then the radial segment PQ will turn out to have shrunk. But the reply is "Why not also similarly correct the length of the allegedly shrunken radial segment PQ, all the more so since it is made of the same solid material as R?" In short, by virtue of its very mensurational function, R effectively becomes as much part of the disk as PQ itself. And how then could there possibly be a destruction of R's pre-rotational coincidence with PQ as a result of the uniform rotational motion?

This objection calls for two sets of comments:

(a) For the sake of specificity, let the end points of the radial segment PQ be $\rho = 0$ and $\rho = 1$. The theory set forth in Section 2 claims via Equation (4) that PQ's pre-rotational length changes from unity to the smaller value

$$(11) \qquad r = \int_0^1 \frac{1}{(1+\rho^2\omega^2/c^2)^{3/2}}\, d\rho$$

Now, it is true but beside the point that if unit rod R had been literally *nailed* to PQ before the onset of rotation so as to coincide with it and had remained nailed in this fashion, then there would have been no destruction of their pre-rotational coincidence. To 'incorporate' R in PQ is *not* to measure PQ by means of R. The relevant question is: What would R reveal if only *one* of its termini were nailed to P or Q? Observe that such *one*-ended nailing in no way assures a concordance with our dynamical constraints on *PQ* qua part of *S* such that the *pre*-rotational coincidence behavior of *R* with *PQ* must be preserved. Also note that *R*'s *uncorrected* findings would depend on its particular constitution. Thus, if the mensurationally-ascertained lengths are *not* to have that kind of dependence, the latter must be eliminated correctionally in such a way that corrected rods of diverse constitution are able to function metrically *in interchangeable fashion* on one and the same disk. Thus, even if a rod *R* that happens to be made of the same material as *S were* to have preserved coincidence with *PQ* under the transition to uniform rotation, this would not suffice to show that the length of *PQ* had remained unchanged. Indeed, as explained by Møller (1972, p. 254), it is a fundamental assumption of the theory that after computational allowance for the distorting effects of Reichenbachian differential forces, the rods on *S* will yield the same metrical findings on *S* as the unit rigid rods of the corresponding co-moving local inertial frames. And, as we explained, the latter *unaccelerated* co-moving rods do physically realize the rotational metric $d\sigma$ of Equation (9).

(b) Moreover, what would it *mean* to 'correct' not just the length of *R* but the length of *PQ* itself? Exactly what would be the physical significance of the latter 'correction'? Being concerned with the interval-lengths and 2-geometry on the disk *during* its uniformly rotating state, it is clearly unavailing to decree that *PQ*'s *pre*-rotational length be regarded as the basis for 'correcting' its rotational length after all! 'Correcting' the rotational lengths of intervals on *S* in this fashion is on a par with 'correcting' the 2-dimensional hyperbolic geometry of a saddle-shaped rubber sheet to be Euclidean 2-space merely because the latter state of the sheet has been obtained by stretching and/or compressing (distorting) a sheet of rubber which was originally in the shape of part of a Euclidean plane with respect to rigid rods! By the same token, even if it were not similarly ill-conceived to 'correct' the rotational lengths of intervals on *S* by decreeing that the lengths in *I* of their respective simultaneity-projections into *I* be their 'true' lengths, this would be unavailing in the case at hand: Since we obtained Equation (11) via Equation (4), we can recall that the radial shrinkage from unity to the *r*-value of Equation (11) is mirrored quantitatively by *PQ*'s respective simultaneity-projections into *I*.

Once the wrong-headed enterprise of 'correcting' the rotational lengths of intervals on S is abandoned, the unaccelerated rigid co-moving rods can be used as a *tertium comparationis* both to specify these lengths and to correct the accelerated rods each one of which is nailed to S at *one* of its termini.

It would seem that the stated allegation of paradox is not cogent.

University of Pittsburgh

NOTES

* The authors are indebted to the National Science Foundation for support of research.

[1] See the valuable bibliographies in Arzeliès (1966, pp. 241–243) and Grøn (1975, p. 876). As a mere example of recent discussions not listed in these bibliographies, see Atwater (1971), Cavalleri (1972), Marsh (1971), Newburg (1972), Noonan (1971), Suzuki (1971), and Whitmire (1972).

[2] The proof is as follows:
The geodesic equation expressed in terms of an arbitrary parameter p takes the form

$$\text{(i)} \qquad \frac{d^2x^\mu}{dp^2}+\begin{Bmatrix}\mu\\ \alpha\beta\end{Bmatrix}\frac{dx^\alpha}{dp}\frac{dx^\beta}{dp}=A\frac{dx^\mu}{dp}$$

for some unspecified A, where the usual summation convention is employed. We wish to show that any path $\theta' =$ constant satisfies Equation (i). If we now let $x^1 = r$ and $x^2 = \theta'$, the path will be such that $dx^2/dp = 0$, but $dx^1/dp \neq 0$. When $\mu = 1$ in Equation (i), we may take the resulting equation to define A. Since for $\mu = 2$ the right-hand side of Equation (i) vanishes, the geodecicity of $\theta' =$ constant will be established by showing that the left-hand side of Equation (i) vanishes for $\mu = 2$. If we use $d\theta'/dp = 0$, the left-hand side of Equation (i) becomes

$$\text{(ii)} \qquad \frac{d^2x^\mu}{dp^2}+\begin{Bmatrix}\mu\\ 11\end{Bmatrix}\left(\frac{dr}{dp}\right)^2.$$

Setting $\mu = 2$ in expression (ii) and using the definition of the Christoffel symbol in terms of the metric tensor, we obtain

$$\text{(iii)} \qquad \tfrac{1}{2}g^{2\alpha}(2g_{1\alpha,1}-g_{11,\alpha}),$$

where a comma denotes partial differentiation. The assumed orthogonality of the (r, θ') coordinate system limits α to the value 2. The first term in the parentheses in expression (iii) then vanishes because of the assumed orthogonality, and the remaining term vanishes because of the assumption that the metric tensor is independent of θ'.

BIBLIOGRAPHY

Arzeliès, H.: 1966, ***Relativistic Kinematics***, Pergamon, New York, ch. IX: 'The Rotating Disc', pp. 204–243. A *valuable* large bibliography is given on pp. 241–243.

Atwater, H. A.: 1971, *Nature Phys. Sci.* **230**, 197–198.
Cavalleri, G.: 1972, *Lett. Nuovo Cim.* **3**, 608.
Einstein, A.: 1905, 'On the Electrodynamics of Moving Bodies', *The Principle of Relativity: A Collection of Original Memoirs*, Dover Publications, New York, 1952.
Einstein, A.: 1916, 'The Foundation of the General Theory of Relativity', *The Principle of Relativity: A Collection of Original Memoirs*, Dover Publications, New York, 1952.
Grøn, Ø.: 1975, 'Relativistic Description of a Rotating Disk', *Amer. J. Phys.* **43**, 869–876.
Grünbaum, A.: 1967, 'Theory of Relativity', in P. Edwards (ed.), *The Encyclopedia of Philosophy*, Macmillan, New York, Vol. 7, pp. 133–140.
Grünbaum, A.: 1973, *Philosophical Problems of Space and Time*, 2nd ed., Dordrecht and Boston, Reidel.
Marsh, G. E.: 1971, *Nature Phys. Sci.* **230**, 197.
Møller, C.: 1972, *The Theory of Relativity*, 2nd ed., Oxford University Press, Oxford.
Newburg, R. G.: 1972, *Lett. Nuovo Cim.* **5**, 387–388.
Noonan, T. W.: 1971, *Nature Phys. Sci.* **230**, 197.
Reichenbach, H.: 1924 (reprinted 1965), *Axiomatik der relativistischen Raum-Zeit-Lehre*, Vieweg, Braunschweig (now Wiesbaden), Part II, ch. IV, § 44, pp. 136–141.
Reichenbach, H.: 1969 (English translation of [Reichenbach, 1924] by Maria Reichenbach with Foreword by Wesley C. Salmon), *Axiomatization of the Theory of Relativity*, Part II, ch. IV, § 44, pp. 172–177.
Strauss, M.: 1974, 'Rotating Frames in Special Relativity', *Int. J. Theoret. Phys.* **11**, 107–123.
Suzuki, M.: 1971, *Nature Phys. Sci.* **230**, 13.
Whitmire, D. P.: 1972, *Nature Phys. Sci.* **235**, 175–176.

Note added in proof: An important early discussion of the geometry of the rotating disk can be found in T. Kaluza, 'Zur Relativitätstheorie', *Physikalische Zeitschrift* **XI** (1910), 977–978. An important recent discussion, John Stachel, 'Einstein and the Rigidly Rotating Disc', will appear in *General Relativity and Gravitation*, an Einstein Centenary volume, Alan Held, *et al.*, eds.

O. COSTA DE BEAUREGARD

TWO LECTURES ON THE DIRECTION OF TIME

I. PHYSICAL IRREVERSIBILITY PROBLEMS*

Since long ago theoretical physicists and/or philosophers of science have been trying to get at the knot of a truly puzzling enigma situated at the very heart of the problem of physical irreversibility. Poincaré [1], for instance, remarks that if the arrow of the Zeroth and Second Laws were reversed, if heat flowed from cooler towards warmer sources, and if viscosity were an accelerating rather than a damping force, no prediction would be possible. It would be, as Grünbaum [4] comments, dangerous to get into a lukewarm bathtub, because one could never know which end is going to boil and which to freeze; and it would be dangerous to play at bowls, because bodies would get in motion by themselves, in unpredictable directions and with unforeseeable velocities. Such a world would be a law-less world, insofar as our thinking conforms to the familiar experience of causes developing after effects. In our familiar world it is 'blind statistical retrodiction', in Watanabe's [2] words, which is impossible. As no records have been preserved, we cannot 'retrotell' the chain of events having generated the quasi-homogeneous swarm of the Little Planets. Neither can we retrotell at which point has been deposited the ink drop now completely diluted inside a glassful of water.

The point is that the paradoxical anti-Carnot world would be a world governed by final causes instead of initial causes. The problem is that perhaps there may be, inside our overall causal world, a small contribution from the anti-Carnot, or final world. Consider, for instance, the biological descent (in Darwin's words) of the horse. Can we safely predict the Horse given the Eohippus? Evidently not. However, 'blind statistical retrodiction' will say that the Eohippus has emerged from a homogeneous molecular soup – which in fact is the very truth. These are, in a short, informal, presentation, the problems we will be concerned with.

When Boltzmann used statistical mechanics for deducing the Second Law, the paradox inherent in extracting time asymmetry from a theory –

W. C. Salmon (ed.), Hans Reichenbach: Logical Empiricist, 341–366.

Newtonian mechanics – that is intrinsically time symmetric was very quickly exposed in the specific forms due to Loschmidt and to Zermelo. If at time instant t all velocities of a mechanical system are exactly reversed, this system will go back through all its previous states so that if, for example, the Boltzmann entropy was increasing in the 'natural' evolution, it will decrease in the reversed evolution, so that there should be just as many entropy decreasing as there are entropy increasing evolutions (Loschmidt). Also, by Poincaré's recurrence theorem, any isolated system with a finite number of degrees of freedom will pass, and has passed, infinitely many times arbitrarily near any specific configuration it has once gone through. And this is true also in the case of improbable configurations (Zermelo).

The so-called refutation of these two famous paradoxes has produced, in vain, flows of writing, from such prominent authors as Boltzmann, Poincaré, Borel, P. and T. Ehrenfest, and certainly quite a few others. These commentaries have uncovered many interesting facets of the problem, such as: occurrence of an 'initial' interaction of the system under study with a larger one, as in the production of the swarm of the Little Planets, by the explosion of a large planet; inevitable weak interaction of any so-called isolated system with the rest of the Universe, if only through Newtonian attraction; inevitable interaction of any system with the measuring apparatus at the moment an observation is made; retarded character of distant interactions, entailing the use of advanced waves in a Loschmidt type reversal; and so on. It is mainly around the sixties that quite a few physicists and/or philosophers of science have independently got the knot of the problem and shown how it is tied [4].

The significant remark is that the common root of the Loschmidt and the Zermelo paradoxes is a deeper one inherent in the probability calculus itself: how is it that, *even if the intrinsic transition probabilities between two states of a system are symmetric*, as for example in card shuffling, or in the exchange of a grain of powder from one square of a chessboard to another one, or in radioactive transitions, how is it that *macroscopic* evolutions always go from lower to higher probability complexions, and not the other way? This is a *fact* we are so used to that it seems a triviality. It is, however, a mathematical paradox, as is immediately made clear if we try to perform 'blind statistical retrodiction' (Watanabe's [2] words) in the same way that we trivially perform 'blind statistical prediction". Starting at time zero with a radionuclide, *blind statistical retrodiction* will tell us that the nuclide has just built up from its elements according to the law $n = n_0 \text{ Exp } (+at)$, which is time symmetric to the decay law $n = n_0 \text{ Exp } (-at)$. And, of course, we will not believe that; at least, *not in physics.* Or: there is nothing unusual if a physicist deposits with a pipette an ink drop inside a glassful of water, when the ink subsequently gets uniformly mixed with the water. But it would be quite unusual *if*, by dipping an empty pipette inside a glassful of mixed water and ink, a physicist induced the ink to concentrate immediately before, so that he could suck it up.

By these examples I purposefully lay aside the well known argument that if a system – consisting for instance of radionuclides and their decay products – were enclosed inside a completely adiabatic box, disintegrations and syntheses would be continuously going on, and fluctuations would continuously occur, even large ones in the long range. Such was the gist in the famous analysis by P. and T. Ehrenfest. However it is definitely not the answer to the question I am raising, this question being: How is it [3] that we can *at will* place a radionuclide inside an adiabatic box and observe later on the trivial decay law, but we cannot *at will* open the box and pick out the atom in its excited state? In other words: *Why* is it that blind statistical prediction is physical while blind statistical retrodiction is not?

For want of a convincing answer the classical probability calculus has at least an operational recipe, stating that, in retrodictive problems (which were termed *problems in the probability of causes* – a very revealing terminology, as we shall see) – one should use Bayes' conditional probability formula implying, besides the intrinsic transition probabilities of the system, also *extrinsic* probabilities describing at best what we know, or what we believe we know, concerning the connection of the system with its surroundings.

So, in this problem, Bayes conditional probabilities describe the *interaction* of the system under study with its surroundings. Our question then assumes the form: Why is it that, in physics, we are allowed to think of an *initial* interaction and not of a *final* one? Or, in other words, why is it that the interaction of our system with its surroundings can be thought of as developing effects after it has ceased, and not before it has begun? If we do not understand *why* things are so, at least *do* we know *that* they are so.

And this, of course, amounts to saying that our physical world is a *causal* rather than a *final* world. Inhomogeneous initial conditions do make sense in it, while inhomogeneous final conditions do not. Something more can be said: the latter statements are essentially macroscopic in character, which means that the concept of a *cause* developing *after effects* (as opposed to an *end* inducing *before effects*) *is a macroscopic, and by no means a microscopic concept.*

This conclusion has been reached more or less explicitly, and almost independently, in the sixties, by quite a few thinkers. As for the Bayesian form of the statistical irreversibility law there is a very often quoted sentence by Willard Gibbs [5]. More specifically, the explicit statement that the statistical deduction of the Second Law implies a time asymmetric use of Bayes conditional probability law, is found in a 1911 paper by Van der Waals [6].

As for the mathematical nature of the prescription stating that, in physics, blind statistical retrodiction is forbidden and interactions only develop after effects, it should be obvious by now that it is, in Mehlberg's [4] words, *factlike* and *not lawlike*. Technically speaking, it is a *boundary condition* for integrating, say, the 'master equation'. It is an *initial condition* as opposed to a *final condition.*

Now, there is in physics another well-known irreversibility principle, also expressed as a time asymmetric boundary condition: the principle of retarded waves, which also excludes one half of the mathematically permissible solutions of the wave equation. Moreover, all sorts of classical waves are derivable, through appropriate averaging procedures, from the quantum mechanical waves, which are, in the present state of our knowledge, the truly fundamental ones. Could it be that the (factlike rather than lawlike) principle of retarded waves is also of a macroscopic nature, and another expression of the physical irreversibility principle?

Physically speaking all sorts of examples support this view. A falling meteorite is slowed down in the Earth's atmosphere, through the emission of retarded ballistic and radiation waves, which provide the mechanism through which the overall entropy is made to increase. More specifically, if, between time instants t_1 and t_2, someone moves a piston in the wall of a vessel containing a gas in equilibrium, *the fact* is that Maxwell's velocity law is disturbed after time t_2, not before time t_1, and, moreover, that the perturbation is in the form of a retarded pressure wave emitted, not of an advanced pressure wave absorbed, by the piston. Also, the mechanism through which the entropy goes up inside an inhomogeneous gas is the mutual scattering of particles; but, according to wave mechanics, this is *also* a mutual scattering of the associated waves. And so on.

Logically speaking, the idea of an essential connection between the two principles of probability increase and of wave retardation has emerged in both the writings of Planck, and the controversy between Einstein and Ritz (1906–1909).

In this bitter controversy, Ritz was maintaining that the principle of retarded waves is implied in the law of increasing entropy, while Einstein was insisting that the law of retarded waves should be deduced from the principle of increasing probability. Both opponents were somewhat overlooking the *de facto* rather than *de jure* nature of the principle they were assuming, and neither seems to have suspected that, far from being contradictory to each other, their approaches were *reciprocal*. The reason why this happened is that, while of course Einstein's photon was known by that time, de Broglie's matter waves were not.

As for Planck's thinking around the turn of the century, Klein, in his book on Ehrenfest (Vol. 2) makes clear that Planck strongly suspected a link between the two irreversibility principles we are speaking of. In any case, the link is made explicit in Planck's formula stating that, even in phase coherent scattering, the entropy increases. But the very concept of scattering implies the principle of wave retardation. If, paradoxically, phase-coherent anti-scattering, or *confusion* instead of *diffusion*, could occur, then the entropy would go down.

I have pointed out quite a few times that, by using a theory where both the wave concept and the probability concept are inherent, namely, wave mechanics, the connection can be made explicit.

Consider first, as a thought experiment, the scattering of a light or of a matter wave by a grating. In order to avoid the subtleties of quantum statistics we may assume that the associated quanta are few, thus coming at long intervals, so that there is no more than one in each wave train; they are thus *discernible*, and Boltzmann statistics is the right one.

Given the wavelength and the grating, to each incident plane wave there corresponds a well defined set of a finite number, say g, of plane diffracted waves. Conversely, to any outgoing plane wave belonging to the preceding set, there corresponds one and the same finite set of g incident plane waves (comprising the one considered first), each able to generate the outgoing wave we are speaking of. Thus the grating induces a quantal transition of the particles, the probability matrix being a $g \times g$ square one. The principle of detailed balance holds, and thus everything is perfectly time symmetric at this fundamental level.

Macroscopically speaking, however, there is a very large *de facto* (if not *de jure*) time asymmetry. Starting with one incident plane wave we generate quite easily, through phase-coherent diffusion, the set of outgoing waves; moreover, their respective intensities can be calculated either through the laws of classical optics, or through the calculus of probabilities using blind statistical prediction. What we cannot do in fact, although we can do it in thought, is to generate the ghost set of incident

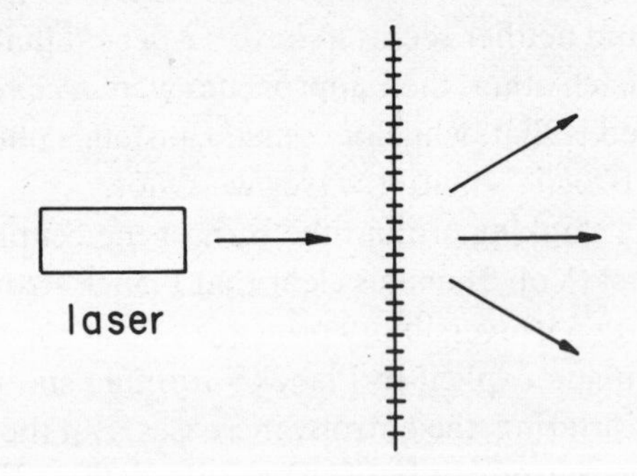

Fig. 1. Retarded waves, increasing probabilities, phase coherent diffusion.

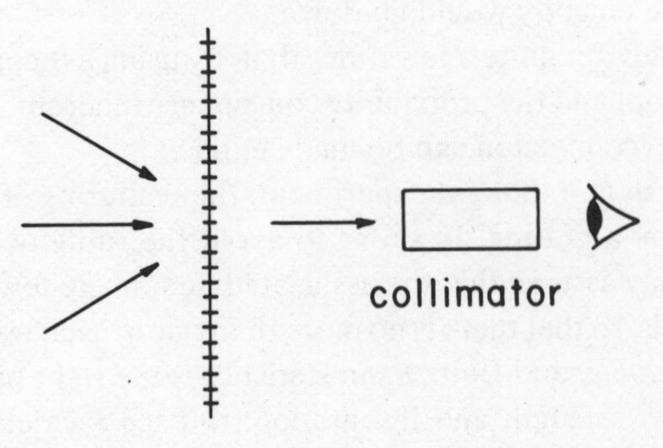

Fig. 2. Advanced waves, decreasing probabilities, phase coherent confusion.

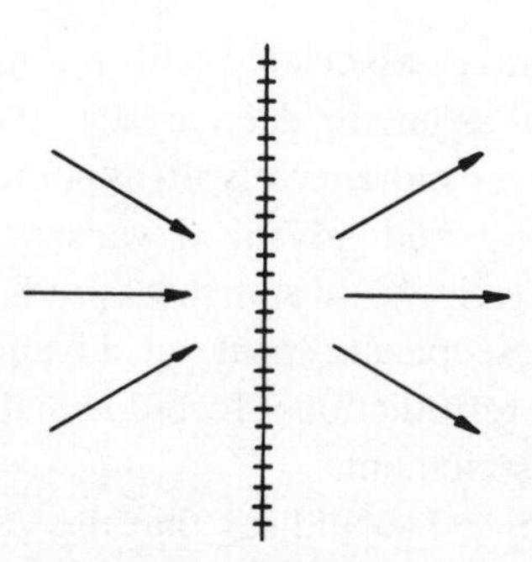

Fig. 3. Intrinsic, or *de Jure*, Symmetry of the Transition Matrix.

plane waves which would produce, in a phase-coherent confusion, the one outgoing plane wave received in a collimator. By this argument the one to one association of the twin principles of retarded waves and of increasing probabilities is made quite obvious – together with the *macroscopic* character of their common nature.

Let me recall that in his Ph. D. thesis M. von Laue has set up an optical apparatus preserving the phases of separate beams and finally producing, through, so to speak, an organized conspiracy, the phenomenon we are speaking of. The point is, however, that (at least in physics) this cannot be done by *merely* requiring the existence of an inhomogeneous final condition imposed upon the macroscopic waves. We know of macroscopic sources emitting waves and particles into the future, but not of macroscopic sinks sucking waves and particles from the past. This amounts to saying that, at first sight, a straightforward analogy drawn from Euclidean steady state hydrodynamics fails to work in Minkowskian wave-mechanical physics. Also, this of course amounts to saying that our physical world is a world of causality as opposed to a world of finality.

A general and abstract version of the preceding argument consists in merely re-reading von Neumann's [8] irreversibility proof in the quantum mechanical measurement process. Von Neumann has shown that the entropy of his ensembles increases whenever the same measurement is performed at time t upon all the members of the ensemble. The proof, however, implies that these measurements are used for blind statistical prediction, by means of the retarded waves produced by the transition at time t. If, instead, a blind statistical retrodiction were performed through use of the advanced waves associated with the measurement, then the entropy would turn out as having decreased.

All this can be said even more tersely. If, as pointed out by Fock [9] and Watanabe [4], retarded and advanced waves are used in quantum mechanics for, respectively, blind statistical prediction and retrodiction, then to say that, macroscopically speaking, advanced waves do not exist or that blind statistical retrodiction is forbidden, are merely two different wordings of the same statement.

Concluding this section I would say that the very *sui generis* form of the probability calculus inherent in quantum mechanics, together with its novel relation to wave amplitudes, making of it some sort of information telegraph through spacetime, is perhaps apt to uncover unexpected fundamentals of the relation between matter and mind.

I come now to discuss the irreversibility problem in terms of the physical equivalence between *neg-entropy* (minus the entropy) and *information*, the equivalence coefficient being $k \ln 2$ when the information is expressed in binary units, and the entropy in the usual thermodynamical units.

First of all, do we have to make our mechanics *statistical* only because our knowledge is incomplete or, (in more up to date words) is *entropy* merely *missing information*? This view was a very accepted one in classical statistical mechanics but, even there, a more refined discussion shows it to be questionable [11]. This problem is very much intertwined with the one of deciding whether probability should be considered as objective or subjective. Arguments in favor of both theses are so cogent that, in my opinion, there is strong reason to believe that *probability is indissolubly objective and subjective, being the hinge around which mind and matter are interacting*. This, of course, implies that we are living inside an irreducibly probabilistic world – namely, the world of quantum mechanics.

My next point, and an extremely important one for the present discussion, is that for Aristotle, the inventor of the concept, *information* was not only knowledge, as it is trivially today – one buys a newspaper one dime or two to find *informations* in it – it was also, symmetrically, an organizing power, the examples given being the craftsman's or the artist's work, and also biological ontogenesis. This second face of Aristotle's *information* seems today rather esoteric, though it is familiar to those few philosophers interested in will and in finality. The reason why the two faces of Aristotle's information are the one trivially obvious and the other one almost hidden is a straightforward corollary of physical irreversibility, as we shall see.

What may seem astounding is that cybernetics, without having searched for it, has hit upon the intrinsic symmetry of the two facets of Aristotle's information. For example, in computers and other information processing machines, the chain

information → *negentropy* → *information*

is considered, meaning that a concept is coded and sent as a message, before being decoded and received. The latter transition is the *learning transition*, where information shows up as *gain in knowledge*, while the former transition is the *willing transition*, where information shows up as an *organizing power*.

Intrinsically, or *de jure*, there is a complete symmetry between the two transitions. *De facto*, however, there is a dissymmetry, because irreversibility creeps in. Misprints in the coding, noise along the line, mistakes in decoding, all loose information, so that *de facto*, if not *de jure*, one has the chain of inequalities

$$I_1 \geqslant N \geqslant I_2$$

which, in cleverly built (and used) systems, may be very close to equalities, but which, in the typical context of classical physics, are extremely large inequalities.

This is made clear (as Brillouin [10] has pointed out) by the very smallness of Boltzmann's constant, k, in the equivalence formula

$$\Delta N = \underbrace{k \operatorname{Ln} 2}_{\simeq 10^{-16}} \Delta I$$

Fig. 4.
Negentropy: 'practical' thermodynamic units.
Information: binary units.

It is obvious that the very choice of the units we find *practical* belongs to what is called today *existentialism*, that is, the way we feel inside the world around us. For instance, the fact that c, the velocity of light in vacuo, is so large in units we find practical (say, meters and seconds) reflects a situation where relativistic effects are not at all obvious, so that, in everyday life, we act and think as Newtonians. I tend to believe that if, for us, the meter and the second are practical as associated units of length and time, this is because the velocity of our nervous influx is of the order of not too many meters per second. If, instead, this velocity was a large fraction of c, relativistic phenomena would become obvious in everyday life.

It is quite striking that the universal constants of the 20th century physics – Einstein's c, Planck's h, Boltzmann's k – are extremely large or small when expressed in practical units. This characterizes a somewhat inhuman physics, the phenomena of which lie quite outside the realm of everyday experience.

Now, it is a time-honored exercise to see what happens when we let go to infinity a very large, or to zero a very small, universal constant – how one thus loses Einstein and recovers Newton, or how one loses de Broglie's waves or Einstein's photons by respectively setting $1/h \to \infty$ or $h \to 0$.

What we lose when letting $k \to 0$ is the esoteric face of Aristotle's information. We describe a situation where gaining knowledge is absolutely costless, and producing order is utterly impossible. Consciousness is thus made totally passive; it registers what goes on without her, and that's all. Such a theory has been famous under the name of *epiphenomenal consciousness.*

The discovery of Cybernetics, however, is as Gabor [11] puts it, that "one cannot have anything for nothing, not even an observation." Cybernetics has made consciousness, as a spectator, pay her ticket – a very cheap price. But consciousness is thus allowed to be an actress also – just as Aristotle had conceived her.

So, the very smallness of Boltzmann's k (as expressed in practical units), reflects a situation where observation is easy, or common, and action painful, or rare on a large scale. In this light the pure Carnot irreversibility, where negentropy goes down uselessly to zero inside an isolated system, is the extreme limit of the 'learning transition', $N \to I$, $N \geq I$, where at least some knowledge is subtracted from the cascading negentropy.

What dawns in this light is that entropy decreasing, and advanced waves processes, are not strictly forbidden, but, instead, very rare in the general physical context. The progressing fluctuations must be (at least potentially) associated, on the subjectivistic side, with willing awareness, just as regressing fluctuations are potentially associated with learning awareness. And this, again, displays the progressing fluctuations as belonging to the category of finality as opposed to causality. This, incidentally, helps understanding why finality in biology (while almost obvious to the naive outlook) is so hard to demonstrate as such. This is, in my opinion, merely because knowing and willing consciousness are symmetrical to each other, the one being aware of causality and the other of finality.

So, the conclusion which all this intrinsic-symmetry sort of reasoning leads me to, is that Aristotle's – and Cybernetics' – symmetry between learning and willing information is reflected in space-time, through the sui generis probability theory of quantum mechanics, as the lawlike symmetry between retarded and advanced waves. This I intend to discuss in Part II because the really painful sting of the Einstein-Podolsky-Rosen paradox will oblige me to do so.

Today, remaining inside the comfortable Carnot state of affairs where causality largely prevails over finality, and passive observation over energetic will, one easily sees how microscopic information goes into macroscopic statistical frequency, and into negentropy considered as something objective; similarly the information aspect, or Copenhagen interpretation, of quantized waves, goes into the so-called objectivity of the various sorts of macroscopic waves. But I need not recall that if everything seems so quiet and familiar, it is only inasmuch as we are unaware of the hidden face of Aristotle's information.

I will end with a quotation from Brillouin

Relativity seemed at the beginning to yield only small corrections to classical mechanics. New applications to nuclear energy now prove the fundamental importance of the mass-energy relation. We may also hope that the entropy-information connection will, sooner or later, come into the foreground, and that we will discover where to use it to its full value [13].

Institut Henri Poincaré, Paris.

* Lecture delivered on 16 April, 1975, at the Massachusetts Institute of Technology (Thermodynamics Group).

BIBLIOGRAPHY

[1] Poincaré, H., *Science et Méthode*, Flammarion, Paris, 1908, Livre I, Chap. IV.
[2] Watanabe, S., *Phys. Rev.* **27** (1975) 179.
[3] See in this respect 'Discussion on Temporal Asymmetry in Thermodynamics and Cosmology', O. Costa de Beauregard, reporter, in *Proceedings of the International Conference on Thermodynamics held in Cardiff* (ed. by P. T. Landsberg), Butterworths, London, 1970.
[4] Lewis, G. N., *Science* **71** (1930) 569.
Weiszacker, C. F. von, *Ann. Phys.* **36** (1939) 275.
Schrödinger, E., *Proc. Roy. Irish Acad.* **53** (1950) A189.
Watanabe, S., *Phys. Rev.* **84** (1951) 1008. 'Réversibilité contre irréversibilité' in *Louis de Broglie, Physicien et Penseur*, A. George, (ed.), Albin Michel, Paris, 1952, p. 385. *Phys. Rev.* **27** (1955) 26, 40, 179. *Transport Processes in Statistical Mechanics*, I. Prigogine (ed.), Interscience, New York 1958, p. 285.
Costa de Beauregard, O., 'Irréversibilité quantique, phénomène macroscopique' in *Louis de Broglie, Physicien et Penseur*, A. George (ed.), Albin Michel, Paris, 1952, p. 401. [Irreversibility Problems, in *Proceedings of the International Congress on Logic, Methodology and Philosophy of Science*, Y. Bar Hillel (ed.), North-Holland Publishing Company, Amsterdam, 1964, p. 313. *Studium Generale* **24** (1971) 10.
Reichenbach, H., *The Direction of Time*, University of California Press, Los Angeles, 1956.
Yanase, N. N., *Ann. Japan. Assoc. Phil. Sci.* **2** (1957) 131.
Adams, E. N., *Phys. Rev.* **120** (1960) 675.
McLennan, J. A., *Phys. Fluids* **3** (1960) 193.
Terletsky, J. O., *J. Phys.* **21** (1970) 680.
Wu, T. Y. and Rivier, D., *Helv. Phys., Acta* **34** (1961) 661.
Mehlberg, H., 'Physical Laws and Time Arrow', in *Current Issues in the Philosophy of Science*, G. Maxwell (ed.), Holt, Rinehart, Winston, New York, 1961, p. 105.
Gold, T., *Amer. J. Phys.* **30** (1962) 403.
Grünbaum, A., *Phil. Sci.* **29** (1962) 146, *Philosophical Problems of Space and Time*, (A. A. Knopp) New York, 1963, p. 209.
Penrose, O. and Percival, I. C., *Proc. Phys. Soc.* **79** (1962) 493.
[5] Gibbs, J. W., *Elementary Principles in Statistical Mechanics*, Yale University Press, New Haven, Conn. 1914, p. 150.
[6] Waals, J. D. von, *Phys. Zeit.* **12** (1911) 547.
[7] Costa De Beauregard, O., *Dialectica* **19** (1965) 280 and **22** (1968) 187. See also [3].
[8] Costa De Beauregard, O., *Cahiers de Physique* **86** (1958) 323. *Studium Generale* **24** (1971) 10.
[9] Fock, V., *Dokl. Akad. Nauk SSSR* **60** (1948) 1157.
[10] Brillouin, L., *Science and Information Theory*, Academic Press, New York, 1956.
[11] Poincaré, H., [1] § I: "Il faut bien que le hasard soit autre chose que le nom que nous donnons à notre ignorance".
[12] Gabor, D., quoted by Brillouin [10], p. 90.
[13] Brillouin, L., [10], p. 294.

II. THE EINSTEIN-PODOLSKY-ROSEN PARADOX*

In my preceding lecture I have expressed the opinion that probability should be said to be neither objective nor subjective, but rather *indissolubly objective – and – subjective*, being the hinge around which mind and matter are interacting; that the sui generis form of probability calculus inherent in quantum mechanics may well be the natural one in this respect; that the way in which it associates wave amplitudes with the expression of probabilities perhaps displays quantum mechanics as the existentialistic sort of space-time telegraph through which living psyches are exchanging *information* in both of Aristotle's senses, namely, *gain in knowledge* and *organizing power*; finally, that the intrinsic symmetry between the *learning transition*, *negentropy → information*, and the *willing transition*, *information → negentropy*, is, so to speak, projected in space-time by quantum mechanics in the form of the intrinsic symmetry between retarded and advanced waves.

I have also stressed that this *de jure* intrinsic symmetry is somewhat obliterated by a large *de facto* asymmetry expressed in the various forms of entropy (or probability) increase, of wave retardation, and of preponderance of information as knowledge over information as will. The mathematical formulation of this *de facto* asymmetry is of the nature of a boundary condition, namely, heterogeneous initial conditions macroscopically permitted, but heterogeneous final conditions macroscopically forbidden. That is, if physical irreversibility is a law of nature, then nature wears in this case a lawyer's robe. In other words, the physical irreversibility statement looks very much like the sign post stating 'one way direction' on highways [1].

There exists in physics another important instance where a *de jure* or intrinsic mathematical symmetry is concealed by an extremely large *de facto* asymmetry. According to Dirac's theory the trivial electron and the very rare positron are perfect twins to each other. Moreover Feynman, by using a time symmetric thinking very much akin to the one I used in Part I, has conferred to this symmetry the most elegant and useful form. I need not say that, by unravelling the consequences of this *de jure* symmetry, that is, by following the Ariadne thread of logic and mathematics, physicists have been introduced to the whole world of antiparticles.

What I am undertaking now is very analogous. I intend to display the nature of the *de jure* symmetry which is almost completely hidden by the *de facto* very large asymmetry expressed by the physical irreversibility principle. Moreover, I intend to show that a typical place where this intrinsic symmetry crops out in physics is the (presently much discussed) Einstein-Podolsky-Rosen Paradox. If, then, we refer to the analogy between our problem and the Dirac and Feynman electron-positron one, and if by definition the Second Law belongs to physics, then the Anti-Second Law we aim at exploring should logically be said to characterize *anti-physics.*

First I will discuss the specific nature which the *stochastic event* inherent in every sort of statistical mechanics, and indeed of probability calculus, assumes in quantum mechanics. It is also called there *quantal transition*, or often *collapse of the state vector* or *reduction of the wave function*. The latter concept has sometimes been questioned, usually by following a frequency (as opposed to information) style of thinking. However the very specificity of quantum mechanics is lost if one bypasses the discussion of the individual stochastic event. And the key which its novelty may well provide for unlocking the secrets of how mind and matter are interacting is also lost. I cannot see what else than a 'collapse of the state vector' the quantal transition could be. And I deem ominous the binding of the two symmetries of Aristotle's twin faces of information, and of retarded and advanced waves, as implied in quantum mechanics.

Von Neumann, and, following him, London and Bauer [2] in a very clear and concise booklet, have drawn what seems to be the only conceivable interpretation of the reduction of the wave function; namely, that it is due to an act of consciousness of the observer. Setting aside for a moment the problem of how this can happen if no physicist is present with his registering apparatus, I will discuss first the nature of that act of consciousness. At this end I look first at the mathematics, and I purposefully do so in a relativistically covariant way.

Figure 1 contains a résumé (in the special case of a free particle of unspecified spin) of the completely relativistic formalism I have derived in a little book [3] for the first quantized wave mechanics. Equations (1) are the Klein-Gordon equation in its space-time and its 4-frequency representations; $\eta(k)=0$ is the mass shell with its positive and negative energy sheets η_+ and η_-; (2) and (3) are the reciprocal Fourier transforms of the wave function, expressed in (5) in the Dirac *bra* and *ket* formalism;

$$(0)\ c=1;\ \hbar=1;\ \lambda,\mu,\nu,\rho=1,2,3,4;\ x^4=ict$$

$$(1)\ (\partial^\lambda_\lambda-m^2)\psi(x)=0\begin{cases}(k_\lambda k^\lambda+m^2)\theta(k)=0\Rightarrow\\ \theta(k)=0 \text{ or } \eta(k)\equiv k^\lambda k_\lambda+m^2=0\end{cases}$$

$$(2)\ \psi(x)=(2\pi)^{-3/2}\iiint_\eta e^{ik_\lambda x^\lambda}\theta(k)\varepsilon(k)\,d\eta$$

$$(3)\ \theta(k)=-\frac{i}{2m}(2\pi)^{-3/2}\iiint_\sigma e^{-ik_\lambda x^\lambda}[\partial_\mu]\psi(x)\,d\sigma^\mu$$

$$(4)\begin{cases}\varepsilon_{\lambda\mu\nu\rho}d\eta^\lambda=-i[dk_\mu d\,dk_\nu\,dk_\rho], & k^\lambda d\eta=-m\,d\eta^\lambda\\ \varepsilon_{\lambda\mu\nu\rho}\,d\sigma^\lambda=-i[dx_\mu\,dx_\nu\,dx_\rho], & k^\lambda\,d\eta_\lambda=m\,d\eta\\ [\partial_\mu]\equiv\underset{\rightarrow}{\partial}_\mu-\underset{\leftarrow}{\partial}_\mu\end{cases}$$

$$\varepsilon(k)=\begin{cases}+1 \text{ on } \eta_+\\ -1 \text{ on } \eta_-\\ 0 \text{ otherwise}\end{cases}$$

$$(5)\ \langle x|a\rangle=\langle x|k\rangle\langle k|a\rangle;\qquad \langle k|a\rangle=\langle k|x\rangle\langle x|a\rangle$$

$$(6)\ \langle x|k\rangle=\langle k|x\rangle^*=\begin{cases}(2\pi)^{-3/2}e^{ik_\lambda}x^\lambda & \text{if}\quad \eta(k)=0\\ 0 & \text{otherwise}\end{cases}$$

$$(7)\ -\frac{i}{2m}\iiint_\sigma\bar\psi_a[\partial_\lambda]\psi_b\,d\sigma^\lambda=\iiint_\eta\bar\theta_a\theta_b\varepsilon(k)\,d\eta=\delta_{ab}$$
$$\equiv\langle a|b\rangle\equiv\langle a|x\rangle\langle x|b\rangle\equiv\langle a|k\rangle\langle k|b\rangle$$

$$(8)\ D(x-x')\equiv\langle x|x'\rangle\equiv(2\pi)^{-3}\iiint_\eta e^{ik_\lambda(x^\lambda-x'^\lambda)}\varepsilon(k)\,d\eta$$
(zero outside of light cone)

$$(9)\ 2D\equiv D_+-D_-=D_r-D_a$$

$$(10)\ \langle x'|x''\rangle=\langle x'|x\rangle\langle x|x''\rangle\quad\text{if}\quad x'\neq x'';\quad \text{valid iff } \langle x'|x''\rangle=0$$

$$(11)\ \langle x|a\rangle=\langle x|x'\rangle\langle x'|a\rangle$$

Fig. 1

Equations (4) define the vector and scalar 3-volume on η, the vector 3-volume on a space like surface σ, and the well known Schrödinger or Gordon current operator; (7) is the Parseval equality, and is used for defining orthonormal bases; (8) defines the well known Jordan-Pauli propagator, which is a covariant generalization of Dirac's δ; it is odd in $x^\lambda - x'^\lambda$ and zero outside the light cone; it can be expressed as the difference between the full retarded and the full advanced propagator; formula (10) is easily shown to be consistent if, and only if, $x' - x''$ is space-like, and then displays the orthogonality of the D propagators having their apexes on a space-like σ; (11) solves formally the Cauchy problem, and can be considered as the expansion of the wave function at any space-time point upon the complete orthogonal set of D's we were just speaking of. (9) displays a *necessary* binding of positive *and* negative energies with retarded *and* advanced waves.

Formula (7), for $a = b$, is the flux of the Gordon current. It gives in its x form the probability distribution for the particle going through any given element on the space-like σ. If the particle goes through an infinitesimal element of σ, then it belongs to the corresponding eigenfunction, or propagator D; that is, it has come inside the past light cone and will go inside the future light cone. This is a relativisitic formulation of a position measurement performed upon a particle, and is an example of the completely relativistic treatment of first quantized wave mechanics I have developed in my book [3].

So far everything is absolutely time symmetric; and this, again, is typical of all single-events of quantum mechanics. How does it happen, then, that the expression *collapse of the wave function*, quite obviously committed to the full retarded waves philosophy, is the accepted one? It is, in my opinion, because the Copenhagen school has forgotten the hidden face of Aristotle's information.

The principle of retarded waves is of a *de facto* and macroscopic nature, implying that one speaks of statistical frequencies. But how can one speak of statistical frequencies in the case of a position measurement performed around the point instant x_0^λ, which by definition is *unique* in space-time? One must, at this end, postulate the existence of a space-time translational invariance allowing for *repetitions* of the measurement by means of so-called *identical* experimental setups. Then all the apexes x_0^λ of our position measurements can be ideally brought into coincidence at one single point instant x_0^λ by appropriate space-time translations, so that we can then speak of the statistical frequency of a subsequent or preceding position measurement, performed inside the future or past light cone of

x_0^λ. It is at this very step that one is willing to believe in blind statistical prediction and unwilling to believe in blind statistical retrodiction – willing to believe in physics but unwilling to believe in anti-physics. However, we are logically bound to explore anti-physics, and this, if only by virtue of the harshness of the Einstein-Podolsky-Rosen Paradox.

Let me first recall how dictionaries define a paradox. For instance Funk and Wagnall's 'Standard Dictionary of the English Language' gives as sense 1, or fundamental: A statement, doctrine, or expression seemingly absurd, or contradictory to common notions, or to what would naturally be believed, but in fact really true. In this sense Copernicus' and Einstein's doctrines have certainly been paradoxical. Tomorrow will say if the logically deduced propositions I will expand now are 'in fact really true'.

Einstein *et al.* [4] have formulated their paradox in non-relativistic quantum mechanics. It is, however, in relativistic quantum mechanics that its full import is made clear. Suppose that, around point-instant O, a composite system Ω in a pure quantum state explodes in two subsystems α and β which fly apart, and are subsequently observed around point instants A and B (Figure 2); both vectors OA and OB are time like, while the vector AB is quite possibly space-like, and eventually quite large. The wave reduction occurs at both places A and B (if both measurements are performed), and the quantum formalism is such that there exist pairs of associated magnitudes which, respectively measured on α and β, are in a one to one correspondence. For instance, if the initial system was a positronium atom at rest around point $x = y = z = 0$ in the laboratory frame, and we make it explode around time $t = 0$ by imparting to it an energy but no momentum, then, if observer A find that the decay electron passes around time T around the point $X_a = R$, $Y_a = Z_a = 0$, then he is sure that if observer B performs a similar measurement upon the decay positron around time T, he is bound to find that it passes around point $X_b = -R$, $Y_b = Z_b = 0$. Of course we have assumed both R and T to be large with respect to the space-time domains at O, A and B. The strict correlation follows from the conservation of linear momentum.

There is nothing unusual in this as long as we can believe in hidden determinism. If our problem were one in classical statistical mechanics, the directions where the electron α and the positron β are ejected would

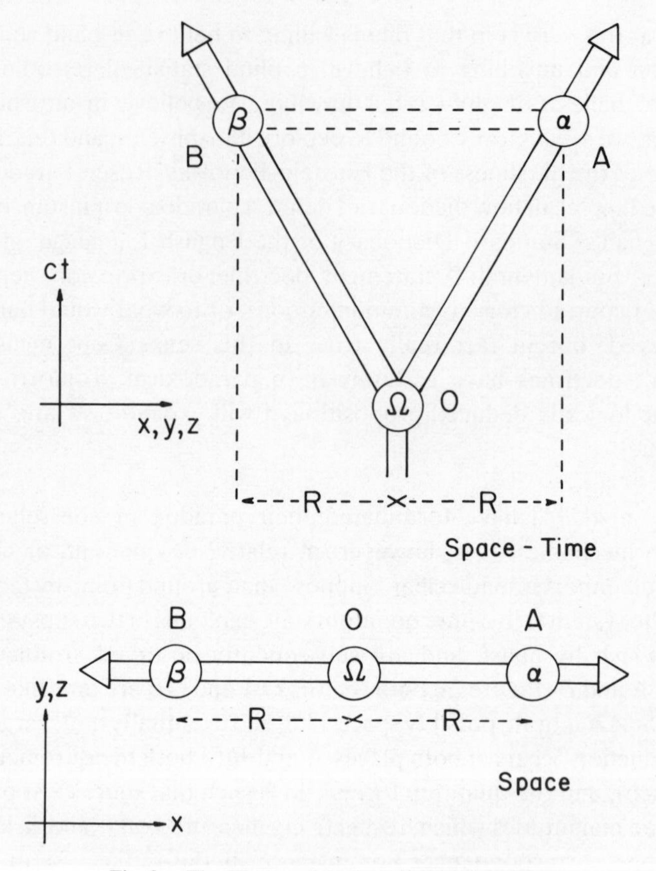

Fig. 2. The Einstein-Podolsky-Rosen Paradox.

be random, but necessarily opposite. There is nothing paradoxical in this, *because, in this scheme of thought, it is at O that the die is cast.* Both the electron and the positron fly apart with velocities each of them *possesses.*

Before we go to quantum mechanics, let us discuss how the logical inference is drawn by observer A as to what observer B finds, and *vice versa.* The very mathematics, duplicating in this the logical thinking, shows that the inference is not telegraphed directly along the space like vector AB (which would be forbidden by Einstein's relativity theory). It is in fact telegraphed along the Feynman style zigzag AOB or BOA, once towards the past, once towards the future, along two timelike vectors. On the whole, it is a space like *telediction*, but it really consists in a timelike retrodiction plus prediction.

It is when we turn towards quantum mechanics that the sting of the paradox is felt and that, up to now, nobody has succeeded in calming the pain. As the remedy I am proposing is a very radical one, I will first give an example of the far-fetched ideas physicists are obliged to ponder by the very physics of the problem – and certainly not by indulging in science fiction.

In a recent article in *Phys. Rev.*, Clauser and Horne [5] draw the conclusion that the mathematics of quantum mechanics is incompatible (in the E.P.R. context) with what they term (and define) as *objective local theories*, except those such that "Systems originate within the intersection of the backward light cones of both analyzers and the source. These propagate into the spatial region of the whole (detecting) apparatus, and simultaneously affect both the experimenter's selections of analyzer orientations and the emissions from the source." The point is that Clauser and Horne are offering no physical interpretation of their mathematics in the place of the understandable 'objective local theories' they have to reject, so that, after all, the paradoxical one they just mention *enpassant* (as being compatible with the mathematics) should perhaps *not* be rejected. In fact, it is the one I am proposing, with the important *proviso* that my version of it bears to Clauser and Horne's the same relationship as does Feynman's *Theory of the Positron* to the former theory of the positron.

Why the sting of the paradox is felt in Quantum Mechanics is that it is not at O that the die is cast, but at A and B (that is, where the observations are performed). Before I discuss this point it should be stressed that precisely *that* paradox occurs not only in sophisticated *ad hoc* setups, but in every quantum measurement whatsoever that more than one observer records independently. As a simple example we can think of the scintillations produced on a ZnS layer by the impact of α particles. If two or more observers are observing the same field, they are in an E.P.R. situation with respect to each other.

It is at this very point that I feel the Copenhagen School is forgetting something important. Everybody agrees that the various observers we are speaking of are bound to find the same result, although they can be very far from each other. This implies that *they are bound to produce strictly correlated reductions of their respective wave functions whenever they are making the same observation.* In other words, this implies that the common reduction of the overall wave function which they produce at the point instant where the α particle flashes occurs *in their past.*

To my cognizance no stronger statement of the reality of advanced waves in a single quantum measurement process has yet been produced. I have been suggesting it in veiled words for years [6], never daring to say it openly. The first time I did it was last year, in Boston, on 19 February. And now I deliberately come out into the open, because, due to the general interest in the E.P.R. paradox, and to the numerous experiments presently in progress, I believe it is the right time to do so.

Before proceeding, however, I will display the connection between this and the Schrödinger cat paradox. Let me recall that the cat, which is a prisoner inside a box, can be either killed (D state) or left alive (L state) according as the decay electron of a radionuclide goes or does not go through a Geiger counter. The amplifying device is of the usual sort, and the choice of the lethal weapon is left open. The L and D states are orthogonal to each other, in the sense that they are faithful macroscopic representations of the two orthogonal waves between which the choice is open.

The point made by Schrödinger [7] is that, as long as no observer has reduced the wave function, the cat is in a coherent superposition of states $\lambda L + \delta D$, λ and δ being known up to a phase exponent. What is wrong in this, in my opinion [8], is that the cat is as good a wave-collapser (or anti-collapser, if we believe in advanced waves) as any other 'observer'. *He* certainly knows when he is alive, and maybe also when he is killed. Therefore, if both the cat and a scientist are observing the decay of the same radionuclide *through parallel channels* (which is quite possible), they must be either cooperating or competing for producing the result. This is because, as soon as we believe in the existence of advanced waves (and we *have to* in this problem), the hidden face of Aristotle's information, namely *will*, shows up beside the familiar face, which is *cognizance.* Things being so, the cat is certainly more motivated in the issue than is the scientist, and one guesses that a normal cat will be in favor of the L issue.

At this point you are certainly suspecting what I have in mind: parapsychology. Well, it so happens that parapsychologists have really performed the Schrödinger cat experiment, not in the form of a death or life dilemma, but in the more sadic form of a punishment or reward experiment (which, in our Carnot-like world, has the advantage of allowing learning). However, as the shameful cheating by W. J. Levy [9] may induce you to question the conclusive results published by other workers [10], I intend to prove now that the problem I am discussing is no science fiction, but a very actual physical problem.

The Einstein-Podolsky-Rosen Paradox has been put in a physically testable form through the successive theoretical efforts of Bohm [11], Bell [12], Clauser *et al.* [13]. The distant correlation that is presently discussed is the one between the spins of two massive particles emitted by a compound system, or between the polarization states of two photons emitted in the cascade transition of an atom. The mathematics has been fairly well worked out, and, as it has now been published quite a few times, I will not go into it. Suffice it to say that, in a typical case of linearly polarized photons, the quantal probabilities for the various answers 'Yes and Yes', 'No and No', 'Yes and No', 'No and Yes' are, α denoting the angle between the polarizers,

$$(Y, Y) = (N, N) = (1/2) \cos^2 \alpha, \qquad (Y, N) = (N, Y) = (1/2) \sin^2 \alpha,$$

and that, for certain combinations of values of α, they can be reproduced by no stochastic theory of the sort characterized by Clauser and Horne [5] as 'objective and local'.

Let me put things this way. As the mathematics of the quantal probability calculus is definitely not identical to that of the classical probability calculus, it should be surmised that certain sophisticated contexts exist where the theoretical predictions are different. Thanks to Einstein-Podolsky-Rosen, Bohm, Bell, Shimony, and others, we now have a very definite instance of this, and a very operational one where crucial experiments are possible.

Now I focus more closely upon the content of Clauser and Horne's [5] recent paper. For one thing, they have deduced the incompatibility we are speaking of from the very reasonable (and remarkably weak) assumption of 'no enhancement of a light beam' by interposition of a linear polarizer. Secondly, they have slightly improved upon the previous interpretations of the mathematics characterizing what they call 'objective local theories'. However, as a mere rejection of a class of mathematical theories does not say much upon the positive side of the non-rejected non-separable theories, as d'Espagnat [8] calls them, my personal feeling is that the truly significant proposal in Clauser and Horne's paper is the one I have quoted, which, according to their own words, is an *objective local theory compatible with the quantal statistics even when the orientations of the polarizers are changed in a time short* in terms of R/c ($2R$ being the distance between the polarizers). *This objective local theory is the only one they can produce at this end, and, in fact, is the only positive physical model they are able to produce.*

However, if accepted, this theory certainly smacks of what Schrödinger [7] termed as 'magic' – precisely in the E.P.R. context. As you may see, it is nothing else than another wording of my own theory.

Concluding this section, I would say that if an E.P.R. experimentation using rapidly changing analyzers still gives results conforming to the accepted quantum mechanics, *there is no other rational choice than the one I am proposing.*

That, though rational, this proposal is radical may be seen through a quotation borrowed from Einstein *et al.* [4]. They wrote: "If, without in any way disturbing a system [say, β], we can by a measurement performed on a distant system, [say, α] predict [I would rather say *teledict*] with certainty ··· the value of a physical quantity [on β], then there exists [a corresponding] element of physical reality." In the jargon of syllogisms, what I am rejecting is the E.P.R. 'major.' I am saying that observers A and B are actually coupled *via* a space-time telegraph, *with the relaying satellite in the past. They have to cooperate or to compete in reducing one and the same wave function. In this way both twin faces of Aristotle's information, cognizance and will, show up as non-separable, like those of a coin.*

By the same token I am *ipso facto* rejecting the often expressed statement [2] that the act of measurement does not react upon the measured system. Truly, this accepted statement is hard to maintain in quantum mechanics, where everybody agrees that the very nature of the question which is asked contributes in producing the answer. Such being the case, the very mathematical structure of the Jordan-Pauli propagator entails that an *individual* measurement performed around the point-instant x^λ is propagated from $-\infty$ to $+\infty$ in time; *there is thus no time asymmetry in the individual quantal measurement.*

Finally, how is my proposal to be tested, and which are its main implications?

I feel that if Freedman and Clauser's [14] very neat experimental verification of quantal predictions in the E.P.R. situation is supported by new experiments now in progress, and thanks to Clauser and Horne's [5] footnote 13, Section C, already quoted, my case is already quite strong.

It would be further strengthened if new experiments, with rapidly changing analyzers, were again in accord with quantum mechanics.

From a different side of the horizon I would quote as highly significant the quite positive Schrödinger cat experiments published by parapsychologists [10]. These so-called psychokinetic experiments, performed with animals as subjects, are not liable to fraud or illusion on their side. Moreover, by using high frequency random outcome generators, tested both before and after the experiments (and then found to be all right), large numbers of tests are easily obtained.

One more proposal would be to use the existing sort of E.P.R. apparatus – for instance a long base one as in Clauser and Horne's experiment with, instead of two passive, impartial, observers, A and B, either one passive observer A and a psychokinetic agent B, or else two psychokinetic agents A and B (with observers looking over their shoulders). In the former case the implication is that, *via* the E.P.R. coupling, the usual predictive frequencies would be disturbed on the observer's line by the action of the agent *on the other line.*

What, now, are the general implications of my philosophy? The first one, as I have already said, is that Einstein's prohibition to telegraph into the past is a macroscopic, not a microscopic one. Everybody should accept the latter statement, which is clearly written in the mathematics, and has been made familiar by Feynman's computational techniques. Even in the macrocosm the prohibition is not a strict one, as exemplified by statistical fluctuations – and, perhaps, by appropriate *ad hoc* contexts: those of parapsychology.

But, if *every* quantum transition needs an act of observation and/or action, what of, say, radioactive disintegrations in the absence of any observer? Should we believe that they are induced by the advanced action of the first onlooker? This, I believe, would be pushing too far one's faith in advanced waves. I rather would assume the existence of many lower psyches inside the cosmos; say, Leibnizian 'monads' that would not be (as Leibniz had them) strictly 'epiphenomenal' in their consciousness, but rather usually very lazy; this would account for the usual exponential decay law.

Due to the assumed existence of these lower psyches, and also to that of numerous interactions inside the universe, a damping of the E.P.R. correlation certainly occurs through time and distance. This is a side aspect of the phenomenon that could probably be tested. The significant point, however, and the one that may prove to be as paradoxically constructive as the observations used by Copernicus and the facts used by Einstein, is that Friedman and Clauser [14] have verified the existence of the E.P.R. correlation over a distance of some 5 meters.

Institut Henri Poincaré, Paris.

* Lecture delivered on 17 April, 1975, at the Massachusetts Institute of Technology (Thermodynamics Group).

NOTE ADDED IN PROOF (24.1.1977)

Had the Freedman-Clauser [14] and the subsequent Clauser and Fry-Thomson [15] experiments been performed in the old quantum theory days, before 1924–27, they would have produced the same stupefaction as did the Michelson experiment disproving the ether wind. They *unambiguously* prove that the photons in each pair *do not possess* polarizations of their own when leaving the source 0, but *borrow one later* by interacting with the analyzers A and B, their coupling occurring nevertheless through the past event at 0!

A technical difference between the experiments [14] and [15] is the parity invariant state they are using, $L_aL_b + R_aR_b$ in the former and $L_aL_b - R_aR_b$ in the latter (where L and R denote left and right circular polarizations of the photons) entailing, in the latter case, an exchange of the values of the probabilities $(Y, Y) = (N, N)$ and $(Y, N) = (N, Y)$ with those of the former.

Consider then the case where $\alpha = \pi/2$ in [14] or $\alpha = 0$ in [15], where $(Y, Y) = 0$: no coincidences. This *definitely* proves that *all* photon pairs are found with parallel (resp. perpendicular) linear polarizations *parallel or perpendicular to those of the analyzers.* If the photons did possess polarizations of their own (compatible of course with the dynamics of the system) when leaving 0, these would be independent of the orientations of the analyzers, and a large proportion of answers Y and Y would be found.

This, of course, is the E.P.R. Paradox, or even the Einstein [16] 1927 Paradox, thrown as an ominous spell upon the very cradle of the New Quantum Mechanics.

An experiment where the orientations of the analyzers are changed while the photons are in flight has also been defined [17] and is being built up.

I have derived the formulas for $(Y, Y) = (N, N)$ and $(Y, N) = (N, Y)$ through an elementary reasoning [18] displaying quite strongly the binding of the Paradox with the neo-quantal rule of squaring the (absolute) sum of partial amplitudes, instead of the paleo-quantal rule of adding partial probabilities; that is, displaying how the paradox belongs to the *new wavelike probability calculus.*

BIBLIOGRAPHY

[1] Mehlberg, H., 'Physical Laws and Time Arrow' in *Current Issues in the Philosophy of Science*, (H. Feigl and G. Maxwell, eds.), Holt, Rinehart, Winston, New York, 1961, p. 105.

[2] London, F. and Bauer, E., *La Théorie de l'Observation an Mécanique Quantique*, Hermann, Paris, 1939.

[3] Costa de Beauregard, O., *Précis de Mécanique Quantique Relativiste*, Dunod, Paris, 1967.

[4] Einstein, A., Podolsky, B. and Rosen, N., *Phys. Rev.* **47** (1935) 777.

[5] Clauser, J. F. and Horne, M. A., *Phys. Rev. D.* **10** (1974) 526. See [13].

[6] Costa de Beauregard, O., *Dialectica* **19** (1965) 280 and **22** (1968) 187. Discussion on temporal asymmetry in thermodynamics and cosmology *in Proc. Intern. Conf. on Thermodynamics*, (P. T. Landsberg ed.), Butterworth, London, 1970, p. 539.

[7] Schrödinger, E., *Naturwiss* **23** (1935) 807, 823 and 844.

[8] Espagnat, B. d', *Conceptual Foundations of Quantum Mechanics*, Benjamin Inc., Menlo Park, California, 1971, makes the same point (p. 302).

[9] See the *New York Times*, Sunday, August 25, 1974.
[10] Duval, P. and Montredon, E., *J. Paraps.* **32** (1968) 153, Schmidt, H., *J. Paraps.* **34** (1970) 255; Watkins, G. K., *Proc. Paraps. Assoc.* **8** (1971) 23.
[11] Bohm, D., *Quantum Theory*, Prentice Hall, Englewood Cliffs, N.J., 1951, p. 164.
[12] Bell, J. S., 1964, 'On the Einstein-Podolsky-Rosen Paradox', *Physics* **1**, 195–200.
[13] Clauser, J. F., Horne, M. A., Shimony, A. and Holt, R. A., *Phys. Rev. Letters* **23** (1969) 880; Shimony, A. in *Foundations of Quantum Mechanics*, (B. d'Espaquat, ed.), Academic Press, New York, 1971, p. 182; Horne, M. A., Ph.D. Thesis, Boston University 1970.
[14] Friedman, S. J. and Clauser, J. F., *Phys. Rev. Letters* **28** (1972) 938; Friedman, S. J., Ph.D. Thesis, University of California, Berkeley, 1972.
[15] Clauser, J. F., *Phys. Rev. Letters* **36** (1976) 1223. Fry, E. S. and Thomson, R. C., *Phys. Rev. Letters* **37** (1976) 405.
[16] Einstein, A., in *Rapports et Discussions du* 5^e *Conseil Solvay*, Gauthier Villars, Paris, 1928, pp. 253–256.
[17] Asped, A., *Phys. Letters* **54A** (1975) 117; *Phys. Rev. D* **14** (1976) 1944.
[18] Costa de Beauregard, O., *Epistemological Letters* (association F. Gouseth) **14** (1976) 1; and **15** (1977), in press.

LAWRENCE SKLAR

WHAT MIGHT BE RIGHT ABOUT THE CAUSAL THEORY OF TIME

I

A causal theory of time, or, more properly, a causal theory of spacetime topology, might merely be the claim that according to some scientific theory (the true theory? our best confirmed theory to date?) some causal relationship among events is, as a matter of law or merely as a matter of physically contingent fact, coextensive with some relationship defined by the concepts of the topology of the spacetime. A strongest version of such a causal theory would be one which demonstrated such a coextensiveness between some causally definable notion and some concept of topology (such as 'open set') sufficient to fully define all other topological notions. Given such a result, one would have demonstrated that for each topological aspect of the spacetime, a causal relationship among events could be found such that that causal relationship held when and only when the appropriate topological relationship held.

I believe, however, that the aim of those philosophers who have espoused causal theories of spacetime topology has been grander than this. There is a general and familiar philosophical program that works like this: In some given area of discourse it is alleged that our total epistemic access to the features of the world described in this discourse is exhausted by access to a set of entities, properties and relations characterized by an apparently proper subset of this conceptual scheme. Then, it is alleged, the full content of propositions framed in this discourse must be characterizable in terms of its totality of observational content, this content being describable in the distinguished concepts of the subset of epistemically basic concepts. So, the program continues, it is up to us to show how all the empirical consequences of the theory really are framable in the distinguished concepts alone. In so far as the theories framed in the totality of concepts appear to make empirical claims outrunning those expressible in the basic concepts, these must be shown to be ultimately

W. C. Salmon (ed.), Hans Reichenbach: Logical Empiricist, 367–383.

reducible to the empirical content expressible in the basic concepts. And if the assertions of the total theory go beyond those reducible in content to assertions of the distinguished vocabulary, then they must be shown to be true merely as a matter of choice or 'convention'.

I believe that, explicitly or implicitly, some such program underlies all those various causal theories of spacetime topology with which we are familiar. So interpreted the philosophical theory is not just the explication of the results of some scientific theorizing, but an attempt at an epistemological and semantic critique which displays initial constraints into which any satisfactory scientific theory must fit. For if the epistemic and semantic analysis proposed by the philosophical theory is correct, then the limits of meaningful assertiveness of the scientific theories are delimited, the distinction between fact asserting and 'mere convention making' elements in the scientific theories is drawn, and the criteria are made clear under which we are entitled to say whether or not two allegedly alternative total theories are or are not properly speaking 'equivalent' to one another and, in Reichenbach's terms, whether they differ merely in descriptive simplicity or in some genuine empirical content.

Taking our total theory to be one which describes the spatiotemporal topological structure on events and that portion of the causal relationships among them which contains reference only to which events are and are not causally connectible with one another, a causal theory of spacetime topology is a philosophical theory which alleges that the total empirical content of this overall theory is exhausted by its causal part; and that any spatiotemporal topological assertion can be reduced to an assertion about causal relatability supplemented, perhaps, by conventional choices of 'ways of speaking'.

But, I believe, we must proceed with some caution here. What counts as 'the full causal structure on events', and, hence, what counts as the allowable empirical basis to which all our theoretical elements are to be reduced may not be the same in all 'causal' theories of spacetime topology. And two philosophical theories even while agreeing as to the nature of the 'epistemic basis', may, as we shall see, disagree as to *why* the elements in question are allowed into the reduction base.

What I wish to suggest here is that, for rather well known reasons, some choices of an epistemological basis in causal theories of spacetime topology won't do the job intended for them. I think that one suggested basis does have some plausibility. But its plausibility, I will suggest, rests on grounds which might lead us to think that the designation 'causal' is inappropriate for this philosophical reductionist account of spacetime topology.

While our ultimate grounds for accepting or rejecting such a 'philosophical' theory of spacetime topology will be 'philosophical' or, more particularly, in terms of an epistemological critique of spacetime theories, it is important and enlightening to see just how particular scientific theories of spacetime fare in the light of one's philosophical account. I will try to show how several recent results in the mathematical foundations of general relativity 'fit in' with the analysis of implausible and more plausible 'causal' theories of topology which I will survey.

Finally, while I believe that the 'causal' theory I end up with is a more plausible candidate for a possibly successful reductionist account of spacetime topology, I do not believe that it itself is uncontrovertibly correct. I will note some reasons for being cautious about accepting it as a philosophical account of spacetime epistemology and semantics.

II

Let us consider first the version of the theory which takes as its 'reduction basis' the relation of causal connectibility among events. As usual, we note, the 'definitions' of the topological notions in terms of the causal are in the mode of possibility. Some topological relationship is alleged to hold if some appropriate causal relationship *could* hold. That topological notions reduce to causal only if we allow causal notions 'in the mode of possibility' is a familiar feature of such philosophical reductionism. I will forego rehearsing the well-known reasons for the necessity of the invocation of such possibility talk and the equally familiar reasons for alleging that such an invocation vitiates at least part of the reductionist program.

In the version we are now exploring the notion of two events being causally connectible, i.e. its being possible for one to be causally related to the other, is taken as our primitive notion. In terms of it we attempt to construct an adequate defining basis for the full gamut of spatiotemporal topological notions. Not surprisingly we do not need to invoke any notion of the direction of time at this point, and causal connectibility, rather than the asymmetric relationship of '*a* could be a cause of *b*' will be sufficient.[1] How plausible is such a causal theory of topology?

There are, I believe, two basic grounds on which a philosophical reductionist thesis is usually challenged:

(1) First it might be claimed that in order to have grounds for believing that a relationship utilized in the reduction basis holds, we must first have some knowledge of relationships holding which are in the portion of the theory to be reduced or defined away. If this is so, it is claimed, then the reduction basis is not a properly epistemically primitive basis.

(2) Second it can be argued that the reduction basis is not adequate to provide 'definitions' for all the concepts of the total theory. That is, even if we can, without knowing any of the defined relationships to hold, establish which relationships in the reduction basis do in fact hold, that information will not be adequate to fully determine all the relationships which are expressed in the total theory.

Our first version of a causal theory of spacetime topology has been challenged on both these grounds.

First, can we really tell which events are causally connectible without already knowing a great deal about their spatiotemporal relationships to one another? In our usual accounts of the notion of causality we find that while causality means falling under a natural law, and a law is a general rule connecting *kinds* of events, knowing which *particular* events are causally related to which other particular events requires knowing a great deal about the spatiotemporal relations of these events to one another. Strikings of matches cause lightings of matches, but to tell that *this* striking caused *that* lighting we must know that this particular striking bears to that particular lighting an appropriate spatiotemporal relationship. Now while I believe that there is some truth to this objection, I think that a possibly viable 'causal' theory of topology is lurking under this first theory despite this initial objection. But rather than pursue the matter here, I will save this discussion for later.[2]

Second, can we really define all our topological notions in terms of the notion of causal connectibility? Here is where some results of the mathematical study of spacetimes become crucial. If the philosophical reductionist theory is correct, it is claimed, then in any possible spacetime we envision the topological notions must be connected to the appropriate causal notions in the way specified by the reductionist account. For the topological notions are supposed to be *defined* by the causal notions. Is there a definition of the appropriate basis for all topological notions (say a definition of 'open set') in terms of causal connectibility which holds in all the spacetimes we take as conceivable?

Not if we allow ourselves the full range of spacetime compatible with general relativity, including those we might view as causally 'pathological'.

In the Minkowski spacetime of special relativity we can indeed causally define (in the present meaning of that term) the open set basis sufficient to fully define the topology. An explicit definition of the open sets in terms of causal connectibility is available in terms of the well-known Alexandroff topology for Minkowski spacetime.

In those general relativistic spacetimes which are *strongly-causal* it can also be shown that the Alexandroff topology and the usual manifold topology will coincide. Indeed, the coincidence of the topology defined by the causally defined open sets of the Alexandroff topology with the usual manifold topology is equivalent to strong causality. When the spacetime is not strongly-causal, however, the manifold topology and the Alexandroff topology will fail to coincide.[3]

Malament has recently shown that in any spacetime which is both *past* and *future distinguishing* (a weaker condition than strong causality) one can at least *implicitly* define the topology in terms of causal connectibility. That is, given any two past and future distinguishing spacetimes, any causal isomorphism between them (bijection preserving causal connectibility) will be a homeomorphism relative to the usual manifold topology. He has also shown that this is a strongest possible result in that one can construct examples of spacetime in which either past or future distinguishing is violated and such that there will be causal isomorphisms between pairs of such spacetimes which fail to be homeomorphisms. If the past and future distinguishing condition is not met the topology cannot be even implicitly defined in terms of causal connectibility. There are, in fact, spacetimes where the failure of causal definition of the topology is made particularly manifest. I refer here to those spacetimes where the topology is non-trivial but where causal connectibility is a relationship which holds between *every* pair of events.[4]

Now we might try to hold that the kinds of spacetimes in which either explicit or implicit definition of the topology by causal connectibility fails are impossible spacetimes. For example, they are, to be sure, infected with the kind of pathological causality which we would not ordinarily expect to find in a causally 'well behaved' universe. But if we can even *understand* what such spacetimes would be like, then the fact that we can understand them (know what it would be like to live in one of them, for example) shows us, I think, that any hopes of establishing that in the actual world our actual concepts of topology are *defined* by notions of causal connectibility are unacceptable. Without going into details here, I believe that such worlds are perfectly intelligible to us and that with a little imagination anyone familiar with the basic concepts of relativistic spacetime can be gotten to understand just what the topological aspects of such a world would be like to an inhabitant of them. Further, we can construct pairs of spacetimes which are causally isomorphic in our present sense and which are such that an inhabitant of them can, by topological exploration, discover in which of the alternative spacetimes he resides.[5]

III

Now the fact that our initial attempt at a causal theory of spacetime topology fails in two diametrically opposite directions might lead us to believe that the prospects for any such theory are dismal indeed. The basis chosen in this original version of the theory is too *weak* to do the job required of it – define the full topological structure in every spacetime we wish to take as intelligible. Yet it is so *strong*, i.e. contains so many elements and of such a kind, that it appears that we already need epistemic access to much spatiotemporal structure on events in order to tell when the relationship which is utilized in the basis, causal connectibility, holds. But we will see that the situation is not quite as hopeless as it looks.

Let us try to remedy the defects of our original theory first by moving to a stronger basis, holding in abeyance for the moment the obvious difficulties this will give rise to. Within the relativistic context two events are causally connectible if there is a continuous causal (timelike or lightlike) path between them. This follows from the standard view that causal interaction is inevitably mediated by the emission, transmission and reception of some 'genidentical' material or lightlike 'particle'. So within this context, we know that the relationship of causal connectibility holds of a pair of events if we know the truth of the existential claim that there is at least one continuous causal path containing the two events.

Now suppose we take as our basis notion not causal connectibility but, instead, the notion of continuous causal path. We then have full knowledge of the basis relationship not when we know merely which pairs of events are connectible by some continuous causal path or other, but when we can tell of any set of events whether or not it constitutes a continuous causal path. Does the introduction of this much stronger basis change the picture significantly?

That is does follows from an important and interesting result of Malament's. Given any two spacetimes (taken to be four-dimensional manifolds with pseudo-Riemannian metrics of Lorentzian signature) any bijection which takes continuous causal curves into continuous causal curves will be a homeomorphism! In other words, the full topology of the spacetime will be fixed by its class of continuous causal curves. So if we can fully determine the latter, then we can fully specify the former as well.[6]

But can we determine the continuous causal curves of a spacetime without already being able to determine its topological features? This obvious objection to the epistemological use of a result like Malament's to establish a causal theory of topology is well known. Let us first present it and then in the next section move on to the deferred task of showing why, despite this argument, something which might be called a causal theory of topology can have a case made out for it that meets this 'epistemological' objection.

In one version, and among philosophers I think the most common one, the picture one has is something like this: A single observer wishes to map out the topology of the spacetime in which he dwells. Equipping himself with an infinite number of material or lightlike particles, he emits them from all points in all directions (we must be generous in such idealizations) so determining the structure of continuous causal paths in his spacetime. Now, by Malament's result, he can pin down uniquely the full topology of the spacetime, including such not-directly-determined features as which spacelike paths are continuous, etc.

But how does our observer tell which classes of events in the life history of a 'genidentical' particle are spatiotemporally continuous portions of its history? Indeed, how does he even tell which events are events in the history of one such particle and not events selected at random from the histories of any number of distinct particles? Only, it is alleged, by already knowing what the continuous spatiotemporal segments really are, i.e. only by already having access to topological features of the spacetime.[7]

IV

But consider the following version of a 'causal' theory of topology: It is true that on the picture just looked at we gain access to the full topology of the spacetime only by already knowing part of it, i.e. only by already knowing which are the continuous timelike and lightlike curves in it. But this is already a 'reductionist' gain. And when we realize the real epistemological importance of the basis to which total topological knowledge has been reduced, we see just how important a gain this is. For what constitutes the continuous causal paths of the spacetime is just what is available to our *direct epistemic access*. Not because we have some wonderful way of spotting continuous segments in the history of genidentical particles without any antecedent knowledge of the topology. That is silly, for what we mean by continuous segment in the history of a genidentical particle is just a continuous set of spacetime locations along a causal path occupied by the same kind of material particle. But, rather, because each causal path can be traversed by a local 'consciousness' who directly and immediately, as a primitive content of his experience, can tell which events in his consciousness, and hence which spacetime locations along the worldline of his history, are 'near' one another!

The idealized picture we have now is of a universe equipped not with one observing 'consciousness' but with a plentitude of them, so that the totality of causal worldlines of the spacetime is covered by consciousnesses sensing immediately and directly which locations of the spacetime along their respective worldlines are 'near', i.e. determining without inference or instrument which are the continuous segments of causal curves in the spacetime. Extravagant as this seems when put this way, the basic idea here has appeared explicitly or implicitly from Robb through Hawking and Malament.[8]

So the causal theory of spacetime topology has now been replaced by a reductionist account which is not really 'causal' in its fundamental epistemic motivation. The reductionist program still takes as the body of concepts to be reduced those characterizing the full topology of the spacetime. The reduction basis consists in those concepts relating to continuity of the one-dimensional causal worldlines of the spacetime. But this set of topological features is not discriminated from the general class because the causal worldlines are the paths of propagation of causal influence throughout the spacetime but, rather, because such curves, being the worldlines of the possible life histories of 'consciousnesses', have their continuity open to 'direct epistemic access' without the intervention of instruments or the necessity of theoretical inference. I will shortly return to the question of just how plausible such a reductionist program might be. I don't wish here to maintain that this is the correct philosophical account of our knowledge of the topology of spacetime and of the semantic analysis of topological concepts in general, but only that it is the most plausible version of a 'causal' theory of topology, combining, as it does, a genuine reduction (from the totality of topological aspects to those along causal paths alone) with an at least *prima facie* case for the epistemic priority (in terms of immediacy of access) of the concepts of the reduction basis. I also believe that at least some readers will agree with me that it is this version of the 'causal' theories which captures the fundamental epistemic 'intuitions' which lay behind the other more 'causal' causal accounts.

V

At this point it is enlightening to introduce some additional results of recent mathematical work on spacetimes. At first glance these new results might seem to militate against the version of the causal theory we ended up with. But on reflection, I believe, they illustrate, rather, the force of our last account of the causal theory.

Suppose we have available to us a full picture of the topology (continuity) of the causal worldlines of the spacetime. Can we then infer immediately the full topology of the spacetime? Malament's result seems to show that we can, for any two spacetimes (in the usual 4-manifold with Lorentz-signature metric sense) which agree on their causal worldline topology will agree in all topological aspects. But it isn't that simple.

Following the work of Zeeman on novel topologies for Minkowski spacetime, recent work has been done on looking for interesting novel topologies for the spacetime of general relativistic worlds. These new topologies, unlike the usual manifold topology, 'code' into themselves the full causal structure of the spacetime. This is true in the sense that any homeomorphism between spacetimes relative to the new topologies are automatically conformal isometries. This is not true of homeomorphism relative to the usual manifold topologies. With these new topologies the spacetimes are not manifolds.[9]

The importance of these new topologies for our present purposes can be seen in the manner of their construction. One starts off imagining a spacetime equipped with the usual manifold topology. One then seeks a new topology which will code the causal structure in the manner described above, and which will then, perforce, differ from the usual manifold topology. But the new topology is so constructed that it *agrees with the manifold topology on the topology induced on the one-dimensional causal curves*. For example, there is the *path topology* of Hawking, King and McCarthy. It is defined as being the finest topology on the spacetime which induces on all continuous causal curves the same topology induced on them by the usual manifold topology.

So, obviously, the standard and the novel topologies will agree on all topological facts about the causal curves. Now aren't we in a position to be skeptical about our 'causal theory of topology'? For if we can fully exhaust our knowledge about the topology along such causal paths and yet still not know the full topology of the spacetime (is it the standard manifold topology or is it, instead, one of the novel topologies?) aren't we in just the same position which caused us to reject our first version of a causal theory of topology? For there we pointed out that in the case of spacetimes which were not both past and future distinguishing, two manifolds could share their causal structure (there could be a causal isomorphism between them) and yet not be topologically identical (i.e. not homeomorphic). This led us to say that the structure determined by causal connectibility was not sufficient, in general, to fully fix the topology of the spacetime. And here, once again, we seem to be saying that once we allow for the possibility of the novel topologies, we can no longer fully fix the topology of the spacetime even given full topological knowledge of the causal paths.

But the situations are not parallel. The trouble with the causal connectibility version of the causal theory of topology is this: Two spacetimes might be causally isomorphic and yet be *empirically distinguishable* in their topological structure. They may very well differ in the structure of continuity along causal paths which is the paradigm of epistemically accessible topological structure. But if two spacetimes share the same topological structure along the causal paths, and if this structure exhausts the empirically determinable topological structure of spacetime, then no empirical observation could tell us which of two incompatible full topologies (say standard vs. the path topology of Hawking *et al.*) is *really* the full topology.

And what that suggests is this (the familiar end product of the problem of theoretical conventionality): If two spacetimes share the same topology induced on the causal paths, then, appearances to the contrary, they are really (topologically) the same spacetime. Their real topological structure may be *expressed* in the standard manifold form (and, as Malament's theorem shows only one of these will be compatible with the topology on the causal paths) or it may be expressed in terms of the path topology of Hawking *et al.*, or in terms of any other non-manifold topology designed to code the causal structure. But these are merely alternative formulations of one and the same set of empirical facts. For (in the usual positivist vein) since the topology on the paths exhausts what can be empirically discovered about the topology in general, the total factual content of the general topology is exhausted by the topological structure it induces on the causal paths.

One awaits at the present time a neat mathematical formulation which eschews the necessity for topological assertions about the spacetime which *appear* to outrun the topological facts about causal curves, and which captures the empirical equivalence among all topological structures which induce the same topology on the causal paths, by offering as the full characterization of the topology of the spacetime the mathematical description of the continuity structure on causal paths alone.

VI

How plausible is a causal theory of spacetime topology, even in its most plausible version in which it becomes rather a 'topology of worldlines traversable by a consciousness' theory of spacetime topology? Now, of course, much of the answer to that question will depend upon just how plausible in general one takes philosophical reductionist theses, with their underlying positivist motivations, to be. While many are skeptical (often without much in the way of argument above and beyond 'realist' dogmatism) of reductionism in general, this is hardly the place to rehearse most of the familiar issues of whether or not 'pure' observation bases exist, of the allowability of modal possibility talk into the reductive definitions, etc. Rather let me focus on just a few issues for discussion which come to mind when one reflects on this particular attempt at philosophical reductionism.[10]

(1) The theory presupposes that at least one aspect of spacetime structure is accessible to consciousness without instrument or inference. But is this belief supportable? Here the reader may remember the discussion of spatiotemporal coincidence in Reichenbach's *Axiomatization of the Theory of Relativity*. Here Reichenbach reflects upon the fact that while coincidence is taken as a primitive of his (and everyone else's) reconstruction of relativity, if this decision to take it as a primitive is based upon the ground that we have immediate non-inferential knowledge of what events in the physical spatiotemporal world are really coincident we are on dangerous ground. Aren't the coincidences of which we are 'immediately' aware *subjective* coincidences; and isn't it *objective* coincidence with which we are concerned in our reconstruction of the theory of physical spacetime?[11]

In the present case, by what right do we identify the continuity of inner experience of some observer, as subjectively experienced, with the 'real' physical continuity of the spacetime locations at which these experiences occur? Imagine, for example, a consciousness instantaneously and discontinuously transported through spacetime, so that the continuity of his 'inner' experiences *misrepresents* the discontinuity of the spatiotemporal locations at which he has them.

What we have here is just one more example of a well known and enormous difficulty for reductionism: If reductionism allows in its basis for reconstruction only the content of the immediately experienced, then how can it ever give us an adequate account of the nature of the objective, physical world? For if we slide down the slippery slope we get on to when we once begin to 'reduce' the inferred to the 'immediately experienced', how can we ever stop before reaching the bottom where the basis is the contents of subjective experience; thereby dropping out of the realm of concepts dealing with the physical 'outer' world altogether?

(2) The theory presupposes that our *only* direct epistemic access is to the continuity of spacetime locations in the lived one-dimensional history of a consciousness traversing a causal worldline. Is this plausible?

First, relying on something quite reminiscent of Kant's distinction between space as the manifold of apperception of outer experience and time as the manifold of apperception of both outer and inner experience, the theory places our experience of time, and continuity in it, in a very special position indeed. All our topological insight into the world is to be grounded in an awareness only of the *temporal* continuity of our experience. Our knowledge of the spatial aspects of the world is derivative from this temporal sense, and, if what we have said about conventionally alternative topologies is correct, really, properly speaking, only conventional. For the only *real* topological facts are the facts about temporal continuity along causal paths.

But is this correct? What about our apparent 'direct' knowledge of the structure of the space around us? Is this to be dismissed? We could, of course, pull the not uncommon move of distinguishing 'perceptual space' from 'physical space' and argue that our direct apprehensions are only of the former, the latter to be 'constructed' out of our temporal experience. But then the peculiar asymmetry of taking most of our perceptual life to be 'merely subjective' while allowing our direct experience of temporal continuity to serve as immediate access to the external world appears in an even more striking light.

Consider, again, what is, on reflection, the rather surprisingly different treatment accorded topological and metrical aspects of spacetime on this view. Nearly everyone who writes about the foundations of our knowledge of spacetime takes metric features of the world, even those along causal paths (elapsed proper time along a causal worldline) to be founded on some physical measuring process. In order to ground our knowledge of just how much proper time elapses between two events on a causal path we must rely, it is almost invariably alleged, either on *clocks* or on a specification of the affine parameter determined by the paths of *freely falling particles*. If one tries to bring forward our immediate experience of the magnitude of duration, this is almost always dismissed as a confusion of 'psychological time' with the objective physical magnitude.

But why should psychological experience of temporal magnitude be irrelevant to founding our knowledge of the magnitude of physical duration and yet the psychological experience of temporal continuity be taken as giving us immediate access to the real facts about continuity along causal paths?

(3) Next consider Reichenbach's well known claim that the topology of spacetime is just as 'conventional' as is its metric. Just as we can save alternative metrics in the light of any experience by a sufficiently rich flexibility in the postulation of universal forces, so any topological thesis can be maintained if we allow ourselves the global flexibility of choosing how to identify or dis-identify events (moving from multiply connected spaces to their simply connected covering spaces, for example) and the local flexibility of tolerating causal anomalies, i.e. spatiotemporally discontinuous causal interaction mediated by the spatiotemporally discontinuous motion of genidentical signals.[12]

But if the continuity of worldlines is determinable directly by consciousness traversing them, aren't we denied at least the second kind of flexibility noted? So aren't at least the local facts about topology non-conventional? Here I think that two (obvious) options present themselves: (1) We could accept the non-conventionality of continuity along causal worldlines, thereby supporting the intuitions which at least some have had that topological facts (or at least some of them) are non-conventional in opposition to the full conventionality of the metric; or (2) We could remember the possibility, noted above, of consciousnesses themselves experiencing as continuous what are actually spatiotemporally discontinuous histories, and allow for the saving of topologies now by contemplation of a sufficiently rich allowance of 'experientially anomalous' worlds.

All of this is just one more way of emphasizing the fundamentally problematic nature of our best causal theory: the fact that it seems to put one kind of experience on our part – the experience of temporal continuity – on a pedestal as the *one* way in which psychological experience gives us 'direct access' to the structure of the objective world.

(4) It is worthwhile to reflect for a moment on the idealized basis of 'directly knowable facts' to which all assertions are to be reduced in our theory of topology. This consists in all facts about the continuity of causal path segments, for each such segment could be traversed by a consciousness able to directly ascertain its topology.

But, of course, no one consciousness, even in the most extreme idealization, could traverse them all. For example, a consciousness which experiences one event is, in principle, excluded from having in its experience, ever, the direct awareness of an event at spacelike separation from this event.

Of course it is a standard problem with philosophical reductionisms of this kind as to whether the reduction basis should consist in the total possible experience of a *single* observer or the total amalgamated experience of *all possible* observers. In the relativistic context this takes on a particularly disturbing aspect. The experiences of some observers simply can't be communicated to some other observers. It is easy to cook up spacetimes such that one has a pair of them which are distinct in their topology and metric structure; whose differences are determinable on the basis of the collective experience of all possible observers; but which are such that no one observer could ever tell – even if he lived a worldline of infinite extent past and future – which of the two spacetimes he lived in. These are the so-called *indistinguishable* spacetimes.[13]

Should a reductionist account of topology allow in its basis of facts everything which could be known to all observers taken collectively, or should the basis consist, rather, in the possible experience of some one observer? The former alternative seems in many ways the more natural, for even on radically positivist grounds it seems unfair to exclude as 'real facts about the world' something which *someone* could ascertain in a direct non-inferential way. On the other hand, the usual pressures toward solipsism encountered in positivist programs may tend to make a sufficiently radical reductionist reject even the basis to which the causal theory reduces topology as allowing too much in, since it takes as factual elements of the world which could never, even in the most 'in principle' way, be known to any single observer trying to ascertain the topology of the world.

VII

Ultimately, deciding on the plausibility of some reductionist account of topology of the kind we have been exploring will require resolving some of the deepest of philosophical questions: questions relating to the existence and nature of an 'observation basis' on which all theorization is to be constructed and questions relating to just what extent one can, on the one hand, support a positivist-reductionist account of 'theoretical' features of the world and, on the other hand, to what extent one can defend a 'realism' which allows for valid inference 'beyond the immediately observable'. A theory of theories which does justice both to our strongly held intuitions of realism and yet at the same time makes coherent sense of the progress of physics in weeding out 'metaphysical' elements by means of epistemic critiques and Ockham's Razorish prunings of theories is not yet before us. But at least I think that we can now see that a causal theory of topology is really one more attempt at such an epistemological critique of theories.

I have been maintaining that 'causal' is really a misnomer for such theories of topology. To be sure even in our last, most plausible, version of a 'causal' theory of topology the 'hard facts' about the topology of the world are reduced to facts about the continuity and discontinuity of causal paths in the spacetime. But not *because* they are the paths of 'genidentical' causal signals, rather because they are the paths which constitute the possible life-histories of experiencing consciousnesses.

One final query: Is the identity of the causal paths with the worldlines of possible consciousnesses just an 'accident', just an artifact, as it were, of the relativistic facts about the lawlike nature of the world? Or is this identity within the relativistic context instead the inevitable result of any physical theory which survives an adequate epistemic critique? If the latter is really the situation, rather than the former, then the misidentification of reductionist topological theories as causal may not be merely a mistake, but rather a symptom leading us to further insight.[14]

The University of Michigan

NOTES

[1] See R. Latzer, 'Nondirected Light Signals and the Structure of Time', *Synthese* **24** (1972), 236–280, and D. Malament, Ph.D. Thesis (Rockefeller University), chap. II, unpublished.

[2] For a statement of this kind of objection to causal theories as epistemologically motivated reductions see A. Grünbaum, *Philosophical Problems of Space and Time*, Knopf, New York, 1963, pp. 190–191.

[3] See S. Hawking and G. Ellis, *The Large Scale Structure of Space-Time*, Cambridge, Cambridge, 1973, pp. 196–197, and S. Hawking, A. King, and P. McCarthy, 'A New Topology for Curved Space-time Which Incorporates the Causal, Differential, and Conformal Structures', *J. Math. Phys.* **17** (1976), 174–181, esp. p. 176.

[4] See Malament's thesis cited, chap. III and also D. Malament, 'The Class of Continuous Timelike Curves Determines the Topology of Spacetime', unpublished at the time this piece is written, p. 11, Theorem 2.
[5] To see this consider the example used by Malament on p. 13 of his 'The Class of Continuous . . .' to show that a failure of past or future distinguishingness is sufficient to generate causal isomorphisms which are not homeomorphisms.
[6] Malament, thesis, chap. III and 'The Class of Continuous . . .', p. 8, Theorem 1.
[7] For this objection see H. Lacey, 'The Causal Theory of Time: A Critique of Grünbaum's Version', *Philos. Sci.* **35** (1968), 332–354. See also L. Sklar, *Space, Time, and Spacetime*, University of California, Berkeley, 1974, sec. IV, E, 3.
[8] See the introductory portions of A. Robb, *A Theory of Time and Space*, Cambridge, Cambridge, 1914 where the epistemic suitability of taking 'after' as the basic primitive for constructing a spacetime theory is discussed. See Hawking, King and McCarthy, *op. cit.*, p. 175. Malament in 'The Class of Continuous . . .', p. 1 says that this result on the implicit definability of the topology be the class of continuous causal curves is ". . . of interest because, in at least some sense, we directly experience whether events on our worldlines are 'close' or not."
[9] See E. Zeeman, 'The Topology of Minkowski Space', *Topology* **6** (1967), 161–170 for the introduction of these non-manifold topologies in special relativity. For the generalization to general relativistic spacetimes see Hawking, King and McCarthy, *op. cit.*
[10] For a discussion of reductionism in general and in the specific context of spacetime theories see Sklar, *op. cit.*, sec. II, H, 4. See also L. Sklar, 'Facts, Conventions and Assumptions in the Theory of Spacetime', forthcoming in *Minnesota Studies in the Philosophy of Science*, vol. 8 (ed. by J. Earman, J. Stachel and C. Glymour).
[11] See H. Reichenbach, *Axiomatization of the Theory of Relativity*, English translation, University of California, Berkeley, 1969, pp. 16–21. See also Sklar, 'Facts, Conventions . . .', sec. IV, B.
[12] See H. Reichenbach, *The Philosophy of Space and Time*, Dover, New York, 1958, p. 65.
[13] On indistinguishable spacetimes see C. Glymour, 'Topology, Cosmology and Convention', *Synthese* **24** (1972), 195–218. See also Glymour's 'Indistinguishable Space-Times and The Fundamental Group' and D. Malament's 'Observationally Indistinguishable Space-Times' forthcoming in vol. 8 of the *Minnesota Studies in the Philosophy of Science*.
[14] Consider a possible world inhabited by 'epiphenomenal' consciousnesses able to experience events at their locations but not capable of exerting causal influence. Now imagine causal signals to be confined to a subclass of paths in this world, as in ours, but these consciousnesses able to traverse what are spacelike paths. Would not the 'epistemic basis' for our spacetime topology in such a world outrun the continuity of causal paths in it that it would include the directly experiencible continuity of the paths traversable by a consciousness but not by a causal signal? Is such a world imaginable? Why not? On the other hand, one is, at least at first, more skeptical of the possibility of a world where the causal paths outrun those traversable by a consciousness, for just how would we 'ultimately' establish the genidentity of the signals marking out these causal paths not open to the direct topological inspection of some idealized local consciousness? Would, if such a world were possible, our epistemic basis for the topology once again be limited to the (now narrower) class of 'directly experiencible' paths?

PHILIP VON BRETZEL

CONCERNING A PROBABILISTIC THEORY OF CAUSATION ADEQUATE FOR THE CAUSAL THEORY OF TIME

The causal theory of time is a species of the relational theory of time which asserts that the temporal order of the events of the universe is given by their causal order. As I construe it this is not a claim about the meanings of terms but a theory about the nature of time. Time is no more than the causal order of the events composing the universe. Hence it should be possible to completely describe the universe without mentioning anything temporal.

Support for the causal theory of time comes primarily from the Special Theory of Relativity. The Special Theory holds that the temporal order of events is invariant if and only if the events are causally connectible; i.e. it would be possible to send a light signal from one to the other. There would appear, then, to be no reason to accept as primitive a special temporal relation in addition to that of causation. The simplest ontology sufficient for physics is free of temporal relations. Consequently, there would be no reason for physics to require such relations, and if physics doesn't require them it is probably the case that nothing does.

If temporal relations are a kind of causal relation it follows that causal relations cannot be defined with the help of temporal ones. Unfortunately, those philosophers whose meta-philosophical and epistemological beliefs dispose them to the causal theory of time also find these same beliefs disposing them to a Humean account of causation, and, as is well known, Hume's theory of causation requires the use of temporal features for its characterization. In fact, examination of most empiricist theories of causation will show that temporal features are so used. For example, both Broad (1914) and Suppes (1970) use the temporal order of events to distinguish, of two causally connected events, which is the cause and which is the effect. Suppes, for example, defines the notion of an event A being a *prima facie* cause of another event B by the following two conditions:

(i) $P(A, B) > P(B)$[1]

W. C. Salmon (ed.), Hans Reichenbach: Logical Empiricist, 385–402.

and

(ii) A occurs before B.

Obviously such a theory of causation cannot be used for the causal theory of time.

We seem to have a problem. Physics requires that temporal relations be defined in terms of causal ones while Humean and empiricist theories of causation have causal relations defined by the use of temporal ones. This could be taken to imply that there is something wrong with the meta-philosophical position underlying both the adoption of the causal theory of time and the Humean theory of causation. While this certainly may be the case I think one should attempt to devise a theory of causation that does not require temporal features and yet is consonant with the requirements of a scientific philosophy. Causation must be defined in such a way that one can discover causal relations by observation. No special nonsensuous intuitions will be allowed; no entailment theories will be accepted. Causation will not be a primitive feature of the universe. This much of Hume must be kept.

1. REICHENBACH'S PROBABILISTIC THEORY OF CAUSATION IN THE DIRECTION OF TIME

In *The Direction of Time* Reichenbach provides some of the materials needed for the construction of a non-temporal theory of causation, yet the view which he presents, at least in Section 22, appears to suffer from the same problem as is found in Suppes. Temporal features are required for the purpose of distinguishing, of two causally connected events, which is the cause and which the effect. Though this clearly makes Reichenbach's view unacceptable it is important that the basic features of his theory be outlined since the view that I will in the end defend is essentially a variant of it.

Reichenbach's theory is a probabilistic one. As is well known he held to the frequency interpretation of probability and it is this interpretation which will be followed here. We start off with individual events x, y, ... classified with respect to various sets A, B, ... Reichenbach places an important restriction on how one may define the sets and hence classify the events. The sets, he says, must be *codefined*. By this he means that one must be able to classify an event x as belonging to a particular set, say A, without making any reference to another event. The definition of the set and thereby the relevant properties of the event must all be one-place predicates. This rules out classifying events in terms of spatial and temporal relations. It also rules out Goodman type classifications, though this should not be taken to imply that restricting the sets to codefined ones provides any sort of solution to paradoxes of the grue type. Now the probabilities that are associated with the individual events are derived from the various codefined sets of which they are members, and it is in terms of these probabilities that Reichenbach attempts to give a purely probabilistic theory of causation.

Reichenbach's first step is to characterize a triadic relation called *causal betweenness* which is used to order the events into a net. The purpose here is to provide a non-linear order to the events so that when the 'direction' of causation is determined for one pair of events one can infer the direction for all others. Figure 1 illustrates such a net. Causal betweenness is defined as follows. Event C is causally between two events A and B, written $btw(ACB)$, if and only if the following hold:

(1) $1 > P(C, B) > P(A, B) > P(B)$

(2) $1 > P(C, A) > P(B, A) > P(A)$

(3) $P(A \,\&\, C, B) = P(C, B)$.

Figure 2 represents this relation.

What do these three propositions mean and why is the relation called one of *causal* betweenness? (1) and (2) express the fact that the cause (or the effect) is positively relevant to the effect (or the cause). This much is common to all empiricist theories of causation. If some principle of density were added to the definition we would have the claim that as one approaches either A or B from either end the conditional probabilities of the event one approaches gets closer to 1. (3) is the relation of *screening off*.[2] Here the presence of C makes A statistically irrelevant to B. C is said to screen off A from B.[3] Reichenbach shows that when $btw(ACB)$ holds not only does C screen off A from B but it also screens off B from A.

Other important properties of causal betweenness are discussed by Reichenbach in Section 22. He shows that if $btw(ACB)$ holds so does $btw(BCA)$. Causal betweenness is symmetrical. Hence there is no preferred direction and it must be supplied by some other means. Causal betweenness also has the property of uniqueness, that is, given any three distinct events only one may be causally between the other two. These properties allow the events to be tied together into a net. That it is a net and not a serial order that is constructed is clear from a third property of the relation; nontransitivity. The proposition $btw(ACB)$ & $btw(CBD) \supset btw(ACD)$ is not always true.

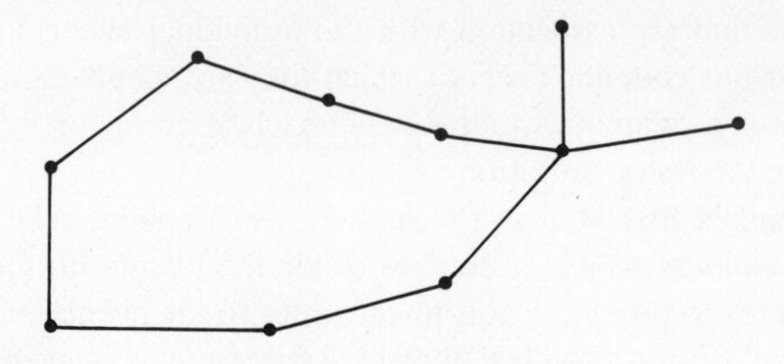

Fig. 1. A Causal Net.

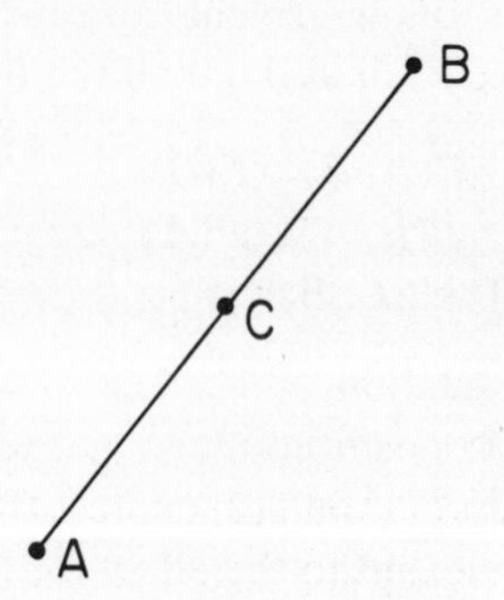

Fig. 2. A Causal Chain.

The problem now facing Reichenbach is to find a way to define the notion of causal direction so that causes can be distinguished from effects. His general strategy is as follows. If $btw(ACB)$ holds and one assumes that a causal direction obtains then there are the following two sets of possibilities; where the arrow indicates the direction of causation.

(a) $A \rightarrow C \rightarrow B$

$A \leftarrow C \leftarrow B.$

Here the directions are all the same. Both cases are instances of *equidirected* causal chains.

(b) $A \rightarrow C \leftarrow B$

$A \leftarrow C \rightarrow B$.

Here the directions are opposed; they are *counterdirected.* Reichenbach holds that if one can provide a means of differentiating those cases where the chains are equidirected from those where they are counterdirected then if one determines the direction of the arrows for one causal chain one can determine it for all.[4]

Reichenbach's procedure then is a three step one. First, events are ordered by the betweenness relation, second, equidirected causal chains are distinguished from counterdirected ones, and finally, the dirction for one or more chains is determined by some means or other. I believe that the first and third steps are adequately carried out by Reichenbach. No temporal features are employed in the definition of causal betweenness or of direction. It is another matter, however, when we come to the second, for in order to distinguish equi- from counterdirected chains Reichenbach invokes what he calls the Principle of the Local Comparability of Time Order.[5] By this he means that one has a means of determining when certain events are approximately coincident in time.

I quote from (Reichenbach, 1956; p. 194).

> Consider an arrangement of the kind illustrated in figure [3]. Assume we find that A_2 is causally between A_3 and A_5, and that A_3 and A_5 are in the neighborhood of A_2 and are still so close together that we can call A_3A_5 an approximate coincidence. We then conclude that the causal lines A_2A_3 and A_2A_5 are counterdirected. Likewise, assume that A_2 is causally between A_1 and A_4, but that the combination A_1A_4 represents an approximate coincidence. We then infer that the causal lines A_2A_1 and A_2A_4 are counterdirected. Furthermore, assume that A_2 is causally between A_1 and A_3, but that A_1 and A_3 do not represent an approximate coincidence. We then conclude that the causal line goes in the direction $A_1A_2A_3$, or in the direction $A_3A_2A_1$, that is, that the lines A_1A_2 and A_2A_3 have the same direction. Similar inferences can be made for the line $A_4A_2A_5$.

The problem with this is obvious. Temporal features, approximate coincidences, are used. We must try, then, to discover a non-temporal way of distinguishing equidirected from counterdirected causal chains.

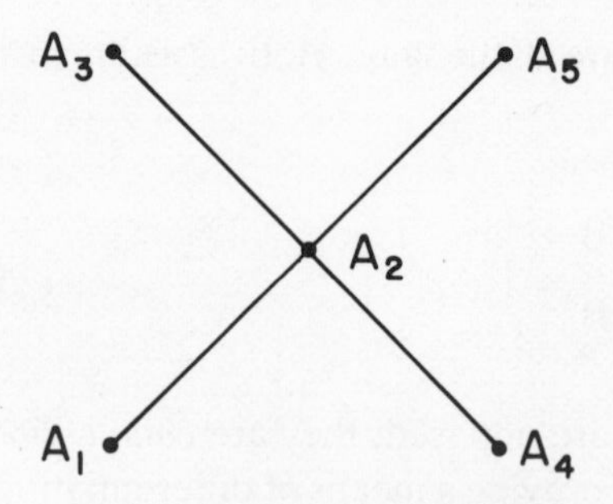

Fig. 3. The Local Comparability of Time Order.

II. CONJUNCTIVE FORKS AND CAUSAL DIRECTION

The problem before us, then, is how to define a non-temporal way of ordering the events of the causal net so that causes may be distinguished from effects. To do this it is necessary to introduce the concept of a conjunctive fork. A conjunctive fork is a state of affairs composed of three distinct events, *A*, *C*, and *B*, such that one of the events, *C*, is either the common cause of the other two or is itself their common effect. For example, a shriek in a library is the common cause of the turns of a number of different heads. Here a fork would be composed of *C*, the shriek, and *A* and *B*, two turnings.[6] The problem is to come up with a purely probabilistic definition of a conjunctive fork and then to provide a means of distinguishing common causes from common effects. Once this is done and it is shown further that when *ACB* is a conjunctive fork, *C* is causally between *A* and *B* the problem of devising a purely probabilistic theory of causality adequate to the demands of the causal theory of time will be solved.

First let us define the notion of a conjunctive fork. Three distinct events, *A*, *C*, *B*, form a conjunctive fork if and only if the following four propositions hold:

$$P(C, A \& B) = P(C, A) \times P(C, B) \tag{4}$$

$$P(\bar{C}, A \& B) = P(\bar{C}, A) \times P(\bar{C}, B) \tag{5}$$

$$P(C, A) > P(\bar{C}, A) \tag{6}$$

$$P(C, B) > P(\bar{C}, B)\,. \tag{7}$$

Some comments on these four propositions are in order. The first thing to note is that from them it is possible to derive $P(A \& B) > P(A) \times P(B)$.[7] This is essential since one important feature of a common cause is that it makes the joint occurrence of A and B more probable than would otherwise be expected. If all the lights in a house go out at the same time it is reasonable to look for a common cause rather than to suppose some fantastic coincidence. In fact the claim that such would be a fantastic coincidence just is the assertion that $P(A \& B) > P(A) \times P(B)$.

The next thing to do is to show that if ACB is a conjunctive fork then C is causally between A and B. Note first that (3) and (4) are equivalent by the Multiplication Theorem. This means that both C screens off A from B and B from A. (1) and (2) are derivable from (5), (6) and (7) as I shall now show. From (6) and (7) the following inequalities follow.[8]

(i) $P(C, B) > P(B) > P(\bar{C}, B)$

(ii) $P(C, A > P(A) > P(\bar{C}, A)$

(iii) $P(A, B) > P(B)$

(iv) $P(B, A) > P(A)$.

(5) is equivalent, again by the Multiplication Theorem, to both

(v) $P(A \& \bar{C}, B) = P(\bar{C}, B)$

and

(vi) $P(B \& \bar{C}, A) = P(\bar{C}, A)$.

These last two propositions express the fact that for a fork not only does C screen off A from B (and B from A) but $\bar{C}$ *does also*. This is important as will be seen in a moment. Now from this mob of inequalities (1) and (2) are easily derivable. $P(A \& \bar{C}, B) < P(A, B)$ follows from (i), (iii) and (v). Furthermore $P(A \& \bar{C}, B) < P(A, B)$ entails $P(A \& C, B) > P(A, B)$. Since we already know that $P(A \& C, B) = P(C, B)$, $P(C, B) > P(A, B)$ immediately follows. Combining this with (iii) and the assumption that $P(C, B) > 1$ gives us (1). An identical proof starting with (vi) gives us (2).

Consider, now, the situation as shown in Figure 4. It is clear that there must be at least one conjunctive fork present. Two are impossible, three unlikely. Assume that ACB is the only fork. These events satisfy, as a

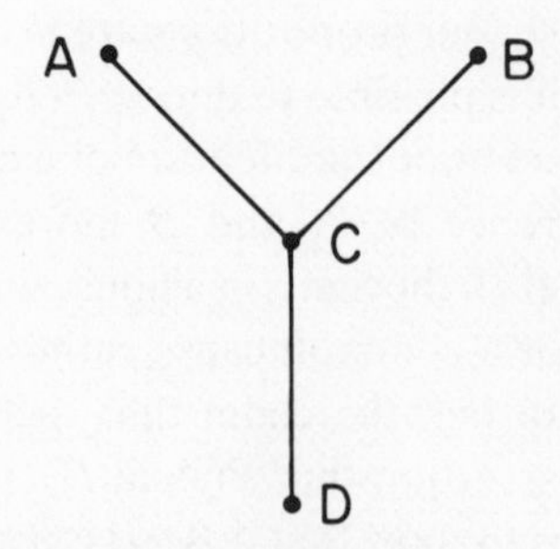

Fig. 4. Here *btw*(ACB), *btw* (DCA) and *btw*(DCB) all hold.

consequence, (4)–(7). The question then becomes which of these four propositions *fail* for the other two triples DCA and DCB. (4) is ruled out since it is equivalent to (3) and it is doubtful that either (6) or (7) can do the job since both should be true of most triples where one event is causally between the other two. One is left, it appears, with (5).

The conclusion to be drawn is that where ACB is a fork *both* C and $\bar{C}$ screen off; but where C is merely causally between the other two then *only* C screens off, $\bar{C}$ does not. A conjunctive fork might then be defined as a triple of events ACB such that (i) *btw*(ACB) holds and (ii) $\bar{C}$ screens off A and B. One could search for forks by going through the net and looking for triples where the middle event and its complement both screen off.

To illustrate this I shall use Reichenbach's theatrical example. The situation is one where both the leading lady and leading man become ill as a result of eating a common meal at a restaurant. (Presumably it is the food and not the company that is responsible.) Let A be the leading lady becomes ill, B be the leading man becomes ill, C be they eat a common meal at some restaurant, and D be they call beforehand to agree on where they shall dine together. These events exhibit the structure of Figure 4. C is the common cause of A and B. ACB forms a fork. DCA, on the other hand, does not. This is shown by the fact that $P(\bar{C}, A) \neq P(D \& \bar{C}, A)$. What does this mean? It allows for some causal 'force' to leak through to A from D by a route other than that via C. For example, the leading lady might have received the call while she was in the tub and as a result of getting out to answer it received a chill which brought about the illness. $\bar{C}$

is then not able to screen off D from A. Not going to the restaurant is not sufficient to make the event of answering the phone irrelevant to becoming ill.

Specification of causal direction at this point is now trivial. Let us define the notion of a conjunctive fork *open on a side* as follows. Forks have two sides. One side is composed of the elements ACB. The fork is said to be closed on this side, or closed at C. The other side would be closed if there were another event E such that AEB formed a conjunctive fork. If no such event exists the fork is said to be open on a side. Figures 5 and 6 illustrate these distinctions. Causal direction is simply defined. Given a fork ACB open on a side then C preceeds A or B. Since, as should now be obvious, counterdirected causal chains are forks and chains which are not forks but which simply exhibit causal betweenness are equidirected, once the direction of one pair of events is determined the direction of the whole net follows.[9]

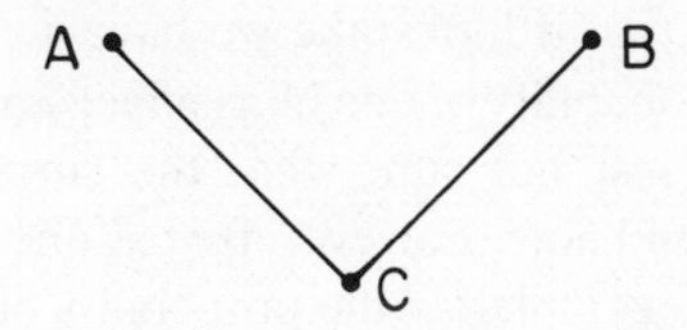

Fig. 5. Closed on the C-side, open on the other.

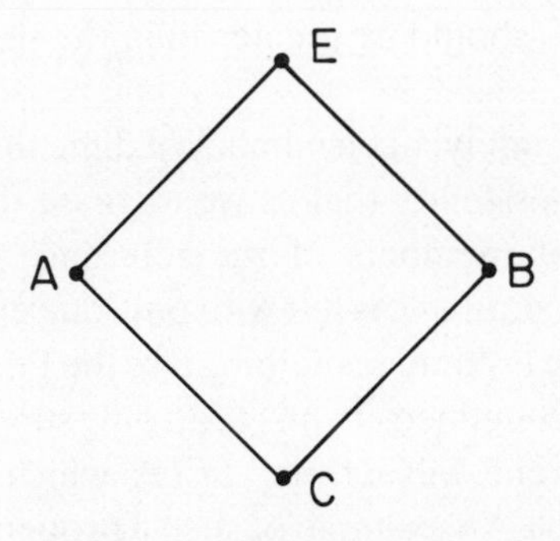

Fig. 6. Closed on both the C-side and the other side.

III. OBJECTIONS AND REPLIES: IMPROBABLE EFFECTS

The theory sketched above is open to two kinds of objections. The first kind involves objections to probabilistic theories of causation; the second to the use of such theories in the construction of the causal theory of time. I shall not spend much time on the first category and will tend to concentrate on the latter. There is an objection mentioned by Suppes that is fairly interesting and may indicate one place where the above theory is flawed. I quote from Suppes (1970; p. 41).

> . . . suppose a golfer makes a shot that hits a limb of a tree close to the green and is thereby deflected into the hole, for a spectacular birdie. Let the event to be explained, $A_{t'}$, to be the event of making a birdie, and let $B_{t'}$ be the event of hitting the limb earlier. If we know something about Mr. Jones' golf we can estimate the probability of his making a birdie on this particular hole. The probability will be low, but the seemingly disturbing thing is that if we estimate the conditional probability of his making a birdie, given that the ball hit the branch, that is, given that $B_{t'}$ occurred, we would ordinarily estimate the probability as being still lower. Yet when we see the event happen, we recognize immediately that hitting the branch in exactly the way it did was essential to the ball's going into the cup.

The problem here is that it looks like we have a case where the cause actually lowers the probability of one of its effects rather than raising it as it should. Now I'm just not sure what the correct response to this objection is. One should note, however, that as one specifies the properties of the event more completely the probability of the effect given this cause, i.e. $P(B, A)$, should increase and hopefully at some point become larger than the probability of the effect alone. If, for example, you know the coefficients of elasticity of the tree limb and the golf ball and the velocity of the ball and its mass the probability of it going in the hole given that it struck the limb should be greater than the simple probability of it going in the hole.

This reply, unfortunately, has a number of difficulties. First off, we have to worry about the possibility that as we increase the description of the event the number of members of the reference classes will become progressively smaller until one is left with only one event, B. This will be a special problem if one holds to something like the Principle of the Identity of Indiscernibles for codefined events. I can only suppose that many of the properties of the event will define classes which will be statistically irrelevant to the effect. An example of such a property might be the color of the golf ball.

Again one might note the fact, as Suppes does, that people who observe the actual course of events (club hitting ball, ball hitting tree, and finally, ball going in the hole) correctly judge that the ball hitting the tree was the cause of it going in the hole and do so without needing any information about masses of balls and elastic properties of tree limbs. This seems to imply that such a person is possessed with sufficient information to judge that the cause of the ball going into the hole was it hitting the tree limb. What is this information? It is, I believe, simply the belief that when balls hit limbs they go where they go because they hit the limb. The problem with this is how is one to state this claim in a properly probabilistic manner. That I think it can be done can be seen by noting the response that might be given if the ball, upon hitting the limb, flew up into the air and into orbit. Most unusual! I suspect one might think some factor other than hitting the limb was relevant. This situation is analogous, for example, to throwing a dart at a large board. Wherever it hits it is unlikely that it would hit there given that it was thrown, but it is not unlikely that it will hit the board. Of course, if one considers the event of the dart with such and such a mass, shape, etc. being thrown with such and such a force, then the probability of it landing where it does with respect to this event increases. If the dart were to fly off and go into orbit we would hesitate to say that the throw was the cause.

The final problem with my reply is that it is just an expression of hope. I have no real reason to suppose that such a move (that is, of specifying further the properties of the cause) will always bring about the desired change in the conditional probability of the effect given the cause. Examples may very well turn up where this is not the case.

IV. OBJECTIONS AND REPLIES: PSEUDO PROCESSES

We come now to a major objection to the attempt to use the above theory of causation for the purpose of constructing a causal theory of time. Special Relativity requires that one distinguish between true causal processes like the propogation of light signals from pseudo causal processes like the motion of a spot of light cast on a screen by a rotating beacon. True causal processes can travel no faster than c while pseudo processes have no upper limit to their speed.[10] Unfortunately it is not clear that our theory of causation has a way of distinguishing pseudo processes from causal ones. This is fairly important since most forms of the causal theory of time introduce a primitive notion of *genidentity* to deal with this problem. Genidentity is a relation that holds between events. Two events are genidentical "if they happen to the same object, if they belong to the history of one and the same object" (Van Fraassen, 1970; p. 34). For example, in Suppes' golf ball example the following events are all genidentical: the hitting of the golf ball by the club, the striking of the limb by the golf ball and the rolling into the hole. All these events are genidentical with each other because all of them 'involve' the same physical object, the golf ball. Now events that comprise a pseudo process such as a moving spot are not genidentical with one another since there is no one physical object which they all involve. Presumably, then, one could restrict his theory of time so that the only events that counted were ones within the field of the genidentity relation. Causal influence then would not propagate at speeds faster then c since causal influence would hold only among genidentical events.

There are, unfortunately, many problems with the notion of genidentity. The major one, as I see it, is that its use commits one to a substance or continuent ontology rather than an event ontology. Event ontologies make more sense out of the Special Theory of Relativity and are more happily married to relational theories of space and time.[11] It would be very nice if we could do away with the concept of genidentity altogether and devise a way of identifying pseudo processes that makes use merely of the notions of causal betweenness and of a conjunctive fork. This would show that the concepts adequate for the construction of the causal net and for its direction are adequate for picking out pseudo processes.

To do this let us start off by calling the series of events that compose the pseudo process of the moving spot the S-series, and the successive states of the rotating beacon the R-series. The events that comprise the R-series, or R-events, will be the total states of the beacon at each moment in time. The excitation of the filament, the forces that the gears apply to each other and the like will all be specified at a particular moment and will count as one element of the R-series. It will then be possible to order the elements of the R-series linearly so that they are ordered as to causal relevance. If these events are, for example, R_1, R_2, R_3... then $btw(R_1R_2R_3)$ will hold with the causal direction going from R_1 to R_2 to R_3. Now with each of these R-events one can associate particular events of the S-series, or S-events, so that the structure of the betweenness relations is that of Figure 7. One will note that we have a series of conjunctive forks with one prong being an S-event and the other being an R-event; the latter being the common cause of the next fork. One should also note that each fork as represented in Figure 7 is *open on a side* and is so in the direction of positive time. This fact of the events of the S-series systematically being prongs of certain conjunctive forks is one important property of pseudo process events.

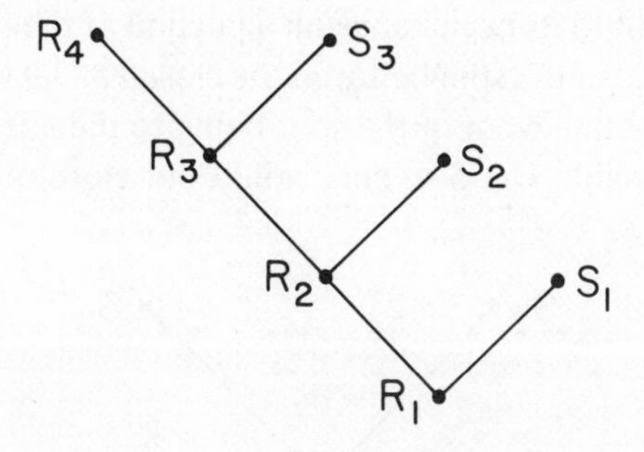

Fig. 7.

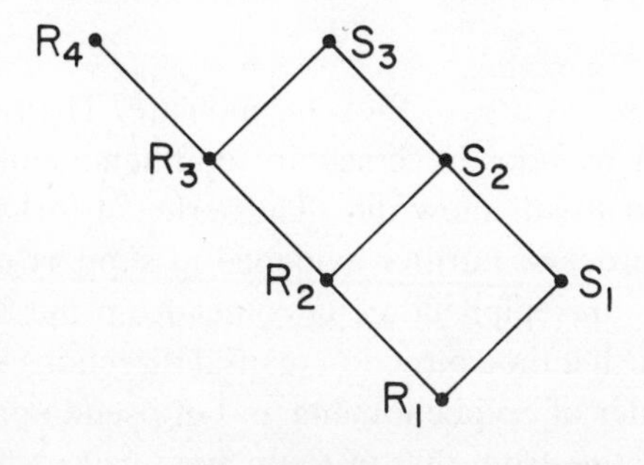

Fig. 8.

Consider the three S-events of Figure 7. Does $btw(S_1S_2S_3)$ hold? It would make things much simpler if it didn't. Unfortunately I suppose that sometimes it does. In such cases the forks $(R_2R_1S_1)$ and $(R_3R_2S_2)$ become *closed on the other side* from the common cause. This situation is depicted in Figure 8. If causal betweenness holds with respect to the members of a pseudo process then one more feature of pseudo processes is revealed, namely, that they systematically close off certain conjunctive forks. These forks have, in our example, a beacon event as a common cause of a spot event for one prong and another beacon event for the other.

Still, the events of the *S*-series don't always exhibit causal betweenness. Consider, for example, a screen with a sliding panel that is open for more times than it is closed. Some of the times when it is closed a spot from a light moves across it. In such cases the occurrence of a spot on the panel is irrelevant to its occurrence on a portion of the screen next to the pannel. The event of the spot being on the closed panel would then not be causally between the event of the spot being to the left of the panel and the spot to the right. The *S*-events would not close off their respective conjunctive forks.

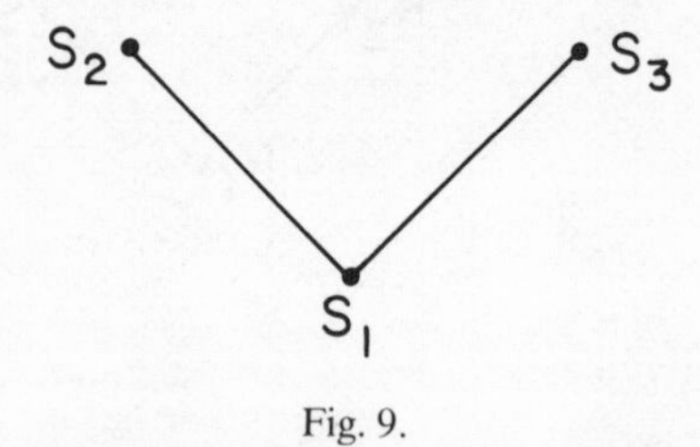

Fig. 9.

What do all these features of *S*-events indicate? They add up, I believe, to the claim that pseudo processes are epiphenomena.[12] A glance at either Figure 7 or 8 will show this. The series of forks is typical of an epiphenomenal process. Further evidence in support of the claim that pseudo processes are epiphenomena comes from the fact that it is not possible to establish a time direction (causal direction) by their sole use. The temporal order of epiphenomena and of pseudo processes must be completely derivative from that of their 'true' causes. What this comes down to is that genuine conjunctive forks do not exist whose members are events of a pseudo process. For example, consider Figure 9. Here we have the splitting of a spot on a screen produced by sending a beam of light through a beam splitter. Both S_2 and S_3 have an *R*-event as their common cause. This implies that S_1 does not screen off S_2 from S_3, and consequently, $(S_2S_1S_3)$ is not a conjunctive fork. This argument presupposes that causal betweenness can be defined for the three events so that *btw*$(S_2S_1S_3)$ holds. As noted there will be cases where this is not the case. In such circumstances there obviously will be no question of the pseudo process events establishing their own temporal direction.

V. CAUSAL INTERACTIONS AND EXPLANATION

The results of the preceeding section allow one to say a few brief things concerning the nature and importance of causal interactions. As is well known, one cannot transmit information by means of a pesudo process. Reichenbach in Section 19 of *The Philosophy of Space and Time* used this feature to distinguish pseudo processes from causal ones. Pseudo processes cannot be marked while causal ones can; consequently, only causal processes can transmit information.

Now, if it is possible to distinguish pseudo processes from causal processes by means of some of the properties of conjunctive forks, then one suspects that some sort of 'explanation' for why pseudo processes cannot be marked could be given. One would be in possession of such an 'explanation' when one has a suitable definition of causal interaction. This follows since any marking activity involves a causal interaction between two processes: the process which marks and the process which is marked. If one cannot interact with a pseudo process then one cannot mark it.

The best way to see what an interaction is is to compare causal interactions with pseudo interactions, i.e. structures that appear to involve interactions between causal processes and pseudo processes. Consider Figures 10 and 11. (Note that the direction of time is toward the top of the page.) In a true interaction the fork (DCE) is conjunctive whether or not it is open to the future while for a pseudo interaction the fork, in this case S_2S_1M is not conjunctive. If it turns out that S_2 and M are statistically relevant to each other then there will be a causal chain leading to S_2 by some other route. That chain will have an event which is the node of a fork with one prong ending in S_2; the other in M. A glance at Figure 11 provides such an example. It is the underlying process, the R-series in the case of the rotating beacon, which provides the explanation for these statistical relevencies.

Again, if the interaction is genuine the fork (ACB) of Figure 10 is conjunctive only when it is closed, since there are no conjunctive forks open toward the past. If the fork is open to the past then it is not conjunctive. It should be remembered that it can be open. In a false or pseudo interaction the fork $(R_2S_2S_1)$ is conjunctive only if it is closed. The difference here is that it cannot be open for otherwise S_1 loses its 'support'.

What is the significance of these distinctions? Basically they show that causal interactions form conjunctive forks, and consequently, common causes. If there were no interactions there would be no common causes, and therefore, no explanations for coincident statistical relevancies and no direction to either causal processes or time. The events which comprise the nodes of interaction between different causal processes are clearly quite basic.[13]

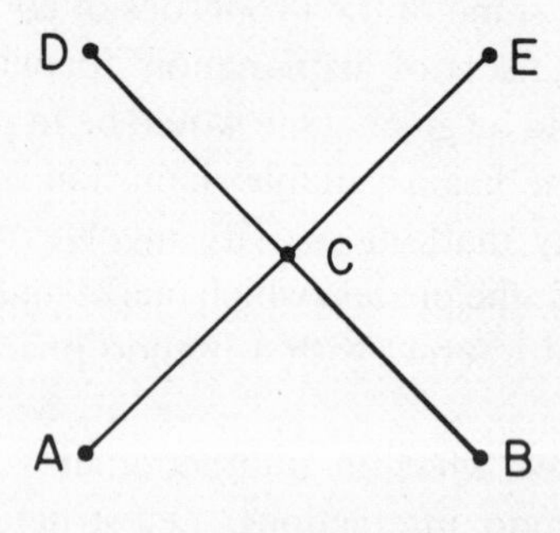

Fig. 10.

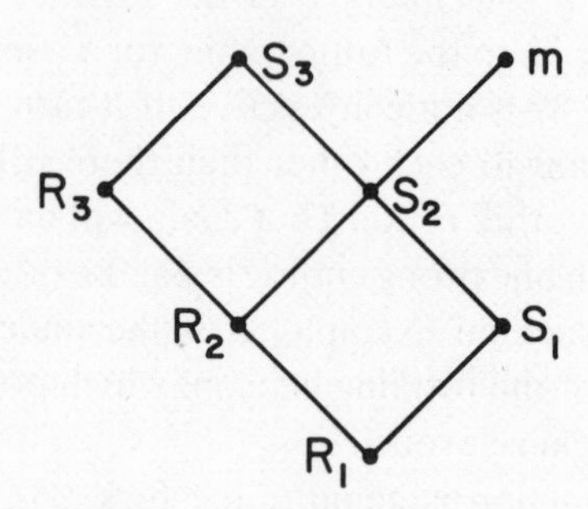

Fig. 11.

VI. SUMMARY

The theory of causal direction presented here is but a slight modification of Reichenbach's. Instead of using the principle of the Local Comparability of Time Order to distinguish counterdirected from equidirected causal chains I use the success (or failure) of $\bar{C}$ in screening off the ends events of those chains (where C is the middle event). This allows one to produce a theory of causality which is adequate for the construction of a causal theory of time free of the troublesome notion of genidentity and true to the strictures of empiricism.[14]

Tucson, Arizona

NOTES

[1] One should note that the notation here is the reverse of the usual. Since most of the paper is a discussion of Reichenbach I thought it best to follow his notation.

[2] For an interesting discussion of the importance of screening off see Salmon (1971); especially Section 6.

[3] One should note that much the same thing is found in Suppes once the explicit reference to time is removed. As has already been noted Suppes' notion of a *prima facie* cause is one of positive relevance. Screening off finds its place in his notion of a spurious cause which he defines as follows: A is a spurious cause of $B =_{\text{df}}$. (i) A is a *prima facia* cause of B, and (ii) there is an event C such that $P(A \;\&\; C, B) = P(C, B)$. That is, C screens off A from B. To say, then, that A is a spurious cause is to say merely that it is positively relevant to the effect and there is some other event C which screens off A from its effect.

[4] Features of the causal net are noted in Section 5 of Reichenbach (1956).

[5] I quote from (Reichenbach, 1956; p. 35).

> ... if two processes occur in spatiotemporal juxtaposition we regard a comparison of their time directions as possible. The statement that, in a given description, both processes have the same time direction, is regarded as observationally verifiable. We speak here of the *local comparability of time order*. (Reichenbach's italics.)

[6] An excellent discussion of the importance of common causes is to be found in Salmon, (1975).

[7] The above definition of a conjunctive fork and a derivation of this inequality are to be found in Section 19 of Reichenbach (1956).

[8] Again the proofs are in Section 19 of Reichenbach (1956).

[9] It should be mentioned that there is a kind of fork which can be open to the past. For example, let C be winning a prize, A be throwing an ace on one die, B, throwing an ace on the other, then where the prize is won when a double ace is thrown, C can be seen as a common effect of A and B. However, such forks are not conjunctive since $P(A \;\&\; B) = P(A) \times P(B)$, and consequently, do not exhibit causal betweenness. This last point

follows from the fact that $P(A \& B) > P(A) \times P(B)$ is equivalent, by the General Multiplication Theorem, to both $P(A, B) > P(B)$ and $P(B, A) > P(A)$. Since these last two inequalities appear in the definition of causal betweenness, the falsity of $P(A \& B) > P(A) \times P(B)$ entails that $btw(ACB)$ does not hold for these non-conjunctive forks.

[10] See Section 22 of Reichenbach (1958) where pseudo processes are called unreal sequences. A good discussion of these is found in Rothman (1960).

[11] This holds, of course, because in the Special Theory it is the space-time interval between events which is invariant not the spatial interval between things.

[12] This suggests the possibility that mental processes are pseudo processes.

[13] A definition of an interaction might go as follows: Two causal chains interact if and only if they have at least one common event which is the node of a conjunctive fork which if open would be open toward the future.

[14] I wish to express my thanks to Wesley Salmon for his help with respect to this paper.

BIBLIOGRAPHY

Broad, C. D.: 1914, *Perception, Physics and Reality*, Cambridge University Press.

Reichenbach, H.: 1956, *The Direction of Time*, University of California Press.

Reichenbach, H.: 1958, *The Philosophy of Space and Time*, Dover, New York.

Rothman, M. A.: 1960, 'Things that Go Faster than Light', *Scientific American* (July), 142–152.

Salmon, W. C.: 1971, 'Statistical Explanation', in *Statistical Explanation and Statistical Relevance*, University of Pittsburgh Press.

Salmon, W. C.: 1975, 'Theoretical Explanation', in S. Körner (ed.), *Explanation*, Basil Blackwell, Oxford.

Suppes, P.: 1970, *A Probabilistic Theory of Causality*, North-Holland, Amsterdam.

van Fraassen, B. C.: 1970, *An Introduction to the Philosophy of Time and Space*, Random House, New York.

WESLEY C. SALMON[1]

WHY ASK, 'WHY?'? AN INQUIRY CONCERNING SCIENTIFIC EXPLANATION

[In 'The Philosophy of Hans Reichenbach' (above), I remarked upon the fertility of his posthumous work, *The Direction of Time,* especially its fourth chapter, which deals with the causal and statistical relations among macrophenomena. I emphasized the value of his ideas in connection with the explication of scientific explanation – in particular, their potential utility for purposes of developing a full-blown theory of *causal* explanation. Since that article was written, I have been devoting considerable attention to just such issues, and I feel that my hopes have been – at least to some extent – vindicated.

The essay which follows is a slightly modified version of the Presidential Address which was presented at the 52nd Annual Meeting of the American Philosophical Association, Pacific Division, held in San Francisco in March, 1978. It embodies my attempt to outline a causal theory of scientific explanation. The resulting account relies heavily upon Bertrand Russell's *The Analysis of Matter* and *Human Knowledge,* as well as Reichenbach's *The Direction of Time.*]

Concerning the first order question 'Why?' I have raised the second order question 'Why ask, "Why?"?' to which you might naturally respond with the third order question 'Why ask, "Why ask, 'Why?'?"?' But this way lies madness, to say nothing of an infinite regress. While an infinite sequence of nested intervals may converge upon a point, the infinite series of nested questions just initiated has no point to it, and so we had better cut it off without delay. The answer to the very natural third order question is this: the question 'Why ask, "Why?"?' expresses a deep philosophical perplexity which I believe to be both significant in its own right and highly relevant to certain current philosophical discussions. I want to share it with you.

The problems I shall be discussing pertain mainly to scientific explanation, but before turning to them, I should remark that I am fully aware

W.C. Salmon (ed.), Hans Reichenbach: Logical Empiricist, 403–425.

that many – perhaps most – why-questions are requests for some sort of *justification* (Why did one employee receive a larger raise than another? Because she had been paid less than a male colleague for doing the same kind of job.) or *consolation* (Why, asked Job, was I singled out for such extraordinary misfortune and suffering?). Since I have neither the time nor the talent to deal with questions of this sort, I shall not pursue them further, except to remark that the seeds of endless philosophical confusion can be sown by failing carefully to distinguish them from requests for scientific explanation.

Let me put the question I do want to discuss to you this way. Suppose you had achieved the epistemic status of Laplace's demon – the hypothetical super-intelligence who knows all of nature's regularities, and the precise state of the universe in full detail at some particular moment (say now, according to some suitable simultaneity slice of the universe). Possessing the requisite logical and mathematical skill, you would be able to predict any future occurrence, and you would be able to retrodict any past event. Given this sort of apparent omniscience, would your scientific knowledge be complete, or would it still leave something to be desired? Laplace asked no more of his demon; should we place further demands upon ourselves? And if so, what should be the nature of the additional demands?

If we look at most contemporary philosophy of science texts, we find an immediate *affirmative* answer to this question. Science, the majority say, has at least two principal aims – prediction (construed broadly enough to include inference from the observed to the unobserved, regardless of temporal relations) and explanation. The first of these provides knowledge of *what* happens; the second is supposed to furnish knowledge of *why* things happen as they do. This is not a new idea. In the *Posterior Analytics* (Book I.2, 71b) Aristotle distinguishes syllogisms which provide scientific understanding from those which do not. In the *Port Royal Logic,* Arnauld distinguishes demonstrations which merely *convince* the mind from those which also *enlighten* the mind.[2]

This view has not been universally adopted. It was not long ago that we often heard statements to the effect that the business of science is to predict, not to explain. Scientific knowledge is descriptive – it tells us *what* and *how*. If we seek explanations – if we want to know *why* – we must go outside of science, perhaps to metaphysics or theology. In his Preface to the Third Edition (1911) of *The Grammar of Science,* Karl Pearson wrote, "Nobody believes now that science *explains* anything; we all look upon it as a shorthand description, as an economy of thought."[3]

This doctrine is not very popular nowadays. It is now fashionable to say that science aims not merely at describing the world – it also provides *understanding, comprehension,* and *enlightenment*. Science presumably accomplishes such high-sounding goals by supplying scientific explanations.

The current attitude leaves us with a deep and perplexing question, namely, if explanation does involve something over and above mere description, just what sort of thing is it? The use of such honorific near-synonyms as 'understanding,' 'comprehension,' and 'enlightenment' makes it sound important and desirable, but helps not at all in the philosophical analysis of explanation – scientific or other. What, over and above its complete descriptive knowledge of the world, would Laplace's demon require in order to achieve understanding? I hope you can see that this is a real problem, especially for those who hold what I shall call 'the inferential view' of scientific explanation, because Laplace's demon can infer every fact about the universe, past, present, and future. If you were to say that the problem does not seem acute, I would make the same remark as Russell made about Zeno's paradox of the flying arrow – "The more the difficulty is meditated, the more real it becomes" (1922, p. 179).

It is not my intention this evening to discuss the details of the various formal models of scientific explanation which have been advanced in the last three decades.[4] Instead, I want to consider the general conceptions which lie beneath the most influential theories of scientific explanation. Two powerful intuitions seem to have guided much of the discussion. Although they have given rise to disparate basic conceptions and considerable controversy, both are, in my opinion, quite sound. Moreover, it seems to me, both can be incorporated into a single overall theory of scientific explanation.

(1) The first of these intuitions is the notion that the explanation of a phenomenon essentially involves *locating and identifying its cause or causes*. This intuition seems to arise rather directly from common sense, and from various contexts in which scientific knowledge is applied to concrete situations. It is strongly supported by a number of paradigms, the most convincing of which are explanations of particular occurrences. To explain a given airplane crash, for example, we seek 'the cause'—a mechanical failure, perhaps, or pilot error. To explain a person's death again we seek the cause – strangulation or drowing, for instance. I shall call the general view of scientific explanation which comes more or less directly from this fundamental intuition *the causal conception*; Michael Scriven (e.g., 1975) has been one of its chief advocates.

(2) The second of these basic intuitions is the notion that all scientific explanation involves *subsumption under laws*. This intuition seems to arise from consideration of developments in theoretical science. It has led to the general 'covering law' conception of explanation, as well as to several formal 'models,' including the well-known deductive-nomological and inductive-statistical models. According to this view, a fact is subsumed under one or more general laws if the assertion of its occurrence follows, either deductively or inductively, from statements of the laws (in conjunction, in some cases, with other premises). Since this view takes explanations to be arguments, I shall call it *the inferential conception*; Carl G. Hempel has been one of its ablest champions.[5]

Although the proponents of this inferential conception have often chosen to illustrate it with explanations of particular occurrences – e.g., why did the bunsen flame turn yellow on this particular occasion? – the paradigms which give it strongest support are explanations of general regularities. When we look to the history of science for the most outstanding cases of scientific explanations, such examples as Newton's explanation of Kepler's laws of planetary motion or Maxwell's electro-magnetic explanation of optical phenomena come immediately to mind.

It is easy to guess how Laplace might have reacted to my question about his demon, and to the two basic intuitions I have just mentioned. The super-intelligence would have everything needed to provide scientific explanations. When, to mention one of Laplace's favorite examples, (1951, pp. 3–6), a seemingly haphazard phenomenon, such as the appearance of a comet, occurs, it can be explained by showing that it actually conforms to natural laws. On Laplace's assumption of determinism, the demon possesses explanations of all happenings in the entire history of the world – past, present, and future. Explanation, for Laplace, seemed to consist in showing how events conform to the laws of nature, and these very laws provide the causal connections among the various states of the world. The Laplacian version of explanation thus seems to conform both to the causal conception and to the inferential conception.

Why, you might well ask, is not the Laplacian view of scientific explanation basically sound? Why do twentieth century philosophers find it necessary to engage in lengthy disputes over this matter? There are, I think, three fundamental reasons: (1) the causal conception faces the difficulty that no adequate treatment of causation has yet been offered; (2) the inferential conception suffers from the fact that it seriously misconstrues the nature of subsumption under laws; and (3) both conceptions have overlooked a central explanatory principle.

The inferential view, as elaborated in detail by Hempel and others, has been the dominant theory of scientific explanation in recent years – indeed, it has become virtually 'the received view.' From that standpoint, anyone who had attained the epistemic status of Laplace's demon could use the known laws and initial conditions to predict a future event, and when the event comes to pass, the argument which enabled us to predict it would ipso facto constitute an explanation of it. If, as Laplace believed, determinism is true, then every *future* event would thus be amenable to deductive-nomological explanation.

When, however, we consider the explanation of past events – events which occurred earlier than our initial conditions – we find a strange disparity. Although, by applying known laws, we can reliably *retrodict* any past occurrence on the basis of facts subsequent to the event, our intuitions rebel at the idea that we can *explain* events in terms of subsequent conditions. Thus, although our inferences to future events qualify as explanations according to the inferential conception, our inferences to the past do not. Laplace's demon can, of course, construct explanations of past events by inferring the existence of still earlier conditions and, with the aid of the known laws, deducing the occurrence of the events to be explained from these conditions which held in the more remote past. But if, as the inferential conception maintains, explanations are essentially inferences, such an approach to past events seems strangely roundabout. Explanations demand an asymmetry not present in inferences.

When we drop the fiction of Laplace's demon, and relinquish the assumption of determinism, the asymmetry becomes even more striking. The demon can predict the future and retrodict the past with complete precision and reliability. We cannot. When we consider the comparative difficulty of prediction vs. retrodiction, it turns out that retrodiction enjoys a tremendous advantage. We have records of the past – tree rings, diaries, fossils – but none of the future. As a result, we can have extensive and detailed knowledge of the past which has no counterpart in knowledge about the future. From a newspaper account of an accident, we can retrodict all sorts of details which could not have been predicted an hour before the collision. But the newspaper story – even though it may *report* the explanation of the accident – surely does not *constitute* the explanation. We see that *inference* has a preferred temporal direction, and that *explanation* also has a preferred temporal direction. The fact that these two are opposite to each other is one thing which makes me seriously doubt that explanations are essentially arguments.[6] As we shall see, however, denying that explanations are arguments does not mean that we must give up the *covering law* conception. Subsumption under laws can take a different form.

Although the Laplacian conception bears strong similarities to the received view, there is a fundamental difference which must be noted. Laplace apparently believed that the explanations provided by his demon would be *casual explanations*, and the laws invoked would be *casual laws*. Hempel's deductive-nomological explanations are often causally called 'causal explanations,' but this is not accurate. Hempel (1965, pp. 352–354) explicitly notes that some laws, such as the ideal gas law,

$$PV = nRT,$$

are non-causal. This law states a mathematical functional relationship among several quantities – pressure P, volume V, temperature T, number of moles of gas n, universal gas constant R – but gives no hint as to how a change in one of the values would lead causally to changes in others. As far as I know, Laplace did not make any distinction between causal and non-causal laws; Hempel has recognized the difference, but he allows non-causal as well as causal laws to function as covering laws in scientific explanations.

This attitude toward non-causal laws is surely too tolerant. If someone inflates an air-mattress of a given size to a certain pressure under conditions which determine the temperature, we can deduce the value of n – the amount of air blown into it. The *subsequent* values of pressure, temperature, and volume are thus taken to explain the quantity of air *previously* introduced. Failure to require covering laws to be causal laws leads to a violation of the temporal requirement on explanations. This is not surprising. The asymmetry of explanation is inherited from the asymmetry of causation – namely, that causes precede their effects. At this point, it seems to me, we experience vividly the force of the intuitions underlying the causal conception of scientific explanation.

There is another reason for maintaining that non-causal laws cannot bear the burden of covering laws in scientific explanations. Non-causal regularities, instead of having explanatory force which enables them to provide understanding of events in the world, cry out to be explained. Mariners, long before Newton, were fully aware of the corrleation between the behavior of the tides and the position and phase of the moon. But inasmuch as they were totally ignorant of the causal relations involved, they rightly made no claim to any understanding of why the tides ebb and flow. When Newton provided the gravitational links, understanding was achieved. Similarly, I should say, the ideal gas law had little or no explanatory power until its causal underpinnings were furnished by the molecular-kinetic theory of gases. Keeping this consideration in mind, we realize that we must give at least as much attention to the explanations of regularities as we do to explanations of particular facts. I will argue, moreover, that these regularities demand causal explanation. Again, we must give the causal conception its due.

Having considered a number of prelininaries, I should now like to turn my attention to an attempt to outline a general theory of causal explanation. I shall not be trying to articulate a formal model; I shall be focusing upon general conceptions and fundamental principles rather than technical details. I am not suggesting, of course, that the technical details are dispensable – merely that this is not the time or place to try to go into them.

Developments in twentieth-century science should prepare us for the eventuality that some of our scientific explanations will have to be statistical – not merely because our knowledge is incomplete (as Laplace would have maintained), but rather, because nature itself is inherently statistical. Some of the laws used in explaining particular events will be statistical, and some of the regularities we wish to explain will also be statistical. I have been urging that causal considerations play a crucial role in explanation; indeed, I have just said that regularities – and this certainly includes statistical regularities – require causal explanation. I do not believe there is any conflict here. It seems to me that, by employing a statistical conception of causation along the lines developed by Patrick Suppes (1970) and Reichenbach (1956, chap. IV), it is possible to fit together harmoniously the causal and statistical factors in explanatory contexts. Let me attempt to illustrate this point by discussing a concrete example.

A good deal of attention has recently been given in the press to cases of leukemia in military personnel who witnessed an atomic bomb test (code name "Smokey") at close range in 1957[7]. Statistical studies of the survivors of the bombings of Hiroshima and Nagasaki have established the fact that exposure to high levels of radiation, such as occur in an atomic blast, is statistically relevant to the occurrence of leukemia – indeed, that the probability of leukemia is closely correlated with the distance from the explosion.[8] A clear pattern of statistical relevance relations is exhibited here. If a particular person contracts leukemia, this fact may be explained by citing the fact that he was, say, 2 kilometers from the hypocenter at the time of the explosion. This relationship is further explained by the fact that individuals located at specific distances from atomic blasts of specified magnitude receive certain high doses of radiation.

This tragic example has several features to which I should like to call special attention:

(1) The location of the individual at the time of the blast is statistically

relevant to the occurrence of leukemia; the probability of leukemia for a person located 2 kilometers from the hypocenter of an atomic blast is radically different from the probability of the disease in the population at large. Notice that the probability of such an individual contracting leukemia is not high; it is much smaller than one-half – indeed, in the case of Smokey it is much less than 1/100. But it is markedly higher than for a random member of the entire human population. It is the *statistical relevance* of exposure to an atomic blast, not a *high probability*, which has explanatory force.[9] Such examples defy explanation according to an inferential view which requires high inductive probability for statistical explanation.[10] The case of leukemia is subsumed under a statistical regularity, but it does not 'follow inductively' from the explanatory facts.

(2) There is a *causal process* which connects the occurrence of the bomb blast with the physiological harm done to people at some distance from the explosion. High energy radiation, released in the nuclear reactions, traverses the space between the blast and the individual. Although some of the details may not yet be known, it is a well-established fact that such radiation does interact with cells in a way which makes them susceptible to leukemia at some later time.

(3) At each end of the causal process – i.e., the transmission of radiation from the bomb to the person – there is a *causal interaction*. The radiation is emitted as a result of a nuclear interaction when the bomb explodes, and it is absorbed by cells in the body of the victim. Each of these interactions may well be irreducibly statistical and indeterministic, but that is no reason to deny that they are causal.

(4) The causal processes begin at a central place, and they travel outward at a finite velocity. A rather complex set of statistical relevance relations is explained by the propagation of a process, or set of processes, from a common central event.

In undertaking a general characterization of causal explanation, we must begin by carefully distinguishing between causal processes and causal interactions. The transmission of light from one place to another, and the motion of a material particle, are obvious examples of causal processes. The collision of two billiard balls, and the emission or absorption of a photon, are standard examples of causal interactions. Interactions are the sorts of things we are inclined to identify as events. Relative to a particular context, an event is comparatively small in its spatial and temporal dimensions; processes typically have much larger durations, and they may be more extended in space as well. A light ray, traveling to earth from a distant star, is a process which covers a large distance and lasts for a long time. What I am calling a 'causal process' is similar to what Russell called a 'causal line' (1948, p. 459).

When we attempt to identify causal processes, it is of crucial importance to distinguish them from such pseudo-processes as a shadow moving across the landscape. This can best be done, I believe, by invoking Reichenbach's *mark criterion*.[11] Causal processes are capable of propagating marks or modifications imposed upon them; pseudo-processes are not. An automobile traveling along a road is an example of a causal process. If a fender is scraped as a result of a collision with a stone wall, the mark of that collision will be carried on by the car long after the interaction with the wall occurred. The shadow of a car moving along the shoulder is a pseudo-process. If it is deformed as it encounters a stone wall, it will immediately resume its former shape as soon as it passes by the wall. It will not transmit a mark or modification. For this reason, we say that a causal process can transmit information or causal influence; a pseudo-process cannot.[12]

When I say that a causal process has the capability of transmitting a causal influence, it might be supposed that I am introducing precisely the sort of mysterious power Hume warned us against. It seems to me that this danger can be circumvented by employing an adaptation of the 'at-at' theory of motion, which Russell used so effectively in dealing with Zeno's paradox of the flying arrow.[13] The flying arrow – which is, by the way, a causal process – gets from one place to another by being *at* the appropriate intermediate points of space *at* the appropriate instants of time. Nothing more is involved in getting *from* one point *to* another. A mark, analogously, can be said to be propagated from the point of interaction at which it is imposed to later stages in the process if it appears *at* the appropriate intermediate stages in the process *at* the appropriate times without additional interactions which regenerate the mark. The precise formulation of this condition is a bit tricky, but I believe the basic idea is simple, and that the details can be worked out.[14]

If this analysis of causal processes is satisfactory, we have an answer to the question, raised by Hume, concerning the connection between cause and effect. If we think of a cause as one event, and of an effect as a distinct event, then the connection between them is simply a spatio-temporally continuous causal process. This sort of answer did not occur to Hume because he did not distinguish between causal processes and causal interactions. When he tried to analyze the connections between distinct events, he treated them as if they were chains of events with discrete links, rather than processes analogous to continuous filaments. I am inclined to attribute considerable philosophical significance to the fact that each link in a chain has adjacent links, while the points in a continuum do not have next-door neighbors. This consideration played an important role in Russell's discussion of Zeno's paradoxes.[15]

After distinguishing between causal interactions and causal processes, and after introducing a criterion by means of which to discriminate the pseudo-processes from the genuine causal processes, we must consider certain configurations of processes which have special explanatory import. Russell noted that we often find similar structures grouped symmetrically about a center – for example, concentric waves moving across an otherwise smooth surface of a pond, or sound waves moving out from a central region, or perceptions of many people viewing a stage from different seats in a theatre. In such cases, Russell (1948, pp. 460–475) postulates the existence of a central event – a pebble dropped into the pond, a starter's gun going off at a race-track, or a play being performed upon the stage – from which the complex array emanates. It is noteworthy that Russell never suggests that the central event is to be explained on the basis of convergence of influences from remote regions upon that locale.

Reichenbach (1956, Sec. 19) articulated a closely-related idea in his *principle of the common cause*. If two or more events of certain types occur at different places, but occur at the same time more frequently than is to be expected if they occurred independently, then this apparent coincidence is to be explained in terms of a common causal antecedent. If, for example, all of the electric lights in a particular area go out simultaneously, we do not believe that they just happened by chance to burn out at the same time. We attribute the coincidence to a common cause such as a blown fuse, a downed transmission line, or trouble at the generating station. If all of the students in a dormitory fall ill on the same night, it is attributed to spoiled food in the meal which all of them ate. Russell's similar structures arranged symmetrically about a center obviously qualify as the sorts of coincidences which require common causes for their explanations.[16]

In order to formulate his common cause principle more precisely, Reichenbach defined what he called a *conjunctive fork*. Suppose we have events of two types, A and B, which happen in conjunction more often than they would if they were statistically independent of one another. For example, let A and B stand for colorblindness in two brothers. There is a certain probability that a male, selected from the population at random, will have that affliction, but since it is often hereditary, occurrences in male siblings are not independent. The probability that both will have it is greater than the product of the two respective probabilities. In cases of such statistical dependencies, we invoke a common cause *C* which accounts for them; in this case, it is a genetic factor carried by the mother. In order to satisfy the conditions for a conjunctive fork, events of the types *A* and *B* must occur independently in the absence of the common cause *C* – that is, for two unrelated males, the probability of both being colorblind is equal to the product of the two separate probabilities. Furthermore, the probabilities of *A* and *B* must each be increased above their overall values if *C* is present. Clearly the probability of colorblindness is greater in sons of mothers carrying the genetic factor than it is among all male children regardless of the genetic make-up of their mothers. Finally, Reichenbach stipulates, the dependency between *A* and *B* is absorbed into the occurrence of the common cause *C*, in the sense that the probability of *A* and *B* given *C* equals the product of the probability of *A* given *C* and the probability of *B* given *C*. This is true in the colorblindness case. Excluding pairs of identical twins, the question of whether a male child inherits colorblindness from the mother who carries the genetic trait depends only upon the genetic relationship between that child and his mother, not upon whether other sons happened to inherit the trait.[17] Note that screening-off occurs here.[18] While the colorblindness of a brother is statistically relevant to colorblindness in a boy, it becomes irrelevant if the genetic factor is known to be present in the mother.

Reichenbach obviously was not the first philosopher to notice that we explain coincidences in terms of common causal antecedents. Leibniz postulated a pre-established harmony for his windowless monads which mirror the same world, and the occasionalists postulated God as the coordinator of mind and body. Reichenbach (1956, pp. 162–163) was, to the best of my knowledge, the first to give a precise characterization of the conjunctive fork, and to formulate the general principle that conjunctive forks are open only to the future, not to the past. The result is that we cannot explain coincidences on the basis of future effects, but only on the basis of antecedent causes. A widespread blackout is explained by a power failure, not by the looting which occurs as a consequence. (*A* common effect *E* may form a conjunctive fork with *A* and *B*, but only if there is also a common cause *C*.) The principle that conjunctive forks are not open to the past accounts for Russell's principle that symmetrical patterns emanate from a central source – they do not converge from afar upon the central point. It is also closely related to the operation of the second law of thermodynamics and the increase of entropy in the physical world.

The common cause principle has, I believe, deep explanatory significance. Bas van Fraassen (1977) has recently subjected it to careful scrutiny, and he has convinced me that Reichenbach's formulation in terms of the conjunctive fork, as he defined it, is faulty. (We do not, however, agree about the nature of the flaw.) There are, it seems, certain sorts of causal *interactions* in which the resulting effects are more strongly correlated with one another than is allowed in Reichenbach's conjunctive forks. If, for example, an energetic photon collides with an electron in a Compton scattering experiment, there is a certain probability that a photon with a given smaller energy will emerge, and there is a certain probability that the electron will be kicked out with a given kinetic energy (see Figure 1). However, because of the law of conservation of energy, there is strong correspondence between the two energies – their sum must be close to the energy of the incident photon. Thus, the probability of getting a photon with energy E_1 and an electron with energy E_2, where $E_1 + E_2$ is approximately equal to E (the energy of the incident photon), is much greater than the product of the probabilities of each energy occurring separately. Assume, for example, that there is a probability of 0.1 that a photon of energy E_1 will emerge if a photon of energy E impinges on a given target, and assume that there is a probability of 0.1 that an electron with kinetic energy E_2 will emerge under the same circumstances (where E, E_1, and E_2 are related as the law of conservation of energy demands). In this case the probability of the joint result is not 0.01, the product of the separate probabilities, but 0.1, for each result will occur if and only if the other does.[19] The same relationships could be illustrated by such macroscopic events as collisions of billiard balls, but I have chosen Compton scattering be-

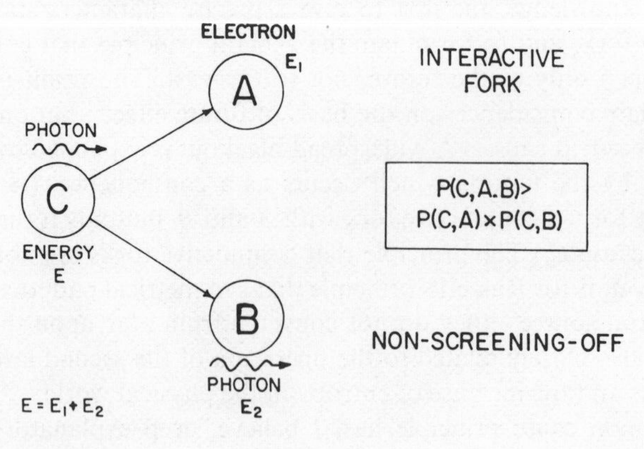

Fig. 1.

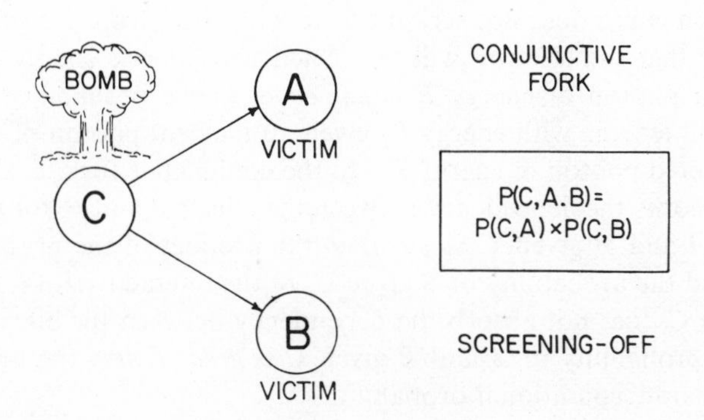

Fig. 2.

cause there is good reason to believe that events of that type are irreducibly statistical. Given a high energy photon impinging upon the electron in a given atom, there is no way, even in principle, of predicting with certainty the energies of the photon and electron which result from the interaction.

This sort of interaction stands in sharp contrast with the sort of statistical dependency we have in the leukemia example (see Figure 2, which also represents the relationships in the colorblindness case). In the absence of a strong source of radiation, such as the atomic blast, we may assume that the probability of next-door neighbors contracting the disease equals the product of the probabilities for each of them separately. If, however, we consider two next-door neighbors who lived at a distance of 2 kilometers from the hypocenter of the atomic explosion, the probability of both of them contracting leukemia is much greater than it would be for any two randomly selected members of the population at large. This apparent dependency between the two leukemia cases is not a direct physical dependency between them; it is merely a statistical result of the fact that the probability for each of them has been enhanced independently of the other by being located in close proximity to the atomic explosion. But the individual photons of radiation which impinge upon the two victims are emitted independently, travel independently, and damage living tissues independently.

It thus appears that there are two kinds of causal forks: (1) Reichenbach's *conjunctive forks*, in which the common cause screens-off the one effect from the other, which are exemplified by the colorblindness and leukemia cases, and (2) *interactive forks*, exemplified by the Compton scattering of a photon and an electron. In forks of the interactive sort, the common cause does not screen-off the one effect from the other. The probability that the electron will be ejected with kinetic energy E_2 given an incident photon of energy E is *not equal to* the probability that the electron will emerge with energy E_2 given an incident photon of energy E *and* a scattered photon of energy E_1. In the conjunctive fork, the common cause C absorbs the dependency between the effects A and B, for the probability of A and B given C is *equal to* the product of the probability A given C and the probability of B given C. In the interactive fork, the common cause C does not absorb the dependency between the effects A and B, for the probability of A and B given C is *greater than* the product of the two separate conditional probabilities.[20]

Recognition and characterization of the interactive fork enables us to fill a serious lacuna in the treatment up to this point. I have discussed causal processes, indicating roughly how they are to be characterized, and I have mentioned causal interactions, but have said nothing about their characterization. Indeed, the criterion by which we distinguished causal processes from pseudo-processes involved the use of marks, and marks are obviously results of causal interactions. Thus, our account stands in serious need of a characterization of causal interactions, and the interactive fork enables us, I believe, to furnish it.

There is a strong temptation to think of events as basic types of entities, and to construe processes – real or pseudo – as collections of events. This viewpoint may be due, at least in part, to the fact that the space-time interval between events is a fundamental invariant of the special theory of relativity, and that events thus enjoy an especially fundamental status. I suggest, nevertheless, that we reverse the approach. Let us begin with processes (which have not yet been sorted out into causal and pseudo) and look at their intersections. We can be reassured about the legitimacy of this new orientation by the fact that the basic space-time structure of both special relativity and general relativity can be built upon processes without direct recourse to events.[21] An electron traveling through space is a process, and so is a photon; if they collide, that is an intersection. A light pulse traveling from a beacon to a screen is a process, and a piece of red glass standing in the path is another; the light passing through the glass is an intersection. Both of these intersections constitute interactions. If two light beams cross one another, we have an intersection without an interaction – except in the extremely unlikely event of a particle-like collision between photons. What we want to say, very roughly, is that when

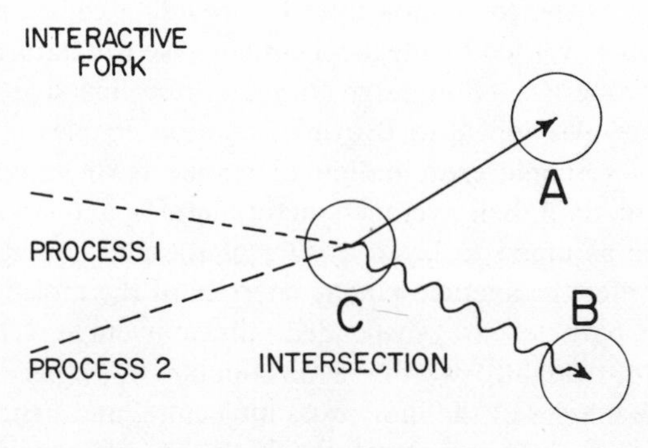

Fig. 3.

two processes intersect, and both are modified in such ways that the changes in one are correlated with changes in the other – in the manner of an interactive fork (see Figure 3) – we have a causal interaction. There are technical details to be worked out before we can claim to have a satisfactory account, but the general idea seems clear enough.[22]

I should like to commend the principle of the common cause – so construed as to make reference to both conjunctive forks and interactive forks – to your serious consideration.[23] Several of its uses have already been mentioned and illustrated. *First,* it supplies a schema for the straightforward explanations of everyday sorts of otherwise improbable coincidences. *Second,* by means of the conjunctive fork, it is the source of the fundamental temporal asymmetry of causality, and it accounts for the temporal asymmetry we impose upon scientific explanations. *Third,* by means of the interactive fork it provides the key to the explication of the concept of causal interaction.[24] These considerations certainly testify to its philosophical importance.

There are, however, two additional applications to which I should like to call attention. *Fourth,* as Russell (1948, pp. 491–492) showed, the principle plays a fundamental role in the causal theory of perception. When various observers (including cameras as well as human beings) arranged around a central region (such as a stage in theatre-in-the-round) have perceptions which correspond systematically with one another in the customary way, we may infer, with reasonable reliability, that they have a common cause – namely, a drama being performed on the stage. This fact has considerable epistemological import.

Fifth, the principle of the common cause can be invoked to support scientific realism.[25] Suppose, going back to a previous example, we have postulated the existence of molecules to provide a causal explanation of the phenomena governed by the ideal gas law. We will naturally be curious about their properties – how large they are, how massive they are, how many there are. An appeal to Brownian motion enables us to infer such things. By microscopic examination of smoke particles suspended in a gas, we can ascertain their average kinetic energies, and since the observed system can be assumed to be in a state of thermal equilibrium, we can immediately infer the average kinetic energies of the molecules of the gas in which the particles are suspended. Since average velocities of the molecules are straightforwardly ascertainable by experiment, we can easily find the masses of the individual molecules, and hence, the number of molecules in a given sample of gas. If the sample consists of precisely one mole (gram molecular weight) of the particular gas, the number of molecules in the sample is Avogadro's number – a fundamental physical constant. Thus, the causal explanation of Brownian motion yields detailed quantitative information about the micro-entities of which the gas is composed.

Now, consider another phenomenon which appears to be of an altogether different sort. If an electric current is passed through an electrolytic solution – for example, one containing a silver salt – a certain amount of metallic silver is deposited on the cathode. The amount deposited is proportional to the amount of electric charge which passes through the solution. In constructing a causal explanation of his phenomenon (known as electrolysis), we postulate that charged ions travel through the solution, and that the amount of charge required to deposit a singly charged ion is equal to the charge on the electron. The magnitude of the electron charge was empirically determined through the work of J. J. Thomson and Robert Millikan. The amount of electric charge required to deposit one mole of a monovalent metal is known as the Faraday, and by experimental determination, it is equal to 96,487 coulombs. When this number is divided by the charge on the electron (-1.602×10^{-19} coulombs), the result is Avogadro's number. Indeed, the Faraday is simply Avogadro's number of electron charges.

The fundamental fact to which I wish to call attention is that the value of Avogadro's number ascertained from the analysis of Brownian motion agrees, within the limits of experimental error, with the value obtained by electrolytic measurement. Without a common causal antecedent, such agreement would constitute a remarkable coincidence. The point may be put in this way. From the molecular-kinetic theory of gases we can derive the statement form, 'The number of molecules in a mole of gas is________.' From the electro-chemical theory of electrolysis, we can derive the statement form, 'The number of electron charges in a Faraday is________.' The astonishing fact is that the same number fills both blanks. In my opinion, the instrumentalist cannot, with impunity, ignore what must be an amazing correspondence between what happens when one scientist is watching smoke particles dancing in a container of gas while another scientist in a different laboratory is observing the electroplating of silver. Without an underlying causal mechanism – of the sort involved in the postulation of atoms, molecules, and ions – the coincidence would be as miraculous as if the number of grapes harvested in California in any given year were equal, up to the limits of observational error, to the number of coffee beans produced in Brazil in the same year. Avogadro's number, I must add, can be ascertained in a variety of other ways as well – e.g., X-ray diffraction from crystals – which also appear to be entirely different unless we postulate the existence of atoms, molecules, and ions. The principle of the common cause thus seems to apply directly to the explanation of observable regularities by appeal to unobservable entities. In this instance, to be sure, the common cause is not some sort of event; it is rather a common constant underlying structure which manifests itself in a variety of different situations.

Let me now summarize the picture of scientific explanation I have tried to outline. If we wish to explain a particular event, such as death by leukemia of GI Joe, we begin by assembling the factors statistically relevant to the occurrence – for example, his distance from the atomic explosion, the magnitude of the blast, and the type of shelter he was in. There will be many others, no doubt, but these will do for purposes of illustration. We must also obtain the probability values associated with the relevancy relations. The statistical relevance relations are statistical regularities, and we proceed to explain them. Although this differs substantially from things I have said previously, I no longer believe that the assemblage of relevant factors provides a complete explanation – or much of anything in the way of an explanation.[26] We do, I believe, have a bona fide explanation of an event if we have a complete set of statistically relevant factors, the pertinent probability values, *and* causal explanations of the relevance relations. Subsumption of a particular occurrence under statistical regularities – which, we recall, does not imply anything about the construction of deductive or inductive arguments – is a necessary part of any adequate explanation of its occurrence, but it is not the whole story. The causal explanation of the regularity is also needed. This claim, it should be noted, is in direct conflict with the received view, according to which the mere subsumption – deductive or inductive – of an event under a lawful regularity constitutes a complete explanation. One can, according to the received view, go on to ask for an explanation of any law used to explain a given event, but that is a different explanation. I am suggesting, on the contrary, that if the regularity invoked is not a causal regularity, then a causal explanation of that very regularity must be made part of the explanation of the event.

If we have events of two types, *A* and *B*, whose respective members are not spatio-temporally contiguous, but whose occurrences are correlated with one another, the causal explanation of this regularity may take either of two forms. Either there is a direct causal connection from *A* to *B* or from *B* to *A*, or there is a common cause *C* which accounts for the statistical dependency. In either case, those events which stand in the cause-effect relation to one another are joined by a causal process.[27] The distinct events *A*, *B*, and *C* which are thus related constitute interactions – as defined in terms of an interactive fork – at the appropriate places in the respective causal processes. The interactions *produce* modifications in the causal processes, and the causal processes *transmit* the modifications. Statistical dependency relations arise out of local interactions – there is no action-at-a-distance (as far as macro-phenomena are concerned, at least) – and they are propagated through the world by causal processes. In our leukemia example, a slow neutron, impinging upon a uranium atom, has a certain probability of inducing nuclear fission, and if fission occurs, gamma radiation is emitted. The gamma ray travels through space, and it may interact with a human cell, producing a modification which may leave the cell open to attack by the virus associated with leukemia. The fact that many such interactions of neutrons with fissionable nuclei are occurring in close spatio-temporal proximity, giving rise to processes which radiate in all directions, produces a pattern of statistical dependency relations. After initiation, these processes go on independently of one another, but they do produce relationships which can be described by means of the conjunctive fork.

Causal processes and causal interactions are, of course, governed by various laws – e.g., conservation of energy and momentum. In a causal process, such as the propagation of a light wave or the free motion of a material particle, energy is being transmitted. The distinction between causal processes and pseudo-processes lies in the distinction between the transmission of energy from one space-time locale to another and the mere appearance of energy at various space-time locations. When causal interactions occur – not merely intersections of processes – we have energy and/or momentum transfer. Such laws as conservation of energy and momentum are causal laws in the sense that they are regularities exhibited by causal processes and interactions.

Near the beginning, I suggested that deduction of a restricted law from a more general law constitutes a paradigm of a certain type of explanation. No theory of scientific explanation can hope to be successful unless it can handle cases of this sort. Lenz's law, for example, which governs the direction of flow of an electric current generated by a changing magnetic field, can be deduced from the law of conservation of energy. But this deductive relation shows that the more restricted regularity is simply part of a more comprehensive physical pattern expressed by the law of conservation of energy. Similarly, Kepler's laws of planetary motion describe a restricted subclass of the class of all motions governed by Newtonian mechanics. The deductive relations *exhibit* what amounts to a part-whole relationship, but it is, in my opinion, the physical relationship between the more comprehensive physical regularity and the less comprehensive physical regularity which has explanatory significance. I should like to put it this way. An explanation may sometimes provide the materials out of which an argument, deductive or inductive, can be constructed; an argument may sometimes exhibit explanatory relations. It does not follow, however, that explanations are arguments.

Earlier in this discussion, I mentioned three shortcomings in the most widely held theories of scientific explanation. I should now like to indicate the ways in which the theory I have been outlining attempts to cope with these problems. (1) The causal conception, I claimed, has lacked an adequate analysis of causation. The foregoing explications of causal processes and causal interactions were intended to fill that gap. (2) The inferential conception, I claimed, had misconstrued the relation of subsumption under law. When we see how statistical relevance relations can be brought to bear upon facts-to-be-explained, we discover that it is possible to have a *covering law conception* of scientific explanation without regarding explanations as arguments. The recognition that subsumption of narrower regularities under broader regularities can be viewed as a part-whole relation reinforces that point. At the same time, the fact that deductive entailment relations mirror these inclusion relations suggests a reason for the tremendous appeal of the inferential conception in the first place. (3) Both of the popular conceptions, I claimed, overlooked a fundamental explanatory principle. That principle, obviously, is the principle of the common cause. I have tried to display its enormous explanatory significance. The theory outlined above is designed to overcome all three of these difficulties.

On the basis of the foregoing characterization of scientific explanation, how should we answer the question posed at the outset? What does Laplace's demon lack, if anything, with respect to the explanatory aim of science? Several items may be mentioned. The demon *may* lack an adequate recognition of the distinction between causal laws and non-causal regularities; it *may* lack adequate knowledge of causal processes and of their ability to *propagate* causal influence; and it *may* lack adequate appreciation of the role of causal interactions in *producing* changes and regularities in the world. None of these capabilities was explicitly demanded by Laplace, for his analysis of causal relations – if he actually had one – was at best rather superficial.

What does scientific explanation offer, over and above the inferential capacity of prediction and retrodiction, at which the Laplacian demon excelled? It provides knowledge of the mechanisms of *production* and *propagation* of structure in the world. That goes some distance beyond mere recognition of regularities, and of the possibility of subsuming particular phenomena thereunder. It is my view that knowledge of the mechanisms of production and propagation of structure in the world yields scientific understanding, and that this is what we seek when we pose explanation-seeking why-questions. The answers are well worth having. That is why we ask, not only 'What?' but 'Why?'

University of Arizona

NOTES

[1]The author wishes to express his gratitude to the National Science Foundation for support of research on scientific explanation. This article is reprinted with the kind permission of the American Philosophical Association.

[2]"Such demonstrations may convince the mind, but they do not enlighten it; and enlightenment ought to be the principal fruit of true knowledge. Our minds are unsatisfied unless they know not only *that* a thing is but *why* it is." p. 330.

[3]P. xi. The first edition appeared in 1892, the second in 1899, and the third was first published in 1911. In the Preface to the Third Edition, Pearson remarked, just before the statement quoted in the text, "Reading the book again after many years, it was surprising to find how the heterodoxy of the 'eighties had become the commonplace and accepted doctrine of to-day." Since the "commonplace and accepted doctrine" of 1911 has again become heterodox, one wonders to what extent such changes in philosophic doctrine are mere matters of changing fashion.

[4]The classic paper by Carl G. Hempel and Paul Oppenheim, 'Studies in the Logic of of Explanation,' which has served as the point of departure for almost all subsequent discussion, was first published just thirty years ago in 1948.
[5]Hempel's conceptions have been most thoroughly elaborated in his monographic essay, 'Aspects of Scientific Explanation.' Hempel, 1965, pp. 331–496.
[6]In my (1977), I have given an extended systematic critique of the thesis (dogma?) that scientific explanations are arguments.
[7]See *Nature*, vol. 271 (2 Feb. 1978), p. 399.
[8]Copi, 1972, pp. 396–397, cites this example from Pauling (1959, pp. 85–91).
[9]According to the article in *Nature* (note 7), "the eight reported cases of leukaemia among 2235 [soldiers] was 'out of the normal range'." Dr. Karl Z. Morgan "had 'no doubt whatever' that [the] radiation had caused the leukaemia now found in those who had taken part in the manoeuvers."
[10]Hempel's inductive-statistical model, as formulated in (1965) embodied such a high-probability requirement, but in 'Nachwort 1976' inserted into a German translation of this article, Hempel (1977), this requirement is retracted.
[11]Reichenbach (1958), sec. 21. Here he offers the mark criterion as a criterion for temporal direction, but as he realized in his (1956), it is not adequate for this purpose. I am using it as a criterion for a symmetric relation of causal connection.
[12]See my (1975), sec. 3, pp. 129–134, for a more detailed discussion of this distinction. It is an unfortunate lacuna in Russell's discussion of causal lines – though one which can easily be repaired – that he does not notice the distinction between causal processes and pseudo-processes.
[13]See Salmon (1970, p. 23) for a description of this 'theory.'
[14]I have made an attempt to elaborate this idea in my (1977a, pp. 215–224). Because of a criticism due to Nancy Cartwright, I now realize that the formulation given in this article is not entirely satisfactory, but I think the difficulty can be repaired.
[15]Russell (1922) Lecture VI. The relevant portions are reprinted in my anthology (1970).
[16]In (1975) I discuss the explanatory import of the common cause principle in greater detail.
[17]Reichenbach (1956, p. 159) offers the following formal definition of a conjunctive fork *ACB*:

$$P(C, A \cdot B) = P(C, A) \times P(C, B)$$
$$P(\bar{C}, A \cdot B) = P(\bar{C}, A) \times P(\bar{C}, B)$$
$$P(C, A) < P(\bar{C}, A)$$
$$P(C, B) > P(\bar{C}, B)$$

[18]C screens-off A from B if

$$P(C \cdot B, A) = P(C, A) \neq P(B, A)$$

[19]The relation between $E_1 + E_2$ and E is an approximate rather than a precise equality because the ejected electron has some energy of its own before scattering, but this energy is so small compared with the energy of the incident X-ray or γ-ray photon that it can be neglected. When I refer to the probability that the scattered photon and electron will have energies E_1 and E_2 respectively, this should be taken to mean that these energies fall within some specified interval, not that they have exact values.
[20]As the boxed formulas in figures 1 and 2 indicate, the difference between a conjunctive fork and an interactive fork lies in the difference between

$$P(C, A \cdot B) = P(C, A) \times P(C, B)$$

and

$$P(C, A \cdot B) > P(C, A) \times P(C, B)$$

The remaining formulas given in note 17 may be incorporated into the definitions of both kinds of forks.

One reason why Reichenbach may have failed to notice the interactive fork is that, in the special case in which

$$P(C, A) = P(C, B) = 1,$$

the conjunctive fork shares a fundamental property of the interactive fork, namely, a perfect correlation between A and B given C. Many of his illustrative examples are instances of this special case.

[21]For the special theory of relativity, this has been shown by John Winnie in (1977), which utilizes much earlier results of A. A. Robb. For general relativity, the approach is discussed under the heading, 'The Geodesic Method' in Grünbaum (1973, pp. 735–750).

[22]The whole idea of characterizing causal interactions in terms of forks was suggested by Philip von Bretzel in 'Concerning a Probabilistic Theory of Causation Adequate for the Causal Theory of Time,' especially note 13, in this volume.

[23]It strikes me as an unfortunate fact that this important principle seems to have gone largely unnoticed by philosophers ever since its publication in Reichenbach's *The Direction of Time* in 1956.

[24]The interactive fork, unlike the conjunctive fork, does not seem to embody a temporal asymmetry. Thus, as seen in Figure 3, the intersection C along with two *previous* stages in the two processes, constitute an interactive fork. This fact is, I believe, closely relared to Reichenbach's analysis of intervention in *The Direction of Time*, Section. 6, where he shows that this concept does not furnish a relation of temporal asymmetry.

[25]Scientific realism is a popular doctrine nowadays, and most contemporary philosophers of science probably do not feel any pressing need for additional arguments to support this view. Although I am thoroughly convinced (in my heart) that scientific realism is correct, I am largely dissatisfied with the arguments usually brought in support of it. The argument I am about to outline seems to me more satisfactory than others.

[26]Compare Salmon, 1971, p. 78, where I ask, "What more could one ask of an explanation?" The present paper attempts to present at least part of the answer.

[27]Reichenbach believed that various causal relations, including conjunctive forks, could be explicated entirely in terms of the statistical relations among the events involved. I do not believe this is possible; it seems to me that we must also establish the appropriate connections via causal processes.

BIBLIOGRAPHY

Aristotle, 1928, *Posterior Analytics* in W. D. Ross, ed. & trans., *The Works of Aristotle*, vol. I, Clarendon Press, Oxford.

Arnauld, Antoine, 1964, *The Art of Thinking*, Bobbs-Merrill, Indianapolis.

Copi, I., 1972, *Introduction to Logic*, 4th ed., Macmillan, New York.

Grünbaum, A., 1973, *Philosophical Problems of Space and Time*, 2nd enlarged ed., D. Reidel, Dordrecht.
Hempel, C., 1965, *Aspects of Scientific Explanation*, Free Press, New York.
Hempel, C., 1977, *Aspekte wissenschaftlicher Erklärung*, Walter de Gruyter, Berlin.
Hempel, C., *et al.*, 1948, 'Studies in the Logic of Explanation,' *Philosophy of Science* **15**, 135–175.
Laplace, P., 1951, *A Philosophical Essay on Probabilities*, Dover Publications, New York.
Pauling, L., 1959, *No More War*, Dodd, Mead & Co., New York.
Pearson, K., 1957, *The Grammar of Science*, 3rd ed., Meridian Books, New York.
Reichenbach, H., 1956, *The Direction of Time*, University of California Press, Berkeley & Los Angeles.
Reichenbach, H., 1958, *The Philosophy of Space and Time*, Dover Publications, New York.
Russell, B., 1922, *Our Knowledge of the External World*, George Allen & Unwin, London.
Russell, B., 1927, *The Analysis of Matter*, George Allen & Unwin, London.
Russell, B., 1948, *Human Knowledge, Its Scope and Limits*, Simon and Schuster, New York.
Salmon, W., ed., 1970, *Zeno's Paradoxes*, Bobbs-Merrill, Indianapolis.
Salmon, W., 1971, *Statistical Explanation and Statistical Relevance*, University of Pittsburgh Press, Pittsburgh.
Salmon, W., 1975, 'Theoretical Explanation' in S. Körner (ed)., *Explanation*, Basil Blackwell, Oxford.
Salmon, W., 1977, 'A Third Dogma of Empiricism' in R. Butts and J. Hintikka (eds.), *Basic Problems in Methodology and Linguistics*, D. Reidel, Dordrecht.
Salmon, W., 1977a, 'An "At-At" Theory of Causal Influence,' *Philosophy of Science* **44**, 215–224.
Scriven, M., 1975, 'Causation as Explanation,' *Nous* **9**, 3–16.
Suppes, P., 1970, *A Probabilistic Theory of Causation*, North-Holland, Amsterdam.
van Fraassen, B., 1977, 'The Pragmatics of Explanation,' *American Philosophical Quarterly* **14**, 143–150.
Winne, J., 1977, 'The Causal Theory of Space–Time' in J. Earman *et al.* (eds.), *Foundations of Space-Time Theories, Minnesota Studies in the Philosophy of Science*, vol. VIII, University of Minnesota Press, Minneapolis.

DONALD RICHARD NILSON

HANS REICHENBACH ON THE LOGIC OF QUANTUM MECHANICS

1. INTRODUCTION

Hans Reichenbach was a philosopher with a truly scientific outlook, who throughout his distinguished career maintained a close working relationship with scientists, keeping abreast of developments in science as they happened. In his work we find a philosophy of science concerned not only with the general epistemological and methodological issues arising from an examination of the practice of science, but also with the detailed, painstaking analysis of the foundations of actual scientific theory as informed by a mastery of them, in particular, of physical theory. Thus, it is not surprising yet laudable to find his deep interest in the two major revolutions in physics which occurred during his lifetime – relativity theory and quantum mechanics[1]—having issued in two major series of works by his hand, each on one of these topics. The first part of this claim is easy to defend, for few would deny the monumental contribution of his works on the philosophy of time, space, and relativistic physics, in particular *Philosophie der Raum-Zeit-Lehre* (1928). Yet granting a comparable status to his studies of the foundations of QM might meet with some reservations. Judged from the standpoint of relative influence there clearly is an asymmetry. The influence of the latter work has certainly been less marked than that of his writings on relativity theory, as other essays in this volume amply demonstrate. However, it is my opinion that Reichenbach's works on QM constitute a major contribution to the philosophy of physics and – when taken in terms of the fundamental point of view maintained – provide insights into the foundations of quantum theory fully comparable to those afforded by his other works. The less than enthusiastic reception of *Philosophic Foundations of QM*[2] (although it was welcomed as 'an important contribution') stemmed less from inherent deficiencies in the book, than from misunderstandings of it and from the understandably unsympathetic reading it received due to the radical thesis concerning logic defended in it. I shall attempt to defend

W. C. Salmon (ed.), Hans Reichenbach: Logical Empiricist, 427–474. *All Rights Reserved.*

this claim and the significance of Professor Reichenbach's foundational studies of QM by showing how frequently many of the critics of that work went awry. Such investigation is valuable not only insofar as it contributes to an understanding of Reichenbach's work, but also because it contributes to the understanding of the general issues surrounding the interpretation and application of non-standard logic, as well as the issues faced in the philosophy of QM itself.

The central theme in Reichenbach's writings on QM with which I wish to deal is his investigation of the *logic* of QM, his contribution to *quantum logic*.[3] Reichenbach has frequently been named along with others as having defended the view that standard logic is to be replaced or revised by reference to facts pertaining to microphysics, and thus as exemplifying Quine's views on the in-principle-revisability of logic. It is much less frequently noted precisely why he defended this view, and how it relates to *his* general philosophical position. Contribution to the discussion of these last mentioned issues is one general aim of the present study.

In the past ten years or so, a constantly increasing volume of writings has appeared on the logic of quantum theory. And while it is still not altogether clear what significance for our understanding of logic these studies hold, it is clear that work on a subject pioneered by Reichenbach is now assuming major proportions.[4] It is to be duly noted that Reichenbach was neither the first nor the only contributor to the subject now known as 'QL'. A brief look at the developments leading up to Reichenbach's studies in QL suffices to demonstrate this.

Attempts earlier than Reichenbach's to apply non-standard logics to quantum physics may be found in works by the Polish logician Z. Zawirski (in 1931) and by the French philosopher Paulette Destouches-Février (in 1937 and later).[5] One of the earliest contributions, that of Garrett Birkhoff and John von Neumann[6] (in 1936), has turned out to be of signal importance in the whole development of QL. However, the relations of influence among the various workers in QL is far from clear. Birkhoff and von Neumann described Mme. Février's work as "logical and philosophical ideas closely related to ours."[7] There is a structural relationship between the systems of Février and Reichenbach.[8] One of the distinctive features of Reichenbach's QL is that it is a three-valued logic.[9] A wide variety of such systems have been developed and investigated in great detail. Modern treatment of formal systems has made it possible to develop systems wherein the principle of excluded middle is not valid. Instead of only the two classical truth-values being available for the valuation of statements in a language, a finite (or in some cases even infinite) spectrum of truth-values is countenanced. Changes in the overall logical structure of the language are then made.

For the topic of this essay, especially important among the variety of many-valued logics so constructed are the systems of E. L. Post and J. Łucasiewicz.[10] Reichenbach made clear and explicit use of their earlier work as will be noted below. Long before enunciating his views on QL, Reichenbach had developed an infinite-valued probability logic. His motivation for this development was to construct a logic with a continuous range of truth-values, to replace standard two-valued logic for certain purposes. No statement would be considered to be 'true' or 'false' *tout court* under this programme, but rather 'true (respectively, false) to the degree **'. One way of interpreting the notion of truth-value (assuming that truth-value is related to information content) is as *degree of assertability*. This is indeed the way Reichenbach interpreted it. But Reichenbach's QL is to be sharply distinguished from his infinite-valued probability logic; the two are conceptually distinct developments. Probability $\frac{1}{2}$ is not what Reichenbach means by truth-value I. It is interesting to note in passing that in the work of C. F. von Weizsäcker, an infinite-valued logic is suggested for use as a logic of QM.[11]

In the context of these developments, Reichenbach set himself to the task of structuring the language of QM by using a three-valued logic (as described in the following section) in such a way as to make explicit the peculiar causal structure of the microcosm.

The present study is structured along the following lines. In Sections 2 and 4, I sketch that part of the framework of Reichenbach's interpretation which is relevant to an understanding of the motivation for his having developed a non-standard logic for the language of QM. There, I also explain in what way Reichenbach believed that a coherent account of microphysics could be provided by adopting a particular QL. Between these sections another is inserted in which I examine the relation of the views of Reichenbach to those of the so-called 'Copenhagen school'. Section 5 covers objections to Reichenbach's QL (methodological and interpretive objections) and includes a defense of his viewpoint. The issue of how to interpret non-standard logic is further considered in the following section; here, too, the intelligibility and cogency of his position is defended. His general interpretation comes under attack in Section 7, where a serious problem at the heart of his interpretation of QM is developed, but this does not immediately bear upon the acceptability of his QL. The penultimate Section 8 elaborates a series of criticisms of the details of his system of logic for QM. His particular system is thus finally held to be untenable, although throughout the study the intelligibility of his view is defended against premature rejection. But, in the closing section, I touch upon what I take to be the lasting contribution of Reichenbach's views on QL.

2. REICHENBACH'S THREE-VALUED QL (I)

As instrumental in incorporating a QL into his interpretation of QM as a central feature of it, Reichenbach in his *Philosophic Foundations of Quantum Mechanics* introduced two fundamental but interrelated distinctions. These distinctions were needed in order for him to precisely articulate what he took to be the central issues and problems which any adequate interpretation of the quantum theory must face. And these distinctions also provided a framework of concepts in terms of which he was able to formulate his own interpretation of QM. The first of these distinctions is that between '*phenomena*' ("... all those occurrences which consist in coincidences, such as coincidences between electrons, or electrons and protons, etc. ..." [Reichenbach, 1944, p. 21] and '*interphenomena*' (occurrences "... introduced by inferential chains ..." [*Ibid.*]). Reichenbach claimed that "the phenomena are ... determinate in the same sense as the unobserved objects of classical physics." He added that "phenomena are connected with macroscopic occurrences by rather short causal chains" [*Ibid.*]. The situation with regard to interphenomena (which may be said to be interpolated between phenomena events) in relation to macroscopic occurrences is a more complex matter involving causal and complex inferential chains which go beyond what is given in the phenomena, in the sense that he has technically redefined this term.

The second distinction used by Reichenbach was introduced in reliance upon the above one. Interpretations of QM will be said to be '*exhaustive*' if and only if in them both interphenomena and phenomena are given a complete and unified interpretation (unified, in the sense that the descriptive categories and laws pertaining to events are the same whether they be phenomena or interphenomena); interpretations are called '*restrictive*' if and only if in them complete interpretation is restricted to phenomena only and is not afforded to interphenomena.

Extensive sections of his book elaborate in detail how exhaustive interpretations (*viz.*, particle and wave interpretations) may be consistently carried through for the quantum mechanical formalism.[12] However, one upshot of all interpretations which are exhaustive, Reichenbach claims, is that *causal anomalies* (such as actions-at-a-distance, causal chains propagated with infinite velocity, or instantaneous field contractions) are clearly present in each of them. That is to say, the principle of action by contact is violated whenever definite values of parameters pertaining to particles or waves are assumed to hold for all phenomena *and* interphenomena. On the other hand, it was apparent to Reichenbach that restrictive interpretations of the formalism were also possible. The interesting consequence of shifting to one or another of the restrictive interpretations, he claimed, was that in so doing causal anomalies may be strictly avoided.

One sort of restrictive interpretation discussed by Reichenbach is the Copenhagen interpretation of Bohr and Heisenberg: an "interpretation by restricted meaning." His own preferred interpretation of QM was yet another restrictive interpretation, distinct from the Bohr–Heisenberg one. He referred to it as an "interpretation by restricted assertability." Statements about interphenomena held to be meaningless according to the Copenhagen view are held to be unassertable, but meaningful in the Reichenbach interpretation. There are some strong affinities between these two interpretations. But differences between them run deeper than these few observations indicate; these differences appear to stem from diverging theories of meaningfulness as we shall have occasion to note again below.

What Reichenbach proposed for use in the description of microphysical events was a neutral language, not the pure wave nor the pure particle languages of exhaustive interpretations. In this language, statements about interphenomena (as well as about phenomena) can be made.[13] Note that this does not hold for the QM object language of the Copenhagen school. Once one opts for "an interpretation by restricted meaning," the resulting language contains only a subset of the set of statements made available within exhaustive interpretations. Thus it is that in the Copenhagen interpretation, statements about those unobserved objects or events termed 'interphenomena' are rendered meaningless and not properly an element in the language of microphysics, whereas in Reichenbach's view such statements *are* admissible elements of the language of QM, $\mathscr{L}_{QM}$, but they are held to be neither true nor false: they have the truth-value *indeterminate*. This last point underscores the relevance of Reichenbach having called his interpretation one of restricted assertability. There is, of course, a problem as to what it *means* to say that a proposition is indeterminate in truth-value. This is a thorny point to which we shall have to address ourselves. But once we commit ourselves – as Hans Reichenbach did – to the advisability of using such a 'tertium quid' as a valuation on $\mathscr{L}_{QM}$, it follows that we must formally structure our neutral, restrictive language with a three-valued propositional logic. The three-valued system that Reichenbach succeeded in formulating we shall refer to as his 'QL'. Before examining this system in any detail, it will be well for us to see why adopting a QL is to be preferred to adopting the Bohr–Heienberg interpretation by restricted meaning.

3. THE BOHR–HEISENBERG RESTRICTIVE INTERPRETATION

Both central figures in the genesis of the quantum theory, Niels Bohr and Werner Heisenberg, offered an alternative to interpreting QM via QL. They did so, moreover, before any system of QL was developed. Their basic move was to place restrictions on what constitutes a legitimate assertion within the language of microphysics. If, for example, an assertation A is made based upon a body of experimental evidence stating that a microobject has momentum p_0 at time t_0, then their restriction on what else is legitimately assertable about this object is embodied in a rule which disallows the assertion of statement B which specifies that (simultaneously) the object has position q_0 and t_0. If A holds, then B is not a meaningful assertion and must be excised from the set of genuine and meaningful quantum theoretic propositions. A similar rule would restrict the meaningfulness of A under the condition that B is determinate in truth-value. And further rules would be added for all other parameter pairs which are incompatible according to Heisenberg's indeterminacy relations.

The philosophical motivation behind the use of 'restricted meaning' by Bohr and Heisenberg apparently was a strongly positivistic strain in their approach to the language of science. While I believe that the case for saying that the Copenhagen interpretation of QM is a positivistic one has been very much overstated and has obscured many of the insights of Bohr and Heisenberg,[14] it is nonetheless hard to see any motivation other than a positivistic one behind *this* aspect of their interpretation. It is widely known that Bohr insisted that the core of scientific language is 'common language'. Indeed, while emphasizing that quantum effects transcended the scope of classical physical analysis, Bohr urged that "the account of the experimental arrangement and the record of the observations must always be expressed in common language supplemented with the terminology of classical physics."[15] The prominence given 'common language' and classical description within scientific language may be further noticed in Bohr's definition of the domain of phenomena proper to the study of microphysics: "one may strongly advocate limitation of the use of the word *phenomenon* to refer exclusively to observations obtained under specified circumstances, including an account of the whole experiment."[16] Taken together with the knowledge that QM description is such that no representation of a state of a system can ever imply the precise specification of both members of a pair of conjugate parameters, the above-mentioned restrictions on the proper scientific domain of phenomena yield the result that at most *one* of a conjugate pair of variables has a precise value. A statement concerning the other member of the conjugate pair is *meaningless* according to this interpretation, because no experimental arrangement could manifest this putative phenomenon (alternatively, we might say that such a proposition is held to be meaningless because no phenomenon, properly-so-called, is described in such a case). If we note (1) divergences between how Bohr and Reichenbach respectively define the domain of microphysical *phenomena*, and (2) the disparity in what they respectively regard as the meaningful use of language, we may see the sources of the differences between these two restrictive interpretations. The most obvious difference to be noted is how much larger the range of the term 'phenomenon' is under Reichenbach's convention than it is under Bohr's.

It must now be clear that Bohr's position regarding the use of a non-standard logic in QM, such as Reichenbach's QL, is simply that such use is misguided, because superfluous or unnecessary.[17] If one properly restricts the application of the term 'phenomenon', one works solely within a framework of 'common language' and so the logic proper to this language – standard logic – will be the only one necessary.[18] If we accept the Copenhagen restrictions on the use of the term 'phenomenon' and on what is to be regarded as meaningful within the language of microphysics, neither Reichenbach's nor any other's QL has any place within $\mathscr{L}_{QM}$. But why accept these restrictions? First of all, the limitation of the use of the term 'phenomena' to cases where actual observation by a classically defined apparatus is performed is an unduly narrow delimitation of use. Perhaps it is to be replied that this linguistic restriction, while not consonant with actual use of the term, is a price we must pay in order to avoid causal anomalies in QM. But, Reichenbach's alternative restrictive interpretation shows this not to be an adequate reply. Secondly, the Bohr–Heisenberg restrictive interpretation makes causal anomalies avoidable only by the introduction of an extensive set of rules, each cutting off this or that 'meaningless' set of sentences from $\mathscr{L}_{QM}$. But one wonders whether this rather unmanageable set of rules is to be avoided solely from considerations of simplicity.

A stronger objection than these is that there are no non-arbitrary grounds for denying the meaningfulness of the propositions in question. *Prima facie*, they have a well-defined use as statements in $\mathscr{L}_{QM}$. If B is a statement held to be meaningless because it asserts a precise momentum value, p_x for Σ (when a precise position value, q_x, for Σ is already known), it follows that $-B$ must also be meaningless. However, the more plausible position to take on such cases, it would seem, is that $-B$ (roughly, 'Σ does not have the precise momentum value, p_x') is true, rather than meaningless. Looking at logically compound sentences, such as $[A \,\&\, B]$ (where A is the statement specifying the precise position value, q_x, for Σ), are these to be held meaningless or false?[19] The charge that interpretation by restricted meaning is arbitrary is a serious one; however, it is clear from a remark of Heisenberg why the restriction in question was held to be justified: the statements held to be meaningless were considered to be contentless and lacking in any consequence.[20] Consider this statement by Heisenberg:

This uncertainty relation $[\Delta q_x \Delta p_x \geqslant h]$ specifies the limits within which the particle picture can be applied. Any use of the words 'position' and 'velocity' with an accuracy exceeding that given by [this equation] . . . is just as meaningless as the use of words whose sense is not defined.[21]

He added the following in a footnote to this passage:

In this connection one should particularly remember that the human language permits the construction of sentences which do not involve any consequences and which therefore have no content at all. . . .

The tenor of this passage is that statements which violate the limits of precision imposed by the uncertainty relations have no content because they have no directly observable consequences. But unless we accept the restrictions in question *a priori*, this argument is an implausible one; for statements like $[A \,\&\, B]$, $[-(A \,\&\, B)]$, etc., *are* in conflict or agreement with the *theory* of QM. And surely consequences of this sort must also bear on whether these statements have content or not.

Furthermore, if one of the purposes of investigating the language of microphysics is to clarify and exhibit the structure of this language, then precisely in regard to this goal the Bohr–Heisenberg approach must be seen as counter-productive. For, the interesting structural features of the language of quantum theory are obfuscated by the move to regard a large portion of that language as being devoid of meaning. This approach certainly does not bring out the intimate relationship of the logical structure of $\mathscr{L}_{QM}$ to the Hilbert space representation of states in the theory. But the appropriateness of this objection is a point of controversy and it may be argued that Reichenbach does not bring out this relationship in an adequate way either. In any case, this point serves to underscore the role of conditions of adequacy on interpretations, in the light of which they are judged to be more or less successful. According to Reichenbach, we are dealing here with a matter of the appropriateness of adopting conventions governing our logical as well as non-logical use of language in physics. It is not a matter of adopting *the true* logic for $\mathscr{L}_{QM}$, but rather of adopting the appropriate logic for $\mathscr{L}_{QM}$.[22] It is not surprising to find that his arguments for the relative advantage of his interpretation *vis-à-vis* the Bohr–Heisenberg one are couched in terms of the goals and commitments of his general philosophical position. We may further bring this out by considering the following four critical remarks of Reichenbach concerning the Copenhagen view:

(i) Under their interpretation QM laws are expressed as semantical rules; this is contrary to the goal of developing a unified object language for QM (Reichenbach, 1944, p. 143).

(ii) In the same spirit as (i), he pointed out that as compared to the Copenhagen interpretation, his view had "the advantage that such statements about interphenomena can be incorporated into the object language of physics, and they can be combined with other statements by logical operations . . ." (Reichenbach, 1944, p. 169, cf. p. 42).

(iii) In a restricted sense, the Bohr–Heisenberg position requires certain expressions which strictly speaking are held by them to be meaningless, nevertheless to be part of the language of physics (Reichenbach, 1944, p. 143).

(iv) Some statements about values of parameters are meaningful under the Copenhagen interpretation depending upon whether or not certain measurements are *in fact* performed; this is not the case in Reichenbach's interpretation.

The force of these critical remarks will move the reader to accept the relative adequacy of Reichenbach's restrictive interpretation as over against that of Bohr–Heisenberg, if the reader acknowledges the import of his implicit conditions of adequacy on an interpretation of a theory. We may note four such conditions, all of which follow from his strongly empiricist position on general philosophical issues, by touching on each of remarks (i)–(iv) above. (i) Reichenbach held that scientific laws are part of the object language of science,[23] and that wherever possible, a unified language of science is to be sought; (ii) wherever possible, the logical relations among and logical combinations of statements in the language of science should be accounted for and clearly exhibited; (iii) an interpretation is to be consistently carried out; (iv) finally, the strictures of a verifiability theory of meaningfulness are to be followed. Of these conditions, perhaps the most questionable is the last one. However, if we look to Reichenbach's works on general epistemological issues, it is immediately clear that he was fundamentally committed to a verifiability theory of meaningfulness.[24] Propositions have truth-value 'true' or 'false' for Reichenbach just in case some physically possible experiment could have decided the proposition had it been performed. "To speak of truth or falsehood is legitimate only when there are ways of finding the truth."[25] Now, under this view, some propositions may never be *known* to be true or *known* to be false, but yet *be* true or *be* false. But if a proposition *could not* be known to be true or false, then it is said to be neither true nor false: it is meaningless or else possesses a third truth-value, *I*. This view is closely tied to Reichenbach's position (mentioned above) on truth as assertability. Now, even if the reader rejects this theory of meaningfulness as untenable, it must be recognized that Reichenbach's version of the theory is superior to the one apparently accepted by the

Copenhagen school. Their account is akin to a verifi*cation* rather than a verifi*ability* theory of meaningfulness in that it requires that measurements be actually performed for language about microobjects to be meaningfully employed.[26] It does not assert, however, that the only meaningful such use of $\mathscr{L}_{QM}$ is in the report of measurement results. As Reichenbach correctly pointed out, the Bohr–Heisenberg position cannot be said to use only verified or verifiable statements. This follows from their acceptance of what he refers to as "Definition 5. The result of a measurement represents the value of the measured entity immediately after the measurement" (Reichenbach, 1944, pp. 140, 142). But all the same, as we saw above, unless an experiment was performed, no truth-value assignment to statements of $\mathscr{L}_{QM}$ could be made in accord with this interpretation. It was this to which Reichenbach objected in point (iv) above.[27]

Whatever insights into QM may be garnered by adopting Bohr's philosophy of complementarity, these are not forthcoming – as it seems Reichenbach was well aware – from that aspect of the Copenhagen view which involves an interpretation of $\mathscr{L}_{QM}$ by restricted meaning.

4. REICHENBACH'S THREE-VALUED QL (II)

After establishing the framework of his 'neutral language' for QM and his commitment to a restrictive interpretation superior to that of the Copenhagen theorists, Reichenbach turned to the task of uncovering a logical structure for the three-valued proposition set of $\mathscr{L}_{QM}$. From among the many alternative sets of matrices which may be used to define the connectives for a three-valued logic, he then selected a particular set of definitions by matrices which, he argued, faithfully reflect the relations among the statements of $\mathscr{L}_{QM}$. Here the systems of Post and Lucasiewicz mentioned earlier were called upon. The details of the system used by Reichenbach are mainly of technical interest, but the complexity of the system (as compared to standard logic) is worth noting; for example, the increased number of truth-values gives the result that more than one form of negation, implication, etc. are found in the system. The new connectives are introduced so as to maintain as much continuity and coherence with standard logic as possible.

For our purposes, what is more pertinent than these technical details is a sense of the overall strategy of Reichenbach's approach to the foundations of the quantum theory. A first desideratum is that one may assume that one of Reichenbach's aims in the philosophy of science was never very far from the goal, characteristic of logical empiricism, of a unified language of science. And if for him it was not required that that language have a logic that was two-valued for every domain covered by it, at least it was required that it be truth-functional in whatever truth-values it made use of, for each domain. It was required that not only should the 'broader' unified language be truth-functional, but that it be related to two-valued languages in a clear and precise way. The goal of improving the language of science was a central motivation for this three-valued QL, hence his remark "... we are not concerned here with questions of truth or falsehood of language systems, but with the superiority of a linguistic technique."[28]

Another aspect of his over-all strategy may be emphasized by noting how Reichenbach investigates to what extent classical conceptions may be applied in microphysics. He inquires as to what extent classical conceptions of causality, logic, relevant physical parameters (such as, momentum, energy, etc.) may be applied in the interpretation of the formalism of QM. When conceptual tensions or contradictions arise, he is concerned with exactly *where* these tensions and contradictions arise, what principles might be restricted or reinterpreted in order to remove them, and so on. It is precisely this line of considerations which led to the insight articulated in the form of his *principle of anomaly*, one of the central contentions of his interpretation of quantum theory.

Examples such as the famous two-slit experiment[29] serve to illustrate this strategy. If we attempt an exhaustive interpretation while postulating orthodox causal behavior of our system, Reichenbach shows how unintelligible consequences result. He goes on, however, to say that if we pinpoint unorthodox causal behavior of the system as the locus of unintelligibility, we can then remove the conceptul tension by positing the existence of causal anomalies. Alternatively, if the demand for an exhaustive interpretation is pinpointed as problematic, this may be given up and a shift made to one of the restrictive interpretations described above. In the latter, there is no assertion of causal anomalies, so again conceptual tension is released. Note that if causal anomalies are posited, we cannot expect that systems will be what Reichenbach calls *normal* systems, that is, ones that behave according to the same natural laws whether observed or not. In classical physics and common sense this postulate – all systems have normal descriptions – is tacitly assumed. One important result of Reichenbach's work was his demonstration that no exhaustive interpretation provides a normal description of the events of microphysics. This was accomplished by showing how causal anomalies raise their ugly heads in exhaustive interpretations of QM. Reichenbach formulated this result less metaphorically and quite succinctly in his *principle of anomaly* ("... no definition of interphenomena may be given which satisfies the requirement of normal causality ..." [Reichenbach, 1944, p. 44]).

This principle has formed a point of strong contention in studies of Reichenbach's work on the foundations of QM. In particular, in a review of Reichenbach's work by Professor P. K. Feyerabend[30] – a review which curiously combines unfortunate misinterpretations with strong and valid criticisms – the view is developed that under one interpretation of the principle of anomaly, it reduces trivially (definitionally) to the claim that the laws of classical physics are different from the laws of QM. He draws this consequence apparently because he misunderstands Reichenbach's statement about (note, *not* definition of) phenomena to the effect that they "... are determinate in the same sense as the unobserved objects of classical physics" (Reichenbach, 1944, p. 21). But, Reichenbach makes this statement in order to (a) point out the kind of inference required to go from 'macroscopic data' to phenomena, and to (b) sidestep (provisionally, and only for the sake of focusing the issue at hand) the problem of unobserved objects in *classical* physics.[31] Feyerabend, on the other hand, takes the statement as a definition such that the laws governing phenomena are classical laws (while correlatively the laws governing interphenomena are defined to be non-classical). But Reichenbach held no such view. The principle of anomaly reduces to no such triviality.

Taking yet another interpretation of the principle of anomaly, Feyerabend concludes that it states that "for any theory which consists of the mathematical formalism of QM together with some E_2 [an exhaustive interpretation] there exist refuting instances."[32] This is an unintelligible inference from Reichenbach's actual view. If Feyerabend means by 'refuting instances' *direct refutation* (which seems to be the case, for it is an adequate rephrasing of his earlier "leads to incorrect predictions"[33]), then he is faced with showing how interphenomena (the only domain where causal anomalies may be exhibited) can be used to directly refute anything. If, on the other hand, he means by 'refuting instances' *indirect refutation*, then he must exhibit how incorrect predictions can be derived from QM with an exhaustive interpretation *without* assuming normal causality for interphenomena. Note that while an exhaustive interpretation does commit one to the modelled entity having precisely defined parameters, it does not commit one to the view that the entity exhibits normal causality.

Considering these attempts to come to grips with the principle of anomaly, it seems clear that Feyerabend did not understand what Reichenbach intended by 'causal anomalies'. Professor Reichenbach explicitly stated that according to his view causal anomalies can "never be used to produce anomalous effects in the world of phenomena" (Reichenbach, 1944, p. 40). Furthermore, he pointed out[34] that causal anomalies are not anomalous physical processes. But this being so, it is hard to see how the principle of anomaly could be said to indicate refutations in any sense of the term.

Causal anomalies and the principle of anomaly were carefully delineated by Reichenbach in order to present us with a fundamental, and perhaps irreducible fact about the microcosm: its structure exhibits acausal features. Causal anomalies are not completely dispensed with, even in his own restrictive interpretation involving QL. In the latter interpretation causal anomalies are eliminated only in the sense that statements of them are strictly unassertable in $\mathscr{L}_{QM}$. These anomalies "cannot be banished from the world as a whole" (Reichenbach, 1944, p. 40). Under the set of conventions adopted by him, the world's acausal structure was reflected in the nonstandardness of the logical structure of $\mathscr{L}_{QM}$. And it was in this regard that Reichenbach amplified Galileo: "the book of quantum phenomena is written in the language of a three-valued logic."[35]

5. OBJECTIONS TO REICHENBACH'S QL

Now it must be clear why under Feyerabend's characterization of the principle of anomaly, Reichenbach's position appeared so repugnant to him. As so characterized, Feyerabend understood Reichenbach's move to render statements of causal anomalies unassertable in the object language of QM as a move to make refuting instances of the theory of QM unassertable. He claimed that the interpretation under consideration violates "one of the most fundamental principles of scientific methodology, namely, the principle to take refutations seriously." The result of this, according to Feyerabend, will be such that "no need is felt to look for a better theory."

It is evident that this sly procedure is only one (the most "modern" one) of the many devices which have been invented for the purpose of saving an incorrect theory in the face of refuting evidence and that, consistently applied, it must lead to the arrest of scientific progress and to stagnation.[36]

Yet if it is still possible to find (and to take seriously!) recalcitrant observations or refutations, it is apparent that QM can clearly be overthrown or revised, just as any other theory may be overthrown or revised. If what Feyerabend is pointing to is 'refuting evidence' gathered from examining interphenomena, then the remarks of Section 4 above again apply. Certainly it does not follow that QM is rendered immune from revision when some statements about interphenomena are restricted by reference to the logic of $\mathscr{L}_{\mathrm{QM}}$. Furthermore, nothing about Reichenbach's approach disallows the postulation of wholly new quantum theories. In this regard it is worth recalling that in commenting on von Neumann's famous 'no hidden variable' proof, he makes the important observation that the proof is conditional upon the assumption of the universal validity of QM, an assumption he of course allows to be open to question.[37] So it is hardly a necessary consequence of the use of a three-valued QL in interpreting current quantum theory, that the stagnation of scientific progress will result.[38]

In addition to this general methodological objection, Feyerabend leveled the special methodological objection that a development of relativistic quantum theory was cut off by the Reichenbach interpretation. But the latter restricts himself to non-relativistic QM only for the sake of simplicity, as have most other philosophers and physicists concerned with the foundations of QM. There appear to be no special barriers to making quantum theory Lorentz-invariant or having any of the other features of a fully relativistic theory, induced by a restriction on the assertability of statements about interphenomena. Feyerabend seems to assume that the only way in which the need for a relativistic version of

QM could manifest itself is through the uncovering of problematic behavior among interphenomena and, in particular, through causal anomalies. But this overlooks the whole range of effects at the level of phenomena, exhibited in systems at 'relativistic velocities'. This domain of phenomena must ultimately be accounted for in any adequate theory of the microworld, and Reichenbach gave no indication of being unaware of this matter. On the contrary, it is to be noted that throughout his works he remained sensitive to the important relations between space–time order and causal connection. So as to preserve the usual topology of space–time order, in his philosophy of space and time, Reichenbach accepts the postulate of no causal anomalies. His particular views on relativistic space–time order, with the emphasis on the principle of the maximal limit of signal velocities, would further motivate acceptance of the postulate of no causal anomalies in QM (in particular, in connection with cases where causal anomalies arise as infinite signal propagation velocities). Allowing for the existence of causal anomalies in the domain of QM would militate against the development of a relativistic QM, but the acceptance of no causal anomalies in either relativistic or quantum physics leads toward, rather than away from, a unified treatment of them. This leaves Feyerabend's special methodological objection with a touch of irony.

Several critics have concentrated their remarks on the Reichenbach three-valued logic for QM on the issue of whether or not three-valued logic as developed in *Philosophic Foundations of QM* is intelligible. The most thorough of these critics, Professors Ernest Nagel and Carl Hempel, have raised the central complaint that Reichenbach has not given sufficient indication of how we are to understand the various truth-values in the proposed three-valued logic. A summary of their objections (which we will investigate in turn) is as follows: (1) 'truth' in $\mathscr{L}_2$ does not mean the same as 'truth' in $\mathscr{L}_3$ [an $\mathscr{L}_2$ is a language having two-valued logic and an $\mathscr{L}_3$ is a language with three-valued logic]; (2) conditions under which a statement is true (or false or indeterminate) in a three-valued sense are left unspecified; (3) Reichenbach's use of a three-valued logic is incompatible with his espousal of a verifiability theory of meaningfulness; (4) 'indeterminate' means the same as 'meaningless'; (5) certain statements in $\mathscr{L}_3$ claimed by Reichenbach to make the same assertion, in fact do not do so; (6) Reichenbach's attempt to move certain statements ordinarily made meta-linguistically into the object language of QM does not work.

(1) Nagel and Hempel were both troubled[39] by the obvious structural differences between $\mathscr{L}_2$ and $\mathscr{L}_3$ and the problem of understanding what it could mean to say that certain statements within $\mathscr{L}_3$ are neither true nor false. Indeed, what do 'true' and 'false' mean in such a language? Surely, they reasoned, 'true' and 'false' must have new meanings if such conditions as stipulated for $\mathscr{L}_3$ are to hold. "Since," Nagel pointed out, "in *his* proposed usage of the word, 'true' has *two* exclusive alternatives and not just *one* (as is the case in its customary meaning), it is clear that 'true' (as used in the three-valued logic) cannot be *identical* in meaning with 'true' (as used in the two-valued logic)."[40] Reichenbach, in his reply to Nagel, was willing to acknowledge the conclusion that, strictly speaking, there is no identity of meaning between the terms.[41] However, one must note that 'truth' in $\mathscr{L}_3$ serves as a through-going analogue of 'truth' in $\mathscr{L}_2$. The terms have the same function in their respective languages and this function may be characterized in Reichenbach's phrase: "... in both cases 'truth' has the meaning of '*assertability*'."[42]

(2) Nagel was not satisfied by this reply,[43] but in any case he claimed that a full account of the conditions under which truth-values are assigned to propositions was not given by Reichenbach.[44] And to Reichenbach's reply[45] that the usage of these terms is specified by the truth-tables, Nagel insisted that granted this does provide a specification of conditions for "a *special* class of *singular* statements in QM," they "do not explain how these predicates are to be employed *in general* – in connection with other singular statements or with universal ones."[46] However, this supposed deficiency turns out to be illusory, for the truth-tables give truth-functions for *arbitrary* singular statements. But what about *universal* statements? This seems to coincide with the objection that no system of quantification is supplied for Reichenbach's QL.[47] No system of quantification to this purpose has been developed, but this can be taken as a provisional difficulty (no one has argued that such a system *cannot* be constructed) and is a difficulty which is peripheral to Nagel's original objection.

(3) and (4) This does not exhaust by any means Nagel's volley of criticisms, for he was puzzled by an apparent conflict between the acceptance of a verifiability theory of meaning (more precisely, meaningfulness) and the use of a three-valued QL – both of which are attributable

to Reichenbach. According to the verifiability theory a statement is meaningful if it is true or false, but there are some statements in Reichenbach's QL which are neither true nor false, yet meaningful. Professor Reichenbach attempted to clarify this point by urging (a) that he now accepted a broader definition of meaningfulness than the one according to which a statement is meaningless if it is not true and not false, and (b) that, according to him, "statements are meaningful if they can be verified as true, false, or indeterminate."[48] But Nagel would not accept this point either, finding the latter clarification "a curious procedure." His misgivings (especially about 'verifying a statement as indeterminate') are spelled out by considering a "strictly parallel reasoning, that if a sentence can be verified as meaningless, the statement is verifiable (and therefore not meaningless!)." "The obvious fault in such reasoning," he goes on, "is that what is being verified is that a given sentence has the *property* of being indeterminate or meaningless, *what* it expresses is not being verified at all."[49] Here the 'parallel reasoning' is carried out without regard for the revised definition of meaningfulness. Perhaps the matter can be crystallized by analyzing 'verified as true, false, or indeterminate'. Assuming some type of correspondence theory of truth and keeping in mind Reichenbach's account of truth as assertability, we may carry out the analysis as follows: 'ϕ' is verified as true means that we have grounds for asserting that 'ϕ' states what is the case; 'ϕ' verified as false implies that we have grounds for asserting that the (cyclic) negation of 'ϕ' ($\sim\phi$) states what is the case; 'ϕ' verified as indeterminate implies that we have grounds for asserting that the double (cyclic) negation of 'ϕ' ($\sim\sim\phi$) states what is the case. It seems clear from this analysis and from the remarks in (2) above that 'indeterminate' is best seen as possessing a structural meaning; that is, the conditions for its applicability are determined by the overall structure of $\mathscr{L}_3$. Nonetheless, something more on the matter of what conditions may obtain which would verify a statement as I may be offered to alleviate the skeptical puzzlement of objection (2) above. An example of U being verified as I is as follows:

We note Reichenbach's statement (1944, p. 158) of the 'Rule of Complementarity'.

(RC) $(V \vee \sim V) \rightarrow \sim\sim U$

Suppose we verify $V =$ true,

then $V \vee \sim V$ holds.

Taking this result together with RC (by modus ponens) yields: $\sim\sim U$. U is thereby verified as indeterminate.

Again considering the above analysis, we conclude that 'verified as indeterminate' indicates a property of statements only in the sense that 'verified as true' also indicates a property of statements. What then of Nagel's 'verified as meaningless'? If we can make sense of this, it is as follows: 'ϕ' verified as meaningless means that we have established grounds indicating (a) that the 'statement' 'ϕ' is unassertable, (b) that it has no logical relation to any statements which are meaningful, and (c) that no 'statement' resulting from the application of logical operations to 'ϕ' is meaningful. But clearly 'verified as I' is quite distinct from 'verified as meaningless', for the former (a) holds, but *not* (b) and (c). This being the case, it seems hardly appropriate to wonder "whether, after all, 'indeterminate' may not be another way of saying 'meaningless'."[50]

(5) Professor Hempel was as unconvinced as Professor Nagel by Reichenbach's exposition of his QL (and whereas their criticisms do overlap, they do not coincide). Hempel leveled the following criticism at Reichenbach's three-valued logic: although Professor Reichenbach claimed that '$-\phi$' and '$\sim\phi$'[51] state the same fact (that 'ϕ' is false), this cannot be so, for the truth tables for these sentences differ. "If two sentences state the same fact, then it would seem that they must have identical truth-tables. This, however, fails to be the case in the example cited as well as in many analogous cases."[52]

In order to further clarify his point, Hempel sets forth the following considerations (owing to ideas developed by Wittgenstein and Carnap):

> In order to know that two statements make the same assertion – i.e., that they determine the same dichotomy of states – it suffices to show that their ranges are identical (for then so will be the complements of their ranges), i.e., that they are made true by exactly the same states. But this is not sufficient in $\mathscr{L}_3$. For here a sentence determines a trichotomy of states, and in order to show that two sentences make the same assertion, it is necessary to ascertain that they make the same trichotomy . . .
>
> Finally the following consideration may be pertinent in this context: For $\mathscr{L}_2$, it can be said that '$\sim a$' asserts the same as "'a' is false"; for as a consequence of the semantical definition of truth, $\sim a$ is the case if and only if '$\sim a$' is true, and hence (by virtue of the truth-table of two-valued negation) if and only if 'a' is false. Thus the basic idea of Reichenbach's device does apply to two-valued logic; but its transfer to $\mathscr{L}_3$ breaks down because neither of the two principles invoked for two-valued logic applies to the three-valued case.
>
> These observations seem to show that it is not generally possible to express assertions about the truth-values of sentences of the neutral language by statements in that language itself.[53]

I submit that Hempel's approach here is misguided and that Reichenbach's view is again defensible. In order to see this, let us restate Reichenbach's original claim [let's call it F] that '$-\phi$' and '$\sim\phi$' both state that 'ϕ' is false. This restatement is as follows:

(F^*) 'ϕ' is false, if $-\phi$ is the case or $\sim\phi$ is the case.

Note that Reichenbach was not claiming in F that '$-\phi$' is equivalent to '$\sim\phi$', which is what Hempel seems to criticize him for in his 'diverging truth-tables' argument. Also note that a semantical definition *is* brought into play. It may be that Hempel's critique goes astray on precisely this point. What I suspect he assumes is that some three-valued use of a semantical definition of truth is required (if it is used at all).[54] However, the above formula, F^*, is a consequence of a semantical definition used in a straightforward and *strictly two-valued* way, taken together with consideration of the truth-tables for $\mathscr{L}_3$. Reichenbach is not committed to the use of a three-valued semantic metalanguage.

(6) The last of the objections to be considered is related to the previous one in that the relation between object language and meta-language is the important desideratum in each case. In remarks *à propos* Reichenbach's formulation of the 'Rule of Complementarity' ('If U and V are noncommutative entities, then $U \vee \sim U \rightarrow \sim\sim V$': 1944, p. 158) Hempel registers the objection that "if this is a sentence in the object language, then the phrase 'if . . . then' occurring in it must express one of the various implications of $\mathscr{L}_3$ rather than the customary meaning of 'if . . . then', and in order to make the meaning of the proposed translation of the complementarity rule clear, it would have to be stated just which implication is meant."[55] It is only the formula '$U \vee \sim U \rightarrow \sim\sim V$', however, that was employed by Reichenbach to capture the Rule of Complementarity. This becomes more apparent from consideration of formulas (34) and (37) (Reichenbach, 1944, pp. 158–159) both of which reformulate the same rule; Reichenbach's wording in the passage cited is unfortunate and somewhat misleading. Although '$U \vee \sim U \rightarrow \sim\sim V$' is a formula in the object language of QM, the above quoted statement (containing the formula as its consequent), strictly speaking, belongs to the meta-language as does its more informal counter-part 'if U is true or false, V is indeterminate' (Reichenbach, 1944, p. 158). Under this interpretation of

Reichenbach's position, Hempel's objection carries no force, for in the meta-language 'if . . . then' has its customary, two-valued sense.

These six objections by no means constitute the whole body of criticism of the Reichenbach approach to QL, but hopefully the present discussion and rectification of these makes the intelligibility of his approach more apparent and paves the way for further considerations.

6. ON THE INTERPRETATION OF THREE-VALUED LOGIC

In the foregoing section, some criticisms of Reichenbach's QL were drawn from the literature and discussed. From that discussion it is readily observed that the criticisms cited there revolve around the problem of understanding what a three-valued logic is and how it functions, in terms available to 'two-valued beings', so to speak. It was in response to this central question that Hilary Putnam's 'Three-Valued Logic'[56] was written. It constitutes an attempt to support Reichenbach's QL both directly (he briefly discusses the relevant portion of Reichenbach (1944)) and indirectly (he tries to provide a general standpoint for viewing alternative logics of a certain variety). Toward the end of providing a framework for viewing non-standard logics,[57] Professor Putnam sets down a set of criteria which must be met by the terms 'true', 'false' and other truth-values in a whole range of languages which are held to be significant alternatives to $\mathscr{L}_2$ (assuming that the point all along is to relate the truth-values in $\mathscr{L}_2$ to those in other languages with more than the customary two values).

The first of these conditions is that the ascription of truth-values must be 'tenseless', that is, truth-value is to be independent of the time at which determination of truth-value is made. A statement cannot be indeterminate at t_0 and become true at t_1. If a statement is false at t_2, it is false at all times.[58] The second condition is that the truth-values must *not* be *defined* in terms of what Putnam calls 'epistemic predicates' (that is, predicates that are ascribed "relative to the evidence at a particular time," *e.g.*, 'verified' or 'falsified'). As he put it, "it makes sense that 'Columbus crosses [tenselessly] the ocean blue in fourteen-hundred and ninety-two' was verified in 1600 and not verified in 1300, but not that it is true in 1600 and false in 1300."[59]

He then adds a condition that will serve to relate the truth-values in $\mathscr{L}_2$ to those in, say, $\mathscr{L}_3$ (without trying to specify necessary and sufficient conditions – a translation – that would hold between them) and in so doing, he determines part of the syntax of $\mathscr{L}_3$. This condition is as follows: "... statements that are accepted as verified are called 'true', and statements that are rejected, that is, whose denials are accepted are called 'false'."[60] "More precisely, *S* is accepted if '*S* is true' is accepted."[61] The consequence of this is that a necessary but not sufficient condition for statements to be 'middle' (*I*) is that they be neither verified nor falsified.

Of course, one way of definitively satisfying those who demand a definition and explication of the meaning of the truth-value *I* or 'middle', would be to give them the necessary and sufficient conditions (if such *were* available) for "'ϕ' is *I*'. But one of Putnam's central points is that *no such translation conditions are available*. A three-valued logic is not something that can be translated into two-valued logic; rather, he says, "... to use a three-valued logic means to adopt a different way of using logical words."[62]

Isaac Levi has taken issue with Putnam on this point.[63] Professor Levi, after registering his request for an explication of '*I*' or 'middle' alongside the requests of Nagel and Hempel mentioned earlier, argued that Putnam would have made his point in urging non-translatability of $\mathscr{L}_3$ plus the across-language relation of truth-values outlined above, if he had also shown that three-valued logic was "... introduced as a means for saying something that could not be said with the aid of a two-valued logic ..."[64] But Reichenbach's use of $\mathscr{L}_3$, which Putnam stands behind, is incompatible with $\mathscr{L}_3$ being more expressive than $\mathscr{L}_2$. So, Levi went on to point out, if an $\mathscr{L}_3$ does *not* have 'greater expressive power' than $\mathscr{L}_2$, then it is clearly appropriate to request a translation between the two 'equally expressive' systems. Either $\mathscr{L}_3$ expresses more than $\mathscr{L}_2$ or there must exist a translation between $\mathscr{L}_2$ and $\mathscr{L}_3$ (assuming also that $\mathscr{L}_2$ does not express more than $\mathscr{L}_3$). Moreover, since it is also held that $\mathscr{L}_3$ and $\mathscr{L}_2$ are equivalently descriptive languages and since any translation of $\mathscr{L}_3$ to $\mathscr{L}_2$ will violate Putnam's criteria on truth-values, the latter's defense of three-valued logic – in view of the above – is threatened.

In one sense, one could say (regarding Levi's first point) that $\mathscr{L}_3$ does provide a framework for saying things that cannot be said in $\mathscr{L}_2$. For example, in $\mathscr{L}_3$ one can assert '$\sim\sim$(*x* has position *q*)' and one cannot directly assert this in $\mathscr{L}_2$, nor does there seem to be a unique, alternative way of saying *the same thing* in $\mathscr{L}_2$. Now (in regard to Levi's second point), it seems appropriate to mention that for Reichenbach two languages $\mathscr{L}_A$, $\mathscr{L}_B$ give equivalent descriptions if *taken as a whole* they can each be used to account equally well for the same observational domain; in the case of QM: two languages $\mathscr{L}_2$, $\mathscr{L}_3$ are equivalent descriptions if nothing that can be said about *phenomena* in $\mathscr{L}_2$ *taken as a whole* can be said about *phenomena* in $\mathscr{L}_3$ *taken as a whole* and vice versa.[65] There is no requirement that there be a non-trivial way of translating a given sentence from the one language into the other.

In another sense, there does exist a 'translation' from $\mathscr{L}_3$ to $\mathscr{L}_2$, for it is possible to map the set {statements of $\mathscr{L}_3$} onto the set {specified sets of statements of $\mathscr{L}_2$}, indeed it is possible to specify an isomorphism for these sets. In Reichenbach's reply to Nagel, he pointed out that a theorem due to Gustin shows that "every three-valued statement may be coordinated to a pair of two-valued statements."[66] Yet, whereas Levi pointed out that a translation of $\mathscr{L}_2$ to $\mathscr{L}_3$ would clearly show that (contrary to Putnam's criteria on truth-values) "epistemic considerations enter into the statement of necessary and sufficient conditions [*i.e.*, the translation] for middle-hood,"[67] no epistemic consideration in fact need enter in determining the above 'translation'. But a 'translation', alas, is not a translation! Showing that one can coordinate $\mathscr{L}_2$ with $\mathscr{L}_3$ does not show that there exists a translation in the sense relevant to Levi's argument. This shows not only the weakness of Levi's critique, but also indicates that this part of the Nagel–Reichenbach dispute was largely a matter of each speaking past the other.

Levi's case against Putnam, thus, is very weak. But Levi might reply that if $\mathscr{L}_3$ does not have more expressive power than $\mathscr{L}_2$ and there exists no translation of $\mathscr{L}_3$ to $\mathscr{L}_2$, then we still don't know what '*I*' or 'middle' means. Now recalling what was said in Section 5 about the definition of the term '*I*' by way of truth tables, we may say that what '*I*' means is given in its *structural meaning*, its rôle in $\mathscr{L}_3$. Furthermore, we can say of indeterminate statements that it is physically impossible to verify them.

If '*I*', and 'true' and 'false' as well, have structural meanings in $\mathscr{L}_3$, it would then seem that 'true' and 'false' should have structural meanings in $\mathscr{L}_2$. This is indeed correct, but we then are readily presented with the question of what relation, if any, exists between the respective truth-values of the two languages. Reichenbach claimed that "the three-valued truth is the *analogue* of the two-valued truth." And he added that "in both cases truth has the meaning of *assertability*."[68] Now, it is this same idea that Putnam tries to clarify by his notion of the 'core meaning' of 'true' and 'false' common to $\mathscr{L}_3$ and $\mathscr{L}_2$.[69]

The task of giving an account of the 'core meaning' of 'true' in $\mathscr{L}_3$ and $\mathscr{L}_2$ would certainly be an easier one if there existed a general theory of meaning change and meaning continuity between successive theories, languages, and so on. Such a theory is notoriously and conspicuously absent from contemporary philosophy of language, and the need for it is, of course, remarkably important. But even without such an account of meaning continuity, Putnam succeeded in the much more narrow task of specifying certain conditions on 'true' and 'false' which remain invariant under transformation from $\mathscr{L}_2$ to at least some other non-standard language structures (the prime case being $\mathscr{L}_3$). One such condition is cited and elaborated above: "statements that are accepted as verified are called 'true' and statements that are rejected, that is, whose denials are accepted are called 'false'." This establishes a relation that holds for all cases in at least $\mathscr{L}_2$ and $\mathscr{L}_3$ (as developed by Putnam) and certainly must also hold for some languages with other alternative logics. One might perhaps add that another condition is further constitutive of the 'core meaning' of 'true': in language, $\mathscr{L}_i$, starting from true premises, if an argument A is valid, then the conclusion of A is also true. Other conditions may or may not be added to the ones stated; what is important is that with the stated conditions, already a core is formed.

The way meaning is then bestowed upon the sentence 'ϕ' is I' parallels the well-known logical empiricist account (in, say, Hempel's latest works) of the way meaning is bestowed upon theoretical terms in the language of science. The 'core meaning' of 'true' and 'false' provide an analogue of 'antecedently understood terms' (the conditions specified on truth-values in $\mathscr{L}_2$ and $\mathscr{L}_3$ being a sort of set of 'bridge principles'), and the structural meaning provided for 'I' via the truth-tables for the logic of $\mathscr{L}_3$ forms an analogue of the indirect meaning afforded to theoretical terms via the 'internal principles' of a theory. Furthermore, it does not seem that a demand for a stronger and narrower specification of meaning for the truth value 'I' than what is provided by the present account is warranted. In the last analysis, we learn what 'I' means by learning how to use $\mathscr{L}_3$.

7. OBJECTIONS TO REICHENBACH'S GENERAL APPROACH TO QUANTUM THEORY

We may distinguish the various criticisms that have been leveled at Reichenbach's use of a three-valued logic in interpreting the quantum theory from various criticisms which have been directed in a more general way to his overall approach to the foundations of QM. There are two main criticisms of the latter sort: (1) the objection that Reichenbach generates a morass of pseudo-problems only because he makes use of a principle which stipulates that microobjects always possess classical parameters (under exhaustive interpretations); (2) the objection that the phenomenon/interphenomenon distinction is indefensible.

(1) Feyerabend argues[70] that many of Reichenbach's problems and solutions thereto in the interpretation of quantum theory are needless exercises arising from his adoption of the (mistaken) 'assumption *C*', which is roughly as follows: any exhaustive interpretation is such that it defines a class of classical categories (parameters) where each entity is assumed to always possess one each of all such categories. Professor Haack has in my opinion decisively refuted this misconception of Reichenbach's position.[71]

(2) The bulk of this section will be an exploration of the most central distinction used in Reichenbach's interpretation of quantum theory: that between phenomena and interphenomena. We shall see that this distinction cannot be adequately drawn along the lines developed by Reichenbach, and so his entire foundational study will be seen to rest on less than secure conceptual ground. This objection to Reichenbach's general approach to QM was developed in one form by Ernest Nagel in his critical study of *Philosophic Foundations of QM*.[72] He held that the distinction is vague and so, inappropriately used to support the other distinctions which Reichenbach builds upon it (viz., exhaustive/restrictive, normal system/non-normal system, etc.). The distinction in question is one of degree, as Professor Reichenbach himself indicated: "the logical difference between the physics of phenomena and the physics of interphenomena is a matter of degree; the latter contains more definitions than the former" (Reichenbach, 1944, p. 42). Clear cases of phenomena

according to Reichenbach are collisions of electrons. But Nagel was unsure of the classification of even such cases, for "'electron' is an empty word until it has been incorporated into a set of statements which ascribe determinate properties to electrons and from which verifiable consequences can be drawn. If this is so, however, it is not evident how one is to distinguish between that part of electronic behavior which is to be counted as a phenomenon and that part which is interphenomenal."[73] From these considerations he concludes that the distinction is vague and so practically useless.

Reichenbach had already specified the basis of the distinction in terms of inferability or non-inferability by classical physical principles from experimental data.[74] Note that both phenomena and interphenomena are inferentially based; they differ on *what* type of inference yields statements about each. Thus the distinction is not one between the macroscopic and the microscopic, neither is it the observed/unobserved distinction.[75] The general point is that there is a class of events which, although strictly-speaking unobservable even when a measurement is made, are still very naturally interpreted in a determinate way and with the use of classical physics. For example, a β-track in a cloud chamber lends itself very naturally to the interpretation that the entity was at the various condensate points at various times. The events in question are 'observable' in broad sense; these are *phenomena*. What is important about this, Reichenbach urges in his reply to Nagel, is that "considering the phenomena as observables means that the problem of unobservable objects can be disregarded for classical physics; only for what I call interphenomena can we not disregard this problem because it is here that the laws of QM intervene."[76] This indicates a less than adequate representation of the relation of the respective domains of classical and quantum physics, but it also throws light on what Reichenbach meant by his statement to the effect that phenomena have the same status as the unobserved objects of classical physics. He emphasized by this that while in classical mechanics there was a simple and natural relation to assume as holding between what was observed and what was unobserved, this happy situation is not preserved in the new physics of the microworld. The phenomena/interphenomena distinction is no more vague according to Reichenbach than the distinction of observed/unobserved encountered in classical physics.

It is sensible to suggest that Nagel's dissatisfaction with this reply originates in his observation of Reichenbach's failure to point out in precisely what way classical physics may be employed to establish statements about phenomena. This point is a good one, for it suggests a possible ambiguity in Reichenbach's claims about phenomena vs. interphenomena (an ambiguity which may also be the source of Feyerabend's perplexities about the 'classical assumption *C*'). He might mean (a) that classical physics may be used in the analysis of experimental data to establish, e.g., the *precise* values of position and momentum parameters of an electron in a collision (but only in the collision and not between it and another event), or he might mean (b) that classical physics may be used in the analysis of data to infer values of parameters within certain intervals (such that the indeterminacy relations are not violated). In (a) it is assumed that an analysis by classical theory yields results about what is in effect an indirectly observed classical entity (the use of the theory in this way requiring the pre-supposition that certain properties always characterize the represented entity); in (b) classical physics may be used as an approximation under certain circumstances without any supposition being made that the measured entities may be precisely represented by exact values of classical parameters so determined.

Reichenbach undoubtedly had conception (b) in mind when he discussed the status of phenomena, but his statement of his position is not always clear. In any case, to consistently work out his position Reichenbach would have held to conception (b) of phenomena.

Now, this clarification having been made, we can ask whether there is any further substance to the type of objection leveled against the phenomena/interphenomena distinction by Professor Nagel. That there is another problem with the distinction will be demonstrated through the example to follow. It will be maintained that some macrooccurrences would be classified as interphenomenal according to Reichenbach's account, and since this is completely unintelligible under that account, serious reservations must be made concerning the tenability of the distinction at the very heart of this interpretation of QM. The objection proceeds from this presupposition about the classificatory purpose of Reichenbach's distinction: while the phenomena/interphenomena distinction does not correspond to the 'macro'/'micro' distinction, it is Reichenbach's intent that no macrooccurrence be classified as interphenomenal (whereas, some microoccurrences are phenomenal, in his sense).

If we focus upon the basis given for the phenomena/interphenomena distinction, we find that it is grounded in the kinds of inference made to the description of events as phenomenal or interphenomenal, respectively. Phenomenal events "are inferred from macroscopic observations by the *sole* use of classical theory."[77] In addition, Reichenbach indicates that he thinks that the laws of QM apply in some sense only to interphenomena. He says, "... it is here that the laws of QM intervene."[76] It is clear from Nagel's rejoinder that he was bothered by this remark and by the fact that Reichenbach did not give an account of how classical theory and how quantum theory can each be used selectively to apply to certain events and not to others.[78]

For most of the macro-domain, Reichenbach's distinction applies in a relatively unproblematic way. This author, as is usually the case, concentrated his discussion of QM on the so-called 'free-particle situation' (isolated electrons, protons, etc.). Most of the interesting questions of interpretation come up in connection with the free-particle case. There are some situations, however, when quantum theoretically derived behavior occurs on a large scale, that is, for large systems of entities. For certain substances, at very low temperatures (where consequently system energy is lowest), rather than the ordinarily large number of states of a broad spectrum involved at high temperatures, one encounters only a very few states, all of them near the ground state. Under these conditions, the quantum-mechanical character of the ground state can manifest itself in large scale observable effects. Here we refer to the well-known properties of super-cooled matter (that is, matter at temperatures near absolute zero) called super-fluidity, super-conductivity (thermal and electrical), etc. Here quantum-mechanical effects are exhibited on an unprecedented scale and a spectacular manner; quantum mechanical considerations 'intervene' not only in an account of interphenomena but also for a variety of observable phenomena as well. Put in another way, we may say that the term 'phenomena' defined as (1) inferrable from data via classical principles alone, and as (2) encompassing macrophysical occurrences which are observable in the usual sense, does not consistently apply to certain domains. An example of such a domain is provided by considering super-cooled helium, to which we now turn.

But here the question must be asked: When Reichenbach claims that the laws of QM intervene only in interphenomena, does he mean that (1) only interphenomenal events are explained by reference to the laws of QM, or that (2) the means of representation provided by QM (i.e., state-vectors) is applicable only to interphenomena? For the reasons given above in showing that the principle of anomaly does not simply state that the laws of classical and quantum physics are distinct, (1) must be rejected. So, we shall interpret Reichenbach as holding that state-vectors or wave functions are applicable as a means of representation only to interphenomena.

Now to the example. For the case of liquified helium, at temperatures below 2.2°K, a range of peculiar effects are observed: viscosity of a vanishingly small magnitude, the famous 'fountain effect', the existence of 'super surface films', and many others. An explanation of these superfluid properties of matter is found in a two-fluid model of liquid helium. It is posited that helium at these temperatures is actually a mixture of two different fluids, one having the properties mentioned, the other a 'normal' fluid. The amount of the superfluid component is said to vary with the temperature, decreasing as the temperature increases until above the 'critical temperature' (2.2°K), there are no superfluid properties of the mixture. There are in turn a number of accounts of the underlying nature of the two fluids. What is perhaps the most successful account provides an understanding of the two fluids in terms of phonons (quantized sound waves, so to speak) in the fluid helium. Liquid helium at absolute zero, would have no phonons present. As heat is introduced, phonons appear, the number a function of the temperature. The phonons collide with each other and act in many ways like gas atoms. In these terms, the super-fluid can be considered as the underlying liquid helium itself. The normal fluid is really a manifestation of phonons. The phonon gas flows around the superfluid, so that what was at first thought to be a mixture of two fluids turns out upon further analysis to be a gas superposed on a background fluid. (cf. the electron gas in conducting metals.) It has been found, furthermore, that the behavior of phonons (in a hard-sphere gas model, i.e., a Bose–Einstein gas) corresponds remarkably well to thermal and dynamic properties of the normal fluid in liquid helium. This oversimplifies the picture somewhat, but will serve our purposes quite well.

The two-fluid model was elaborated again in a theory due to Onsager and Feynman. It is well known that one cannot obtain a precise solution for the Schrödinger equation applied to systems as complex as we are now considering. Feynman, however, succeeded in developing a theory whereby "a macroscopic wave function, ψ, interpreted analogously to the ordinary quantum mechanics,"[79] is used to account for the superfluid component of liquid helium, and with remarkable results. The upshot of this is that we cannot apply classical physics to data pertaining to super-cooled matter in order to specify all and only *phenomena* (in Reichenbach's sense) for this domain. Even the macroscopic characteristics of super-cooled helium are to be accounted for by use of the means of representation of, and the principles of, quantum theory. This shows that the phenomena/interphenomena distinction, if it is to be drawn at all, requires a more subtle account than Reichenbach offers.

8. LASTING DIFFICULTIES WITH REICHENBACH'S QL

While we have seen that many of the problems posed for, and alleged difficulties inherent in, Reichenbach's interpretation of QM by way of a three-valued QL are of no lasting consequence, we have also found that there are serious problems with the central distinction used in that interpretation. But it may be that there remains to be found a way of suitably modifying and reformulating the phenomena/interphenomena distinction so as to avoid the objection of the previous section.[80] Assuming that this could be accomplished, it would remain for us to raise a further question. Is the QL developed by Reichenbach fully adequate with respect to his conditions of adequacy and to the task of providing a solid and illuminating account of the language of quantum physics? The answer to this question, we shall see, is decidedly in the negative, and this even though there is much to be found in Reichenbach's view that is helpful and illuminating concerning the interpretation of QM. The answer is in the negative because of some four final and lasting objections to Professor Reichenbach's QL.

(1) The first and strongest of these is one suggested and elaborated by Paul Feyerabend. It is distinct from his other criticisms of Reichenbach which were discussed earlier and rejected. It is duly to be noted that Reichenbach put two requirements on the laws of QM as formulated in

his reconstruction of $\mathscr{L}_{QM}$. One is that the laws of QM (including the physical laws which under some interpretations have been stated as *rules* in the metalanguage of QM, such as the 'Rule of Complementarity') are to be formulated in the object language of QM. The second is that we "... put into the true-false class [that is, into the subset of statements of $\mathscr{L}_3$ which have as truth-values either true or false, but not I] those statements which we call quantum mechanical *laws*" (Reichenbach, 1944, p. 160). Thus, according to this set of requirements no law of QM should appear either in the metalanguage pertaining to $\mathscr{L}_{QM}$ (that is, *used* in the metalanguage rather than simply mentioned there) or as an indeterminate statement. Clearly there are some cases – for example, ones discussed by Reichenbach – where laws of QM meet these requirements. But do *all* laws of QM meet them?

Feyerabend raises this question as follows:

Consider for that purpose the law of conservation of energy and assume that it is formulated as saying that the sum of the potential energy and the kinetic energy (both taken in their classical sense) is a constant. Now according to (a) [(a) = statements expressing anomalies are to have truth-value I] either the statement that the first part of the sum has a definite value is indeterminate, or the statement that the second part has a definite value is indeterminate, or both statements are indeterminate from whence it follows that the law of the conservation of energy will itself be a statement which has always the value 'indeterminate'. The same results if we use the statement in the form in which it appears in QM.[81]

We observe that Reichenbach himself noted this. He claimed that it "follows that the principle of conservation of energy is eliminated by the restrictive interpretation, from the domain of true statements, without being transformed into a false statement; it is an indeterminate statement" (Reichenbach, 1944, p. 166).

But Feyerabend then makes the even more telling objection that:

... every quantum-mechanical statement containing noncommuting operators can only possess the truth value 'indeterminate'. This implies that *the commutation rules* which range among the basic laws of QM *as well as the equations of motion* (consider them in Heisenberg form) *will be indeterminate* ...[82]

He thus shows that Reichenbach's QL does not satisfy the conditions of adequacy on the statement of laws set forth by Reichenbach himself. This criticism meets the interpretation on its own terms.

Professors B. C. van Fraassen and K. Lambert in their reconstructions of Reichenbach's QL[83] both note this problem as a serious one for Reichenbach and one to be overcome in their reconstructions. The same approach is taken by each of them to alleviate the problem: remove the laws from the object language of science. While the problem *does* disappear under this alteration, there is the further question of to what extent the modification does violence to Reichenbach's overall intended interpretation of QM. Since he did require – presumably for reasons of coherence, simplicity, and so on – that laws be part of the object language of QM, one wonders how sympathetic with the van Fraassen/Lambert reconstruction Reichenbach would have been. In any case, making the revision in question amounts to a renunciation of a central tenet of Reichenbach's interpretation of QM.

(2) A second objection to Reichenbach's QL which invites revision rather than reply is one showing that the statement used by Reichenbach to state the 'Rule of Complementarity' within the language $\mathscr{L}_{QM}$;

(RC) $U \vee {\sim} U \rightarrow {\sim}{\sim} V$

is satisfied by certain parameters which are *not* incompatible. Van Fraassen has demonstrated this result by first pointing out that the relation of incompatibility expressed in the 'Rule of Complementarity' may be telescoped into a single connective, ♁ ('apple'), such that

$$U \text{ ♁ } V \quad \text{iff} \quad U \vee {\sim} U \rightarrow {\sim}{\sim} V.$$

A truth-table for this connective may be stated as follows:

♁	*T*	*F*	*I*
T	*F*	*F*	*T*
F	*F*	*F*	*T*
I	*T*	*T*	*T*

He then argues that

. . .it could happen that *U* is true in a model, and *V* does not have a truth-value in that model, although *U* and *V* do not correspond to incompatible magnitudes. For example, let *Q* and *Q′* be the operators corresponding to the *X* and *Y* coordinates of position respectively and *P* the operator corresponding to the *X* coordinate of momentum. Then suppose that in $K = \langle X, f \rangle$, $x = f(X)$ is such that $Qx = rx$ and $Q'x = r'x$ [x, eigenvectors of *Q*, *Q′* corresponding to eigenvalues r, r']. The former entails that *x* is not an eigenvector of *P* [because *Px*, *Qx* are incompatible]. Hence $U = U(q', r')$ is true in *U*[sic; I believe that this should read *K*, not *U*] and $V = U\,(p, r'')$ is indeterminate in *K*; by the above table we see that *U* ♁ *V* is true in *K*. Yet *U* and *V* do not correspond to incompatible magnitudes.[84]

Van Fraassen's response[85] to this problem was to show how to revise Reichenbach's QL so as to allow the use of a semantic metalanguage and the machinery of semantic entailment for the statement of the relation of incompatibility and hence of the above 'Rule of Complementarity' (RC). This move parallels the one mentioned in (1) above.[86]

(3) We may add to the above objections one which we touched upon in considering the Bohr–Heisenberg restrictive interpretation of QM, namely that Reichenbach's QL is not as adequate to the purpose of clarifying the foundations of QM as certain other rival QLs (such as that of Birkhoff–von Neumann, 1936), because the latter provide richer insight into the logico-algebraic structure of the theory whereas the Reichenbach approach is much more indirectly related to the theoretical structure of QM. Thus, in the Birkhoff–von Neumann QL we find an obvious expression of the relation of 'experimental propositions' to the theoretical structure formed by the closed subspaces of the Hilbert space used to represent quantities in QM. Some glimpse of this structure is available in Reichenbach's work if we consult his (1944), Part II, where he outlines the mathematics of quantum theory.

Professor Patrick Suppes reproved Reichenbach's approach to QL because it "seems to have little relevance to quantum-mechanical statements of either a theoretical or experimental nature." He added, what he "particularly fails to show is how the three-valued logic he proposes has any functional role in the theoretical development of QM."[87] In the final analysis, it "has little if anything to do with the underlying logic required for quantum mechanical probability spaces. . . . ."[88] Suppes' argument is that the logic of QM, by which he means the logic of QM events, has been shown to be non-truth functional,[89] and since Reichenbach's QL *is* truth functional, it is thus defective. Elsewhere[90] Suppes expressed sympathy with Reichenbach's search for a truth functional QL. He urged that one may be developed which (a) reflects the logic of events and the theoretical structure of QM, and (b) provides the underlying structure for a countably additive probability measure. While the resulting logic is truth functional, it has non-standard truth-values. Instead of Reichenbach's T, F, and I, Suppes uses ordered pairs as truth-values (T, R), (F, R) and (μ, R) where "R is the set of coordinates of the cylinder set corresponding to the proposition in question."[91]

If Suppes is right on either of these two counts, reasons are provided for choosing one or another of the alternative QLs to that of Reichenbach. But one cannot turn attention away from Suppes' truth functional QL without wondering what Levi and others who found such profound difficulty in attaching meaning to Reichenbach's truth-value I, would say for example, about the meanings of (F, R) and (μ, R) as truth-values.

(4) Reichenbach claimed to have removed all causal anomalies from QM, in the sense of rendering them unassertable in $\mathscr{L}_{QM}$, once his restrictive interpretation of QM was adopted. We have brought out a number of problems with his interpretation, but we have not asked whether or not he was right in saying that all causal anomalies are indeed removed under his reconstruction of $\mathscr{L}_{QM}$. The two main attempts to show otherwise are due to Professors M. Strauss and M. Gardner. Both of these authors agree that Reichenbach succeeded in resolving the frequently discussed causal anomalies posed by the two-slit thought experiment. However, Strauss maintained that while the usual statements of causal anomalies cannot be derived in Reichenbach's system, yet another causal anomaly arises under his account. Gardner, on the other hand, sees no further problems in connection with the two-slit experiment, but holds that in *other contexts* causal anomalies are unresolved by Reichenbach's QL.

Professor Strauss points out that "... in Reichenbach's language it is *not possible to infer from the observed flash on the screen* [recording the distribution of 'particle hits' behind the diaphragm in the two-slit experiment] *that the particle causing the flash has passed through the slits in the diaphragm*."[92] This, he remarked, "is a causal anomaly even worse than the one suppressed."[93] But, is it? We may observe that no statement of an action at a distance, or of instantaneous contraction of a wave, or of any other type of causal anomaly may be *asserted* in $\mathscr{L}_{QM}$ in connection with Strauss' example. Strictly speaking, in Reichenbach's terms, do we then have an anomaly at all? Secondly, given either the definition of the restrictive interpretation due to Reichenbach or the one due to Bohr and Heisenberg, would one expect to retrodict from the occurrence of a flash on the screen that some statement about prior behavior of interphenomena (such as that a particle passed through one or another of the slits) is true? I submit that one would not. Given their definitions, it is only statements about what immediately follows a measurement which may be deductively drawn from the statement of measurement outcome.[94] Strauss' result is perplexing all the same, but it would seem that the way to show Reichenbach's account inadequate in this regard would be to develop an alternative to Reichenbach's QL in which one can derive that a particle went through either one slit or the other, given that it has registered a flash on the detection screen placed behind the diaphragm.[95]

Professor Gardner has set forth a series of criticisms in a recent article, which, if correct, bear both directly and indirectly upon the tenability of Reichenbach's QL. The direct critique is an effort to show that "... while his theory seems to resolve the two-slit paradox, it cannot resolve the tunneling and orbital electron paradoxes ...,"[96] and thus that not all causal anomalies are resolved by use of Reichenbach's QL. He shows that it is not at all clear that the QL approach is at all helpful in dealing with *some* paradoxes of QM. The indirect critique proceeds by casting doubt upon Reichenbach's (implicit) theory of measurement and continues by showing that under Reichenbach's view, we cannot deal in a satisfactory way with either the Schrödinger cat paradox or a version of the Einstein–Podolsky–Rosen (EPR) paradox.[97] The first of these critiques, the one which most directly bears upon the line of argument of this paper, seems to me to be entirely unsuccessful as it stands, but it must be admitted that it *might* be developed into a successful criticism of not just Reichenbach's, but systems of QL in general. The second of the critiques is more telling and points to genuine problems for Reichenbach's interpretation of quantum theory.

The reason for rejecting Gardner's criticism of Reichenbach's solution to the tunneling and orbital electron paradoxes is that it does not meet Reichenbach on his own ground, as it were. His criticism is that Reichenbach (in his argument blocking the above mentioned paradoxes) interprets the uncertainty principle as referring to individual systems rather than to ensembles, and that Reichenbach holds that simultaneous values of p_x and x (where 'x' is to be read 'q_x' in the notation used earlier) cannot be measured. In short, Gardner accepts the (minimal) statistical interpretation of QM and asserts that Reichenbach cannot resolve the two paradoxes in question without rejecting two central tenets of the statistical interpretation. But clearly, Reichenbach rejected the statistical interpretation in the form it is now held by Gardner, Ballentine, and others. He held that the state-function, strictly speaking, does not apply to individual systems,[98] but the uncertainty principle, he claimed, *does*. Consider his statement:

Within such an interpretation [a restrictive interpretation such as his own] we must say that if we regard a situation for which q is known to a smaller or greater degree of exactness Δq, p is completely unknown for this situation and cannot even be said to be at least within the interval Δp coordinated to Δq by the Heisenberg inequality.[99]

Now it may turn out that the accounts according to which such simultaneous measurability is possible *are* correct, but then the centrel tenet upon which all QLs (at least those interpreted *as logics*) are founded would be mistaken, and so this objection *might* be developed into an attack against the QL approach to interpreting QM in general. Thus, given his presupposition of meaningful comeasurability of simultaneous values of some conjugate parameters (under special conditions), it is not surprising to find an incompatibility between Gardner's point of view and that of the quantum logicians. He might have pointed out that some statements formulating the results of such experiments are *true* and therefore not indeterminate as required by Reichenbach of *all* such statements.[100] No special examination of, for example, the orbital electron paradox would be needed to refute Reichenbach. But those who take little stock in the statistical interpretation of QM and the possibility of comeasurability would hardly be moved by this first criticism of Reichenbach's QL.

Professor Gardner's second critique is designed to show that Reichenbach's interpretation of QM cannot be used to solve the Schrödinger cat paradox and a version of the EPR paradox. Here Gardner's remarks concerning Reichenbach do not make use of any unfair presuppositions deriving from the statistical interpretation and his argument that Reichenbach does not adequately dispense with the EPR paradox seems to me to be entirely cogent; no restatement of his argument will be attempted here for sake of brevity.[101] This is the fourth telling objection to Reichenbach's system of QL.

His objection to Reichenbach's solution to the Schrödinger cat paradox we must also take as basically cogent, if we take Reichenbach literally. However, what is of most interest in this regard is what is suggested about the quantum theory of measurement by Reichenbach's attempted solution.

In the Schrödinger cat paradox, the justification for assuming that parameters pertaining to macrosystems always retain sharp values is questioned by showing that QM does not predict that a macrosystem will always be in a definite state. In the case in point, the macrosystem consists of a cat in a box, whose death is contingent upon the decay process of a radioactive atom (also in the box) which has a likelihood of 0.5 of disintegrating during a specified time interval, and the relevant macro-observable states are 'cat alive' (α) and 'cat dead' (δ). But since we assume that a cat must be alive or dead, our common sense view of the world demands that the system be in a definite macrostate.[102]

Reichenbach believed, as Gardner points out, that this problem could be resolved by adopting what he calls the 'ignorance interpretation of mixtures', i.e., that a system represented as a mixture of eigenstates of the relevant observable is actually in *one* of these eigenstates although we are ignorant (to the extent specified by the probabilities given to states in the density matrix for the mixture) as to *which* one it is in. This view of mixtures would allow us to maintain that, as Reichenbach says, "... the status of the box [i.e., atom + cat as an isolated system] is a mixture and should be interpreted not as the living and the dead cat being smeared out over the state, but as a status describable in terms of an ≪ either-or ≫ and certain degrees of probability. There is either a living or a dead cat...."[103] But there are problems with taking this way out of the paradox, as Gardner tells us, because there are problems with the 'ignorance interpretation', specifically because there is no unique decomposition of a mixed state representation, although this is required under Reichenbach's view of the matter.

A further objection is added by Gardner, which we state in somewhat more precise terms than the above discussion. If the initial state of the atom + cat system is

$$\Psi(0) = \alpha \otimes \phi\,,$$

then after the specified interval, *t*, according to the Schrödinger equation we expect the system to be in the superposition

$$\Psi(t) = c_1(t)\alpha \otimes \phi + c_2(t)\delta \otimes \theta$$

such that the density matrix is $\rho = P\Psi(t)$. (Here the c_i are weighting factors, P is a projection operator, and ϕ, θ are initial and decay states of the atom.) But the theory also specifies that the system's final state is to be represented by a mixture, such that the density matrix is

$$\rho' = |c_1(t)|^2 P\alpha \otimes \phi + |c_2(t)|^2 P\delta \otimes \theta .$$

To say that the theory separately implies each of these results, where $\rho \neq \rho'$ is one way of raising the measurement problem in QM. Reichenbach's approach here is clear: (1) ρ doesn't represent the system, ρ' does, and moreover (2) one of the states (alive or dead) 'in' ρ' represents the actual state of the cat, whether we know which it is or not. We have already pointed out Gardner's objection to (2), but (1) is a dubious claim as well, and Gardner makes the plausible counter claim that "... the statement '$\rho = P\Psi(t)$', since it follows from a well-confirmed theory, is true."[104] However, to leave the matter at this point would be to overlook a significant alternative approach to quantum measurement which is naturally suggested by Gardner's critique of Reichenbach's theory of measurement. Professor Gardner points out that Feyerabend (at one time), Jauch, and others have maintained an approximationist quantum theory of measurement according to which no physically possible measurement could distinguish the state represented by ρ from the state represented by ρ'. We may then in some sense identify these states in order to resolve the measurement problem and cat paradox. A defender of Reichenbach could then require only an 'approximate eigenstate' as sufficient for a system to have a certain observable value.

I believe that this is more than a vague suggestion as to how the measurement problem might be resolved, but is rather in line with the most promising approach to quantum measurement now being investigated.[105] First of all, from a qualitative point of view, would one not expect a macrostate such as a cat-being-alive to correspond to not one, but to a whole class of macroobservationally indistinguishable states or an 'approximate eigenstate'. From this vantage point, we need not strictly identify the members of the classes of states discussed, but rather group them into classes of states which are mutually indistinguishable. The approximationist quantum theory of measurement thus attempts to explain what it means for a system to be in what Gardner is gesturing toward when he refers to 'approximate eigenstates' and why it is appropriate to represent a system after measurement (or cat-testing) by ρ', even if we hold – with Professor Gardner – that ρ is true of the system. To Gardner's suggestion, however, that such a condition on observables obtaining for a system is invariably vague and diffuse or precise and arbitrary, we must reply that the approximationists, insofar as they give a well-defined *theory* of measurement, specify a non-arbitrary correspondence between the obtaining of macroobservables and approximate 'underlying' eigenstates. A fully adequate treatment of this problem awaits the further elaboration and defense of the approximationist theory. But what I wish to emphasize here is that the interest of Reichenbach's treatment of Schrödinger's cat paradox is of great contemporary relevance not so much for the details of his attempted resolution as for the directions suggested by it.

9. CONCLUSION

Central to Reichenbach's concerns in interpreting QM were the problems posed by attempting exhaustive interpretations of the theory. Ultimately this forceful set of considerations resulted in his *principle of anomaly*. Whatever genuine problems for this interpretation of QM arise as a result of the critique of the phenomena/interphenomena distinction developed in this paper, it is doubtful whether this critique raises any real problems for the principle of anomaly, for it may be stated equally forcefully in terms independent of the distinction in question.

This principle, as Reichenbach saw the matter, pertains to restrictive as well as exhaustive interpretations. The development of his three-valued QL was not an attempt on Reichenbach's part to avoid the consequences of the principle of anomaly, but rather served to bear out the principle. The anomalous causal structure asserted to hold within the context of exhaustive interpretations is reflected within restrictive interpretations by corresponding anomalous semantic or logical structures.

> The physical status of the quantum mechanical world, expressed through a restrictive interpretation, is the same status expressed through exhaustive interpretations with causal anomalies which can be transformed away locally. The restrictive interpretations do not *say* the causal anomalies, but they do not *remove* them.[106]

Thus rather than saying that Reichenbach's QL was designed to banish causal anomalies from the microworld, it would be more propitious to point out that QL served to remove the assertions of causal anomalies from $\mathscr{L}_{QM}$.

The fact that Reichenbach held that the causal structure of microphysics could be represented equally well either by assertions about extraordinary processes within a language with an ordinary logical structure, or by the conspicuous absence of such assertions within a language with an extraordinary logical structure, here indicates the presence of the same pervasive interest in *equivalent descriptions* and the role of *conventions* in physical theory as we find in his studies of alternative physical geometries and the conventionality of distant simultaneity. Reichenbach held that the decision between the two above mentioned means of representation in QM was a matter of which conventions to adopt. He was thus a conventionalist in his philosophy of logic, but a subtle one. He held that "the subject matter of a science, without the addition of further qualifications, does not determine a particular form of logic" (Reichenbach, 1944, p. 43). But, of course, this was not to say that adopting a particular set of conventions is a *purely* conventional matter. Choice of a logic, according to his point of view, is to be made neither from *a priori* reasons alone, nor is it dictated by matters of empirical fact. Putnam's question 'Is Logic Empirical?' he would have taken to be somewhat ambiguous, however provocatively phrased. The more fundamental way of putting the question for Reichenbach is, I would suggest, 'do matters of fact bear upon the choice of a logic?' And to this question, Reichenbach's answer was a resounding, if controversial, *yes*.

> I don't believe that one can speak of the truth of a logic. . . . One can consider the consequences of the choice of a logic for a language; and, to the extent that these consequences depend . . . upon physical law, they reflect properties of the physical world. It is thus the combination of logic and physics which indicates the structure of the reality that concerns the physicist.[107]

His conventionalist view of logic was intimately bound to his general commitments to conventionalism. It seems to me then, that if we wish to see Reichenbach's study of QL in relation to his general philosophical position, we must note that his general views on the conventionality of the abstract structures used in modern physical theory are exemplified in both his studies in the philosophy of space–time and also in his studies of the logic of QM.

Whether it be particular concrete proposals presented, or suggestions pertinent to current research, Reichenbach's writings on the foundations of quantum theory provide us with a wealth of themes worthy of detailed study. In the context of the present investigation, we have been able to touch upon only a very few of these themes. Our task has been to single out and follow a leading thread in his works on QM, the development of a non-standard logic tailored to remove the paradoxes which have conceptually plagued the theory from its origins. In our exploration of his interpretation, we found that Reichenbach's QL is based upon much more secure ground than his early critics would have it. However, we also found that an examination of another line of criticism showed the inability of Reichenbach's QL to satisfy the conditions of adequacy which he had established for it. We also saw that the most central distinction used by him in his interpretation – that between phenomena and interphenomena – was not adequately drawn by Reichenbach, and so the entirety of his foundational study is seen to rest on less than secure support. Finally, we suggested that there is a close relation between Reichenbach's views on measurement in QM and certain more recent developments in this hotly contested subject area. In particular, the import of approximations and the relation of the macro- to microphysical (which is also explored in his study of QL) are topics immediately brought out by a reading of his writings on QM. It is hoped that the present study has in large part vindicated my earlier claim of the great value and relative merit of Reichenbach's works in the philosophy of QM. They remain as magisterial but neglected writings by a magisterial but neglected philosopher.

Emory University

NOTES

[1] Hereafter, I shall use the abbreviation 'QM' for this expression.
[2] Reichenbach (1944); I shall adopt the practice of including most references to this work in brackets or parentheses in the text.
[3] Hereafter, the abbreviation 'QL' will be used for this expression. Also, '$\mathscr{L}_i$' will be used to pick out a language indexed by *i*. Thus, '$\mathscr{L}_2$' will stand for a bivalent language, '$\mathscr{L}_3$' will stand for a trivalent language, and '$\mathscr{L}_{QM}$' will stand for the language of quantum mechanics.
[4] See for example Hooker (1975) for a collection of works on QL; a second volume is forthcoming.
[5] See Jammer (1974), Chapter 8, for references to these and other early writers on QL.
[6] Birkhoff–von Neumann (1936), reprinted in Hooker (1975).
[7] Birkhoff (1961), p. 157.
[8] Törnebohm (1957) brings this out very nicely.
[9] For a most illuminating study in the relations of Reichenbach's QL to that of Birkhoff–von Neumann, see the reconstruction of van Fraassen (1974), also reprinted in Hooker (1975).
[10] Post (1921); Łucasiewicz (1920). For a more recent work on many-valued logic containing a very helpful and complete bibliography, see Rescher (1969).

[11] Von Weizsäcker (1955).
[12] Reichenbach (1944), Sections 25 and 27.
[13] This is so unless it is held that 'making' a statement which has indeterminate truth-value is not *making* a statement at all. It is clear that this is a semantic point which is either trivial or question-begging.
[14] For illuminating studies which avoid such overemphasis and which separate the views of Bohr from those of Heisenberg (something which did not seem necessary in the context of the present study), see Hooker (1972) and D'Espagnat (1971), Chapter 18.
[15] Bohr (1948), p. 313.
[16] *Ibid.*, p. 317.
[17] W. Pauli has also taken this position. In his (1964), pp. 1403–1405, Pauli claims that Reichenbach's QL is unnecessary because physicists regard the Bohr approach and the QL approach as 'completely equivalent'.
[18] Bohr (1963), pp. 5–6.
[19] Note that $[A \& B]$ is held to be false and not meaningless or indeterminate in some of the systems of QL not discussed in this paper. See, for example, Birkhoff and von Neumann (1936).
[20] *Cf.* this remark by Reichenbach in (1944), p. 142: "The only justification [for adopting the definition which delimits the class of meaningful QM statements] . . . is that it eliminates the causal anomalies." Here Reichenbach overstates the point. We must object that while considerations of causal anomalies may have motivated the decision of the Copenhagen school to use the definition mentioned above, it is clear from their writings that considerations of what is verified or verifiable was also highly relevant to them in this decision. See notes 21 and 27.
[21] Heisenberg (1930), p. 15. Interpolations are mine.
[22] Reichenbach (1953), p. 136.
[23] Of course, one may be an empiricist and hold that laws are rules of inference and not part of an object language. The point is, however, that Reichenbach did not hold to that view.
[24] Reichenbach (1938). We must take care not to confuse theories of meaningfulness with theories of meaning. *Cf.* Salmon (1966) on this point.
[25] Reichenbach (1946a), p. 8
[26] *Cf.*, Reichenbach (1948), pp. 347–348.
[27] This point was also made in an excellent but most infrequently discussed article by H. Mehlberg (1949/50), pp. 203–204. He goes on to argue that this same view also commits the Copenhagen theorist to ontological idealism.
[28] *Cf.* Reichenbach (1948), p. 346.
[29] See Reichenbach (1944), Section 7, or a QM text such as Feynman (1965), Chapter 1. Briefly stated, the experimental arrangement in the two-slit experiment is as follows. We have a beam of electrons which pass through a diaphragm with two slits (which may be open or closed separately) and strike a screen where their impact is registered on a photographic plate. The patterns of data resulting from trials run with a single slit open may be compared with trials run with both slits open. The pattern formed when both slits are open differs from a pattern formed as the composite sum of the patterns resulting when each of the slits are open individually.
[30] Feyerabend (1958), see esp. p. 51 and pp. 58–59, n. 4. The paperback edition of Reichenbach (1944) received a somewhat less complete critique by the same author in *British J. Philosophy Sci.* **17** (1966/67), 326–328.
[31] Reichenbach (1944), Sections 5 and 6, esp. p. 21.

[32] Feyerabend (1958), p. 52, italics removed; the interpolation is mine.
[33] *Ibid.*, p. 52.
[34] Reichenbach (1944), p. 40. Feyerabend in (1958) p. 51 and p. 59, n. 5, himself observed this.
[35] Reichenbach (1946a), p. 9.
[36] Feyerabend (1958), p. 50. This comment, of course, is that of *early* Feyerabend. One wonders whether now, after his writing of 'Against Method' (Feyerabend, 1970), he has come perforce to admire 'this sly procedure' along with those of Galileo, Einstein, and others.
[37] Reichenbach (1944), p. 14, n.
[38] Cf. the supporting critique of Feyerabend in Haack (1974), pp. 153–155, which I discovered after writing this paper.
[39] Nagel (1945), p. 443; Hempel (1945), p. 99.
[40] Nagel (1945), p. 443.
[41] Reichenbach (1946b), p. 244.
[42] *Ibid.*
[43] This dissatisfaction seems to have been based upon Nagel's restriction of the sense of 'assertability' to languages taken separately, that is, this term is taken to have distinct senses in each of $\mathscr{L}_2$ and $\mathscr{L}_3$ – hence his unwillingness to accept 'assertability' as an explicans. However, is there not a sense of this term which may be taken to be common to at least $\mathscr{L}_2$ and $\mathscr{L}_3$, and perhaps to other languages as well? It is upon this sense that Reichenbach's view depends.
[44] Nagel (1945), p. 444.
[45] Reichenbach (1946b), p. 246.
[46] Nagel (1946), p. 249.
[47] Hempel and Turquette both made the 'quantifier objection' as well. See Hempel (1945), p. 100 and Turquette (1945), pp. 513–516.
[48] Reichenbach (1946b), p. 244; see also Reichenbach (1951), p. 58.
[49] Nagel (1946), pp. 249–250. Professor Max Jammer in his excellent recent introduction (1974) accepts this objection, but with no further argument than the one given by Nagel.
[50] Nagel (1945), p. 444.
[51] Consult p. 151 of Reichenbach (1944) for the definition of these distinct connectives.
[52] Hempel (1945), p. 99.
[53] *Ibid.*, pp. 99–100.
[54] That is, it seems to be assumed that since the object language is non-bivalent, so (because the semantic definition of truth is used) the metalanguage must be non-bivalent also. Problems along these lines might have arisen if Reichenbach had tried to specify necessary and sufficient conditions for ''ϕ' is false' and ''ϕ' is indeterminate', but this he does not do nor is it clear that he is required to do so in order to make his view intelligible. See van Fraassen (1971), pp. 163ff., for problems that arise when a transition from ''ϕ' is true iff ϕ is the case' to '... iff ϕ is *not* the case' is attempted for a non-bivalent language.
[55] Hempel (1945), p. 100.
[56] Putnam (1957), also reprinted in Hooker (1975).
[57] Putnam concentrated his attention on three-valued logics, but his results clearly are generalizable. Be this as it may, we will focus on three-valued logic.
[58] The statements that are mentioned here will themselves include reference to a time in their internal structure. The treatment of time and tense presupposed here is close to that of W. V. O. Quine in (1960).

[59] See Putnam (1957), Section I, for both of these conditions.
[60] *Ibid.*, p. 74.
[61] *Ibid.*, p. 80, n. 2.
[62] *Ibid.*, p. 76.
[63] Levi (1959), pp. 65–69.
[64] *Ibid.*, p. 67.
[65] Reichenbach (1944), p. 19.
[66] Also M. Strauss (1971), Chapter XXIV, Section I, shows how a many-one correspondence between $\mathscr{L}_2$ and $\mathscr{L}_3$ may be charted.
[67] Levi (1959), p. 68.
[68] Reichenbach (1946b), p. 244.
[69] Putnam (1959), p. 79. One can find the same underlying idea that there is continuity of meaning between analogous connectives in alternative logics in Farber (1942), p. 52, where it is argued that the law of excluded middle analogues in non-standard logics 'have a central core'.
[70] Feyerabend (1958), pp. 50–52.
[71] Haack (1974), pp. 157–159.
[72] Nagel (1945).
[73] *Ibid.*, p. 443.
[74] Reichenbach (1953), p. 126. "Such quantum phenomena are inferred from macroscopic observations by the sole use of classical theory; this is why they can be taken on a par with macroscopic phenomena." I am grateful to Mr. P. Drew for advice on translation.
[75] Gardner (1972), pp. 102–103, however, mistakenly conflates the distinction of 'phenomena'/'interphenomena' with those of 'macro'/'micro' and 'observable'/ 'unobservable'. Gardner asks (p. 103): "The distinction between 'macro' and 'micro' is vague and arbitrary, whereas that between 'true' and 'indeterminate' is not. How, then, can the distinctions be correlative?" The first reply would be that they are not correlative, as Reichenbach emphasized in remarking that some statements are held to be true of micro-objects (even though they are not undergoing measurement), specifically when his Definition 4 [Reichenbach (1944), p. 140] applies. Secondly, Reichenbach did maintain a sharp distinction between 'true' and 'indeterminate' (*Cf.* n. 78), but this distinction is correlative with that between cases where a corresponding measurement on a system can be performed and cases where it cannot – this presumably being a sharp distinction.
[76] Reichenbach (1946b), p. 243.
[77] Reichenbach (1953), p. 126. The italics are mine.
[78] From his discussion of phenomena in his (1944) and of the Schrödinger cat paradox in his (1948), it seems plausible to infer that Reichenbach saw that the approximation of microsystems to macrosystems (in the sense of having macroproperties *approximately* applicable to microsystems) was a means of dealing with fundamental issues in the interpretation of QM. Moreover, Reichenbach held that "the use of the 'sharp' categories *true* and *false* must be considered in both cases classical and quantum physics as an idealization applicable only in the sense of an approximation." [(1944), pp. 147–148; the interpolation is mine.] This was intimately bound to his concern with probabilistic logic. To the same passage he added "[t]he quantum mechanical truth-value *indeterminate*, however, represents a topologically different category." *Cf.*, the remarks on Gardner (1972) in n. 75.
[79] Putterman and Rudnick (1971), p. 40; this paper gives a good survey of recent experimental results in low temperature physics. For an account of the relevant theory see

Galasiewicz (1971), especially the editor's introduction and a paper by R. P. Feynman, 'Application of QM to Liquid Helium', pp. 268–313.

[80] Perhaps, for example, one might wish to define the events in super-cooled matter discussed above to be interphenomenal, however paradoxical this might seem.

[81] Feyerabend (1958), pp. 53–54. Interpolation is mine.

[82] *Ibid.*, p. 54.

[83] Van Fraassen (1974) and Lambert (1969). A difference (which one might want to argue is of no consequence) between Reichenbach and van Fraassen/Lambert is that '*I*' for the former is an assignable truth-value, but for the latter pair it indicates a 'truth-value gap'. The difference is that one assigns a truth-value, the others assign no truth-value at all. The question thus arises of the faithfulness of the van Fraassen/Lambert reconstruction of Reichenbach.

[84] Van Fraassen (1974), pp. 232–233.

[85] *Ibid.*, pp. 233–235. Interpolations are mine.

[86] In an unpublished paper, Mr. G. Hardegree has carried out important work pertinent to Reichenbach's explication of incompatibility. He argues for a generalization of the notion of compatibility, according to which propositions may be absolutely compatible, absolutely incompatible, or compatible relative to certain states. See Hardegree (1975). Also see Hardegree's essay in this volume.

[87] Suppes (1966), p. 14.

[88] *Ibid.*, p. 20. We find a closely related objection suggested by M. Strauss in (1972), p. 296. He points out that Reichenbach did not develop a probability calculus defined on three-valued language. (This essay was written however in 1945.)

[89] Suppes (1966), p. 20.

[90] Suppes (1965).

[91] *Ibid.*, pp. 369–375.

[92] Strauss (1971), p. 296.

[93] *Ibid.*

[94] See Reichenbach (1944), esp. Definition 4, p. 140. Reichenbach *was* aware of Strauss' 'causal anomaly'. He had pointed out *Ibid.* p. 41: "Thus it is hard to abandon a statement like the one concerning the particle's going through one slit or the other. All that can be said against this statement is that it leads to undesired consequences." Cf. Putnam (1970), p. 186, for an alternative account.

[95] For Strauss' own attempt to deal with the two-slit paradox, see his (1971), Chapter XVII, 'The Paradoxes of Quantum Physics and the Complementary Mode of Description'.

[96] Gardner (1972), p. 102.

[97] These two paradoxes are carefully discussed in Gardner (1972).

[98] Reichenbach (1944), p. 95.

[99] *Ibid.*, p. 170; the interpolation is mine. *Cf.* p. 98 and Reichenbach (1948), pp. 344–345.

[100] And any QL in which such statements must be held to be *false* would be equally overturned. This is the intent of Park and Margenau (1968) and (1971), as I read them.

[101] See Gardner (1972), pp. 105–107; *cf.* Reichenbach (1944), Section 36.

[102] See Gardner (1972), pp. 103–105. (Some forty-odd years have passed since Schrödinger proposed this experiment, and there has not been one complaint lodged by the S.P.C.A.!)

[103] Reichenbach (1948), p. 344.

[104] Gardner (1972), p. 105.

[105] I have in mind the work of the Italian group, most prominently, A. Daneri – A. Loinger – G. M. Prosperi; see, for example, their 'Quantum Theory of Measurement and Ergodicity Conditions', *Nuclear Phys.* **33** (1962): 297–319, and later works.

It should go without saying that while I am in basic agreement with the QL approach to the foundations of quantum physics, I also regard the attempt of J. Bub to solve the measurement problem using QL in Chapter XI of his *The Interpretation of QM*, D. Reidel Publishing Co., Dordrecht, 1974, as well as his (independent) attempts to refute the work of the Italian group, as unsuccessful. These claims cannot be defended here, however.

[106] Reichenbach (1944), p. 44.

[107] Reichenbach (1953), p. 136; based upon a translation by Professor L. Beauregard.

BIBLIOGRAPHY

Birkhoff, G., 1961, 'Lattices in Applied Mathematics', in *American Mathematical Society Proceedings of Symposia in Pure Mathematics* **2**, American Mathematical Society, Providence, R.I.

Birkhoff, G. and von Neumann, J., 1936, 'The Logic of Quantum Mechanics', *Ann. Mathematics* **37**, 823–843.

Bohr, N., 1948, 'On the Notions of Causality and Complementarity', *Dialectica* **2**, 312–319.

Bohr, N., 1963, 'Quantum Physics and Philosophy: Causality and Complementarity', in N. Bohr, *Essays 1958–1962 on Atomic Physics and Human Knowledge*, Interscience, New York.

D'Espagnat, B., 1971, *Conceptual Foundations of Quantum Mechanics*, W. A. Benjamin, Menlo Park, Calif.

Farber, M., 1942, 'Logical Systems and the Principles of Logic', *Philosophy Sci.* **9**, 40–54.

Feyerabend, P. K., 1958, 'Reichenbach's Interpretation of Quantum Mechanics', *Philosophical Stud.* **9**, 47–59.

Feyerabend, P. K., 1970, 'Against Method' in M. Radner and S. Winokur (eds.), *Minnesota Studies in the Philosophy of Science*, Vol. IV, University of Minnesota Press, Minneapolis, pp. 17–130.

Feynman, R. P., 1965, *Lectures on Physics*, Vol. III, Addison-Wesley, Reading, Mass.

van Fraassen, B. C., 1971, *Formal Semantics and Logic*, Macmillan, New York.

van Fraassen, B. C., 1974, 'The Labyrinth of Quantum Logic', in R. S. Cohen and M. Wartofsky (eds.), *Logical and Epistemological Studies in Contemporary Physics, Boston Studies in the Philosophy of Science*, Vol. 13, D. Reidel Publishing Company, Dordrecht.

Galasiewicz, Z. M. (ed.), 1971, *Helium 4*, Pergamon Press, London.

Gardner, M., 1972, 'Two Deviant Logics for Quantum Theory: Bohr and Reichenbach', *British J. Philosophy Sci.* **23**, 89–109.

Haack, S., 1974, *Deviant Logic*, Cambridge University Press, Cambridge.

Hardegree, G., 1975, 'Compatibility and Relative Compatibility in Quantum Mechanics', an unpublished paper read at the International Congress of Logic, Methodology and Philosophy of Science, London, Ontario, August, 1975.

Heisenberg, W., 1930, *The Physical Principles of Quantum Theory*, Dover Publications, New York.

Hempel, C. G., 1945, 'Review of H. Reichenbach, *Philosophic Foundations of Quantum Mechanics*', *J. Symbolic Logic* **10**, 97–100.

Henkin, L., 1960, 'Review of H. Putnam, 'Three-Valued Logic'; P. K. Feyerabend, 'Reichenbach's Interpretation of Quantum Mechanics'; and I. Levi, 'Putnam's Three Truth-Values'', *J. Symbolic Logic* **25**, 289–291.

Hooker, C. A., 1972, 'The Nature of Quantum Mechanical Reality: Einstein Versus Bohr', in R. Colodny (ed.), *Paradigms & Paradoxes: The Philosophical Challenge of the Quantum Domain*, University of Pittsburgh Press, Pittsburgh.

Hooker, C. A., 1975, *The Logico-Algebraic Approach to Quantum Mechanics*, D. Reidel Publishing Company, Dordrecht, Vol. I.

Jammer, M., 1974, *The Philosophy of Quantum Mechanics*, McGraw-Hill, New York.

Jauch, J. M., 1968, *The Conceptual Development of Quantum Mechanics*, Addison-Wesley, Reading, Mass.

Lambert, K., 1969, 'Logical Truth and Microphysics', in K. Lambert, *The Logical Way of Doing Things*, Yale University Press, New Haven.

Levi, I., 1959, 'Putnam's Three Truth Values', *Philosophical Stud.* **10**, 65–69.

Łucasiewicz, J., 1920, 'On Three-Valued Logic' (in Polish), *Ruch Filozoficzny* **5**, 169–170.

Mehlberg, H., 1949/50, 'The Idealistic Interpretation of Atomic Physics', *Studia Philosophica* **4**, 171–235.

Nagel, E., 1945, 'Book Review: *Philosophical Foundations of Quantum Mechanics* by H. Reichenbach', *J. Philosophy* **42**, 437–444.

Nagel, E., 1946, 'Professor Reichenbach on Quantum Mechanics: A Rejoinder', *J. Philosophy* **43**, 247–250.

Park, J. L. and Margenau, H., 1968, 'Simultaneous Measurability in Quantum Theory', *Int. J. Theoret. Phys.* **1**, 211–283.

Park, J. L. and Margenau, H., 1971, 'The Logic of Noncommutability of Quantum Mechanical Operators and Its Empirical Consequences', in W. Yourgrau and A. van der Merwe (eds.), *Perspectives in Quantum Theory*, MIT Press, Cambridge, pp. 37–70.

Pauli, W., 1964, 'Reviewing Study of Hans Reichenbach's *Philosophical Foundations of Quantum Mechanics*', in R. Kronig and V. F. Weisskopf (eds.), *Collected Scientific Papers*, Vol. 2, Interscience Publishers, New York.

Post, E. L., 1921, 'Introduction to a General Theory of Elementary Propositions', *Amer. J. Math.* **XLIII**, 163ff.

Putnam, H., 1957, 'Three-Valued Logic', *Philosophical Stud.* **8**, 73–80.

Putnam, H., 1970, 'Is Logic Empirical?', in R. S. Cohen and M. Wartofsky (eds.), *Boston Studies in the Philosophy of Science*, Vol. 5, D. Reidel Publishing Company, Dordrecht.

Putterman, S. J. and Rudnick, I., 1971, 'Quantum Nature of Superfluid Helium', *Phys. Today* **24** (Aug.), 40ff.

Quine, W. V. O., 1960, *Word and Object*, MIT Press, Cambridge, Mass.

Reichenbach, H., 1938, *Experience and Prediction*, University of Chicago Press, Chicago.

Reichenbach, H., 1944, *Philosophic Foundations of Quantum Mechanics*, University of California Press, Los Angeles and Berkeley.

Reichenbach, H., 1946a, 'Philosophy and Physics', in *Faculty Research Lectures – University of California*, #19, University of California Press, Berkeley and Los Angeles.

Reichenbach, H., 1946b, 'Reply to Ernest Nagel's Criticism of My Views on Quantum Mechanics', *J. Philosophy* **43**, 239–247.

Reichenbach, H., 1948, 'The Principle of Anomaly in Quantum Mechanics', *Dialectica* **2**, 337–350.

Reichenbach, H., 1951a, 'The Verifiability Theory of Meaning', *Contributions to the Analysis and Synthesis of Knowledge: Proceedings of the American Association of Arts and Sciences* **80**: 1, 46–60.

Reichenbach, H., 1951b, 'Über die erkenntnistheoretische Problemlage und den Gebrauch einer dreiwertigen Logik in der Quantenmechanik', *Z. Naturforschung* **6a**, 569–575.

Reichenbach, H., 1953, 'Les fondements logiques de la mécanique des quanta', *Ann. Inst. Henri Poincaré* **XIII**, 109–158.

Rescher, N., 1969, *Many-Valued Logic*, McGraw-Hill, New York.

Salmon, W. C., 1966, 'Verifiability and Logic', in P. K. Feyerabend and G. Maxwell (eds.), *Mind, Matter and Method*, University of Minnesota Press, Minneapolis.

Strauss, M., 1971, *Modern Physics and Its Philosophy*, D. Reidel Publishing Company, Dordrecht.

Suppes, P., 1965, 'Logics Appropriate to Empirical Theories', in J. Addison *et. al.* (eds.), *The Theory of Models*, North-Holland Publishing Company, Amsterdam.

Suppes, P., 1966, 'The Probabilistic Argument for a Non-Classical Logic in Quantum Mechanics', *Philosophy Sci.* **23**, 14–21.

Törnebohm, H., 1957, 'On Two Logical Systems Proposed in the Philosophy of Quantum Mechanics', *Theoria* **23**, 84–101.

Turquette, A. R., 1945, 'Review of Reichenbach's *Philosophical Foundations of Quantum Mechanics*', *Philosophical Rev.* **54**, 513–516.

von Weizsäcker, C. F., 1955, 'Komplementarität und Logik', *Naturwissenschaften* **42**, 521–529 and 545–555.

GARY M. HARDEGREE

REICHENBACH AND THE LOGIC OF QUANTUM MECHANICS

1. INTRODUCTION

It is generally agreed that quantum mechanics (QM) constitutes a revolutionary physical theory, and it has been suggested that the revolutionary character of QM penetrates even to the level of logic. This suggestion stems from at least two sources. First of all, there are the various foundational approaches to QM which propose to formulate quantum theory on the basis of 'quantum logic'.[1] This approach to the postulational foundations of QM originates with the seminal work of Birkhoff and von Neumann (1936), which is based on the Hilbert space formulation of quantum theory expounded by von Neumann in his classic treatise (1932) on the mathematical foundations of QM. A parallel and closely related treatment of QM was proposed by Martin Strauss (1937–38) and developed primarily by Kochen and Specker (see, e.g., 1967). Together these approaches comprise what may be called mainstream (or conventional) quantum logic.[2] Secondly, there are a number of non-mainstream quantum logics which have been proposed,[3] most notably perhaps by Hans Reichenbach. In his treatise on the philosophical foundations of QM (1944, henceforth PhF), Reichenbach proposed a three-valued truth-functional logic as a means of dealing with certain conceptual tensions arising in the quantum mechanical description of the world.

In the present article, I wish to compare mainstream quantum logic (MQL) with Reichenbach's quantum logic (RQL). This has been done previously, for example, by van Fraassen (1974). However, in attempting to provide a unified account of the 'labyrinth of quantum logics', van Fraassen stresses purely formal similarities and differences between the various quantum logics, and consequently fails to make clear what I maintain to be a fundamental difference between MQL and RQL. Specifically, whereas van Fraassen treats RQL and MQL as if they

W. C. Salmon (ed.), Hans Reichenbach: Logical Empiricist, 475–512.

pertain to exactly the same linguistic domain, I wish to argue that their intended domains are significantly different. In particular, I wish to argue that whereas RQL was intended to provide the logico-linguistic framework for an alternative formulation of quantum theory,[4] MQL constitutes merely a particular formal analysis of the empirical postulates of quantum theory, regarded as formulated in a language with a classical logical structure.

In order to clarify my position within the context of conventional (Hilbert space) quantum theory, in Sections 3 and 4, I distinguish two empirical languages to be examined in any logical analysis of a kinematic theory K: the *theory formulation language* TL(K) and the *observation language* OL(K). Whereas the laws of K occur (are formulated) in TL(K), sentences employed to make observation reports relevant to K – for example, 'quantity q has value r' – occur in OL(K). I argue that according to the conventional construal of quantum logic, these two languages are categorically distinct or disjoint. On the other hand, the names of expressions (sentences) of OL(K) occur in TL(K), which is to say that TL(K) provides a metalanguage for OL(K).

This provides the touchstone for the investigation in Sections 5–7 of the logical structure of OL(K), in particular OL(QM). The fundamental idea is quite simple: in terms of the descriptive vocabulary of TL(K), employing the resources of first order logic, one may define various 'logical predicates' (not to be confused with the logical vocabulary of TL(K)), which pertain to the sentences occurring in OL(K). One then deduces, by purely classical methods, the logical structure of OL(K). In Section 7, we briefly examine the logical structure of the observation language OL(QM) of kinematic quantum theory, which is seen to be isomorphic to the family of subspaces of a separable Hilbert space – the conclusion of Birkhoff and von Neumann.

A major point emphasized in Part I is the disjointness of the two languages OL(QM) and TL(QM). Similarly, it is emphasized that, whereas the logical structure of OL(QM) is non-classical, the logical structure in terms of which quantum theory is formulated as a system of empirical postulates is classical. In particular it is stressed that MQL pertains exclusively to OL(QM).

Having characterized the structure of MQL as well as its relation to the formulation of conventional quantum theory, in Part II we turn to the logic proposed by Reichenbach. A summary is presented in the preface to Part II, so I confine myself here to a few general remarks. Reichenbach criticized the Bohr-Heisenberg interpretation of QM because it involved formulating laws of QM – for example, the law of complementarity – in the metalanguage rather than the object language of QM. By the object language of QM, I interpret Reichenbach to mean the observation language OL(QM), possibly augmented by truth-functional connectives. Thus, I see his criticism of Bohr-Heisenberg as tantamount to rejecting the disjointness of TL(QM) and OL(QM), and I interpret Reichenbach's program primarily as an attempt to replace TL(QM) and OL(QM) by a unified language RL(QM), which is basically OL(QM) augmented by a variety of three-valued truth-functional connectives. Furthermore, since Reichenbach proposed a unified quantum language, RQL must be understood as pertaining to both the formulation of observation reports and the formulation of the laws of QM. The upshot of Part II is that the unification of OL(QM) and TL(QM) envisioned by Reichenbach cannot succeed in light of a fundamental criticism first presented by van Fraassen (1974).

I. CONVENTIONAL QUANTUM LOGIC

2. FORMAL SEMANTIC PRELIMINARIES

Formal semantically speaking (cf., e.g., Thomason, 1973), a logic L may be characterized as an ordered pair ⟨SYN, SEM⟩, where SYN is the underlying *syntax* or *language* of L, and SEM is the *semantics* of L, which consists of a class of semantic assignments on SYN. A *semantic assignment* on SYN is a function which assigns various entities to well-formed expressions of SYN (individual terms, predicate terms, etc.), and additionally assigns truth values (however construed) to at least certain sentences and formulas of SYN. In examining various quantum logics, we are interested primarily in sentential logics; a *sentential logic* is an ordered pair $\langle S, V \rangle$, where S is a class of sentences (closed formulas) and V is a class of valuations (partially) defined on S. A *valuation* (partially) defined on S is a map v whose domain is (some subset of) S, and whose codomain is some superset of $\{0, 1\}$ ({False, True}). Observe that we permit both the possibility of more than two truth values and the possibility of 'truth value gaps' (cf., e.g., Lambert, 1969).

These ideas are made explicit as follows. A valuation v on S is said to be *complete* relative to S exactly if v is fully defined on S, that is, exactly if $\text{dom}(v) = S$. A valuation v is said to be *bi-valued* exactly if its codomain is $\{0, 1\}$. Finally, a valuation v is said to be *bivalent* exactly if it is both bi-valued and complete. Analogous definitions may be given for a semantics V. A semantics V is complete relative to S (resp., bi-valued, bivalent) exactly if every valuation $v \in V$ is complete relative to S (resp., bi-valued, bivalent). On the basis of these definitions, we see that the failure of bivalence may be attributed either to the failure of bi-valuedness or to the failure of completeness. We consider both cases in this work.

We next define various formal semantic notions: satisfaction, validity, (weak and strong) entailment, and (weak and strong) exclusion. We say that a valuation v *satisfies* a sentence P exactly if $v(P) = 1$; we say that P is *valid* in V exactly if P is satisfied by every $v \in V$. Where P and Q are sentences, P *entails* Q in the weak sense (relative to V) exactly if for all $v \in V$, $v(P) = 1$ only if $v(Q) = 1$; P *excludes* Q in the weak sense exactly if for all $v \in V$, $v(P) = 1$ only if $v(Q) = 0$. In order to define the strong versions of entailment and exclusion, we require that the codomain of every valuation be the closed unit interval $[0, 1]$ of real numbers. Then, P *entails* Q in the strong sense exactly if for all $v \in V$, $v(P) \leq v(Q)$, and P *excludes* Q in the strong sense exactly if for all $v \in V$, $v(P) \leqslant 1 - v(Q)$. Observe that, whereas P strongly entails (excludes) Q only if P weakly entails (excludes) Q, these concepts may not be expected automatically to be identical except in the case that V is bi-valued.[5]

3. THE FORMULATION LANGUAGE OF A KINEMATIC THEORY

In order to clarify the role of quantum logic within conventional QM as well as the role of Reichenbach's quantum logic within his proposed interpretation of quantum theory, it is important to specify formally exactly what the language of a physical theory looks like.

Ideally a physical theory can be formulated in a manner completely analogous to the formulation of a mathematical theory, for example, group theory. However, there are different approaches to the exact formulation of mathematical theories, including most prominently the linguistic (syntactic) approach and the 'set-theoretic predicate' or semantic approach. These two approaches yield corresponding views of scientific theories: the standard or received view (cf., e.g., Hempel, 1970) and the 'semantic view of theories' variously proposed by Suppes (1967), Sneed (1971), van Fraassen (1970, 1974), and Suppe (1967, 1972). Previously (1974, 1975, 1976a), I have employed the semantic view of theories in describing quantum logic. However, in order to make my viewpoint as clear as possible, in the present work I adopt the purely linguistic approach, although the languages utilized are sufficiently rich to render the two approaches largely equivalent.

Fundamental to the linguistic approach to theories is the thesis that theories, either mathematical or physical, are individuated entirely with respect to the claims they make, that is, the theorems they postulate, and accordingly a physical theory may simply be identified with its class of theorems. Thus, to analyze a physical theory T is first of all to determine the specific language $L(T)$ in terms of which it is formulated, and secondly to ascertain which among the well-formed expressions of $L(T)$ are in fact theorems postulated by T. The straightforwardness of the enterprise, so stated, is perhaps the chief philosophical virtue of the linguistic construal of theories.

As a result of modern foundational studies in quantum theory, it is currently believed (cf., e.g., Jauch, 1968 and Stein, 1972) that the fundamental differences between QM and classical mechanics (CM) are reducible to differences at the purely kinematic (static) level. Therefore, for the sake of simplicity, we restrict ourselves to purely kinematic theories, or to the kinematic components of physical theories.

Within the framework of the linguistic construal of theories, a kinematic theory K can be formulated in a four-sorted language $L(K)$ described as follows. First of all, sort 1 is intended to be (designate) a class M of *physical magnitudes* (observables); sort 2 is intended to be the class $B(R)$ of Borel subsets of the real line R; sort 3 is intended to be a class W of (*statistical*) *states*; finally, sort 4 is intended to be the closed unit interval $[0, 1]$ of real numbers. Logical signs include the appropriately sorted variables ranging over each of these four sorts, generically denoted m, X, w, r, as well as the corresponding quantifiers. The universal quantifiers are generically denoted (m), (X), (w), (r); the existential quantifiers are generically denoted $(\exists m)$, $(\exists X)$, $(\exists w)$, $(\exists r)$. The logical signs of $L(K)$ also include a two-place predicate '=' (equality) as well as a collection of sentential connectives, including negation ($\neg$), conjunction (&), disjunction (*or*), and conditional ($\Rightarrow$). Concerning the non-logical signs, we suppose that $L(K)$ is sufficiently rich to have individual names for all elements of each sort; indulging in set theory, we may simply employ the objects themselves as their own names. The only other non-logical sign is a four-place predicate letter S of type (1, 2, 3, 4). Thus, the atomic sentences of $L(K)$ are quite simple; they all have either the form of simple equality formulas ('$s = t$') or the form $S(m, X, w, r)$, where $m \in M$, $X \in B(R)$, $w \in W$, $r \in R$ (remember the naming convention!), or m, X, w, r are variables of the appropriate sort.

On the inferential or logical side of things, in addition to all the inference rules validated by classical first order logic, $L(K)$ contains four special infinitary inference rules, one for each sort, for introducing universal quantifiers. For example, let x be a variable ranging over sort 3 (the class W of statistical states), and let $\Phi(x)$ be a formula with one free variable of sort 3. Then the sort 3 rule reads: if $\Phi(w)$ is a thesis of K for all $w \in W$, then $(x)\Phi(x)$ is also a thesis of K (note: here K is any theory formulated in terms of $L(K)$).

Concerning the intended interpretation of $L(K)$, the atomic formula $S(m, X, w, r)$ is intended to be read as follows: "for any physical system s, if s is in state w, then the probability that magnitude m lies in X is r." That the intended reading of $S(m, X, w, r)$ involves a universal quantifier suggests that we are simplifying our account somewhat; a kinematic theory might be expected in general to yield quite different sets of postulates for different (types of) physical systems. To accommodate this generality our language would have to be enlarged to include variables and names to denote (types of) physical systems. However, for the sake of simplicity, we restrict our attention to a single (type of) physical system, so as to avoid complicating the language.

Now, a kinematic theory K may be specified simply by 'listing'[6] all its basic postulates of the form $S(m, X, w, r)$, and accordingly K may be identified with the deductive closure of this basic postulate set. For example, if the basic formulas have their intended probabilistic meaning, then among the deductive consequences of K one would expect theorems such as $(m)(w)S(m, R, w, 1)$ and $(m)(w)S(m, \varnothing, w, 0)$. The first theorem states that for any magnitude m and any state w the probability that m lies in R is one; the second theorem states that the probability that m lies in the empty set $\varnothing$ is zero. There are numerous other probabilistically interpretable consequences of K which we will consider presently.

At this point it must be emphasized that, except for the infinitary character of $L(K)$, there is nothing unusual about the logic in terms of which K is formulated; rather, the logic is a simple extension of classical first order logic. Furthermore, this linguistic scheme is fully adequate to formulate both classical and quantum kinematics. One might therefore wonder where quantum logic appears in this formulation of quantum theory. The important point I wish to emphasize is that, however (conventional) quantum logic is construed, it should in no case be understood as the logic underlying the formulation of quantum theory, in the sense that classical first order logic underlies the formulation of, say, group theory. Such a logic I wish to call the *formulation logic* of quantum theory, which is basically the set of logical or inferential principles of the formulation language of the theory, which one employs in generating the set of deductive consequences. In other words, the formulation logic of a theory is constituted by the closure conditions on the theory regarded as a deductively closed system of postulates. There simply are no grounds for supposing that the formulation logic or language of quantum theory is in any way significantly different from the formulation logic/language of group theory or classical mechanics.

Of course, although quantum logic is not explicitly employed in formulating kinematic quantum theory and deducing its various logical consequences, it is nevertheless lurking about, as it were, not within the formulation logic of QM, but rather within the theory's descriptive ($\neq$ logical) content. To be more specific, quantum logic is part of the (classical logical) deductive consequences of kinematic quantum theory, obtained by the introduction of appropriate derived terms defined in terms of the theory's descriptive and logical vocabulary (see below). This point has been previously emphasized, from the viewpoint of the semantic view of theories, by van Fraassen, who writes the following (1974, p. 247).

> ... from our point of view a logic of quantum mechanics is simply an attempt to give a systematic account of the semantic relations among the elementary statements of that theory. And these semantic relations are to be *deduced from* quantum theory – that is the sense in which this logic is a quantum logic. It is not meant to be the basis for a formalization of the theory, or for a new, non-standard *Principia*.

4. THE OBSERVATION LANGUAGE OF A KINEMATIC THEORY

Having made our point of view clear, let us now consider how one proceeds to ascertain the logic *inherent to* a kinematic theory such as QM, formulated according to the scheme presented in the previous section. Recall that the basic formulas of a kinematic theory K all have the form $S(m, X, w, r)$, or restoring the suppressed name of the physical system s, they all have the form $S(s, m, X, w, r)$. By a process of extracting the first three components, we obtain the ordered triple (s, m, X), which may be regarded as an *elementary descriptive* or *observation sentence* (compare with van Fraassen's expression 'elementary statement' in the above quote), to be read 'system s is such that m lies in X'. This elementary sentence may be regarded as ascribing to system s an elementary monadic predicate (m, X), which in turn denotes the property of m lying in X. For example, if m is position, then (m, X) corresponds to the property of being located in a certain spatial region denoted by X. It is customary in the quantum logic literature to drop reference to system s and call the ordered pair (m, X) an elementary sentence, although 'elementary predicate' is clearly more appropriate. Insofar as confusion is unlikely to arise we follow this custom.[7]

It is crucial to recognize that the elementary sentences of the form (m, X), or even of the form (s, m, X), have a character entirely different from the theoretical sentences of the form $S(m, X, w, r)$. An elementary sentence (m, X) makes a genuine assertion (statement) only relative to a specific *context of utterance*, in reference to a particular physical system s, at a particular time t, and so forth. In other words, elementary sentences of the form (m, X) are *indexical expressions* (i.e., their meaning is *context-dependent*), where the indexical coordinates include a system coordinate as well as a temporal coordinate (cf. Montague, 1968, 1970 and Lewis, 1970).[8] Let us suppose for simplicity that we are always speaking of a single physical system, say, Oscar the electron (supposing the genidentity of electrons); we can therefore concentrate on the temporal index. Now, from our viewpoint the role of time in kinematics (statics) is not particularly important, except insofar as it serves to parameterize the set of physically possible states. From the theoretical viewpoint it is pointless to add a temporal index t if it only denotes actual historical moments, for theoretical physics is presumably not concerned with historical reports such as "Oscar's position lies in region Delta at 18:00 G.M.T. July 14, 1789." Rather, in kinematics time only enters as an indirect indexical coordinate, the physically important index being the possible states (configurations) of the system. Thus, rather than indexing the elementary sentences by time, we index them relative to the class of physically possible states. Also, since the theory under consideration may describe the world only probabilistically, we may also include a 'weighting factor' as well. Making these indices explicit, we arrive at sentences of the form (m, X, w, r), where w is a physical state, and r is an appropriate weighting factor. We thus arrive at formulas very much like the theoretical assertions of kinematic theory K.

Before continuing, it might be useful to note the following analogy. The relation between the context-free theoretical sentences of K and the indexical or context-dependent observation sentences is completely analogous to the corresponding relation between the theoretical assertions of meteorology and the sentences employed to report the contingent state of the weather, for example, 'it is raining'. It is evident that 'it is raining' and 'm lies in X' are on a par with respect to indexicality; at the same time, sentences like 'whenever it rains, it pours' and 'whenever m lies in X, n lies in Y' are similarly on a par.[9] Sentences of the latter sort are non-indexical (context-free) and are therefore of the correct category to appear in the theory formulation language, and are candidates to be law statements. On the other hand, sentences of the former sort, being indexical or context-dependent, are not even of the correct category to be law statements.

In order to maintain the categorical distinction between these two sorts of empirical sentences, I propose to distinguish between the *theory formulation language* TL(K) and the corresponding *observation language* OL(K) of a kinematic theory. In our specific formulation, sentences of TL(K) are generated from the basic sentences of the form $S(m, X, w, r)$, and sentences of OL(K) are all of the basic form (m, X) where $m \in M$ and $X \in B(R)$ (in other words, $\text{OL}(K) = M \times B(R)$). In examining Reichenbach's interpretation of QM, a very important question concerns the precise formal relation between these two languages (in the case of QM, they are denoted TL(QM) and OL(QM)). This question is discussed in the next three sections in relation to conventional quantum logic, and in Part II in relation to Reichenbach's quantum logic.

5. THE LOGICAL STRUCTURE OF OL(K): FIRST FORMULATION

We have argued that the logic of the theory formulation language TL(K) of a kinematic theory K is a simple extension of classical first order logic. In the next three sections, I describe the conventional answer concerning the logical structure of the observation language OL(K), in particular OL(QM).

We have isolated certain 'extracts' in the basic formulas of TL(K) and identified them as elementary sentences in the corresponding observation language OL(K). This identification permits trivially reformulating TL(K) in such a way that the basic predicate S is a three-place predicate of type (0, 3, 4) (0 is the sort designating the elementary sentences, that is $M \times B(R)$), whose arguments therefore include an elementary sentence $P = (m, X)$ together with a state w and a probability value r. We employ P, Q as variables ranging over the class of elementary sentences. Thus formulated, the formulas of the form $S(P, w, r)$ in TL(QM) are evidently assertions about (mentioning) elementary sentences, and are accordingly metalinguistic with respect to the sentences of OL(K). As we will observe later, Reichenbach viewed a bifurcation of the language of QM of this sort as an unacceptable disunification of the language of physics, and accordingly sought to reformulate quantum theory so as to avoid it. On the other hand, it is evident that this 'disunification' of the quantum mechanical language amounts to little more than the fact that certain sentences (viz., TL(K)) are context-free, and other sentences (viz., OL(K)) are context-dependent or indexical.

We now formulate definitions of various relations holding among elementary sentences and between elementary sentences and states. First of all, we define three two-place relations – T, F, I – each of type (0, 3) as follows (note: '$=_{\text{df}}$' may be read 'is shorthand for' and holds between schematic formulas of TL(K)).[10]

(T) $\quad T(P, w) =_{\text{df}} S(P, w, 1)$

(F) $\quad F(P, w) =_{\text{df}} S(P, w, 0)$

(I) $\quad I(P, w) =_{\text{df}} \neg\, T(P, w)\, \&\, \neg\, F(P, w)\,.$

In relation to the theory formulation language TL(K), these three predicates are simply two-place descriptive predicates definable on the basis of the descriptive vocabulary of TL(K) employing the resources of first order logic. In relation to the observation language OL(K), these predicates act as *semantic predicates* True, False, Indeterminate, *all relative to W*. For example, T is the predicate 'is true at w'.

Besides these, one can define the semantic predicates ent and exc of *entailment* and *exclusion* (both of type (0, 0)) as follows.

(ent) $P \text{ ent } Q =_{\text{df}} (w)[T(P, w) \Rightarrow T(Q, w)]$

(exc) $P \text{ exc } Q =_{\text{df}} (w)[T(P, w) \Rightarrow F(Q, w)]$.

Finally, one can define the one-place predicate val (type (0)) of validity as follows.

(val) $\text{val}(P) =_{\text{df}} (w)T(P, w)$.

Each of these expressions has a simple reading. 'P ent Q' reads 'whenever P is true, so is Q'; 'P exc Q' reads 'whenever P is true, Q is false'; 'val(P)' reads 'P is always true'. Note that the temporal modalities are colloquial ways of reading the corresponding expressions employing quantifiers over possible states; for example, 'whenever' is universally quantified 'if . . . then'.

Given these specific definitions of the various semantic concepts, we see how a kinematic theory K can be interpreted as providing a formal semantics (or logic) for its associated observation language OL(K). In particular, a kinematic theory K specifies which elementary sentences are valid and which sentences entail (exclude) one another.[11]

We are particularly interested in the algebraic structure of the observation language OL(K) with respect to the two semantic relations ent and exc. This structure may be characterized by first defining a binary relation $\simeq$ on OL(K) so that $P \simeq Q$ iff P ent Q and Q ent P; one can show that $\simeq$ is an equivalence relation on OL(K). Next, a binary relation $\leqslant$ can be defined on the family $S/\simeq$ of equivalence classes of sentences so that: $[P] \leqslant [Q]$ iff P ent Q; one can show that $\leqslant$ is a partial order relation. (Note: here $[P] = \{Q \in S: P \simeq Q\}$; $S/\simeq = \{[P]: P \in S\}$.) One also shows that with respect to $\leqslant$, $S/\simeq$ is bounded above by $[(m, R)]$ and below by $[(m, \varnothing)]$; in other words, $\langle S/\simeq, \leqslant \rangle$ is a *bounded partially ordered set* (poset).

Next, one defines an orthocomplementation operation * so that $[P]^* =_{\text{df}} [P']$, where $P' = (m, R - X)$ whenever $P = (m, X)$. One can show that P exc Q iff P ent Q' iff $[P] \leq [Q]^*$; one can furthermore show that $\langle S/\simeq, \leq, ^* \rangle$ is an *orthocomplemented partially ordered set* (ortho-poset). This is about as far as we can proceed, however, without knowing more about the actual kinematic theory. In Section 7 we discuss the concrete case of quantum theory.

6. THE LOGICAL STRUCTURE OF OL(K): SECOND FORMULATION

Recall that in Section 2, we distinguished between weak and strong entailment and exclusion. Whereas in the previous section we exclusively employed the weak versions of entailment and exclusion, in the present section I wish to examine the logical structure of OL(K) in terms of strong entailment and exclusion.

We begin with a kinematic theory K and its associated observation language OL(K). Given such a theory K and given a statistical state $w \in W$, we can define a function p_w from $S(= M \times B(R))$ into $[0, 1]$ in the following way.

$$p_w(m, W) = r \quad \text{iff} \quad S(m, X, w, r) \text{ is a thesis of } K.$$

We assume that K is constructed so that p_w is well-defined. Furthermore, since K is intended to be probabilistically interpreted, we suppose for each magnitude $m \in M$ that the induced map p^m_w from $B(R)$ into $[0, 1]$ – defined so that $p^m_w(X) = p_w(m, X)$ – is a *probability measure* on $\langle R, B(R) \rangle$. Recall that a probability measure over $\langle Z, F \rangle$ – where Z is any set, and F is a Borel field (sigma-algebra) of subsets of Z – is a function p from F into $[0, 1]$ which is countably additive over disjoint sets, and which is such that $p(Z) = 1$. We call a function p_w from $M \times B(R)$ into $[0, 1]$, satisfying the condition that each p^m_w is a probability measure over $\langle R, B(R) \rangle$, a *probability assignment* on $S = M \times B(R)$. Given this definition of probability assignment, we can then construct for any kinematic theory K the associated *simple probability logic* (see my 1976c, d) as the ordered pair $\langle S, W \rangle$, where W is the class of probability assignments of the form p_w determined by K in the manner specified above.

Every simple probability logic $\langle S, W \rangle$ determines a logical structure by specifying a class of *propositions* and a relation of *probabilistic implication*, a species of strong entailment. It is customary to identify the proposition associated with (expressed by) a sentence P with the set of possible worlds (states, situations, contexts, states of affairs, etc.) which satisfy P. In this context, a possible world is an admissible valuation, and a valuation v satisfies a sentence P exactly if $v(P) = 1$. Now, the underlying assumption in individuating propositions in this way is that two sentences P, Q make the same assertion exactly if they are true under precisely the same circumstances (possible worlds), or in our terms, exactly if for all $v \in V$ $v(P) = 1$ iff $v(Q) = 1$. In the language of weak and strong entailment, two sentences P, Q are propositionally equivalent in V under this construal exactly if they weakly entail each other.

We have already remarked that weak entailment does not always coincide with strong entailment. Accordingly, in the most general non-bivalent contexts this particular construal of propositional equivalence is not entirely adequate, the more adequate construal employing not weak but strong entailment. Specifically, we define strong equivalence $\equiv$ so that $P \equiv Q$ iff P and Q strongly entail each other, which in terms of our general semantics is to say that for all $v \in V$ $v(P) = v(Q)$. Then the proposition associated with P is identified, not with a set of possible valuations, but with an equivalence class of sentences, in terms of which the logico-algebraic structure of OL(K) is defined as follows.

SD1. $P \equiv Q =_{\text{df}}$ for all $w \in W$ $w(P) = w(Q)$

SD2. $|P| =_{\text{df}} \{Q \in S: P \equiv Q\}$

SD3. $P(S, W) =_{\text{df}} \{|P|: P \in S\} = S/\equiv$

SD4. $|P| \leqslant |Q| =_{\text{df}}$ for all $w \in W$ $w(P) \leqslant w(Q)$

SD5. $|P|^* =_{\text{df}} |P'|$ where $P' = (m, R - X)$ whenever $P = (m, X)$

SD6. $1(W) =_{\text{df}} |(m, R)|$; $0(W) =_{\text{df}} |(m, \varnothing)|$

SD7. $P(m, W) =_{\text{df}} \{|P|: P = (m, X)$ for some $X \in B(R)\}$.

Terminology: we read '$P \equiv Q$' as 'P is *statistically equivalent* to Q' and '$|P| \leqslant |Q|$' as '*P probabilistically implies Q*'. $P(m, W)$ is called the *statistical propositional range* of m.

Now, supposing that for each state w and each magnitude m the induced map p_w^m is a probability measure over $\langle R, B(R)\rangle$, we can prove the following theorem concerning the statistical propositional structure

$P(S, W)$: $P(S, W)$ is partially ordered by $\leqslant$, with respect to which $1(W)$ and $0(W)$ are greatest and least elements, and with respect to which * is an orthocomplementation. Moreover, for each magnitude m the propositional range $P(m, W)$ is a Boolean (sub)lattice of $P(S, W)$. What additional structure $P(S, W)$ has is not presently determined, for example, whether it is a partial Boolean algebra, or whether it is a lattice. This requires additional information concerning the specific kinematic theory.

7. THE LOGICAL STRUCTURE OF KINEMATIC QUANTUM THEORY

The concrete case which interests us is of course quantum theory. According to conventional QM, each physical system s is associated with a separable Hilbert space H. Then, each physical magnitude m pertaining to s is associated with a self-adjoint (s.a.) operator M on H, and the statistical states are generated by the statistical operators (positive definite trace class s.a. operators) on H. Specifically, the probability assignment w associated with the statistical operator W is given as follows: $w(m, X) = \mathrm{Tr}(WE)/\mathrm{Tr}(W)$, where Tr is trace and E is short for $E(m, X)$, which is the projection operator (idempotent s.a. operator) on H uniquely associated with Borel set X in the spectral decomposition of the s.a. operator M.

These particular identifications yield the following interesting and important consequence: two elementary sentences (m, X) and (n, Y) are statistically equivalent in the simple probability logic SPL(QM) associated with kinematic quantum theory exactly if the corresponding projections $E(m, X)$ and $E(n, Y)$ are identical. Consequently, the statistical propositions of SPL(QM) are individuated in accordance with their natural 1–1 correspondence with the projections on H, which in turn are in a natural 1–1 correspondence with the subspaces (closed linear manifolds) of H. Furthermore, the probabilistic implication (strong entailment) relation $\leqslant$ is completely determined by the partial ordering among projections, equivalently the set-inclusion relation among subspaces.

$$|(m, X)| \leqslant |(n, Y)| \quad \text{iff} \quad E(m, X) \leqslant E(n, Y) \quad \text{iff}$$
$$\mathrm{SE}(m, X) \subseteq \mathrm{SE}(n, Y).$$

SE is the unique subspace of H associated with projection E. Also, the orthocomplementation operation * on $P(S, W)$ is completely determined by the function $^{\perp}$ ("perp") on the family of subspaces of H which maps each subspace S onto its orthogonal complement $S^{\perp}$.

Thus, the ortho-poset of statistical propositions of K(QM) is isomorphic to the ortho-poset of subspaces of H, equivalently the ortho-poset of projections on H. Now, the family $S(H)$ of subspaces (projections) on H is not merely an ortho-poset; $S(H)$ also has the structure of an orthomodular semi-modular (modular if H is finite-dimensional) atomic complete lattice. Furthermore, being an orthomodular lattice, $S(H)$ also has the structure of a partial Boolean algebra in the sense of Kochen and Specker (1967). Beyond its structure as a partial Boolean algebra, whether the additional mathematical structure – latticity, complete latticity, atomicity, etc. – carries any genuine physical significance is a matter of dispute in current research on the foundations of quantum theory.

We have thus far analyzed the probabilistic structure of QM exclusively in terms of the relation of strong entailment, whereas in Section 5 we observed that a logical structure also arises by considering instead the relation of weak entailment. One may therefore ask what relation this logical structure has to the structure presented in this section; the answer is that they are identical. This follows from the fact that in conventional QM weak and strong entailment (exclusion) coincide.[12]

We conclude Part I by reiterating a point made earlier. From the viewpoint of conventional quantum logic, the 'logic of QM', properly understood, is simply the 'logical' structure postulated by QM to pertain to the physical world. It is not the logic on the basis of which quantum theory is formulated, the deductive structure of quantum theory being completely classical. The claim that QM *employs* a non-standard logic is misleading, for QM employs exactly the same logic that classical mechanics employs. What is novel about QM is not *its* logic, which is quite pedestrian, but the world's 'logic' as described by QM.[13]

Even the term 'quantum logic' is misleading and tends to obscure both the important similarities and the important differences between QM and its classical counterpart. One of the first writers to complain about the expression 'quantum logic', as chosen by Birkhoff and von Neumann, was Hermann Weyl (1940) who suggested the alternative expression 'quantum geometry' (cf. Varadarajan, 1968). Along the same lines I would like to suggest that the area of physics investigated by conventional quantum logic be called *proto-kinematics*, since in particular conventional quantum logic is currently regarded as the structure underlying quantum kinematics (cf., e.g., Jauch, 1968 and Stein, 1972). It might be noted that this term is completely in keeping with Weyl's suggestion, since kinematics has traditionally been regarded as a sort of descriptive geometry.

II. AN EXAMINATION OF REICHENBACH'S PROPOSED LOGIC FOR QUANTUM MECHANICS

In Part I, I argued that conventional (mainstream) quantum logic is not correctly understood as the *formulation logic* of QM, regarded as a deductively closed system of empirical postulates, but should be understood as the 'logical' ('geometric' or proto-kinematical) structure postulated by QM to pertain to the physical world. When we consider the quantum logic proposed by Reichenbach, we are presented with a logic prima facie of an entirely different category. Specifically, whereas mainstream quantum logic (MQL) is not intended to replace classical logic as the formulation logic of QM, but rather represents a particular analysis of the *descriptive* postulates of quantum theory, it appears that Reichenbach's quantum logic (RQL) was *intended* to form the logico-linguistic basis of an alternative *formulation* of quantum theory.

In Part II, I examine certain features of the logic proposed by Reichenbach as the basis for the reformulation of quantum theory, assessing the success achieved (achievable) in such a program. The first two sections concern two objections raised by Reichenbach against the Bohr-Heisenberg interpretation of QM(BHI). The first objection, reiterated by numerous commentators – that BHI severely restricts the language of QM, that it involves a basic formal awkwardness – is seen to be founded on a misunderstanding of the expression 'meaningless' as it appears in BHI. The second objection – that the laws of QM, for example, the law of complementarity, must be stated in the metalanguage and not the object language of quantum theory – is seen to amount to the 'objection' that the laws of QM are formulated in TL(QM) rather than OL(QM).

Since Reichenbach apparently regarded as unacceptable the bifurcation of the language of QM into TL(QM) and OL(QM), we review Reichenbach's proposed program to reformulate QM in a unified language – denoted RL(QM) – consisting of OL(QM) augmented by a collection of three-valued truth-functional connectives. We consider two objections to this program. First, we consider Hempel's (1945) objection that Reichenbach's translation of the semantic expressions 'is true', etc. into the object language is flawed inasmuch as the proposed object language expressions are not semantically (strongly) equivalent to the metalinguistic expressions for which they are intended to serve as translations. This objection is easily circumvented, however, by reformulating RQL and the proposed translation function. The second objection, raised by van Fraassen (1974), is more telling, however; it maintains that the law of complementarity (or any law of physics) cannot be adequately formulated in Reichenbach's object language RL(QM). In order to circumvent van Fraassen's objection, the reformulation of the law of complementarity must occur at the level of TL(QM) rather than OL(QM) or even RL(QM).

But this formulation of the law of complementarity is subject to exactly the same objection raised by Reichenbach against BHI, the objection that it formulates a law of physics in the metalanguage rather than in the object language. However, this objection is seen to be insubstantial, being based on a misunderstanding concerning exactly what constitutes the object language of quantum theory. If we understand quantum theory as a body of law-like assertions, then the object language of quantum theory is the language in which these assertions are formulated, which is TL(QM) not OL(QM). Neither OL(QM) nor RL(QM) is capable of formulating physical laws, but only indexical observation statements. Having reformulated the law of complementarity in TL(QM) so as to circumvent van Fraassen's objection, we see that the deployment of multi-valued connectives is superfluous at least to the formulation of the *laws* of QM, if not to the formulation of observation reports. Although the observation language of QM might be regarded as displaying a non-classical logical structure, as for example investigated in Part I, there are no grounds to suppose that there is any need to implement non-classical logic at the level of the formulation language.

8. REICHENBACH VERSUS BOHR AND HEISENBERG: 1

Motivating Reichenbach's development of a three-valued logic for quantum theory is the idea that on the Bohr-Heisenberg interpretation of

QM(BHI), as Reichenbach construes it, because of 'in-principle' limitations of measurement imposed by QM, certain observation statements are meaningless under certain circumstances. For example, whenever any sentence (p, r) ascribing an exact value of momentum to a system s is true, any sentence (q, r) ascribing to s an exact value of position is meaningless, and conversely.

This formulation of the law of complementarity was objectionable to Reichenbach and later commentators, who understand BHI to be an incoherent position so formulated because it makes syntactic features of the observation language of QM peculiarly dependent on contingent empirical circumstances. In PhF (1944) Reichenbach writes the following (p. 143).

> The law [of complementarity] can be stated only by reference to a class of linguistic expressions which includes both meaningful and meaningless expressions; with this law, therefore, meaningless expressions are included, in a certain sense, in the language of physics.
>
> The latter fact is also illustrated by the following consideration. Let $U(t)$ be the propositional function 'the entity has the value u at time t'. Whether $U(t)$ is meaningful at a given time t depends on whether a measurement m_u is made at that time. We therefore have in this interpretation propositional functions which are meaningful for some values of the variable t, and meaningless for other values of t.

In his review of PhF, Hempel comments on this passage as follows (1945, p. 98): "Thus, the syntactical rules of sentence formation for the language of quantum mechanics have here to refer to certain empirical criteria, a situation which seems highly undesirable." Similarly, Gardner (1972, p. 93) in comparing Bohr and Reichenbach writes that BHI "embodies a certain formal awkwardness – specifically, that well-formedness is not a purely syntactic property [– which] makes it impossible to formulate Bohr's theory as anything remotely resembling a formal system with a recursive class of well-formed formulas." Haack (1974, p. 150) makes a completely analogous remark, objecting to the cumbersomeness of BHI, "since it makes contingent information about what measurements have been made relevant to whether or not an expression counts as well-formed." Finally, in this volume, Nilson similarly interprets the expression 'meaningless' in BHI: "Once one opts for 'an interpretation by restricted meaning', the resulting language contains only a subset of the set of statements made available with exhaustive interpretations.

... statements about those unobserved objects or events termed 'inter-phenomena' are rendered meaningless and not properly in the language of microphysics ..."

It thus seems that all these writers construe the term 'meaningless' in BHI, not as a semantic predicate analogous to 'true' and 'false', but as a syntactic predicate, something like 'ill-formed'. Now, in philosophical circles at least, a meaningless expression E of a language L is meaningless simpliciter, which is to say that E is not even a potential argument term of any semantic assignment of L (recall Section 2). But clearly, Bohr and Heisenberg do not have this in mind in employing the term 'meaningless', for they obviously countenance the same *linguistic expression* – for example, 'momentum has value r' – being meaningful (= true or false) under certain circumstances, but at the same time being meaningless (= neither true nor false) under others. In other words, the expression 'meaningless' does not play a syntactic role, but a semantic role; in particular, 'meaningless' does not mean 'syntactically ill-formed' in BHI.[14]

What is meaningless is not an otherwise well-formed *expression* (*sentence*)'momentum has value r', but rather a contextually conditioned *utterance* (*statement*) roughly like '*this* system's momentum has value r *now*' (see Section 4 and below concerning indexicals).[15] According to BHI, whether the particular utterance is meaningful may be dependent on what *this* system is and what is additionally going on *now*. BHI should not be understood as advocating as suggested by Reichenbach (PhF, p. 41), removing any linguistic expressions from the *language* of physics. What is 'removed' is not the expression 'momentum has value r' but rather the corresponding indexical utterance, and this only relative to particular circumstances.

The envisioned circumstance is not particularly unusual, since an indexical expression E might generally be expected not to have a well-defined denotation relative to every index point i. For example, E might be the definite description 'this year's volume of *Erkenntnis*', where the index is obviously temporal, or E might be a sentence, and the presupposition (cf. van Fraassen, 1968, 1970) on which E is predicated might fail to obtain at a given index point. From the viewpoint of formal semantics, more particularly formal pragmatics (see Montague, 1968, 1970 and Lewis, 1970), the indexical features of all linguistic expressions are formally characterized in the metalanguage by an ordered tuple of 'contextual coordinates', including indices for time, space, speaker, audience, etc. Formally the entire tuple can be lumped together and regarded as a (pragmatic) 'possible world' or context. In any case, to say that a sentential expression P may give rise to a meaningless statement is merely to say that relative to certain relevant indices, the truth conditions of P are not well-defined. Thus, it seems natural to interpret the expression 'meaningless' in BHI as a semantic predicate completely on a par with 'true' and 'false'. Specifically, according to BHI, relative to every 'possible world' (physically possible state of affairs), certain observation sentences are true, others are false, and still others are neither true nor false, that is, they are 'meaningless'. Once we see that 'meaningless' is a semantic predicate completely on a par with 'true' and 'false', it is easy to dispose of the above-mentioned criticisms of BHI. For example, it is no more mysterious that $U(t)$ can be meaningless for certain values of t (obviously an indexical coordinate in modern parlance) but meaningful for others, than it is mysterious that $U(t)$ can be true for some values of t and false for others.

Nevertheless, even granting that 'meaningless' is a semantic predicate, we cannot simply go on to say that 'meaningless' and the semantic expression 'indeterminate' introduced by Reichenbach are completely equivalent. The difference, however, is not between a syntactic and semantic predicate, but between a semantic expression (indeterminate) which with respect to compound sentential formulas behaves in a completely determinate and straightforward (i.e., truth-functional) manner, and a semantic expression (meaningless) whose semantic characteristics may be exceedingly involved. For example, meaningful components may form meaningless compounds, or meaningless components may form meaningful compounds under certain circumstances and meaningless compounds under others.

A plausible explication of the term 'meaningless', as it appears in the Bohr-Heisenberg interpretation, construes it first of all as serving as a merely notational placeholder for 'neither true nor false', and secondly as serving to warn us that 'meaningless' is strictly forbidden to serve as an argument in any (classical) truth function. Certain of Bohr's remarks may be interpreted as supporting this construal of BHI; for example in (1948, p. 317), Bohr writes as follows.

... one may strongly advocate limitation of the use of the word *phenomenon* to refer exclusively to observations obtained under specified circumstances, including an account of the whole experiment .... Incidentally, it would seem that the recourse to three-valued logic, sometimes proposed as means for dealing with the paradoxical features of quantum theory, is not suited to give a clearer account of the situation, since all well-defined experimental evidence, even if it cannot be analyzed in terms of classical physics, must be expressed in ordinary language making use of common logic.

We may read this as follows: in each particular circumstance, if we are careful to restrict ourselves to phenomena as defined by quantum theory, that is, to statements which are meaningful or to sentences which are either true or false, then there is simply no occasion to employ non-standard logic. All truth-functional compounds of meaningful sentences are likewise meaningful, their truth values being completely determined by classical sentential logic.[16]

This idea corresponds to what I have previously (1976a) called *quasi-truth-functionality*. A semantics V is quasi-truth-functional exactly if for any two valuations $v, v' \in V$, if v and v' agree and assign either T or F to every atomic formula p in Δ, then they also agree and assign either T or F to every compound formed out of elements of Δ. It may be noted that conventional quantum logic, construed in accordance with the Copenhagen modal interpretation, is quasi-truth-functional (see my, 1975, 1976a). Thus, the difference between BHI and RQL is that, whereas Bohr and Heisenberg employ the semantic value 'meaningless' and opt for classical logic and quasi-truth-functionality. Reichenbach pursues a different line of attack, employing the semantic value 'indeterminate' and opting for multi-valued logic and complete truth-functionality. The resulting differences between BHI and RQL are important; they are not equivalent manners of speaking (see, e.g., Pauli, 1964), for in particular, they treat compounds of non-true-false sentences from entirely different points of view.

At the same time, no fundamental advantage of RQL over BHI exists, notwithstanding the remarks of Reichenbach, who maintains that the language of QM is significantly richer under the RQL interpretation. However, this claim is based on the misunderstanding exposed earlier, that calling a *statement* meaningless relative to certain circumstances is tantamount to excising the corresponding linguistic expression from the associated language. At least relative to the class of atomic sentences of the form (m, X), the observation language of QM according to BHI and according to RQL are identical (see Section 9 concerning differences with regard to molecular sentences). The key difference is not the syntax, but the semantics which is reflected in the logics respectively advocated by Bohr-Heisenberg and Reichenbach.

9. REICHENBACH VERSUS BOHR AND HEISENBERG: 2

We have disposed of the initial objection against BHI raised by Reichenbach and reiterated by commentators, the objection that BHI severely restricts the language of QM and renders syntactic features of the language dependent on empirical circumstances. Reichenbach's second major criticism of BHI maintains that, according to BHI many of the theorems of QM, for example, the law of complementarity, must be stated in the metalanguage of QM rather than in the object language. According to Reichenbach (PhF, p. 143), "This is unsatisfactory, since, usually, physical laws are expressed in the object language, not in the metalanguage." Let us first examine the exact nature of Reichenbach's accusation, after which we will examine the alternative object language formulation of physical laws such as complementarity proposed by Reichenbach.

Recall the distinction (Sections 3 and 4) between the theory formulation language TL(QM), which is generated from the class of basic formulas of the form $S(m, X, w, r)$, and the corresponding observation language OL(QM), which consists of elementary (indexical) sentences of the form (m, X). We have already remarked that with respect to the observation language OL(QM) the sentences of TL(QM) are metalinguistic. For example, the sentence $S(m, X, w, r)$ mentions (m, X), stating that its probability in state w is r. Also recall the derived predicate T of TL(QM): $T(P, w) =_{\mathrm{df}} S(P, w, 1)$. Now, the law of complementarity between magnitudes p and q may be regarded (cf. PhF, p. 143) as asserting that whenever any sentence (p, r) ascribing an exact value to system s is true, any sentence (q, r) ascribing an exact value of q to s is meaningless. We can write this in terms of TL(QM) as follows.[17]

$$\text{(Comp)} \quad (w)[(\exists r)T(p, r, w) \Rightarrow \neg\, (\exists r)T(q, r, w)]\,.$$

We translate 'meaningless' so that the sentence (p, r_0) – r_0 being fixed – is meaningless at state w exactly if the open formula (p, r) – r being free – is not satisfied by *w* relative to any value of *r*. Ignoring the complexities involved in dealing with physical magnitudes, such as position and momentum, which do not have pure discrete spectra, we suppose that (Comp) is a non-trivial theorem of kinematic quantum theory, whenever *p* and *q* are complementary.

Concerning Reichenbach's objection, it is evident that since (Comp) appears in the theory formulation language TL(QM), it is metalinguistic in relation to the observation language OL(QM); it states in effect that whenever (p, r) is true for any value of *r*, (q, s) is not true for any value of *s* (note: as usual, 'whenever' is universally quantified 'if . . . then'). Being a (context-free) physical law statement and not a (context-dependent) observation statement or compound of such, the law of complementarity is formulated in the theory formulation language and not in the observation language. This is the precise sense in which the law of complementarity, or any physical law, is formulated in the 'metalanguage' rather than in the 'object language'. Nevertheless, this bifurcation of the language of QM into TL(QM) and OL(QM) appears to have been objectionable to Reichenbach, so let us examine how he proposed to replace these categorically distinct languages by a unified language for QM.

Although Reichenbach does not explicitly display his syntax apart from the various sentential connectives, it is evident that the basic (atomic) sentences of his language RL(QM) are precisely the basic sentences of OL(QM), although Reichenbach concentrates on the special sentences of the form (m, r). The chief difference between RL(QM) and OL(QM) is that Reichenbach's language employs a host of three-valued truth-functional connectives, and consequently has myriads of proper molecular formulas, whereas the sentences of OL(QM) are all atomic, having the form (m, X) (see my 1975b). (This difference was first remarked by van Fraassen (1974a) in his comparison of RQL with MQL.) Specifically, RL(QM) is equipped with conjunction, disjunction, three negations, three conditionals, and two biconditionals, which are explicitly displayed (PhF, p. 151).

Actually, not all of these connectives are required in providing a minimal explication of RQL; furthermore, Reichenbach provides little or no motivation for introducing his non-normal connectives: cyclical negation, complete negation, and quasi-implication. As defined by Rescher (1969, p. 196) a connective is normal exactly if its associated truth function yields a classical truth value (T or F) whenever all its arguments are classical truth values; otherwise it is non-normal. The idea that three-valued logic merely extends classical two-valued logic depends crucially upon restricting oneself to normal connectives. I accordingly wish to recast RQL in purely normal terms, thereby eliminating a number of Reichenbach's connectives. In place of all Reichenbach's explicitly defined connectives, I propose to employ exactly two primitive independent connectives: Łukasiewicz negation ($-$) and Łukasiewicz conditional ($\supset$), which correspond to Reichenbach's 'diametrical negation' and 'standard implication'. They are semantically characterized by the following truth tables ('I' stands for 'indeterminate').

$\supset$	T	I	F	$-$
T	T	I	F	F
I	T	T	I	I
F	T	T	T	T

Every connective I wish to employ is definable in terms of these two; since both $\supset$ and $-$ are normal connectives, every such connective is likewise normal. The most important ones are defined syncategorimatically as follows; the corresponding truth tables follow.

(exclusion negation) $\neg A =_{df} (A \supset -A)$

(disjunction) $(A \vee B) =_{df} ((A \supset B) \supset B)$

(conjunction) $(A \wedge B) =_{df} -(-A \vee -B)$

(assertion) $\mathbf{T}A =_{df} -(\neg A)$

(denial) $\mathbf{F}\mathrm{A} =_{df} \mathbf{T}(-A)$

(indeterminacy) $\mathbf{I}A =_{df} (-\mathbf{T}A \wedge -\mathbf{F}A)$

$\vee$	T	I	F
T	T	T	T
I	T	I	I
F	T	I	F

$\wedge$	T	I	F
T	T	I	F
I	I	I	F
F	F	F	F

A	$\neg A$	$\mathbf{T}A$	$\mathbf{F}A$	$\mathbf{I}A$
T	F	T	F	F
I	T	F	F	T .
F	T	F	T	F

The three operators **T**, **F**, **I** may be regarded intuitively as the object language counterparts of the semantic predicates 'true', 'false', and 'indeterminate'. To be precise, from the vantage point of the meta-metalanguage, the expression 'p is true' (resp., 'p is false' and 'p is indeterminate') in the metalanguage (presumed to be classical; cf. PhF, p. 150) and the expression '$\mathbf{T}p$' (resp., '$\mathbf{F}p$' and '$\mathbf{I}p$') in the object language have exactly the same truth conditions; they are true or false under exactly the same circumstances. Also, concerning indexicality, these expressions are completely on a par; for the actual expression occurring in the metalanguage is not 'p is true (simpliciter)' but 'p is true at w'. Thus, the expression 'p is true' should be understood, not as a *sentence* of the metalanguage, but as an *open formula*. In particular, the object language expressions '$\mathbf{T}p$', '$\mathbf{F}p$', and '$\mathbf{I}p$' respectively correspond to the metalinguistic expressions '$T(p, x)$', '$F(p, x)$', and '$I(p, x)$', where x is a free variable ranging over the index set of possible states of affairs. This must constantly be kept in mind translating between the metalanguage and the object language of RL(QM).

10. REICHENBACH VERSUS HEMPEL

Reichenbach is not entirely clear as to what can and what cannot be translated between the object language and metalanguage of RL(QM), nor is he clear concerning which particular object language expressions correspond to which particular metalinguistic expressions. For example, on page 153 of PhF he writes the following:

Our truth values are so defined that only a statement having the value T can be asserted. When we wish to state that a statement has a truth value other than T, this can be done by means of negations. Thus the assertion

$$\sim\sim A \qquad (2)$$

states that A is indeterminate. Similarly, either one of the assertions

$$\sim A \qquad -A \tag{3}$$

states that A is false.

This use of negations enables us to eliminate statements in the metalanguage about truth values. Thus the statement of the object language *next-next-A* takes the place of the semantical statement 'A is indeterminate'. Similarly, the statement of the metalanguage 'A is false' is translated into one of the statements (3) of the object language.

Note: Reichenbach's connective $\sim$ is called 'cyclical negation', which is associated with the truth function which maps (F, I, T) onto (T, F, I) (i.e., which cyclically permutes the truth values).

The above remarks of Reichenbach prompted criticism by Hempel (1945) who argues that Reichenbach's translation of the metalinguistic predicates 'is true', etc., is inadequate because, in the case of non-bivalent logics, it is not correct to say that two sentences make the same assertion or express the same proposition merely on the grounds that they are both true under the same circumstances (recall discussion in Section 6). Specifically, Hempel writes as follows (p. 99).

A third difficulty of interpretation arises in connection with the author's method of expressing assertions about the truth-value of a sentence of a language by a statement in that language itself. Consider, for example, Reichenbach's assertion that '$\sim A$' as well as '$-A$' state that 'A' has the truth-value F. If two sentences state the same fact, then it would seem that they must have identical truth-tables. This, however, fails to be the case in the example cited as well as in many analogous cases.

In accordance with ideas expressed by Wittgenstein and Carnap, Hempel goes on to note the following (p. 99).

. . . a statement in two-valued logic divides all possible states of affairs into two classes: the class of those which make the statement true – these constitute its range – and the complement of this class. In order to know that two statements make the same assertion – i.e., that they determine the same dichotomy of states – it suffices therefore to know that their ranges are identical. . . . But this is not sufficient in L_3. For here, a sentence determines a trichotomy of all states, and in order to show that two sentences make the same assertion, it is necessary to ascertain that they establish the same trichotomy . . . the fact that two sentences in L_3 have the truth value T in exactly the same cases is not sufficient to show that they make the same assertion.

In light of these remarks, we see that in order for any object language expression E^* to be the precise counterpart of the metalinguistic expression E, since the metalanguage is classical or two-valued (PhF, p. 150), E^* must be an expression of the true-false category, for E is of that category. However, none of Reichenbach's object language expressions mentioned above – 'A', '$-A$', '$\sim A$', '$\sim\sim A$' – are of this category, and so none of them can perform the task required of them. On the other hand, it is clear that the expressions '$\mathbf{T}p$', '$\mathbf{F}p$', and '$\mathbf{I}p$' are secure against Hempel's criticism; these expressions are all true-false sentences, and are furthermore in exact accord with their associated semantic formulas 'p is true', 'p is false', and 'p is indeterminate'. Thus, in light of Hempel's objection to RQL, we henceforth employ the expressions '$\mathbf{T}p$', '$\mathbf{F}p$', and '$\mathbf{I}p$' in place of Reichenbach's 'p', '$\sim p$', and '$\sim\sim p$'.

Having reformulated RQL so as to avoid non-normal connectives, and so as to circumvent the objection raised by Hempel mentioned above, we now consider the manner in which Reichenbach proposed to reconstruct quantum theory entirely in the object (observation) language. Reichenbach's chief illustration of his general approach is the law of complementarity, whose semantic formulation he construed as follows (PhF, p. 157): "We call two statements *complementary* if they satisfy the relation

(28) $A \vee \sim A \rightarrow \sim\sim B$

...[which]...can be read: If A is true or false, B is indeterminate." Note: $\sim$ is cyclical negation as above; $\rightarrow$ is 'alternative implication'. Employing the resources of our reformulation of RQL, we can directly translate the metalinguistic expression from which (28) derives as follows (changing to lower case letters).

(28a) $(\mathbf{T}a \vee \mathbf{F}a) \supset \mathbf{I}b$

(RC1) is straightforwardly equivalent to the following.

(28b) $(\mathbf{I}a \vee \mathbf{I}b)$.

The reader may easily verify that (28), (28a), and (28b) all have exactly the same truth tables.

11. REICHENBACH VERSUS VAN FRAASSEN

That none of these object language expressions can adequately formulate the notion of complementarity was pointed out by van Fraassen (1974), who writes that (p. 233), "... the semantic relation of incompatibility is not adequately expressed by any three-valued truth-functional connective . . ." Fundamental to van Fraassen's criticism is the simple fact that the formula (28) – equivalently, the formulas (28a) and (28b) – being a truth-functional compound of observation sentences is itself of precisely the same category. Specifically, it is a context-dependent or indexical sentence, which is true at an index point i exactly if A is indeterminate at i or B is indeterminate at i, and false otherwise. On the basis of (28), we can conclude, for example, that whenever a sentence is indeterminate in truth value it is incompatible with itself!

What happened in the translation? First of all, we began with a misleading formulation of the metalinguistic statement of complementarity. It is misleading because it suppresses the fact that the semantic predicates 'true', 'false', and 'indeterminate', are two-place predicates in the metalanguage, such that the expression '$p$is true' is merely shorthand for the open formula 'p is true at i', where i ranges over the possible indices or possible states of affairs. Therefore, the original formulation – 'If A is true or false, B is indeterminate' – surreptitiously eludes the logical requirement that in order for the rule of complementarity to make any statement whatsoever, the implicit variable must be bound, in this case by a universal quantifier. The formulation of complementarity should have originally been stated as follows: *Whenever* A is true or false, B is indeterminate. As usual, 'whenever' is universally quantified 'if . . . then', where the variable of quantification ranges over possible indices or possible states.

By way of illustration of the contrast between 'whenever' and simple 'if . . . then', let us consider an example from rudimentary meteorology. Specifically, let us consider the simple observation sentences 'it is raining' and 'it is pouring'. With these two elementary sentences, we can construct the truth-functional compound 'if it is raining, then it is pouring', which might be employed by someone who hears a thundering noise on the roof, but is unable to look out the window. In any case, it is an observation sentence just like 'it is raining' and 'it is pouring', and accordingly is an indexical expression with at least a spatial and temporal index. Now consider instead the sentence 'in Spokane, whenever it rains, it pours'. This latter sentence is of an entirely different character from 'if it is raining, then it is pouring', since in particular all the indexical slots are filled appropriately, and is accordingly at least of the correct category to be a law statement (candidate) of meteorology. This must be contrasted with the observation statement 'if it is raining, then it is pouring', which is not even a potential candidate to be a law statement.

Let us now consider a second illustration, the relation of semantic entailment, which is completely analogous to the relation of complementarity. To say that P entails Q is to say that whenever P is true so is Q. Unravelling the suppressed quantifier in the colloquial 'whenever', this should be read as follows: for all $v \in V$, $v(P) = 1$ only if $v(Q) = 1$; alternatively, for all $i \in I$, P is true at i only if Q is true at i. We cannot simply read semantic entailment as: if P is true, then Q is true, for whereas the former expression 'P entails Q' is context-free (the implicit indexical variable is bound), the latter expression is not.

Now, consider once again the semantic relation of complementarity, which is completely analogous to semantic entailment. Correctly stated, two sentences P and Q are complementary exactly if whenever P is true or false, Q is indeterminate, equivalently, exactly if under all possible circumstances either P is indeterminate or Q is indeterminate. In order to obtain the object language formulation of complementarity provided by Reichenbach in the form of (28), reformulated as (28a) and (28b), it is sufficient merely to change 'whenever' to 'if . . . then' and then translate the resulting expression, which is basically the open formula 'if A is true or false at i, then B is indeterminate at i', into the corresponding object language expression '$(\mathbf{T}a \vee \mathbf{F}a) \supset \mathbf{I}b$'. As remarked by van Fraassen (1974a, p. 233), Reichenbach's object language formulation of complementarity stands to the semantic relation of complementarity as the material conditional stands to the semantic relation of entailment.

12. REICHENBACH'S SECOND OBJECTION TO BHI RECONSIDERED

Reichenbach evidently recognized the crucial need to employ quantifiers in formulating the rule of complementarity, so as to avoid the sort of criticism raised by van Fraassen. On page 159 of PhF, he reformulates complementarity as follows.

(37) $(u)(v)(t)[(v1(U, t) = u) \vee \sim(v1(U, t) = u) \rightarrow \sim\sim(v1(V, t) = v)]$.

Reformulating this in accordance with our notation, we obtain the following.

(37a) $(r)(s)(t)[\mathbf{T}(u, r, t) \vee \mathbf{F}(u, r, t). \supset \mathbf{I}(v, s, t)]$

Note carefully that **T**, **F**, and **I** are here intended to be sentential operators, not semantic predicates, and apply to sentences of the form (u, r, t) to form compound sentences at the same linguistic level.

As noted earlier, time may be regarded simply as a parameterization of the physically possible states. Therefore, formulas (37) and (37a) should not be understood as quantifying directly over time, that is, the actual historical moments of the universe. Rather, they should be understood as implicitly quantifying over quantum mechanically possible states, which yields the following reformulation of complementarity.

(37b) $(r)(s)(w)[\mathbf{T}(u, r, w) \vee \mathbf{F}(u, r, w). \supset \mathbf{I}(v, s, w)]$

But observe that in order to introduce the variable over possible states, in virtue of our analysis of indexical coordinates, we must ascend to the metalanguage. In this case, (37b) should be written as a statement in the theory formulation language TL(QM), in which case **T**, **F**, and **I** are understood as metalinguistic semantic predicates rather than object language operators. (37b) is thus rewritten as follows.

(37c) $(r)(s)(w)[T(u, r, w) \textit{ or } F(u, r, w). \Rightarrow I(v, r, w)]$.

With formula (37c) we have formulated the rule of complementarity so as to circumvent van Fraassen's objection, and it is also fairly clear that Reichenbach had such a formulation in mind. On the other hand, it is clear that (37c) is completely on a par with the Bohr-Heisenberg formulation of complementarity given earlier by formula (Comp). This formulation is therefore subject to the same objection what Reichenbach raised against BHI, the objection that it states a law of physics, not in the object language, but in the metalanguage.

But is this particular objection to BHI substantial? I think not. For ultimately we must ask: exactly what is the object language of quantum theory? If we follow Reichenbach, then the object language is simply the observation language OL(QM) augmented by a variety of three-valued connectives. If this is the case, then no law of physics can be adequately formulated in the object language, for every sentence occurring in

OL(QM), as well as every truth-functional compound, is an indexical expression. Accordingly, no such expression is even of the correct category to be a law of physics. Equivalently stated, any putative law statement formulated in the observation language is in principle subject to the van Fraassen objection to Reichenbach's formulation of the law of complementarity.

On the other hand, we noted in Part I that the language of kinematic quantum theory is TL(QM), whereas OL(QM) is merely a theoretical extract from TL(QM) employed for the purpose of heuristics. In other words, if we must identify either TL(QM) or OL(QM) as the unique object language of quantum theory, we must choose the language in which the theory as a body of empirical claims is formulated, which is TL(QM). Given this identification of the object language of QM, we see that the laws of QM, including the law of complementarity as formulated by Bohr and Heisenberg, is not in the metalanguage of QM. Thus Reichenbach's objection is without substance.

I now wish to consider the adequacy of formula (37c), as well as the earlier formula (Comp), to express the notions of incompatibility and complementarity. Recall that we defined the semantic predicates T, F, and I in TL(QM) so that: $T(m, r, w) =_{\text{df}} S(m, r, w, 1)$; $F(m, r, w) =_{\text{df}} S(m, r, w, 0)$; $I(m, r, w) =_{\text{df}} \neg T(m, r, w) \,\&\, \neg F(m, r, w)$. Under this construal of truth and falsity, which is the conventional quantum logical construal, (37c) does not correctly formulate the relation of complementarity. Consider two magnitudes p and q, both with pure discrete spectra, and suppose that they are completely incompatible, which is to say that the corresponding operators share no common eigenvectors. Then it seems that p and q are complementary; however, under the construal of truth and falsity, two elementary sentences (p, r) and (q, s) can be simultaneously false, which is to say that they can simultaneously receive probability 0 in some state (see my (1976b)). Thus, they are not complementary according to (37c).

However, this is not the way that Reichenbach construed falsity in RQL; a plausible explication of Reichenbach's notion of falsity in RQL has been proposed by van Fraassen (1974). We restrict ourselves to pure statistical states, which can be represented by vectors in H, and we also restrict ourselves to magnitudes which have pure discrete spectra; then van Fraassen's explication is given as follows (M is the s.a. operator associated with magnitude m; x is regarded as a vector in H).[18]

(T) (m, r) is true at x iff $mx = rx$

(F) (m, r) is false at x iff $Mx = r'x$ for some $r' \neq r$

(I) (m, r) is indeterminate at x iff $Mx \neq rx$ for all x

Thus in RQL, just as in conventional quantum logic, (m, r) is true at state w exactly if the probability of (m, r) is 1. On the other hand, whereas in conventional quantum logic falsity is identified with zero probability, in RQL (m, r) is false at w exactly if (m, r') is true at w for some $r' \neq r$. Consequently, the sentence (m, r) is true or false at state w just in case w is an eigenstate of the observable m (presuming pure discrete spectra); otherwise it is indeterminate. We may thus reformulate (37c) in accordance with this construal of truth and falsity as follows.

(37d) $(w)[(\exists r)T(u, r, w) \Rightarrow \neg(\exists r)T(v, r, w)]$

It may be noted that (37d) is identical to the earlier formula (Comp), which represents the BHI formulation of the law of complementarity.[19]

It is also important to note that formula (37d) or (Comp) occurs in the theory formulation language TL(QM), and as we have previously emphasized, the logic appropriate to TL(QM) is an extension of classical first order logic. Thus, there is no need to formulate quantum theory in a language with a trivalent semantics; what can be expressed in RL(QM) is more perspicuously rendered in TL(QM) – compare (37) and (37d). Of course, deciding the logic of TL(QM) does not settle the logic appropriate to OL(QM).

13. CONCLUDING REMARKS

There seem to be (at least) two major viewpoints concerning the nature of quantum logic. The first is represented in the passage from van Fraassen quoted earlier (Section 3), which claims that quantum logic exclusively concerns the semantic relations holding among the elementary sentences of QM, or in our terminology, it exclusively pertains to OL(QM). The second major viewpoint is described by Jammer in the following passage (1974, p. 341).

> Certain developments in mathematics and philosophy, however, have led to the idea that . . . a fourth component . . . could also be the object of inquiry in the search for an interpretation: the formal structure of the deductive reasoning applied in formulating *T*. If a certain theory *T* leads to an impasse, it was claimed, it is not necessarily its mathematical formalism as such nor the meaning of its extralogical concepts that may have to be modified; it may equally well be the logic underlying the formulation of *T* which has to be revised.

In this article I have suggested that neither of these general descriptions of quantum logic adequately accounts for all proposed quantum logics. In Part I, I argued that van Fraassen's description correctly accounts for mainstream quantum logic, but in Part II, I have argued that his account does not adequately subsume Reichenbach's quantum logic, which seems to fit more readily into Jammer's scheme. I have interpreted Reichenbach's program as primarily an attempt to unify the two languages TL(QM) and OL(QM) into a language RL(QM) with a trivalent sementics. However, I have pointed out that, due to inherent limitations of RL(QM), this program (as we have described it) is unsuccessful.

Although it appears that RL(QM) is inadequate to formulate the *law statements* of quantum theory, this does not necessarily mean that RL(QM) and RQL do not adequately describe certain features of the observation language relevant to QM. Specifically, we may restructure the observation language of QM so that it includes not only the syntactic atomic formulas of OL(QM), but also includes proper syntactic molecular formulas, formed by the application of three-valued connectives. Van Fraassen noted the possibility of combining logics such as MQL and RQL in (1974, p. 245), and he also noted in (1973, p. 99) the limitations of MQL in distinguishing between mixtures and superpositions. I propose that these ideas may be combined in the following way: whereas the MQL disjunction is to be employed in speaking of superpositions (for example, the interference pattern in the two-slit experiment), the RQL disjunction is to be employed in speaking of mixtures (for example, the additive pattern in the two-slit experiment). This furthermore suggests that, as an analysis of the observation language relevant to QM, one need not reject RQL merely on the basis of accepting MQL; they do not necessarily exclude one another. However, this approach to the general problems of quantum logic requires an investigation of its own, which we leave for future work.

Indiana University

NOTES

[1] Major contributors in this area include Mackey (1963), Jauch (1968, 1974), Piron (1964, 1972), and Varadarajan (1968).

[2] For a detailed sourcebook on the development of the quantum logic approach to QM, C. A. Hooker's anthology (1975) is highly recommended (a second volume on current research is in press). Also recommended is Jammer (1974), Chapter 8.

[3] Other logics for QM include Destouches-Fevrier (1951) and von Weizsacker (1955).

[4] See Jammer (1974, p. 341) and the passage quoted in Section 13.

[5] Given the definitions of weak and strong entailment, one can define the analogous notions of weak and strong equivalence; specifically, P and Q are strongly (weakly) equivalent just in case P and Q strongly (weakly) entail each other.

[6] We are using the notion of listing somewhat loosely here, since the number of basic postulates is expected to be non-denumerable.

[7] One way to interpret the expression (m, X) as a fullfledged sentence is to regard it as an indexical sentence with an indexical slot for the physical system in addition to the usual indexical slots for time, place, etc. See note 8.

[8] The investigation of sentences (expressions) whose truth or falsity (denotation) must be referred to a particular context of use is customarily referred to as pragmatics. For example, Montague (1970, p. 68) writes that, "... [Bar-Hillel suggested] that pragmatics [should] concern itself with what C. S. Peirce has in the last century called *indexical expressions*. An indexical word or sentence is one of which the reference cannot be determined without knowledge of the context of use ...". Montague goes on to say that formal pragmatics should follow the lead of formal semantics in which the key notions are truth and satisfaction in a model or interpretation; he writes, "pragmatics, then, should employ similar notions, though here we should speak about truth and satisfaction with respect not only to an interpretation but also to a context of use." It may be noted that the observation language of a kinematic theory may be regarded as an interpreted pragmatic language in the sense of Montague (1968, 1970), where the kinematically possible states comprise the set of index points. It may also be noted that Reichenbach in (1947) pioneered a pragmatic theory of indexical expressions – calling then 'token reflexive expressions' – in particular the class of tensed sentences.

[9] In this particular context I use 'whenever' as universally quantified 'if ... then', where the variable of quantification ranges over all possible indices. One can also use 'whenever' for quantified 'if ... then' where the variable ranges over only the temporal indices.

[10] At this point the reader should be warned that our construal of falsity of elementary sentences does not coincide with Reichenbach's, although it agrees with the conventional quantum logical construal. See Section 12.

[11] One should note that the notions of validity and entailment might be more accurately termed *pragmatic validity* and *pragmatic entailment* relative to a particular pragmatic language. In his introduction to *Formal Philosophy: Selected Papers of Richard Montague*, Thomason writes the following (1974, p. 65): "... the introduction of contexts of use also gives rise to a new species of validity that does not arise in semantics. It is natural, for instance, to treat 'I exist' as pragmatically valid, since with respect to any context of use in which it is uttered it will be true. But certainly it is not semantically valid – that is, it need not express a necessary proposition."

[12] The reader should note that our treatment of the logical structure of QM is not as general as it should be. For the logical structure of QM may be regarded as dividing into the

statistical structure and the semantic structure (see my 1976c, d). The statistical structure, which is examined in Part I, is determined by the statistical states of quantum theory and is straightforwardly derivable from the extant formalism. The semantic structure, on the other hand, is not explicitly displayed in the extant formalism, but intimately depends on one's proposed interpretation of QM. For example, if one opts for an 'orthodox' interpretation, such as van Fraassen's Copenhagen modal interpretation (1973) or Bub and Demopoulos' logical interpretation (1973, 1974a, b, 1976), then the semantic and statistical structures are equivalent. However, if one opts for a 'hidden variable' interpretation, such as Fine's (1973, 1974) or van Fraassen's anti-Copenhagen modal interpretation (1973), then the semantic structure is quite different from the statistical structure, being embeddable into a Boolean algebra.

[13] The position described in this paragraph is referred to as the logical interpretation of QM, which was expounded by Bub and Demopoulos (1973, 1974a, b, 1976). The chief tenet of this interpretation is that the fundamental postulates of QM concern the logical structure of micro-events or micro-properties of the form '*m* lies in *X*', what we have called elementary sentences.

[14] This interpretation of Bohr has been advanced also by Hooker and van Fraassen (1976), who distinguish between 'syntactically ill-formed' and 'physically ill-formed'.

[15] Reichenbach makes an analogous distinction on page 141 of PhF, writing the following: "In this definition, we are using the term 'statement' in a sense somewhat wider than usual, since a statement is usually defined as having meaning. Let us use the term 'proposition' in this narrower sense as including meaning. Then, definition 5 states that not every statement of the form 'the value of the entity is *u*', is a proposition . . .".

[16] It may be noted that Lambert has reached a similar conclusion, writing as follows (1969, p. 93): "My main purpose is to show that certain microphysical statements can be construed as neither true nor false, while yet retaining the classical codification of statement logic. The major philosophical implication of this thesis is that whether microphysics requires a nonclassical codification of statement logic is independent of the question of what truth-value, if any, is to be assigned certain microphysical statements."

[17] In the case of magnitudes with pure continuous spectra (e.g., position and momentum), one can show that for all quantum states w there is no r such that $T(m, r, w)$ is true. Thus, for such magnitudes, (Comp) is trivially satisfied: if p has a pure continuous spectrum, then p is complementary to every magnitude m, according to (Comp). See Section 12.

[18] These definitions are not entirely adequate in the case of magnitudes with pure continuous spectra, which have no eigen*vectors* (i.e., in H) whatsoever.

[19] We have already remarked (n. 17) that (37d) is inadequate to formulate the notion of complementarity in the case of magnitudes with pure continuous spectra. We now observe that, even with respect to magnitudes with pure discrete spectra, (37d) does not adequately characterize the notion of incompatibility (non-commuting). If A and B both have pure discrete spectra, then they commute just in case their common eigenvectors (eigenstates) form a basis for H, that is, just in case they share all their eigenvectors in common. It follows that A and B are non-commuting just in case they do not share *all* their eigenvectors in common. Thus, while failing to commute, A and B may nevertheless share a significantly large class of eigenvectors in common. It has been previously argued (1976b) that measurements of A and B can be performed upon any system s which is in a superposition (or mixture) of common eigenstates of A and B. Thus, that A and B are non-compatible (non-commuting) merely means that there exists states relative to which they cannot be simultaneously measured. However, it is clearly invalid to conclude, from the fact that A

and B fail to commute, that they cannot be simultaneously measured relative to *any* states. For it is invalid to suppose, because A and B do not share all their eigenvectors in common, that they share no common eigenvectors. In light of these distinctions it seems reasonable to distinguish between 'merely non-commuting' and 'strongly non-commuting', where the latter corresponds to the notion of complementarity (see my 1976b). Reichenbach evidently identified 'non-commuting' with 'complementary'; on page 159 of PhF he refers to the various components of angular momentum L_x, L_y, L_z as complementary. However, although these magnitudes do not commute pairwise, they nevertheless share common eigenstates. Consider the ground state of Hydrogen; then all three of the following sentences are simultaneously true: $(L_x, 0)$,$(L_y, 0)$, $(L_z, 0)$. Therefore, they are not complementary according to formula (37d). Indeed, for magnitudes with pure discrete spectra, formula (37d) is true of a pair of magnitudes just in case they share *no* common eigenstates; on the other hand, a pair of magnitudes is non-commuting just in case they do not share *all* their eigenstates in common.

BIBLIOGRAPHY

Birkhoff, G. and von Neumann, J.: 1936, 'The Logic of Quantum Mechanics', *Ann. Math.* **37**, 823–43.

Bohr, N.: 1948, 'On the Notions of Causality and Complementarity', *Dialectica* **2**, 312–19.

Bub, J.: 1973, 'On the Completeness of Quantum Mechanics', in C. A. Hooker (ed.), *Contemporary Research in the Foundations and Philosophy of Quantum Theory*, D. Reidel, Dordrecht.

Bub, J.: 1974 , *The Interpretation of Quantum Mechanics*, D. Reidel, Dordrecht.

Bub, J. and Demopoulos, W.: 1974b, 'The Interpretation of Quantum Mechanics', in R. Cohen and M. Wartofsky (eds.), *Boston Studies in the Philosophy of Science*, Vol. 13, D. Reidel, Dordrecht.

Demopoulos, W.: 1976, 'The Possibility Structure of Physical Systems', in W. Harper and C. A. Hooker (eds.), *Foundations of Probability Theory, Statistical Inference, and Statistical Theories of Science*, Vol. 3, D. Reidel, Dordrecht, pp. 55–80.

Destouches-Fevrier, P.: 1951, *Structure des Theories Physiques*, Paris, 1951.

Fine, A.: 1973, "Probability and the Interpretation of Quantum Mechanics', *British J. Philos. Sci.* **24**, 1–37.

Fine, A.: 1974, 'On the Completeness of Quantum Mechanics', *Synthese* **29**, 257–89. Reprinted in P. A. Suppes (ed.), *Logic and Probability in Quantum Mechanics*, D. Reidel, Dordrecht, 1976, pp. 249–81.

Gardner, M.: 1972, 'Two Deviant Logics for Quantum Theory: Bohr and Reichenbach', *British J. Philos. Sci.* **23**, 89–109.

Haack, S.: 1974, *Deviant Logic*, Cambridge Press, Cambridge.

Hardegree, G. M.: 1974, 'The Conditional in Quantum Logic', *Synthese* **29**, 63–80, reprinted in P. A. Suppes (ed.), *Logic and Probability in Quantum Mechanics*, D. Reidel, Dordrecht, 1976, pp. 55–72.

Hardegree, G. M., 1975, 'Stalnaker Conditionals and Quantum *Logic*', *J. Philos. Logic* **4**, 399–421.

Hardegree, G. M.: 1976a, 'The Conditional in Abstract and Concrete Quantum Logic', in C. A. Hooker (ed.), *The Logico-Algebraic Approach to Quantum Mechanics*, Vol. 2, D. Reidel, Dordrecht, in press.

Hardegree, G. M.: 1976b, 'Relative Compatibility in Conventional Quantum Mechanics', *Foundations of Physics*, in press.

Hardegree, G. M.: 1976c, 'The Modal Interpretation of Quantum Mechanics', *Proceedings of the Fifth Bienial Meeting of the Philosophy of Science Association.*

Hardegree, G. M.: 1976d, 'Semantics and the Interpretation of Quantum Mechanics. Part I', unpublished ms.

Hempel. C. G.: 1945, 'Review of Reichenbach's *Philosophic Foundations of Quantum Mechanics*', *J. Symbolic Logic* **10**, 97–100.

Hempel, C. G.: 1970, 'On the "Standard Conception" of Scientific Theories', in M. Radner and S. Winokur (eds.), *Minnesota Studies in the Philosophy of Science*, Minnesota Press, Minneapolis.

Hooker, C. A. (ed.): 1975, *The Logico-Algebraic Approach to Quantum Mechanics, Vol. 1: Historical Evolution*, D. Reidel, Dordrecht.

Hooker, C. A. and van Fraassen, B. C.: 1976, 'A Semantic Analysis of Niels Bohr's Philosophy of Quantum Theory', in W. L. Harper and C. A. Hooker (eds.), *Foundations of Probability Theory, Statistical Inference, and Statistical Theories of Science*, Vol. 3, D. Reidel, Dordrecht, pp. 221–241.

Jammer, M.: 1974, *The Philosophy of Quantum Mechanics*, Wiley, New York.

Jauch, J. M.: 1968, *Foundations of Quantum Mechanics*, Addison-Wesley, New York.

Jauch, J. M.: 1974, 'The Quantum Probability Calculus', *Synthese* **29**, 131–154, reprinted in P. A. Suppes (ed.), *Logic and Probability in Quantum Mechanics*, D. Reidel, Dordrecht, 1976, pp. 123–46.

Kochen, S. and Specker, E. P.: 1967, 'The Problem of Hidden Variables in Quantum Mechanics', *J. Math. Mech.* **17**, 59–87.

Lambert, K.: 1969, 'Logical Truth and Microphysics', in K. Lambert (ed.), *The Logical Way of Doing Things*, Yale Press, New Haven.

Lewis, D.: 1970, 'General Semantics', *Synthese* **22**, 18–67.

Mackey, G. W.: 1963, *Mathematical Foundations of Quantum Mechanics*, Benjamin, New York.

Montague, R.: 1968, 'Pragmatics' in R. Klibansky (ed.), *Contemporary Philosophy: A Survey*, 102–22, La Nuova Italia Editrice, Florence, reprinted in R. H. Thomason (1974).

Montague, R.: 1970, 'Pragmatics and Intensional Logic', *Synthese* **22**, 68–94.

Nilson, D. R.: 'Hans Reichenbach on the Logic of Quantum Mechanics', *Synthese* **34**, 313–360.

Pauli, W.: 1964, 'Reviewing Study of Hans Reichenbach's *Philosophic Foundations of Quantum Mechanics*, in *Collected Scientific Papers*, Vol. 2 (ed. by R. Kronig and V. Weisskopf), Interscience, New York.

Piron, C.: 1964, 'Axiomatique Quantique', *Helv. Phys. Acta* **37**, 439–68.

Piron, C.: 1972, 'Survey of General Quantum Physics', *Found. Phys.* **2**, 287–314.

Reichenbach, H.: 1944, *Philosophic Foundations of Quantum Mechanics*, Univ. of California Press, Berkeley.

Reichenbach, H.: 1947, *Elements of Symbolic Logic*, Free Press, New York.

Rescher, N.: 1969, *Many-Valued Logic*, McGraw-Hill, New York.

Sneed, J.: 1971, *The Logical Structure of Mathematical Physics*, D. Reidel, Dordrecht.

Stein, H.: 1972, 'On the Conceptual Structure of Quantum Mechanics', in R. G. Colodny (ed.), *Paradigms and Paradoxes*, Pittsburgh Press, Pittsburgh.

Strauss, M.: 1937–38, 'Mathematics as Logical Syntax – a Method to Formalize the Language of a Physical Theory', *Erkenntnis* **7**, 147–53.

Suppe, F.: 1967, *Meaning and Use of Models in Mathematics and the Exact Sciences*, Doctoral Dissertation, Department of Philosophy, University of Michigan.

Suppe, F.: 1972, 'Theories, Their Formulations, and the Operational Imperative', *Synthese* **25**, 129–64.

Suppes, P. A.: 1967, 'What is a Scientific Theory?', in S. Morgenbesser (ed.), *Philosophy of Science Today*, Basic Books, New York.

Thomason, R. H.: 1973, 'Philosophy and Formal Semantics', in H. Leblanc (ed.), *Truth, Syntax and Modality*, North-Holland, Amsterdam.

Thomason, R. H.: (ed.), 1974, *Formal Philosophy: Selected Papers of Richard Montague*, Yale Press, New Haven.

van Fraassen, B. C.: 1970, 'On the Extension of Beth's Semantics of Physical Theories', *Philos. Sci.* **37**, 325–39.

van Fraassen, B. C.: 1968, 'Presupposition, Implication, and Self-Reference', *J. Philos.* **65**, 136–52.

van Fraassen, B. C.: 1969, 'Presuppositions, Supervaluations, and Free Logic', in K. Lambert (ed.), *The Logical Way of Doing Things*, Yale Press, New Haven.

van Fraassen, B. C.: 1973, 'Semantic Analysis of Quantum Logic', in C. A. Hooker (ed.), *Contemporary Research in the Foundations and Philosophy of Quantum Theory*, D. Reidel, Dordrecht.

van Fraassen, B. C.: 1974, 'The Ladyrinth of Quantum Logics', in R. Cohen and M. Wartofsky (eds.), *Boston Studies in the Philosophy of Science*, Vol. 13, D. Reidel, Dordrecht.

Varadarajan, V. S: 1968, *Geometry of Quantum Theory* (two vols.), Van Nostrand, Princeton.

von Neumann, J.: 1932, *The Mathematical Foundations of Quantum Mechanics* (tr. by R. T. Beyer), Princeton Press, 1955 (originally published 1932).

von Weizsäcker, C. F.: 1955, 'Komplementarität und Logik', *Naturwissenschaften* **42**, 521–29 and 545–55.

Weyl, H.: 1940, 'The Ghost of Modality', in M. Farber (ed.), *Philosophical Essays in Memory of Edmund Husserl*, Harvard Press, Cambridge.

GARY M. HARDEGREE

REICHENBACH AND THE INTERPRETATION OF QUANTUM MECHANICS

1. INTRODUCTION

The present work, which is a sequel to 'Reichenbach and the Logic of Quantum Mechanics' (this volume; henceforth referred to as Part I), endeavors to place Reichenbach's proposed quantum logic RQL within the context of his overall interpretation of quantum mechanics (QM). Although it is argued in Part I that Reichenbach intended RQL to provide an alternative logico-linguistic framework for the *formulation* of the quantum *theory*, in the present work we interpret RQL as being entirely on a par with mainstream quantum logic (MQL). In particular, we regard RQL as pertaining, not to the theory formulation language TL(QM), but rather exclusively to the elementary (observation) language RL(QM), which is obtained from the elementary language OL(QM), associated with MQL, by adding a variety of three-valued truth-functional connectives. Also, like Part I, the present work is essentially comparative in nature, the purpose being to analyze Reichenbach's interpretation of QM in the light of more recently proposed interpretations. We take the chief themes of Reichenbach's interpretation to include the following: the distinction between phenomena and interphenomena; the distinction between exhaus-

W.C. Salmon (ed.), Hans Reichenbach: Logical Empiricist, 513–566. *All Rights Reserved.*

tive and restrictive interpretations of QM; the Principle of Anomaly; the theory of equivalent descriptions; the proposal of a non-bivalent semantics for the observation language of QM.

Section 2 examines two widely different conceptions of generalized (non-classical) probability theory. As noted in Section 3, whereas the former underlies the standard explication of the conventional Hilbert space formulation of quantum theory, the latter offers significant insights into the *fine structure* of quantum probability. The key difference between the two interpretations of quantum probability concerns the nature of quantum events, and hence quantum magnitudes (observables). Whereas the conventional (partial Boolean) interpretation construes quantum events as being represented by the *subspaces* of the relevant Hilbert space $\mathscr{H}$, the non-conventional (semi-Boolean) interpretation construes quantum events as being represented by the *subframes* of $\mathscr{H}$. Whereas the subspaces of $\mathscr{H}$ form a partial Boolean algebra, the subframes of $\mathscr{H}$ form a semi-Boolean algebra; hence, our terminology.

Although the differences between the partial Boolean and the semi-Boolean conception of quantum probability are not reflected at the level of *statistical states*, they are reflected at the level of *micro-states*, which is the subject of Section 4. Two radically different conceptions of quantum micro-states are presented, based respectively on the Copenhagen variant and the anti-Copenhagen variant of van Fraassen's (1973) *modal interpretation* of QM. It is argued that whereas the Copenhagen modal interpretation (CMI) identifies quantum events in the partial Boolean fashion with subspaces, the anti-Copenhagen interpretation (ACMI) identifies quantum events in the semi-Boolean fashion with subframes. The different characterizations of quantum events yield correspondingly different characterizations of quantum magnitudes and quantum micro-states. Whereas Copenhagen magnitudes are identified with self-adjoint operators on $\mathscr{H}$, anti-Copenhagen magnitudes are identified with semi-Boolean random variables on $\mathscr{H}$; whereas every anti-Copenhagen micro-state assigns an exact value to every (anti-Copenhagen) magnitude, no Copenhagen micro-state does.

Section 5 describes the salient features of Reichenbach's interpretation of QM. Section 6 presents a formal semantic reconstruction of a *restrictive interpretation*, RI, arguably attributable to Reichenbach or Bohr-Heisenberg. The reconstruction is based on translating each elementary sentence (m, X), which is read '*m* has value in *X*', into the conditional $[m] \ni\!\!- [m, X]$, where $[m]$ is read '*m* is measured', $[m, X]$ is read '*m* is observed to have value in *X*', and the conditional connective is Reichenbach's quasi-implication, which has the property that $A \ni\!- B$ is indeterminate if A is false or indeterminate, and otherwise has the truth value of B. The upshot of the investigation of Section 6 is that, according to RI, (m, X) is true if m is actually observed to have a value in X, it is false if m is actually observed to have a value in R-X, and it is indeterminate otherwise.

Whereas the restrictive interpretation RI assigns (exact) values only to a particular maximal compatible subset of magnitudes (in any given situation), the Copenhagen modal interpretation generally assigns (exact) values to infinitely many mutually incompatible magnitudes. In Section 7, this difference is traced to the respective manners in which the elementary sentence (m, r) ('m has value r') is translated. Whereas RI translates (m, r) as the truth-functional conditional $[m] \ni\!- [m, r]$, CMI translates (m, r) as the *counterfactual conditional* $[m] \rightarrow [m, r]$, and evaluates (m, r) as true precisely when the probability of $[m] \rightarrow [m, r]$, which is identical to the conditional probability $p([m, r]/[m])$, is one. CMI is not a restrictive interpretation, since in any given situation numerous unmeasured (interphenomenal) magnitudes are assigned exact values, nor is CMI an exhaustive interpretation, since numerous magnitudes are not assigned exact values. Section 7 concludes with a discussion of the anti-Copenhagen modal interpretation, which is an exhaustive interpretation. It is argued that, whereas CMI asserts 'm has value r' (as true) whenever a measurement of m *would certainly* yield the value r, ACMI asserts 'm has value r' whenever a measurement of m *would actually* yield the value r, where in both cases the underlying conditional connective is a Stalnaker (1968) conditional.

Section 8 describes Reichenbach's Principle of Anomaly, which poses serious conceptual problems for any exhaustive interpretation. The exposition of this principle is based on a theorem of Gleason (1957) and Kochen and Specker (1967), which entails the following. Insofar as we identify quantum magnitudes in the orthodox fashion with self-adjoint operators, every assignment of exact values to all quantum magnitudes violates the functional relations among these magnitudes. On the other hand, it is pointed out that, if we identify quantum magnitudes in the non-orthodox fashion with semi-Boolean random variables, then no valuation associated with an anti-Copenhagen micro-state involves functional anomalies of this sort. It is nevertheless argued that ACMI does not circumvent the Principle of Anomaly, for the *functional* anomalies in the context of individual systems reappear as *causal* anomalies in the context of *composite* systems.

Section 9 presents a minimal modification of ACMI designed to remove the functional (causal) anomalies. The modified interpretation, CACMI, is called the Copenhagen variant of ACMI, and is intermediate between ACMI and CMI; although it assigns values to *all maximal* magnitudes, it does not assign (exact) values to *all* magnitudes. CACMI is obtained by replacing the Stalnaker conditional of ACMI with a Lewis (1973) conditional; as van Fraassen (1974) has shown, this may be accomplished by the method of *supervaluations*. This method is extended in a more or less natural way to magnitudes, so that we speak of the 'supervalue' of a magnitude m, which is a non-empty set of real numbers representing the *various* exact (but anomalous!) values that m *may* take.

2. GENERALIZED PROBABILITY THEORY

Since QM is a probabilistic theory, we begin by examining various alternative probability theories or models of probabilistic phenomena. First of all, the classical model of probability (Kolmogorov, 1933) is formulated in terms of sample spaces; a *sample space* is a structure $\langle Z, F \rangle$, where Z is a set of *outcomes*, and F is a set of *events*, which are subsets of Z postulated to form a (Borel) field. A *probability measure* over $\langle Z, F \rangle$ is any function w from F into the closed unit interval $[0, 1]$, which is (countably) additive over disjoint events (subsets of Z), and which is such that $w(Z) = 1$. A *random variable* over $\langle Z, F \rangle$ is any real-valued measurable function m over $\langle Z, F \rangle$, that is, any function m from Z into R subject to the requirement that the inverse image $m^{-1}(X)$ of every Borel subset X of R is an event (i.e., an element of F). Henceforth, R is the set of real numbers, and $B(R)$ is the set of Borel subsets of R. Given any probability measure w and random variable m over $\langle Z, F \rangle$, one can define an associated probability measure p_m^w over $\langle R, B(R) \rangle$, where $p_m^w(X) = w(m^{-1}(X))$ for all $X \in B(R)$.[1]

Although the Kolmogorov model of probability lies at the heart of present day statistical practice, the peculiarities of QM have strongly suggested that this model is not sufficiently general to subsume all probabilistic phenomena, and accordingly a number of generalizations have been proposed. In the present article, we concentrate on two generalized theories of probability, which will respectively be called *partial Boolean probability theory* and *semi-Boolean probability theory*.

However, before examining these, it is useful first to examine *Boolean probability theory*, which serves as a bridge between classical and partial Boolean probability theory. According to Boolean probability theory, the basic objects are not outcomes, as in the classical model, but rather events, which are postulated to form a (sigma) Boolean algebra. Since every (Borel) field of subsets of a set is a (sigma) Boolean algebra, Boolean probability subsumes classical probability. On the other hand, in virtue of the Stone (Loomis) representation theorem, nothing essential is gained in moving from classical to Boolean probability structures. In fact, it may be plausibly argued that something important is lost, namely, the 'fine structure' provided by the outcomes. A *probability measure* over a (sigma) Boolean algebra B is any function w from B into $[0, 1]$ which is (countably) additive over disjoint elements of B, and which is such that $w(1) = 1$ (here, '1' also denotes the unit element of B). Whereas Boolean probability measures are defined just like classical probability measures, Boolean random variables are defined very differently from their classical counterparts. Specifically a *random variable* over a (sigma) Boolean algebra B is any function m from $B(R)$ into B which is a (sigma) homomorphism. As with the classical model, given any Boolean probability measure w and random variable m, one can define an associated (classical) probability measure p^w_m over $\langle R, B(R)\rangle$; the definition is different, however: $p^w_m(X) = w(m(X))$. This is because a Boolean random variable does not correspond directly to a classical random variable, but rather to its associated inverse image function. Whereas a classical random variable m over $\langle Z, F\rangle$ assigns a real number $m(z)$ to each outcome $z \in Z$, and the associated inverse image function assigns an event $m^{-1}(X)$ to each $X \in B(R)$, a Boolean random variable m over B is more direct assigning an event $m(X)$ to each $X \in B(R)$. In each case there is an event $e(m, X)$ associated with 'm has value in X', and the probability that m has value in X is simply the probability of $e(m, X)$; the difference is that in the classical case $e(m, X) = m^{-1}(X)$, whereas in the Boolean case $e(m, X) = m(X)$. Henceforth, in order to avoid terminological confusion, we use the term 'random variable' to refer to functions that assign real numbers to outcomes, and we use the term 'spectral measure' to refer to functions that assign events to Borel subsets of R (cf., e.g., Jauch, 1968). Thus, for example, under this terminology, we can say that every random variable m gives rise to an associated spectral measure m^{-1}. We can also say that the Boolean counterparts of classical random variables are spectral measures.

Partial Boolean probability theory[2] is a natural generalization of Boolean probability theory obtained by enlarging the relevant category of event structures from Boolean algebras to partial Boolean algebras. In the context of this work, a *partial Boolean algebra* (PBA) is a family $(B(i) : i \in I)$ of Boolean algebras 'pasted together' so as to satisfy the following (redundant) restrictions. (1) The unit elements of the various $B(i)$ are all identical; i.e., $1_i = 1_j$ for all $i, j \in I$. (2) The zero elements of the various $B(i)$ are all identical; i.e., $0_i = 0_j$ for all $i, j \in I$. (3) The meet, join, and complement operations agree on all overlapping regions; i.e., if $x, y \in B(i) \cap B(j)$, then $M_i(x, y) = M_j(x, y)$, $J_i(x, y) = J_j(x, y)$, $C_i(x) = C_j(x)$; M_i, J_i, C_i respectively denote the meet, join, and complement operations on $B(i)$.

Where $\mathbf{B} = (B(i) : i \in I)$ is a PBA, the set of elements that belong to some $B(i)$ or other is also denoted $\mathbf{B}$; the ambiguity is harmless. Two elements $x, y \in \mathbf{B}$ are said to be *compatible* if they belong to a common Boolean algebra $B(i)$ for some $i \in I$. A subset S of $\mathbf{B}$ is said to be *pairwise compatible* if every pair $x, y \in S$ is compatible; S is said to be *jointly compatible* if *all* the elements of S belong to a common Boolean algebra $B(i)$, i.e., if $S \subseteq B(i)$ for some $i \in I$. A PBA $\mathbf{B}$ is said to be a *compatible* PBA if every finite pairwise compatible subset S of $\mathbf{B}$ is also jointly compatible. What we call a compatible PBA is what is ordinarily called simply a PBA (see, e.g, Kochen and Specker, 1967, Demopoulos, 1976); our definition is chosen for terminological ease (see below).

Next, one can define a relation $\leqslant$ on $\mathbf{B}$ as follows: $x \leqslant y$ iff $M_i(x, y) = x$ for some $i \in I$. This relation is always reflexive and anti-symmetric, although it need not be transitive. If $\leqslant$ is transitive, and hence a *partial ordering,* then $\mathbf{B}$ is called a *transitive PBA*. Every transitive PBA $\mathbf{B}$ gives rise to a *partially ordered set* (poset) $\langle \mathbf{B}, \leqslant \rangle$, which is moreover *orthocomplemented* by the map that takes each element $x \in \mathbf{B}$ to $C(x)$, where $C(x) = C_i(x)$ for *any* $i \in I$. One can also define a relation $\perp$ on $\mathbf{B}$ as follows: $x \perp y$ iff for some $i \in I$, $x, y \in B(i)$ and $M_i(x, y) = 0_i$. A transitive PBA is said to be *coherent* if the following holds: for all $x, y \in \mathbf{B}$, if $x \perp y$, then the join $J_i(x, y)$ of x, y *relative to* $B(i)$ is also the join of x, y *relative to* the poset $\langle \mathbf{B}, \leqslant \rangle$. Every compatible transitive PBA is a coherent transitive PBA, but not conversely. According to a theorem of Finch (1969), every coherent transitive PBA gives rise to an *orthomodular poset,* and conversely. According to a theorem of Gudder (1972), every compatible transitive PBA gives rise to a compatible orthomodular poset, and conversely. Finally, we note that in the special case that the partial ordering $\leqslant$ on a transitive PBA $\mathbf{B}$ is a *lattice ordering* (i.e., meets and joins exist for *arbitrary* pairs of elements of $\mathbf{B}$), the associated poset $\langle \mathbf{B}, \leqslant \rangle$ is an *orthomodular lattice,* in which case the conditions of compatibility and coherence are automatically satisfied.

Turning to probabilistic notions, a *probability measure* on a PBA $(B(i): i \in I)$ is any function w from $\mathbf{B}$ into $[0, 1]$ subject to the requirement that

w restricted to any $B(i)$ is a probability measure over $B(i)$. In other words, a partial Boolean probability measure may be described as a *locally Boolean* probability measure. A 'random variable' on $(B(i) : i \in I)$ is any function m from $B(R)$ into $B(i)$, for some $i \in I$, which is a Boolean 'random variable' (i.e., a spectral measure) relative to $B(i)$; by extension of terminology, we call such functions spectral measures. As in the Boolean case associated with every partial Boolean probability measure w and spectral measure m, there is a classical probability measure p^w_m over $\langle R, B(R) \rangle$, where $p^w_m(X) = w(m(X))$.

We have now given a very broad and general outline of the more or less standard formulation of generalized probability, as enunciated (for example) by Jauch (1968), Varadarajan (1962, 1968), and Greechie and Gudder (1973).

There is an alternative model of generalized probability, developed by Foulis and Randall (1972, 1973, 1974, 1976, 1978), which they alternately call *operational statistics* and *empirical logic,* and which we call *semi-Boolean probability theory*. Unlike the Boolean and partial Boolean models, and like the classical model, the Foulis-Randall (i.e., semi-Boolean) model of probability takes *outcomes* as the basic objects. In particular, the outcomes form what they call a manual; a *manual* is a non-empty family $(Z(i) : i \in I)$ of non-empty sets $Z(i)$ subject to the condition of *irredundancy* (if $i \neq j$, then $Z(i) \not\subseteq Z(j)$ for all j, $j \in I$) and the condition of *coherence* (see below). The elements of a manual $\mathscr{M} = (Z(i)$: $i \in I)$ are called *operations*; each $Z(i)$ is regarded as representing a physical (e.g., measurement) operation, where the elements of $Z(i)$ represents the abstractly possible outcomes of performing this operation. If x is an element of $Z(i)$, then it is said to be an outcome of $Z(i)$; if x is an element of $Z(i)$ for some $i \in I$, then x is said to be an *outcome of* $\mathscr{M}$; we denote the set of outcomes of $\mathscr{M}$: $O(\mathscr{M})$.

Since the outcomes of a manual $\mathscr{M}$ are grouped into various operations, not just any subset of $O(\mathscr{M})$ constitutes an event. Rather a subset S of $O(\mathscr{M})$ is an event if and only if S is an event relative to some particular operation $Z(i)$. An event on $Z(i)$ is *any* subset of $Z(i)$; the set of events on $Z(i)$ is denoted $\mathbb{E}(i)$. Thus, letting $\mathbb{E}(\mathscr{M})$ denote the class of events on M, we have: $S \in \mathbb{E}(\mathscr{M})$ iff $S \in \mathbb{E}(i)$ for some $i \in I$. The set $\mathbb{E}(\mathscr{M})$ of events on a manual $\mathscr{M}$ is partially ordered by set-inclusion; with respect to this ordering, $\mathbb{E}(\mathscr{M})$ has a least element, the empty set, but generally it does not have a greatest element. If E, F are events in $\mathbb{E}(\mathscr{M})$, then so is their intersection $E \cap F$, although their union $E \cup F$ need not be in $\mathbb{E}(\mathscr{M})$. Given an event $F \in \mathbb{E}(\mathscr{M})$, the set $\{E \in \mathbb{E}(\mathscr{M})$: $E \subseteq F\}$ forms a Boolean lattice, being simply the set of all subsets of F. In other words, the class $\mathbb{E}(\mathscr{M})$ of events of any manual M form what Abbott (1967, 1969) calls a *semi-Boolean algebra*. Whereas every manual gives rise to a semi-Boolean algebra, not every semi-Boolean algebra arises in this manner. Nevertheless, I wish to call the Foulis-Randall model semi-Boolean probability theory, in order to render our terminology parallel. Thus, we have the Boolean, partial Boolean, and semi-Boolean models of probability.

Concerning the relevant probabilistic notions, a *probability measure*[3] on a manual $\mathscr{M} = (Z(i): i \in I)$ is any function w from $\mathbb{E}(\mathscr{M})$ into [0, 1] subject to the requirement that w restricted to any $\mathbb{E}(i)$ is a classical probability measure over $\langle Z(i), \mathbb{E}(i)\rangle$. The class of probability measures over $\mathscr{M}$ is denoted $P(\mathscr{M})$. The definition of random variables on a manual is very intricate in the general case (Foulis and Randall, 1974, 1978), but in the special case of random variables with a *countable spectrum*[4] (i.e., *discrete* random variables), a completely straightforward definition can be given (cf. Finch, 1976). Specifically, every discrete random variable over $(Z(i): i \in I)$ may be defined to be a classical random variable over $\langle Z(i), \mathbb{E}(i)\rangle$ (that is, a real-valued function over $Z(i)$) for some $i \in I$. In the present work, in order to avoid unnecessary measure-theoretic subtleties, we deal exclusively with discrete random variables (magnitudes, observables). We accordingly adopt the simplified, although not completely general, definition of random variables as real-valued functions on the various $Z(i)$ of a manual.

Besides events, which form a semi-Boolean algebra, there are also what Foulis and Randall call *propositions*, which in the 'nice' case form a partial Boolean algebra. First of all, outcomes $x, y \in O(\mathscr{M})$ are said to be orthogonal (written $x \perp y$) if they are distinct and belong to a common operation: $x \perp y$ iff $x \neq y$ and $x, y \in Z(i)$ for some $i \in I$. Two events $E, F \in \mathbb{E}(\mathscr{M})$ are derivatively said to be orthogonal (written $E \perp F$) if every outcome in E is orthogonal to every outcome in F: $E \perp F$ iff $e \perp f$ for all $e \in E$ and all $f \in F$. We can now define the condition of *coherence*: for all $E, F \in \mathbb{E}(M)$, if $E \perp F$, then $E \cup F \in \mathbb{E}(M)$. The orthogonality relation $\perp$ gives rise to the 'perp' operation, denoted $^{\perp}$, defined on the subsets of $O(\mathscr{M})$ so that $S^{\perp}$ is the set of all outcomes that are orthogonal to every outcome in S: $S^{\perp} = \{x \in O(\mathscr{M}): \{x\} \perp S\}$. The *proposition* $p(E)$ associated with event E is defined to be the set $(E^{\perp})^{\perp}$; that is, $p(E)$ consists of all outcomes that are orthogonal to every outcome orthogonal to E.[5] The set $\{p(E): E \in \mathbb{E}(\mathscr{M})\}$ of propositions associated with a manual $\mathscr{M}$ is denoted $\mathbb{P}(\mathscr{M})$. $\mathbb{P}(\mathscr{M})$ is partially ordered by set-inclusion, but this poset is generally very different from the poset $\mathbb{E}(\mathscr{M})$ of events. In the special (and natural) case that $\mathscr{M}$ is a *Dacey manual*, $\mathbb{P}(\mathscr{M})$ forms an orthomodular poset, and hence a coherent (but not necessarily compatible) partial Boolean algebra.

Besides propositions in the above sense, one might also wish to define *statistical propositions* to be equivalence classes of events, where the equivalence relation $\equiv$ is *statistical equivalence*: $E \equiv F$ iff $w(E) = w(F)$ for all $w \in P(M)$. The *statistical proposition* $sp(E)$ associated with the event E is defined to be the set $\{F \in \mathbb{E}(M): E \equiv F\}$. In the most general case, statistical propositions do not correspond biuniquely to ordinary propositions; whereas $p(E) = p(F)$ only if $sp(E) = sp(F)$ (i.e., only if $E \equiv F$), the converse need not be true. If the manual $\mathscr{M}$ has a set $P(\mathscr{M})$ of probability measures that separate (individuate) the outcomes (i.e., for all $x, y \in O(\mathscr{M})$, if $x \neq y$, then there is a $w \in P(M)$ such that $w(\{x\}) \neq w(\{x\})$), in which case $\mathscr{M}$ is automatically a Dacey manual, we have $p(E) = p(F)$ if *and* only if $sp(E) = sp(F)$. In this case, we have $p(E) = \{x \in O(\mathscr{M}): x \in F$ for some $F \in sp(E)\}$, and $sp(E) = \{F \in \mathbb{E}(\mathscr{M}): p(E) = p(F)\}$.

Since $p(E) = p(F)$ only if $E \equiv F$, every probability measure w on the class $\mathbb{E}(\mathscr{M})$ of events on $\mathscr{M}$ can be transferred to a function w° on the class $\mathbb{P}(\mathscr{M})$ of propositions on $\mathscr{M}$, defined so that $w^\circ(p(E)) = w(E)$ for all $E \in \mathbb{E}(M)$. If $\mathbb{P}(M)$ forms an orthomodular poset, and hence a partial Boolean algebra, then w° is a partial Boolean probability measure; conversely, every partial Boolean probability measure w on $\mathbb{P}(\mathscr{M})$ can be transferred to a probability measure w^* on $\mathbb{E}(\mathscr{M})$, defined so that $w^*(E) = w(p(E))$ for all $E \in \mathbb{E}(\mathscr{M})$. Thus, to the extent that the semi-Boolean propositions form a partial Boolean algebra, the partial Boolean probability measures coincide with the semi-Boolean probability measures. The difference between the two approaches does not concern the class of probability measures, but rather the *underlying structure*; whereas the partial Boolean model postulates only propositions (i.e., partial Boolean events), the semi-Boolean model postulates events and outcomes in addition to propositions. As we will see in subsequent sections, this additional 'fine structure' is a valuable tool in the analysis and interpretation of quantum probability.

3. THE PROBABILISTIC FORMULATION OF QUANTUM THEORY

With the various models of generalized probability at our disposal, we turn our attention to the probabilistic formulation of quantum theory. According to the conventional model of quantum theory, as originally enunciated by von Neumann (1932), every physical system $\mathscr{sys}$ is characterized by a separable complex Hilbert space $\mathscr{H}$. The *magnitudes* (observables) pertaining to $\mathscr{sys}$ are represented by self-adjoint (s.a.) operators on $\mathscr{H}$, and the *statistical states* pertaining to $\mathscr{sys}$ are represented by density operators on $\mathscr{H}$ (i.e., positive definite s.a. operators with unit trace). Given any magnitude (s.a. operator) A and statistical state (density operator) W, the probability in state W that the value of A lies in Borel set X is $Tr(WE)$, where E is the projection (idempotent s.a.) operator associated with X according to the spectral measure determined by A.

The density operators on a Hilbert space $\mathscr{H}$ form a convex set, the extremal points of which are interpreted as *pure statistical states*, the remainder of which are interpreted as *mixed statistical states*. Every density operator W has a range $ran(W)$, which is a sub(vector)space of $\mathscr{H}$; W is a pure statistical state if and only if $ran(W)$ is a one-dimensional subspace (ray) of $\mathscr{H}$[6]. Indeed, the pure statistical states (density operators) on $\mathscr{H}$ stand in a natural 1-1 correspondence with the rays of $\mathscr{H}$, which may accordingly be used as alternative representatives of pure statistical states. In actual physical practice, every pure statistical state is specified, not by a ray r in $\mathscr{H}$, but rather by a unit vector in r; which among the infinitely many unit vectors is chosen is determined by the 'phase' of the system.[7]

There are at least two explications, or interpretations, of the von Neumann model of quantum theory, corresponding respectively to the two generalized probability theories discussed in Section 2. The orthodox explication has been variously proposed by Birkhoff and von Neumann (1936), Mackey (1963), Piron (1964), Jauch (1968), and Varadarajan (1968), among others (cf. Stein, 1972, for a lucid exposition). According to this interpretation, the underlying structure postulated by QM is an ordered triple $\langle \mathbf{E}, \mathbf{M}, \mathbf{S} \rangle$; $\mathbf{E}$ is a set of *events* (also called properties, attributes, qualities, propositions, questions, and eventualities); $\mathbf{M}$ is a set of *magnitudes* (also called observables, quantities, variables, random variables, and statistical variables); $\mathbf{S}$ is a set of *statistical states* (also called states). What distinguishes the orthodox explication of the von Neumann model from the non-orthodox explication, discussed later, is the proposal that the events band together to form some sort of partial Boolean algebra. I accordingly refer to the orthodox explication as the *partial Boolean interpretation* of quantum probability.

According to this interpretation of quantum probability, the events pertaining to a physical system $\mathscr{sys}$ with Hilbert space $\mathscr{H}$ are represented by the *subspaces* of $\mathscr{H}$. As is well-known, the subspaces of a Hilbert space form a PBA; indeed, they form an orthomodular lattice, but the exact physical significance of the lattice property remains unclear. In any case, given the partial Boolean structure of the class $E(\mathscr{H})$ of events pertaining to system sys, one can provide completely general and abstract definitions of the associated class of partial Boolean spectral measures (random variables), which will comprise the class **M** of magnitudes, and the class of partial Boolean probability measures, which will comprise the class **S** of statistical states. As it turns out, the spectral measures definable over $E(\mathscr{H})$ correspond biuniquely to the s.a. operators on $\mathscr{H}$, and in the case $\dim(H) \geqslant 3$, the probability measures definable over $E(\mathscr{H})$ correspond biuniquely to the density operators on $\mathscr{H}$ (Gleason, 1957)[8]. In other words, once we identify the relevant quantum events as the subspaces of $\mathscr{H}$, the remaining objects in the von Neumann model – the magnitudes and the statistical states – do not require an independent specification. Rather, their identities are uniquely determined by the definition of magnitudes as spectral measures and the definition of statistical states as probability measures. Thus, according to this explication, the empirically significant feature of the von Neumann model of quantum theory is that it postulates an event structure represented by the subspaces of a Hilbert space.

Having considered the conventional explication of quantum theory based on partial Boolean probability theory, we now examine a non-conventional explication based on semi-Boolean probability theory. According to the semi-Boolean interpretation of quantum probability, the fundamental mathematjcal object associated with a system sys with Hilbert space $\mathscr{H}$ is the manual $\mathscr{M}(\mathscr{H})$, which is the family $(Z(i): i \in I)$, where each $Z(i)$ is a *frame* on $\mathscr{H}$, that is, a maximal collection of mutually orthogonal rays of $\mathscr{H}$ (a complete orthonormal system). Accordingly, each event $E \in \mathbb{E}(\mathscr{M}(\mathscr{H}))$ is a subset of some frame $Z(i)$ on $\mathscr{H}$, that is, an orthogonal collection of rays of $\mathscr{H}$ – what we will call a *subframe*. Thus, whereas the quantum events are *subspaces* according to the partial Boolean interpretation, they are *subframes* according to the semi-Boolean interpretation. If the subspace generated by a subframe F is S, then F is said to *resolve* S, or S is said to *resolve into* F. Except in the case of one-dimensional subspaces (minimal events) every subspace (partial Boolean event) resolves into infinitely many subframes (semi-Boolean events).

Whereas partial Boolean events do not correspond to semi-Boolean events, they do correspond to semi-Boolean *propositions*.[9] If F is a semi-Boolean event, then the associated proposition $p(F)$, which is the set $(F\perp)\perp$, consists of precisely those rays contained in the subspace $S(F)$ generated by F. Furthermore, the statistical proposition $sp(F)$ associated with F, which consists of those events (subframes) that are statistically equivalent to F, consists of precisely those subframes that resolve $S(F)$. As noted in Section 2, there is a direct correspondence between probability measures over the semi-Boolean algebra of events and probability measures over the associated poset of propositions, which depends crucially upon the fact that every proposition resolves into events all of which are statistically equivalent. In the case of QM, the associated poset of propositions is the partial Boolean algebra (orthomodular lattice) of partial Boolean events. As an immediate consequence, even though the partial Boolean interpretation and the semi-Boolean interpretation characterize the class of quantum events in radically different ways, they nevertheless agree in their characterizations of the associated class of quantum statistical states.[10]

Just as the transition from the conventional partial Boolean interpretation of quantum probability to the non-conventional semi-Boolean interpretation involves resolving all non-minimal partial Boolean events into a multiplicity of (statistically equivalent) semi-Boolean events, this transition involves a similar resolution of magnitudes, such that the semi-Boolean construal of quantum magnitudes is radically different from the conventional construal. Recall that a (discrete) random variable over a manual $\mathscr{M} = (Z(i): i \in I)$ is any real-valued function over $Z(i)$ for some $i \in I$. In the case of a Hilbert space manual $\mathscr{M}(\mathscr{H})$, if **m** is a *maximal* random variable on $\mathscr{M}(\mathscr{H})$ (i.e., a 1-1 function), then **m** is biuniquely associated with a s.a. operator on $\mathscr{H}$, and hence a partial Boolean spectral measure.[11] On the other hand, whereas every non-maximal random variable (non 1-1 function) on $\mathscr{M}(\mathscr{H})$ determines a unique s.a. operator on $\mathscr{H}$, the converse is not true; every non-maximal s.a. operator on $\mathscr{H}$ has associated with it infinitely many random variables on $\mathscr{M}(\mathscr{H})$. A random variable **m** on $\mathscr{M}(\mathscr{H})$ is associated with (resolves) s.a. operator A on $\mathscr{H}$ just in case the following holds: where $\text{dom}(\mathbf{m}) = Z(i[\mathbf{m}])$, every ray r in $Z(i[\mathbf{m}])$ is contained in the α-eigenspace $A(\alpha)$ of A for some $\alpha \in R$, and moreover, $r \in A(\alpha)$ iff $\mathbf{m}(r) = \alpha$, for all $r \in Z(i[\mathbf{m}])$. A (discrete) s.a. operator is maximal just in case every eigenspace $A(\alpha)$ of A is one-dimensional. Therefore, if A is a maximal s.a. operator on $\mathscr{H}$, then there is exactly one frame, call it $Z(a)$, every ray of which is contained in an eigenspace of A, and accordingly there is exactly one random variable **a** that *resolves* A, defined so that $\mathbf{a}(A(\alpha)) = \alpha$ for all $A(\alpha) \in Z(a)$. In other words, the random variable **a**, whose domain consists of the eigenspaces of A, assigns to each eigenspace $A(\alpha)$ its associated eigenvalue α. On the other hand, if A is non-maximal (degenerate), then not every eigenspace of A is one-dimensional, in which case there are infinitely many frames contained in the eigenspaces of A, and so there are infinitely many random variables that resolve A. Specifically, if every ray in $Z(a_k)$ is contained in an eigenspace of A, one can

b3
a3
b2
a2
a1 b1

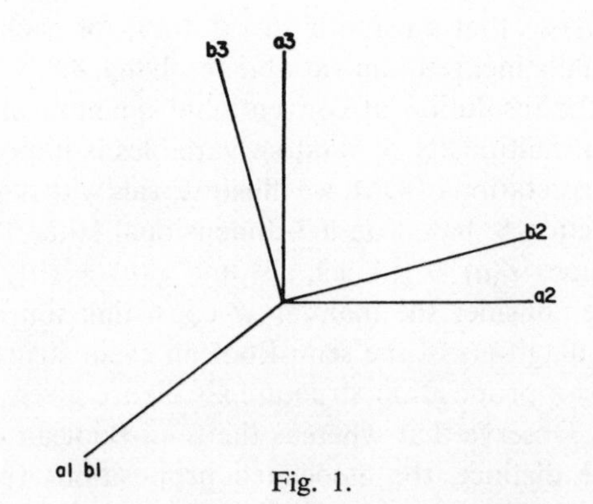

Fig. 1.

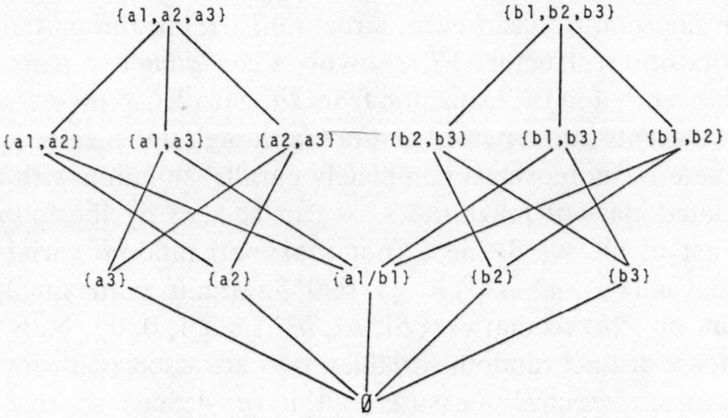

{a1,a2,a3}
{b1,b2,b3}
{a1,a2}
{a1,a3}
{a2,a3}
{b2,b3}
{b1,b3}
{b1,b2}
{a3}
{a2}
{a1/b1}
{b2}
{b3}
Ø

Fig. 2

p({a1,a2,a3})
p({b1,b2,b3})
p({a1,a2})
p({a1,a3})
p({a2,a3})
p({b2,b3})
p({b1,b3})
p({b1,b2})
p({a3})
p({a2})
p({a1/b1})
p({b2})
p({b3})
p(Ø)

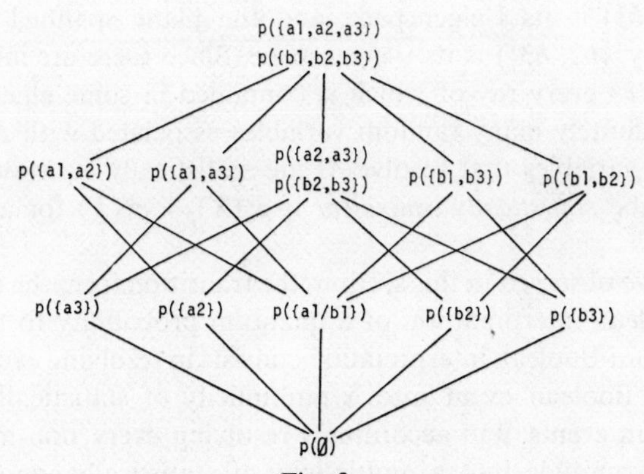

Fig. 3.

define $\mathbf{a}_k$ on $Z(a_k)$ so that $\mathbf{a}_k(r) = \alpha$ iff $r \in A(\alpha)$; for each distinct frame $Z(a_k)$, $\mathbf{a}_k$ will be a distinct random variable resolving A.

Inasmuch as the resolution of conventional quantum magnitudes (s.a. operators) into a multiplicity of random variables is important to understanding the interpretation of QM, we illustrate this with a simple example, used again in Section 8, based on a 3-dimensional Hilbert space $H3$. We consider two frames $Z(a) = \{a1, a2, a3\}$ and $Z(b) = \{b1, b2, b3\}$, where $a1 = b1$, and we consider the manual $\mathscr{M}$ consisting simply of $Z(a)$ and $Z(b)$. The Hasse diagrams of the semi-Boolean event structure $\mathbb{E}(\mathscr{M})$ and the partial Boolean proposition structure $\mathbb{P}(\mathscr{M})$ are given respectively by Figures 2 and 3. Observe that whereas the semi-Boolean events $\{a2, a3\}$ and $\{b2, b3\}$ are distinct, the associated propositions (partial Boolean events) $p(\{a2, a3\})$ and $p(\{b2, b3\})$ are identical; similarly, although $\{a1, a2, a3\} \neq \{b1, b2, b3\}$, $p(\{a1, a2, a3\}) = p(\{b1, b2, b3\})$. Thus, the transition from the semi-Boolean event structure $\mathbb{E}(\mathscr{M})$ to the partial Boolean event (proposition) structure $\mathbb{P}(\mathscr{M})$ involves *conflating* two pairs of events into single events, and the transition from $\mathbb{P}(\mathscr{M})$ to $\mathbb{E}(\mathscr{M})$ involves *resolving* each of two events into a pair of distinct (although statistically equivalent) events. There is, moreover, a completely parallel situation with regard to the associated class of magnitudes, as can be seen by the following example. First of all, we define a (non-maximal) random variable $\mathbf{m}_a$ on $Z(a)$ so that $\mathbf{m}_a(\langle a1, a2, a3\rangle) = \langle 1, 0, 0\rangle$, and a (non-maximal) random variable $\mathbf{m}_b$ on $Z(b)$ so that $\mathbf{m}_b(\langle b1, b2, b3\rangle) = \langle 1, 0, 0\rangle$. Now although $\mathbf{m}_a$ and $\mathbf{m}_b$ are distinct random variables, they are associated with the same partial Boolean spectral measure, call it m, defined so that $m(X) = p(\mathbf{m}_a^{-1}(X))$, or equivalently, $m(X) = p(\mathbf{m}_b^{-1}(X))$. By the same token, $\mathbf{m}_a$ and $\mathbf{m}_b$ are associated with the same s.a. operator on $H3$, call it M, defined so that $a1$ $(= b1)$ is its 1-eigenspace, and the plane spanned by $\{a2, a3\}$ (equivalently, $\{b2, b3\}$) is its 0-eigenspace. Since there are infinitely many frames on $H3$ every ray of which is contained in some eigenspace of A, there are infinitely many random variables associated with M. However, all random variables that resolve M are statistically equivalent; $\mathbf{a}$ and $\mathbf{b}$ are said to be *statistically equivalent* if $p_{\mathbf{a}}^{w}(X) = p_{\mathbf{b}}^{w}(X)$ for all $w \in P(\mathscr{M})$, $X \in B(R)$.

As we have observed in this section, the transition from the conventional partial Boolean interpretation of a quantum probability to the non-conventional semi-Boolean interpretation consists in resolving every non-minimal partial Boolean event into a multiplicity of statistically equivalent semi-Boolean events, and accordingly resolving every non-maximal conventional magnitude into a multiplicity of statistically equivalent (semi-

Boolean) random variables. The obvious philosophical issue concerns the individuation of quantum events, and hence the individuation of quantum magnitudes. According to the conventional interpretation, statistically indistinguishable events (magnitudes) are identical, whereas according to the non-conventional interpretation, statistically indistinguishable events (magnitudes) may nevertheless be physically distinct. The existence and nature of non-statistical criteria by which to individuate events (magnitudes) presently falls within the province of the *interpretation of QM*, which is the subject of the remainder of this article.

4. THE INTERPRETATION OF QUANTUM MECHANICS

Each of the interpretations of quantum probability presented in Section 3 provides an abstract account of the exact manner in which the statistical states of QM arise, either as probability measures over a partial Boolean event structure, or as probability measures over a semi-Boolean event structure. Although the partial Boolean and semi-Boolean construals of quantum events are radically different, they nevertheless give rise to the same class of quantum statistical states; that is, although they disagree concerning the fine structure of quantum probability, they agree concerning its gross structure. Although the fine structure does not play a crucial role in the determination of the quantum statistical states, it does play a crucial role in the interpretation of QM, as demonstrated in the present section.

The statistical states of QM provide information of a purely probabilistic nature; each statistical state w determines for each magnitude m and Borel set X the probability $p^w_m(X)$ that the value of m lies in X. However, except in the case that $p^w_m(X) = 1$ or 0, w does not disclose whether the value of m *actually* lies in X. The interpretative problem for QM has been understood for the most part as the question whether (or to what extent, or in what sense) the description provided by the quantum statistical state is *complete* (cf., e.g., Bub, 1974, Fine, 1974). The question of completeness in turn has been analyzed largely in terms of the question whether it is possible to augment the quantum mechanical description in a way that is consistent with the principles of quantum theory. The possible augmentations of the quantum statistical states assume various forms, but they may all be understood as complementing the class of quantum statistical states with a class of *micro-states*. Presumably, whereas the former describe the (hypothetical) long-run relative frequency characteristics of quantum magnitudes, the latter describe their 'trial-by-trial' characteristics.

From the formal semantic viewpoint, an interpretation in this sense provides an abstract account of the contingent truth conditions of the elementary sentences of the form 'm has value in X', that is, it provides a semantics for the elementary (observation) language OL(QM) of QM (see Part I). Previously (1976), I have described a general formal semantic scheme for the interpretation of QM, which is based on the notion of an *interpreted probability logic* (IPL). An IPL is a structure $\langle S, W, V \rangle$, where S is a class of *elementary sentences*, W is a class of *probability assignments* on S, and V is a class of *valuations* on S. First of all, the class S of elementary sentences should include at least all atomic sentences of the form (m, X), which is read 'm has value in X', where m is a magnitude and X is a Borel subset of R. S may additionally include molecular sentences as in RL(QM) discussed in Part I and Section 6, but for the moment at least we simply identify S with OL(QM), which consists exclusively of atomic sentences. Next, a *probability assignment* on S is any function p from S into [0, 1] subject to the requirement that for each $m \in M$, the relativized function p_m is a probability measure over $\langle R, B(R) \rangle$; $p_m(X) = p(m, X)$ for all $X \in B(R)$. Finally, a valuation on S is any function v partially defined on S taking values in $\{0,1\}$ $(= \{F, T\})$ subject to the following conditions for each $m \in M$.

(v1) $v(m, R) = 1$;
(v2) $v(m, R - X) = 1 - v(m, X)$;
(v3) $v(m, X \cap Y) = v(m, X) \cdot v(m, Y)$.

Note that (v2) and (v3) are intended conditionally to mean that if either side is defined, the other side is defined and is related in the specified manner.

Every IPL$\langle S, W, V \rangle$ divides into a *statistical component* $\langle S, W \rangle$ and a *semantic component* $\langle S, V \rangle$, and associated with these two components are two propositional structures, respectively called the *statistical propositional structure* and the *semantic propositional structure*. A proposition is understood to be an equivalence class of sentences of S, where the equivalence relation is either statistical or semantic equivalence. Two sentences P, Q are said to be *statistically equivalent* in $\langle S, W, V \rangle$ if $p(P) = p(Q)$ for all $p \in W$; P, Q are said to be *semantically equivalent* in $\langle S, W, V \rangle$ if $v(P) = v(Q)$ for all $v \in V$.

It is plausible to require that any two semantically equivalent sentences of S are automatically statistically equivalent as well. Interpreted physically, this amounts to saying that if two events E, F are physically equivalent in the sense that they always *co-occur*, then they always have the same long-run relative frequency. On the other hand, it does not seem plausible to require that statistically equivalent sentences are automatically semantically equivalent, for it seems possible that two events E, F may always occur with the same long-run relative frequency, and yet E and F may not always co-occur (cf. Fine, 1973).

In order to illustrate these ideas within a relatively simple and non-controversial context, we consider classical statistical mechanics (CSM). According to CSM, every physical system $\mathscr{sys}$ is characterized by a *classical phase space* Z together with the set $B(Z)$ of Borel subsets of Z; Z is a $6n$-dimensional Euclidean space, where n is the number of degrees of freedom of $\mathscr{sys}$. Thus, $\langle Z, B(Z)\rangle$ form a classical sample space, where Z is the set of outcomes, and $B(Z)$ is the set of events. The random variables definable over $\langle Z, B(Z)\rangle$ are identified as the magnitudes pertaining to $\mathscr{sys}$, and the probability measures definable $\langle Z, B(Z)\rangle$ are identified as the statistical states pertaining to $\mathscr{sys}$. This enables us to define the statistical component $\langle S, W\rangle$ of the associated IPL $\langle S, W, V\rangle$ as follows: $S = \mathbf{M} \times B(R)$, where $\mathbf{M}$ is the class of random variables over $\langle Z, B(Z)\rangle$; each probability assignment $p \in W$ is of the form p_w for some probability measure w over $\langle Z, B(Z)\rangle$, where $p_w(m, X) = w(m^{-1}(X))$. In other words, associated with every elementary sentence (m, X), there is an event $e(m, X)$ in $B(Z)$, which is the subset $m^{-1}(X)$ of Z that the random variable m maps into X, and the probability of (m, X) according to p_w is simply the measure of the associated event $e(m, X)$ according to w.

Whereas the statistical component of the IPL associated with CSM is determined by the class of statistical states, the semantic component $\langle S, V\rangle$ is determined by the class of *micro-states,* which in CSM are identified with phase space points. Each micro-state $z \in Z$ demarcates the events into occurrent and non-occurrent according to whether or not $z \in e(m, X)$ Formal semantically speaking, each micro-state $z \in Z$ determines a *complete* valuation v_z on S according to the following rule.

$$v_z(m, X) = T \text{ if } z \in e(m, X); = F \text{ otherwise.}$$

As an immediate consequence of this particular identification of the micro-states and associated valuations, the semantic and statistical propositional structures of the IPL associated with CSM are identical. Two elementary sentences P, Q are statistically equivalent if and only if they are semantically equivalent, which in turn is true if and only if the associated events $e(P)$ and $e(Q)$ are identical.

Concerning how we interpret the two sorts of states, we regard each physical system as having both a statistical state and a micro-state, which together form what we call a *total state.* A classical total state, in particular, is an ordered pair (w, z), where w is a classical statistical state, and

z is a classical micro-state, which is an element of every countable measurable subset E for which $w(E) = 1$. Formal semantically construed, each total state is an ordered pair (p, v), where p is a probability assignment on S, and v is a valuation on S. Assigning a particular total state (w, z) to a system $\mathscr{sys}$ consists in placing $\mathscr{sys}$ in an (idealized) ensemble of similar systems, described by w, in addition to attributing to $\mathscr{sys}$ a number of individual properties, described by z. Whereas the statistical state is customarily associated with the thermodynamic characteristics of $\mathscr{sys}$, the micro-state is customarily associated with the mechanical characteristics of $\mathscr{sys}$.

Turning now to the problem of defining the IPL associated with QM, we first note that the statistical component $\langle S, W\rangle$, which is described in Sections 6 and 7 of Part I, is completely straightforward and non-controversial. Briefly, for a system with Hilbert space $\mathscr{H}$, the set W of probability assignments is generated by the class of statistical states (density operators) on $\mathscr{H}$, as a consequence of which the statistical propositions are biuniquely associated with the subspaces of $\mathscr{H}$. Whereas the statistical compoment of the IPL associated with QM is straightforwardly derivable from the extant quantum formalism, the semantic component is not. This is because the nature of quantum micro-states is not explicitly postulated by the conventional formalism, and accordingly must be specified by a particular interpretation of QM. In order to make comparisons with Reichenbach's interpretation of QM, in the remainder of this section, we examine two possible characterizations of quantum micro-states, which are based respectively on the Copenhagen variant and the anti-Copenhagen variant of van Fraassen's (1973) modal interpretation of QM, which I have previously examined in (1976).

According to the Copenhagen modal interpretation (CMI), which is intended to reflect orthodoxy, there is a strict analogy between the points in a classical phase space and the rays in a Hilbert space. In particular, every micro-state pertaining to a system with Hilbert space $\mathscr{H}$ is represented by a ray r in $\mathscr{H}$, or equivalently, it is represented by the pure statistical state w with range r. A total state is then defined to be an ordered pair (w, r), where w is a statistical state, and r is a ray contained in the range of w. Each Copenhagen micro-state r determines an associated valuation v_r *partially* defined on S as follows, where $\text{ran}(w) = r$.

$$v_r(m, X) = \begin{cases} T \text{ if } p_m^w(X) = 1 \\ F \text{ if } p_m^w(X) = 0 \\ \text{undefined otherwise} \end{cases}$$

As an immediate consequence of identifying micro-states with rays in $\mathscr{H}$, the semantic propositions of the quantum IPL, like the statistical propositions, are biuniquely associated with the subspaces of $\mathscr{H}$, which is to say that according to CMI, the semantic propositional structure and the statistical propositional structure of the quantum IPL are identical. It follows that the quantum events (magnitudes) are individuated entirely with respect to their statistical behavior, and may accordingly be identified with the subspaces of (s.a. operators on) $\mathscr{H}$, which is what one would expect from orthodoxy. The identification of quantum events with subspaces of $\mathscr{H}$ enables us to provide an alternative characterization of the valuation v_r determined by the micro-state r, as follows.

$$v_r(m, X) = \begin{cases} T \text{ if } r \in e(m, X) \\ F \text{ if } r \perp e(m, X) \\ \text{undefined otherwise} \end{cases}$$

Here, $e(m, X)$ is the (collection of rays contained in the) subspace assigned to X by the spectral measure associated with m.

Whereas CMI maintains the identity of the semantic and statistical propositional structures, equivalently CMI maintains that the quantum events (magnitudes) are individuated by the statistical states, this is explicitly denied by the anti-Copenhagen modal interpretation (ACMI), which is intended to be a 'hidden variable' interpretation of QM. The ACMI is best understood, I believe, from the viewpoint of semi-Boolean probability theory, according to which the fundamental structure associated with a system with Hilbert space $\mathscr{H}$ is the manual $\mathscr{M}(\mathscr{H})$, which is identified with the family $(Z(i): i \in I)$ of frames on $\mathscr{H}$. In particular, ACMI does not identify quantum events with partial Boolean events (subspaces of $\mathscr{H}$), but rather with semi-Boolean events (subframes of $\mathscr{H}$), and accordingly identifies each quantum magnitude with a (semi-Boolean) random variable on $\mathscr{M}(\mathscr{H})$, that is, a function $\mathbf{m}$ from $Z(i)$ into R for some $i \in I$.

Whereas CMI regards the quantum 'phase space' simply to be the set of all rays of $\mathscr{H}$, and accordingly construes quantum micro-states as *individual* rays in $\mathscr{H}$, ACMI regards the quantum 'phase space' to be a collection $(Z(i): i \in I)$ of classical phase spaces (frames), and accordingly construes quantum micro-states to be *collections* of rays of $\mathscr{H}$. To be more precise, each anti-Copenhagen (a.c.) micro-state is a choice function λ on $(Z(i): i \in I)$ which selects from each classical phase space (frame) $Z(i)$ a classical micro-state (ray) $\lambda(i) \in Z(i)$[12] Next, a total state according to ACMI is an ordered pair (w, λ), where w is a statistical state (density operator), and

λ is a choice function on $(Z(i) : i \in I)$ subject to the restriction that no $\lambda(i)$ is orthogonal to $ran(w)$.

Unlike their Copenhagen counterparts, every a.c. micro-state assigns an exact value to every *anti-Copenhagen* magnitude (i.e., random variable over $\mathcal{M}(\mathcal{H})$). Every random variable over $\mathcal{M}(\mathcal{H})$ is a real-valued function $\mathbf{m}$ with domain $Z(i)$ for some $i \in I$; henceforth, $i[\mathbf{m}]$ denotes the *unique* index point $i \in I$ such that $dom(\mathbf{m}) = Z(i)$ (we assume the indexing is 1–1). Now, every a.c. micro-state λ selects a ray $\lambda(i[\mathbf{m}]) \in Z(i[\mathbf{m}])$, and accordingly assigns a value $val_\lambda(\mathbf{m})$ to $\mathbf{m}$ in a completely classical manner. Specifically, $val_\lambda(\mathbf{m})$ is the value of the random variable $\mathbf{m}$ at the selected point $\lambda(i[m])$: $val_\lambda(\mathbf{m}) = \mathbf{m}(\lambda(i[\mathbf{m}]))$. Accordingly, every a.c. micro-state λ determines an associated *complete* valuation v_λ on $S^\circ = \mathbf{M}^\circ \times B(R)$, where $\mathbf{M}^\circ$ is the class of a.c. magnitudes, according to the following rule.

$$v_\lambda(\mathbf{m}, X) = T \text{ if } \mathrm{val}_\lambda(\mathbf{m}) \in X; = F \text{ otherwise.}$$

As a consequence of the ACMI identification of quantum micro-states and the associated valuations, the semantic propositional structure of the IPL of QM is considerably different from the statistical propositional structure. Whereas the latter is a partial Boolean algebra $(\mathbf{B}(i) : i \in I)$ in which there is a considerable amount of overlap among the various $B(i)$, the former is a partial Boolean algebra $(B(k) : k \in K)$ in which the overlap among the various $B(k)$ is absolutely minimal; specifically, for all $j, k \in K$ $(j \neq k)$, $B(j) \cap B(k) = \{1, 0\}$.

5. SYNOPSIS OF REICHENBACH'S INTERPRETATION OF QUANTUM MECHANICS

In the present section, we outline the salient features of Reichenbach's interpretation of QM, indicating how they are *inter*related, and in later sections we examine how they are related to the general problem of interpreting QM, as described in the previous sections. I take Reichenbach's chief tenets to include the following: the distinction between phenomena and interphenomena; the distinction between restrictive and exhaustive interpretations; the distinction between normal and non-normal systems of descriptions; the theory of equivalent descriptions; the proposal of a non-bivalent semantics for the elementary language of QM.

Fundamental to Reichenbach's general interpretation of QM is the distinction between phenomena and interphenomena, which he describes as follows (1944; henceforth PhF, 20, 21).

. . . There is, however, a class of occurrences which are so easily inferable from macroscopic data that they may be considered as observable in a wider sense. We mean all those occurrences which consist in coincidences, such as coincidences between electrons, or electrons and protons, etc. We shall call occurrences of this kind *phenomena*. The phenomena are connected with macroscopic occurrences by rather short causal chains; we therefore say that they can be 'directly' verified by such devices as the Geiger counter, a photographic film, a Wilson cloud chamber, etc.

We then shall consider as unobservable all those occurrences which happen between coincidences, such as the movement of an electron, of a light ray from its source to a collision with matter. We call this class of occurrences the *interphenomena*. Occurrences of this kind are introduced by inferential chains of a much more complicated sort; they are constructed in the form of an *interpolation* within the world of phenomena, and we can therefore consider the distinction between phenomena and interphenomena as the quantum mechanical analogue of the distinction between observed and unobserved things.

Intimately related to the phenomena-interphenomena distinction is the distinction between *normal systems* of description and *non-normal systems* of description, and the distinction between *exhaustive* interpretations and *restrictive interpretations*. According to Reichenbach (PhF, p. 19), a normal system of description is one in which "the laws of nature are the same whether or not the objects are observed [, and] the state of the objects is the same whether or not the objects are observed." He also states somewhat more precisely (PhF, pp. 23–34) that a normal descriptive system is one which is free from *causal anomalies*, that is, non-local effects. It is the claim of Reichenbach's *Principle of Anomaly* (PhF, p. 33) that "the class of descriptions of interphenomena contains no normal system." (Also see Reichenbach (1956), pp. 217, 218.) That no *complete* description of interphenomena is free from causal anomalies suggests the usefulness of restrictive interpretations, which Reichenbach describes as follows (PhF, p. 33).

. . . Since no normal description of interphenomena exists, it has been suggested we should renounce any description of interphenomena; we should restrict quantum mechanics to statements about phenomena—then no difficulties of causality will arise. The impossibility of a normal system is construed, in this conception, as a reason for abandoning all descriptions of interphenomena. We shall call conceptions of this kind *restrictive interpretations*, since they restrict the assertions of quantum mechanics to statements about phenomena. The rule expressing this restriction can assume various forms, and we shall therefore have several restrictive interpretations. Interpretations which do not use restrictions, like the corpuscle and wave interpretation, will be called *exhaustive interpretations*, since they include a complete description of interphenomena.

Whereas a restrictive interpretation provides a description only of phenomena, or phenomenal characteristics of a system, an exhaustive interpretation provides a description of interphenomenal characteristics as well as phenomenal characteristics of a system. However, in any given physical situation, among the various abstractly possible interpolations of interphenomenal characteristics, there are, according to Reichenbach, a number which are equally true or equally faithful to the objective physical facts. This is intimately related to Reichenbach's conception of *equivalent* descriptions, which he describes as follows (PhF, p. 23).

> . . . Given the world of phenomena, we can introduce the world of interphenomena in different ways; we then shall obtain a class of equivalent descriptions of interphenomena, each of which is equally true, and all of which belong to the same world of phenomena. In other words: In the class of equivalent descriptions of the world the interphenomena vary with the descriptions, whereas the phenomena constitute the invariants of the class . . . Nowhere do we find an unambiguous supplementation of observations; interpolation of unobserved values can be given only by a class of equivalent descriptions.

Although it is not clear that Reichenbach had this in mind in proposing his three-valued logic for QM, the existence of distinct equivalent descriptions leads in a natural way to a non-bivalent semantics for the elementary quantum language OL(QM). If two descriptions D_1 and D_2 are formulated in the same language, say OL(QM), then they are distinct exactly to the extent that they disagree concerning the truth value of at least one elementary sentence. On the other hand, if D_1 and D_2 are equivalent in the sense that they both *faithfully reflect* the physical situation S, then they cannot disagree concerning any sentence whose truth value is uniquely determined by S, they can only disagree concerning sentences whose truth value is *arbitrary* with respect to S. If S admits a class K of equivalent descriptions formulated in the same language OL(QM), then there are sentences of OL(QM) whose truth value is arbitrary with respect to S. In order to remove the arbitrariness from K, we simply take what is *common to all* descriptions in K, that is, we construct the associated *supervaluation* v_k; v_k evaluates a sentence A as true (false) just in case *every* description in K evaluates A as true (false); otherwise, v_k does not evaluate A, or it evaluates A as 'indeterminate' (cf. van Fraassen, 1966, 1969, 1971; see Section 6).

Let us summarize. Among the class of complete (exhaustive) descriptions of micro-systems, according to Reichenbach's Principle of Anomaly, none are free from causal anomalies. It is additionally claimed that the anomalies can be avoided by deleting the *statements* that concern interphenomena; the deletion can be accomplished by assigning a third truth value (or no truth value) to all such *statements*. The proposal is supported by the claim that, in each physical situation, there is not a unique correct description, but rather a whole class of alternative equivalent descriptions; each complete description extends the restrictive description, but in an arbitrary manner.

6. A RESTRICTIVE INTERPRETATION OF QUANTUM MECHANICS

According to Reichenbach, the events pertaining to a physical system divide into phenomena and interphenomena. This distinction, which we call the *P-I* distinction, has both a dynamical and statical interpretation. On the one hand, a system sys can be observed at one instant of time t_1, but not at another instant t_2; we might therefore call the t_1-events phenomenal, and the t_2-events interphenomenal. Thus, the dynamical version of the *P-I* distinction applies not only to quantum mechanical objects, but also to classical and everyday objects. On the other hand, it is possible that certain properties (magnitudes) pertaining to sys are *in fact* tested (measured) at a particular instant t_1, while *at the same time* other properties (magnitudes) pertaining to sys are not *in fact* tested (measured). We might therefore call the tested properties (measured magnitudes) phenomenal, and the untested properties (unmeasured magnitudes) interphenomenal. In particular, according to the statical interpretation of the *P-I* distinction, at any given instant, the properties (magnitudes) pertaining to a physical system divide into those which are phenomenal and those which are interphenomenal.

That Reichenbach regarded the statical interpretation to be an admissible construal of the *P-I* distinction is evidenced on page 40 of PhF, where he writes that according to the Bohr-Heisenberg interpretation (see Part I), "only statements about measured entities, i.e., phenomena, are admissible; statements about unmeasured entities, or interphenomena, are called meaningless." Note that Reichenbach uses the term 'entity' in the same way we use the term 'magnitude'. He goes on to say, "This has the immediate consequence that statements about simultaneous values of complementary entities cannot be made." It is difficult to see how this latter claim follows unless 'phenomena' and 'interphenomena' are themselves simultaneously applicable notions. Since the statical interpretation of the *P-I* distinction is the version that bears crucially upon the conceptual issues of QM, we adopt the statical version rather than the dynamical version in the remainder of this work.

The *P-I* distinction gives rise to a classification of interpretations into those which are exhaustive and those which are restrictive. Whereas an *exhaustive* interpretation provides (an abstract schema for) a description of *all* events (properties, magnitudes), both phenomenal and interphenomenal, a *restrictive* interpretation provides a description of only phenomenal events. Since this classification is not exhaustive, we add a third category of *intermediate* interpretations, which provide a description of all phenomenal events, and *some but not all* interphenomenal events. Whereas the anti-Copenhagen modal interpretation is an exhaustive interpretation, the Copenhagen modal interpretation is an intermediate interpretation (see Section 7). In the present section, we examine a restrictive interpretation.

Although there are a large variety of non-exhaustive interpretations, they may all be regarded as proposing a non-bivalent[13] semantics for the elementary quantum language OL(QM). We suppose that an event E is 'described' exactly insofar as the associated elementary sentence $s(E)$ is *classically evaluated*, that is, assigned a classical truth value T or F, and that E is 'not described' exactly insofar as $s(E)$ is not classically evaluated. Thus, for example, according to a restrictive interpretation, in any particular situation, only elementary sentences referring to phenomenal events are classically evaluated, whereas according to an exhaustive interpretation, in every situation, every elementary sentence is classically evaluated. There are basically two methods by which a non-bivalent semantics can be implemented; either one can opt for truth-value gaps (see, e.g., van Fraassen, 1966, Lambert, 1969), allowing valuations that assign *no* truth value to certain sentences, or one can opt for additional non-classical truth-values, which are assigned to those sentences not classically evaluated. In Part I, I argue that whereas the Bohr-Heisenberg interpretation (as presented by Reichenbach in PhF) can be understood as adopting truth-value gaps, using the term 'meaningless' to mark the gaps, Reichenbach adopts a genuine three-valued truth-functional semantics, using the third truth value I ('indeterminate') much in the same way that Bohr-Heisenberg use the term 'meaningless'.

In the remainder of this section, we examine a restrictive interpretation, denoted RI, which could plausibly be attributed to Reichenbach. Both in his reply to Nagel (1946) and in his paper (1948) on the Principle of Anomaly (see Section 8), Reichenbach insists that the intermediate truth value I may be explicated by Table 2 on page 146 of PhF. This table concerns a special observation language, which contains sentences of the form 'm is measured' and 'm is observed to have value r', and it correlates the elementary sentence 'm has value r' (denoted (m, r)) with the ordered pair consisting of 'm is measured' (denoted $[m]$) and 'm is observed to have value r' (denoted $[m, r]$).[14] In particular, (m, r) is indeterminate in truth value if $[m]$ is false, and otherwise has the truth value of $[m, r]$. Note that since $[m]$

and $[m, r]$ pertain to laboratory operations and results, they are of the true-false category. Furthermore, in Section 34 of PhF, Reichenbach discusses conditionals of the form $[m] \ni [m, r]$,[15] where the conditional connective $\ni$ is called *quasi-implication*, which is a three-valued connective having the property that $A \ni B$ has the truth value of B if A is true, and is otherwise indeterminate.[16] Combining these two suggestions in the natural way leads to the following reconstruction of Reichenbach's table.

(m, r)	$[m]$	$[m, r]$	$[m] \ni [m, r]$
T	T	T	T
F	T	F	F
I	F	T	I
I	F	F	I

Since it is more natural to translate each elementary sentence into a sentence rather than an ordered pair of sentences, and since the above table constitutes a plausible construal of Reichenbach's table, we will interpret Reichenbach as having intended the elementary sentence (m, r) to be translated as the conditional $[m] \ni [m, r]$.

In order to determine the extent to which Reichenbach's proposal succeeds in explicating the third truth value, we first show how the proposed translation provides a very broad and general account of the third truth value, and then we show how the specific details of the restrictive interpretation can be filled in by reference to the Hilbert space formalism. In particular, the quantum formalism is consulted in order to determine the 'meaning postulates' by which the semantics of *abstractly admissible* valuations is narrowed down to the class of *quantum physically admissible* valuations.

Consider first the language employed by Reichenbach, denoted RL(QM), which consists of the elementary (observation) language OL(QM) augmented by a collection of three-valued truth-functional connectives (see Part I). Like OL(QM), RL(QM) consists of atomic formulas of the form (m, X), although Reichenbach concentrates on those of the form (m, r), short for $(m, \{r\})$. However, unlike OL(QM), RL(QM) consists additionally of a host of proper molecular formulas generated from the atomic formulas by the application of three-valued truth-functional connectives. In Part I, we employ two (independent) primitive connectives: negation and the conditional. In the present section, we employ a functionally equivalent set of five (non-independent) primitive connectives: negation ($\sim$), conjunction (&), disjunction ($\vee$), the assertion operator

(**T**), and the denial operator (**F**), whose truth tables appear in Part I.[17] Concerning the semantics of RL(QM), each valuation v is determined by a trivalent assignment of truth values T, I, F to the atomic formulas (atoms) of RL(QM), and each trivalent assignment is extended to an (abstractly) admissible valuation on RL(QM), in accordance with the truth tables of the connectives of RL(QM).

We next consider the special observation language, denoted SOL, which is intended to explicate RL(QM). The atomic formulas of SOL are of two sorts: if m is a magnitude, then the expression $[m]$, to be read 'm is measured', is an atomic formula, and if X is a Borel subset of R, then the expression $[m, X]$, to be read 'm is observed to have value in X', is an atomic formula. The primitive connectives of SOL include negation, conjunction, disjunction (with truth tables just like RL(QM)), in addition to the quasi-implication connective. Concerning the semantics of SOL, each valuation on SOL is determined by a *bivalent* assignment of classical truth values to the atoms of SOL, and each bivalent assignment on the atoms of SOL is extended to an (abstractly) admissible valuation on SOL, in accordance with the truth tables of the connectives of SOL.

We now show how RL(QM) can be translated into SOL in a way that preserves the semantics. The translation of each formula A of RL(QM) involves two steps. First of all, A is mapped by the function n into its normal form equivalent $n(A)$; a formula B in RL(QM) is in normal form, in this sense, just in case no molecular subformula of B is within the scope of an occurrence of either **T** or **F**. Next, the formula $n(A)$ is mapped into SOL by the function t, recursively defined on the class of normal form formulas of RL(QM) as follows.

Base clauses:

$$t(m, X) = [m] \ni\!\!- [m, X]$$
$$t(\mathbf{T}(m, X)) = [m] \,\&\, [m, X]$$
$$t(\mathbf{F}(m, X)) = [m] \,\&\, {\sim}[m, X]$$

Inductive clauses:

$$t({\sim}A) = {\sim}t(A)$$
$$t(A \,\&\, B) = t(A) \,\&\, t(B)$$
$$t(A \vee B) = t(A) \vee t(B)$$

The translation function is simply the composition of t with n, and is denoted tn. That this function is an adequate translation of RL(QM) into SOL amounts to the following facts, which may be shown routinely by induction. (1) For every admissible valuation v on SOL, the function v^*

on RL(QM), defined so that $v^*(A) = v(tn(A))$, is an admissible valuation on RL(QM). (2) For every admissible valuation v on RL(QM), there is an admissible valuation v° on SOL such that $v(A) = v^\circ(tn(A))$.

That the translation *tn* of RL(QM) into SOL provides an *explication* of RL(QM) rests primiarily on the fact that, unlike RL(QM), SOL has no irreducible non-true-false formulas; rather, each non-true-false formula is molecular, containing at least one occurrence of the quasi-implication connective. Whereas molecules of this sort may be neither true nor false, every atom is always either true or false in a completely classical manner. In particular, every physical (experimental) situation is formally characterized by a bivalent assignment on the atoms of SOL, which is then extended to a trivalent valuation on the entire class of formulas of SOL. However, the extension is purely formal in character, since no further reference to the world is required once the atoms are evaluated, and it is only in the extension that the failure of bivalence arises. Thus, inasmuch as SOL has no irreducible non-true-false formulas, and inasmuch as *tn* is an adequate and natural translation of RL(QM) into SOL, the occurrence of the third truth value in RL(QM) is in some sense explained.

Nevertheless, our explication of RL(QM) on the basis of SOL is incomplete as it stands, since the class of admissible valuations on SOL, and hence RL(QM), is not sufficiently narrow to be of any physical significance. What is required is a set of "meaning postulates" that delimit the meanings of the formulas of SOL (in effect) by narrowing down the class of *abstractly admissible* valuations to the class of *quantum physically admissible* valuations. For example, the physically significant valuations should reflect the following facts: (1) $[m, r]$ and $[m, r^*]$ cannot be simultaneously true unless $r = r^*$. (2) If $X \subseteq Y$, then $[m, X]$ can be true only if $[m, Y]$ is also true. (3) $[m, X]$ can be true only if $[m]$ is true. Furthermore, the quantum physically admissible valuations should reflect the various relations among the magnitudes, most importantly perhaps the relation of complementarity. In order to delimit the quantum physically admissible valuations on SOL, and hence RL(QM), in the remainder of this section, we propose a semantics which is based directly on the Hilbert space formalism of QM.

In the proposed semantics, denoted RS, every valuation on SOL is determined by what we call a *state of affairs*, which is identified with an ordered pair consisting of a *micro-state*, representing the physical system, and a *macro-context*, representing the experimental environment. We suppose that each macro-context is maximal and classically describable (à la Bohr), which leads us to identify each macro-context with a frame $Z(i)$ in the relevant Hilbert space manual $(Z(i): i \in I)$. Actually, as a purely technical matter, we identify each macro-context with an index point $i \in I$, which in turn determines a particular frame $Z(i)$. Although it is unnecessary in the context of semantics RS, in order to facilitate the transition to non-restrictive semantics, we define each micro-state in the anti-Copenhagen fashion as a choice function λ on $(Z(i): i \in I)$. Thus, each state of affairs is an ordered pair (λ, i), where $\lambda \in \mathbf{X}(Z(i): i \in I)$, and $i \in I$.

For the sake of completeness, and for the sake of later comparison, we give the truth conditions for the formulas of SOL according to RS both with respect to the conventional construal and with respect to the non-conventional construal of quantum magnitudes. We identify each semi-Boolean magnitude with a random variable $\mathbf{m}$ over the Hilbert space manual $\mathcal{M}(\mathcal{H}) = (Z(i): i \in I)$, and we identify each conventional magnitude with an equivalence class M of random variables over $\mathcal{M}(\mathcal{H})$. We continue to use lower case boldface letters to denote random variables; we use upper case italic letters primarily to denote equivalence classes of random variables and secondarily to denote the associated partial Boolean spectral measures or s.a. operators; finally, we use lower case italic letters to denote 'generic' magnitudes.

Insofar as we identify conventional magnitudes with equivalence classes of random variables, to say that $\mathbf{m}$ resolves M is simply to say that $\mathbf{m} \in M$. A conventional magnitude M is said to be *definable* on a frame Z if there exists an $\mathbf{m} \in M$ that is *defined* on Z (i.e., $\text{dom}(\mathbf{m}) = Z$). The set of frames over which m is defined (definable) is denoted $DOM(m)$; in particular, $DOM(M) = \{Z \in \mathcal{M}(\mathcal{H}): \text{dom}(\mathbf{m}) = Z \text{ for some } \mathbf{m} \in M\}$, and $DOM(\mathbf{m}) = \{\text{dom}(\mathbf{m})\}$. Recall that associated with every magnitude m and Borel set X, there is an event $e(m, X)$; specifically, $e(\mathbf{m}, X)$ is the set $\mathbf{m}^{-1}(X)$ of rays that are mapped by $\mathbf{m}$ into X, and $e(M, X)$ is the set of rays contained in the subspace assigned to X by the spectral measure associated with M; equivalently, $e(M, X)$ is the set of rays contained in the subspace generated by $\mathbf{m}^{-1}(X)$ for *any* $\mathbf{m} \in M$.

The generic definitions of $DOM(m)$ and $e(m, X)$ permit us to give the truth conditions for the formulas of SOL without explicit reference to the exact character of the magnitudes. Where $s = (\lambda, i)$ is a state of affairs, the asociated valuation v_s on SOL is determined by a bivalent assignment on the atoms of SOL, also denoted v_s, defined as follows.

$$v_s([m]) = T \text{ if } Z(i) \in DOM(m); = F \text{ otherwise.}$$
$$v_s([m, X]) = T \text{ if } Z(i) \in DOM(m) \text{ and } \lambda(i) \in e(m, X);$$
$$= F \text{ otherwise.}$$

As noted earlier, every valuation v on SOL induces an associated valution v^* on RL(QM), where $v^*(A) = v(tn(A))$ for all $A \in$ RL(QM). For example, the formula (m, X) is translated by tn into the conditional $[m] \ni\!\!- [m, X]$,[18] so $v^*(m, X) = v([m] \ni\!\!- [m, X])$, but according to the definition of quasi-implication, $[m] \ni\!\!- [m, X]$ is indeterminate if $[m]$ is false, and otherwise has the truth value of $[m, X]$. Thus, the valuation v_s^* on RL(QM) induced by v_s ($s = (\lambda, i)$) is determined by a trivalent assignment on the atoms of RL(QM), also denoted v_s^*, which may be defined as follows.

$$v_s^*(m, X) = \begin{cases} T \text{ if } Z(i) \in \mathit{DOM}(m) \text{ and } \lambda(i) \in e(m, X) \\ F \text{ if } Z(i) \in \mathit{DOM}(m) \text{ and } \lambda(i) \notin e(m, X) \\ I \text{ otherwise} \end{cases}$$

In other words, the elementary sentence (m, X) is true or false at (λ, i) according to RS if and only if m is defined (definable) over $Z(i)$, and is otherwise indeterminate at (λ, i).

The set of valuations over OL(QM) (RL(QM)) induced by valuations of RS is denoted RI, which constitutes a restrictive interpretation of QM.[19] In particular, in any given situation (state of affairs), *only* elementary sentences referring to phenomenal events are classically evaluated. Each state of affairs (λ, i) divides the quantum magnitudes into those which are phenomenal and those which are interpehenomenal; magnitude m is phenomenal at (λ, i) if m is defined (definable) over $Z(i)$ (i.e., $Z(i) \in \mathit{DOM}(m)$); otherwise, m is interpehenomenal. In the case of events, an event e is phenomenal if $e \in \mathbb{E}(i)$, in the case of semi-Boolean events, or $e = p(f)$ for some $f \in \mathbb{E}(i)$, in the case of partial Boolean events (semi-Boolean propositions); otherwise, e is interphenomenal. Note carefully that, although (m, X) is classically evaluated only if the associated event $e(m, X)$ is phenomenal, the converse need not be true. Consider the example in Section 3 given by Figures 1–3, and consider the state of affairs (λ, a), where $\lambda(a) = a1$ $(= b1)$; then $\mathbf{m}_a$ is phenomenal, so $(\mathbf{m}_a, 1)$ is classically evaluated (as true), whereas $\mathbf{m}_b$ is interphenomenal, so $(\mathbf{m}_b, 1)$ is evaluated as indeterminate. On the other hand, $e(\mathbf{m}_b, 1) = e(\mathbf{m}_a, 1)$, so $e(\mathbf{m}_b, 1)$ is phenomenal according to our definition.

7. INTERPOLATING INTERPHENOMENA WITH STALNAKER CONDITIONALS

The restrictive semantics RS (RI) proposed in the previous section has a serious shortcoming: merely knowing the actual truth values of all the formulas of SOL (RL(QM)) does not provide as much information about the physical system as seems *physicaally* vailable. Although every valuation on SOL (RL(QM)) discloses the value of every phenomenal magnitude, it provides no information whatsoever concerning the interphenomenal magnitudes, except of course that they are not actually measured. Now, the choice of which particular maximal compatible class of magnitudes to measure is presumably a matter of volition on the part of the experimenter; in any given situation, although we may *in fact* measure maximal magnitude m, we *could* have chosen *instead* to measure a complementary magnitude m^*, and we might therefore wonder what *would* the result *have been if* we had instead measured m^*. Answers to questions of this sort may be expressed as *counterfactual conditionals* of the form 'if m^* *were* (had been) measured, then m^* *would* be (would have been) observed to have value r.' Concerning sentences of this sort, the restrictive semantics RI provides no information; if a magnitude m is not measured, that is, if m is interphenomenal, then every sentence pertaining to m is simply evaluated as indeterminate, even 'm has value in R'.

In Section 4, we examined two *non-restrictive* interpretations of QM: the Copenhagen modal interpretation (CMI) and the anti-Copenhagen modal interpretation (ACMI). CMI is an intermediate interpretation, and assigns (exact) values to some but not all interphenomenal magnitudes (even pairs of incompatible magnitudes), whereas ACMI is an exhaustive interpretation, and assigns (exact) values to all magnitudes. In the present section, we examine these two interpretations in the light of the restrictive interpretation RI; in particular, we will view both of them as interpolating interphenomenal events by means of counterfactual conditionals. Whereas CMI will be seen as evaluating (m, X) as true precisely when a measurement of m *would necessarily* yield a value in S, ACMI will be seen as evaluating (m, X) as true precisely when a measurement of m *would actually* yield a value in X.

Recall that according to CMI, every micro-state pertaining to a system with Hilbert space $\mathcal{H}$ is identified with a ray r in $\mathcal{H}$, equivalently with a pure statistical state w with range r. An elementary sentence (m, X) is then true (false) at r if and only if the associated event $e(m, X)$, a subspace of $\mathcal{H}$ according to CMI, contains (is orthogonal to) r; equivalently, (m, X) is true (false) at r if and only if the probability according to w that m lies in X is one (zero). Henceforth, we say that an event e is true (false) if $e = e(m, X)$ for some (m, X) that is true (false), and we say that an event e is a *T-F* event if e is either true or false, in this sense. Now, the subspaces containing (orthogonal to) a given ray r form a maximal filter (ideal) on the lattice $L(\mathcal{H})$ of subspaces of $\mathcal{H}$. Thus, according to CMI, in any given situation, there are considerably more *T-F* events than phenomenal events, which may be illustrated by a 3-dimensional Hilbert space *H*3. Every frame on *H*3 consists of three rays, and accordingly each frame generates a eight element Boolean sublattice of *L*(*H*3). It follows that, in any given situation, there are only eight phenomenal events, and hence only eight *T-F* events according to RI.[20] On the other hand, every maximal filter (ideal) on *L*(*H*3) has infinitely many elements, being isomorphic to *L*(*H*2). Thus, the number of *T-F* events according to CMI is vastly larger than the number of phenomenal events. The contrast between RI and CMI may be described by saying that, whereas the true (false) events according to RI form a maximal filter (ideal) on some particular Boolean sublattice of *L*(*H*), the true (false) events according to CMI form a maximal filter (ideal) on the entire lattice *L*(*H*).

Whereas the *T-F* events according to RI are always jointly compatible, since they form a Boolean sublattice, the *T-F* events according to CMI are generally not jointly compatible. To state things somewhat more dramatically, every Copenhagen micro-state assigns (exact) values to infinitely many mutually incompatible magnitudes, although it does not assign (exact) values to all magnitudes. To the extent that we uphold the more or less standard view that incompatible magnitudes cannot be *actually* observed simultaneously,[21] and insofar as CMI does assign simultaneous exact values to incompatible magnitudes, we must provide an account of non-phenomenal *T-F* events.

In order to reconcile the two opposing versions of 'orthodoxy', I propose to analyze quantum events (properties) as dispositions, where phenomenal events are *manifest* dispositions and interphenomenal events are *latent* dispositions. A disposition *D* with 'triggering' event *T* is said to be *manifest* (relative to a particular individual system *sys* and at a particular instant of time *t*) if *T* is present (i.e., 'imposed on' *sys* at time *t*); *D* is said to be *latent* if *T* is not present. By way of illustration, if at time *t* a specimen *s* of sugar is immersed in water, and accordingly is in solution, then the *solubility* of *s* is manifest at time *t*. On the other hand, the *flammability* of *s* is latent at time *t*, since the relevant triggering conditions are not (indeed, cannot be) present for *s* at time *t*. Indeed, the solubility and flammability of *s* are *mutually incompatible* dispositions, insofar as one can be manifest only if the other is latent.

According to the dispositional account, quantum events (properties) are dispositions where the relevant triggering events are measurements. The chief complication is that they are *probabilistic dispositions*, or what are often called *propensities* (following Popper, 1957, 1959). The quantum events first of all divide, at any time, into those which are manifest (tested, phenomenal) and those which are latent (untested, interphenomenal). The latent quantum dispositions in turn divide into those which have 'universal strength', which correspond to the *T-F* interphenomenal events, and those which have 'non-universal strength', which correspond to non-*T-F* interphenomenal events. A propensity is said to have universal strength if it has strength (probability) 1 or 0; otherwise, it is said to have non-universal strength. Viewing quantum events as dispositions, we can say that (m, r) and (m^*, r^*) are simultaneously true of a system $\mathscr{sys}$ even when m and m^* are incompatible, and we can say that m and m^* cannot be measured simultaneously. For to say that (m, r) is true of $\mathscr{sys}$ is not to say that m is *actually observed* to have value r, but rather that m *would* be observed to have value r *if m were* measured, and similarly to say that (m^*, r^*) is true is to say that m^* *would* be observed to have value r^* *if* m^* *were* measured. This is completely analogous to saying that the specimen *s would* dissolve *if* immersed in water, and that *s would* burn *if* ignited, which does not entail that one can *simultaneously* dissolve and burn *s*.

There are two ways that the quantum propensities can be analyzed, either in terms of conditional probability or in terms of counterfactual conditionals.[22] In the former case, we speak of the probability that a magnitude m will/would be observed to have value in X *on the condition* that m is measured. Employing our earlier notation in a slightly more general way, we use $[m]$ to denote the measurement *event* and $[m, X]$ to denote the observation *event*, so the conditional probability is written $p([m, X]/[m])$. The interpretation of quantum probability as being basically conditional in nature was suggested by Sneed (1970), who developed this idea in the finite case. The detailed technical machinery required to implement this idea in the general case was developed by Hooker and van Fraassen (1976). The chief difficulty is that according to the conventional Kolmogorov model, the conditional probability $p(B/A)$ of event B relative to event A is undefined if $p(A) = 0$, and is otherwise equal to $p(A \cap B)/p(A)$. It is therefore impossible within the Kolmogorov framework to define more than denumerably many conditional probability functions of the form $p(./A)$, whereas there are non-denumerably many distinct complementary magnitudes in QM. The framework adopted by Hooker and van Fraassen is the theory of *Popper probability measures*, which represents an axiomatization in which probability is taken as fundamentally a two-place function (something to which Reichenbach was obviously sympathetic!) The upshot of their investigation, which is intended to explicate Bohr's interpretation of QM, is that quantum probability can consistently be interpreted as conditional probability where the the conditioning events are measurements. The Hooker-van-Fraassen analysis enables us to reinterpret the Copenhagen modal interpretation as saying that (m, X) is true (false) just in case the probability that m will/would be observed to have value in X *on the condition that* m is measured is one (zero), and that (m, X) is truth-valueless (indeterminate) otherwise.

The alternative analysis of quantum propensities is in terms of counterfactual conditionals of the form 'if m *were* measured, then m *would* be observed to have value in X', which we symbolize $[m] \rightarrow [m, X]$. There is an intimate relation between counterfactual conditionals and conditional probability; according to Stalnaker's thesis (Stalnaker, 1970, van Fraassen, 1976), the absolute probability of the conditional $A \rightarrow B$ is the conditional probability of B relative to A: $p(A \rightarrow B) = p(B/A)$. Whether or not this thesis can be upheld in general, it can be upheld in the context of QM, as shown by Hooker and van Fraassen, who construct a conditional connective, which is a Stalnaker conditional (see below) for which the absolute probability $p([m] \rightarrow [m, X])$ equals the conditional probability $p([m, X]/[m])$. This enables us to provide yet another analysis of CMI: (m, X) is true (false) just in case the probability of $[m] \rightarrow [m, X]$ is one (zero), and is otherwise truth-valueless (indeterminate).

Both of these analyses of CMI suggest that CMI may be understood as translating the elementary sentence (m, X) as the conditional 'if m were measured, then m *would certainly* (necessarily) be observed to have value in X'. We are therefore tempted to ask what sort of interpretation of QM might result if (m, X) is instead translated as 'if m were measured, then m *would actually* be observed to have value in X'. There are at least two alternative answers to this question, which are discussed in Section 9 and in the remainder of this section.

Numerous explications of counterfactual conditionals have been proposed in recent years, most notably perhaps by Stalnaker (1968, 1970)[23] and Lewis (1973). Van Fraassen (1974) has constructed a general scheme subsuming both Stalnaker conditionals (of the $C2$ variety) and Lewis conditionals (of the C1 variety), which we adopt (and adapt) here. According to this scheme, every *model* is characterized by a class W of *possible worlds* together with a *selection function* f. Each possible world $w \in W$ determines a valuation v_w on the associated language $\mathscr{L}$. We say that w *satisfies* a formula A of $\mathscr{L}$ if $v_w(A) = T$; equivalently, we say that w is an *A-world*; the set of A-worlds is denoted $|A|$. The selection function f assigns a (possibly empty) subset $f(A, w)$ to each admissible antecedent formula A (see below) and world w, which is interpreted as the set of A-worlds that are *minimally different* (distant) from w; $f(A, w) = \varnothing$ iff $|A| = \varnothing$. The truth conditions for conditional formulas are specified as follows; w satisfies $A \rightarrow B$ if and only if every world in $f(A, w)$ satisfies B: $w \in |A \rightarrow B|$ iff $f(A, w) \subseteq |B|$. The basic idea is that in evaluating $A \rightarrow B$ at a given world w, we consider (envisage) all the minimal alterations of w that make A true, and we determine whether every such minimal alteration also makes B true.

The selection function f is required to satisfy a number of conditions.

(f1) $f(A, w) \subseteq |A|$
(f2) if $w \in |A|$, then $f(A, w) = \{w\}$
(f3) if $f(A, w) \subseteq |B|$ and $f(B, w) \subseteq |A|$, then $f(A, w) = f(B, w)$
(f4) $f(A \vee B, w) \subseteq |A|$ or $f(A \vee B, w) \subseteq |B|$ or $f(A \vee B, w) = f(A, w) \cup f(B, w)$
(f5) $f(A, w) = \{x\}$ for some $x \in W$ or $f(A, w) = \varnothing$

If f satisfies conditions (f1)–(f4), then the model is called a *Lewis model*; a Lewis model that does not satisfy (f5) is called a *proper Lewis model*, and a Lewis model that satisfies (f5) is called a *Stalnaker model*. In all cases, these notions are defined relative to the class of admissible antecedent formulas of $\mathscr{L}$; for example, $\mathscr{L}$ may not have any non-first-degree conditionals, in which case a conditional formula $A \rightarrow B$ is not an admissible antecedent formula of $\mathscr{L}$, and accordingly $f(A \rightarrow B, w)$ is not defined. In the languages that interest us, the only admissible antecedent (A) formulas are measurement formulas.

The difference between Stalnaker models and (proper) Lewis models is that in the former, if there are any A-worlds, then there is a unique *closest* A-world to any given world w, whereas in the latter, if there are any A-worlds, then there is generally a set of A-worlds *minimally distant* from any given world w. This yields an important difference in the respective logics of Stalnaker conditionals and Lewis conditionals: whereas *conditional excluded middle* (CEM) – $(A \rightarrow B) \vee (A \rightarrow {\sim} B)$ – is validated by every Stalnaker model, it is not validated by every Lewis model. In a Stalnaker model, the conditional $A \rightarrow B$ is true at w just in case no A-worlds exist or the unique A-world w^* closest to w is also a B-world. If w^* is not a B-world, then it is a ${\sim}B$-world, in which case the alternate conditional $A \rightarrow {\sim} B$ is true at w. Thus, *at least one* of $A \rightarrow B$, $A \rightarrow {\sim} B$ is true at w, for any $w \in W$, which is to say that CEM is valid. On the other hand, in a proper Lewis model, $A \rightarrow B$ is true at w just in case no A-world exists or *every* A-world minimally distant from w is a B-world, and $A \rightarrow {\sim} B$ is true at w just in case *no* A-world minimally distant from w is a B-world. Thus, in the non-extreme cases in which some but not all words in $\mathscr{f}(A, w)$ satisfy B, neither $A \rightarrow B$ nor $A \rightarrow \sim B$ is true at w. Circumstances such as this prompt Lewis (1973) to define an additional counterfactual conditional connective, denoted $\diamond\!\!\rightarrow$; whereas the official reading of $A \rightarrow B$ is 'if it were the case that A, then it *would* be the case that B', the official reading of $A \diamond\!\!\rightarrow B$ is 'if it were the case that A, then it *might* be the case that B.' In particular, $A \diamond\!\!\rightarrow B$ is equivalent to $\sim(A \rightarrow \sim B)$, and accordingly is true at w just in case there exists at least one world in $\mathscr{f}(A, w)$ that satisfies B. Henceforth, let us say that a conditional $A \rightarrow B$ is *determinate* at w if exactly one of $A \rightarrow B$, $A \rightarrow \sim B$ is true at w, and let us say that $A \rightarrow B$ is *indeterminate* at w if neither $A \rightarrow B$ nor $A \rightarrow \sim B$ is true at w, that is, if both $A \diamond\!\!\rightarrow B$ and $A \diamond\!\!\rightarrow \sim B$ are true at w.

In applying these ideas to the interpretation of QM, we employ a language (syntax) $\mathscr{CL}$, which is described as follows. For every magnitude m and Borel set X, the expressions $[m]$ and $[m, X]$ are atomic formulas of $\mathscr{CL}$. The primitive connectives of $\mathscr{CL}$ include negation ($\sim$), conjunction (&), and the conditional ($\rightarrow$), all of which have restricted domains, made precise as follows. First of all, there are measurement (M) formulas: $[m]$ is an M-formula; if A, B are M-formulas, then so is A & B; nothing else is an M-formula. Next, there are observation (O) formulas; $[m, X]$ is an O-formula; if A, B are O-formulas, then so are $\sim A$ and A & B; nothing else is an O-formula. Next, there are conditional (C) formulas: if A is an M-formula and B is an O-formula, then $A \rightarrow B$ is a C-formula; nothing else is a C-formula. Finally there are the proper (P) formulas of $\mathscr{CL}$: every M-formula is a P-formula; every O-formula is a P-formula; every C-formula is a P-formula; if A, B are P-formulas, then so are $\sim A$ and A & B; nothing else is a P-formula.

The semantics for $\mathscr{CL}$ may be outlined as follows, the details requiring that the nature of the quantum magnitudes be specified. We suppose we have a system $\mathscr{sys}$ characterized by Hilbert space $\mathscr{H}$ and manual $(Z(i)$: $i \in I)$; in the associated model $(W, \mathscr{f})$, each world $w \in W$ is a state of affairs (λ, i) as in Section 6; that is, $\lambda \in \mathbf{X}(Z(i)\colon i \in I)$ and $i \in I$. The selection function $\mathscr{f}$ is defined as follows.

$$\mathscr{f}(A, (\lambda, i)) = \begin{cases} \{(\lambda, i)\} \text{ if } (\lambda, i) \in |A| \\ \{(\mu, j)\colon \lambda = \mu \text{ and } (\mu, j) \in |A|\} \text{ otherwise.} \end{cases}$$

In other words, in considering alterations of the reference world (λ, i), we only consider alterations of the macro-context, keeping the micro-state fixed. This might be regarded as taking the same 'object' and looking at it from different vantage points, represented by the macro-contexts.

We next observe that the selection function $\mathscr{f}$ satisfies conditions ($\mathscr{f}$1)–($\mathscr{f}$4) for *any* formula A of $\mathscr{CL}$, as a straightforward consequence of the definition of $\mathscr{f}$. Whether $\mathscr{f}$ additionally satisfies condition ($\mathscr{f}$5) for *all* formulas of $\mathscr{CL}$ is immaterial, since the only admissible antecedent formulas of $\mathscr{CL}$ are M-formulas. The relevant question is whether $\mathscr{f}$ restricted to M-formulas satisfies ($\mathscr{f}$5), and here the answer depends on how we construe quantum magnitudes. If we adopt the non-conventional definition of quantum magnitudes as random variables, then $\mathscr{f}$ satisfies ($\mathscr{f}$5), whereas if we adopt the conventional definition of quantum magnitudes as equivalence classes of random variables, then $\mathscr{f}$ does not satisfy ($\mathscr{f}$5). Thus, depending on our construal of quantum magnitudes, we obtain either a Stalnaker model or a proper Lewis model.

Leaving the Lewis model for Section 9, we conclude this section by examining the Stalnaker model and its associated semantics, denoted SCS. Consider the conditional $[\mathbf{m}] \to [\mathbf{m}, X]$, which is intended to translate the elementary sentence $(\mathbf{m}, X)$, and consider the state of affairs (world) (λ, i). If (λ, i) satisfies $[\mathbf{m}]$, then it satisfies $[\mathbf{m}] \to [\mathbf{m}, X]$ if and only if it satisfies $[\mathbf{m}, x]$. But (λ, i) satisfies $[\mathbf{m}, X]$ if and only if $\mathrm{dom}(\mathbf{m}) = Z(i)$ and $\lambda(i) \in e(\mathbf{m}, X)$. Recall that $\mathrm{dom}(\mathbf{m}) = Z(i)$ iff $i = i[\mathbf{m}]$. If (λ, i) does not satisfy $[\mathbf{m}]$, then it satisfies $[\mathbf{m}] \to [\mathbf{m}, X]$ if and only if every (λ, j) that satisfies $[\mathbf{m}]$ also satisfies $[\mathbf{m}, X]$. But (λ, j) satisfies $[\mathbf{m}]$ if and only if $\mathrm{dom}(\mathbf{m}) = Z(j)$ (i.e., $j = i[\mathbf{m}]$), and (λ, j) satisfies $[\mathbf{m}, X]$ if and only if $\mathrm{dom}(\mathbf{m}) = Z(j)$ and $\lambda(j) \in e(\mathbf{m}, X)$. Therefore, in either case (λ, i) satisfies $[\mathbf{m}] \to [\mathbf{m}, X]$ if and only if $\lambda(i[\mathbf{m}]) \in e(\mathbf{m}, X)$, which is to say that the truth of $[\mathbf{m}] \to [\mathbf{m}, X]$ is independent of the macro-context. We can accordingly speak of the truth of $[\mathbf{m}] \to [\mathbf{m}, X]$ relative to a micro-state λ, and since $[\mathbf{m}] \to [\mathbf{m}, X]$ is intended to be the translation of the elementary sentence $(\mathbf{m}, X)$, we can derivatively speak of the truth of $(\mathbf{m}, X)$ at λ. Specifically, $v_\lambda(\mathbf{m}, X) = T$ iff $(\lambda, i) \in \|[\mathbf{m}] \to [\mathbf{m}, X]\|$ for any $i \in I$. Recall that, according to ACMI, $v_\lambda(\mathbf{m}, X) = T$ iff $\lambda(i[\mathbf{m}]) \in e(\mathbf{m}, X)$. Thus, the Stalnaker conditional semantics SCS gives rise to precisely the same class of valuations on OL(QM) (RL(QM)) as ACMI.

Let us review what has transpired. In Section 6, we examined a semantics RS, which is intended to explicate Reichenbach's restrictive interpretation. According to our explication, the elementary sentence (m, X) is translated as the conditional $[m] \ni\!\!- [m, X]$, where the connective is Reichenbach's quasi-implication, which is truth-functional. According to RS, $[\mathbf{m}]$ is true at state of affairs (λ, i) exactly if m is defined (definable) over $Z(i)$, and $[m, X]$ is true at (λ, i) exactly if m is defined (definable) over $Z(i)$ and the value of m according to $\lambda(i)$ lies in X. Since the conditional connective $\ni\!\!-$ is truth-functional, RS gives no information except concerning what magnitudes are actually measured and what values are actually observed, and accordingly only phenomenal magnitudes are assigned values. In the semantics SCS presented in Section 7, which explicates ACMI in a way that renders it consistent with the phenomenon-interphenomenon distinction, the truth-functional quasi-implication is replaced by a non-truth-functional connective, which for semi-Boolean magnitudes is a Stalnaker conditional. Translating each elementary sentence $(\mathbf{m}, X)$ as a conditional $[\mathbf{m}] \rightarrow [\mathbf{m}, X]$ enables us to interpolate *all* interphenomenal events, at least in a dispositional sense; to say that $(\mathbf{m}, X)$ is true is to say that a measurement of **m** *would actually* yield a value in X. That the proposed interpolation, or any complete interpolation, poses serious conceptual problems is the subject of Section 8.

8. THE PRINCIPLE OF ANOMALY

We have now examined three interpretations of QM. According to the restrictive interpretation RI, in any given situation, only phenomenal events are *T-F*; according to the intermediate interpretation CMI, all phenomenal, and some but not all interphenomenal, events are *T-F*; according to the exhaustive interpretation ACMI, all events are *T-F*. Equivalently, in any given situation, according to RI only phenomenal magnitudes are assigned (exact) values, according to CMI all phenomenal, and some but not all interphenomenal, magnitudes are assigned values, and according to ACMI all magnitudes are assigned values. According to Reichenbach's Principle of Anomaly, no exhaustive interpretation of QM is consistent with normal causality; that is, no complete description of quantum events, or complete assignment of values to quantum magnitudes, is free from causal anomalies. Rather than pursue the details of Reichenbach's presentation of the Principle of Anomaly, which is discussed by Jones in this volume, we provide a rather different exposition of this principle based on the work of Gleason (1957) and Kochen and Specker (1967).

Let us call any function *val* that assigns a real number $val(m)$ to *every* quantum magnitude m a *valuation* (not to be confused with a semantic valuation). Let us say that a valuation *val* is *anomalous* if it has the following property: magnitude n *functionally related* by f to magnitude m – written $n = f(m)$ – but $val(n) \neq f(val(m))$, for *some* magnitudes m, n. In order for the definition of 'anomalous' not to be circular, we require a characterization of the functional relations among quantum magnitudes that is independent of the values they take. If we identify quantum magnitudes with random variables, then '$\mathbf{n} = f(\mathbf{m})$' means that $\mathbf{m}$ and $\mathbf{n}$ are defined over a common frame Z, and moreover, $\mathbf{n}(r) = f(\mathbf{m}(r))$ for all $r \in Z$. If we identify quantum magnitudes with s.a. operators, then '$N = f(M)$' means that $N = f(M)$ according to the operator calculus; equivalently, if we identify quantum magnitudes with equivalence classes of random variables, then '$N = f(M)$' means that there exist random variables $\mathbf{n} \in N$, $\mathbf{m} \in M$ such $\mathbf{n} = f(\mathbf{m})$ in the sense of random variables.

If we adopt the anti-Copenhagen (semi-Boolean) construal of quantum magnitudes as random variables, then no valuation val_λ associated with a micro-state λ is anomalous, which may be seen as follows. As observed in Section 4, the valuation val_λ is defined so that $val_\lambda(\mathbf{m}) = \mathbf{m}(\lambda(i[\mathbf{m}]))$, where $\text{dom}(\mathbf{m}) = Z(i[\mathbf{m}])$. Now, $\mathbf{n} = f(\mathbf{m})$ if and only if $\text{dom}(\mathbf{m}) = \text{dom}(\mathbf{n})$ – call it, say, $Z(j)$ – and $\mathbf{n}(r) = f(\mathbf{m}(r))$ for all $r \in Z(j)$. So if $\mathbf{n} = f(\mathbf{m})$, then $val\,(\mathbf{n}) = \mathbf{n}(\lambda(j)) = f(\mathbf{m}(\lambda(j))) = f(val_\lambda(\mathbf{m}))$. Thus, insofar as we identify quantum magnitudes with *individual* random variables, no valuation val_λ associated with a micro-state λ is anomalous.

If, on the other hand, we adopt the conventional (partial Boolean) construal of quantum magnitudes as s.a. operators, equivalently, equivalence classes of random variables, then the functional relations among the magnitudes are considerably more complex. Regarded as an equivalence class of random variables, conventional magnitude N is related by f to conventional magnitude M just in case there are random variables $\mathbf{n} \in N$, $\mathbf{m} \in M$ such that $\mathbf{n} = f(\mathbf{m})$. Characterizing functional relations among magnitudes this way immediately yields a complication not arising under the semi-Boolean construal of magnitudes: every non-maximal magnitude M is functionally related to infinitely many mutually incompatible maximal magnitudes. This places serious restrictions on the class of non-anomalous valuations; in fact, if the relevant Hilbert space $\mathscr{H}$ is such that $\dim(\mathscr{H}) \geqslant 3$, then the class of non-anomalous valuations of conventional magnitudes is empty!

In order to show this, we begin by noting that every valuation *val* determines a unique micro-state λ according to the values assigned by *val* to the maximal magnitudes; assigning a value to each maximal magnitude is tantamount to fixing a point in the associated Cartesian product space, which is set-theoretically identical to a choice function on $(Z(i): i \in I)$.[24] Next, every micro-state λ determines a partial valuation val^*_λ on the class $\mathscr{M}ax$ of maximal magnitudes; every $M \in \mathscr{M}ax$ is a singleton $\{\mathbf{m}\}$, so $val^*_\lambda(M) = val_\lambda(\mathbf{m}) = \mathbf{m}(\lambda(i[\mathbf{m}]))$. Each valuation val^*_λ on $\mathscr{M}ax$ in turn determines an associated partial function from $\mathbf{M} \times \mathscr{M}_{ax}$ into R, denoted val_λ, defined as follows: where $M \in \mathbf{M}$ and $A \in \mathscr{M}_{ax}$, if M and A are compatible, in which case there is a function f such that $M = f(A)$, then $val_\lambda(M/A) = f(val^*_\lambda(A))$; if M and A are not compatible, then $val_\lambda(M/A)$ is not defined. We call $val_\lambda(M/A)$ the value (at λ) of M *relative to* A. Next, we assign a 'supervalue' $supval_\lambda(M)$ to each magnitude M, where $supval_\lambda(M) =_{df} \{val_\lambda(M/A) : A \in \mathscr{M}_{ax}\}$, which is a non-empty subset of real numbers. We can equivalently define relative values of magnitudes relative to the class I of macro-contexts (frames); specifically, if M is definable on $Z(j)$, then $val_\lambda(M/j) = \mathbf{m}_j(\lambda(j))$, where $\mathbf{m}_j \in M$ and $\text{dom}(\mathbf{m}_j) = Z(j)$, and if M is not definable on $Z(j)$, then $val_\lambda(M/j)$ is not defined. Then $supval_\lambda(M) =_{df} \{val_\lambda(M/i) : i \in I\}$. A magnitude M is said to be *determinate* at a micro-state λ if $supval(M) = \{r\}$ for some $r \in R$; otherwise, M is said to be *indeterminate* at λ. A micro-state λ is said to be *determinate* if every magnitude M is determinate at λ; otherwise, λ is said to be *indeterminate*.

We now show that if the micro-state λ induced by a valuation val is indeterminate, then val is anomalous. First of all, in order for val not to be anomalous, it is necessary that $val(M) \in supval_\lambda(M)$ for all $M \in \mathbf{M}$; otherwise, $val(M)$ is not appropriately related to $val(A)$ for *any* $A \in \mathscr{M}_{ax}$ to which M is functionally related; for if $M = f(A)$, then $f(val_\lambda(A)) \in supval_\lambda(M)$. It is furthermore necessary that $supval_\lambda(M) = \{r\}$ for some $r \in R$, for all $M \in \mathbf{M}$, that is, it is necessary that λ is determinate. For suppose that magnitude M is not determinate at λ; then there exist maximal magnitudes A, B and functions f, g such that $M = f(A)$ and $M = g(B)$, and such that $val_\lambda(M/A) \neq val_\lambda(M/B)$, where $val_\lambda(M/A) = f(val_\lambda(A))$ and $val_\lambda(M/B) = g(val_\lambda(B))$. But for all maximal magnitudes C we have $val_\lambda(C) = val(C)$, and so $f(val(A)) \neq g(val(B))$. Consequently, either $val(M) \neq f(val(A))$ or $val(M) \neq g(val(B))$, but $M = f(A)$ and $M = g(B)$, so val is anomalous.

Thus, in order to show that every valuation on the class of conventional magnitudes is anomalous, it is sufficient to show that every micro-state λ is indeterminate. This cannot be shown in complete generality, inasmuch as *no* micro-state defined on a two-dimensional Hilbert space is indeterminate, but it can be shown in nearly complete generality – for all Hilbert spaces $\mathscr{H}$ for which $\dim(\mathscr{H}) \geqslant 3$. As a consequence of a theorem proved by Gleason (1957) and Kochen and Specker (1967) (henceforth GKS), if $\dim(\mathscr{H}) \geqslant 3$, then every choice function λ on the associated manual $(Z(i) : i \in I)$ has the following property: λ selects at least two rays $\lambda(j)$, $\lambda(k)$ that are orthogonal; equivalently, for some $j, k \in I$, $\lambda(j)$, $\lambda(k) \in Z(j)$ but

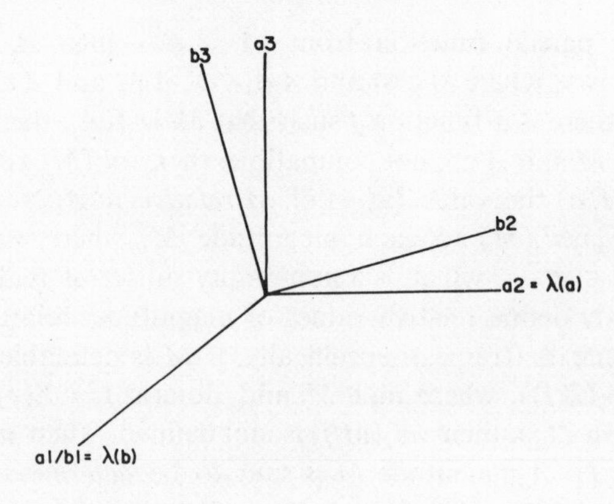

Fig. 4.

$\lambda(j) \neq \lambda(k)$. In other words, no choice function λ on $(Z(i): i \in I)$ selects *exactly one* ray $\lambda(i)$ from each $Z(i)$; rather, there is inevitably a frame $Z(j)$ with an 'extraneous' ray $\lambda(k)$ different from $\lambda(j)$.

That this implies that λ is indeterminate may be illustrated by reference to Hilbert space $H3$, and frames $Z(a)$ and $Z(b)$, as in Section 3, which may be constructed as follows. We begin with any frame $Z(a)$ that contains an extraneous ray $\lambda(b)$, and then add frame $Z(b)$ by rotating through the appropriate angle around $\lambda(b)$, so that $\lambda(b) \in Z(a) \cap Z(b)$. This yields Figure 4. Here, $Z(a) = \{a1, a2, a3\}$, and $Z(b) = \{b1, b2, b3\}$, where $a1 = b1 = \lambda(b)$; we may set $\lambda(a) = a2$ without any loss of generality. Next, we define two maximal conventional magnitudes A and B, both with spectrum $\{1, 2, 3\}$, with eigenspaces $\{a1, a2, a3\}$ and $\{b1, b2, b3\}$ respectively. Since they are maximal, A and B determine unique random variables, denoted $\mathbf{a}$ and $\mathbf{b}$ respectively, defined so that $\mathbf{a}(\langle a1, a2, a3\rangle) = \langle 1, 2, 3\rangle$ and $\mathbf{b}(\langle b1, b2, b3\rangle) = \langle 1, 2, 3\rangle$. We also define a non-maximal conventional magnitude M with spectrum $\{0, 1\}$, where the 1-eigenspace of M is $a1$ $(= b1)$ and the 0-eigenspace of M is the plane spanned by $\{a2, a3\}$ (equivalently, $\{b2, b3\}$). Now, M resolves into two random variables $\mathbf{m}_a$ and $\mathbf{m}_b$, among others, defined respectively on $Z(a)$ and $Z(b)$, where $\mathbf{m}_a(\langle a1, a2, a3\rangle) = \langle 1, 0, 0\rangle$, and $\mathbf{m}_b(\langle b1, b2, b3\rangle) = \langle 1, 0, 0\rangle$. Thus, $\mathbf{m}_a = f(\mathbf{a})$, and $\mathbf{m}_b = f(\mathbf{b})$, where $f(\langle 1, 2, 3\rangle) = \langle 1, 0, 0\rangle$. It follows that M is functionally related to A and to B, in both cases by the same function f, so M is indeterminate at λ unless $f(val_\lambda(A)) = f(val_\lambda(B))$, that is, unless $val_\lambda(M/A) = val_\lambda(M/B)$. Now, $\lambda(a) = a2$ and $\mathbf{a}(a2) = 2$, so $val_\lambda(A) = 2$, and so $val_\lambda(M/A) = f(val_\lambda(A)) = f(2) = 0$. On the other hand, $\lambda(b) = b1$ and

$\mathbf{b}(b1) = 1$, so $val_\lambda(B) = 1$, and so $val_\lambda(M/B) = f(val_\lambda(B)) = f(1) = 1$. Thus, $val_\lambda(M/A) \neq val_\lambda(M/B)$, and $supval_\lambda(M) \supseteq \{0, 1\}$, which is to say that M is indeterminate at λ, which in turn entails λ is indeterminate.

As a consequence of the GKS theorem, one cannot assign exact values to all conventional quantum magnitudes without violating certain of their functional relations; every complete valuation is anomalous. The problem of anomalous valuations is circumvented by the anti-Copenhagen modal interpretation by resolving each non-maximal conventional quantum magnitude into infinitely many anti-Copenhagen magnitudes, which are identified with (semi-Boolean) random variables. As a result no anti-Copenhagen magnitude is functionally related to more than one maximal magnitude. According to ACMI, a non-maximal conventional magnitude M is indeterminate at a microstate λ simply because M represents several physically distinct (though statistically equivalent) magnitudes, whose values are different at λ. In the example described above, M is assigned two different relative values, $val(M/A)$ and $val(M/B)$; according to ACMI, this is not anomalous, because $val(M/A)$ is $val(\mathbf{m}_a)$ and $val(M/B)$ is $val(\mathbf{m}_b)$, but $val(\mathbf{m}_a)$ and $val(\mathbf{m}_b)$ may be different since $\mathbf{m}_a$ and $\mathbf{m}_b$ are physically distinct magnitudes.

Although the anti-Copenhagen multiplication of magnitudes manages to evade the functional anomalies, it does not in fact circumvent the Principle of Anomaly, for the *functional* anomalies in the context of *individual* systems become genuine *causal* anomalies in the context of *composite* systems. This may be illustrated with the spin states of two spin-1 systems. The Hilbert spaces pertaining to $\mathscr{sys}(1)$ and $\mathscr{sys}(2)$ are $\mathscr{H}(1)$ and $\mathscr{H}(2)$, respectively, which are both 3-dimensional, and the Hilbert space pertaining to the composite system $\mathscr{sys}(1 + 2)$ is the tensor product $\mathscr{H}(1) \otimes \mathscr{H}(2)$, which is 9-dimensional. If a magnitude m pertaining to $\mathscr{sys}(1)$ is represented by operator A on $\mathscr{H}(1)$, then it is represented by the operator $A \otimes I$ on $\mathscr{H}(1) \otimes \mathscr{H}(2)$, where I is the identity on $\mathscr{H}(2)$. Similarly, if magnitude n is represented by operator B on $\mathscr{H}(2)$, then it is represented by operator $I \otimes B$ on $\mathscr{H}(1) \otimes \mathscr{H}(2)$, where in this case I is the identity on $\mathscr{H}(1)$. Thus depending on whether we view, say, $\mathscr{sys}(1)$ as an individual or as a component of the composite system $\mathscr{sys}(1 + 2)$, a given magnitude will be represented by two different s.a. operators, A and $A \otimes I$. Henceforth, we concentrate on magnitudes that are maximal *relative to* one of the component systems, $\mathscr{sys}(1)$ or $\mathscr{sys}(2)$, that is, magnitudes that are represented by maximal s.a. operators either on $\mathscr{H}(1)$ or on $\mathscr{H}(2)$; following Bub (1976), we call these magnitudes *locally maximal*. The problem is this: although A is maximal relative to $\mathscr{sys}(1)$, and accordingly can only be resolved into one random variable **a** over $\mathscr{H}(1)$, $A \otimes I$, which represents the same physical magnitude, is non-maximal relative to $\mathscr{sys}(1 + 2)$, and accordingly can be resolved into infinitely many random variables $(\mathbf{a}_k : k \in K)$ over $\mathscr{H}(1) \otimes \mathscr{H}(2)$.

In dealing with locally maximal magnitudes, there are two plausible alternatives. Either we can adopt the *extreme form* of ACMI, asserting in effect that as soon as $\mathscr{sys}(1)$ comes into 'contact' (see below) with $\mathscr{sys}(2)$, every one of its maximal magnitudes splits into infinitely many physically distinct magnitudes; or we can adopt the *moderate form* of ACMI, asserting in effect that locally maximal magnitudes do not split. Suppose we adopt the extreme position. Then since every conventional magnitude $A \otimes I$ is non-maximal, it splits into infinitely many random variables over $\mathscr{H}(1) \otimes \mathscr{H}(2)$, each associated with a particular maximal magnitude on $\mathscr{H}(1) \otimes \mathscr{H}(2)$ compatible with $A \otimes I$. Therefore, in order to measure $A \otimes I$ on $\mathscr{sys}(1 + 2)$, which is the same as measuring A on $\mathscr{sys}(1)$, we must in fact measure a particular maximal magnitude compatible with $A \otimes I$, or else we do not know which particular anti-Copenhagen magnitude we are in fact measuring. But every such magnitude M pertains to $\mathscr{sys}(2)$, although not exclusively, so in order to measure M we must perform some measurement or set of measurements on $\mathscr{sys}(2)$. In other words, it is a consequence of the extreme position that we cannot perform a measurement of A on $\mathscr{sys}(1)$ without performing a measurement of some magnitude on $\mathscr{sys}(2)$. This is odd, since $\mathscr{sys}(1)$ and $\mathscr{sys}(2)$ may be in separate galaxies! Note carefully that in moving to the tensor product representetion, we need not postulate that $\mathscr{sys}(1)$ and $\mathscr{sys}(2)$ are *physically related*; rather, we may simply be *describing* them in the *same* '*picture*'.

Suppose, on the other hand, that we adopt the moderate position, according to which locally maximal magnitudes do not split. Then A relative to $\mathscr{sys}(1)$ will be represented by a single random variable on $\mathscr{H}(1)$, and $A \otimes I$ relative to $\mathscr{sys}(1 + 2)$ will be represented by an equivalence class $(\mathbf{a}_k : k \in K)$ of random variables on $\mathscr{H}(1) \otimes \mathscr{H}(2)$. Since we do not allow splitting in moving from one representation to the other, we restrict the class of admissible micro-states on $\mathscr{H}(1) \otimes \mathscr{H}(2)$ to those that assign the same value to each $\mathbf{a}_k$; a micro-state λ is said to be *locally determinate* if every locally maximal magnitude is determinate at λ. The problem is that there are *no* locally determinate micro-states, at least if $\mathscr{sys}(1 + 2)$ is characterized by an $n \times n$-dimensional Hilbert space, where $n \geqslant 3$, which follows from the GKS theorem (cf. Bub, 1976, n.18). Thus, the only way to have any micro-states is to permit locally maximal magnitudes to split, but as we have already seen, this introduces causal anomalies.

9. DESCRIBING ANOMALIES WITH LEWIS CONDITIONALS

According to our formulation of the Principle of Anomaly, every (anti-Copenhagen) micro-state λ is indeterminate (for composite systems, no micro-state is locally determinate); in other words, for every micro-state λ, there is at least one (non-maximal) magnitude M whose value is not uniquely determined by λ; rather, the outcome of measuring M depends on which particular maximal magnitude M is measured with respect to. As Reichenbach noted, we can avoid anomalous *descriptions* of micro-systems by adopting a restrictive interpretation of QM, according to which no statement about interphenomena is asserted. Although a restrictive interpretation certainly succeeds in *evading* anomalies, it does so at the expense of making any counterfactual assertions: a restrictive interpretation describes the value of each measured (phenomenal) magnitude, but it says nothing about unmeasured (interphenomenal) magnitudes; in particular, it says nothing about the value that *would*, or the values that *might*, be observed if a given unmeasured magnitude were in fact measured. A more liberal semantics is represented by the Copenhagen modal interpretation, according to which the elementary sentence (m, X) may be asserted (as true) whenever a measurement of m *would certainly* (necessarily) yield a value in X. A still more liberal semantics is represented by the anti-Copenhagen modal interpretation, according to which (m, X) may be asserted merely whenever a measurement of m *would actually* yield a value in X. Unfortunately, however, according to the Principle of Anomaly, causal anomalies arise when this semantics is implemented in the context of composite quantum systems.

In the present section, we examine an interpretation which is referred to as the Copenhagen variant of the anti-Copenhagen variant of the modal interpretation (CACMI). This interpretation, which is intermediate between CMI and ACMI, preserves the anti-Copenhagen intuition concerning "the value that *would actually* be observed *if m were* measured", but it also preserves the conventional identification of quantum magnitudes. Specifically, whereas micro-states are identified in an anti-Copenhagen fashion with choice functions on the quantum manual $\mathscr{M}(\mathscr{H})$, magnitudes are identified in a Copenhagen fashion with s.a. operators on $\mathscr{H}$, equivalently equivalence classes of random variables on $\mathscr{M}(\mathscr{H})$. The interpretation CACMI can be given directly in terms of the micro-states as follows; every micro-state λ determines an associated valuation v_λ partially defined on $S = \mathbf{M} \times B(R)$, where $\mathbf{M}$ is the class of conventional quantum magnitudes, as follows.

$$v_\lambda(M, X) = \begin{cases} T \text{ if } \mathit{supval}_\lambda(M) \subseteq X \\ F \text{ if } \mathit{supval}_\lambda(M) \subseteq R\text{-}X \\ \text{undefined otherwise} \end{cases}$$

In virtue of the Principle of Anomaly, no v_λ is completely defined on S; for every λ, there is a $P \in S$ such that $v_\lambda(P)$ is not defined.

An alternative, more illuminating, characterization of CACMI involves translating OL(QM) into the conditional language $\mathscr{CL}$, as in Section 7, so that the elementary sentence (M, X) is translated as the conditional $[M] \rightarrow [M, X]$. Whereas the underlying syntax $\mathscr{CL}$ is the same, and the overall semantic framework is the same, the details of the semantics, which is denoted LCS (Lewis conditional semantics), are radically different because the quantum magnitudes are identified, not with individual random variables, but with equivalence classes of random variables. In particular, the model is a proper Lewis model, not a Stalnaker model, and accordingly the associated conditional connective is a Lewis conditional rather than a Stalnaker conditional.

As in the semantics SCS of Section 7, every possible world is an ordered pair (λ, i) consisting of a micro-state λ and a macro-context i, and the selection function is defined similarly: $\mathscr{f}(A, (\lambda, i)) = (\lambda, i)$ if $(\lambda, i) \in |A|$; otherwise, $\mathscr{f}(A, (\lambda, i)) = \{(\mu, j) : \mu = \lambda \text{ and } (\mu, j) \in |A|\}$. As in Sections 6 and 7, the formula $[M]$ is true at (λ, i) just in case M is definable over $Z(i)$ (i.e., $Z(i) \in \mathit{DOM}(M)$), and $[M, X]$ is true at (λ, j) just in case $[M]$ is true at (λ, i) and $\lambda(i) \in e(M, X)$. As usual $w \in |A \rightarrow B|$ iff $\mathscr{f}(A, w) \subseteq |B|$, where all the relevant A-formulas are measurement (M) formulas. Since magnitudes are identified with equivalence classes of random variables, there is not in general a unique closest A-world, which opens up the possibility that certain conditionals are indeterminate. Recall that $A \rightarrow B$ is indeterminate at w just in case neither $A \rightarrow B$ nor $A \rightarrow \sim B$ is true at w.

In order to illustrate this, consider the conditional $[M] \rightarrow [M, X]$ and the state of affairs (world) (λ, i). If $(\lambda, i) \in |[M]|$ (i.e., $Z(i) \in \mathit{DOM}(M)$), then $(\lambda, i) \in |[M] \rightarrow [M, X]|$ iff $(\lambda, i) \in |[M, X]|$ (i.e., $\lambda(i) \in e(M, X)$). On the other hand, if $(\lambda, i) \notin |[M]|$, then $(\lambda, i) \in |[M] \rightarrow [M, X]|$ iff the following holds: every (λ, j) that satisfies $[M]$ also satisfies $[M, X]$, that is, $\lambda(j) \in e(M, X)$ for every j such that $Z(j) \in \mathit{DOM}(M)$. To say that $\lambda(j) \in e(M, X)$ is just to say that $\mathit{val}_\lambda(\mathbf{m}_j) \in X$, where $\mathbf{m}_j$ is the unique random variable defined on $Z(j)$ resolving M. But $\mathit{val}_\lambda(\mathbf{m}_j) \in X$ for every j such that $Z(j) \in \mathit{DOM}(M)$ if and only if $\mathit{supval}_\lambda(M) \subseteq X$; recall that $\mathit{supval}_\lambda(M) = \{\mathit{val}_\lambda(\mathbf{m}) : \mathbf{m} \in M\}$. Thus, if $(\lambda, i) \notin |[M]|$, then $(\lambda, i) \in |[M] \rightarrow [M, X]|$ iff $\mathit{supval}_\lambda(M) \subseteq X$. It immediately follows that $(\lambda, i) \in |[M] \rightarrow [M, X]|$ for *all* $i \in I$ iff $\mathit{supval}_\lambda(M) \subseteq X$.

Let us say that a micro-state λ satisfies a formula A of $\mathscr{CL}$ just in case (λ, i) satisfies A for all $i \in I$. Thus, we may say that λ satisfies the conditional $[M] \rightarrow [M, X]$ if and only if $supval_\lambda(M) \subseteq X$. This enables us to define a valuation v_λ on OL(QM) so that: $v_\lambda(M, X) = T$ if λ satisfies $[M] \rightarrow [M, X]$; $v_\lambda(M, X) = F$ if $v_\lambda(M, R\text{-}X) = T$; $v_\lambda(M, X)$ is undefined otherwise. Evidently, this characterization of v_λ agrees with the original one associated with CACMI.

Recall that a conventional magnitude M is determinate at micro-state λ if and only if $supval_\lambda(M) = \{r\}$ for some $r \in R$. Thus, if M is determinate at λ, then every conditional $[M] \rightarrow [M, X]$ is determinate at (λ, i) for all $i \in I$. The converse is also true, for if M is *not* determinate at λ, then $supval_\lambda(M) \supseteq \{r, r^*\}$ for some $r, r^* \in R$ $(r \neq r^*)$, in which case it is false that $[M] \rightarrow [M, r]$ is determinate at (λ, i) for all $i \in I$; similarly with $[M] \rightarrow [M, r^*]$. If we consider a world (λ, j) not in $\|[M]\|$, in which case $[M] \rightarrow [M, X]$ is a genuine counterfactual, then $[M] \rightarrow [M, r]$ is not true because a measurement of M *might* yield r^*, and $[M] \rightarrow [M, r^*]$ is not true because a measurement of M *might* yield r. This can in fact be expressed in the object language $\mathscr{CL}$ by means of 'might' counterfactuals: $[M] \diamond\!\!\rightarrow [M, r]$, $[M] \diamond\!\!\rightarrow [M, r^*]$. A micro-state λ satisfies $[M] \diamond\!\!\rightarrow [M, r]$ just in case $r \in supval_\lambda(M)$, that is, just in case for *some* $A \in \mathscr{M}ax$ $val_\lambda(M/A) = r$, just as λ satisfies $[M] \rightarrow [M, r]$ just in case $\{r\} = supval_\lambda(M)$, that is, just in case for *every* $A \in \mathscr{M}ax$ $val_\lambda(M/A) = r$.

The expressive power of $\mathscr{CL}$ is not essentially greater than that of OL(QM) if we consider only conditionals of the form $[M] \rightarrow [M, X]$ or $[M] \diamond\!\!\rightarrow [M, X]$. The advantage of $\mathscr{CL}$ over OL(QM) rather concerns more complex conditional sentences such as $[M] \,\&\, [N]. \rightarrow [M, X]$, which may be read, 'if M were measured *in conjunction with* N, then M would be observed to have value in X'. Now, a micro-state λ satisfies $[M] \,\&\, [N]. \rightarrow [M, X]$ just in case every world (λ, i) that satisfies $[M] \,\&\, [N]$ also satisfies $[M, X]$. A given world (λ, i) satisfies $[M] \,\&\, [N]$ just in case both M and N are definable over $Z(i)$, that is, just in case $Z(i) \in \mathsf{DOM}(M) \cap \mathsf{DOM}(N)$. Since two conventional quantum magnitudes M, N are compatible (commute) if and only if they are definable over at least one common frame (i.e., $\mathsf{DOM}(M) \cap \mathsf{DOM}(N) \neq \varnothing$), no world (λ, i) satisfies $[M] \,\&\, [N]$ unless M and N are compatible. Thus, if M and N are not compatible, in which case $\mathsf{DOM}(M) \cap \mathsf{DOM}(N) = \varnothing$, every micro-state trivially satisfies *every* conditional of the form $[M] \,\&\, [N]. \rightarrow [M, X]$. On the other hand, if M and N are compatible, then $\mathsf{DOM}(M) \cap \mathsf{DOM}(N) \neq \varnothing$, so there is at least one world (λ, i) that satisfies $[M] \,\&\, [N]$, and if M and N form what is customarily called a complete commuting set of magnitudes, then there is also at most one such world. A set S of mutually compatible magnitudes is said to be *complete* if there is exactly one frame over which all the magnitudes of S are jointly definable. One way to ensure that M and N form a complete commuting set is to choose N from the set of maximal magnitudes compatible with M, in which case $M = f(N)$ for some function f.

Let us consider the conditional $[M] \& [A]. \rightarrow [M, X]$, where A is maximal and compatible with M. In this case, there is a unique frame over which M and A are definable, call it $Z(a)$, and accordingly for any micro-state λ there is a unique world, (λ, a), that satisfies $[M] \& [A]$. Consequently, λ satisfies $[M] \& [A]. \rightarrow [M, X]$ just in case (λ, a) satisfies $[M, X]$, that is, just in case $\lambda(a) \in e(M, X)$. Recall that $\lambda(a) \in e(M, X)$ if and only if $val_\lambda(\mathbf{m}_a) \in X$, where $\mathbf{m}_a \in M$ and $\mathrm{dom}(\mathbf{m}_a) = Z(a)$, and also recall that $val_\lambda(\mathbf{m}_a) = val_\lambda(M/a)$. Thus, λ satisfies $[M] \& [A]. = [M, X]$ just in case the value at λ of M relative to A is an element of X. In other words, the notion of the value of M relative to A can be expressed in the object language $\mathscr{CL}$, as one would hope, by conditionals of the form 'if M were measured *in conjunction with* A, then M would be observed to have value X,'

With compound conditionals in hand, we can show how causal anomalies can be described in the language $\mathscr{CL}$. Basically, every causal anomaly is a situation in which there are three magnitudes A, B, M, where A and B are both maximal and compatible with M, such that $val(M/A) \neq val(M/B)$. This may be expressed in $\mathscr{CL}$ by a conjunction of the form $([M] \& [A]. \rightarrow [M, r]) \& ([M] \& [A]. \rightarrow [M, r^*])$, which may be read, 'if M were measured *in conjunction with* A, then M would be observed to have value r, *but* if M were measured *in conjunction with* B, then M would be observed to have value r^*'. Since A and B are incompatible with one another, the two measurements – M in conjunction with A, and M in conjunction with B – involve different experimental arrangements, and so the fact that different values might be observed is not paradoxical as it stands. The paradoxical character of anomalies of this sort rather arises in the context of composite systems, such as described in Section 8. In the case of composite systems, M may pertain to $\mathscr{sys}(1)$, but A and B must pertain also to $\mathscr{sys}(2)$, which may be light years away. It is paradoxical that the result of measuring M on $\mathscr{sys}(1)$ should depend in any way on what measurement is performed on $\mathscr{sys}(2)$, since these measurements cannot *causally* influence one another. Any dependency of this sort must be regarded as a causal anomaly.

We conclude this section by showing how the Copenhagen modal interpretation (CMI) may be construed as translating the elementary sentence (m, X) by the conditional $[m] \rightarrow [m, X]$, where the conditional connective is a Lewis conditional, although not the same one as used by CACMI. In both cases, we can regard a possible world as an ordered triple (w, λ, i), where w is a Copenhagen micro-state (pure statistical state), λ is an anti-Copenhagen micro-state subject to the restriction that no $\lambda(j)$ is orthogonal to the range of w, and i is an index point which represents as usual the macro-context. A possible world (w, λ, i) satisfies an atomic sentence A of $\mathscr{CL}$ just in case (λ, i) satisfies A in the manner specified by CACMI. The difference between CMI and CACMI concerns conditional formulas, which is based on the difference in the respective manners in which they define the selection function f. In the case of CACMI, f is defined in terms of an 'accessibility' relation R defined so that: $(w, \lambda, i)\ R\ (w', \lambda', i')$ iff $w = w'$ and $\lambda = \lambda'$. Specifically, letting x, y range over possible worlds, $f(A, x) = \{x\}$ if $x \in |A|$, and otherwise $f(A, x) = \{y : xRy \text{ and } y \in |A|\}$. In the case of CMI, the relevant accessibility relation T is defined so that: $(w, \lambda, i)\ T\ (w', \lambda', i')$ iff $w = w'$, and the associated selection function f^* is defined just like f, except that R is replaced by T. Concerning the conditional defined in terms of f^*, one can prove the following: (w, λ, i) satisifes $[m] \rightarrow [m, X]$ just in case either (w, λ, i) satisfies $[m]$ & $[m, X]$ or $p^w_m(X) = 1$.[25] In other words the counterfactual conditional $[m] \rightarrow [m, X]$ is true at (w, λ, i) just in case either m is actually observed to have value in X or m is not actually measured but the probability, according to w, that a measurement of m would result in a value in X is one.

10. CONCLUDING REMARKS

In the present work, we have discussed four very different interpretations of QM: the restrictive interpretation RI, the Copenhagen modal interpretation CMI, the anti-Copenhagen modal interpretation ACMI, and the Copenhagen variant of the anti-Copenhagen modal interpretation CACMI. Each interpretation assigns in each situation values to certain quantum magnitudes. At one extreme, the restrictive intetpretation assigns values only to phenomenal magnitudes, which in each situation form a maximal set of mutually compatible magnitudes. At the opposite extreme, the anti-Copenhagen modal interpretation assigns values to all magnitudes. Between these two extremes, the two intermediate interpretations, CMI and CACMI, assign values to all phenomenal magnitudes and some but not all interphenomenal magnitudes. In each interpretation, the value assignments can be understood in terms of conditionals involving sentences of the form 'm is measured' and 'm is observed to have value in X'.

In the case of RI, the conditional connective is truth-functional, and a magnitude m is assigned the value r just in case m is observed to have that value. In all non-restrictive interpretations, the conditional connective is intensional, and m is assigned the value r just in case a measurement of m would yield that value.

In a given 'possible world' ω, what *would*, and correspondingly what *might*, result if a given measurement were performed is determined by the set of possible worlds that are 'accessible' to ω. As noted that the end of Section 9, the two intermediate interpretations, CMI and CACMI, involve two different construals of accessibility among possible worlds, that is, they involve two different construals of physical possibility. According to CMI, any world that is not excluded by the pure statistical state (Copenhagen micro-state) of the system $\mathscr{sys}$ is regarded as (objectively) physically possible. On the other hand, CACMI adopts a less liberal stance, maintaining that the physical possibilities pertaining to $\mathscr{sys}$ are further constrained by 'hidden variables' specified by the anti-Copenhagen micro-state of $\mathscr{sys}$. Insofar as the physical possibilities according to the anti-Copenhagen modal interpretation are further restricted, what *might* occur is likewise restricted, which permits a more precise specification of $\mathscr{sys}$ than given by the Copenhagen modal interpretation.

Whether the more precise specification of $\mathscr{sys}$ provided by the anti-Copenhagen modal interpretation carries any physical significance is a difficult question. For example, Reichenbach might have regarded the relation between the orthodox CMI and the non-orthodox ACMI and CACMI as follows. Each extension of the orthodox description of $\mathscr{sys}$ contains *arbitrary* elements, that is, elements that do not reflect *objective* features of $\mathscr{sys}$. Rather than admitting a unique anti-Copenhagen description, $\mathscr{sys}$ admits a class K of *equivalent descriptions*, no one of which is (objectively) preferable to any other in K. In particular, each admissible anti-Copenhagen extension (w, λ, i) of the description of $\mathscr{sys}$ provided by the pure statistical state w is equivalent to every other admissible extension of w (see Note 25). Mathematically speaking, this amounts to partitioning the class of possible worlds of the form (w, λ, i) into equivalence classes $K(w)$, one for each pure statistical state w. Each equivalence class K in turn determines a *super*-valuation s_K, in the sense of van Fraassen; s_K valuates a sentence A as true (false) just in case every $v \in K$ valuates A as true (false). By supervaluating over K, we remove the arbitrary features while preserving the objective (i.e., invariant) features of the various descriptions in K. As can easily be shown, the conditional $[m] \rightarrow [m, X]$ is supervaluated by $s_{K(w)}$ as true (false) just in case $p_m^w(X) = 1\ (0)$. In other words, CMI can be straightforwardly recovered from ACMI (CACMI) by the method of supervaluation.

But the supervaluation coin has two sides, as noted by van Fraassen (1974). Whereas the exponent of orthodoxy sees each anti-Copenhagen extension of the Copenhagen description as incorporating arbitrary (non-objective) elements, his counterpart or 'dual', the exponent of non-orthodoxy, sees things quite differently. Specifically, the latter maintains that, whereas the anti-Copenhagen interpretation assigns values when measurements would *actually* yield those values, the Copenhagen interpretation is willing to assign values only when measurements would *certainly (necessarily)* yield those values. In virtue of the duality of these two views, formal logic by itself cannot resolve this dispute; as usual, it can only delineate our alternatives.

University of Massachusetts, Amherst

NOTES

[1]Our description of classical and Boolean probability includes both finite probability and infinite probability structures. The difference concerns, not the size of the sample (outcome) space or event space, but rather whether the events are closed under countable union or under finite union, and whether the probability measures are countably additive or finitely additive.

[2]In the context of partial Boolean and semi-Boolean probability, all the definitions are for *finite probability*: the events are closed under finite union (join), and the probability measures are finitely additive.

[3]There are also *regular probability measures,* which are determined by *weights*. A *weight* is a function w on the class $O(\mathscr{M})$ of outcomes of $\mathscr{M}$ taking values in [0, 1], subject to the following restraint: the unordered sum $\sum(w(z): z \in Z) = 1$ for all $Z \in \mathscr{M}$. The (regular) probability measure p_w determined by w is defined on $\mathbb{E}(\mathscr{M})$ so that $p_w(E) = \sum(w(z): z \in E)$. Not every probability measure is a regular probability measure.

[4]Strictly speaking, a discrete random variable is a function that takes values in a countable Borel subset of R, not just any countable subset of R.

[5]According to the general definition in the Foulis-Randall framework, a proposition is an ordered pair $(c(p), r(p))$, where $c(p), r(p) \in O(\mathscr{M})$; $c(p)$ is called the *confirmation set* of p, and $r(p)$ is called the *refutation set* of p. The operational proposition $p(E)$ associated with an event E is the ordered pair $((E^{\perp})^{\perp}, E^{\perp})$, but since the dsecon component is uniquely determined by the first component, we may simply regard $p(E)$ to be $(E^{\perp})^{\perp}$.

[6]By a *subspace* of a Hilbert space $\mathscr{H}$ is often meant a sub(vector) space of $\mathscr{H}$. Then a *closed subspace* of $\mathscr{H}$ is a sub(vector)space of $\mathscr{H}$ which is topologically closed, and hence a Hilbert space in its own right. In other words, a closed subspace of $\mathscr{H}$ is a sub(Hilbert) space of $\mathscr{H}$. In applications to QM, there is little motivation to deal with sub(vector) spaces which are not also sub(Hilbert)spaces, so we use the term 'subspace' to mean sub(Hilbert)space, rather than sub(vector)space (cf. Jauch, 1968).

[7]In the case of a Hilbert space over the *complex numbers,* although each ray is *algebrai-*

cally one-dimensional, it is *topologically* two-dimensional. Accordingly, each ray contains a circle of unit vectors, so that picking a particular one fixes the 'phase' of the pure statistical state.

[8]In the context of Gleason's theorem, a probability measure is defined to be *countably* additive.

[9]Although technically speaking, a subspace is a subset of *vectors* of $\mathscr{H}$, since rays are the smallest unit with which we deal, we may equivalently regard each subspace as a subset of *rays* of $\mathscr{H}$. Thus, if we have a subframe F of $\mathscr{H}$, then although its associated proposition $p(F)$ $(=(F^{\perp})^{\perp})$ is not strictly speaking a subspace of $\mathscr{H}$, but rather a collection of rays, we shall nevertheless regard the subspace $S(F)$ generated by F and the proposition $p(F)$ as interchangeable. We will therefore speak of a ray r as being an *element* of a subspace S, although strictly speaking r is a *subset* of S.

[10]In this context, we mean *countably* additive probability measures.

[11]Henceforth, I use lower case boldface letters exclusively to denote (semi-Boolean) random variables.

[12]Technically, the Cartesian product of a family $(Z(i)\colon i \in I)$ of sets – denoted $\mathbf{X}(Z(i)\colon i \in I)$ – is the set of chocie functions on $(Z(i)\colon i \in I)$. The ordinary finite Cartesian product is a special case in which the indexing set I is some natural number n. In CSM, the phase space is a finite Cartesian product of the real line with itself, and every micro-state is a a choice function on this Cartesian product, that is, an ordered n-tuple of real numbers. Anti-Copenhagen micro-states are formally similar to classical micro-states, with the following differences: the Cartesian product is over a non-denumerably infinite rather than a finite indexing set; the various $Z(i)$ are all distinct (though not all disjoint) from each other.

[13]We regard a semantics V to be *genuinely bivalent* if and only if for every formula A, for every valuation $v \in V$ $v(A) = T$ or $v(\sim A) = T$. In the context of the elementary language OL(QM) of QM, the negation of (m, X) is $(m, R\text{-}X)$. Every valuation v on OL(QM) that we consider is defined so that $v(m, X) = F$ if and only if $v(m, R\text{-}X) = T$. In this case genuine bivalence obtains precisely if every formula is either true or false.

[14]These are not exactly Reichenbach's readings; he would read $[m]$ as "a measurement of m is made" and $[m, r]$ as "the indication of the measuring instrument shows the value r"; see PhF, p. 141.

[15]All logical expressions appearing in this work are *names* of the corresponding expressions in the object language, no part of which is ever explicitly displayed. Two conventions are adopted. First, outer parentheses are always dropped. Second, the juxtaposition of names is the name of the corresponding juxtaposition of named objects. Thus, for example, it is admissible to say the following. A & B is a sentence in the object language, and the 'word' at the beginning of this sentence is its name.

[16]It might be useful to note that Reichenbach's quasi-implication connective $\ni\!\!-$ is analogous to Belnap's (1970) conditional assertion connective; also see Dunn (1975), who axiomatizes, it, and van Fraassen (1975) who discusses a variant called the quasi-Belnap conditional, and who discusses it in reference to quantum logic. A Belnap conditional A/B, read 'if A, then B', is *unassertive* if A is false, and otherwise states (in effect) that B is true. For sentences A, B which are themselves assertive (true or false), A/B is true (false, unassertive) just in case $A \ni\!\!- B$ is true (false, indeterminate). For van Fraassen's quasi-Belnap conditional, this is true *also* if A, B are unassertive (indeterminate).

[17]The symbols (names) we use to denote conjunction and negation differ from Part I. Specifically, '&' denotes the same connective as '$\wedge$', and '$\sim$' denotes the same connective as '$-$'.
[18]Other examples of translations: $\mathbf{T}(m, X) \vee \mathbf{F}(m, X)$ is translated by $([m]\ \&\ [m, X]) \vee ([m]\&\sim [m, X])$, which is equivalent to $[m]$; accordingly the sentence $\mathbf{I}(m, X)$ is translated by a formula equivalent to $\sim[m]$, and the complementarity formula (see Part I) (Comp) – $(\mathbf{T}(m, r) \vee \mathbf{F}(m, r)) \supset \mathbf{I}(n, s)$ – is translated as a formula equivalent to $\sim([m]\ \&\ [n])$. As noted in Part I, to say that m and n are complementary is not merely to assert (Comp) but to assert that it is valid. It may be noted that $\sim([m]\ \&\ [n])$ is valid according to the semantics RS just in case $DOM(m) \cap DOM(n) = \emptyset$, that is, just in case m and n are incompatible. Thus, although (Comp) does not correspond to incompatibility with respect to other semantics considered (see Part I, esp. n. 19), it does correspond to incompatibility for the restrictive semantics RS.
[19]Note that the truth conditions for OL(QM) (RL(QM)) given by RI do not correspond to those attributed to Reichenbach in Section 12 of Part I. According to the latter, (m, r) is ture at *pure statistical state* w if $p_m^w(r) = 1$; (m, r) is false at w if $p_m^w(r^*) = 1$ for some $r^* \neq r$; otherwise (m, r) is indeterminate. Concerning truth, but not falsity, the truth conditions of Part I coincide with those for the Copenhagen modal interpretation.
[20]Note that an event e may be true (false) without every elementary sentence referring to e being true (false) as noted at the end of Section 6.
[21]There are a number of interpretations of the notion of incompatibility in QM. According to one interpretation, which we adopt in the present work, if two magnitudes A and B are incompatible (non-commuting), then measurements of A and B simply exclude one another, that is, they involve mutually conflicting experimental arrangements. According to an alternative interpretation, A and B can be measured simultaneously to any degree of *precision*, but the resulting 'numbers' provide little or no *information*; rather, they represent 'noise'. Now, whether the results of measuring A and B are informative depends to a certain extent on the statistical state w of the system prior to measurement. If w generates a joint distribution for A and B, then a measurement of A and B will be informative; w generates a joint distribution for A and B just in case A and B are *compatible relative* to w; in the discrete case, this is true just in case w is a superposition (or mixture) of *common eigenstates* of A and B. See my (1977) for an examination of the notion of relative compatibility.
[22]The role of counterfactual conditionals in QM has been examined previously in my (1974, 1975, 1978), which stems from van Fraassen (1973, Section 6), which stems from Pool (1968), which stems from Foulis (1960); the latter two are reprinted in Hooker (1975). In this particular context, the sentences connected by the conditional connective are all elementary sentences; a typical conditional reads, 'if m *were* to have value in X, then n *would* have value in Y'. In particular, I show that there is a natural ordering among 'possible worlds', which are taken to be pure statistical states, determined by the Hilbert space metric, which gives rise in a completely natural way to a Stalnaker conditional.
[23]See also Stalnaker and Thomason (1970).
[24]Everything in this section is predicated on identifying any two magnitudes that can be obtained from one another by a permutation of the real numbers (1–1 function from R onto R). That is, we regard two magnitudes m, n to be equivalent modulo permutation exactly if $m = f(n)$, where f is a permutation of R. For example, two maximal random

variables **m**, **n** are equivalent modulo permutation just in case *dom*(**m**) = *dom*(**n**), from which it follows that there is a 1–1 correspondence between frames and maximal magnitudes modulo permutation.

[25]If one has an accessibility relation R defined on the set of possible worlds, then one can define a necessity connective $\Box$ in the associated language L so that: $\omega \in |\Box A|$ iff $x \in |A|$ for all x such that $\omega\ Rx$. In other words, ω satisfies 'necessarily A' just in case every possible world accessible to ω satisfies A. Also, given any accessibility relation R, one can define a selection function f as follows: $f(A,\omega) = \{\omega\}$ if $\omega \in |A|$; otherwise, $f(A, \omega) = \{x{:}\omega\ Rx \text{ and } x \in |A|\}$. It is routine to show that f is a Lewis selection function. The relation between the associated counterfactual conditional $\rightarrow$ and the necessity connective $\Box$ is moreover quite simple: $\omega \in |A \rightarrow B|$ iff $\omega \in |(A \,\&\, B) \vee (\Box(A \supset B))|$. In the case of CMI, and the accessibility relation R defined so that $(w, \lambda, i)\ \mathrm{R}\ (w', \lambda', i')$ iff $w = w'$, a possible world (w, λ, i) satisfies $\Box([m] \supset [m, X])$ just in case $p^{w}_{m}(X) = 1$.

REFERENCES

Abbott, J. C., 1967, 'Semi-Boolean Algebra', *Matematicki Vesnik* **4**, 177–98.

Abbott, J. C., 1969, *Sets, Lattices, and Boolean Algebras,* Allyn and Bacon, Boston.

Belnap, N. D., Jr., 1970, 'Conditional Assertion and Restricted Quantification', *Nous* **4**, 1–12.

Birkhoff, G., and von Neumann, J., 1936, 'The Logic of Quantum Mechanics', *Annals of Mathematics* **37**, 823–43.

Bub, J., 1974, *The Interpretation of Quantum Mechanics,* D. Reidel, Dordrecht, Holland.

Bub, J., 1976, 'Randomness and Locality in Quantum Mechanics', in P. A. Suppes, ed. (1976), 397–420.

Demopoulos, W., 1976, 'The Possibility Structure of Physical Systems', in W. Harper and C. A. Hooker, eds. (1976) Vol. III, 55–80.

Dunn, J. M., 1975, 'Axiomatizing Belnap's Conditional Assertion', *Journal of Philosophical Logic* **4**, 383–97.

Finch, P. D., 1969, 'On the Structure of Quantum Logic', *Journal of Symbolic Locic* **34**, 275–82.

Finch, P. D., 1976, 'Quantum Mechanical Physical Quantities as Random Variables', in W. Harper and C. A. Hooker, eds. (1976), Vol. III, 81–103.

Fine, A., 1973, 'Probability and the Interpretation of Quantum Mechanics', *British Journal for the Philosophy of Science* **24**, 1–37.

Fine, A., 1974, 'On the Completeness of Quantum Mechanics', *Synthese* **29**, 257–289, reprinted in Suppes, ed. (1976), 249–281.

Foulis, D. J., 1960, 'Baer *-Semigroups', *Proceedings of the American Mathematical Society* **11**, 648–54.

Foulis, D. J., and Randall, C. H., 1972, 'Operational Statistics, I. Basic Concepts', *Journal of Mathematical Physics* **13**, 1667–75.

Foulis, D. J., and Randall, C. H., 1974, 'Empirical Logic and Quantum Mechanics', *Synthese* **29**, 81–111, reprinted in Suppes, ed. (1976), 73–103.

Gleason, A. M., 1957, 'Measures on the Closed Subspaces of Hilbert Space', *Journal of Mathematics and Mechanics* **6**, 885–93.

Greechie, R. J., and Gudder, S. P., 1973, 'Quantum Logics', in C. A. Hooker, ed. (1973), 143–173.

Gudder, S. P., 1972, 'Partial Algebraic Structures Associated with Orthomodular Posets', *Pacific Journal of Mathematics* **41**, 717–30.

Hardegree, G. M., 1974, 'The Conditional in Quantum Logic', *Synthese* **29**, 63–80, reprinted in Suppes, ed. (1976), 55–72.

Hardegree, G. M., 'Stalnaker Conditionals and Quantum Logic', *Journal of Philosophical Logic* **4**, 399–421.

Hardegree, G. M., 1976, 'The Modal Interpretation of Quantum Mechanics', *Proceedings of the Fifth Biennial Meeting of the Philosophy of Science Association,* October, 1976, Chicago.

Hardegree, G. M., 1977, 'Relative Compatibility in Conventional Quantum Mechanics', *Foundations of Physics* **7**, 495–510.

Hardegree, G. M., 1978, 'The Conditional in Abstract and Concrete Quantum Logic', in C. A. Hooker (ed.), *The Logico-Algebraic Approach to Quantum Mechanics,* Vol. II, D. Reidel, Dordrecht, Holland, 49–108.

Harper, W. L., and Hooker, C. A. (eds.), 1976, *Foundations of Probability Theory, Statistical Inference, and Statistical Theories of Science,* D. Reidel, Dordrecht, Holland.

Hooker, C. A. (ed.), 1973, *Contemporary Research in the Foundations and Philosophy of Quantum Theory,* D. Reidel, Dordrecht, Holland.

Hooker, C. A. (ed.), 1975, *The Logico-Algebraic Approach to Quantum Mechanics, Volume I: Historical Evolution,* D. Reidel, Dordrecht, Holland.

Hooker, C. A., and van Fraassen, B. C., 1976, 'A Semantic Analysis of Niels Bohr's Philosophy of Quantum Theory', in W. L. Harper and C. A. Hooker (1976), Vol. 3, 221–241.

Jauch, J. M., 1968, *Foundations of Quantum Mechanics,* Addison-Wesley, New York.

Kochen, S., and Specker, E. P., 1967, 'The Problem of Hidden Variables in Quantum Mechanics', *Journal of Mathematics and Mechanics* **17**, 59–87.

Kolmogorov, A. N., 1933, *Foundations of the Theory of Probability* (tr. Nathan Morrison), Chelsea Publishing Co., New York, 1950 (orig. 1933).

Lambert, Karel, 1969, 'Logical Truth and Microphysics' in Lambert, K. (ed.) *The Logical Way of Doing Things,* Yale University Press, New Haven.

Lewis, D., 1973, *Counterfactuals,* Harvard University Press, Cambridge, Mass.

Mackey, G. W., 1963, *Mathematical Foundations of Quantum Mechanics,* Benjamin, New York.

Nagel, E., 1945, 'Review of Hans Reichenbach's *Philosophic Foundations of Quantum Mechanics', Journal of Philosophy* **42**, 437–44.

Piron, C., 1964, 'Axiomatique Quantique', *Helvetica Physica Acta* **37**, 339–68.

Pool, J. C. T., 1968, 'Baer *-Semigroups and the Logic of Quantum Mechanics', *Communications in Mathematical Physics* **9**, 118–41.

Popper, K. R., 1957, 'The Propensity Interpretation of the Calculus of Probability, and the Quantum Theory', in S. Körner (ed.), 1957, *Observation and Interpretation in Philosophy of Physics, Butterworth, London,* 65–70.

Popper, K.R., 1959, 'The Propensity Interpretation of Probability', *British Journal for the Philsophy of Science* **10**, 25–42.

Randall, C. H., and Foulis, D. J., 1973, 'Operational Statistics, II. Manuals of Operations and their Logics', *Journal of Mathematical Physics* **14**, 1472–80.

Randall, C. H., and Foulis, D. J., 1976, 'A Mathematical Setting for Inductive Reasoning', in W. Harper and C. A Hooker (1976) Vol III., 169–205.

Randall, C. H., and Foulis, D. J., 1978, 'The Operational Approach to Quantum Mechanics', in C. A. Hooker (ed.), *Physics as a Logico-Operational Structure,* D. Reidel, Dordrecht, Holland, in press.

Reichenbach, H., 1944, *Philosophic Foundations of Quantum Mechanics,* University of California Press, Los Angeles.

Reichenbach, H., 1946, 'Reply to Ernest Nagel's Criticism of my View on Quantum Mechanics', *Journal of Philosophy* **43**, 239–47.

Reichenbach, H., 1948 'The Principle of Anomaly in Quantum Mechanics', *Dialectica* **2**, 337–350.

Reichenbach, H., 1956, *The Direction of Time,* University of California Press, Berkeley.

Sneed, J. D., 1970, 'Quantum Mechanics and Classical Probability Theory', *Synthese* **21**, 34–64.

Stalnaker, R. C., 1968, 'A Theory of Conditionals', in N. Rescher (ed.), *Studies in Logical Theory,* Blackwell, Oxford.

Stalnaker, R. C., 1970, 'Probability and Conditionals', *Philosophy of Science* **37**, 68–80.

Stalnaker, R., and Thomason, R., 1970, 'A Semantic Analysis of Conditional Logic', *Theoria* **36**, 23–42.

Stein, H., 1972, 'On the Conceptual Structure of Quantum Mechanics', in R. G. Colodny (ed.), *Paradigms and Paradoxes,* Pittsburgh Series in the Philosophy of Science, Vol. 5, University of Pittsburgh Press, Pittsburgh, 367–438.

Suppes, P. A. (ed.) (1976), *Logic and Probability in Quantum Mechanics,* D. Reidel, Dordrecht, Holland.

van Fraassen, B. C., 1966, 'Singular Terms, Truth-Value Gaps, and Free Logic', *Journal of Philosophy* **63**, 481–95.

van Fraassen, B. C., 1969, 'Presuppositions, Supervaluations, and Free Logic', in K. Lambert (ed.), *The Logical Way of Doing Things,* Yale University Press, New Haven, 67–91.

van Fraassen, B. C., 1971, *Formal Semantics and Logic,* Macmillan, New York.

van Fraassen, B. C., 1973, 'Semantic Analysis of Quantum Logic', in C. A. Hooker (1973).

van Fraassen, B. C., 1974, 'Hidden Variables in Conditional Logic', *Theoria* **40**, 176–90.

van Fraassen, B. C., 1975, 'Incomplete Assertion and Belnap Connectives', in Hockney, Harper, Freed (eds.), *Contemporary Research in Philosophical Logic and Linguistic Semantics,* D. Reidel, Dordrecht, Holland, 43–70.

van Fraassen, B. C., 1976, 'Probabilities of Conditionals', in W. Harper and C. A. Hooker (1976), Vol. 1, 261–301.

Vardarajan, V. S., 1962, 'Probability in Physics and a Theorem on Simultaneous Observability', *Communications in Pure and Applied Mathematics* **15**, 189–217.

Varadarajan, V. S., 1968, *Geometry of Quantum Theory,* Vol. I, Van Nostrand, Princeton.

von Neumann, J., 1932, *The Mathematical Foundations of Quantum Mechanics* (tr. Robert T. Beyer), Princeton University Press, Princeton, 1955. (orig. 1932).

ROGER JONES

CAUSAL ANOMALIES AND THE COMPLETENESS OF QUANTUM THEORY*

I. INTRODUCTION

Hans Reichenbach's *Philosophic Foundations of Quantum Mechanics* (1944) has attracted the attention of philosophers because of its proposal for the interpretation of the theory in terms of a non-standard, three-valued logic.[1] Such an interpretation is, as Reichenbach calls it, a 'restrictive' one; that is, it leaves certain propositions formulable in the language of the theory without standard truth conditions.[2] These propositions can be neither verified nor falsified, and the facts they express thus neither predicted nor explained by the theory. A typical such proposition assigns a determinate position and momentum to a particle at some instant. Clearly, embracing such an interpretation involves renouncing a space-time description of the world, and is thus not to be entered into lightly.

Reichenbach did not enter into it lightly, but only under the duress of a conviction of the unassailability of what he considered to be the two great principles of quantum mechanics – the Principle of Indeterminacy and the Principle of Anomaly. Much has been written about the indeterminacy principle, but Reichenbach's Principle of Anomaly is not widely discussed, or, at least, not knowingly so. I argue here that the validity of a form of the Principle is at issue in much of the recent work on hidden variable theories. I give a compact statement of the Principle in the form of a theorem, provide a concise version of Reichenbach's general proof, and illustrate the application of the theorem with several examples.

An up-to-date form of the Principle can be subjected to up-to-date criticism, and I consider, in particular, recent critical discussions of the hidden variable program by Arthur Fine. These considerations lead to a re-evaluation of the relationship between the theorem and the Principle (a relationship Reichenbach assumed to be one of equivalence), and of

* I would like to thank Arthur Fine for his encouragement and critical appraisal at all stages in the preparation of this work.

W. C. Salmon (ed.), Hans Reichenbach: Logical Empiricist, 567–604.

the validity of the Principle itself, thus of the grounds for accepting *any* restrictive interpretation of quantum mechanics.

II. THE COMPLETENESS OF QUANTUM THEORY

One reason that the Principle of Anomaly has not been more widely discussed in the thirty years since the publication of the *Philosophic Foundations* must surely be the fragmented form in which it is presented there. Reichenbach did discuss the Principle in two other major papers (1948, 1952), and though the presentation in these is more direct, it is also considerably less thorough. I will be concerned exclusively with the book. My concern is not to give a detailed and faithful critical exegesis of Reichenbach's derivation of the Principle from general considerations of the 'logical nature of interpretations' of physical theories, but rather to get on with a bare minimum of such remarks to a straightforward statement of the Principle in the form of a theorem and the presentation of a proof.

Reichenbach considers the 'basic principles of the quantum mechanical method' to be as follows (109 ff):

(1) The characterization of physical entities in a given context by operators, eigenfunctions, and eigenvalues, determined by the first Schrödinger equation.

I will write a typical such equation $Q\xi_i = q_i\xi_i$, where Q is the operator corresponding to the physical quantity, ξ_i is its ith eigenfunction, and q_i is its ith eigenvalue.[3]

(2) The characterization of the physical situation by a function which . . . determines the probability distributions of the values of the physical entities in (1), and is inversely determined by these distributions.

I will write $P_Q^{\psi}(q_i)$ for the probability of occurrence of the ith eigenvalue of the quantity Q in a physical situation characterized by the state function ψ.[4]

(3) The law of time dependence of ψ, given in the second Schrödinger equation.

The equation (also called the 'time-dependent Schrödinger equation') is usually written $H\psi(t) = (ih/2\pi)(d\psi(t)/dt)$, where H is the 'Hamiltonian' operator of the system (representing its energy) and h is Planck's constant. This establishes the version of the time-evolution of the physical system usually called the 'Schrödinger representation': the physical quantities of the system are viewed as time-dependent operators acting on state functions evolving according to the equation $\psi(t) = U(t, t_0)\psi(t_0)$, where $U(t, t_0)$ is a unitary operator.[5]

An alternative, but completely equivalent version of the time-evolution of the system is given by the 'Heisenberg representation', in which the physical quantities become time-dependent operators acting on a time-dependent state function. This representation is obtained if we perform upon the state functions and operators of the Schrödinger representation the unitary, time-dependent transformation $U^{-1}(t, t_0)$, yielding

$$\psi_H = U^{-1}(t, t_0)\psi_S(t) = U^{-1}(t, t_0)[U(t, t_0)\psi_S(t_0)] = \psi_S(t_0)$$

and

$$Q_H(t) = U^{-1}(t, t_0)Q_S U(t, t_0)$$

where the subscripts H and S refer respectively to the elements of the respective representations. Note in particular that the quantities $Q_H(t)$ and $Q_H(t')$ do not commute unless Q_H is conserved, that is, unless $Q_H(t) = Q_H(t')$. I shall make frequent use of the Heisenberg representation in discussing the examples in Section IV.

(4) The quantum mechanical definition of measurement.

For Reichenbach, "A measurement of an entity u is a physical operation relative to which the ψ-function of the physical system is represented by one of the eigenfunctions of u" (97). This will be recognized as a statement of what is now called the 'projection postulate'. The definition arises from the following requirement: a quantity Q can be said to have the value q_i in some physical situation characterized by state function ψ if $P^{\psi}_{Q}(q_i) = 1$. And, of course, this holds precisely when $\psi = \xi_i$.

So this is, for Reichenbach, the quantum mechanical view of the world: physical situations are characterized by state functions which determine the probability distributions of the eigenvalues of the various operators corresponding to physical quantities associated with the situations. We know how the situations evolve in time, and we have sufficient conditions for associating particular eigenvalues with physical quantities. What then is the interpretive problem of quantum mechanics? It is precisely that these conditions are *sufficient*, and not necessary. That is, we have criteria for establishing the values only of quantities for which the state ψ of the system is an eigenfunction (measured quantities, for Reichenbach), for which the probabilities that particular eigenvalues occur are 0 and 1. What are we to say about the values of quantities for which the state function is not an eigenfunction, but rather a superposition of eigenfunctions? We must be able to complete the assignment of values to such quantities, or give up a space-time picture of the world, since without such a completion we can never, for example, assign both a determinate position and a determinate momentum to any particle at any instant.

Arthur Fine puts the problem very neatly in his recent paper, 'On the Completeness of Quantum Theory' (1974).

> Just what is the problem over completeness? It is that the states of the theory underdetermine the values of the various quantities, leaving value gaps. (The quantities are only partial functions on the set of states.) The problem, then, is whether one can consistently interpolate values so as to fill the gaps. One wants to do this in a way consistent with the assignments of values quantum theory does make, consistent with experiment, and consistent with other plausible constraints (whether derived from physics or from the metaphysics embedded in one or another 'interpretation' of the theory).[6]

III. THE PRINCIPLE OF ANOMALY

Reichenbach terms the completeness problem that of providing an 'exhaustive interpretation' for quantum mechanics, and he does wish to introduce such a 'plausible constraint'. The problem, as he puts it, is 'to assign definite values to the unmeasured, or not exactly measured entities, in such a way that the observed results appear as the causal consequences of the values introduced by our assumption' (5). He continues, rounding out a remark very much like Fine's, "Even if we wish to follow such a procedure however, we must answer the question of whether such a *causal supplementation of observable data by interpolation of unobserved values* can be consistently done" (5). And it can be consistently done" as he says further on, "if only it establishes a connection between observable distributions corresponding to the numerical results of quantum mechanics" (117).

Reichenbach characterizes his notion of 'causal law' by two conditions: "First, it is required that the cause determine the effect univocally;

second, it is demanded that the effect spread continuously through space, following the principle of action by contact" (117). Subsequently, he weakens the first condition to admit "probability relations between cause and effect". Thus, Reichenbach's constraint on exhaustive interpretations is that they satisfy 'normal causality'; only what we might today call 'local stochastic causes' are to be admitted. Failure to satisfy this constraint constitutes a 'causal anomaly'. For instance, if one can alter some property (e.g., the probability of observing some value of some quantity) associated with a set of particles by altering the positions of arbitrarily distant particles, a causal anomaly exists.

The interpretive problem of quantum mechanics can thus be stated concisely in Reichenbach's terms: Is there an exhaustive interpretation of quantum mechanics that satisfies the requirements of normal causality? The Principle of Anomaly states that there is not. In positive terms the Principle states that an exhaustive interpretation of quantum mechanics is possible only if causal anomalies are admitted.

This is the intuitive idea of the Principle, but clearly, no definitive judgments can be made about it in this form. What is needed is some formal characterization of an 'exhaustive interpretation' and a formal condition appropriate to such a characterization sufficient to identify it as admitting a causal anomaly. Given these, we can provide a concrete mathematical statement of the Principle of Anomaly, a statement which can be evaluated by a proof 'based on a general theory of the relations between quantum mechanical entities' (32). Reichenbach does make such a statement of the Principle in §25 and §26, but the statement is a bit mixed up with the proof in a non-perspicuous fashion. I shall try to clear things up.

An exhaustive interpretation provides assignments of values to all physical quantities in all physical states. Each assignment, for a given state, assigns one value to each quantity, and there is such a set of assignments, a subset of the whole set, associated with each state. In each assignment in the stateset, all the quantities for which the state is an eigenstate are assigned the same value – the appropriate eigenvalue – and some value from the possible values of each non-eigenstate quantity is assigned. Let us assume in general that the possible values of some quantum mechanical quantity Q in state ψ are just those Q-eigenvalues whose quantum mechanical probability if ψ is different from zero.[7] Let us assume moreover that the distributions of these values among the assignments associated with state ψ coincides with the probabilities assigned by quantum mechanics. That is, in the ψ-set of assignments the eigenvalue q_i for Q occurs with a relative frequency corresponding to the quantum mechanically assigned probability $P_Q^\psi(q_i)$.[8] We make these assumptions for all quantities in all states. It seems natural to suppose likewise that the probability for some *combination of values* of the quantities in state ψ is related to the relative frequency with which this *combination of values* appears in the set of assignments associated with that state. For quantities P and Q, these 'joint probabilities' will be denoted $P_{Q,P}(q_i, p_j)$.

It is, in fact, not necessary for our purposes that we investigate the individual assignments of eigenvalues in all states that provide an exhaustive interpretation. We will be satisfied to consider merely the relative frequencies with which combinations of values occur. As Reichenbach says, "It is sufficient for our purposes if we define for the general case, not the value of an unobserved entity, but the probability of such a value" (120). It is sufficient because, if it can be shown, no matter what the distributions of the values (or combinations of values) of unobserved quantities in non-eigenstate situations, that the statistical results of quantum mechanics cannot be reproduced without admitting causal anomalies, then the Principle of Anomaly will be vindicated. We yet need, however, a formal condition sufficient to characterize the distribution of values (or combinations of values) arising in an exhaustive interpretation as causally anomalous. Let me try to motivate Reichenbach's condition.

First a brief aside. By the formal rule Reichenbach refers to as the "general rule of multiplication of probabilities", or what is now called the "rule of conditionalization", the joint probability distribution $P_{Q,P}(q_i, p_j)$ associated with a system characterized by state ψ must satisfy the following identity:

$$P_{Q,P}(q_i, p_j) = P_Q^\psi(q_i) \cdot P(P = p_j \mid Q = q_i)$$

$P(P = p_j \mid Q = q_i)$ is related to the relative frequency of appearance of p_j values for P among assignments in which Q has value q_i; it is the 'conditional probability' for $P = p_j$ given that $Q = q_i$. So the existence of joint probabilities is equivalent to the existence of conditional probabilities.[9]

Consider now some quantum mechanical system S, which I will take to be a collection of similar particles (e.g., all electrons) present all at once in some experimental arrangement. Let Q be an (approximate) position quantity describing S at time t, with eigenvalue equation $Q\xi_i = q_i\xi_i$. The probability distribution of eigenvalues of Q can be taken to be related to the frequency of appearance of these values among the particles of the collection.[10] For convenience we can consider an arrangement involving only two possible widely separated positions of particles, q_1 and q_2. Assume that we can select the subset of the particles with position value q_1, and consider the distribution among these particles of the eigenvalues of some non-commuting quantity describing S at t, say (approximate) momentum P with eigenvalue equation $P\pi_i = p_i\pi_i$. Elements of this distribution are denoted by $P(P = p_j \mid Q = q_1)$ – the probability that momentum value p_j appears among the particles with position q_1, or the conditional probability for $P = p_j$ given that $Q = q_1$.

Now if I suddenly obliterate all particles at position q_2, I am left with a system S' consisting only of particles at q_1. Such a system is characterized by ξ_1, the q_1-eigenstate of Q, and for it the observed results of quantum mechanics are that $P'(P = p_j \mid Q = q_1) = P_P^{\xi_1}(p_j)$. If $P(P = p_j \mid Q = q_1) \neq P'(P = p_j \mid Q = q_1)$, we have precisely the sort of causal anomaly mentioned on page 45: I can alter the distribution of momentum values among particles at position q_1 (essentially instantaneously) by altering the distribution of positions of particles at a great distance from q_1.

Reichenbach's *sufficient* condition for the presence of causal anomalies is precisely this one, generalized to all non-commuting quantities A and B (with eigenvalue equations $A\alpha_i = a_i\alpha_i$ and $B\beta_j = b_j\beta_j$). Whenever an exhaustive interpretation features joint distributions $P_{A,B}(a_i, b_j)$ (for all i, j) for which the conditional probabilities $P(B = b_j \mid A = a_i)$ are different in two physical situations in which the distribution of particle locations has been altered, causal anomalies are admitted. Since the observed results of quantum mechanics (the assigned probabilities which the exhaustive interpretation must reproduce) require that $P(B = b_j \mid A = a_i) = P_B^{\alpha_i}(b_j)$ in a system in which all particles have a_i values for A, this probability value fixes the probability for all particle distributions – for all states. Reichenbach's 'normal causality' condition (which I will call the RCA condition) can thus be expressed so:

An exhaustive interpretation admits causal anomalies if it assigns values to two (or more) non-commuting quantities A and B (with eigenvalue equations $A\alpha_i = a_i\alpha_i$, $B\beta_j = b_j\beta_j$) *in such a way that the resulting joint distributions* $P_{A,B}(a_i, b_j)$ *(for any state* ψ *and all i, j) are such that*

$$P(B = b_j \mid A = a_i) \neq P_B^{\alpha_i}(b_j)\,. \qquad (RCA)$$

It is interesting to note that Reichenbach attempts to use a motivating argument of the same form as that above to show that any assignment of values to some quantum mechanical quantity Q in state ψ *other than* those Q-eigenvalues whose probability in ψ is different from zero admits causal anomalies. Let t denote the time of a measurement of Q. Denote Q after the measurement as Q_0 and the observed eigenvalue of position q_0. Denote Q before t as Q_b, and a value before measurement q_b, where q_b need *not* be an eigenvalue of Q. Now assume that we can select the subset of particles in S all giving the same value of Q_0, that is, all of which have, on measurement, the position eigenvalue q_0. Consider the distribution of pre-measurement values q_b among the particles of this subset, a distribution whose elements are denoted by $P(Q_b = q_b \mid Q_0 = q_0)$ – the relative frequency with which the unmeasured value q_b appears among the particles with measured value q_0.

Now if I suddenly obliterate all particles with unmeasured values $q_b \neq q_0$, I am left with a system S' consisting only of particles with unmeasured values $q_b = q_0$, a system characterized by the q_0-eigenstate of Q. In such a system the observed results of quantum mechanics are that $P'(Q_b = q_b \mid Q_0 = q_0)$ is unity for $q_b = q_0$ and zero for all other values of Q_b. Reichenbach wishes to argue that if $P(Q_b = q_b \mid Q_0 = q_0) \neq P'(Q_b = q_b \mid Q_0 = q_0)$, then we have a causal anomaly as above: the distribution in *pre-measured* values among particles which interact with the experimental apparatus to give *measured* values q_0 has been altered (essentially instantaneously) by altering the distribution of *measured* positions of particles possibly far distant from q_0. (Stated more graphically, the measurement interaction leading to the observation of eigenvalue q_0 has been altered by altering the interactions leading to the observation of particles at possibly far distant position eigenvalues.) But this argument does not succeed here. It is certainly the case that we have altered the distribution of measured positions of particles (or the interactions leading to these observations) other than those at q_0, in analogy with the situation in the argument above, but here we have altered more: by obliterating those particles with *non-eigenvalue* positions which yet give on measurement the position value q_0, we have, in a perfectly local way, altered precisely the distribution in question. So we cannot, at least by the kind of argument Reichenbach wishes to give, demonstrate that non-eigenvalue assignments of values to unmeasured quantities involve causal anomalies, and our restrictions on assignments remain but reasonable assumptions.

Let us summarize then where we have come, following Reichenbach, in an effort to provide a concrete mathematical statement of the Principle of Anomaly. An exhaustive interpretation of quantum mechanics provides assignments of eigenvalues to all physical quantities in all physical states. Joint probability distributions related to the frequency of appearance of combinations of eigenvalues of non-commuting quantities arise naturally from such assignments, as consequently do conditional probabilities. The conditional probabilities must, however, satisfy a causality condition, RCA, or the exhaustive interpretation admits causal anomalies. Thus, a ncecessary condition for the existence of an exhaustive interpretation that reproduces the probability assignments of quantum mechanics while satisfying the requirements of normal causality is that there exist, for any two non-commuting quantities and any physical state, a joint probability distribution that meets these two criteria. It is the content of the Principle of Anomaly, so Reichenbach claims, that there does not exist any such distribution. It is this statement that yields a theorem.

IV. A THEOREM ON JOINT DISTRIBUTIONS

THEOREM: No probability distribution for the joint occurrence of eigenvalues of two non-commuting quantities describing a quantum mechanical system characterized by some state ψ exists which satisfies the following two criteria:

(1) No contradictions with the probabilities quantum mechanics does assign are derivable.

(2) No causal anomalies are admitted.

Proof: Let A and B be non-commuting quantities (with eigenvalue equations $A\alpha_i = a_i\alpha_i$, $B\beta_j = b_j\beta_j$) describing a quantum mechanical system in a physical situation characterized by state ψ. Assume that a joint probability distribution $P_{A,B}(a_i, b_j)$ (for all i, j) exists. By the rule of conditionalization

$$P_{A,B}(a_i, b_j) = P^{\psi}_{A}(a_i) \cdot P(B = b_j \mid A = a_i) .$$

We can classify exhaustively all joint distributions as those for which either

(a) $\quad P(B = b_j \mid A = a_i) \neq P^{\alpha_i}_{B}(b_j)$

or

(b) $\quad P(B = b_j \mid A = a_i) = P^{\alpha_i}_{B}(b_j) .$

By Reichenbach's sufficient condition for the existence of causal anomalies (RCA), all these distributions in the (a) class fail to satisfy criterion 2.

Consider now the second class of distributions. We wish to show that this class of conditionals violates criterion 1, that is, that it leads to assignments of probabilities contradicting those of quantum mechanics. First of all, since

$$P_{A,B}(a_i, b_j) = P^{\psi}_{A}(a_i) \cdot P^{\alpha_i}_{B}(b_j)$$

the joint distributions in this class are clearly state functionals, and we will denote this by writing $P^{\psi}_{A,B}(a_i, b_j)$.[11] Applying what Reichenbach calls the 'rule of elimination', or, as it is now more often called, computing the 'marginal probabilities' from the joint distribution, we recover the probabilities for the appearance of eigenvalues of the individual quantities. That is,

$$\sum_i P^{\psi}_{A,B}(a_i, b_j) = P^{\psi}_{B}(b_j)$$

Thus, from the conditionals of the (b) class we obtain

$$P^{\psi}_{B}(b_j) = \sum_i P^{\psi}_{A}(a_i) \cdot P^{\alpha_i}_{B}(b_j) .$$

Since the component probabilities in this equation are all assigned by quantum mechanics, it can be established directly whether the equality holds, that is, whether criterion 1 of the theorem is satisfied.

To choose a simple case for checking, let $\psi = c_1\alpha_1 + c_2\alpha_2$ with $|c_1|^2 + |c_2|^2 = 1$. Then

$$P^{\psi}_{B}(b_j) = |(\beta_j, \psi)|^2 = |c_1(\beta_j, \alpha_1) + c_2(\beta_j, \alpha_2)^2.$$

But

$$\sum_i P^{\psi}_{A}(a_i) \cdot P^{\alpha_i}_{B}(b_j) = |c_1|^2|(\beta_j, \alpha_1)|^2 + |c_2|^2|(\beta_j, \alpha_2)|^2.$$

Clearly, these two expressions are not equal in general, and, in fact, their inequality is a notorious fact in quantum mechanics: the probability in the first expression is that for the occurrence of value b_j for B in a system described by the so-called 'pure state' ψ; that in the second is the probability for the occurrence of value b_j for B in a 'mixture' of pure states α_1 and α_2, themselves occurring with probabilities $|c_1|^2$ and $c_2|^2$. Thus, a joint distribution of the (b) class violates criterion 1 of the theorem.

This exhausts the classes of joint distributions, and the theorem is proved.

A few comments on the theorem are in order before we consider examples of its application. Reichenbach's idea for such a theorem is the first in what has now become a sequence of 'no-joint-distribution' theorems – theorems which show the incompatibility of joint distributions of non-commuting quantities with the requirements of a 'classical' probabilistic reconstruction of quantum mechanics. But whereas Reichenbach interprets 'classical' to reflect causality conditions, more recent authors impose conditions of other sorts. Cohen (1966), considering joint position-momentum distributions, requires (besides the obvious marginal conditions) that one be able to derive from them 'in the classical manner' the quantum mechanical probabilities not only for every *other* quantity, but for every function of every other quantity as well. Nelson (1967) requires from the probabilities and joint distributions of any set of quantities that one be able to reproduce the quantum mechanical probabilities for any quantity formed by taking a real linear combination of the quantities in the set. Fine (1973) requires that the quantum mechanical probabilities be reproduced for quantities formed from any measurable function of the quantities in such a set.

Reichenbach's causality condition (RCA) on joint distributions is stronger than these other conditions, and excludes at a stroke all joint distributions of non-commuting quantities satisfying even the minimal marginal conditions. Thus, for instance, Margenau's frequently used example (1963) of joint distributions for stochastically independent quantities A and B is immediately rejected. That is,

$$P^{\psi}_{A,B}(a_i, b_j) = P^{\psi}_{A}(a_i) \cdot P^{\psi}_{B}(b_j)$$

must admit causal anomalies since $P(B = b_j | A = a_i) = P^{\psi}_{B}(b_j)$ in this case and $P^{\psi}_{B}(b_j) \neq P^{\alpha_i}_{B}(b_j)$ in general, e.g., when $\psi = \beta_j$.[12] Of course, this joint distribution does satisfy the marginal conditions, e.g.,

$$\begin{aligned}\sum_i P^{\psi}_{A,B}(a_i, b_j) = P^{\psi}_{B}(b_j) &= \sum_i P^{\psi}_{A}(a_i) \cdot P^{\psi}_{B}(b_j) \\ &= P^{\psi}_{B}(b_j) \cdot \sum_i P^{\psi}_{A}(a_i) \\ &= P^{\psi}_{B}(b_j) \cdot 1\,.\end{aligned}$$

Moreover, the more radical claim made by Margenau (Park and Margenau, 1968), and by Cartwright (1974) that joint distributions need *not* be considered as state functionals, i.e., that $P_{A,B} = P'_{A,B}$ for *all* quantities A and B describing some quantum mechanical system characterized by states ψ and ψ' does not imply that $\psi = \psi'$, is rejected as well, as was pointed out explicitly.

V. EXAMPLES

I would like to give several examples of the application of the theorem. In each case I will describe some (considerably idealized) 'experimental arrangement' and identify the 'observed quantities'. I will then give an informal 'exhaustive interpretation' of the quantum mechanical description of the arrangement (a 'space-time' account) followed by a more formal account that identifies explicitly the quantities for which values must be provided by an exhaustive interpretation. Following Reichenbach, I will consider only the probability distributions of the values of the quantities provided by the exhaustive interpretation and, proceeding roughly according to the method of the proof, show that no distribution reproduces the statistical results of quantum mechanics without admitting causal anomalies.

Example 1. In the 'two-slit' experimental arrangement pictured in Figure 1, electrons leave a source with a common energy and encounter successively a plate containing two holes (spaced equidistantly from the source) and a sensitized screen on which their impacts are recorded. The

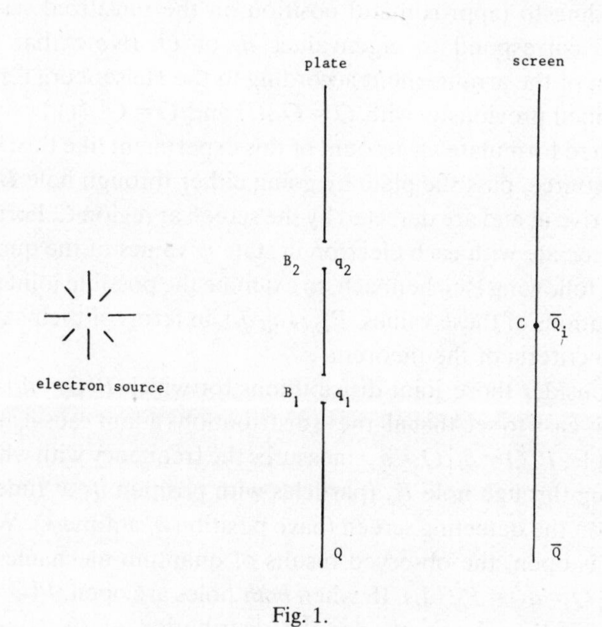

Fig. 1.

pattern of electron impacts on the screen observed in this arrangement is usually called the 'interference pattern'. The pattern observed when hole B_2 is blocked and hole B_1 is open will be called the 'B_1 pattern'; the pattern with B_1 blocked and B_2 open, the 'B_2 pattern'; and the superposition of the B_1 and B_2 patterns (as would be obtained if the two single hole experiments were run one after another for equal times), the 'additive pattern'. The additive and interference patterns are, of course, markedly different.

The distribution of electron impacts resulting from a given arrangement gives rise to a probability distribution for the arrival of electrons at locations on the sensitized screen. If C is a region on the screen, the probability of arrival at C should be proportional to the relative number of electron impacts recorded at C.

Let ψ describe the state of electrons emitted from the source toward the two holes. Q is the quantity corresponding to (approximate) position on the plate. We are concerned only with the eigenvalues q_1 and q_2 of Q corresponding to the locations of holes B_1 and B_2. $\bar{Q}$ is the quantity corresponding to (approximate) position on the sensitized screen, and regions C correspond to eigenvalues $\bar{q}_i$ of $\bar{Q}$. (Note that this is a description of the arrangement according to the Heisenberg representation described previously, with $Q = Q_H(t)$ and $\bar{Q} = Q_H(\bar{t})$.)

We wish to formulate an account of this experiment like this: electrons leave the source, pass the plate by going either through hole B_1 or hole B_2, and arrive at and are detected by the screen at region C. Formally, we wish to associate with each electron in state ψ values of the quantities Q and $\bar{Q}$, or, following Reichenbach, to examine the possible joint probability distributions of these values, $P^{\psi}_{Q,\bar{Q}}(q_i, \bar{q}_j)$, in terms of their satisfaction of the two criteria of the theorem.

First, consider those joint distributions for which $P(\bar{Q} = \bar{q}_j \,|\, Q = q_i) \neq P^{\xi_i}_{\bar{Q}}(\bar{q}_j)$. It is easy to see that all these distributions admit causal anomalies. For example, $P(\bar{Q} = \bar{q}_j \,|\, Q = q_1)$ measures the frequency with which particles passing through hole B_1 (particles with position q_1 at time t) strike region C on the detecting screen (have position $\bar{q}_j$ at time $\bar{t}$). When hole B_1 *alone* is open, the observed results of quantum mechanics are that $P(\bar{Q} = \bar{q}_j \,|\, Q = q_1) = P^{\xi_1}_{\bar{Q}}(\bar{q}_j)$. If, when *both* holes are open, $P(\bar{Q} = \bar{q}_j \,|\, Q = q_1) \neq P^{\xi_1}_{\bar{Q}}(\bar{q}_j)$, then I can *change* the distribution of $\bar{Q}$ values among particles with Q-value q_1 by opening or shutting hole B_2, that is, by changing the distribution of particle positions at time t at locations *other than* q_1. This is just the sort of causal anomaly we discussed on page 45, a non-local effect, an effect, as Reichenbach says, that does not "spread continuously through space, following the principle of action by contact" (117). This can be seen, he says, "from the fact that changes in the material of the diaphragm between B_1 and B_2, or changes in its shape such as would result from corrugating the diaphragm, would not influence the effect" (29).

Of course, if such causal anomalies are admitted, joint distributions do exist which at least give the proper quantum mechanical marginals, as we saw in the previous section. For stochastically independent Q, $\bar{Q}$

$$P^{\psi}_{Q,\bar{Q}}(q_i, \bar{q}_j) = P^{\psi}_{Q}(q_i) \cdot P^{\psi}_{\bar{Q}}(\bar{q}_j)$$

and computing the marginal probabilities just yields identities:

$$\sum_i P^{\psi}_{Q,\bar{Q}}(q_i, \bar{q}_j) = P^{\psi}_{\bar{Q}}(\bar{q}_j) = \sum_i P^{\psi}_{Q}(q_i) \cdot P^{\psi}_{\bar{Q}}(\bar{q}_j)$$

$$= P^{\psi}_{\bar{Q}}(\bar{q}_j) \cdot \sum_i P^{\psi}_{Q}(q_i)$$

$$= P^{\psi}_{\bar{Q}}(\bar{q}_j) \cdot 1 .$$

Causal anomalies are not admitted by those interpretations involving joint distributions for which $P(\bar{Q} = \bar{q}_j \mid Q = q_i) = P^{\xi_i}_{\bar{Q}}(\bar{q}_j)$. In this case

$$P^{\psi}_{Q,\bar{Q}}(q_i, \bar{q}_j) = P^{\psi}_{Q}(q_i) \cdot P^{\xi_i}_{\bar{Q}}(\bar{q}_j) .$$

Computing the marginals, we obtain

$$\sum_i P^{\psi}_{Q,\bar{Q}}(q_i, \bar{q}_j) = P^{\psi}_{\bar{Q}}(\bar{q}_j) \neq P^{\psi}_{Q}(q_1) \cdot P^{\xi_1}_{\bar{Q}}(\bar{q}_j) + P^{\psi}_{Q}(q_2) \cdot P^{\xi_1}_{\bar{Q}}(\bar{q}_j) .$$

Thus, the statistical results of quantum mechanics are not reproduced. The failure of equality in the expression above shows up in the experiment as follows: the first term on the right gives the relative number of electron impacts on the detecting screen observed at some region C (corresponding to $\bar{q}_j$) when hole B_1 alone is open; the second term, the relative number when hole B_2 alone is open. But their sum does not give the relative number of impacts at C observed when both holes are open throughout, the term on the left. That is, the 'interference pattern' is not identical to the 'additive pattern'.

Thus, any joint probability distribution of the values of Q and $\bar{Q}$ that reproduces the statistical results of quantum mechanics must admit causal anomalies.

Example 2. The 'Stern-Gerlach' experiment pictured in Figure 2 involves passing a collimated beam of electrons (or other spin-$\frac{1}{2}$ particles) through a non-uniform magnetic field and recording the pattern of electron impacts on a sensitized screen located downstream of the field region.

It is found in such an arrangement that equal numbers of electron impacts on the screen are observed at two regions evenly spaced above and below the axis of the beam in a line parallel to the field gradient direction.

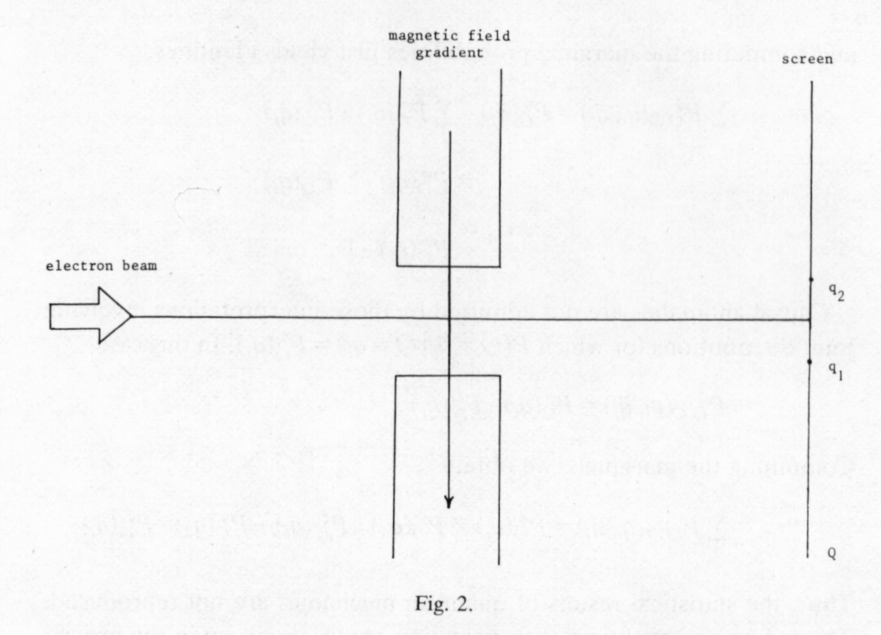

Fig. 2.

Consider the apparatus also to be equipped with sliding plates within the magnets, as pictured in Figure 3, that can be extended from either side of the beam to near the axis. If the upper plate is slid down, inserted into the beam, electron impacts on the screen are observed only in the lower position, and *vice versa*. I shall call this a 'type-I apparatus'.

Let ψ describe the state of electrons travelling down the beam axis. Q is the quantity corresponding to (approximate) position on the screen, with eigenvalues q_1 and q_2. We wish to formulate an account of this experiment like this: each electron in the collimated beam has its spin vector $\boldsymbol{\sigma}$ oriented in a particular direction. Let the magnetic field gradient direction be given by vector $\boldsymbol{\mu}$. Then Q can be considered proportional to the spin component of an electron in direction $\boldsymbol{\mu}$, i.e., $Q \propto \boldsymbol{\sigma} \cdot \boldsymbol{\mu}$. The electrons with spin oriented 'positively' with respect to $\boldsymbol{\mu}$ are described by eigenvalue q_2 of Q; those oriented 'negatively', by eigenvalue q_1.

It will be convenient to imagine also an arrangement slightly different from that above, one in which the single magnet pair of Figure 3 is replaced with three pairs arranged as pictured in Figure 4.[13] The first and third magnets are precisely the same sort as those in a type-I apparatus. The second is twice as long, and the polarity of its magnetic field is

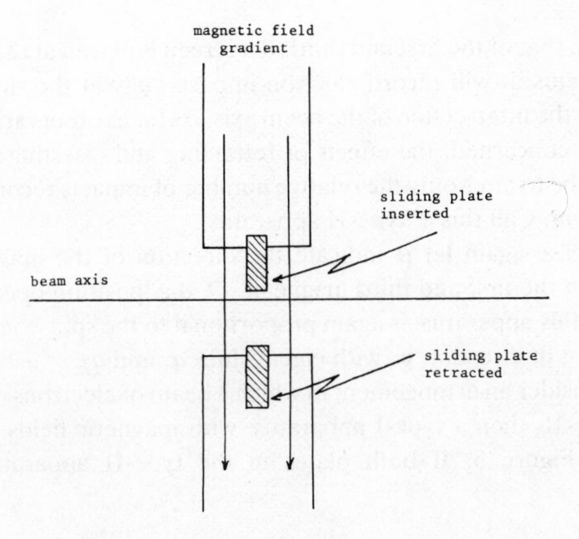
magnetic field
gradient
sliding plate
inserted
beam axis
sliding plate
retracted

Fig. 3.

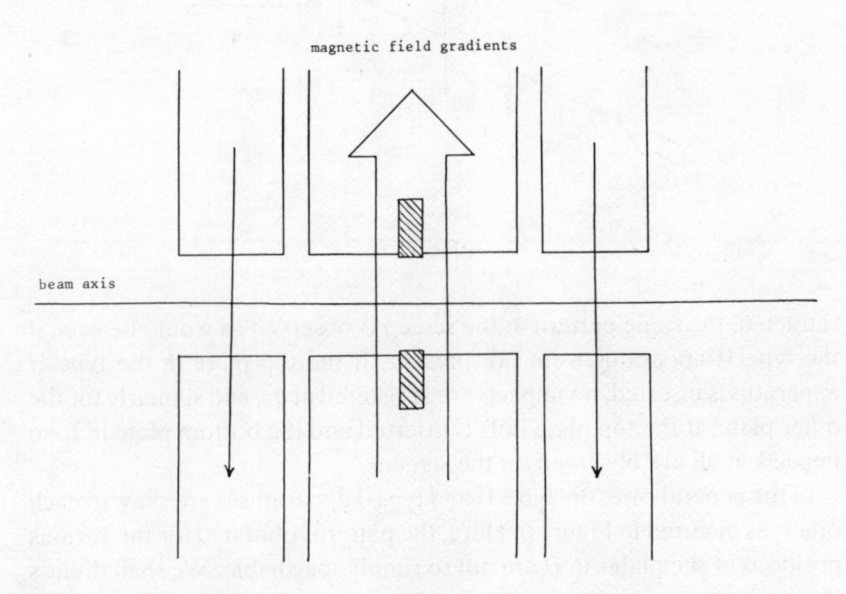
magnetic field gradients
beam axis

Fig. 4.

opposite to that of the first and third. If a screen is placed at the far end of this apparatus, it will record electron impacts only at the single region marked by the intersection of the beam axis. As far as observations on this screen are concerned, the effects of retracting and inserting the sliding plates will be to alter only the relative number of impacts recorded at this single region. Call this a 'type-II apparatus'.

In this case again let $\boldsymbol{\mu}$ indicate the direction of the magnetic field gradient (in the first and third magnets). Q, the 'position operator' with respect to this apparatus, is again proportional to the spin-component of the electron in direction $\boldsymbol{\mu}$, with eigenvalues q_1 and q_2.

Now consider an arrangement in which a beam of electrons encounters first a type-II, then a type-I apparatus, with magnetic fields aligned as shown in Figure 5. If both plates in the type-II apparatus remain

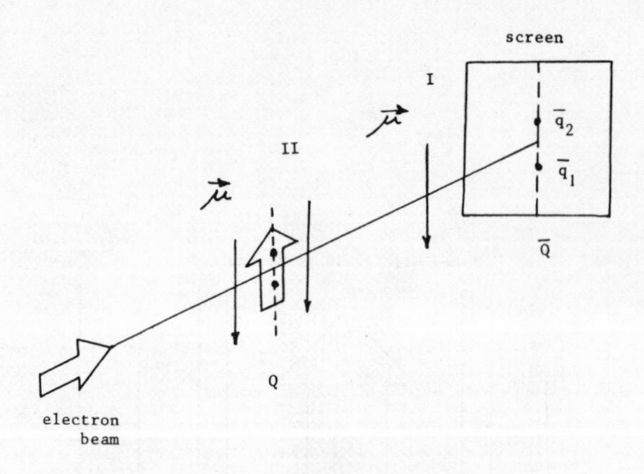

Fig. 5.

retracted, the same pattern at the screen is observed as would be even if the type-II apparatus were not present. If the top plate in the type-II apparatus is inserted, no impacts are registered at $\bar{q}_2$, and similarly for the other plate. If the top plate in II is inserted and the bottom plate in I, no impacts at all are observed on the screen.

In the general case, the type-II and type-I apparatuses are skew to each other, as pictured in Figure 6. Here, the patterns obtained for the various positions of the plates in II are not so simply specifiable. We shall discuss them quantitatively shortly. But first let us formulate our account of the

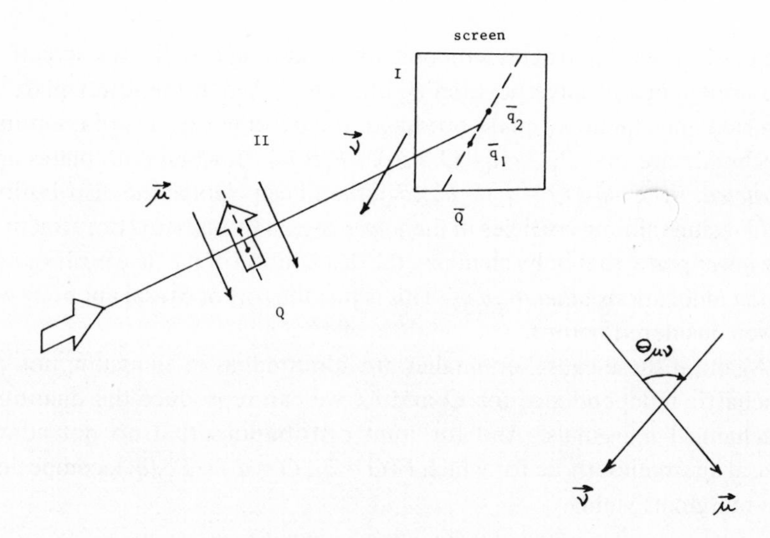

Fig. 6.

experiment: each electron in the collimated beam (with a particular orientation of its spin vector $\boldsymbol{\sigma}$) passes through the type-II apparatus either in the upper or lower region (with respect to the field gradient direction), but not both, continues through the type-I apparatus, again either in the upper or lower region with respect to the field gradient direction, and is detected in the appropriate region on the screen. (Henceforth, whenever I refer to an 'upper region' or a 'lower region', it is to be understood that these regions are defined with respect to the associated magnetic field gradient orientation.) Formally, we wish to associate with each electron in the collimated beam (hence described by ψ) values of the quantities Q and $\bar{Q}$. or, as we have done before, to examine the possible joint probability distributions of these values, $P^{\psi}_{Q,\bar{Q}}(q_i, \bar{q}_j)$, in terms of their satisfaction of the two criteria of the theorem.

This formal description is identical to that just considered in the 'two-slit' case, and, clearly, this example has been set up to be subject to a treatment analogous to that of the two-slit case throughout. Thus, it is easy to see that all distributions for which $P(\bar{Q}=\bar{q}_j \mid Q=q_i) \neq P^{\xi_i}_{\bar{Q}}(\bar{q}_j)$ admit causal anomalies. For example, $P(\bar{Q}=\bar{q}_2 \mid Q=q_2)$ measures the frequency with which particles passing through the type-II apparatus in the upper region (particles with position q_2 at time t) strike the screen in the upper region (have position $\bar{q}_2$ at time $\bar{t}$). When the lower plate is inserted and the upper plate retracted, the observed results of quantum mechanics are that $P(\bar{Q}=\bar{q}_2 \mid Q=q_2)=P^{\xi_2}_{\bar{Q}}(\bar{q}_2)$. If, when both plates are *retracted*, $P(\bar{Q}=\bar{q}_2 \mid Q=q_2) \neq P^{\xi_2}_{\bar{Q}}(\bar{q}_2)$, then I can *change* the distribution of $\bar{Q}$-values among particles in the *upper* region by inserting or retracting the *lower* plate, that is, by changing the distribution of particle positions at time t at locations *other than* q_2. This is just the sort of causal anomaly we have considered before.

Again, if these causal anomalies are admitted, as in an assumption of stochastic independence for Q and $\bar{Q}$, we can reproduce the quantum mechanical marginals. And for joint distributions that do not admit causal anomalies, those for which $P(\bar{Q}=\bar{q}_j \mid Q=q_i)=P^{\xi_i}_{\bar{Q}}(\bar{q}_j)$, computing the marginals yields

$$\sum_i P^{\psi}_{Q,\bar{Q}}(q_i, \bar{q}_j)=P^{\psi}_{\bar{Q}}(\bar{q}_j) \neq P^{\psi}_{Q}(q_1) \cdot P^{\xi_1}_{\bar{Q}}(\bar{q}_j)+P^{\psi}_{Q}(q_2) \cdot P^{\xi_2}_{\bar{Q}}(\bar{q}_j)$$

and the statistical results of quantum mechanics are not reproduced. The first term on the right gives the relative number of electron impacts in the jth (i.e., upper or lower) region of the detecting screen when the upper plate of the type-II apparatus is inserted, the lower plate retracted. The second term gives the relative number when the lower plate is inserted, the upper plate retracted. The term on the left gives the relative number of impacts in the jth region when both plates are retracted throughout.

Now, in fact, when the magnetic field gradients are aligned (as in Figure 5) or anti-aligned (i.e., exactly oppositely directed), the value of the sum of the two terms on the right is equal to the value of the term on the left. But in the general case pictured in Figure 6, this is certainly not the case. For example, if $\theta_{\mu\nu}=\pi/2$, and $\psi=(1/\sqrt{2})\xi_1+(1/\sqrt{2})\xi_2$, we find that

$$P^{\psi}_{\bar{Q}}(\bar{q}_1)=1 \quad \text{and} \quad P^{\psi}_{\bar{Q}}(\bar{q}_2)=0$$

while

$$P^{\psi}_{Q}(q_1) \cdot P^{\xi_1}_{\bar{Q}}(\bar{q}_j)+P^{\psi}_{Q}(q_2) \cdot P^{\xi_2}_{\bar{Q}}(\bar{q}_j)=\tfrac{1}{2}$$

for $j=1, 2$.[14]

Example 3. Having carried through an analysis of the arrangement of Example 2, we can deal in a precisely analogous way with any number of type-II apparatuses placed upstream of the one in Figure 6. It will be useful to consider only one more, as pictured in Figure 7.[15] The notation has been adjusted slightly to make it more concise in this enlarged case.

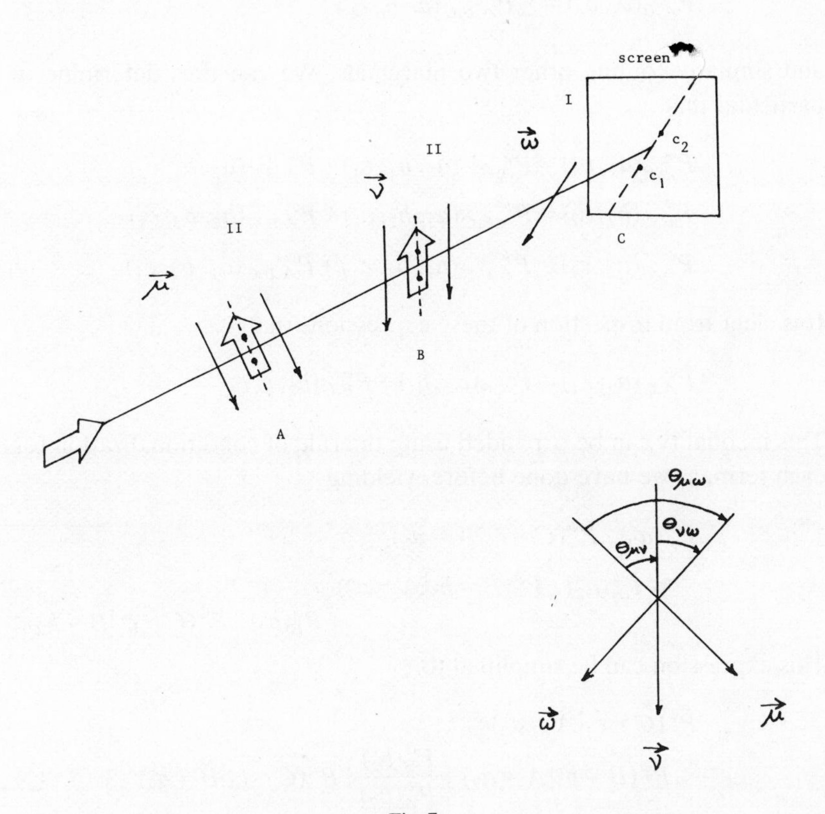

Fig. 7.

There are now three 'position operators' with respect to the appropriate apparatuses, in each case proportional to the spin component of the electron in the direction of the appropriate magnetic field gradient: $A \propto \boldsymbol{\sigma} \cdot \boldsymbol{\mu}$, $B \propto \boldsymbol{\sigma} \cdot \boldsymbol{\nu}$, $C \propto \boldsymbol{\sigma} \cdot \boldsymbol{\omega}$. Their respective eigenvalue equations are $A\alpha_i = a_i\alpha_i$, $B\beta_j = b_j\beta_j$, $C\alpha_k = c_k\gamma_k$. We will in this case be concerned to investigate the properties of possible probability distributions of the three quantities, $P^{\psi}_{A,B,C}(a_i, b_j, c_k)$, in terms of their satisfaction of the two criteria of the theorem.

It follows, of course, from the 'rule of elimination', i.e., from the marginal probability requirements, that, e.g.,

$$P^{\psi}_{A,B}(a_i, b_j) = \sum_k P^{\psi}_{A,B,C}(a_i, b_j, c_k)$$

and similarly for the other two marginals. We can thus determine in particular that

$$P^{\psi}_{A,B}(a_2, b_1) = P^{\psi}_{A,B,C}(a_2, b_1, c_1) + P^{\psi}_{A,B,C}(a_2, b_1, c_2)$$
$$P^{\psi}_{B,C}(b_2, c_1) = P^{\psi}_{A,B,C}(a_1, b_2, c_1) + P^{\psi}_{A,B,C}(a_2, b_2, c_1)$$
$$P^{\psi}_{A,G}(a_2, c_1) = P^{\psi}_{A,B,C}(a_2, b_1, c_1) + P^{\psi}_{A,B,C}(a_2, b_2, c_1).$$

It is clear from inspection of these expressions that

$$P^{\psi}_{A,C}(a_2, c_1) \leqslant P^{\psi}_{A,B}(a_2, b_1) + P^{\psi}_{B,C}(b_2, c_1)\,.$$

This inequality can be expanded using the rule of conditionalization for each term, as we have done before, yielding

$$\begin{aligned} &P^{\psi}_{A}(a_2) \cdot P^{\psi}(C = c_1 | A = a_2) \\ &\quad \leqslant P^{\psi}_{A}(a_2) \cdot P^{\psi}(B = b_1 | A = a_2) \\ &\qquad\qquad + P^{\psi}_{B}(b_2) \cdot P^{\psi}(C = c_1 | B = b_2)\,. \end{aligned}$$

This expression can be simplified to

$$\begin{aligned} &P^{\psi}(C = c_1 | A = a_2) \\ &\quad \leqslant P^{\psi}(B = b_1 | A = a_2) + \frac{P^{\psi}_{B}(b_2)}{P^{\psi}_{A}(a_2)} \cdot P^{\psi}(C = c_1 | B = b_2) \end{aligned} \qquad (*)$$

We thus arrive at the familiar stage of the calculation when we must examine the implications of the various choices of joint distributions. To make this examination particularly simple in this case, assume that $\psi = \alpha_2$. For this choice of ψ, we can write immediately

$$\begin{aligned} P^{\alpha_2}(C = c_1 | A = a_2) &= P^{\alpha_2}_{C}(c_1) \\ P^{\alpha_2}(B = b_1 | A = a_2) &= P^{\alpha_2}_{B}(b_1) \end{aligned} \quad \text{and} \quad \frac{P^{\alpha_2}_{B}(b_2)}{P^{\alpha_2}_{A}(a_2)} \leqslant 1$$

since these are just the observed results of quantum mechanics. The burden of the examination thus falls on $P^{\alpha_2}(C = c_1 | B = b_2)$. If we assume that B and C are stochastically independent, admitting causal anomalies of the sort described previously, then (*) is indeed valid:[16]

$$P_C^{\alpha_2}(c_1) \leqslant P_B^{\alpha_2}(b_1) + P_C^{\alpha_2}(c_1) \quad \text{or} \quad 0 \leqslant P_B^{\alpha_2}(b_1).$$

If we assume that $P(C = c_1 | B = b_2) = P_C^{\beta_2}(c_1)$, preserving local causality, we obtain for (*)

$$P_C^{\alpha_2}(c_1) \leqslant P_B^{\alpha_2}(b_1) + P_C^{\beta_2}(c_1)$$

an inequality not, in general, in accord with the quantum mechanically assigned values. In particular, for the values appropriate to the arrangement of Figure 7, we obtain

$$\sin^2 \theta_{\mu\omega}/2 \leqslant \sin^2 \theta_{\mu\nu}/2 + \sin^2 \theta_{\nu\omega}/2$$

which is false in general.

Thus, for the class of experimental arrangements considered in this example and the previous one, the only joint probability distributions of the values of the non-commuting quantities involved that reproduce the statistical results of quantum mechanics admit causal anomalies.

Example 4. As a last example of the application of the theorem, let us consider an experimental arrangement of the 'EPR' type (Einstein *et al.*, 1935), an arrangement involving systems described by quantum mechanical quantities whose values are correlated due to some past interaction of the systems. In this case we shall consider two beams of electrons moving in opposite directions, each of which passes through an apparatus of the type considered in Example 2, Figure 6, as pictured below. We shall stipulate that the state of the overall system (composed of the two beams) be described by the *spin singlet* state ψ. That is, we should like to consider the two beams to be composed of electrons that emerge from the source in pairs, their spins correlated in such a way that the net spin of each pair is identically zero.

The observations in this arrangement are distributions of recorded electron impacts on the two screens, but it will be helpful for the purposes of this example to think of the experiment as being run one electron pair at a time, in which case we can consider the observations to consist of coincidences of electron impacts at various pairs of points on the two screens.

The account we wish to implement of this arrangement goes like this: electrons leave the source and travel along the beam axis in pairs, with the particular orientations of their spin vectors correlated in such a way that $\boldsymbol{\sigma}^1 = -\boldsymbol{\sigma}^2$. Both electrons pass through their respective type-II apparatuses (at the same time) either in the upper or lower regions, but not in both, continue through their respective type-I apparatuses, again either in the upper or lower regions, and are detected at the appropriate regions of their respective screens. Formally, we wish to associate with *pairs* of electrons that leave the source (hence are described by ψ) values of the (position) quantities Q and $\bar{Q}$. In the arrangement pictured in Figure 8, these are A^1C^2 and B^1B^2, with four eigenvalues in each case, e.g., $a_1^1c_1^2 = q_1$, $a_2^1c_1^2 = q_2$, $a_1^1c_2^2 = q_3$, $a_2^1c_2^2 = q_4$.[17] In a manner completely analogous to the examples already considered, we examine the various forms of the joint probability distribution, $P^{\psi}_{Q,\bar{Q}}(q_i, \bar{q}_j)$, in terms of their satisfaction of the two criteria of the theorem.

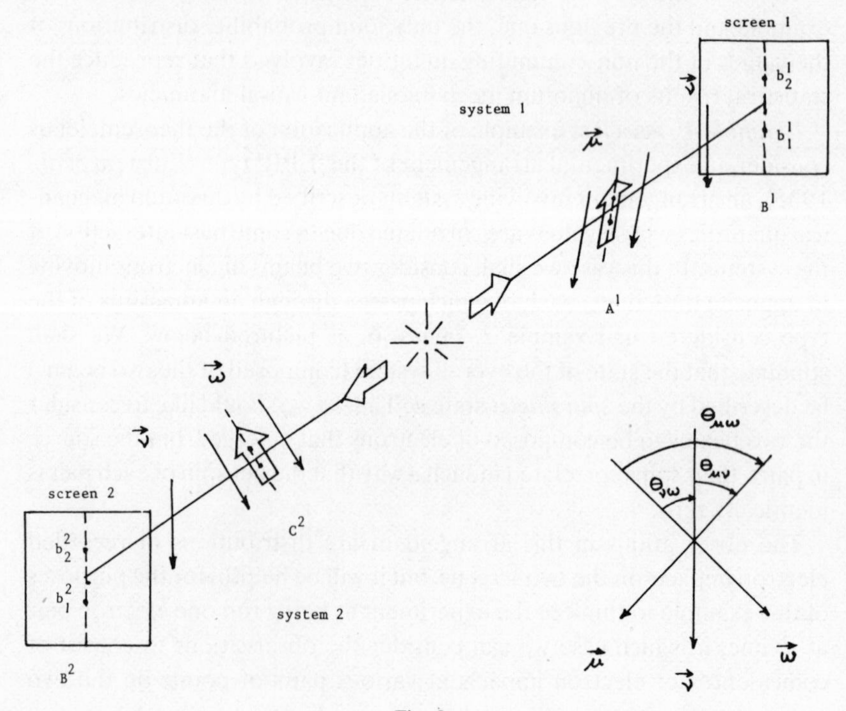

Fig. 8.

By the rule of conditionalization,

$$P^{\psi}_{Q,\bar{Q}}(q_i, \bar{q}_j) = P^{\psi}_{Q}(q_i) \cdot P^{\psi}(\bar{Q} = \bar{q}_j \mid Q = q_i)$$

and we expect that all distributions for which $P(\bar{Q} = \bar{q}_j | Q = q_i) \neq P^{\xi_i}_{\bar{Q}}(\bar{q}_j)$ admit causal anomalies of the kind we have seen before. It is easy to see that this is so. For example, $P(B^1B^2 = b_1^1b_1^2 | A^1C^2 = a_1^1c_1^2)$ measures the frequency with which pairs of electrons, both passing through the lower regions of their type-II apparatuses, are detected at the lower regions of their respective screens. When (either or both of) the upper plates of the type-II apparatuses are inserted, and the lower plates retracted, the observed results of quantum mechanics are that $P(B^1B^2 = b_1^1b_1^2 | A^1C^2 = a_1^1c_1^2) = P^{\alpha_1^1\gamma_1^2}_{B^1B^2}(b_1^1b_1^2)$. If, when both pairs are retracted, $P(B^1B^2 = b_1^1b_1^2 | A^1C^2 = a_1^1c_1^2) \neq P^{\alpha_1^1\gamma_1^2}_{B^1B^2}(b_1^1b_1^2)$, then I can *change* the distribution of coincident particle impacts on the screens (B^1B^2 values) among particles passing through the *lower* regions of their type-II apparatuses (those with $a_1^1c_1^2$ values of A^1C^2) by inserting or retracting the *upper* plates of these apparatuses, that is, by changing the distribution of particle positions at time t at locations *other than* a_1^1 and/or c_1^2. This is just the familiar sort of causal anomaly.[18]

Again, if these causal anomalies are admitted, as for stochastically independent Q, $\bar{Q}$, we can reproduce the quantum mechanical marginals. And for joint distributions that do not admit causal anomalies, those for which, as above, $P(B^1B^2 = b_1^1b_1^2 | A^1C^2 = a_i^1c_l^2) = P^{\alpha_i^1\gamma_l^2}_{B^1B^2}(b_1^1b_1^2)$, we find, analogously to our previous results, that

$$\sum_{i,l} P^{\psi}_{A^1C^2,B^1B^2}(a_i^1c_l^2, b_1^1b_1^2)$$

$$= P^{\psi}_{B^1B^2}(b_1^1b_1^2) \neq \sum_{i,l} P^{\psi}_{A^1C^2}(a_i^1c_l^2) \cdot P^{\alpha_i^1\gamma_l^2}_{B^1B^2}(b_1^1b_1^2)$$

where the sum involves four terms. For example, the $i = l = 1$ term gives the relative number of coincident impacts at the lower regions of screens 1 and 2 (at b_1^1 and b_1^2) when the upper plates of both type-II apparatuses are inserted (or only one is, as explained in the previous footnote). The sum of four such terms (corresponding to the four possible configurations of plates in the type-II apparatuses) is not generally equal to the value of the term on the left, corresponding, in this example, to the relative number of coincidences observed at b_1^1 and b_1^2 when all plates are retracted throughout. In fact, for this particular example, the value of the term on the left, the correct quantum mechanical marginal, is identically zero. (This follows immediately from the spherical symmetry of the singlet state ψ and expresses the requirement that the spins of the paired electrons be oppositely directed.) The sum of the terms on the right is easily evaluated in a case where $\theta_{\mu\omega} = \pi$, and $\theta_{\mu\nu} = \theta_{\nu\omega} = \pi/2$, and is equal to $\frac{1}{4}$ in this case. Thus, as in previous examples, joint distributions for which $P(\bar{Q} = \bar{q}_j \mid Q = q_i) = P^{\xi i}_{\bar{Q}}(\bar{q}_j)$, those that do not admit causal anomalies, do not reproduce the statistical results of quantum mechanics.

Let us now return to the original account of this arrangement. How would the account be different if, instead of the orientations of magnets pictured in Figure 8, the orientations of the type-I and type-II system 2 magnets were interchanged? Clearly, we would associate with each electron pair run through this apparatus particular values of the quantities A^1B^2 (at time t) and B^1C^2 (at time $\bar{t}$), consider joint probability distributions $P^{\psi}_{A^1,B^2,B^1C^2}(a^1_i b^2_k, b^1_j c^2_l)$, and draw the same conclusions about the various distributions for quantities A^1B^2, B^1C^2 as we have for the quantities A^1C^2, B^1B^2.

These two arrangements are, in some sense, merely alternative adjustments of the same apparatus: the system 1 magnets have possible (relative) orientations A^1 and B^1, and those in system 2, B^2 and C^2. Consider then all assignments to these four quantities *separately*, or, as we have done before, all joint probability distributions $P^{\psi}_{A^1,B^1,B^2,C^2}(a^1_i, b^1_j, b^2_k, c^2_l)$, in terms of their satisfaction of the two criteria of the theorem. By the rule of conditionalization, $P^{\psi}_{A^1,B^1,B^2,C^2}(a^1_i, b^1_j, b^2_k, c^2_l)$ can be expressed either as

$$P^{\psi}_{A^1C^2}(a^1_i c^2_l) \cdot P^{\psi}(B^1B^2 = b^1_j b^2_k \mid A^1C^2 = a^1_i c^2_l)$$

or as

$$P^{\psi}_{A^1B^2}(a^1_i b^2_k) \cdot P^{\psi}(B^1C^2 = b^1_j c^2_l \mid A^1B^2 = a^1_i b^2_k) .$$

These, of course, are just the distributions considered already in the two separate arrangements, and we know precisely what to say about them: no distributions of this type exist which reproduce the results of quantum mechanics (give the proper marginals) without admitting causal anomalies. Distributions admitting causal anomalies which do reproduce these results do exist, however, e.g., those for stochastically independent

$Q, \bar{Q}$. Thus any distribution $P^{\psi}_{A^1,B^1,B^2,C^2}$ which gives the proper quantum mechanical marginals, e.g., for which $\sum_{i,l} P^{\psi}_{A^1,B^1,B^2,C^2}(a^1_i, b^1_j, b^2_k, c^2_l) = P^{\psi}_{B^1B^2}(b^1_j b^2_k)$, must admit the familiar causal anomalies.

Consider such distributions for a moment. We have already seen that $P^{\psi}_{B^1B^2}(b^1_j b^2_j) = 0, j = 1, 2$, for all singlet states ψ. Thus all terms in these distributions of the form $P^{\psi}_{A^1,B^1,B^2,C^2}(a^1_i, b^1_j, b^2_j, c^2_l)$ are identically zero for all i, j, and l. (This is, of course, eight terms.) Using this fact, we determine that

$$\begin{aligned} P^{\psi}_{A^1B^2}(a^1_1 b^2_1) &= P^{\psi}_{A^1,B^1,B^2,C^2}(a^1_1, b^1_2, b^2_1, c^2_1) \\ &\quad + P^{\psi}_{A^1,B^1,B^2,C^2}(a^1_1, b^1_2, b^2_1, c^2_2) \\ P^{\psi}_{B^1C^2}(b^1_1 c^2_1) &= P^{\psi}_{A^1,B^1,B^2,C^2}(a^1_2, b^1_1, b^2_2, c^2_1) \\ &\quad + P^{\psi}_{A^1,B^1,B^2,C^2}(a^1_1, b^1_1, b^2_2, c^2_1) \\ P^{\psi}_{A^1C^2}(a^1_1 c^2_1) &= P^{\psi}_{A^1,B^1,B^2,C^2}(a^1_1, b^1_2, b^2_1, c^2_1) \\ &\quad + P^{\psi}_{A^1,B^1,B^2,C^2}(a^1_1, b^1_1, b^2_2, c^2_1)\,. \end{aligned}$$

It is clear from inspection of these expressions that

$$P^{\psi}_{A^1C^2}(a^1_1 c^2_1) \leq P^{\psi}_{A^1B^2}(a^1_1 b^2_1) + P^{\psi}_{B^1C^2}(b^1_1 c^2_1)\,.$$

But since there are all probabilities quantum mechanics does assign, we can substitute the appropriate values, obtaining

$$\tfrac{1}{2}\sin^2 \theta_{\mu\omega}/2 \leq \tfrac{1}{2}\sin^2 \theta_{\mu\nu}/2 + \tfrac{1}{2}\sin^2 \theta_{\nu\omega}/2$$

which is certainly false in general. Thus, even for those distributions $P^{\psi}_{A^1,B^1,B^2,C^2}$ that admit the familiar causal anomalies, those that give the proper quantum mechanical marginals, we obtain other contradictions with the results of quantum mechanics.

This conclusion can be drawn without a calculation of the above sort for the special case of distributions involving stochastically independent quantities $Q, \bar{Q}$. As we saw, for any $P^{\psi}_{A^1,B^1,B^2,C^2}$,

$$\begin{aligned} P^{\psi}_{A^1C^2}(a^1_i c^2_l) \cdot P^{\psi}(B^1B^2 = b^1_j b^2_k \mid A^1C^2 = a^1_i c^2_l) &= P^{\psi}_{A^1B^2}(a^1_i b^2_k) \\ &\quad \cdot P^{\psi}(B^1C^2 = b^1_j c^2_l \mid A^1B^2 = a^1_i b^2_k) \end{aligned}$$

and for stochastically independent $Q, \bar{Q}$'s, this gives

$$P^{\psi}_{A^1C^2}(a^1_i c^2_l) \cdot P^{\psi}_{B^1B^2}(b^1_j b^2_k) = P^{\psi}_{A^1B^2}(a^1_i b^2_k) \cdot P^{\psi}_{B^1C^2}(b^1_j c^2_l)$$

which itself is in contradiction with the probabilities quantum mechanics does assign. Of course, this result can be obtained in such an obvious way only for stochastically independent $Q, \bar{Q}$. The general result above is stronger: any distribution $P^{\psi}_{A^1,B^1,B^2,C^2}$ which gives the proper quantum mechanical marginals admits the familiar causal anomalies. But any distribution of this form which gives the quantum mechanical marginals leads to the above results in contradiction to those of quantum mechanics.

Thus, in some sense, this EPR situation seems to be even in worse interpretive shape than the previous examples. At least in these cases joint distributions exist which satisfy one of the criteria of the theorem. Here it seems we can satisfy neither. But is this conclusion justified? Have we not in fact described the *separate* arrangements characterized by A^1C^2, B^1B^2 and A^1B^2, B^1C^2 precisely analogously to the previous examples? We have, of course, but in such a way that $P^{\psi}_{A^1C^2,B^1B^2}(a^1_i c^2_l, b^1_j b^2_k) \neq P^{\psi}_{A^1B^2,B^1C^2}(a^1_i b^2_k, b^1_j c^2_l)$ in general, and if these two distributions are not equivalent, no distribution of the form $P^{\psi}_{A^1,B^1,B^2,C^2}(a^1_i, b^1_j, b^2_k, c^2_l)$ exists.[19]

Now is has been claimed that a distribution $P^{\psi}_{A^1,B^1,B^2,C^2}$ exists, rather than merely inequivalent $P^{\psi}_{A^1C^2,B^1B^2}$ and $P^{\psi}_{A^1B^2,B^1C^2}$, only for interpretations in which the value *a* quantity *in one system* takes (at some time) does *not* depend on the choice of quantity with respect to which *the other system* is described (at the same time).[20] For example, the value of A^1 in such an interpretation is independent of whether B^2 or C^2 is considered in the second system. In more physical terms, the value at time *t* of the spin component is some direction of one electron of a (spin-correlated) pair must be independent of the *orientation* of the magnetic field gradient through which the other electron is passing at time *t*. Thus, an interpretation for which inequivalent $P^{\psi}_{A^1B^2,B^1C^2}$ and $P^{\psi}_{A^1C^2,B^1B^2}$ exist, but $P^{\psi}_{A^1,B^1,B^2,C^2}$ does not, has the following feature: I can change the position of a *system 1* electron (and hence the distribution of position eigenvalues for a collection of electrons) at time *t* by changing the orientation of the *system 2* magnets at time *t*. Such an interpretation clearly admits causal anomalies, and anomalies of a different kind from those 'intra-system' anomalies considered already in this example and previous ones; they are 'inter-system' anomalies.

Let me summarize the conclusions of this example. We have seen that exhaustive interpretations of an arrangement of the sort pictured in Figure 8 with possible relative orientations of system 1 and system 2 magnets A^1 and B^1, B^2 and C^2 respectively, lead to joint distributions with the following features:

(1) If both inter-system and intra-system anomalies are admitted, inequivalent joint distributions of the form $P^{\psi}_{A^1C^2,B^1B^2}$ and $P^{\psi}_{A^1B^2,B^1C^2}$ exist which reproduce the quantum mechanical results considered.

(2) If only intra-system anomalies are admitted, joint distributions of the form $P^{\psi}_{A^1,B^1,B^2,C^2}$ exist (from which *equivalent* distributions of the above sort follow automatically) which give the correct quantum mechanical marginals, but which lead to other results in contradiction to those of quantum mechanics.

(3) If no causal anomalies are admitted, the only joint distributions of the form $P^{\psi}_{A^1,B^1,B^2,C^2}$ that exist do not even give the correct quantum mechanical marginals.

VI. DISCUSSION

With this consideration of the EPR class of experimental arrangements we bring our discussion of the Principle of Anomaly up to date, for many of the arguments presented in Example 4 are just those of Wigner (1970), clarifying a result of Bell (1964, 1971). Recently, Arthur Fine (1973, 1974) has raised important objections to the Wigner-Bell conclusions, and I would like now to consider these objections.

Recall yet again (from Section III) the intuitive statement of the Principle of Anomaly and the connection of this intuitive statement with the theorem of Section IV. First the Principle: no exhaustive interpretation of quantum mechanics satisfies the requirements of normal causality. To get to the theorem we have, according to Reichenbach, only to consider carefully of what an exhaustive interpretation consists, namely, assignments of values to all physical quantities in all physical states. The notion of "the probability for some particular value of an unobserved entity in a state" arises naturally from the frequency with which this value appears in the set of assignments associated with that state. And it seems natural to suppose likewise that the probability for some particular combination of values of the unobserved quantities in some state is related to the frequency with which this combination of values appears in the set of assignments associated with that state.

If we do suppose that the probabilities for such combinations of values are well-defined, then we can go on immediately to investigate whether any such distributions can reproduce the statistical results of quantum mechanics without admitting causal anomalies. In particular, we can consider distributions for two non-commuting quantities. I have shown that no such distributions satisfying these two criteria exist; this is the theorem of Section IV. Does this result then vindicate the Principle of Anomaly?

Fine would claim that it does not. As he points out (1973), the assumption that the probabilities of combinations of values for *all* relevant quantum mechanical quantities describing some physical system are well-defined in all states is a *stronger* assumption than that required by an adequate exhaustive interpretation which, recall, requires only that we "assign definite values to the unmeasured, or not exactly measured entities, in such a way that the observed results appear as the causal consequences of the values introduced by our assumption" (5). But the 'observed results' are never distributions of combinations of values of *non-commuting* quantities; the only combinations of values for which quantum mechanics does provide probability assignments are those of *commuting* quantities. Thus Fine would reject the necessity of the existence of joint distributions of non-commuting quantities for the existence of exhaustive interpretations satisfying the requirements of normal causality.

In fact, Fine takes all the more recent 'no-joint-distribution' proofs as the very best grounds for rejecting the existence of these joint distributions altogether. He claims (1973) that results of this kind simply show that the quantum mechanical description of the world is one in which

(i) individual events are well-defined;
(ii) the probabilities of individual events are well-defined;
(iii) compound events are well-defined;
(iv) but the probabilities of compound events are *not*, in general, well-defined.

The situation, as Fine says, is "like the situation that obtains in the limiting relative frequency account of probability. There one may have a pair of zero-one sequences in each of which there is a limit to the relative frequencies for 1, but where in the compound sequence (in which a 1 occurs in just those places where a 1 occurs in both original sequences) there is no limit." He goes on to argue that this particular sort of situation, in which there are 'gaps' in the assignments of probabilities to well-defined sets of events, does not reflect anything 'non-classical or aberrant' in the use being made of probability. "The situation does reflect – in the sense that it results from – those special features of the microcosm that are built into quantum mechanics." And these features are, generally, that the sort of compound events indicated by the occurrence of simultaneous values of non-commuting quantities are simply "too erratic to admit of probability assessment, though the component events themselves are well-behaved."

This is Fine's response to the more recent 'no-joint-distributions' proofs. But as we saw in Section IV, these proofs all impose conditions on joint distributions that are weaker than the RCA condition. And though RCA as it stands is inapplicable to exhaustive interpretations in which joint distributions of non-commuting quantities fail, in general, to exist, it seems likely that such interpretations violate Reichenbach's intuitive notions of normal causality. In fact I think Reichenbach would claim that they admit the same sort of anomalies described before. Let us see why.

The standard sort of anomaly, recall, arose in consideration of joint distributions $P^{\psi}_{Q,\bar{Q}}(q_i, \bar{q}_j)$ of non-commuting $Q, \bar{Q}$. Consider for the moment exhaustive interpretations of quantum mechanics in which such distributions do not exist, in general. Now while they may not be well-defined *generally* (i.e., in *all* states), they are certainly well-defined in *some* states, e.g., in eigenstates of Q. Thus, in a q_i-eigenstate ξ_i, the probability assignments of quantum mechanics are such that

$$\begin{aligned}\sum_k P^{\xi_i}_{Q,\bar{Q}}(q_k, \bar{q}_j) = P^{\xi_i}_{\bar{Q}}(\bar{q}_j) &= \sum_k P^{\xi_i}_{Q}(q_k) \cdot P^{\xi_i}(\bar{Q} = \bar{q}_j \mid Q = q_k) \\ &= P^{\xi_i}_{Q}(q_i) \cdot P^{\xi_i}(\bar{Q} = \bar{q}_j \mid Q = q_i) \\ &= 1 \cdot P^{\xi_i}(\bar{Q} = \bar{q}_j \mid Q = q_i)\,.\end{aligned}$$

For instance, in the two-slit arrangement of Example 1, with one hole alone (called simply B) open, as Reichenbach says,

> We have a certain probability . . . that a particle leaving the source A will arrive at B, and a probability . . . that a particle leaving A and passing through B will arrive at C. Both values can be statistically determined by counting on a surrounding screen all particles leaving A, then all those particles arriving on the screen (the number arrived at meaning the number of particles passing through slit (B) and then all particles arriving at C. (25)[21]

But causal anomalies are avoided in cases such as these, as we saw, only if the conditional probability $P(\bar{Q} = \bar{q}_j \mid Q = q_i)$ is independent of the arrangement of hole openings (the state of the electrons downstream from the plate); that is, the probability that a particle leaving source A and passing through hole B_1 strikes region C on the detecting screen is the same whether or not hole B_2 is open. If it is not, then, as Reichenbach would say, I can change the distribution of $\bar{Q}$-values among the particles with Q-value q_1 by opening or closing hole B_2, that is, by changing the distribution of particle positions at locations other than q_1. What does it mean, then, to say that the conditional probability $P(\bar{Q} = \bar{q}_j \mid Q = q_i)$ for arbitrary arrangements of open holes (i.e., states expressed as superpositions of ξ_i's) is not only not *identical* to the probability when hole B_i alone is open (i.e., in state ξ_i), but is not even *well-defined*?

Fine says it indicates that the sort of compound events involved (e.g., passing through hole B_1 and striking region C on the screen) are 'too erratic to admit of probability assessment'. Reichenbach would presumably claim, however, that since the compound event of passing through B_1 and striking C is *not* too erratic to admit of probability assessment when B_1 *alone* is open, and *is* too erratic to admit of probability assessment when B_1 and B_2 are *both* open, I can so alter the distribution of $\bar{Q}$-values among particles with Q-value q_1 by closing or opening hole B_2 as to prevent their characterization by probabilities altogether in the latter case – a more dramatic causal anomaly than before!

What is at issue here is the very nature of the probablilistic character of the physical processes described. For Reichenbach, the only cogent 'corpuscular interpretation' of the two-slit experiment is in terms of interactions of individual particles with individual holes. Thus, at hole B in a plate (a diaphragm, as Reichenbach calls it),

> particles are subject to impacts or other forms of interaction imparted to them by the particles of which the substance of the diaphragm is composed. They thus deviate from their paths. These impacts follow statistical laws in such a way that some parts of the screen are frequently hit, others less frequently. The interference pattern of the photographic film indicates, therefore, the probability distribution of the impacts given in B to the passing particles. (25)

Any change in the distribution of impacts on the screen from particles that pass through B when *other* holes are opened thus indicates a change in the interactions at B – a non-local effect.

Fine understands the probabilities in a different way. If the arrangement consists of a collection of particles passing singly through the plate, on his view the probabilities are related to the character of the plate *as a whole*, not merely to interactions of individual particles with individual holes in it.[22] After all, the state description of the particles considered is such that the sum of the probabilities of particles striking the screen is always the same value – usually normalized to unity. (See note 21.) That is, when only a single hole is open, the state description is such that the probability that a particle which strikes the screen passes through the plate at any other region is zero. When a second hole is opened, the probability that a particle passes through this region is changed from zero to some positive value, and the probability at the initial hole is decreased by this amount. The same holds for any number of holes. So the opening and closing of holes in the plate changes much more than the distribution of $\bar{Q}$-values among particles with particular Q-values; it can change the probability distributions for whole sets of Q-values as well. In general, since probabilities depend on features of the plate as a whole, *any* alteration in the array of holes in the plate may change *all* the distributions of probability in question, and might even therefore render the probability for various compound events no longer well-defined.[23]

It appears then, that in Fine's view, probabilities are relational properties that depend in special ways on physical situations as a whole. Presumably he would deny that, in general, locality conditions of any sort are appropriate for such probabilities.

Thus the question of whether the effects present in all these examples violate the requirements of local causality remains. A firm conclusion would require a more thorough analysis of both Fine's and Reichenbach's views on probability, and this is a job for another paper. What their difference of viewpoint shows, however, is the richness which general questions about the nature of probability take on when they are instantiated in the context of a detailed physical theory.

VII. CONCLUDING REMARKS

Reichenbach proposed an interpretation of quantum mechanics in terms of a three-valued logic due to his conviction in the impossibility of there being any adequate bivalent semantics for the theory – a condition he elevated to the status of a fundamental principle, the Principle of Anomaly. Such a semantics, or exhaustive interpretation of the theory, would assign values to all physical quantities in all states in such a way as to reproduce the results of quantum mechanics – both the eigenvalues observed in measurements and the statistical distributions of these values in the states – while satisfying the requirements of classical causality.

Reichenbach took, as a necessary condition for the existence of such an interpretation, the existence, for any two non-commuting quantities and any state, of a joint probability distribution which would reproduce the quantum mechanical probability assignments while satisfying a particular causality condition – RCA as I have called it. Following Reichenbach's general line of reasoning, I have established as a theorem that no such joint distributions exist.

But the issue of whether the existence of such joint distributions is in fact necessary for the existence of an adequate exhaustive interpretation, and thus whether the theorem establishes the Principle of Anomaly, is not as clear-cut as Reichenbach thought. Arthur Fine, in particular, has discussed exhaustive interpretations in which such joint distributions are not well-defined in general. It seems clear that Reichenbach would reject such interpretations as not satisfying the requirements of classical causality, and thus would consider the Principle of Anomaly established. Fine, however, with a view of the character of probabilities in quantum mechanics quite different from Reichenbach's, does not see any obvious violations of classical causality in these interpretations.

Thus I would say that the status of the Principle of Anomaly is still not resolved, and the necessity for a restrictive interpretation of quantum mechanics not demonstrated.

University of Tennessee, Knoxville

NOTES

1. See, for example, van Fraassen (1974) and Gardner (1972).
2. Reichenbach's interpretation restricts the class of propositions to which standard truth values are assigned, and thus is a restriction on 'assertability'. (See Reichenbach, 1944, p. 145. Subsequent page references, except where noted, are to this text.) He classifies the 'Bohr-Heisenberg' interpretation as one of 'restricted meaning': "The *rule of restriction* states that only statements about measured entities . . . are admissable; statements about unmeasured entities . . . are called meaningless." (40).

3. Throughout this paper I will consider only operators with discrete (and non-degenerate as well) sets of eigenvalues. When 'position' and 'momentum' quantities are mentioned, they are generally to be understood in the 'approximate' sense, that is, as 'coarse-grained' position and momentum. None of the conclusions drawn depend on this restriction, however, and in fact most of Reichenbach's arguments in the book are given in 'continuous' terms. There are many fewer technical problems to be considered in the discrete case though, and the formalism is easier to write down and talk about.
4. For the simple cases we will be considering (see previous note), $P^{\psi}_{Q}(q_i) = |(\xi_i, \psi)|^2$, where ξ_i is the ith eigenfunction of Q and (,) denotes the inner product on the Hilbert space of states characterizing the system.
5. See Messiah (1958), p. 310 ff.
6. This problem of completeness can also be presented as one of providing a complete semantics for quantum mechanics, i.e., of providing truth conditions for *all* statements of the form 'the value of quantity Q is in set S'. Fine goes on immediately to cast the problem in this form, and, in fact, his characterization of the 'hidden parameters' ("The hidden parameters are nothing more than the ways of assigning definite values to each of the physically meaningful quantities already employed by the quantum theory. The parameters are logical constructs which complete the definition of these quantities so as to make them well-defined in very state.") would probably appeal very much to Reichenbach, who did not see his work as bearing on the issue of whether "the laws of quantum mechanics, perhaps, hold only for a certain kind of parameter; that at a later stage of science other parameters may be found for which the relation of uncertainty does not hold; and that the new parameters may enable us to make strict predictions" (13), but was vitally concerned with just the 'logical constructs' Fine mentions. Much of Reichenbach's discussion does take a semantical approach (almost exclusively after § 28) and, of course, his 'restrictive interpretation' is a modification of the standard bivalent truth value assignment scheme. Strictly for the purposes of elucidating and criticizing the Principle of Anomaly however, such considerations are unnecessary, and I prefer to stick to the problem as formulated above.
7. This is certainly a reasonable assumption. (See Fine (1973) for an argument.) Reichenbach, in fact, offers a 'proof' that it is only these assignments that satisfy the requirements of normal causality. I find his argument unconvincing though, and I will criticize it after I have formulated a formal causality condition.
8. As Fine (1973) points out, there will in general be too many assignments in this set to measure by simple proportions. Thus, what one does here is to form a 'classical probability space' out of these assignments and then transfer the measure on this space to the quantum mechanical probability for the event in question. Rather than introduce these notions formally however, I refer the reader to this paper for details.
9. It should be pointed out that this equivalence is denied by Popper (1968, Appendices *IV and *V) who considers the conditionals more fundamental than the joint probabilities. (See also van Fraassen, 1975). Actually, this equivalence is not crucial for any of the arguments in this paper; all could be carried through in terms of conditionals only. I shall treat the sorts of distributions as equivalent, however, to make things simpler.
10. Alternatively, S can be taken to be a collection of individual particles each in an identical experimental arrangement, or a collection of passes of similar particles through the same experimental arrangement, with what is meant by Q and t specified appropriately. The probability distributions of values of all quantities are generally to be related to the frequency of appearance of these values in the given collection.

11. That is, if, for all pairs of quantum mechanical quantities A, B describing some physical system characterized by states ψ and ψ', $P^{\psi}_{A,B} = P^{\psi'}_{A,B}$, then $\psi = \psi'$.
12. Reichenbach specifically argues that this joint distribution admits anomalies, for A the (approximate) position operator on pp. 127–128.
13. This sort of arrangement is described in Feynman *et al.* (1965), Vol. III, Section 5.
14. Note that this result is precisely what one would expect intuitively in this case. ψ is actually the state that would be prepared, for example, by placing another type-II apparatus upstream of the one in Figure 6, oriented in the same direction as the type-I apparatus there, with the lower plate retracted and the upper plate inserted, that is, a $\bar{q}_1$-eigenstate, $\bar{\xi}_1$.
15. The idea of this example is taken from Bub (1973). Aspects of the formal treatment of it follow Bub's treatment closely.
16. Of course, this special case does not suffice to demonstrate that (*) is always satisfied for any state ψ. But it is easy to check the other cases: ψ can be an eigenstate of *at most* one of the quantities A, B, and C (since they are non-commuting) or an eigenstate of none.
17. Note that ψ is an element of the Hilbert space formed by taking the tensor product of that associated with particle 1, H^1, and that associated with particle 2, H^2, e.g., $\psi = \xi_1 = \alpha^1_1 \otimes \gamma^2_1 \in H^1 \otimes H^2$. Likewise, A^1C^2, as an operator on $H^1 \otimes H^2$, is usually expressed as $A^1 \otimes C^2$. The notation here is cumbersome enough as it is, however, so I will just use, e.g., $\alpha^1_1\gamma^2_1$ and A^1C^2 with the tensor product understood.
18. Note that inserting either upper plate, or both, leads to precisely the same probability of *coincidences* at the two screens $P^{\alpha^1_1\gamma^2_1}_{B^1B^2}(b^1_1 b^2_1)$. Of course, the *number* of impacts, or the relative probability of impacts on, say, screen 1 *alone*, is not affected by the position of the plate on the system 2 type-II apparatus, and *vice-versa*. The non-locality in this case is only with respect to electrons passing through the single apparatus. The same considerations apply if we are considering beams of electrons passing through together. In this case, though ψ determines the distributions of positions (or of spins) of *all* electrons at time t, the non-local interactions need only be admitted among the electrons of a single beam.
19. In this case the inequality above cannot be derived. If we try to repeat the calculation that leads to it in the case of stochastically independent Q, $\bar{Q}$, we obtain

$$P^{\psi}_{A^1B^2}(a^1_1 b^2_1) = \sum_{j,l} P^{\psi}_{B^1C^2}(b^1_j c^2_l) \cdot P^{\psi}_{A^1B^2}(a^1_1 b^2_1)$$

$$P^{\psi}_{B^1C^2}(b^1_1 c^2_1) = \sum_{i,k} P^{\psi}_{A^1B^2}(a^1_i b^2_k) \cdot P^{\psi}_{B^1C^2}(b^1_1 c^2_l)$$

from the second arrangement of magnets considered, and

$$P^{\psi}_{A^1C^2}(a^1_1 c^2_1) = \sum_{j,k} P^{\psi}_{B^1B^2}(b^1_j b^2_k) \cdot P^{\psi}_{A^1C^2}(a^1_1 c^2_1)$$

from the first. The inequality can be derived in this case only if

$$P^{\psi}_{A^1C^2}(a^1_i c^2_l) \cdot P^{\psi}_{B^1B^2}(b^1_j b^2_k) = P^{\psi}_{A^1B^2}(a^1_i b^2_k) \cdot P^{\psi}_{B^1C^2}(b^1_j c^2_l)$$

and, as we noted above, this expression itself contradicts the assigned probabilities of quantum mechanics.
20. The claim is made explicitly in this form by Wigner (1970) and Bub (1973). Bell's claim (1964, 1971) is equivalent, and is stated in more detail by, e.g., Shimony (1973) and Clauser and Horne (1974).

21. Note that the little calculation above involves a 'normalization' to unity of the probability that a particle leaving the source A will arrive at B. Reichenbach's treatment has the advantage of making explicit the fact that the arrangement of open holes in the plate has nothing to do with the emission of particles from the source and hence with the state of the particles up stream of the plate. The advantage of the above treatment is that one considers only those electrons that pass *through* the plate, ignoring those that strike away from the open holes, or never get to the plate at all.

22. Michael Audi (1973) presents an analysis of the origin of the characteristic statistical results of quantum mechanics which draws somewhat this same conclusion. But his views and Fine's are very different. Audi attempts to analyze the two-slit experiment along the lines of the usual treatment of particle scattering by crystal lattices: the collection of particles is treated as a plane wave scattering off a 'quasi-periodic potential'. This leads to 'selective momentum transfer' and the characteristic diffraction patterns. Since "The space distribution of the perturbing potential represented by the slit arrangement is quite different in the two [single slit and double slit] cases," the conditional probabilities derived from these two arrangements are different as well, and no non-local effects are involved. (See p. 118).

I think there are two kinds of difficulties with Audi's analysis. First, it does not adequately describe the physics involved. It is a standard result in classical diffraction theory (Jackson, 1962, p. 262) that the diffraction pattern in the region on the order of wavelengths away from the plate (the Fresnel zone) is quite different from that at distances many orders of magnitude greater (the Fraunhofer zone). Audi's calculation is designed to exhibit only the Fraunhofer pattern (which is, of course, reasonable, since that is the pattern usually measured). With different approximations, perhaps the Fresnel pattern could be obtained as well. It would be interesting to see.

The other objection is more fundamental. It follows from Audi's treatment that the joint distributions $P_{Q,\bar{Q}}(c_i, \bar{q}_j)$ are well-defined in all states. But as I have mentioned obliquely before (in Section IV), in any interpretation in which such distributions exist generally (with the possible exception of interpretations involving non-state functional distributions proposed by Margenau and Cartwright), results can be obtained which contradict those of quantum mechanics. Cohen (1966), Nelson (1967), and Fine (1973, 1974) provide examples of such results. Thus, whatever the success of Audi's treatment of the 'locality' issue, it cannot escape criticism on this basis.

23. It should be clear from the preceding paragraph how Fine would eliminate the issue of 'locality' from the two-slit case, and all the other analogous cases of the examples, including the 'intra-system' locality of the EPR arrangement. He need not even adopt the radical conclusion that the joint distributions do not exist in general to obtain the 'observed results' in these cases – the proper quantum mechanical marginals. In showing that the contradictions with quantum mechanical probabilities derived by Bell and Wigner for the EPR case do not depend on any sort of 'locality' assumptions, Fine (1974) eliminates the 'inter-system' locality issue from this example as well. (In terms of the treatment I give here of this example, Fine's result is that any exhaustive interpretation of the EPR arrangement in which joint distributions are well-defined (as state-functionals) generally must admit distributions of the form P_{A^1,B^1,B^2,C^2}, and thus must give the contradictions with the results of quantum mechanics derived on p. 67.) The example thus becomes a particularly nice way to display the rather more subtle (than the marginals) sort of contradiction with the results of quantum mechanics that, by the theorems of Cohen (1966), Nelson (1967), and Fine himself (1973), any interpretation admitting joint distributions of non-commuting quantities must involve.

BIBLIOGRAPHY

Audi, M., *The Interpretation of Quantum Mechanics*, University of Chicago Press, Chicago, 1973.

Bell, J. S., 'On the Einstein-Podolsky-Rosen Paradox', *Physics* **1** (1964), 195–200.

Bell, J. S., 'Introduction to the Hidden Variable Question', in B. d'Espagnat (ed.), *Foundations of Quantum Mechanics*, Academic Press, New York, 1971.

Bub, J., 'On the Possibility of a Phase-Space Reconstruction of Quantum Statistics: A Refutation of the Bell-Wigner Locality Argument', *Foundations Phys.* **3** (1973), 29–44.

Cartwright, N., 'Correlations without Distributions in Quantum Mechanics', *Foundations Phys.* **4** (1974), 127–136.

Clauser, J. F. and Horne, M. A., 'Experimental Consequences of Objective Local Theories', *Phys. Rev.* **D10** (1974), 526–535.

Cohen, L., 'Can Quantum Mechanics be Formulated as a Classical Probability Theory', *Philos. Sci.* **33** (1966), 317–322.

Einstein, A., Podolsky, B., and Rosen, N., 'Can Quantum Mechanical Description of Physical Reality Be Considered Complete', *Phys. Rev.* **47** (1935), 770–780.

Feynman, R. P., Leighton, R., and Sands, M., *The Feynman Lectures on Physics*, Addison-Wesley, Reading, Mass., 1965.

Fine, A., 'Probability and the Interpretation of Quantum Mechanics', *Brit. J. Philos. Sci.* **24** (1973), 1–37.

Fine, A., 'On the Completeness of Quantum Theory', *Synthese* **29** (1974), 257–289.

Gardner, M., 'Two Deviant Logics for Quantum Theory: Bohr and Reichenbach', *Brit. J. Philo. Sci.* **23** (1972), 89–109.

Jackson, J. D., *Classical Electrodynamics*, John Wiley & Sons, New York, 1962.

Margenau, H., 'Measurements and Quantum States', I and II, *Philos. Sci.* **30** (1963), 1–16, 138–157.

Messiah, A., *Quantum Mechanics*, North-Holland, Amsterdam, 1958 (trans. by G. M. Temmer).

Nelson, E., *Dynamical Theories of Brownian Motion*, Princeton University Press, 1967.

Park, J. L. and Margenau, H., 'Simultaneous Measurability in Quantum Theory', in W. Yourgrau and A. van der Merwe (eds.), *Perspectives in Quantum Theory: Essays in Honor of Alfred Landé*, MIT Press, Cambridge, Mass., 1971.

Popper, K., *The Logic of Scientific Discovery* (2nd ed.), Harper Torchbooks, NY, 1968.

Reichenbach, H., *Philosophic Foundations of Quantum Mechanics*, University of California Press, Berkeley and Los Angeles, 1944.

Reichenbach, H., 'The Principle of Anomaly in Quantum Mechanics', *Dialectica* **2** (1948), 337–348.

Reichenbach, H., 'Les Fondements Logiques de La Mecanique des Quanta', *Ann. Inst. Henri Poincaré* **13** (1952), 109–158.

Shimony, A., 'The Status of Hidden Variable Theories', in P. Suppes, L. Henken, Gr. C. Moisil, and A. Joja (eds.), *Logic, Methodology and Philosophy of Science* IV, North-Holland, Amsterdam, 1973.

van Fraassen, B., 'The Labyrinth of Quantum Logics', in R. S. Cohen and M. W. Wartofsky (eds.), *Boston Studies in the Philosophy of Science XIII*, Humanities Press, NY, 1974.

van Fraassen, B., 'Construction of Popper Probability Functions', (preprint) (1975).

Wigner, E. P., 'On Hidden Variables and Quantum Mechanical Probabilities', *Amer. J. Phys.* **38** (1970), 1005–1009.

NEAL GROSSMAN

METAPHYSICAL IMPLICATIONS OF THE QUANTUM THEORY

According to Arthur Danto[1] one of the distinguishing features of philosophy is that, unlike any other discipline, questions about its own nature are internal to the discipline; questions like 'What is a philosophical question?' and 'What is philosophy?' are themselves philosophical questions. Conceptions of the scope and nature of philosophy have varied markedly throughout the history of philosophy, and vary today from one 'school' to the next. The conception of the nature of philosophy naturally affects the way philosophy is done, which of course affects the content of philosophy. Hence, in order to appreciate the content of any philosopher's writings it is important to understand his general attitude towards philosophy and how the content is structured by that attitude. Reichenbach is exceptionally clear about his attitude towards philosophy:

> I do not believe that there are philosophical questions as such, exempt from scientific treatment. If philosophy is to deal with problems of the structure of the world, it should abandon the idea of deriving this structure from an intuition alleged to be based on eternal truths. The philosophical *truths* of yesterday have become the *errors* of today. The philosopher who wants to contribute towards making the universe intelligible can aid the physicist only in searching for the correct form of a question, but not in searching for the answer. That is to say, his contribution will consist in a *logical analysis* of the problems which separate the physical content of a theory from the additions occurring in the form of definition, and which clarifies the meanings of terms instead of prescribing the ways of thought.[2] (Italics mine.)

This expresses a rather unusual conception of what philosophy is – a conception with which, I am sure, hardly any philosopher would agree today. Nevertheless, even though I regard speculative metaphysics as perhaps the most important part of philosophy – a view diametrically opposed to Reichenbach's – I find myself deeply sympathetic towards certain aspects of Reichenbach's conception of philosophy. Although I do not believe that philosophy is parasitical on the sciences, I do believe that philosophers ought to be sensitive to the content of the fundamental scientific theories, and to appreciate the relevance of that content to their

W. C. Salmon (ed.), Hans Reichenbach: Logical Empiricist, 605–623.

own work. In particular, I want to argue, in this paper, for a view which other philosophers may find as outrageous as Reichenbach's: namely, that the quantum theory, when interpreted *realistically* has *radical* implications for *all* of philosophy, especially metaphysics.[3] Specifically, I shall argue that *atomism* – the view which conceives of reality as constituted by independently existing entities – is at variance with a realistic[4] interpretation of quantum theory. I shall develop my argument using Reichenbach's principle of causal anomaly,[5] and then discuss some ontological implications of the fact (at least, I am taking it to be a fact) that there are no exhaustive interpretations of quantum mechanics free from causal anomaly.

The principle of anomaly is best discussed in the context of a familiar example. Consider an arrangement consisting of a diaphragm with two movable shutters, and a screen. A beam of electrons is incident on the diaphragm from the left. We want to know how the electrons distribute themselves on the screen if (i) one slit is open and the other is closed, and (ii) both slits are open. Let '$\langle x|\phi_1\rangle$' mean 'the amplitude for the electron to arrive at x if it passes through the upper slit'. If the lower slit is closed then the distribution of electrons on the screen is given by $P_1(x) = |\langle x|\phi_1\rangle|^2$. If the upper slit is closed, the distribution is given by $P_2(x) = |\langle x|\phi_2\rangle|^2$. If both slits are open, the distribution is given by

$$P_{12}(x) = |\langle x|\phi_1 + \phi_2\rangle|^2 = P_1(x) + P_2(x) + 2\langle x|\phi_1\rangle\langle x|\phi_2\rangle\,.$$

'$P_{12}(x)$' is usually called the *interference pattern*; '$P_1(x) + P_2(x)$' is usually called the additive pattern. The experimental fact that one gets the interference pattern and *not* the additive pattern is the crux of all attempts to make 'sense' out of quantum mechanics. It is this fact which makes it impossible[6] (or at least extremely difficult) to maintain that each electron goes through either the upper or the lower slit, for if that *were* the case, one would expect to get the additive pattern. And indeed, if observations are made to determine which of the two slits the electrons traverse, then one does get the additive pattern.

For Reichenbach, as for all the logical empiricists, the 'real' is to be identified with actual observations (phenomena). We are free to tell any consistent story we like about what we are not observing (inter-phenomena) as long as the story accounts for the phenomena. All such stories (same phenomena, different interphenomena) are descriptively equivalent because, according to Reichenbach, they all agree with respect to what counts, namely, the phenomena. Any interpretation of a body of data which talks only about what is actually observed is called a *restrictive interpretation*; an interpretation which goes beyond what is observed is called *exhaustive*. Thus 'all observed metals conduct electricity' and 'all metals conduct electricity' are, respectively, restrictive and exhaustive interpretations of the same body of data. Moreover 'all metals conduct electricity' attributes to metals the same properties as the restrictive interpretation; Reichenbach calls it the *normal* description. Any other exhaustive interpretation (e.g. 'all observed metals conduct electricity and all metals not observed on Sundays do not conduct electricity'), although descriptively equivalent to the normal interpretation, involves *causal anomalies*.

Applying these concepts to the two slit example, let us consider a single electron incident upon the diaphragm (both slits open) and arriving at the point x on the screen. Since we do not actually observe the electron passing through the diaphragm, that event is an interphenomenon, and we may say anything we like about it consistent with the phenomenon (arrival at x). Suppose we say that it went through one of the two slits. There are, then, two experimentally distinguishable situations:

(i) the electron goes through one of the slits and the other slit is closed.

(ii) the electron goes through one of the two slits and the other slit is open.

(i) yields the additive pattern; (ii) yields the interference pattern. Thus it makes a difference, to an electron about to pass through one of the two slits, whether or not the other slit is open or closed. The electron, as it is about to pass through the upper slit, say, must 'know' that the lower slit is open. This involves a causal anomaly, since the 'information' would have to be transmitted instantaneously.

If, on the other hand, we represent the interphenomena as a wave process, so that we can say the electron goes through both slits in the form of a wave, then we have a different causal anomaly: namely, the collapse of the wave to a single point at x. Reichenbach's conclusion is that any exhaustive interpretation of quantum mechanics (in our example, any attempt to say what is happening at the two slits when we are not actually performing measurements there) must involve causal anomalies. Quantum mechanics does not have a normal interpretation.

I think Reichenbach is entirely correct as far as he goes, but I think his positivist ideology prevented him from asking the obvious question, namely, why is it the case that quantum mechanics does not have a normal interpretation? After all, *something real is happening at the two slits.* We cannot intelligibly represent what is happening in terms of localized particles, and if we try to represent it in terms of waves, then we cannot also intelligibly represent what happens at the screen. Does this mean that we must give up? Does this mean that reality is unintelligible? Unfortunately, Reichenbach's general philosophical beliefs allowed him to ignore such questions. For his identification of what is real with what is observed led to the distinction between the phenomena and the interphenomena, and implicit in that distinction is the belief that the interphenomena are not that important, and perhaps not altogether real. And this is why we are free to say anything at all about the interphenomena, or to refrain from saying anything at all. I find this attitude somewhat anthropocentric. It is perhaps epistemologically useful to distinguish what we observe from what we do not observe, especially if, like Reichenbach, we want to base our theories upon observation of the world. But it seems to me that the content of our theories, once arrived at, ought not to confer special status to man. From a realist's perspective, what there is is what the theory says there is, whether we make observations or not. The interphenomena are every bit as real as the phenomena. The problem is to give a coherent, intelligible account of reality, that is, of both the phenomena *and* the interphenomena. It is thus important to realize that Reichenbach's principle of anomaly is by no means an *interpretation* of the quantum theory; but rather, it is a statement of the problem.

At present, there is no satisfactory solution of this problem, although there are numerous re-statements of it. Unfortunately, re-statements have often been mistaken for solutions. For example, it is sometimes claimed that the problem does not arise if non-standard logics are used. But I think it is easy to see that the very fact that the microworld is amenable to non-standard logics is itself a paradox which needs to be understood, and I also think that Reichenbach himself was aware of this. Reichenbach's proposal is for a three-valued logic in which a disjunction is true but the disjuncts have indeterminate truth-value. In his own words:

> If no observation is made at the slits, one would say, using two valued logic, that an individual particle observed on the screen has passed either through one slit, or through the other. This statement, which belongs to an exhaustive description, leads to causal anomalies. The *restricted description, in the form of a three-valued logic*, replaces this disjunction with another ... it has the following properties: if the particle has been observed in the vicinity of one of the slits, the disjunction is true, and if no observation has been made at the slits and the statement of passage is indeterminate, the disjunction is true just the same. By employing this meaning for the word "or" we can thus say that the particle passes through one slit or the other, without arriving at causal anomalies.[7] (Italics mine.)

I want to make two related points here: firstly, Reichenbach is explicitly clear in recognizing that a three-valued logic is a form of the restricted description. The restricted description is the one which confines itself solely to the phenomena, and which, consequently, does not provide an exhaustive description of reality (where reality includes the inter-phenomena). To say that quantum mechanics *requires* a three-valued logic is just another way of saying that we *must* settle for the restricted description – that there is no exhaustive description free from causal anomaly. And this, of course, is precisely what the problem is.

Secondly, any proposal to change the meaning of logical connectives must be examined with great care, especially when the connectives are 'interpreted', that is, used to connect physical objects and events, not merely mathematical objects. In terms of the two-slit example, Reichenbach's proposal to change the meaning of 'or' has the following consequences: when the particle is not being observed at the slits, we have

'ϕ_1' – indeterminate
'ϕ_2' – indeterminate
'ϕ_1 or ϕ_2' – true .

In words: 'The particle passed through the upper slit' is indeterminate, 'the particle passed through the lower slit' is indeterminate, but 'either the particle passed through the upper slit or the particle passed through the lower slit' is true. But although one can *say* 'ϕ_1 or ϕ_2', using this different sense of 'or', we are no closer to understanding what is going on at the slits. If 'ϕ_1' were true, I would understand that the particle traversed the upper slit; if 'ϕ_1' were false, I would understand that the particle traversed the lower slit; but I haven't the slightest idea what to understand about the actual physical situation when 'ϕ_1' is indeterminate. To say that 'ϕ_1 or ϕ_2' is true also fails to bring understanding because (given that 'ϕ_1 or ϕ_2' is true does not entail either 'ϕ_1' is true or 'ϕ_2' is true), I do not know what sort of physical situation 'ϕ_1 or ϕ_2' is true of. That is, I do not understand what is going on at the slits. What is needed is "to understand *why* the [micro-world] is associated with some particular logic",[8] and this understanding can come about (if at all) only through conceptual (metaphysical) analysis *and speculation*.

Consider the general question: can one alter the meanings of the logical connectives, or the rules which the connectives obey,[9] without also altering the nature of the entities so connected? If not, then the entities (objects, events) which non-standard logics connect are not like the entities which ordinary bivalent logic connects. Hence what is called for are *new concepts* to grasp the non-standard entities of quantum mechanics. ". . . the criticism is that quantum logic, pursued as *normative* rather than simply as an important quantum algebraic structure, diverts our eyes from the *physical situations with which we are confronted in the micro-domain*, and the present *inability of our conceptual structures to comprehend them adequately*."[10]

I wish now to reformulate what I take to be 'the problem' in very general, metaphysical terms. I shall then develop an analogy which I believe both exemplifies the relation between conceptual structures and logic and shows why the logic is secondary to the concepts. Suppose a given conceptual structure prescribes a set of fundamental concepts, or categories (e.g. mind and matter, substance and property, wave and particle, etc.), which it claims is both *exhaustive* (any given thing must fall under one of the categories) and *exclusive* (no entity can fall under more than one category). A 'problem' for such a conceptual structure would be generated if there were good reason to believe that there exist entities which either do not fall neatly into one of the prescribed categories or are able to cross categories. That is, even if at a given time every observed entity (the 'phenomena') fit neatly into the conceptual structures, there would still be a major problem with the conceptual structure if there were good reason to believe that the basic entities can cross the fundamental categories. Why? Because (i) one would need a concept of cross-categorical *identity* for the basic entities, and this concept could not be formulated in the original conceptual structure, and (ii) no account of *how* the entities change categories could be given within the original conceptual structure.

This is exactly the situation in quantum mechanics. According to quantum mechanics, there are two basic categories for dealing with the events of the micro-world: particle and wave. These two categories are exclusive in the sense that nothing can be both a particle and a wave at the same time. However, it appears to be the case that there exist entities (electrons, photons, etc.) which under certain experimental conditions, C_1, behave like waves and under other experimental conditions, C_2, (where C_1 and C_2 are not simultaneously realizable) behave like particles. Thus we have two ontological categories, wave and particle, and entities which are able to cross those categories (e.g. as in the 'reduction of the wave-packet'). How does one make sense of this? It seems to me there are two alternatives: (a) one can cling to the belief that the wave/particle category system is ultimate, in which case the most that one can do is to try to make *logical sense* of the situation by means of rules (logical, semantical, pragmatic or otherwise) according to which we can employ the concepts, wave and particle, without contradiction, or (b) one can abandon the belief that the wave/particle category system is ultimate, and attempt to make *conceptual sense* of the situation by means of new categories. The following analogy develops this point of view.

Let us imagine that we are back at a time prior to the general acceptance of the atomic theory of matter. We have a set of three categories: solid, liquid, gas. We believe this set to be exhaustive (any given thing must be solid, liquid, or gas) and exclusive (no given thing can be both solid and liquid, etc.) This classification system works perfectly well until it is realized that there are things (like water) which fall under different categories at different times. There are two ways of dealing with this situation, corresponding to the two above mentioned paradigms: either (i) one clings to the belief that the classification system, solid-liquid-gas, is ontologically ultimate or (ii) one gives up this belief and begins searching for a more fundamental category. The most that can be achieved under (i) is the formulation of rules which enable us to apply the solid-liquid-gas system to water without contradiction. Fortunately, it was (ii) which was adopted, and this led both to the *discovery* of a new ontological category (the atom, or molecule) and to an *explanation* of the old categories in terms of the new one. However, I think that it will be both instructive and amusing to outline how the water example might be handled under (i).

Let us assume, then, that the three categories, S (solid), L (liquid), and G (gas) are fundamental. Thus, for any particular thing x, exactly one of the following three statements must be true: (1) S_x, (2) L_x, (3) G_x. Each category has associated with it a set of properties, P_S, P_L, and P_G. Now although these property sets are not disjoint [e.g. *weight* belongs to all three sets], there are certainly *some* properties which belong exclusively to one set. The property of having an intrinsic shape belongs only to P_S; the property of having a certain surface tension belongs only to P_L. It is clearly inappropriate to apply a property, which belongs only to P_S, to anything which falls under the categories L or G. Thus any formal language game would want to stipulate that questions such as 'what is the intrinsic shape of this batch of krypton gas?' are inadmissible. Now, in the case of water, although it was formerly thought that water belonged to the L category, it is now discovered that only under certain conditions C_2 does water behave like a liquid, whereas under other conditions C_1 it behaves like a solid and under still other conditions C_3 it behaves like a gas. Hence, instead of being able to assert categorically that water is a liquid, we must now make essential reference to the *conditions* under which water is a liquid. We thus have the following three propositions:

$$(x) \quad [(W_x \cdot C_1) \rightarrow S_x],$$

$$(x) \quad [(W_x \cdot C_2) \rightarrow L_x],$$

and

$$(x) \quad [(W_x \cdot C_3) \rightarrow G_x].$$

Let us ask some experimental questions about the properties of water. Suppose I have a batch of water k and I want to determine, experimentally, the value of the two properties: *hardness* (h) and *surface tension* (t). Since h belongs *only to* P_S and t belongs only to P_L, and since k cannot be both S and L, we cannot answer both questions at the same time. For the conditions C_1 under which h is measurable are incompatible with the conditions C_2 under which t is measurable. It is not *merely* that under conditions C_1 we are unable to measure the value of k's surface tension; it is rather that under conditions C_1 the very property of having a surface tension is undefined, and under conditions C_2, the very property of having a hardness is undefined. If k is initially in its liquid state, and we want to measure its hardness we must first create the conditions, C_1, under which k appears in its solid state. One might even say that, in creating the conditions C_1, we are creating the very property whose value we are trying to ascertain!

The basis for constructing a three-valued logic is quite obvious. Let z be a particular real number, let 'T', 'F' and 'N' stand for 'true', 'false' and 'neuter' respectively. Consider any proposition which could be used to express the outcome of some possible experiment designed to measure some property of k, say (to be specified) k's hardness: 'The value of h for k is z.' = 'a'. The semantics for 'a' is:

(1) Assign 'T' to 'a' iff C_1 obtains and z is the case

(2) Assign 'F' to 'a' iff C_1 obtains and z', $(z' \neq z)$, is the case

(3) Assign 'N' to 'a' iff C_1 does not obtain .

A quantum-mechanical analogue of (3) would be a proposition asserting some value for an electron's position while the electron is in a momentum eigenstate, or, in terms of the two slit example, a proposition asserting that the electron went through one of the two slits. The quantum-logician's claim that the discovery of the electron's properties 'proves' that the 'logic' of the world is non-standard is no more viable than the 'water logician's' claim that the discovery of water's properties 'proves' that the logic of the world is non-standard. The 'logic' of the situation is, I think, clear: given a set of categories which are believed to be ontologically basic, and given the discovery of an entity which belongs exclusively to none of those categories, we can either (i) retain our belief in the ontological fundamentalness of our present categories or (ii) search for new categories. It is only with respect to choosing alternative (i) that we are stuck with non-standard logics. And of course, to boldly proclaim that the logic of the micro-world is non-standard is to naively presuppose that the concepts which we *now* have at our disposal for describing the micro-world are ultimate. But a simple induction on the history of physics should yield the conclusion that there is no reason to expect that physics ever will arrive at anything like 'ultimate' categories. Indeed, *the fact that a given physical theory requires a non-standard logic could be, and I think ought to be, taken as indicating that the categories used and/or presupposed by that theory are not fundamental.* Or, in Reichenbach's terminology, the fact that quantum theory admits of no exhaustive description free from causal anomaly indicates that the categories used and/or presupposed by that theory are not fundamental.

Although I cannot say what the new category is, I think I can say something about what it is not. Specifically, I think the new category will have to be (i) non-atomistic and (ii) non-spatio-temporal. The need for these conditions has already been hinted at above. A concept which purports to explain *how* a micro-object crosses categories (wave to particle, as in the collapse of the wave packet) must be of greater generality than the wave/particle concepts, in the same way that the category of a molecule is more general than the categories solid/liquid/gas. Indeed, if, in each case, the former concept were not more general than the latter, then the former concept could not be used to explain how entities can transform across the latter categories. In the remainder of this paper, I wish to discuss three related features of the micro-world which indicate the need for a non-atomistic ontology, and which hence support the thesis that, to the extent that quantum mechanics is true, atomism is false.

1. TRANSFORMATION OF MICRO-OBJECTS

Not only do micro-objects cross categories, but also they transform into one another. Thus a neutron can transform into a proton and an electron, and photons can be transformed into electron-positron pairs. Now it seems to me that any 'thing' which can be transformed into some other 'thing' ought not to be regarded as basic in an ontological sense. For to regard such 'things' as ontologically basic commits one to a magical picture of reality in which the fundamental building blocks of the world can pop in and out of existence.

The history of physics seems to indicate the belief that the presence of change must be explained on the basis of something which does *not* change. That which remains invariant under a given transformation is always regarded as belonging to a deeper level of reality than that which undergoes the transformation. Thus, if we observe that A changes to B, then we postulate the existence of something else, x, and functions (or rather 'real definitions') f and g, such that $A = f(x)$ and $B = g(x)$. This rule not only gives sense to the expression 'x is ontologically more fundamental than A', but also serves as a principle of intelligibility. This rule expresses the belief that the world is intelligible; not to use this rule is to have given up the search for an intelligible physical (or metaphysical) account of the structure of reality.

Consider our water example: let A be a batch of ice which changes into B, a batch of water. The change from A to B is explained by postulating the existence of something else, x, (molecules), and defining both A and B in terms of x. That is, A is defined as a collection of molecules bound in a crystalline lattice structure; B is defined as a collection of molecules having freedom of mobility among their neighboring molecules. It is in virtue of the fact that the molecules which constitute the batch of ice, A, are the *same* molecules which constitute the batch of water, B, that we can explain, or render intelligible, the cross-categorical (solid→liquid) transformation. Another illustration of the above rule (or principle of intelligibility) which is more relevant to the case of particle transformation, is molecular transformation. Consider the transformation of water into hydrogen and oxygen: $2H_2O \rightarrow 2H_2 + O_2$. Neither the *number* of molecules (two before, three after) nor the *kind* of molecules (water before, hydrogen and oxygen after) are invariant under this transformation. The atomic theory of matter *explains* (i.e. *makes intelligible*) this change by postulating the existence of entities in terms of which the molecules are defined [e.g. 1 molecule of water $= H_2O$] and which *do* remain invariant both in kind and in number [there are 4 atoms of hydrogen and 2 atoms of oxygen, both before and after the transformation]. Similarly, the fact that atoms can themselves be transformed into one another indicates (applying the principle of intelligibility) that there is something still more fundamental than atoms, and in terms of which the atom must be defined.[11] Considering now the so-called elementary particles "it is clear that they are not the immutable atoms of Democritus, since neither their number nor their kind remains fixed with time. The only things which do remain constant in the transformations that the particles undergo are certain quantitative properties that the particles carry, such as charge, energy, and momentum. *It is as if these properties were the reality, while the particles are just transitory manifestations* of them." (Italics mine.)[12] It should be noted that properties are neither atomistic or spatio-temporal. Perhaps (and this is a very speculative 'perhaps') the problem of explaining how these properties manifest themselves in the form of particles is not unlike the problem of explaining how Platonic Universals manifest themselves in the form of particular things.

2. IDENTITY

In virtue of *what* are we able to say that the very same entity which passes through the slits in the form of a wave strikes the screen in the form of a particle? In terms of the water analogy, this question would be, in virtue of what are we able to say that the very same thing which was in solid form (ice) is now in liquid form? The 'what' in this latter case consists in discovering a new concept (the molecule) and defining water in terms of the new concept (as a collection of molecules) in such a way that the *identity* of a given batch of water is independent of *which* of the solid/liquid/gas categories the given batch happens to belong. And because the identity of water is independent of the solid/liquid/gas categories we are able to say that the *same* batch of water which formerly was solid is now liquid. Hence what is required is a new category, and a definition of micro-entities in terms of this new category in such a way that the identity of a given micro-object is independent of the wave/particle categories. Only in this way can sense be made out of saying that the same entity which formerly was a wave is now a particle. Note that one cannot identify this new category with the micro-objects themselves, since, as discussed above, the micro-entities transform among themselves.

These considerations, although pointing to the need for new categories of physical existence, do not by themselves demonstrate the likelihood that the new categories will be non-atomistic. What does, I think, indicate the need for a non-atomistic category is the *fact* that the micro entities obey non-standard statistics. It is a tribute to Reichenbach's genius that he recognized the importance of this fact. I shall use the same example Reichenbach[13] uses to illustrate the peculiarities of the Bose-Einstein statistics. Suppose we are given two fair coins, which we label '1' and '2'. The probability of getting 'heads-up' on coin 1 is one-half. In symbols $P(H_1) = \frac{1}{2}$. Similarly $P(H_2) = \frac{1}{2}$, $P(T_1) = \frac{1}{2}$, and $P(T_2) = \frac{1}{2}$. If we now flip both at the same time there are *four* possible outcomes: H_1H_2, H_1T_2, T_1H_2, T_1T_2, with probabilities $P(H_1H_2) = P(H_1)P(H_2) = \frac{1}{4}$, $P(H_1T_2) = \frac{1}{4}$, $P(T_1H_2) = \frac{1}{4}$ and $P(T_1T_2) = \frac{1}{4}$. Suppose, however, we are interested not in *which coins* come up heads or tails, but only in *how many* come up heads or tails. That is, we want to know the probabilities of getting two heads, one head and one tail (without regard for which coin is heads and which is tails), and two tails. The probabilities for these outcomes are, respectively, $P(H_1H_2) = \frac{1}{4}$, $P(H_1T_2 \text{ or } T_1H_2) = \frac{1}{2}$, $P(T_1T_2) = \frac{1}{4}$. Note: it is *because* H_1T_2 and T_1H_2 are *distinct* outcomes that we *count* them both to get P (one heads and one tails) $= P(H_1T_2 \text{ or } T_1H_2) = P(H_1T_2) + P(T_1H_2) = \frac{1}{4} + \frac{1}{4} = \frac{1}{2}$. Assume now that the coins obeyed the Bose-Einstein statistics. Then the empirically verified probability distribution is equivalent to the assignment P (both heads) $= \frac{1}{3}$, P (both tails) $= \frac{1}{3}$, and P (one heads and one tails) $= \frac{1}{3}$. The physical situation here is *as if* 'H_1T_2' and 'H_2T_1' did not refer to distinct outcomes, but were merely different names for the same outcome.

At any rate, the tautology '(i) either the outcomes are distinct (from one another) or (ii) they are not distinct' is certainly true. Reichenbach spelled out the consequences of adopting (i). If one assumes that the outcomes are indeed distinct, then since P (one heads and one tails) $= P(H_1T_2 \text{ or } T_1H_2) = \frac{1}{3}$, the outcomes themselves must have probabilities $P(H_1T_2) = P(T_1H_2) = \frac{1}{6}$. Recall that, for ordinary non-quantum coins, the probability of getting say, tails on the second coin is $\frac{1}{2}$, regardless of whether the first coin came up heads or tails. This is no longer true for quantum coins. For assuming that $P(H_1) = P(T_1) = \frac{1}{2}$, and given that $P(H_1T_2) = \frac{1}{6}$ and $P(T_1T_2) = \frac{1}{3}$, getting tails on the second coin is more likely if the first coin came up tails than if it came up heads. Thus if H_1T_2 and T_1H_2 are distinct outcomes we must conclude that what happens to one of the coins is *not independent* of what happens to the other. A consideration of the nature of this dependency leads quickly to Reichenbach's concept of causal anomaly; for example, the dependency would still exist even if the coins were tossed millions of miles apart. In Reichenbach's words

> If we are to put [the fact of quantum statistics] in accord with the idea of distinguishable particles, we will have to introduce interparticle forces which attract or repel them in such a way that their motions are no longer mutually independent. Each particle would depend not only on each other particle, but also on the totality of positions of the other particles. These forces would have a rather strange nature because they would be transmitted instantaneously across space and could be observed only by their effects on the statistics.[14]

In the next paragraph Reichenbach dramatically spells out what would result from the attempt to reconcile the concept of distinguishable particles, each of which has a definite trajectory, with the fact that these particles obey quantum statistics:

> But what kind of physics would we then have? It would be a physics of mystical forces, far from the accepted idea of force in classical physics. In fact, this physics would be further removed from the physics of common sense than the indeterministic physics accepted today.[15]

Returning to our coin example, if we hold (i) that 'H_1T_2' and 'T_1H_2' refer to distinct outcomes, then we must conclude that the individual coins depend on one another in ways so strange that the only word Reichenbach can think of to describe this dependency is 'mystical'. And since, for Reichenbach, 'mystical' means 'unintelligible' (or worse), alternative (i) is unintelligible in that it leads to unintelligible consequences.

Before discussing (ii) above, I want to relate what we have concluded thus far to metaphysical atomism. I take atomism to be the thesis that reality is constituted by distinct, independently existing entities. Now the phrase 'independently existing' is of crucial importance. For if something does not have independent existence, then that upon which its existence depends is (ontologically) more fundamental. Now, the consequence of adopting (i) is that the particles are not independent. On what do they depend? "... on each other ... and also on the totality of positions of the other particles." Now the essence of atomism consists, I believe, in the attitude that any composite thing (like a collection of particles) is constituted by its parts, and that the behavior of the thing is determined by the properties of its parts. But it looks as if, in the quantum domain, the behavior of the parts (particles) depends on the whole ('the totality of positions of the other particles') which those parts supposedly constitute. We no longer have atomism, or anything like atomism. For how can the parts be regarded as constituting the whole if the properties of the parts depend on the whole? Thus adopting (i) above leads to a rejection of atomism.

Atomism fares just as badly if we adopt the view (ii) that the outcomes are not distinct. For what does it mean, from the realists' perspective, to claim that the expression 'coin number one lands heads-up and coin number two lands tails-up' and 'coin number one lands tails-up and coin number two lands heads-up' do not refer to distinct physical situations? It means, I think, that not only is there no way of telling which coin is which, but also, there is, ontologically speaking, nothing in the world in virtue of which one could attach the labels 'coin number one' and 'coin number two'. Nature can *count* (we can tell *how many* are heads and how many are tails) but she cannot *individuate* (we cannot say *which coins* are heads and *which coins* are tails). Thus to deny that the outcomes are distinct is to deny that there is anything *in the world* which could distinguish them. But this is to admit that there is no principle of individuation for the fundamental entities. What kind of atomism is it in which the basic entities are not individuated?

Let us look at this closely: in the classical case, we explain the fact that getting one heads and one tails is twice as probable as getting two heads (or two tails) by pointing out that the outcome, one head and one tail, can occur in two distinct ways [H_1T_2, T_1H_2], whereas the outcome, two heads, can occur in only one way [H_1H_2]. But in the quantum case, the observed fact is that the three outcomes, one heads and one tails, two heads, two tails, are equi-probable. Now, to preserve atomism, it is essential that the labels '1' and '2' have physical significance, but this, as we have seen, leads to what Reichenbach has called "... a physics of mystic forces...." However, it is possible to account for the observed distribution by denying physical significance to the labels '1' and '2'. Thus one would have three outcomes, *HH*, *HT*, and *TT*, and only one way for each of the outcomes to occur (*HT* is identical to *TH*). But, since the labels '1' and '2' are the only way of individuating which coins are which, *denying physical significance to the labels is to deny that there are in the world distinct entities* which could be referred to by '1' and '2' – and this is tantamount to a rejection of atomistic ontology. For when the labels are removed from H_1T_2, what is left are just the properties *H* and *T*; there is no room for any 'entity' to act as a 'carrier' of those properties. So it is incorrect to attribute the property, heads, to an already existing coin. To quote Feinberg again, "It is as if these properties were the reality, while the particles are just transitory manifestations of them."[17]

3. COMPOSITE SYSTEMS

The power, beauty and elegance of atomism lies in its ability to explain a wide range of diverse phenomena by assuming that the phenomena are constituted by independently existing, qualitatively similar, entities. For if the world is really made up of parts, then it is natural to expect that the behavior of the observable phenomena is determined by the nature of the parts. For example, because a gas is nothing over and above a collection of molecules, the behavior of a gas can be completely explained in terms of the nature and properties of molecules. Extending this 'reductionist' attitude, one attempts to explain the behavior of molecules in terms of *its* parts, namely atoms, and then attempts to explain the behavior of atoms in terms of *its* parts, namely particles. Unfortunately for atomism, the application of quantum mechanics to composite systems demonstrates the impossibility of regarding the whole as constituted by independently existing parts.

Consider a composite system consisting of two identical particles, each of which has spin $\frac{1}{2}$,[16] but which are combined in such a way that net spin of the two-particle system is zero. It would seem (forgetting, for the moment, the discussion in the previous section) that there are two ways for two spinning particles to arrange themselves so that their spins cancel out: If we let '$\psi_-(1)$' mean 'particle #1 is in the spin-up state' then the empirically verified wave function for the composite system is:

$$\psi(1, 2) = \psi_+(1)\psi_-(2) - \psi_-(1)\psi_+(2) = \psi^a - \psi^b,$$

where

$$\psi^a \equiv \psi_+(1)\psi_-(2); \psi^b = \psi_-(1)\psi_+(2)$$

The analysis of this situation follows the same pattern as the earlier analysis of the two slit experiment. Recall that any attempt to construe the particle as actually passing through one of the two slits must involve unpalatable causal anomalies. Similarly, because experimentally, one gets the interference distribution, $\psi^2(1, 2) = (\psi^a - \psi^b)^2 = (\psi^a)^2 + (\psi^b)^2 - 2\psi^a\psi^b$, and not the additive distribution, $(\psi^a)^2 + (\psi^b)^2$, one cannot construe $\psi(1, 2) = \psi^a - \psi^b$ to mean that either ψ^a or ψ^b is the case, without involving oneself in the most horrendous sorts of causal anomalies.

To be sure, if one performs a measurement to ascertain whether ψ^a or ψ^b is the case, one will always get either ψ^a or ψ^b; but then, the physical situation after the measurement is no longer described by $\psi(1, 2)$. One gets, in fact, the additive pattern, not the interference pattern. This is completely analogous to the two-slit case, where a measurement to determine through which slit the particle passes will always find the particle at one of the slits, but the measurement will alter the physical situation in such a way that the interference pattern no longer obtains. So, consistent with the physical situation (both slits open) in which the interference pattern obtains, one cannot conceive of the particle as passing through one of the slits. Similarly, consistent with the physical situation represented by $\psi(1, 2)$, one cannot hold that either ψ^a or ψ^b is the case [if ψ^a or ψ^b *were* the case, the physical situation could no longer be represented by $\psi(1, 2)$].[17] But to deny that either ψ^a or ψ^b is the case is *to deny that there is anything in the world represented by 'ψ^a' or 'ψ^b'*. This is the *coup de grace* as far as atomism goes. For the question before us is: can the physical situation of a whole, (1, 2), represented by '$\psi(1, 2)$', be explained or reduced to its parts, 1 and 2, represented by $\psi_{\pm}(1)$ and $\psi_{\pm}(2)$. The answer, according to the quantum theory is 'no'. Thus, *given* that '$\psi(1, 2)$' correctly describes the physical situation, there exists *nothing in the world* which could be described by either '$\psi_{+}(1)$' or '$\psi_{-}(1)$ or '$\psi_{+}(2)$' or '$\psi_{-}(2)$'. Therefore, because there is absolutely no way for the so-called 'parts' to arrange themselves to form the composite whole, it is no longer meaningful to conceive of the whole as being constituted by parts. The spin-zero composite system can *not* be conceived of as consisting of two particles spinning in opposite directions. For not only does the whole exhibit properties (interference effects) not explainable in terms of parts, but also, these properties are not compatible with the assumption that there exist 'parts' at all.

CONCLUSION

In this paper I have argued that in the quantum domain we have reached a level where atomistic concepts are no longer applicable. From the perspective of theoretical realism, what there is is what our best theories say there is. Now quantum mechanics describes a most fundamental level of physical existence. After all, do we not believe that the phenomenal level (macro-objects) depends on the molecular level which in turn depends on the atomic level . . . ? Does it not then follow that atomism, construed as a thesis about the nature of physical existence, is false? It seems so to me. It also seems to me that the consequences of this for philosophy are very very important.

Philosophy is concerned with the nature of reality, including ourselves and our knowledge of reality as a 'part' of reality. Because atomism fails at the most fundamental level of physical existence, it follows[18] that reality, including ourselves, our knowledge, our language, etc. can no longer be understood from within the framework of an atomistic metaphysics. Thus, in the same way that one cannot regard the world as constituted by independently existing entities, one cannot regard say, knowledge as ultimately consisting of or reducible to independently existing 'sense impressions'; one cannot regard beliefs as 'justified' by preceptions which are assumed to be independent of the beliefs; one cannot regard society as consisting of persons who exist independently of the society; one cannot regard action as independent of perception or desire. This is not to deny that atomism, in any area of philosophy, does at times give an approximate and useful analysis of things. But as philosophers concerned with understanding the nature of reality, we must not be satisfied with approximations. We must take our cue from physics, and begin the very difficult task of examining seriously the non-atomistic philosophies of the past (Plato, Spinoza, Hegel, Whitehead, etc., and even including non-Western philosophies) and, should these prove inadequate, inventing new ones. Of one thing I am certain: if we don't do this job, the physicists will!

University of Illinois at Chicago Circle

NOTES

[1] *What Philosophy Is*, Arthur Danto, Harper & Row, 1968.
[2] Reichenbach, 'Logical Foundations of Quantum Mechanics', *Ann. Inst. Henri Poincaré* **XIII**, fascicule II, 1953.
[3] It has always seemed to me that physics is where *real* (as opposed to *linguistic*) metaphysical questions are discussed, such as causation, the nature of space and time, and the fundamental building blocks (ontology) of physical reality. Plato required that students undertake ten years of mathematics before coming to philosophy; perhaps five or six years of physics ought to be added to this requirement.
[4] I take realism to be "the view that if a scientific theory is in fact true, then there is in the world exactly those entities which the theory says there is, having exactly those characteristics which the terms of the theory describe them as having." Hooker, 'Systematic Realism', *Synthese* **26** (1974) 409.
[5] See the paper by Roger Jones, in this volume, for a precise definition of the principle of anomaly.

[6] See my paper 'Q.M. and Int. of Prob.', *Philosophy of Science*, December 1972.
[7] Reichenbach, *op. cit.*
[8] Hooker, Clifford A, 'The Nature of Quantum Mechanical Reality', in *Paradigms and Paradoxes* (ed. by Colodny), p. 181.
[9] Putnam, for example, advocates abandoning the distributive law:

$$(\phi_1 \vee \phi_2) \cdot x \not\equiv (\phi_1 \cdot x) \vee (\phi_2 \cdot x)$$

in words, 'the electron goes through either the upper slit or the lower slit, and arrives at the screen at point x' is not equivalent to "either the electron goes through the upper slit and arrives at point x or the electron goes through the lower slit and arrives at point x."
[10] Hooker, *op. cit.*, p. 181.
[11] Historically, it was the other way around: the discovery that atoms have a structure led to the realization that they could be transformed.
[12] Feinberg, G., 'Philosophical Implications of Contemporary Particle Physics', in *Paradigms and Paradoxes* (ed. by Colodny).
[13] Reichenbach, *op. cit.*, p. 8.
[14] Reichenbach, *op. cit.*, p. 8.
[15] *Ibid.*
[16] The reader who is unfamiliar with spin may think of it simply as an intrinsic property of the particle.
[17] For a more detailed discussion of this point, see my paper 'The Ignorance Interpretation Defended', *Philosophy of Science*, December 1974.
[18] It *certainly* follows, if one is a materialist; for in that case our knowledge, perceptions, language, etc., depend on the atomic level in the same way that tables and chairs do.

MERRILEE H. SALMON

CONSISTENCY PROOFS FOR APPLIED MATHEMATICS

In 1948, Hans Reichenbach distributed a mimeographed set of notes, entitled 'Theory of Series and Gödel's Theorems', to members of his advanced logic class at UCLA. A part of this material has been published for the first time (Reichenbach, 1948). This portion contains Reichenbach's highly interesting and provocative discussion of the importance of consistency proofs for pure and applied mathematics.

In spite of wide acceptance of the view that consistency proofs for systems of applied mathematics are unnecessary, it is difficult to find arguments which support this position in the literature which deals with consistency proofs. If writers discuss the problem at all, they usually just refer to the history of mathematics, claiming that, for example, inconsistencies in the infinitesimal calculus caused no difficulties for its application, even before Weierstrass and others repaired its foundations. It is not always clear whether these writers are claiming that systems of applied mathematics are consistent even though they may be embedded in inconsistent theories, or whether they are maintaining that any inconsistencies in applied mathematics itself never lead to difficulty in actual practice. For example, Karl Menger claims that all the difficulties in the foundations of calculus involved the notion of a limit, and so did not hinder mathematicians from using the calculus to ascertain the areas bounded by various curves. Such calculations, which depend only on the fundamental theorem, do not require a consistent account of limits (Menger, 1952, p. xx). In quantum mechanics, on the other hand, the Dirac delta function was used although its author explicitly acknowledged his inability to give a consistent account of it.[1]

Sometimes historical facts of this sort are offered in support of the position that formal systems are unnecessary or undesirable in applied mathematics. Although traditionally there have been close connections between the development of formalism and the interest in consistency proofs, there are some formalists who believe that consistency proofs have minor import, at best, even for systems of pure mathematics. This

W. C. Salmon (ed.), Hans Reichenbach: Logical Empiricist, 625–636.

results from their view that inconsistent systems may be interesting from a purely formal standpoint, particularly while the inconsistency remains undetected (Curry, 1963, p. 16).

Even formalists who admit the value of consistency proofs recognize that such proofs leave important questions unanswered. Reichenbach describes their dilemma: A consistency proof of an object language theory within that theory is worthless because of circularity. For suppose the statement of the theory's consistency can be expressed as a well formed formula in the object language, and that this formula can be derived as a theorem. Such a procedure is helpful only if we already know that the theory is consistent. If it is inconsistent, then any well formed formula may be derived, including the formula which states in its intended interpretation that the theory is consistent. But the alternative is to argue in a metalanguage that the theory is consistent. Then we may ask whether the metalinguistic theory in which the proof is given is consistent. Thus, the attempt to formalize all proofs of consistency leads either to a vicious circle or to an infinite regress (Reichenbach, 1947, §33).

As a prelude to further discussion of the importance of consistency proofs, some remarks about applied mathematics and its relation to formal systems are in order. A formal system suitable for mathematics may be interpreted in different ways. One sort of interpretation involves correlating the symbols of the formalism with ordinary middle-sized physical objects and with physical operations performed upon these objects. Such an interpretation gives us an instance of applied mathematics.

Whether physical objects and operations can correspond in the requisite fashion to the objects and operations of the formal system is itself an empirical question, and must be settled accordingly. Indeed, we know that some objects and operations upon them, such as the juxtaposition of hydrogen and oxygen molecules at very high temperatures, do not provide a suitable model for the statements of applied arithmetic. Questions of truth and reliability arise only in connection with a particular interpretation. In applied mathematics the interpretations in which we are interested are the ones which correlate the formalism with (parts of) the physical world (Reichenbach, 1946b, pp. 36–37).

The interpreted formulas which constitute applied mathematics have the status of well confirmed empirical statements. Verification of individual statements of applied mathematics like '$2+2=4$' can be obtained by identifying individual physical objects, such as apples, interpreting '+' as a physical operation such as putting apples into the same container, and interpreting '=' as some physical one-one correspondence. If a set of axioms of some formal system of mathematics can be interpreted and verified in this way, we may speak of the interpreted axioms as axioms of a (formalized) system of applied mathematics.

The certainty of such a system of applied mathematics is similar to the certainty of any highly confirmed empirical theory. The value of formal systems of applied mathematics, like formal systems of other good physical theories, lies in their confining uncertainty to the truth of the interpreted axioms and to the truth preserving character of the rules of inference of the system. In addition, physical evidence for derived theorems can be added to the evidence for the axioms, increasing the reliability of the system as a whole. Every statement which is deduced in the system then shares in this increased reliability.

This explains why applied mathematicians prefer theorems which are deduced in a system to mathematical statements which have been established merely by direct empirical means. An applied mathematician is happier if he can *calculate*, for example, the reinforcement required for a bridge to support a given weight than he is if he knows only that bridges built a certain way have not collapsed under the given weight so far.

Since, however, theorems may be derived in a formal system with no reference to the interpretation, and since the use of uninterpreted axioms from the original formal system (the system of which the interpreted axioms form a subset) may simplify deductions, the question of the consistency of the formal system may arise. Hao Wang provides a clear statement of the problem:

> It is known that we can derive the differential and integral calculus in some system of set theory which is not known to be consistent. Suppose the system is found to be inconsistent. It follows that we can derive all sorts of false and absurd consequences in this system, some of them having to do with differential and integral calculus.
>
> Since the calculus can be applied in constructing bridges, we may be able to prove that a pillar whose diameter is three feet long is strong enough although actually we need a pillar whose diameter is seven feet long. Hence, it might be argued, bridges may collapse because of the inconsistency of the particular system in which we developed the calculus. (Wang, 1961, pp. 345–346.)

Wang argues that no such thing will happen. Some of his reasons seem irrelevant, e.g., the claim that applied mathematicians are usually not engaged in work on the axiomatic foundations of the calculus, and would not be likely to derive their calculations from any version of axiomatic set theory. More to the point, Wang says that the applied mathematician does not use the whole complicated apparatus of set theory, but only performs such calculations as could be justified in a consistent system.

Wang concludes, "It is not necessary to formalize mathematics nor to prove the consistency of formal systems if the problem is that bridges shall not collapse unexpectedly. There are many things which are more pertinent in so far as bridges are concerned."[2]

Wang does not say what any of these considerations are. Nor does he say anything about the complicated turns the mathematician must side-step in order to avoid inconsistency. Moreover, mere complexity is neither a necessary nor a sufficient condition for inconsistency. There is nothing particularly complicated about the moves in naïve set theory which lead to a contradiction. And anyone who has studied advanced texts in applied mathematics is aware of the complexity there.

Reichenbach presents a far more careful analysis of the dispensability of consistency proofs for formal systems of applied mathematics. Even if there is a contradiction in the system, he argues, it does not lower the probability that a deduced theorem is true. His proof is provided within the framework of the probability calculus (Reichenbach, 1948).

We know that there is no interpretation which simultaneously satisfies any inconsistent set of axioms. The claim that applied mathematics (the intended interpretation of a part of the formal system which is used for both pure and applied mathematics) is an admissible interpretation, i.e., one which renders all the axioms true, is only a 'posit'. As such, it is open to correction or revision as a result of further observations or new information. It is an 'appraised posit' since it is based on good physical evidence, but it lacks absolute certainty. However, since a set of sentences which are all true must be consistent, no *additional* posit regarding the consistency of the axioms is required. If the axioms are all true in the intended interpretation then the axioms are consistent.

We wish to know, however, not only whether the *axioms* are consistent, but also whether it is possible by use of the *rules* of the language to derive any contradiction from the axioms. This point can be covered by positing that the rules are truth preserving. Such posits cannot be supported by the physical evidence which supports the truth of the interpreted axioms. Reichenbach believes, however, that they can be supported by *logical evidence* (Reichenbach, 1947, §34).

Logical evidence is the logical constraint we feel in using simple tautologies and applying truth-preserving rules. Consider, for example, our acceptance of modus ponens as a truth-preserving rule of inference. In a metalanguage we can construct a truth table in the usual way, and using perceptual evidence, convince ourselves that formulas of the form 'q' are assigned the value 'T' whenever formulas of the form '$p \supset q$' and 'p' are both assigned 'T'. Perceptual evidence is the simplest type of physical evidence. It refers to the sense observations we make in order to convince ourselves, for example, that certain symbols occur in various positions in a truth table. But perceptual evidence cannot justify the claim that *truth* is preserved when we infer sentences with particular structural properties from sentences with other structual properties. Logical evidence is the guide to which we appeal in connecting 'T' with truth and in constructing truth tables. For sentences which are either true or false, it is 'self-evident' that there are only four possible combinations of truth values, represented by 'T' and 'F', in the initial columns of a truth table for molecular sentences with two distinct components.[3] Our belief that these possibilities result in the value 'true' for whatever sentence 'q' represents whenever the sentences represented by 'p' and '$p \supset q$' are assigned truth is based on our understanding of the logical operation of implication, represented by '$\supset$'. Reichenbach does not deny that such considerations, which he calls 'material thinking', can be formalized. But he points out that they can be formalized only by using another metalanguage, and in this new metalanguage other principles – or perhaps even the same principles – will be used without being formalized. In order to stop a regress we must admit that we have immediate knowledge of some principles.

Reichenbach's claims should not be construed as arguments against formalization. His point is merely that we cannot entirely eliminate dependence on logical evidence if we are to avoid a regress. Even though logical evidence cannot be eliminated, we prefer to minimize it. Formal systems are an aid to this minimization. They limit our reliance on logical evidence by confining its support to a small number of statements and rules of inference.

Logical evidence provides no guarantee of reliability. Statements based on logical evidence, like those based on perceptual evidence, are empirical. This claim does not obliterate the distinction between logical truth and empirical truth. Tautologies are logically true, i.e., any substitution instance of a formal tautology is a true statement regardless of the content of the non-logical vocabulary employed. However, the claim that a particular formula is a tautology is an empirical claim about the structural properties of the formula. Sometimes we feel very certain that a particular formula has a structure which would make it tautologous, but since this is an empirical claim, it cannot be absolutely certain. Statements based on logical evidence are posits, which may be dealt with as true so long as they have not been falsified. If they are falsified they are replaced by new posits. In this area – as in the rest of empirical knowledge – the method of trial, error, and new trial is appropriate. Reichenbach's view is strikingly similar to Popper's and Lakatos's account of the method of conjecture, proof, and refutations as the road to mathematical knowledge. Like them, he honestly admits mathematical fallibility.

Logical evidence, for Reichenbach, is a psychological criterion. Our belief that a proposition cannot be both true and not true, for example, comes from our understanding of words like 'true', 'and', and 'not'. We cannot conceive the possibility of a proposition's being true and failing to be true, given our understanding of these terms. To say that the criterion is psychological is not to say that we feel a kind of twinge when faced with a logical truth, but rather that a systematic search for counter-examples has been unproductive, that no alternative seems feasible. This brings out an important difference between perceptual evidence and logical evidence, though both types of evidence result in empirical claims. Whenever we use perceptual evidence we can always imagine that the situation we believe to obtain could actually be otherwise, even in very simple cases. Anyone who has read proof knows how easy it is to be mistaken about the occurrence of such simple things as type symbols, even when a page has been checked and rechecked. It is not possible in the same way to understand how statements based on logical evidence could be wrong. We do know, however, from the history of mathematics, that proofs which appeared to be evident in the past no longer seem evident to us, even though it was not possible at the time to conceive their incorrectness. An interesting example of this is presented in Lakatos (1963).

If we can use logical evidence to convince ourselves that the rules of a language are truth preserving, then this posit will assure us that the rules cannot be the source of any inconsistency. Reichenbach calls a language which consists only of interpreted axioms, tautologies and truth preserving rules an *acceptable empirical language*.[4] In an acceptable empirical language we can say that the probability that the axioms are true is equal to the probability that they are true and that the system is consistent. Remembering that any deduced theorem is also meant to be interpreted, and so lacks absolute certainty, we can see that the probability that the theorem is true is greater than or equal to the probability that the language is consistent and that the theorem is true. Moreover, if the language is consistent, the probability that the theorem is true is greater than or equal to the probability that the axioms are true and that the language is consistent. But since this latter probability is equal to the probability that the axioms are true, we can see that the probability of the truth of a deduced theorem is greater than or at least equal to the probability that the axioms are true.

The preceding paragraph sketches Reichenbach's argument which shows that it is not necessary to prove the consistency of an acceptable empirical language theory, such as we believe the language of applied mathematics to be. For details see Reichenbach (1948). Any uncertainty concerning the truth of a theorem which *might appear* to arise from lack of a consistency proof is actually absorbed into the question of the truth of the interpreted axioms and the truth-preserving character of the rules. We cannot be absolutely certain that the theorem is true, but that uncertainty can be traced to the possibility that an error of interpretation was made.

Reichenbach's account does not consider an important feature of applied mathematics as it is actually practiced. In order to simplify his work the applied mathematician may use any part of pure mathematics. But the entire system of pure mathematics is not an empirical language. The system cannot be given a complete observational physical interpretation since there are no physical observations of the infinite. This objection can be mitigated by noticing that the only theorems in which the applied mathematician is interested are interpreted theorems, and so he could derive these, although inconveniently, from the interpreted axioms alone. The applied mathematician could get along with a subsystem of mathematics, a subsystem which was capable of complete physical interpretation, and no consistency proof would be required for such a system[5] Nevertheless, the fact that the applied mathematician actually uses systems which are incapable of complete interpretation is one of the considerations which makes a consistency proof for all of mathematics so desirable.

There is some inductive evidence that the entire system of mathematics is consistent – no contradictions are known to exist. But this is only the weakest sort of inductive evidence, comparable to the evidence for a statement of applied mathematics which has not been shown to be false, but for which there is no deductive proof from the interpreted axioms of the system. And, it must be added, this evidence is somewhat counterbalanced by the fact that formal systems which were regarded as consistent in the past have subsequently spawned explicit contradictions.

Reichenbach recognized the value of deductive proofs of the consistency of parts of pure mathematics, such as Gentzen's proof of the consistency of first-order number theory, and hoped for the discovery of consistency proofs of other parts of pure mathematics as well. A deductive proof of consistency increases the reliability of the statement of consistency in the same way that the deductive proof of an interpreted theorem increases its reliability. Furthermore, Reichenbach believed that a regress with such proofs might be avoided. The metalanguage within which such a proof is given would not need to be proved consistent if we could use perceptual evidence and logical evidence to show that the metalanguage fulfills the conditions of what Reichenbach calls an empirical language, i.e., a theory consisting only of interpreted axioms, tautologies, and truth-preserving rules of inference.

Reichenbach claimed that the metalanguage in which a consistency proof for pure mathematics may be given is an empirical language. In saying that the metalanguage is subject to a complete interpretation, Reichenbach envisioned what he called a *semantical* interpretation of the metalanguage rather than a physical object interpretation. In the metalanguage we understand the meaning of such expressions as 'the same sign', 'constant', and 'variable'. We understand the rules controlling the operations of substitution and inference. The semantic interpretation of a derivation tells us that when various operations are performed on signs this results in the occurrence of other signs. The semantical interpretation of the statement of consistency, for example, tells us that if a particular formula is derivable by operations on the axiom signs then the same formula with the sign for negation preceding it is not derivable. Put this way, we can see that the statement of consistency is to be regarded as an empirical statement that a particular pair of formulas shall not occur as a result of derivations in the system. Since this is an empirical statement it is not absolutely reliable. But when it is deductively proved, the inductive support for it is stronger than the weak support provided by the fact that no contradictions have been discovered. It is based on a 'concatenated' induction, one running through the axioms of the system, rather than on a direct induction.

This attempt to stop the regress is very appealing. It is, however, doubtful whether the metalanguage used, for example, in Gentzen's proof of the consistency of formal number theory fulfills Reichenbach's requirements for an empirical language. The problem seems to lie with the condition that logical evidence convince us that the rules of the metalanguage in which the proof is given are truth-preserving. Whether there is sufficient logical evidence to guarantee this for a rule such as transfinite induction up to ε_0 is questionable. Perhaps the most we can say is that nothing but logical evidence could convince us of such a rule, for it cannot be derived from simpler, more evident rules. However, then we must allow for degrees of logical evidence, for we cannot claim that this rule is as evident as modus ponens.

Perhaps Reichenbach would admit such degrees of evidence. He does draw on the analogy between perceptual evidence and logical evidence, and clearly there are degrees of perceptual evidence. The ability to recognize subtle color distinctions, for example, varies among perceivers, even when color-blindness is not a factor. Such ability may depend on native talent, as well as training or practice. Furthermore, recognition of subtle perceptual distinctions may not occur at first glance. Careful and extended looking, and perhaps guidance in what to look for may be necessary. Similar considerations may apply to logical evidence.[6]

But even if we admit that there are degrees of logical evidence, we still must face the issue of whether we can use logical evidence to convince ourselves that a rule such as the rule of transfinite induction is truth-preserving. Reichenbach introduced the semantic interpretation to avoid certain difficulties with the infinite in the physical interpretation of formulas in the system. But while the formulas can be given a finite semantic interpretation, in the really interesting metalanguages strong non-finitistic rules are required. The difficulty of supporting such rules on the basis of logical evidence seens to count against Reichenbach's claim that the regress with consistency proofs can be halted in these cases.

Reichenbach has provided a thorough and coherent account of why the absence of consistency proofs does not affect the reliability of applied mathematics. At the same time he has shown why formal systems of applied mathematics are so valuable. As in good physical theories, formalizations localize the problem of truth, and enable us to see how the degree of support for a theory can be greater than for its miscellaneous parts.

These results for applied mathematics are important for pure mathematics as well. Although the language of pure mathematics is not a completely interpreted language, Reichenbach believed that a metalanguage used to talk about pure mathematics is subject to a complete semantical interpretation. The consistency problem for pure mathematics is formulated in the metalanguage, and so it can be seen as a problem in applied mathematics. Even when consistency is proved deductively, the statement that a system is consistent is an interpreted statement. Its reliability, like that of any empirical statement, is a problem for inductive rather than deductive logic.

The success of Reichenbach's approach clearly depends on the acceptance of his account of logical evidence. His proof that applied mathematics needs no consistency proof depends on the consistency of the probability calculus, which is a part of pure mathematics. But the portion of the probability calculus which he uses is capable of a finite interpretation. Logical evidence and perceptual evidence support this interpreted calculus.

If Reichenbach is right, then ultimately the correctness of deductive reasoning depends upon empirical claims based on logical and perceptual evidence. Not only empirical science, but mathematics as well, must be concerned with the development of a satisfactory theory of inductive inference. Those who hold that mathematical reasoning deserves a different – or 'better' – foundation bear the burden of providing a detailed formulation of a more palatable alternative.[7]

University of Arizona

NOTES

[1] "$\delta(x)$ is not a function of x according to the usual mathematical definition of a function, which requires a function to have a definite value for each point in its domain, but is something more general, which we may call an improper function to show up its difference from a function defined by the usual definition. Thus $\delta(x)$ is not a quantity which can be generally used in mathematical analysis like an ordinary function, but its use must be confined to certain simple types of expressions for which it is obvious that no inconsistency can arise." (Dirac, 1958, p. 58) J. von Neumann severely criticized the Dirac method for its lack of rigor, and presented an alternative to it (von Neumann, 1935, p. ix.) Even after this the Dirac function was commonly used. Still later, Laurent Schwartz gave a mathematically consistent refinement and explanation of the Dirac method.

[2] Another view of why consistency proofs for applied mathematics are superfluous is discussed in Goodstein (1965, p. 137.) "In the application of arithmetic to commerce or to physics we retrace the evolutionary development, replacing theorems by statements of observation, and so long as these observed regularities persist arithmetic will be capable of application." But since Goodstein admits, in the same work, that we know nothing about the origins of arithmetic, and can offer only a speculative account of the evolutionary development which we are to retrace, his suggestion is of limited value.

[3] As anyone familiar with Reichenbach's work knows, he does not intend reliance on logical evidence to commit us to a system of two-valued logic. Tautological formulas and truth-preserving rules will be different in systems which are not two-valued, but one can still apply the criterion of logical evidence for determining whether the formulas and rules have the structure which would make them tautologous or truth-preserving, given the different background assumptions. We might say that logical evidence convinces us that truth is an exclusive concept, i.e., that no proposition can be both true and not true. But whether to identify lack of truth with falsity, or to allow other values, such as 'indeterminate', or to assign a system of 'weights', may be a question which can be settled only by taking into account pragmatic considerations. See Reichenbach (1946a, pp. 139–166).

[4] Reichenbach uses 'language' here in the sense of 'theory' or 'axiomatic system'. I shall follow this usage in the rest of the paper.

[5] This is a strong assumption. But for a mathematician's defense of it see Klein (1924, pp. 35–36).

[6] The analogy between perceptual evidence and logical evidence must be used with some caution. K. Gödel (1964, pp. 271–272) employs it in ways which Reichenbach would approve, i.e., to explain why the axioms of set theory "force themselves upon us as being true." But Gödel also draws platonistic conclusions about the objective existence of mathematical entities from the analogy, comparing their existence to the objective existence of the external world.

[7] An earlier version of this paper was read at the 1974 Philosophy of Science Association meeting. I am grateful to A. W. Burks, H. C. Byerly, and W. C. Salmon for their helpful comments.

BIBLIOGRAPHY

Curry, H., 1963, *Foundations of Mathematical Logic*, McGraw-Hill, New York.
Dirac, P., 1958, *Principles of Quantum Mechanics*, 4th ed., Oxford University, Oxford.

Gödel, K., 1964, 'What is Cantor's Continuum Problem?', in *Philosophy of Mathematics* (ed. by P. Benacerraf and H. Putnam), Prentice-Hall, Englewood Cliffs, pp. 258–273.

Goodstein, R., 1965, *Essays in the Philosophy of Mathematics*, Leicester University, Leicester.

Klein, F., 1924, *Elementary Mathematics from an Advanced Standpoint*, Dover, New York.

Lakatos, I., 1963, 'Proofs and Refutations', *Brit. J. Philos. Sci.* **14**, 1–25.

Menger, K., 1952, *Calculus, A Modern Approach*, University of Chicago, Chicago.

Reichenbach, H., 1946a, *The Philosophical Foundations of Quantum Mechanics*, University of California, Berkely and Los Angeles.

Reichenbach, H., 1946b, 'Bertrand Russell's Logic', in *The Philosophy of Bertrand Russell*, (ed. by P. A. Schillp), Library of Living Philosophers, Inc., Evanston, pp. 21–54.

Reichenbach, H., 1947, *Elements of Symbolic Logic*, Macmillan, New York.

Reichenbach, H., 1948, 'Theory of Series and Gödel's Theorems', in *Vienna Circle Collection*, Vol. 4, *Hans Reichenbach: Selected Writings, 1909–1953* (ed. by Maria Reichenbach and Robert S. Cohen), D. Reidel, Dordrecht, 1978.

von Neumann, J., 1935, *Mathematical Foundations of Quantum Mechanics*, Princeton, Princeton.

Wang, H., 1961, 'Process and Existence in Mathematics', in *Essays on the Foundations of Mathematics*, (ed. by Y. Bar-Hillel *et al.*), The Magnes Press, Jerusalem, pp. 328–351.

WILLIAM E. McMAHON

A GENERATIVE MODEL FOR TRANSLATING FROM ORDINARY LANGUAGE INTO SYMBOLIC NOTATION

A key issue for 20th century logical theory has been the thesis that one can design a symbolic language which expresses all that is cognitively significant within a conversational language. This paper will examine this contention of the 'Carnapian' school of philosophy in the light of recent findings in the field of linguistics. It is impossible in this space to consider all the aspects of the problem,[1] so what we want to do here is consider the type of linguistic system required for the task of translating from conversational language (CL) into a symbolic language (SL) and then make a judgment as to the efficacy of the translations.

In doing these things we would like to try to reverse what has been a common practice. It has become a standard procedure since Russell to apply logical theory to analyses of conversational language and linguistic structure. This even has resulted in a 'Copernican revolution' in the field of linguistics, with Chomsky's development of the theory of generative grammar. But given the theory of generative grammar, which was itself the product of the application of logical theory to linguistics, let us see what it can tell us about the nature of those artificial symbolic languages which may be used to express the character of natural languages. In other words, we shall be employing grammatical theory to understand the nature of certain kinds of logistic languages.

Probably the most ambitious project for constructing an SL-model for CL was that of Hans Reichenbach.[2] Reichenbach's work on this has been noted by many people, but insufficient attention has been paid to it.[3] So here we shall attempt to characterize Reichenbach's conception of a logistic model of CL, as it exemplifies other efforts to achieve the same sort of thing. Reichenbach believed that a 'logistic grammar' for CL could be constructed by using the concepts of Russell–Whitehead logic. In order for this to be done CL-terms have to be put in correspondence with SL-terms and translations must be made which preserve the cognitive significance of the CL-terms. Prior to the development of generative grammar it was difficult to see what this project amounted to.[4] But now,

W. C. Salmon (ed.), Hans Reichenbach: Logical Empiricist, 637–654.

we believe, not only the nature of the project but the problems involved therein may be made clear.

As is noted above, a 'logistic grammar' for CL will be that of a correlative SL. This requires that an appropriate SL be constructed and CL-terms be translated into it. A wide variety of CL-terms must be so translatable, i.e. it will not suffice just to do the standard translations of noun phrases, verb phrases, etc. Reichenbach's SL is essentially the Russell–Whitehead predicate calculus, with the addition of such symbols as indexed quantifiers and descriptional and pragmatic operators.[5] If SL is to serve as the model for the grammar of CL, CL-sentences must then be generable from SL. So what we propose to do is to take a sample CL-sentence of a moderate degree of complexity and try to construct a grammar which can generate it. The sentence we have chosen is one of Reichenbach's favorites, 'the present king of England is ceremoniously crowned at Westminster Abbey'.[6] Our procedure will consist in (1) constructing a phrase-marker for the SL-version of this sentence, (2) deriving the SL-form of the sentence from its structural description, and (3) deriving the CL-version from the SL-version. The aim of all this is to see whether CL-sentences are generable from their putative SL-correlates, and if so, in what way. The results obtained will enable us to make some judgments on the feasibility of constructing SL-models for CL.

At each step in the procedure we shall employ certain kinds of rules. We assume the reader has some familiarity with the theory of generative grammar as well as with Chapter VII of Reichenbach's *Elements of Symbolic Logic*.

I

The first step, then, is to construct a phrase-marker to represent the deep structure of 'the present king of England is ceremoniously crowned at Westminster Abbey.' This requires a set of 'rewriting' rules; the rules to be used and the glossary of category symbols are given in Table I. These rules have been constructed from Reichenbach's suggested 'IC'[7] analysis of SL-sentences and his implicit formation rules for SL.[8] These rules are 'branching rules' according to linguistic terminology,[9] for they break down constructions into expressions which are ultimate constituents in Reichenbach's grammar.

TABLE I

Categories and base rules for logistic grammar

Categories

MS – molecular sentence
S – sentence
C_1 – one-place connective
C_2 – two-place connective
CF – complex function
A – argument
F – function
Adv – adverbial function
PN – proper name
DO – descriptional operator
Def. O – definite descriptional operator
Indef. O – indefinite descriptional operator
PN_T – proper name of a thing
PN_E – proper name of an event
PN_V – name variable
PN_{E-PL} – proper name of a place-event
PN_{E-T} – proper name of a time-event
F_A – function in active voice
F_P – function in passive voice

Base Rules

(i) $MS \rightarrow \begin{Bmatrix} S \\ C_1 {}^\frown S \\ S {}^\frown C_2 {}^\frown S \end{Bmatrix}$

(ii) $S \rightarrow \begin{Bmatrix} CF {}^\frown A {}^\frown A {}^\frown A \\ CF {}^\frown A {}^\frown A \\ CF {}^\frown A \end{Bmatrix}$

(iii) $CF \rightarrow \begin{Bmatrix} F \\ \text{Adv} {}^\frown F \end{Bmatrix}$

(iv) $A \rightarrow \begin{Bmatrix} PN \\ DO {}^\frown S \end{Bmatrix}$

(v) $DO \rightarrow \begin{Bmatrix} \text{Def. } O \\ \text{Indef. } O \end{Bmatrix}$

(vi) $PN \rightarrow \begin{Bmatrix} PN_T \\ PN_E \\ PN_V \end{Bmatrix}$

(vii) $PN_E \rightarrow \begin{Bmatrix} PN_{E-PL} \\ PN_{E-T} \end{Bmatrix}$

(viii) $F \rightarrow \begin{Bmatrix} F_A \\ F_P \end{Bmatrix}$

We shall gloss over a few peculiarities of these categories and rules.[10] A minor point concerns the distinction between names of place-events and names of time-events; Reichenbach makes this distinction in the text of *Elements of Symbolic Logic*[11] but does not include it in his table of ultimate constituents on pp. 352–354. A more important issue centers around the distinction between active and passive functions (Rule viii), which is not in Reichenbach. We made it in order to generate 'the present king of England is crowned at Westminster Abbey' but not 'the present king of England crowns Westminster Abbey'. Similar considerations have led us to ignore the problem of tenses for the moment and to take the sentence tenselessly. We disregarded these matters to keep this grammar simple (but still adequate to serve our purposes). However, the complexities that would be raised in trying to carry out Reichenbach's suggestions for the analysis of functions and tenses do raise important questions about the adequacy of his grammar, and we will come back to these questions later. Furthermore, Reichenbach includes within his category of 'logical terms' such diverse types as connectives, quantifiers, and descriptional operators. We did not set up the rewriting rules of Table I so that all logical terms are generated from some common class LT, for different 'logical terms' do not have analogous syntactic functions.

Having made these qualifications, we now turn to rules (i–viii) and see that they generate the following phrase-marker:

- *MS*
  - *S*
    - *CF*
      - Adv
      - *F*
        - F_P
    - *A*
      - *DO*
        - Def. *O*
      - *S*
        - *F*
          - F_A
        - *A*
          - *PN*
            - PN_V
        - *A*
          - *PN*
            - PN_E
              - PN_{E-PL}
        - *A*
          - *PN*
            - PN_E
              - PN_{E-T}
    - *A*
      - *PN*
        - PN_E
          - PN_{E-PL}

The above *P*-marker lays out what, from the standpoint of generative grammar, could be called the 'deep structure' of the sentence, 'the present king of England is ceremoniously crowned at Westminster Abbey'. We do not have, as yet, a full explication of the logical structure[12] (in Reichenbach's sense) of the sentence. To get the logical structure the pre-terminal string so far generated ($\text{Adv}\frown F_P\frown((\text{Def. } O\frown F_A\frown$-$(PN_V\frown PN_{E-PL}\frown PN_{E-T}))\frown PN_{E-PL}))$ must be mapped onto a string of SL-symbols. This will be the next step in the derivation. At present, as can be seen, only PS-rewriting rules have been employed. No context-conditioned rules have been used, nor have any explicitly 'transformational' rules come into play. The meta-rule for substitution of the symbols to the right of the '→' for the symbol on the left has employed. This is a 'transformation' rule in the Carnapian but not in the Chomskyan sense.[13] The bracketing of expressions is not achieved by a transformation rule but simply by means of the rewriting rules. That is, the tree-diagram is one way of writing what we might equally as well write using labelled brackets.[14]

II

Step two in the generative process converts the pre-terminal string into the SL-correlate of 'the present king of England is ceremoniously crowned at Westminster Abbey,' which will be the SL-sentence

$$`\varphi_1\{\text{ICR}([(\iota x)RL(x, \text{en}, \text{now})], \text{wa})\}'.$$

The most obvious technique for achieving this would be to construct a lexicon of SL-terms and assume a transformation rule permitting the substitution into a *P*-marker of a formative (SL-term) having the same features as a branch of the *P*-marker,[15] so we shall do this to see what happens. Assuming the necessary transformation rule, we shall posit the following sample lexicon (giving also the CL-correlates of the SL-terms):

SL-LEXICON

$(wa, [+PN, +PN_E, +PN_{E-PL}, \ldots])$	(*Westminster Abbey*)
$(now, [+PN, +PN_E + PN_{E-T}, \ldots])$	(*now*)[16]
$(en, [+PN, +PN_E + PN_{E-PL}, \ldots])$	(*England*)
$(\varphi_1[+\text{Adv}, \ldots])$	(*ceremoniously*)
$(ICR, [+F, +F_p, \ldots])$	(*is crowned*)
$(RL, [+F, +F_A, \ldots])$	(*rules*)

By means of the lexicon and the substitution rule, the *P*-marker can be expanded in the following way:

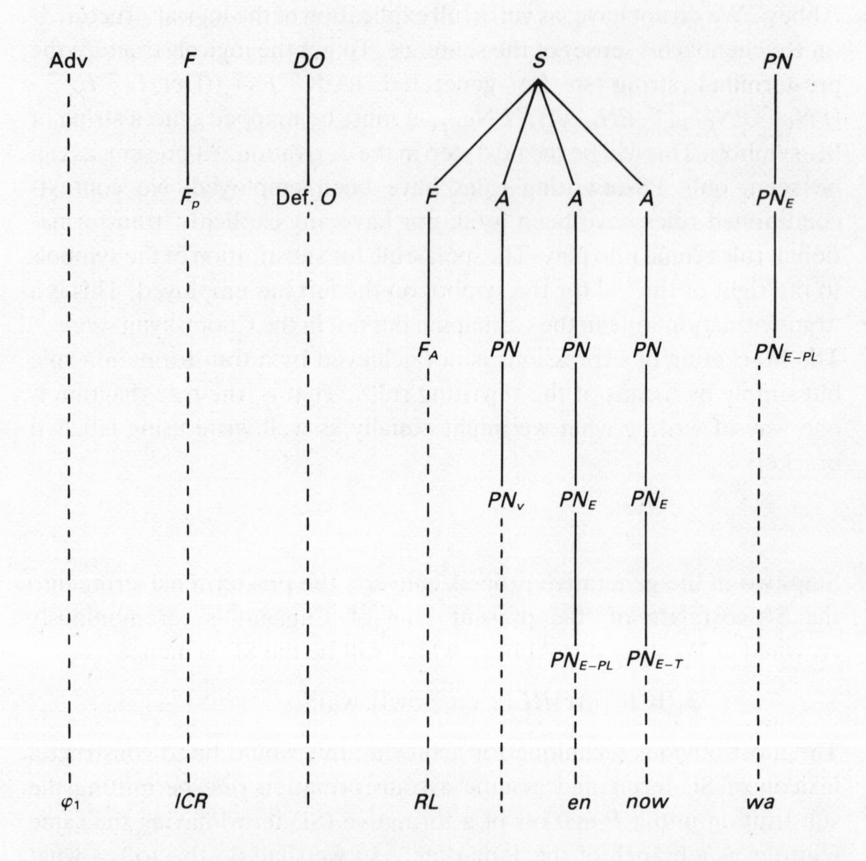

Note that two lexical items, '(ιx)' and 'x', have not been entered into the phrase marker. There is a good reason for this omission. Such entries cannot be made without context-conditioned rules. Allowance must be made for the scope of operators in the predicate calculus. In the case at hand, 'x' is to be bound by the iota-operator. Suppose the expression preceding the description contained a quantifier binding a PN_V 'y' and including the description in its scope.[17] Context-free rules would then allow the substitution of '(ιx)' and 'y' for 'Def. *O*' and 'PN_V.' This would yield an SL-sentence with a vacuous descriptional operator (and in which '*y*' would be bound by the wrong operator). Although vacuous operators may occur in some logistic systems, they certainly cannot be used to symbolize sentences of CL. Reichenbach does not discuss the problem of vacuous operators and quantifiers, but we think it safe to postulate that they cannot occur in logistic grammar.

If our stipulation forbidding vacuous quantification is correct, then Reichenbach does need context sensitive rules in order to generate the desired sentence and to avoid generating unacceptable forms. We shall propose the following two rules for the sentence at hand:

$$\text{(ix)} \qquad \text{Def. } O \rightarrow \left\{ \begin{array}{l} (\iota x)/X \underline{\quad\quad} F(A_1, \ldots) \text{ where } x \text{ does not appear in } X. \\ (\iota y)/X \underline{\quad\quad} F(A_1, \ldots) \text{ where } x \text{ but not } y \text{ appears in } X \end{array} \right\}$$

$$\vdots$$

$$\text{(x)} \qquad PN_V \rightarrow \left\{ \begin{array}{l} x/(\iota x)\, F(A_1 \underline{\quad\quad} A_2) \text{ where } A_1 \text{ and/or } A_2 \text{ may be null} \\ y/(\iota y)\, F(A_1 \underline{\quad\quad} A_2) \text{ where } A_1 \text{ and/or } A_2 \text{ may be null} \end{array} \right\}$$

$$\vdots$$

These rules state contextual limitations on the substitution of a formative for a lexical category-symbol. They are ordered, i.e., (ix) must be applied before (x). These rules may be taken as addenda to the lexicon.[18] So whenever we have an operator (such as a quantifier or descriptional operator) and a PN_V in a phrase-marker, we shall have to apply rules like (ix) and (x) to generate an SL-sentence. By using (ix) and (x) to substitute '(ιx)' and 'x' into the above phrase-marker, we have now generated the following sentence:

$$\varphi_1\{\text{ICR}([(\iota x)\text{RL}(x, \text{en}, \text{now})], \text{wa})\}$$

This is the SL-formulation of 'the present king of England is ceremoniously crowned at Westminster Abbey'. Let us call it the *SL-terminal string*. It is still, however, pre-terminal in regard to CL. What we have achieved by means of the derivations so far is an explication of the *logical structure* of the given CL-sentence. Our interpretation of Reichenbach's conception of the *logical structure* of a CL-sentence, is that it consists of its SL-correlate plus the structural analysis of the SL-sentence.[19] Reichenbach's conception of 'logical structure' thus differs from Chomsky's conception of 'deep structure' as follows: For Chomsky, the 'deep structure' of a CL-sentence is a CL-structure as such; it must, however, be correlated to the surface structure via *transformation* rules. On the other hand, for Reichenbach the 'logical structure' of a CL-sentence is the structure of an SL-sentence; it is shown to be the structure of the CL-sentence via *translation* rules. For Reichenbach the 'deep structure' of a CL-sentence *S* is the structure of its SL-correlate; for him it would make no sense to talk of *S* as having a deep structure independently of its correlation to some *S'* in SL.[20] So the notion of 'logical structure' posits a mediating factor between deep structure and CL, i.e., SL. A phrase-marker must be generated for *S'* before one can talk about the grammatical structure of *S*. Then, given that *S'* translates into *S*, the *logical structure of S* is the *grammatical structure of S'*.

So far we have shown how the description of *S'* may be generated and mapped onto the sentence *S'*. But it remains to be shown how the structure of *S'* may be said to be the structure of *S*. This can be shown only if rules can be constructed so as to generate *S* from *S'*. The construction and application of these rules thus is the final step of this analysis of the underlying principles of Reichenbach's grammar.

Before undertaking the generation of *S* from *S'*, let us comment briefly on the rules used to generate *S'*. Note that we had to abandon the methods of context-free phrase structure grammar to get *S'*. Transformation rules had to be brought in, and context-sensitive rules were necessary to generate the desired sentence and to avoid undesirable kinds of sentences. Hence to achieve what he wants Reichenbach would have to use more powerful methods than those of which he apparently was aware.

III

Now for the final step in the derivation of a CL-sentence from the structural analysis of SL. The string of SL-symbols must be correlated by means of the appropriate rules to the string of CL-formatives written as, 'the present king of England is ceremoniously crowned at Westminster Abbey'. A lexicon and rules sufficient to do the job can be constructed, but, as we shall see, once again the process has very interesting implications.

The easiest way to construct the required lexicon would be to have an SL-CL lexicon consisting of rewriting rules such as the following:

SL-CL LEXICON

φ_1[21]	→	*ceremoniously*
ICR	→	*is crowned (at)*[22]
RL	→	*rules*, *is the king of*[23] / *is the queen of*
en	→	*England*
now	→	*now, present*$_1$[24]
wa	→	*Westminster Abbey*
(℩*x*)	→	*the x such that, the only x such that*[25]

Here we have followed Chomsky's earlier technique of lexical rewriting rules rather than his later method of defining lexical items as sets of features.[26] We could have made the second lexicon like the first, with entries like '(ceremoniously, [+φ_1, . . .])', but we think the way we have chosen is simpler for the purposes of explication. The second lexicon, then, consists of simple rewriting rules which map an SL-term onto one or more CL-terms. SL would have a narrower vocabulary than CL; furthermore, for Reichenbach terms in different conversational languages would correspond to one SL-term. Again for purposes of simplicity we have restricted CL to English, realizing, however, that Reichenbach's full project would entail either having one such lexicon for all languages or different SL-CL lexicons for each natural language.

There are some problems as to the accurate CL-translation of some SL-terms (e.g., the '$\supset$' symbol). By working from SL to CL instead of conversely we have eliminated some such problems, e.g., the ambiguity of 'the'. Such problems fall under the issue of the justification of the SL-CL lexicon. Although this is an important issue, space does not allow us to do more than note it here.[27] Finally, it should be pointed out that we have allowed name-variables (and probably other kinds) to be legitimate CL-expressions. We find it also simpler for the purposes of the lexicon to do this. As we shall see, the elimination of such terms as 'x' will be achieved via transformation rules.

Before inserting CL-formatives from the lexicon into the *P*-marker, we need to use the following two transformation rules:

$$T_1\colon\ DO^\frown F^\frown(A_1{}^\frown P_V{}^\frown A_2)$$
$$\Rightarrow DO^\frown P_V{}^\frown F^\frown(A_1{}^\frown A_2)$$

where A_1 and/or A_2 may be null

$$T_2\colon\ CF^\frown[(DO^\frown X)^\frown(A_1{}^\frown\ldots{}^\frown A_n)]$$
$$\Rightarrow (DO^\frown X)^\frown CF^\frown(A_1{}^\frown\ldots{}^\frown A_n)$$

where $n \geqslant 1$

T_1 and T_2 are not necessarily ordered in regard to one another. Either one may be applied before the other, although we have chosen to use T_1 first. By applying T_1 to the last line of the *P*-marker, we get the string:

$$\varphi_1\{\text{ICR}[(\iota x)xRL(\text{en, now})], \text{wa}\}.$$

Thus the variable '*x*' has been pulled out from among the arguments and placed between the descriptional operator and the function. If we apply T_2 to this result, we get:

$$[(\iota x)xRL(\text{en, now})]\ \varphi_1\ (\text{ICR wa}).$$

Here the adverb-main function complex has been placed between the two arguments of the main function. It should not be disturbing that what we now have no longer looks like a typical sentence of the predicate calculus. Recalling Chomsky's view that grammatical analysis brings into play forms that would not be suggested by intuitive considerations, we would like to point out that we are at present at an intermediate point between the surface structures of SL and CL. Hence the intermediate forms in the derivation may not closely resemble forms of either SL or CL.

Now the lexical rules can be employed, so as to arrive at:

[(the *x* such that) *x* rules (England, now)] ceremoniously
(is crowned at Westminster Abbey)

In order to put 'ceremoniously' in the right spot, we shall posit the following rule:

$$T_3\colon\ \text{Adv}^\frown(\text{is}^\frown X) \Rightarrow \text{is}^\frown\text{Adv}^\frown X$$

T_3 is a transformation rule, but it may be considered as an addition to the second lexicon, as it refers to an expression which is a formative in CL.[28]

T_3 would appear to be an optional rule, as the adverb could occur before 'is'. As we have set up the derivation, T_3 must be employed after T_1 and T_2, as T_3 requires the prior application of lexical rewriting rules. Using T_3 and the rule from the lexicon which permits the substitution of 'is the king of' for 'rules',[29] the derived string now becomes:

[(the x such that) $\tilde{x}$ is the king of (England, now)] (is ceremoniously crowned at Westminster Abbey)

The parentheses and brackets may now be dropped. This could be done by a rule which is to be used at some specific stage in the derivation. We shall assume such a rule without attempting to construct it. The lexicon now needs another transformation rule to permit the deletion of 'the x such that x is'. Let us then posit the following rule:

T_4: the x such that x is$\frown DO \frown X \Rightarrow DO \frown X$

which enables us to derive:

the king of England now is ceremoniously crowned at Westminster Abbey

There may be some question as to whether T_4 is sufficiently general to be of much value. Even if we disregard the relatively minor problem that a rule about 'x' does not take into account expressions containing other variables, there still is the serious problem of whether the derivation of each new CL-sentence would require rules invented *ad hoc* to handle the intricacies of expression in CL. Let us suppose T_4 is sufficiently general, however, for the problem may be brought into sharper focus by considering how we must treat the word 'now'.

It appears that we can only put 'now' in the proper place for being translated into 'present' by means of special rules designed to apply only to the contextual occurrence of that one formative. That is, 'now' would have to have its own special transformation rules. Let me suggest adding the following two rules to the lexicon:

T_5: $DO \frown P_V \frown F \frown (A_1 \frown \text{now} \frown A_2)$
$\Rightarrow \text{DO} \frown \text{P}_\text{V} \frown \text{now} \frown F \frown (A_1 \frown A_2)$ where A_1 and/or A_2 may be null.
T_6: $X_1 \frown \text{now} \frown \text{is the} \frown X_2 \Rightarrow X_1 \frown \text{is the} \frown \text{now} \frown X_2$

These two rules would have to be employed after T_1 but before T_4. T_5 would convert 'the x such that x is the king of England now' into 'the x such that x now is the king of England'. Thus it takes 'now' out from among the arguments and places it between the $DO^\frown P_V$ complex and the function. T_6 converts the result of T_5 into 'the x such that x is the now king of England'. It splits the components of the function 'is the king of', putting 'now' between 'is the' and 'king of'. The result of T_5 and T_6 is the expression 'is the now king of', which can, by means of the second lexicon, be altered to 'is the present_1 king of'.

Now the derivation is complete. It illustrates what is involved in what Reichenbach called a 'logistic grammar', i.e., a set of categories and rules for translating from any CL into an appropriately chosen SL. Since the structure of SL constitutes the 'deep structure' of CL, in order for the project of logistic grammar to make sense the sentences of CL must be generable from their putative SL-correlates. We have shown that they are and how they are, thus making explicit what is implied by the suggestion that logistic languages may serve as models for conversational languages.

It would surely be otiose to deny that logistic languages may be employed to represent certain aspects of natural languages. Successful formulations of such things as inference patterns enable us to discard that contention out of hand, for new ground is continually being broken in such areas as chronological propositions and modal logics. So what we are really concerned with here is how far we can push the use of logic for 'meaning analysis'. Reichenbach had as ambitious a project possible for capturing the cognitive significance of CL via an SL. His logistic grammar was intended to cover many (if not all) conversational languages, and he tried to make a logistic analysis of virtually every kind of CL-expression.[30] What we have done here is to examine the project from a new perspective, and now we can draw some conclusions as to its feasibility.

Although Chomsky claims that "the grammars of the 'artificial languages' of logic or theory of programming are, apparently without exception, simple phrase structure grammars in most significant respects,"[31] it is clear that the grammar Reichenbach had in mind cannot be such. The grammar required for Reichenbach's derivations must be both context-sensitive and transformational. So the process of generating CL from an SL is far more complicated than has been customarily thought.

In assessing the conception of a logistic grammar, we grant that CL-sentences can be generated from a sufficiently rich SL. But that does not *per se* guarantee that the grammar in question is an *adequate* grammar. Chomsky has formulated criteria for the adequacy of grammars,[32] and it will suffice for our purposes to consider whether a logistic grammar can be 'descriptively adequate'.

One need not belabor the point that Reichenbach, *de facto*, did not formulate his theory in the rigorous manner required by recent grammarians. What Reichenbach actually gave us is only a loosely constructed taxonomy which does not even meet good structuralist standards. So what we are testing is not an actual product of Reichenbach's thinking but an attempt at the rigorous formulation of logistic grammar desired by Reichenbach. A formulation of this sort would seem to represent the optimum achievable by someone taking the Reichenbachian approach. An evaluation of it in terms of Chomsky's criteria for adequacy enables us to assess the descriptive or explanatory potential of such a theory.

For the purposes of this discussion we shall beg the question of observational adequacy, recognizing that Reichenbach did not really pay attention to it.[33] The interesting question for us is, assuming that a logistic grammarian can make observationally adequate statements about natural languages, can he construct a descriptively adequate grammar? Another way to put this is, what is the generative capacity of 'Reichenbachian' grammar for deriving CL-sentences?

In viewing the situation, we conclude that logistic grammar could not be descriptively adequate, as it would fail the tests of both strong and weak generative capacity. Such a grammar could not generate very many sentences or structural descriptions in any particular language. Our reason for this conclusion is that, taking English as a sample language, the derivations above show that only sentences which are relatively simple in structure are generable from a Reichenbachian grammar. One could, for example, generate such sentences as 'the president of the United States is wise' or 'John loves Mary'. But, we have seen, in order to derive a sentence of a moderate degree of complexity, such as 'the present king of England is ceremoniously crowned at Westminster Abbey', it would be necessary to introduce *ad hoc* rules applying only to single cases. Even to derive that sentence we disregarded such problems as the analysis of tenses and token-reflexives, and the rules still had to be manipulated to handle the word 'present'. Suppose we tried to supply a structural description for a sentence like 'the present king of England, who was ceremoniously crowned recently at Westminster Abbey, today told the British parliament that before he made the proposed trip to Germany, he would have to be given convincing reasons as to why they desired him to go'. Even assuming a Reichenbachian grammar can handle the high degree of embedding in such a sentence, an enormous number of rules would still be required to handle the sentence. To get the structural description of the SL-version, intricate permutation rules would be necessary for the tenses and token-reflexives. Expressions like 'who', 'that', 'would have to be given', 'why', etc., would be very tricky to deal with. Then consider the problem of mapping the SL-formulation onto CL. The SL-version would be so different from its CL-correlate in structure that a whole new set of rules would be required to put the sentence into final form.

We don't want to contend that the above sentence cannot be analyzed by Reichenbachian methods; what we are saying is that such an analysis would be most unwieldy. The rules for sentences which are structurally complex would not square with any criteria of simplicity whatever. Adequate rules must be sufficiently general that a number of sentences and structural descriptions can be derived from them. Reichenbach's rules would neither be economical nor sufficiently general, so their generative capacity would be extremely limited. Hence Chomsky would say that as a theory of CL, Reichenbach's grammatical theory "can apply only clumsily . . . any grammar that can be constructed in terms of this theory will be extremely complex, *ad hoc*, and 'unrevealing' . . . certain very simple ways of describing grammatical sentences cannot be accommodated . . . certain fundamental formal properties of natural language cannot be utilized to simplify grammars."[34] In short, Reichenbach's logistic theory of grammar cannot produce a descriptively adequate grammar for English. And one could say the same for any other natural language. Measuring Reichenbach's program by what Chomsky has achieved, the value of logistic grammar is virtually nil.

And since it cannot describe the structures of English, Reichenbachian grammar certainly would be unable to *explain* them. Hence the contention of Carnap and other 'ideal language' philosophers that an SL is needed to explicate the structure of CL is seen to fail in one important respect. The grammar of a CL is not to be identified with that of a 'perspicuous' SL. An SL *per se* does not supply the categories and structures of a CL or the meanings for CL-terms. However, as Katz has pointed out,[35] the ideal language program was not completely wrong-headed. Contemporary grammar and semantics owe much to ideal language philosophy and its conception of rigor. The idea that a theory must be constructed rigorously, i.e., in accord with the canons of logic, was appropriated by Chomsky. So although logic does not supply the substance of grammatical theory, it does supply the form. Similar considerations apply to semantics. Just as the grammatical category of a formative is not that of its putative SL-correlate, the meaning of a CL-term is not simply that of one in an SL. The point is that, given notions of CL-structure and of the meanings of CL-terms, we may employ logistic languages for achieving rigorous formulations of these notions. The cognitive (descriptive, literal) meanings of CL-terms are not derivative from those of SL-terms, and men like Reichenbach erred in suggesting we need procedures for translating from SL into CL in order to make CL intelligible. But since Aristotle logicians have performed a most valuable service in suggesting means of making our CL-concepts precise and testing inferences from CL-sentences. The present work on modal logics, for example, is illuminating the semantics of value terms, tense expressions, necessity and possibility, and intensions. So logic is far from being irrelevant to the grammarian and semanticist.

In conclusion let me say that although the project of logistic grammar, as conceived by Reichenbach, was bound to fail as an analysis of CL-structure, it is of considerable historical interest. The idea of a universal grammar based on logic has occupied thinkers since the Middle Ages. Although Reichenbach wrongly posited a theory of substantive linguistic universals, he also suggested one about formal universals, and a minor alteration in some of his assumptions could have led to the earlier development of the theory of generative grammar.

The University of Akron

NOTES

[1] For a more detailed treatment see McMahon, *Hans Reichenbach's Philosophy of Grammar*, Mouton, the Hague, 1976.

[2] Hans Reichenbach, *Elements of Symbolic Logic*, MacMillan, New York, 1947, especially Chapter VII.

[3] Prior to the study cited in Note 1 the most extensive discussion of Reichenbach's logistic grammar is in R. M. Martin, *Logic, Language and Metaphysics*, New York U. Press, New York, 1971, pp. 75–99. Cf. also Uriel Weinreich, 'On the Semantic Structure of Language', in Joseph H. Greenberg (ed.), *Universals of Language*, MIT Press, Cambridge, Mass., 1963, pp. 114–115; Yehoshua Bar-Hillel, *Language and Information*, Addison-Wesley, Reading, Mass., 1964, p. 4; George D. W. Berry, 'Review: Hans Reichenbach, *Elements of Symbolic Logic*', *J. Symbol. Logic* **XIV** (March, 1949), 50–52; Milka Ivič, *Trends in Linguistics*, Mouton, the Hague, 1965, pp. 189–190.

[4] The confusion on this issue is illustrated by the debate between Bar-Hillel and Chomsky in the mid-1950's. Cf. Yehoshua Bar-Hillel, 'Logical Syntax and Semantics', *Language* **XXX** (1954), 230–237, and Noam Chomsky, 'Logical Syntax and Semantics, their Linguistic Relevance', *Language* **XXXI** (Jan.–March, 1955), 36–45.

[5] Reichenbach's SL is not a first-order predicate calculus, as he employs predicates of higher levels, but he does try to operate within a truth-functional system. See McMahon, *Hans Reichenbach's Philosophy of Grammar*, Chapter 2, for an extensive discussion of Reichenbach's 'logistic' analysis of CL.

[6] Cf. Reichenbach, *Elements of Symbolic Logic*, pp. 419, 436. Also pp. 257, 260, 268–269, 350–351.

[7] Cf. Reichenbach, *Elements of Symbolic Logic*, pp. 349–352. Also McMahon, *Hans Reichenbach's Philosophy of Grammar*, pp. 41–45. 'IC' is an abbreviation for 'immediate constituent'.

[8] Reichenbach's suggestions on formation rules for SL are scattered throughout *Elements of Symbolic Logic*. In McMahon, *Hans Reichenbach's Philosophy of Grammar*, they are treated in Chapter 2 and presented in tabular form on pp. 54–55.

[9] Cf. Chomsky, *Aspects of the Theory of Syntax*, MIT Press, Cambridge, Mass., 1965, pp. 112–120.

[10] For elaboration see *Hans Reichenbach's Philosophy of Grammar*, pp. 256–258.

[11] Reichenbach, *Elements of Symbolic Logic*, pp. 259–260.

[12] Although Reichenbach does not explicitly use the term, we have chosen to call his conception of the underlying structure of CL the 'logical structure' of CL. The structure of an SL can be called a syntactic structure of CL only if the SL-terms can be mapped onto CL-terms. Thus the analysis of the logical structure of CL involves rules for generating SL-sentences from their structural descriptions and rules for translating from SL into CL. For further discussion of this see *Hans Reichenbach's Philosophy of Grammar*, pp. 28–34, 234-248.

[13] For Carnap transformation rules are meta-linguistic statements which allow inferences from axioms to theorems within a logical system. Cf. *The Logical Syntax of Language*, Littlefield, Adams, and Co., Paterson, N. J., 1959, pp. 1–6. Hence from a Carnapian viewpoint there would be at least one transformation rule in a PS (phrase structure) grammar, the one allowing the substitution of the expression on the right of the '→' symbol for the expression to the left. However, for Chomsky the rewriting rules function as axioms in a transformational grammar, and the 'transformation rules', instead of being mere substitution rules, enable one to perform alterations on the rewriting rules. Cf., e.g., Chomsky, *Syntactic Structures*, Mouton, the Hague, 1957, p. 111.

[14] Cf. e.g., Chomsky, 'Current Issues in Linguistic Theory', in Fodor and Katz (eds.), *The Structure of Language*, Prentice-Hall, Englewood Cliffs, N. J., 1964, pp. 53, 82.

[15] Another technique is to have rewriting rules which list the members of formative-classes,

as in *Syntactic Structures*, p. 26. However, here we are following Chomsky's later method of constructing a lexicon. Cf. *Aspects of the Theory of Syntax*, p. 84.

[16] How this term corresponds to 'present' will be explained shortly. Since 'now' is a token-reflexive, for Reichenbach it would also require detailed analysis. But as in the case of tenses, we shall forego analysis in detail.

[17] An example of this would be the sentence, 'Everybody loves the president of the United States', which can be symbolized as '$(y)\{L[y, (\iota x)P(x, us)]\}$'. Quantifiers and the variables bound by them would also have to be introduced via context-sensitive rules. Where a quantifier or operator precedes a sentence, we could have a rule stipulating that it be assigned a specific variable-symbol. Then rules would have to be designed to handle the embedded quantifiers and bound variables. To take another example, an expression like 'the man who is loved by everybody' ['$(\iota x)(y)(Mx \cdot Lyx)$'] would require special rules for the universal quantifier, e.g.,

$$UQ \rightarrow \left\{ \begin{array}{l} (x)/X \_\_\_ Y \text{ where } x \text{ does not appear in } X \\ (y)/X \_\_\_ Y \text{ where } x \text{ but not } y \text{ appears in } X \end{array} \right\}$$

[18] Another way that '(ιx)' and 'x' could be entered into the *P*-marker in the desired way would be to have rewriting rules which introduce a dummy symbol such as '$\triangle$' into the *P*-marker, e.g.:

$$DO \rightarrow (\text{Def. } O \frown \triangle)$$

$$PN_v \rightarrow \triangle$$

Then the lexicon could contain a general context-sensitive rule for the binding of variables by operators, e.g., something like:

$$\triangle \rightarrow \left\{ \begin{array}{l} x/ \left\{ \begin{array}{l} X \frown (DO \_\_\_) \frown F \frown (A_1 \_\_\_ A_2) \\ X \frown (Q \_\_\_) \frown F \frown (A_1 \_\_\_ A_2) \\ \end{array} \right. \quad \left. \begin{array}{l} \text{where } x \text{ does not appear} \\ \text{in } X, \text{ and } X, A_1 \text{ and/or} \\ A_2 \text{ may be null} \end{array} \right\} \\ \\ y/ \left\{ \begin{array}{l} X \frown (DO \_\_\_) \frown F \frown (A_1 \_\_\_ A_2) \\ \vdots \end{array} \right. \quad \left. \begin{array}{l} \text{where } x \text{ but not } y \\ \text{appears in } X, \text{ etc.} \end{array} \right\} \end{array} \right\}$$

Here 'Q' stands for 'quantifier', and the lexicon could contain the simple entry $(\iota, [+DO, +\text{Def. } O, \ldots])$ for the descriptional operator. This method would have greater generality than the one proposed, but the other method is simpler for the purposes of exposition. The ideas here were suggested by remarks in Chomsky, *Aspects of the Theory of Syntax*, pp. 122, 128–137.

[19] An interesting question is whether a distinction between the deep and surface structures of the SL-sentence should be made. In order to make this distinction one would have to give primitive structural descriptions for SL along with transformation rules for conversions into derivative structures. The procedure that we have followed contains no transformation rules of that sort, so our analysis of the SL-sentence could be called a direct analysis of its surface structure. Reichenbach does suggest, however, that there can be transformation rules for converting certain SL-forms into others. A particular case mentioned is the transformation of statements about events. Cf. *Elements of Symbolic Logic*, pp. 267–272. Let us make it clear, however, that 'rules of inference' in logistic systems are not to be equated with the kind of transformation rules we are speaking of here. The formulas generated by rules of inference can be analyzed by means of the same phrase-structure rules used for the analysis of axioms in logistic systems.

[20] The key texts for Reichenbach's conception of the structure of CL and the need for translation into SL are *Elements of Symbolic Logic*, pp. 251–255, 301–302; *The Rise of Scientific Philosophy*, University of California Press, Berkeley, 1956, p. 220. Reichenbach's views are akin to those expressed in Carnap, *The Logical Syntax of Language*, pp. 2, 8–9, 168, 312. For further discussion of why, for Reichenbach, doing a grammar of CL requires translation into SL, see McMahon, *Hans Reichenbach's Philosophy of Grammar*, pp. 234–248.

[21] The symbol for 'ceremoniously' probably would not be 'φ_1'. Here we are just following Reichenbach's use of Greek letters for adverbs and taking an arbitrary number for a subscript.

[22] Another tricky problem for Reichenbach is whether to take 'at' as belonging to 'ICR' ('is crowned at') or to 'wa' ('at Westminster Abbey'). Either way leads to further complexities. 'Is crowned at' should be distinguished from 'is crowned on', 'is crowned in', etc., and 'at Westminster Abbey' differs from 'in Westminster Abbey' or 'Westminster Abbey' (taken as a subject).

[23] 'Rules' doesn't mean the same as 'is the king of', but we shall disregard this relatively minor issue.

[24] We subscripted 'present' here to differentiate it from its CL-homonyms, all of which would probably belong to some other category, e.g., the triadic functions 'is a present from . . . to . . .' and '(I) presént . . . to . . .'.

[25] Once again we have the problem of handling different variables. Here the question is whether we should have separate entries for $(\neg y)$', '$(\neg z)$', etc., or whether we should have a general rule to cover all such cases.

[26] See Note 15 above.

[27] Cf. McMahon, *Hans Reichenbach's Philosophy of Grammar*, pp. 179, 206–221, 243–247, 269–272.

[28] Actually, the 'is' which occurs in the rule is part of a formative in Reichenbach's grammar.

[29] Strictly speaking, 'is the king of' is not synonymous with 'rules'. We have noted that 'rules' in some cases could mean 'is the queen of', and there are, of course, other possibilities ('is the emperor of', 'is the dictator of'). But for the present purposes we have chosen to ignore such intricacies.

[30] In Chapter VII of *Elements of Symbolic Logic* Reichenbach goes beyond his predecessors in proposing analyses of such expressions as pronouns, adverbs, prepositions, interjections, contour morphemes, and tenses, and he also considers, besides English, expressions in German, French, Latin, Greek, and one non-Indo-European language, Turkish.

[31] Chomsky, *Aspects of the Theory of Syntax*, p. 136.

[32] Cf. Chomsky, 'Current Issues in Linguistic Theory', in Fodor and Katz, pp. 62–77; *Aspects of the Theory of Syntax*, pp. 24–27, 30–46.

[33] Reichenbach was not really concerned with the collection of CL-data nor with distinguishing sentences which belong to a specific CL from expressions which do not. He merely considered a few sample sentences from different CL's. And although his examples are valid sentences of the various languages considered, he was not concerned with why this is the case. His technique was to start with theoretical premises and to find data to substantiate his theory rather than to work from data to theory.

[34] Chomsky, *Syntactic Structures*, p. 34.

[35] Cf. Jerrold J. Katz, *The Philosophy of Language*, Harper and Row, New York, 1966, pp. 18–97.

WESLEY C. SALMON

LAWS, MODALITIES AND COUNTERFACTUALS*

There is a close-knit complex of philosophical problems which has been with us since antiquity. These problems involve such concepts as potentiality and disposition; necessity, actuality, and possibility; and even material and subjunctive implication. They are philosophically ubiquitous, cropping up in a range of fields broad enough to include metaphysics, epistemology, philosophy of science, logic, philosophy of language, and ethics. They seem, moreover, to be strangely recalcitrant. It is to this set of perennial philosophical problems that Hans Reichenbach's 1954 monograph – originally published under the title *Nomological Statements and Admissible Operations*, but reissued in 1976 under the title *Laws, Modalities, and Counterfactuals* – is addressed. The new title has been adopted to provide the philosophical community with a clearer idea of the subject matter of the book (the original pagination is retained).

Republication of this work, which has been out of print for many years, requires a few words of explanation. First, these problems are still with us. They have not gone away in the last twenty years. Philosophers are actively working and publishing on these topics, and no resolutions which are widely accepted as adequate are anywhere in sight.

Second, Reichenbach's monograph attracted little attention when it was first published, and it is almost completely ignored today. This lack of attention is not due to any demonstrated inadequacy in his results. Only a few philosophers seem to have been aware of the existence of this monograph, or what it is about, and those who were cognizant of it may have been repelled by its admitted complexity.

Third, Reichenbach's treatment of these problems does, I firmly believe, merit careful attention. While I do not intend to argue that he has found definitive solutions to all of these problems, I do wish to urge consideration of the major concepts he develops in the course of his analysis. Even if his answers are not fully adequate, the central concepts will be enlightening to those who are working in this area. Although they embody seminal ideas, these concepts have not been generally assimilated by contemporary philosophers.

W. C. Salmon (ed.), Hans Reichenbach: Logical Empiricist, 655–696.

Fourth, Reichenbach's approach has philosophical virtures to recommend it. He offers a systematic account of laws, modalities, and counterfactuals which is thoroughly consonant with his empiricism. He eschews appeal to such devices as mysterious 'connections' of the sort criticized by Hume, or to realms of reality reminiscent of Meinong's jungle. According to Occam's razor, postulation of such entities is to be avoided if possible. It is important to ascertain whether a satisfactory resolution of these problems can be achieved without them.

Finally, Reichenbach did not seek to evade these problems. He clearly recognized their central role in philosophy of science, as well as other areas. He spared no effort in attempting to provide answers which are adequate to the problems, and which do not violate the spirit of empiricism which characterized all of his philosophical enterprises.

This essay, which will be devoted to elucidation and justification of the foregoing claims, will be divided into three main parts. In Section I, I shall make some general remarks about the problems, about their interrelations, and about the ways in which modern discussions have developed. This Section will make it evident that Reichenbach's book is highly relevant to current philosophical concerns. In Section II, I shall discuss some basic aspects of Reichenbach's earlier attempt, in *Elements of Symbolic Logic*, to deal with the same set of problems. Although the 1954 monograph is logically self-contained, the earlier treatment has, I believe, considerable heuristic value for anyone attempting to understand his subsequent, more technical, treatment. In Section III, I shall offer some hints which are intended to facilitate the understanding of *Laws, Modalities, and Counterfactuals*. Reichenbach does not give us easy answers to these problems, but he has provided a sustained and penetrating attack upon them. His work deseserves serious study.

I

The most distinctive feature of twentieth-century philosophy is, perhaps, the development and widespread use of formal logic. Leibniz had envisioned the day when philosophical problems would be definitively settled by such methods, but it was not until Whitehead and Russell had produced *Principia Mathematica* that tools which might be adequate to the tasks became widely known. With the application of these resources to a variety of philosophical problems, some philosophers seemed to believe that the dream of Leibniz had come true. Logical positivists and logical empiricists, for example, sought to emphasize their reliance upon logistic techniques in the very names with which they baptized their movements.

Looking back at the results, one can hardly doubt the impact of formal logic in twentieth-century philosophy, even if the Leibnizian millenium has not as yet been realized. In the foundations of mathematics, Russell's work on paradoxes, Gödel's work on consistency and completeness, and Cohen's work on the continuum hypothesis are obvious landmarks. In semantics, Tarski's work on truth and Robinson's work on model theory stand out conspicuously. Carnap's monumental studies in inductive logic and confirmation theory rest solidly upon the foundation of formal logic. The influential studies of scientific explanation by Braithwaite, Hempel, and others stand upon the same foundations. The list is easily extended. I am not trying to give an exhaustive catalogue of applications of modern logic, but only to provide a brief reminder to those of us who might now take these achievements for granted, forgetting the vast changes wrought in philosophy since the beginning of this century.

Yet, strangely enough, the problems mentioned in the opening paragraph have appeared to be amazingly refractory against the best efforts of contemporary philosophers using a variety of logical techniques. This phenomenon is all the more surprising, because the problems of modalities, lawlike generalizations, counterfactual conditionals, and dispositional predicates would seem at first blush to be precisely the sort that ought to be most amenable to formal logical analysis. The current philosophical literature, nevertheless, gives ample testimony to the absence of generally accepted solutions.

As one of the outstanding leaders of logical empiricism, Reichenbach was squarely within the tradition which places great emphasis upon logistic methods. He often reiterated his confidence in the power of formal logic to resolve philosophical problems. In his work on probability and induction, on quantum mechanics, on space and time, and on the analysis of conversational language, he attempted to exemplify the applicability of these methods. His 1954 monograph constitutes his final effort to make formal methods yield answers to the puzzles associated with laws, modalities, and counterfactuals. Let us take a look at some of the background of these problems.

1. *Modalities*

The rigorous formal treatment of this ancient topic began, shortly after the publication of *Principia Mathematica*, with the work of C. I. Lewis on strict implication. C. I. Lewis developed several systems of modal logic, and other logicians have subsequently created a multitude of others. Such modal logics were formal systems containing a primitive symbol (for necessity or possibility) over and above the primitive symbols of some standard logical system. In the absence of one or more clear interpretations of the modal primitive, there was considerable room for skepticism regarding the philosophical value and import of modal logics.

Beginning about 1959, Saul Kripke and others developed rigorous semantics for modal logics. It is essential to recognize, however, that the interpretations thereby provided are abstract – that is, the entities invoked in these interpretations are mathematical constructions of one sort or another. Although the informal chat about such interpretations has involved much talk of such things as 'possible worlds', the entities are, strictly speaking, entirely set-theoretical in character. A strong analogy can be drawn to analytic geometry – that is, the interpretation of the primitive terms of Euclidean geometry in terms of such extra-geometrical *mathematical* entities as algebraic equations and triples of numbers. Important as the existence of such abstract interpretations is, they do not provide any enlightenment whatever to the philosopher who has questions about the structure of *physical* space. Likewise, valuable as it is to have abstract interpretations by means of which to prove completeness or consistency of modal systems, the formal systems with their abstract interpretations do not tell us anything about what is necessary or possible in the real world.[1]

In his recent book, *Counterfactuals* (1973), David Lewis has taken the bull by the horns and asserted the reality of possible worlds. Among the real worlds one is considered actual because we happen to be living in it, but the others are no less real because of being merely possible and not actual. Denying that possible worlds are somehow to be identified with either linguistic or mathematical structures, he writes:

> When I profess realism about possible worlds, I mean to be taken literally. Possible worlds are what they are, and not some other thing. If asked what sort of thing they are, I cannot give the kind of reply my questioner probably expects: that is, a proposal to reduce possible worlds to something else.

I can only ask him to admit that he knows what sort of thing our actual world is, and then explain that other worlds are more things of *that* sort, differing not in kind but only in what goes on at them. Our actual world is only one world among others. We call it alone actual not because it differs in kind from all the rest but because it is the world we inhabit. (p. 85.)

Such shocking metaphysical proliferation requires some attempt at justification. David Lewis continues,

So it is throughout metaphysics; and so it is with my doctrine of realism about possible worlds. Among my common opinions that philosophy must respect (if it is to deserve credence) are not only my naive belief in tables and chairs, but also my naive belief that these tables and chairs might have been otherwise arranged. Realism about possible worlds is an attempt, the only successful attempt I know of, to systematize these preexisting modal opinions. (p. 88.)

We must agree, of course, that it is sensible (and probably true) to say that tables and chairs might have been differently arranged. To those who have appreciated Russell's beautiful theory of definite descriptions as a way to make sense of statements about golden mountains, without becoming enmeshed in Meinong's jungle, David Lewis's views on the reality of possible worlds will appear to be a giant step backward. Reichenbach offered a well-developed theory of logical and physical modalities which does not indulge in such ontological excesses, but it is not cited in David Lewis's book.

The laws of nature, it might be said, delineate the realm of physical possibility; indeed, laws are often expressed quite explicitly in this form. A basic principle of relativity theory states that it is impossible to propagate any signal with a velocity greater than that of light. The first and second laws of thermodynamics, respectively, are often expressed by asserting the impossibility of perpetual motion machines of the first and second kinds. Conservation laws assert the impossibility of creating or destroying an entity or quantity of some sort. It would, of course, be foolhardy to expect modal logic (or any other kind of logic) to determine which statements express laws of nature and which statements do not.[2] But there is a more modest task – the problem of delineating the characteristics of 'law-like statements', that is, the problem of specifying the characteristics a statement must have *over and above truth* if it is to express a law of nature.

If we ask whether the study of modal logic has been of any significant help in clarifying the nature of physical necessity, I think the answer must be negative. Little of the work done in modal logic has been primarily motivated by a concern with the analysis of physics (or any other empirical science).[3] The situation is just the reverse. If we had a clear grasp of the concept of physical necessity, we might be able to find a modal logic which would embody its structure. In the absence of such a prior understanding, it is unlikely that modal logic by itself will provide any insight into the nature of physical necessity.[4] Reichenbach's analysis of synthetic nomological statements – law statements – was designed to furnish clear concepts of physical modalities.

The investigations of C. I. Lewis were primarily directed, not at physical modality, but rather to logical modalities. There has been considerable subsequent discussion of whether such modal systems as he constructed can capture the concept of *entailment.* Some philosophers, noting that C. I. Lewis' systems give rise to 'paradoxes of strict implication', have sought what they consider a more adequate formalization of the entailment relation. The *magnum opus* on this topic (Anderson *et al.*, 1975) has just been published. In his analysis of *analytic* nomological statements (of the narrower kind), Reichenbach was attacking precisely the same sort of problem.

Philosophers have, of course, looked for other kinds of interpretations of modal logics. Some have found modal systems useful in their attempts to construct 'deontic logic'.[5] Others have used modal systems in an effort to characterize the nature of time; these are the 'tense logics'.[6] There may be many other applications of the formal structures that have been constructed under the general heading of 'modal logics'.[7] Interesting as they are in their own right, they do not bear upon the problems of physical modalities and laws of nature, or upon logical modalities and entailment.

2. *Laws*

Concern with the nature of laws stems from a variety of sources. We have just seen that it is intimately connected with the problem of physical modalities. Another source is the extensive work – done over the past three decades – on scientific explanation. C. G. Hempel and Paul Oppenheim's classic (1948) article formulated a basic schema of scientific explanation which has come to be known as the 'deductive-nomological model'. One of the main requirements imposed upon such explanations is that the explanans must include at least one law-statement in an essential way. This is one among several models – elaborated by Hempel and others – which, because they require an appeal to laws of nature, are called 'covering laws models' of scientific explanation. The problem of providing an accurate and precise characterization of the concept of a law has been a recurrent problem in this approach to scientific explanation.[8]

Nelson Goodman's influential book, *Fact, Fiction, and Forecast* (1965; 1st ed., 1955) has posed the problem of laws in a somewhat different way. By introducing certain peculiar predicates, he shows that some generalizations can apparently be supported inductively by their instances, while others cannot. Goodman defines the expression, 'x is grue at time t' to mean 'x is green and $t \leq 2000$ A.D. or x is blue and $t > 2000$ A.D.'. The expression 'x is bleen at t' is defined in the same manner with 'blue' and 'green' interchanged in the foregoing definition. Goodman then presents the 'new riddle of induction' as the problem of showing why observed instances can confirm the generalization, 'All emeralds are green', while precisely the same instances cannot confirm the generalization, 'All emeralds are grue'.[9] Those who would seek to resolve the problem by admitting only 'purely qualitative' predicates like 'green' and 'blue' into the basic scientific vocabulary – refusing to grant inductive confirmation to generalizations containing 'positional' predicates which, like 'grue' and 'bleen', involve an explicit reference to a particular time t – should consider carefully Goodman's potent argument to the effect that 'green' and 'blue' are the positional predicates in a language in which 'grue' and 'bleen' are taken as primitive.[10] In order to answer this argument, it would seem necessary to show that the *properties* (as opposed to predicates) *grue* and *bleen* are actually *time dependent* in a way which is relevant to the possibility of inductively confirming generalizations making reference to them.[11]

If generalizations containing Goodmanesque predicates are suspect from the standpoint of confirmability, they are at least equally dubious candidates for the status of explanatory law.[12] One might wonder whether they are unconfirmable because they are non-lawlike, or whether they are non-lawlike because they are unconfirmable. Reichenbach, as we shall see, takes the latter alternative, making inductive verifiability *the* primary hallmark of law-statements. Given an adequate analysis of the predicates 'grue' and 'bleen', Reichenbach's requirement of universality will dispatch such statements as 'All emeralds are grue'.

In a popular introduction to philosophy of science, Hempel (1966, p. 55) cites the generalization, 'All bodies consisting of pure gold have a mass of less than 100 000 kg,' as presumably true, but non-lawful.[13] It would be agreed, in contrast, that the generalization, 'No signal travels faster than light', is a lawful statement. The problem is to specify the characteristics upon which this distinction rests. Reichenbach considers a similar pair of examples; I shall discuss them in some detail below.

One immediate temptation might be to say that, although perhaps no golden object of 100 000 kg has ever existed, it is possible in principle to fabricate one. It is, on the other hand, impossible in principle to make a signal go faster than light. This answer, while true, is of no help, for it depends upon the physical modalities. As we saw in the preceding section, physical possibility and impossibility are determined by the laws of nature. The present answer thus takes us around a very neat little circle.

3. *Counterfactuals*

The material conditional was explicitly defined in antiquity; Sextus Empiricus presents the truth table definition, though not in a tabular form. One wonders whether ancient logicians might have drawn truth tables in the sand – after the fashion of geometers and their diagrams – in order to present the definition more concisely. Evidently there was considerable controversy among ancient philosophers concerning the nature of the conditional. Callimachus is quoted as saying, "Even the crows on rooftops are cawing over the question as to which conditionals are true" (Mates, 1972, pp. 212–214). The ancient logicians were fully aware of the 'paradoxes of material implication'.

As the beginning logic student learns while being introduced to truth tables, a material conditional with a false antecedent or with a true consequent is *ipso facto* true, regardless of any relation of meaning between the antecedent and consequent. The perceptive student also quickly realizes that it is very difficult to find examples of material conditionals in everyday life. 'If Smith wins the election, then I'll be a monkey's uncle', which is merely a rhetorical way of expressing the speaker's confidence in the falsity of the antecedent, is about the best we can do. The vast majority of conditional statements are of the sort, 'If the fuse blows, the lights will go out', or 'If the fuse *were to* blow, the lights *would* go out'. Because of the grammatical form of the latter sentence, the designation, 'problem of subjunctive conditionals', is sometimes preferred. As long as the fuse does not blow, the conditional is counterfactual. If it were treated as a material conditional, it would be true simply because of its false antecedent. The difficulty is that the statement, 'If the fuse blows, the lights will not go out', is also true for precisely the same reason. The patent unacceptability of this result shows that counterfactual conditionals have an import quite different from that of the material conditional. The problem, therefore, is to provide some adequate analysis of counterfactuals in particular, and more generally, of subjunctive conditionals.

Much of the contemporary discussion of counterfactual conditionals stems from two classic papers, Chisholm (1946) and Goodman (1947). These two essays state the basic problem in clear and engaging terms, showing that all sorts of rather obvious moves toward resolution will not work. Chisholm's article convincingly demonstrates the degrees to which this problem reaches into a wide variety of philosophical contexts, while Goodman's article exhibits the intimate relations of this problem with that of laws, modalities, and dispositional properties. Goodman treats these as problems within philosophy of science and remarks, " . . . if we set aside all the problems of dispositions, possibility, scientific law, confirmation, and the like, we virtually abandon philosophy of science" (1965, p. x). Goodman's essay was reprinted in 1955 and again in 1965. In the Introduction to the 1965 edition he says, "The scores of articles that have been published since then have made so little progress toward settling the matter that current opinion varies all the way from the view that the problem is no problem at all to the view that is insoluble" (1965, pp. ix–x). Philosophers are still working actively on the problem of counterfactuals; as we have seen above, one of them, David Lewis, has adopted heroic measures to resolve it.[14]

It has often been noted that acceptable counterfactual conditionals can be 'supported' by laws of nature. By examining the circuitry of the house, and by applying laws of electricity, we can confidently predict that the lights will go off and will not remain on if a fuse blows. As we have already noted, however, it is no easy task to state the criteria a generalization must fulfil in order to qualify as a law. Goodman offers as an example of a (presumed true) non-lawlike generalization the statement, "All of the coins in my pocket are silver", pointing out that it does not support the unacceptable counterfactual, "If this penny were in my pocket it would be made of silver." As a contrasting example of a lawlike generalization – which now brings a twinge of nostalgia – he offers the statement, 'All dimes are silver', which presumably does support the counterfactual conditional, 'If this coin (which happens to be a penny) were a dime, it would be made of silver.' The difficulty is that a lawlike generalization is just the sort of generalization that could support counterfactuals, while a non-lawlike generalization cannot. Again, we have gone in a circle. A reasonable counterfactual is one which is supported by a law, while a generalization is a law provided it can lend support to counterfactual assertions.

4. *Dispositions*

One of the most pervasive philosophical influences in modern science – due more to scientists than to philosophers – is the viewpoint known as operationism. The operationist thesis is, briefly, that a scientific concept cannot be meaningful unless its applicability is associated with some sort of physical operation. A metal rod is one meter long if its ends coincide with the ends of a suitably situated meter stick. A piece of iron is magnetic if it attracts nearby iron filings. Such 'operational definitions' clearly have a conditional form. To say that the piece of iron is magnetic *means* iron filings will be attracted *if* they are in the neighborhood. If we want to say, as most scientists presumably would, that the rod is one meter long whether it is actually being measured at the moment or not, and that the piece of iron is magnetic whether there happen to be any iron filings in the vicinity at the moment or not, then obviously these operational definitions embody counterfactual conditionals. Operationists, by and large, have not concerned themselves with the analysis of counterfactual conditionals, apparently believing that our common sense understanding of this type of statement is adequate for all practical purposes. Yet, if one were to try to analyze these definitions in terms of material conditionals, absurd results would ensue. A piece of glass which had never been placed near any iron filings would qualify as magnetic. Operationists, who pride themselves on their precision, would presumably applaud efforts to find a clear analysis of such conditionals.

The classic attempt to provide a precise logical analysis of operational definitions, and definitions of dispositional predicates in general, is Carnap's 'Testability and Meaning' (1936–37). In order to avoid problems surrounding the material conditional, Carnap constructs so-called *reduction sentences* which are to serve as partial definitions of scientific terms. A typical reduction sentence might read,

> If x has iron filings in its vicinity, then x is magnetic if and only if the iron filings are attracted to x.

In this statement, the conditional and the biconditional are to be construed as material. The import of the partial definition is that x is judged magnetic or non-magnetic only if iron filings are nearby; if there are no iron filings in the vicinity the reduction sentence remains true, but no conclusion whatever can be drawn about the magnetic character of x. Another reduction sentence could be introduced to provide a further definition of the term 'magnetic'.

> If x is moved through a coil of conducting wire, then x is magnetic if and only if a current flows in the coil.

By furnishing additional reduction sentences, we can more fully specify the meaning of a dispositional concept.

It is now generally recognized, I believe, that Carnap's reduction sentences, ingenious as they are, do not handle the problem of dispositional concepts, for the dispositional predicate can only be meaningfully applied if one or another of the test conditions – e.g., having iron filings in the vicinity or being moved through a conducting coil – is satisfied. The problem of dispositional terms is precisely that of specifying conditions for their applicability when *no* test conditions happen to obtain.[15] It appears that a subjunctive conditional is what we need for this purpose. Thus, the problem of dispositional concepts falls neatly into place along with the problems of counterfactuals, laws, and modalities.

It is worth noting that an increasing number of philosophers have recently been adopting the 'propensity interpretation' of probability. According to this interpretation, a probability is a *dispositional property* of a 'chance set-up'. Unlike such traditional dispositions as inflammability or solubility, propensities are statistical dispositions to manifest a certain outcome, not in all cases in which the conditions are satisfied, but in a certain percentage of such cases. Like the standard dispositions, the statistical dispositions also have a counterfactual import. For example, a penny which was placed on a railroad track and flattened by a passing train before it was ever flipped could still have been said to have had a propensity of one-half for showing heads *if it had been flipped in the prescribed manner.* This concept of probability has been the subject of a large and growing literature within the last few years, and it constitutes still another locus of the problems of dispositions and counterfactuals.

The cluster of problems concerning modalities, laws, counterfactuals, and dispositions remains a matter of serious concern to philosophers interested in modal logic, entailment, scientific explanation, inductive logic ('projectability'), probability, operational definitions, counterfactual conditionals, determinism and free will ('avoidability'), and a host of other major issues. Let us now take a look at Reichenbach's earlier attempt to provide a systematic resolution.

II

In 1947, Hans Reichenbach published *Elements of Symbolic Logic.* Its final chapter is devoted to the problems of laws, modalities, and counterfactuals. One of Reichenbach's major concerns in this book was to show how symbolic logic could be used for the analysis of conversational language. When he first introduces the truth table definitions of the truth-functional operations, he is careful to emphasize the profound differences between their strictly truth-functional interpretation and the interpretations of such connectives when they occur in conversational language – including, it should be emphasized, the ordinary language of science. He attempts to explain this difference in terms of two distinct ways of reading truth tables.[16] One way of reading yields what he calls "adjunctive operations", the other what he calls "connective operations" (1947, pp. 27–34). The material implication is adjunctive; the subjunctive implication is a connective operation.

This distinction between adjunctive and connective interpretations obviously applies to equivalences just as it does to implications – the material equivalence is simply a conjunction of two material implications. Reichenbach realized that a similar distinction applies to two ways of interpreting a disjunction. When a gun-brandishing bandit says, 'If you do not give me your money, I'll blow your head off', his statement would normally be understood in the subjunctive or connective sense. If, after handing over your money, you notice that the gun he holds is a toy, you might chide him for issuing a false threat. If he has studied symbolic logic he could reply that his implication was true – since you handed over your money, it has a false antecedent. The same threat could obviously have been couched in the form of a disjunction: 'You will give me your money or I will blow your head off.' The ensuing dialogue can easily be constructed. Recognizing clearly that the problem at hand was more general than the interpretation of conditional statements, Reichenbach posed it in terms of 'connective operations' in general.

I have never found Reichenbach's explanations of the relations between adjunctive and connective operations in terms of two ways of reading the truth tables particularly helpful. The main reason is that this explanation relies heavily upon the use of 'connective operations' in the metalanguage – a point that Reichenbach recognized explicitly and, with characteristic philosophic candor, commented upon. These preliminary discussions of the truth tables are not question begging; they are merely informal attempts at clarification. The actual analysis of the connective operations is reserved for the last chapter, where it is treated in the context of logical and physical laws, and logical and physical modalities. Although Reichenbach's subsequent treatment of these problems in his 1954 monograph is logically self-contained, I believe an understanding of some of the material from *Elements of Symbolic Logic* may be heuristically valuable. Most of the main thrust of his later treatment is contained in this earlier discussion. *Laws, Modalities, and Counterfactuals* (1976 reprint of 1954) embodies important revisions and refinements, but the earlier account provides an excellent point of departure.

Reichenbach's first major task is to define the class of *original nomological statements.* This class is intended to contain both the fundamental laws of logic and the fundamental laws of nature. The qualifying term 'fundamental' is important, for he maintains that there are *derivative nomological statements*; these include both laws of logic and laws of nature which do not qualify as fundamental. Original nomological statements, as defined in *Elements of Symbolic Logic*, must fulfill four requirements. In this earlier work, Reichenbach was not very careful to distinguish criteria which depend upon the form in which a statement is given from those which depend solely upon its content. Since I am not offering a critique of the earlier treatment, I shall attempt to capture his intent, even if his explicit statement sometimes falls a bit short. The criteria are:

(1) Original nomological statements must be *all-statements*. An all-statement is simply a statement that begins with an all-operator which has the entire statement as its scope. (Reichenbach used the term 'all-operator' for what is now usually called a 'universal quantifier'; I shall follow his usage.) Clearly, this requirement relates partly to the form of the statement. It would seem to capture Reichenbach's intent, however, to stipulate that the statement first be transformed into a prenex form – an equivalent statement with operators initially placed and having the entire formula as scope – and then to apply the test to see whether it qualifies as an all-statement. If it does, then the given statement, even if it is not originally formulated with an initial all-operator whose scope is the whole statement, would qualify as an all-statement. Being equivalent to a statement which has the requisite form, the given statement would certainly possess the type of generality we want to demand of fundamental laws.

(2) Original nomological statements must be *universal*. A statement is defined as universal if it contains no individual-signs – e.g., proper names, definite descriptions, designators of particular space-time locations, etc. Again, the criterion makes reference to the form of the statement, for a statement which contains an individual-sign may be transformed into an equivalent statement which does not, and vice-versa. In *Laws, Modalities, and Counterfactuals*, Reichenbach is very careful to spell out the requirement in detail, but in the earlier treatment he was not as precise. The intent, nevertheless, is straightforward; a fundamental law must not make essential reference to any individual entity. It is worth noting that, because of this criterion, neither Galileo's law of falling bodies (which makes essential reference to the Earth) or Kepler's laws of planetary motion (which make essential reference to our Solar System) would qualify as fundamental laws. With suitable corrections, they would, however, qualify as relative nomological statements (1954/1976, chap. VIII).

(3) Original nomological statements must be *fully exhaustive*. Requirements of the first two types are familiar from practically every discussion of the nature of laws. This third requirement is not generally

mentioned; it demands fuller explanation. If a given statement is transformed into a prenex form, with all operators standing in front and having scopes that extend to the end of the formula, then that portion of the formula which does not contain the operators – what Reichenbach calls the 'operand' – can be expanded in either of two ways. (i) *Expansion in major terms*. The operand has a truth-functional structure, and as such, it has a major operation. If, for example, the operand has the form $A \supset B$, regardless of the internal structures of the components A and B, then it can be expanded into a formula with the form $A \cdot B \vee \bar{A} \cdot B \vee \bar{A} \cdot \bar{B}$.[17] It is a disjunctive normal form where A and B are treated as unanalyzed units; A and B are the *terms* of the major operation. Reichenbach now defines a *residual statement in major terms* as a statement which results by dropping one or more of the disjuncts of the expanded operand (along with a suitable number of disjunction symbols). The residual statement thus consists of the set of prefixed quantifiers followed by an operand expanded in major terms, less one or more of the disjuncts of the expanded operand.

A true statement is said to be *exhaustive in major terms* if none of its residual statements in major terms is true. The concept of exhaustiveness is not defined for false statements. As the fourth requirement will stipulate, original nomological statements must be true, so we have no need to apply the concept of exhaustiveness of false statements.

The requirement of exhaustiveness in major terms disqualifies universal statements with empty antecedents from the class of original nomological statements. Consider, for example, the statement, 'All unicorns are four-legged'. Expanded in major terms, it becomes

$$(x)[u(x) \cdot f(x) \vee \overline{u(x)} \cdot f(x) \vee \overline{u(x)} \cdot \overline{f(x)}].$$

Since there are no unicorns, the following residual statement is true:

$$(x)[\overline{u(x)} \cdot f(x) \vee \overline{u(x)} \cdot \overline{f(x)}].$$

Somewhat surprisingly, perhaps, it also excludes from the class of original nomological statements such statements as 'All animals which (naturally) have hearts also (naturally) have kidneys', which, expanded in major terms, becomes,

$$(x)[h(x) \cdot k(x) \vee \overline{h(x)} \cdot k(x) \vee \overline{h(x)} \cdot \overline{k(x)}].$$

Since (I believe) only animals with hearts have kidneys, the following residual is true:

$$(x)[h(x) \cdot k(x) \vee \overline{h(x)} \cdot \overline{k(x)}].$$

This statement does, however, qualify as derivative nomological (in the narrower as well as the wider sense). Reichenbach offers the following justification for invoking the exhaustiveness requirement: "If an all-statement is not exhaustive in its major terms, its major operation is used out of place, so to speak; the operation then suggests connections which do not exist" (1947, p. 363).

(ii) *Expansion in elementary terms.* If a quantified statement, in prenex form, has an operand with more than a single truth-functional operation, that operand can be expanded into its full disjunctive normal form. Consider, for example, an operand of the form $A \supset (B \supset C)$, where A, B, and C have no internal truth-functional structure. Its disjunctive normal form would be

$$A \cdot B \cdot C \vee A \cdot \bar{B} \cdot C \vee A \cdot \bar{B} \cdot \bar{C} \vee \bar{A} \cdot B \cdot C$$
$$\vee \bar{A} \cdot B \cdot \bar{C} \vee \bar{A} \cdot \bar{B} \cdot C \vee \bar{A} \cdot \bar{B} \cdot \bar{C}.$$

A residual statement in elementary terms is defined in the same fashion as a residual statement in major terms; it is the statement that results from deleting one or more of the disjuncts of the expansion in elementary terms. A true statement is *exhaustive in elementary terms* if none of its residual statements in elementary terms is true. Again, there is no need to define this concept with respect to false all-statements.

Finally, a statement is *fully exhaustive* if it is exhaustive both in major terms and in elementary terms. This is the property original nomological statements must possess, according to requirement 3. This requirement is a stringent prohibition against many types of vacuousness in statements that are to qualify as original nomological. It is important to note that the residual is stronger, not weaker than the given statement. Statements which fail the test of exhaustiveness do so because they say too little, not too much.

(4) Original nomological statements must be *demonstrable as true*. This requirement is, in a sense, the crucial one, and it introduces a number of serious problems. As a first step, it is essential to distinguish between

analytic and *synthetic* original nomological statements. Those of the analytic variety are the basic laws of logic, and their truth is demonstrated by showing them to be theorems of the logical system to which they belong.[18] Nothing more need be said about them at this juncture.

Synthetic original nomological statements are more problematic; they must be demonstrated inductively, or, in the terms Reichenbach adopted in (1954/1976), verified inductively. Such demonstration or verification cannot be conclusive in the way that deductive demonstration is; it must consist in strong inductive confirmation or great weight of inductive evidence. Reichenbach says such statements are "verified as *practically true* at some time during the past, present, or future history of mankind" (1954, p. 11). This strikes me as a rather awkward way of saying that synthetic original nomological statements must be true, and that evidence exists which would provide warrant for asserting them, not with absolute certainty, but with a high degree of inductive reliability. Reichenbach carefully avoids saying that it must be *possible* to obtain inductive evidence, for he does not want to risk the circularity of using the concept of possibility which is yet to be explicated in his ensuing theory of physical modalities.

The fundamental aim of Reichenbach's whole theory of synthetic original nomological statements is to insure that "the assertion of such statements is necessarily left to the methods of induction" (1947, p. 360). The foregoing three requirements imposed upon synthetic original nomological statements are designed to guarantee that they cannot be verified by any other methods. (1) 'Some swans are black', for example, violates the requirement of being an all-statement. It can be verified by direct observation, without involving any process of inductive generalization. (2) 'All coins in my pocket at present are silver' violates the requirement of universality. It can be verified by complete enumeration, without any use of inductive generalization. (3) 'All unicorns are four-legged' violates the requirement of exhaustiveness. This violation indicates that the statement in question does not have the full generality that the inductive evidence warrants. Consider its contrapositive, 'All non-four-legged animals are non-unicorns'. The available inductive evidence obviously justifies a stronger generalization, namely, 'All animals are non-unicorns'. Since the latter statement *does* fulfil all requirements, *it* should be taken as original nomological. The former statement is not verified by direct inductive generalization; it is asserted on the basis of our knowledge of the emptiness of its antecedent, which follows from the latter generalization.

Requirement (4), demonstrability as true, is designed to insure that statements, whose truth cannot be certified by other means, can be established (to some suitable degree) by inductive methods. Reichenbach offers a concrete example to illustrate this condition:

Thus the statement, 'all gold cubes are smaller than one cubic mile' may be true; but since we cannot prove it to be true it is no nomological statement. We thus exclude statements that are true by chance. Such exclusion is in accordance with our maxim of reducing the definition . . to the methods of inductive evidence. . . . (1947, p. 368.)

This attempt to explain the force of the requirement of demonstrability as true is singularly unilluminating. It appears that Reichenbach is saying that the statement about gold cubes is not a nomological statement because it is 'true by chance', which seems to mean that, if true, it is not lawful – that is, it is not nomological. All this passage seems to say is that 'All gold cubes are smaller than one cubic mile' is not nomological because it is not nomological. The discussion in *Elements of Symbolic Logic* offers no further help in understanding this requirement. Reichenbach seems to be allowing his account of lawfulness to rest upon our unanalyzed ability to distinguish those generalizations that can be supported by inductive evidence from those that cannot. As we saw above, in the discussion of Goodman's 'grue-bleen paradox', one important way of formulating the fundamental problem of laws, counterfactuals, dispositions, etc., is in terms of the problem of analyzing the difference between those generalizations which can be supported by inductive evidence and those which cannot. Reichenbach's treatment of *Elements of Symbolic Logic* seems to beg this very question.

Reichenbach published his logic book in 1947, the year immediately following Goodman's first publication of this famous problem. Reichenbach was evidently unaware of it at the time he wrote the foregoing passage on the statement about gold cubes. By 1949, when he published *The Theory of Probability* (the English translation and revised edition of *Wahrscheinlichkeitslehre*), he was aware of Goodman's puzzle, and he offered what he considered an appropriate resolution of it (pp. 448–449). It is my impression that he did not take Goodman's problem very seriously, and I am not satisfied that his solution is adequate. This was, of course, earlier than the posthumous publication of *Nomological Statements and Admissible Operations* (1954).

It is my view, as mentioned above, that Reichenbach's theory of nomological statements should handle Goodman's example, 'All emeralds are grue', by saying that it violates his condition of universality on account of the positional character of the predicate 'grue'. This answer would certainly seem appropriate if we have an adequate way of establishing the fundamental positionality of 'grue' and 'bleen' which does not fall victim to Goodman's symmetry argument. I believe I have furnished the essentials of such an analysis (Salmon, 1967). Goodman's argument, and the widespread lack of consensus on its appropriate resolution should, nevertheless, make us strongly suspicious of any claim that there is an easy or obvious distinction between generalizations that can be supported by inductive evidence and generalizations that cannot.

In *Laws, Modalities, and Counterfactuals*, Reichenbach offers some further illumination. (1) He points out the obvious fact that this book is not intended to provide an analysis of inductive evidence. For answers to questions about the nature of inductive evidence we must consult his other writings on the subject, especially *The Theory of Probability*. (2) He offers some further remarks about the gold cube example:

> When we reject a statement of this kind as not expressing a law of nature, we mean to say that observable facts do not require any such statement for their interpretation and thus do not confer any truth, or any degree of probability, on it. If they did, if we had good inductive evidence for the statement, we would be willing to accept it. For instance, the statement, 'all signals are slower than or equally fast as light signals', is accepted as a law of nature because observable facts confer a high probability upon it. It is the inductive verification, not mere truth, which makes an all-statement a law of nature. In fact, if we could prove that gold cubes of giant size would condense under gravitational pressure into a sun-like gas ball whose atoms were all disintegrated, we would be willing also to accept the statement about gold cubes among the laws of nature (1954, pp. 11–12).

This passage is, if not explicitly enlightening, at least suggestive. For one thing, it serves to remind us that the statement, 'No signal travels faster than light', has been tested experimentally to some extent. The attempt to accelerate electrons to super-light velocities has resulted in velocities which asymptotically approach that of light, but never reach or exceed it. Moreover, all of the experimental evidence – of which an enormous amount exists – supporting the special theory of relativity at least indirectly supports the assertion that no signals exceed the speed of light.

The statement about gold cubes, in contrast, has never – to the best of our knowledge – been subjected to experimental test. This certainly represents a crucial difference between the two generalizations. If asked why the statement about gold cubes has never been tested, we could answer that there is no theoretical interest in performing what would inevitably be an extraordinarily expensive experiment. This calls attention to another – closely related – difference. Even if we were to verify the statement about gold cubes, it would not have any explanatory import that we can foresee.

(3) In the Appendix of *Laws, Modalities, and Counterfactuals*, Reichenbach does deal with certain important features of the inductive verification of all-statements. The most significant point is the vast difference between the establishing of a probability – even a probability approximately equal to one – by induction by enumeration, on the one hand, and the inductive verification of an original nomological statement, on the other. This distinction is signalled unmistakably by the fact that limiting frequencies (probabilities) can be posited in *primitive knowledge*, while generalizations can only be established in *advanced knowledge. There are no synthetic nomological statements in primitive knowledge.*

If one approaches both statements, 'All gold cubes are smaller than one cubic mile' and 'All signals have velocities less than or equal to that of light', from the standpoint of simple enumeration in primitive knowledge, it is hard to discern any difference. We can equally assert that all observed gold cubes fall below the size mentioned and that all observed signals have velocities that fall below the mentioned limit. Thus we might posit that the probability of a gold cube being less than one cubic mile in size is one, and the probability that a signal will have a velocity not exceeding that of light is also one. It is, of course, ludicrous – given the considerable theoretical content of both of these statements – to suppose that either could be treated in primitive knowledge. But let us keep up the fiction for a moment in order to see what more is needed.

Reichenbach considers the question of what kind of evidence is required to establish a generalization of the form 'All A's are B'. There are three main conditions to be satisfied. First, we must have evidence that $P(A, B)$ – the probability that an A is a B – is high. That is the kind of evidence we have already mentioned, and we are assuming that it exists for both of the generalizations we are discussing. Second, we must consider evidence for the contrapositive 'All non-B's are non-A'. This evidence must be considered independently, for we cannot infer that the probability $P(\bar{B}, \bar{A})$ is high from the fact that the probability $P(A, B)$ is high. From the fact that most kind people are non-philosophers (because most people are non-philosophers), we obviously cannot infer that most philosophers are unkind. Since contraposition is a valid transformation when applied to universal generalizations, but not when applied to probability relations, we must first establish inductively the high probability of the contrapositive, if we want to make a transition from a statement of high probability to an all-statement.

Let us return to the examples we were considering. There is considerable inductive evidence to support the view that all things with velocities greater than light are non-signals. This point is elaborated in some detail in Reichenbach's *Philosophy of Space and Time* (1958, sec. 23, Unreal Sequences). Although we may have some inductive evidence for the assertion that all things larger than one cubic mile are not gold cubes, one can hardly claim that any systematic effort has ever been made to find a gold cube among the objects of that size.

The third requirement – which is probably the most important in this context – deals with partitions or restrictions of the reference class. Given an all statement, 'All A's are B', we can immediately deduce the restricted statement 'All A's which are C's are B'. From the statement, 'All men are mortal', we can obviously conclude, 'All men over six feet tall are mortal'. Again – as with contraposition – this relation has no analogue for high probability statements. From the fact that it is very probable that a man of 40 will survive for another year we obviously cannot conclude that it is probable that a 40 year old man with an advanced case of lung cancer will survive for another year. Again, if we want to make the transition from a high probability statement to an all-statement, we must consider independently whether we have evidence to support the claim that the high probability will be sustained under restrictions on the reference class.[19]

Let us indulge in a little science fiction. Suppose that on a planet orbiting a star in a distant galaxy there is a race of highly intelligent, technologically sophisticated beings with an intense interest in rapid communication. Can we say with confidence that all signals sent by such beings travel with velocities not exceeding that of light? I believe an affirmative answer is justified. Suppose that such beings also have intense aesthetic admiration for gold cubes – the bigger the better. Suppose, further, that the planet these beings inhabit is richly endowed with gold. Can we say with comparable confidence that all gold cubes fabricated by these beings are smaller than one cubic mile? The answer, I believe, must be negative. This third consideration at least lends a bit of plausibility to the claim that the all-statement about signals differs from the all-statement about gold cubes in specifiable ways which bear upon their nomological status.

An additional point deserves explicit mention. In his theory of probability, Reichenbach attempted to defend the claim that Bayes's theorem provides the schema for the inductive support of scientific hypotheses by observational evidence. Although his discussion of this thesis was not easy to understand, its main outline can be discerned.[20] When one considers an all-statement from the standpoint of the requirement of demonstrability as true, it is *incorrect* to approach it in terms of induction by enumeration. Rather, the statement must be taken as one which has to satisfy the three above-mentioned requirements for transition from a high-probability-statement to an all-statement, and its evidence has to be evaluated within the schema provided by Bayes's theorem. When we ask whether an all-statement is inductively verified, we must consider the degree to which it has been subjected to observational and/or experimental testing within the bayesian framework, and how strongly all of the accumulated evidence supports it. This schema involves reference to prior probabilities, and to the possibility of explaining the data by means of alternative hypotheses, as well as any direct instantial confirming evidence.

The purpose of my discussion of Reichenbach's requirement of *demonstrability as true* has not been to argue that he has proved an entirely satisfactory account of the inductive support of scientific hypotheses. I have merely been trying to show that this requirement – Requirement 1.1 in *Laws, Modalities, and Counterfactuals* – is not trivially question-begging. I urge the reader to turn to the Appendix upon encountering the term, 'verifiably true', in this book. It is especially important not to allow the difficulties surrounding Requirement 1.1 to constitute a major obstacle to digging into the detail of Reichenbach's deep but highly technical treatment of laws, modalities, and counterfactuals.

As we approach Reichenbach's later treatment of original nomological statements, we shall be able to make good use of his explication of that concept from *Elements of Symbolic Logic*:

An original nomological statement is an all-statement that is demonstrably true, fully exhaustive, and universal (1947, p. 369).

It will be advisable to keep clearly in mind the fact that his main purpose in these discussions of original nomological statements is to lay down requirements which will *exclude* generalizations which can be verified in any way other than by inductive generalization – not in the crude sense of primitive induction by enumeration, but in the sophisticated schema furnished by Bayes's theorem. His extensive discussions of inductive verification are given in other works – mainly *The Theory of Probability* – and they are not repeated in either of these treatments of original nomological statements.

III

Having discussed some basic aspects of Reichenbach's treatment of original nomological statements in *Elements of Symbolic Logic*, I shall now offer some remarks on the content of his later monograph (1954/1976) which, it is hoped, will make the reader's job in tackling this complex body of material somewhat easier. The principal features of the earlier treatment are carried over into this later work, but the details had to be modified considerably to meet certain objections raised by critics, and other objections Reichenbach himself discovered. In this book, for example, Reichenbach takes pains to distinguish I-terms and I-requirements (terms and requirements which do not depend upon the form in which the statement is given, and which are therefore *invariant* with respect to the linguistic form) from V-terms and V-requirements (terms and requirements which do depend upon the linguistic form, and which are therefore *variant*). I commented above upon his failure to handle this distinction adequately in his earlier treatment. There are many other revisions as well; altogether they result in a great increase in the complexity of the theory.

Reichenbach's initial goal is again to define the class of original nomological statements. This class contains both analytic and synthetic statements, and in the ensuing development as well, Reichenbach attempts to construct a unified theory in which *logical* laws and modalities are treated along with *physical* laws and modalities. While there are parallels between the two types of laws and modalities, it seems to me that it would be heuristically advantageous to construct two distinct theories – pointing out interesting parallels where they exist – rather than combining the two into one. The desirability of this approach is indicated by the fact that certain terms are I-terms for synthetic statements but V-terms for analytic statements, as well as by the fact that some requirements apply to synthetic statements but not to analytic statements.[21] Throughout chapter II, in which the basic definitions are given, frequent exceptions must be introducted on account of analytic statements, and in subsequent elaborations the reader is frequently distracted by further distinctions and exceptions. In the discussion which is to follow, I shall separate analytic from synthetic statements.

1. *Analytic Statements and Operations*

Original nomological statements of the analytic type constitute basic logical laws, and they provide the source from which logical necessity originates. The definition of analytic original nomological statements in this book contains many refinements on that of *Elements of Symbolic Logic*, but no basic changes in substance. Given the definition of original nomological statements, Reichenbach then goes on to define the class of *derivative nomological statements*. This class contains all statements that can be deduced from the analytic original nomological statements. The class of derivative nomological statements *in the wider sense* contains all statements deducible in any way from the original nomological statements. Since all statements in the class of analytic original nomological statements are logically necessary, and since any statement which follows necessarily from a logically necessary statement is itself logically necessary, the class of analytic derivative nomological statements (in conjunction with the class of analytic original nomological statements) defines the concept of logical necessity. The other logical modalities can, of course, be defined in terms of logical necessity.

By imposing certain restrictions (to be discussed below) upon the class of nomological statements, Reichenbach defines the class of analytic derivative nomological statements *in the narrower sense*. These restrictions are designed to block such 'unreasonable' analytic implications as

$$a \cdot \bar{a} \supset b\,.$$

Reichenbach's aim is evidently to deal with such problems as the 'paradoxes of strict implication'. He is also attempting to define a set of 'admissible operations' which, it seems to me, might capture some of the relations which have been of interest to logicians concerned with relevant entailment. Within the realm of analytic statements – with his analysis of original nomological statements, derivative nomological statements in the broader sense, and derivative nomological statements in the narrower sense – Reichenbach claims to have provided an adequate explication of the concepts of logical law, logical modality, and admissible logical operation.

2. *Synthetic Statements and Operations*

Reichenbach explicitly remarks that his main interest lies in synthetic statements; his primary motivation is to explicate the concept of a law of nature, along with the related concepts of physical modalities and reasonable synthetic operations. From this point on, I shall confine my discussion to synthetic nomological statements, and I shall omit the qualifying adjective 'synthetic'.

In chapter III of *Laws, Modalities, and Counterfactuals*, Reichenbach presents his analysis of original nomological statements. This explication is preceded by 26 definitions, some rather abstruse, set out in chapter II. Since the reader is apt to get bogged down in a morass of terminology before getting to the requirements which must be satisfied by a statement *p*, if it is to qualify as original nomological, it might be helpful if I were to attempt an overview of the theory Reichenbach is developing. I shall therefore discuss the requirements on original nomological statements, informally showing what each of them is driving at, and giving some indication of the reasons for the preceding set of definitions. The requirements are divided into two groups; there are five I-requirements (requirements which are invariant with respect to the particular linguistic formulation of a statement) and four V-requirements (requirements which do depend upon the linguistic form in which the statement is given). I shall begin with the five I-requirements.

The I-requirements in the present book strongly resemble the requirements for original nomological statements set forth in *Elements of Symbolic Logic*. There are three important differences, and I shall comment upon them in due course. According to Requirement 1.1, the statement *p* must be verifiably true. This requirement is the counterpart of the earlier requirement that the statement be demonstrably true; I have discussed the import of this requirement above. Reichenbach's theory of nomological statements – laws of nature – is, I believe, unintelligible if this fundamental requirement is not clearly understood.

The remaining four I-requirements are designed to capture four distinct types of generality original nomological statements must possess. Three of these types of generality were mentioned in the requirements in *Elements of Symbolic logic*, namely, the statement *p* must be universal, an all-statement, and exhaustive. Requirements 1.2, 1.3, and 1.4, respectively, state these requirements. The fourth generality condition, Requirement 1.5 – that *p* must be general in self-contained factors – is new. The addition of this requirement is the first of the important differences mentioned above. In order to articulate these conditions of generality, Reichenbach introduces several distinct kinds of normal forms. The basic function of the definitions of chapter II is to characterize these normal forms and relate them to his generality requirements. A number of these definitions clarify the meanings of concepts used without explicit definition in *Elements of Symbolic Logic*.

As we saw above, the requirement of universality – Requirement 1.2 – prohibits the statement *p* from making explicit or implicit reference to individuals or restricted space-time regions. In order to carry through this requirement, Reichenbach has to introduce the notion of a *reduced statement* – roughly, one that contains no redundant parts. Chapter II contains a definition of that concept, and a procedure for carrying out a reduction. Requirements 1.2–1.5 do not require *p* to be a reduced statement, but universality does demand that *p* not be logically equivalent to any reduced statement that contains an individual-term. The basic aim is to insure that, if *p* contains an individual-term, it occurs inessentially or redundantly, and moreover, that *p* not contain any implicit reference to particular individuals or restricted space-time regions. In the new version of *Laws, Modalities, and Counterfactuals*, Requirement 1.4 (exhaustiveness) helps to carry some of this burden. But the intent of the new requirements is to formulate more precisely the condition of universality that was articulated in the earlier explication of original nomological statements.

Requirement 1.3 demands that p be an all-statement. This requirement reads exactly like one of the earlier requirements, but its meaning has been changed. This change is unmentioned by Reichenbach, but it seems rather significant; this is the second of three above-mentioned changes. In order to apply the requirement that p be an all-statement, it is necessary to transform p into prenex form – i.e., an equivalent formula whose operators are all initially placed with scopes extending to the end. In *Elements of Symbolic Logic*, p was an all-statement if and only if the *initial* operator in its prenex form was an all-operator (1947, p. 368). In *Laws, Modalities, and Counterfactuals*, the requirement is considerably weaker; it is enough that the set of operators in the prenex form contain an all-operator (Def. 13). Thus, for both versions of the requirement,

$$(x)(\exists y)[f(x) \supset g(y) \cdot h(x, y)],$$

qualifies as an all-statement, while

$$(\exists y)(x)[f(x) \supset g(y) \cdot h(x, y)]$$

would not have qualified as an all-statement under the earlier requirement, but does qualify under the present requirement. Examining the contexts, I find it hard to believe that this change was made inadvertantly.[22] At the same time, it is unlike Reichenbach to make such a significant change without comment.

Regardless of Reichenbach's intent, this modification of the all-statement requirement demands serious consideration. It seems clear that certain statements with both universal and existential operators – e.g., every atom contains at least one proton – qualify as laws. This is unproblematic because the universal operator precedes the existential operator. It is more difficult to decide whether a statement with an existential operator preceding all universal operators should be allowed to qualify as a fundamental law. I do not know of any general argument against admitting some such statements as laws, and several plausible examples come to mind. Perhaps such statements as 'There is a field of 3 °K black body radiation in which every object in the universe is immersed', 'There is a quantum of action which must occur in integral amounts in every physical interaction', or 'There is a universal constant of gravitation G which determines the force of attraction between any two bodies of given masses at a given distance from one another', might qualify as original nomological. It might be argued, of course, that some or all of these statements should be relegated to the status of derivative nomological; at the same time, the fact that the last example is the denial of 'Mach's principle' suggests that it enjoys a fundamental status. The question of whether statements which, in prenex form, have an initial existential operator should be allowed to qualify as original nomological statements is an issue to which philosophers should, I think, devote some careful deliberation.[23]

Having transformed a statement into a prenex form, we must go one step farther and transform its operand into a disjunctive normal form (D-form) in order to apply the condition of exhaustiveness – Requirement 1.4. The further concept of a residual (discussed above) is also needed. Although the present formulation of the requirement of exhaustiveness differs in detail from the previous version, the intent is clearly the same. It will be recalled from our previous discussion that the requirement of exhaustiveness is designed to insure that the general statements which qualify as original nomological should be as fully general as the inductive evidence permits. This requirement is designed to rule out certain kinds of vacuously true statements, such as those with empty antecedents, but it also blocks other types of vacuousness which are similar in principle. The requirement applies to statements whose major operations are not implications; it also applies to statements, with both kinds of operators, which suffer from vacuousness related to the existential operator rather than the universal operator. Reichenbach shows, for example, that statements of the form

$$(x)(\exists y)[f(x, y) \supset g(x, y)]$$

may be vacuous, but not because of an always-false antecedent (1954, p. 49).

Requirement 1.5, demanding that p be *general in self-contained factors*, did not appear in the earlier treatment of original nomological statements. This requirement is motivated by the fact that a statement which consists, for example, of a conjunction of an all-statement and an existential statement, can be transformed into a prenex form that fulfills the preceding three generality requirements; because of the purely existential part such statements should not however, be taken as original nomological. Consider, for instance, a statement of the form

$$(x)(\exists y)\{[f(x) \supset g(x)] \cdot h(y)\},$$

which is equivalent to

$$(x)[f(x) \supset g(x)] \cdot (\exists y)h(y).$$

'All metal objects expand when heated and whales exist' exemplifies this form. The first conjunct may certainly qualify as an original nomological statement, but the second conjunct cannot, for it is not an all-statement. We surely do not want to allow the conjunction of an original nomological statement with a purely existential statement to qualify as original nomological. This statement violates requirement 1.5.

In order to explain what is meant by 'being general in self-contained factors', Reichenbach has to introduce still another type of normal form, the conjunctive normal form or C-form. A residual of a C-form is defined analogously to a residual of a D-form. Requirement 1.5 simply demands that the residuals of the C-forms be all-statements.

We can now give a rough summary of the I-requirements for original nomological statements:

> An original nomological statement is a true statement that is inductively verifiable and possesses the following four types of generality: (1) it is universally quantified; (2) it contains no implicit or explicit reference to particular individuals or restricted regions of space-time; (3) it is as fully general as the inductive evidence permits; and (4) it cannot be decomposed into independent conjunctive parts with any part failing to be universally quantified.

This set of conditions, which is, of course, spelled out much more fully and precisely in *Laws, Modalities, and Counterfactuals*, seems to me to deserve careful consideration as a reasonable set of qualifications for statements that purport to state fundamental laws of nature.

Reichenbach goes on, however, to impose four additional V-requirements. It would be natural to ask why he bothers with such conditions. Would it not be adequate to spell out the I-requirements (as he has done) and leave it at that, admitting that any statement, regardless of the form in which it is stated, is original nomological if it fulfills the I-requirements? This would be tantamount to saying that any statement which is logically equivalent to an original nomological statement also qualifies as original nomological. Why is he unwilling to accept this approach?

The answer lies in the fact that Reichenbach is trying to characterize not only basic law statements but also reasonable propositional operations. These operations, we recall, are needed to deal with such problems as the nature of the subjunctive conditional statement. Near the end of chapter III, Reichenbach states his aim with regard to reasonable operations: "...propositional operations can be called reasonable when they stand as major operations of original nomological statements" (1954, p. 56). It is obvious that two logically equivalent statements can have different major operations, so having a particular major operation is not an I-property of statements. It is also apparent that, if one is free to perform any equivalence transformation, redundancies can be introduced in such a way as to render the major operation 'unreasonable'. Thus, a statement of the form

$$(x)[f(x) \supset g(x)]$$

is logically equivalent to one having the form

$$(x)[h(x) \vee \overline{h(x)} \supset \overline{f(x)} \vee g(x)]\,;$$

for example, 'All metals expand if heated' is logically equivalent to 'Anything which is either a raven or not a raven is either not a heated metal or it expands'. Reichenbach will obviously want to exclude the implication which is the major operation of the latter statement from the class of reasonable operations even if the implication in the former statement is considered completely reasonable. In order to secure this sort of result, Reichenbach imposes the first V-requirement, 1.6, that p must be reduced. The additional three V-requirements, 1.7–1.9, serve to strengthen the reduction requirement in ways that need not be discussed in here.

I must now mention the third important difference between Reichenbach's new I-requirements for original nomological statements and those given in *Elements of Symbolic Logic*. Although the earlier explication required that original nomological statements be fully exhaustive, the later treatment relaxes that condition somewhat. The net result is that, in some cases, an original nomological statement may fail to be exhaustive in major terms. This makes it possible to say, in some cases, that an implication is reasonable when an equivalence can be asserted as original nomological, and that an inclusive disjunction is reasonable when an exclusive disjunction can be asserted as original nomological. This weakening seems to conform to common usage.

Having defined the class of original nomological statements in terms of the nine I- and V-requirements, Reichenbach can immediately define the class of *nomological statements* (or *nomological statements in the wider sense*) as the class of statements deducible from original nomological statements. Evidently the class of nomological statements contains the class of original nomological statements, since any statement is deducible from itself. As mentioned above, the class of nomological statements defines the modality of physical necessity. The modalities of physical possibility and physical impossibility are easily definable in terms of physical necessity. Thus, the theory of nomological statements has provided an explication of the physical modalities.

The remaining task is to explicate the concept of an *admissible statement* to serve as a basis for a definition of a 'reasonable operation'. Original nomological statements are admissible. Some, but not all, statements deducible from original nomological statements are admissible. We must see what restrictions are needed to define the class of admissible statements (or *nomological statements in the narrower sense*). This definition, it turns out, will have two parts; Reichenbach introduces both *fully admissible statements and semi-admissible statements*. The class of admissible statements contains both types.

The key to an understanding of the treatment of admissible statements is a hierarchy of orders of truth and falsity. Tautologies (logically or analytically true statements) are true of order 3. Synthetic nomological statements (original or derivative) are true of order 2. Verifiably true statements which are not nomological are true of order 1. The order of falsity of a statement $\bar{p}$ is equal to the order of truth of its contradictory p. Reichenbach's general strategy is to rule out, as inadmissible or 'unreasonable', counterfactuals whose order of truth is not higher than the order of falsity of their antecedents.

Consider, for example, the statement, 'If this table salt were placed in water, it would dissolve'. The conditional statement is true of order 2, for it follows from law-statements. The falsity of the antecedent – that the salt was placed in water – is clearly of order 1, for its falsity is not a consequence of any law of nature. Similarly, the statement, 'If light were not wave-like, then such diffraction phenomena as the Poisson bright spot would not occur', is admissible. The wave-like character of light is certainly nomological, so the antecedent of the conditional is false of order 2, but the conditional – which expresses a logical relationship – is true of order 3. The statement, 'If Ohm's law were not true, then Newton's law of gravitation would not be true', is not an acceptable subjunctive conditional because its indicative counterpart, 'If Ohm's law is not true then Newton's law of gravitation is not true', violates the condition on orders of truth and falsity. The conditional itself is true of order 2, for its truth follows from the truth of Ohm's law. The antecedent is also false of order 2, for its falsity follows from the same law. This example shows once more that the class of nomological statements (in the wider sense) is too broad to provide a satisfactory explication of admissible statements and admissible operations.

In this context, as in the treatment of original nomological statements, Reichenbach is concerned with all types of propositional operations, not just implications. In order to exploit the concept of orders of truth in defining a class of reasonable operations, he extends the concept of exhaustiveness. A general statement, it will be recalled, can be transformed into a prenex form in which all of the operators stand at the front of the formula and have scopes extending to the end. The operand of all these operators can then be transformed into a disjunctive normal form. A residual of such a statement was defined as the statement that results from deletion of one or more terms of the disjunction (along with a suitable number of disjunction-symbols). The original statement was said to be exhaustive (in major terms, or in elementary terms, depending on which type of disjunctive normal form was used) if none of its residuals is true. Reichenbach now offers the following definition:

A statement p which is true of order k is *quasi-exhaustive* (in major or in elementary terms) if none of its disjunctive residuals (in major or in elementary terms) is true of an order $\geqslant k$ (1954, p. 67).

Since a statement which has no true residuals obviously has no residuals true of an order $\geqslant k$, every exhaustive statement is also quasi-exhaustive.

Fully admissible statements are now defined in terms of a set of I-requirements and a set of V-requirements. The I-requirements are:

2.1 The statement p must be deductively derivable from a set of original nomological statements i.e., p must be a nomological statement (in the wider sense).

2.2 The statement p must be quasi-exhaustive in elementary terms. (*Ibid.*)

Requirement 2.2 obviously rules out such unreasonable statements as 'If Newton's law of gravitation holds, then Ohm's law of electric circuits is true', but it admits such reasonable conditionals as 'If Newton's law of gravitation is true, then Kepler's laws of planetary motion are true'. The V-requirements for fully admissible statements are precisely the same as the V-requirements for original nomological statements. These, it will be recalled, had essentially the force of demanding that such statements be reduced – i.e., free of redundancies.

Given the set of fully admissible statements – i.e., statements which are nomological, quasi-exhaustive, and reduced – we might naturally ask whether these do not provide an adequate basis for the characterization of admissible operations. The answer, it turns out, is negative, as we can see by consideration of an example introduced earlier. If we assume that the statement, 'All and only those animals which have hearts have kidneys', is original nomological, then 'All animals with hearts have kidneys' is not original nomological, for it fails the test of exhaustiveness.[24] This latter statement is, of course, nomological in the wider sense, but that is not enough to insure that its major operation is reasonable. If this statement were admissible, the job would be done, but we can see immediately that it does not qualify as fully admissible because it is not quasi-exhaustive. The residual which results from the deletion of the term 'x is an animal and x does not have a heart and x has a kidney' is true of order 2, which is the same order as the statement itself (1954, pp. 69–70). A definition of admissible statements which ruled out this statement would be unsatisfactory, for we would want to allow it to support the counterfactual, 'If x were an animal with a heart then x would be an animal with a kidney'. Thus we need to define the class of semi-admissible statements to include such examples among the class of admissible statements.

In providing the explication of semi-admissible statements, Reichenbach introduces the further concept of *supplementability*. A statement is supplementable, roughly, if it can be made a factor in a reduced conjunction which is equivalent to a fully admissible statement. The foregoing example obviously fulfills this condition. 'All animals with hearts have kidneys' can be conjoined to 'All animals with kidneys have hearts', and this conjunction (which is reduced) is equivalent to the fully admissible statement, 'All and only animals with hearts have kidneys'. The relationship between the inclusive and exclusive disjunction can be handled in the same way. A disjunction of the form $A \vee B$ which is in fact exclusive can be written as a conjunction of the form $(A \vee B) \cdot (\bar{A} \vee \bar{B})$.

The admissible statements include those which are fully admissible or semi-admissible. The major operation of an admissible statement is an admissible operation. Such operations may occur in subjunctive or counterfactual conditionals as well as other types of statements, such as disjunctions, containing 'reasonable' operations. Reichenbach is able to prove that his admissible operations, especially the implications, have a variety of desirable characteristics. In particular, if a pair of contrary implications of the forms $A \supset B$ and $A \supset \bar{B}$ are both true neither of them can be admissible (theorem 20) (1954, p. 81).

It should be emphatically reemphasized that the foregoing characterizations of Reichenbach's concepts of original nomological statements, nomological statements, and admissible statements are rough, inaccurate, and inexact. I have been trying merely to bring out what I take to be the most significant and intuitive features of his treatment. The reader who goes through the first five chapters of *Laws, Modalities, and Counterfactuals* (the chapters that contain these analyses) will encounter 36 definitions, 13 requirements, and 21 theorems – many of which are complex and unfamiliar. It seems to me, however, that anyone who has a fairly clear grasp of the following concepts can understand with some degree of adequacy what Reichenbach is attempting to accomplish and how he goes about it.

For original nomological statements:

1. Verifiably true statement
2. Universal statement

3. All-statement
4. Exhaustive statement
5. Statement general in self-contained factors
6. Reduced statement

For admissible statements:

7. Order of truth (or falsity)
8. Quasi-exhaustive statement
9. Supplementable statement.

I have tried to say enough about these concepts to give an intuitive notion of their content and their function. Anyone who wants to achieve genuine understanding of Reichenbach's theory must – needless to say – study his precise formulations in detail and with care. My aim has been partly to provide some guidance and partly to furnish some motivation for this rather formidable task.

With the completion of chapter V, the fundamental concepts have all been introduced and explicated. The remaining chapters contain some interesting extensions, but the reader who has mastered the content of the first five chapters will have no difficulty with the remaining ones. Two topics covered in these latter chapters are worthy of mention, however, because they bear upon certain misgivings that might have been aroused concerning Reichenbach's use of his hierarchy of orders of truth. In the first place, there seem to be cases in which one wishes to deal with counterfactuals in which some, but not all, laws of nature are suspended. Such cases could give rise to counterfactuals whose truth is of order 2, and whose antecedent is also false of order 2. Chapter VI shows how the present theory can be extended to handle such cases. In the second place, it would seem that many counterfactuals which we encounter in everyday life, whose antecedents are false of order 1, are themselves true only of order 1. 'If it had not rained I would have gone on a picnic', for example, does not follow from any nomological statement alone. In chapter VIII, Reichenbach develops in detail his theory of *relative nomological statements*, within which counterfactuals of this sort are straightforwardly handled.

Any reader who does take the pains to work through this complex treatment of laws, modalities, and reasonable operations (including counterfactuals) may well react with the protest, 'But surely there must be an easier way!' Whether this is true, I do not know.[25] It seems plausible to suppose that someone with a good deal of logical acumen and a fair amount of patience could significantly streamline Reichenbach's account, without materially changing its content. I believe this is an enterprise eminently worth undertaking by someone possessing the requisite skills. I do not know of any other approach to this set of problems which gives as much promise of success as does Reichenbach's. And whether Reichenbach's analysis turns out, upon careful examination and reflection, to be satisfactory or unsatisfactory, the major concepts listed above are surely important to anyone attempting to grapple with this set of problems.

Near the beginning of chapter VIII of *Elements of Symbolic Logic*, Reichenbach sets out the program he intends to implement:

> We agree with Hume that physical necessity is translatable into statements about repeated concurrences, including the prediction that the same combination will occur in the future, without exception. 'Physically necessary' is expressible in terms of 'always'. However, we have to find out how this definition can be given without misinterpretation of the language of physics (1947, p. 356.)

Similar considerations apply to law statements and counterfactuals. Reichenbach carries through his analysis without invoking any mysterious 'connections' between properties of the sort that Hume's critique so persuasively proscribed. Chisholm, in his classic article on counterfactuals (1946), seems to suggest that the problems cannot be resolved without reinstating such 'connections'.

Reichenbach also avoids the postulation of myriad 'possible worlds' of the sort David Lewis (1973) resorts to. Although such worlds may be 'accessible' in the abstract set-theoretic sense of the term, they are forever inaccessible to us for purposes of observation. Reichenbach would have abhorred such metaphysical fantasies. Unlike many modal logicians, Reichenbach tried to provide more than abstract set-theoretical constructions; he was attempting to construct concepts that would be applicable to physics. Moreover, Reichenbach's concept of *verifiability as true* rests upon much deeper and more secure inductive foundations than does Goodman's concept of *projectability* (1965, chap. IV) with its merely linguistic analysis in terms of 'entrenchment'.

It is not my intention to try to pass judgment upon the adequacy of Reichenbach's results. I am inclined to think, as hinted above, that the whole theory could be considerably simplified. I also believe that his scale of degrees of truth is probably too crude; he, himself, suggested (1954,

p. 83) that a finer scale might be needed. What I am attempting to show is that Reichenbach's treatment deserves serious study, in spite of its difficulty and complexity. If there were a simple alternative that manifestly provides an adequate account of laws, modalities, counterfactuals, and related concepts, then we might be justified in neglecting Reichenbach's theory. Since no such alternative account does exist, it seems reasonable to take the trouble to learn what we can from Reichenbach's penetrating work. It certainly does not deserve the virtually total neglect it has received from authors currently writing on these topics.

IV

POSTSCRIPT: LAWS IN DEDUCTIVE-NOMOLOGICAL EXPLANATION – AN APPLICATION OF THE THEORY OF NOMOLOGICAL STATEMENTS

More than a dozen years ago, Henry Kyburg (1965) offered the following example of a deductive-nomological explanation:

> This sample of table salt dissolves in water, for it has a dissolving spell cast upon it, and all samples of table salt that have had dissolving spells cast upon them dissolve in water.

During the next few years, I collected a number of examples of a similar sort; here is my favorite:

> John Jones avoided becoming pregnant during the past year, for he has taken his wife's birth control pills regularly, and every man who regularly takes birth control pills avoids pregnancy. (Salmon, 1971, p. 34.)

Although one might suppose that such examples would make defenders of the D–N model of scientific explanation a bit queasy, they have not, to the best of my knowledge, dealt explicitly with them.[26]

D–N theorists have not shown that examples of the foregoing sort violate any of the requirements normally imposed upon D–N explanations. Each one is a valid deductive argument, and each has (we may assume) true premises. Each explanans embodies a true universal generalization which seems to fulfill all of the widely accepted requirements necessary to qualify it as a law.

If these examples do fulfill the traditional requirements of the D–N model, then it would seem reasonable to seek further restrictions in order to block them. D–N theorists have not, as far as I know, taken that tack either. In fact, Carl G. Hempel has explicitly rejected a suggestion I offered in connection with statistical explanation, which could easily have been incorporated into the D–N model.[27]

Since this earlier suggestion has been found unacceptable, perhaps for very sound reasons, I should like to offer another. This one has to do with the delineation of laws. In his theory of nomological statements, Reichenbach sought to define the class of "original nomological statements" as a class of basic law-statements. His requirements are more stringent than those found in more familiar treatments, for original nomological statements must fulfill a condition of *exhaustiveness* (pp. 668–670 above). Let us apply it to the generalization, "All men who regularly take birth control pills avoid becoming pregnant," which I shall symbolize as

$$(1) \qquad (x)[M(x) \cdot C(x) \supset \overline{P(x)}]$$

where '$M(x)$' means 'x is a man,' '$C(x)$' means 'x takes oral contraceptives regularly,' and '$P(x)$' means 'x becomes pregnant,' In order to apply the requirement of exhaustiveness, we transform the statement into prenex form and then transform the truth-functional part into disjunctive normal form:

$$\begin{aligned}(2) \qquad (x)[&M(x) \cdot C(x) \cdot \overline{P(x)} \vee M(x) \cdot \overline{C(x)} \cdot \overline{P(x)} \\ &\vee \overline{M(x)} \cdot C(x) \cdot \overline{P(x)} \vee \overline{M(x)} \cdot \overline{C(x)} \cdot \overline{P(x)} \\ &\vee \overline{M(x)} \cdot C(x) \cdot P(x) \vee M(x) \cdot \overline{C(x)} \cdot P(x) \\ &\vee \overline{M(x)} \cdot \overline{C(x)} \cdot P(x)]\end{aligned}$$

A *residual* of this formula is any formula which results from the deletion of one or more disjuncts along with a suitable number of adjacent disjunction symbols. A true statement is exhaustive if none of its residuals is true. Statement (1) is not exhaustive. Since there are no men who do not consume birth control pills and do become pregnant, the next to last disjunct "$M(x) \cdot \overline{C(x)} \cdot P(x)$" of (2) is always false. It can therefore be deleted without impairing the truth of the disjunction. Thus the statement, "All men who regularly take birth control pills avoid becoming pregnant," is not an original nomological statement.

The class of original nomological statements does not encompass all law-statements. After defining the class of original nomological statements, Reichenbach goes on to define the class of *nomological statements* (or *nomological statements in the wider sense*) as the class consisting of all statements which can be deduced from the class of original nomological statements. This class obviously includes all original nomological statements, since any statement entails itself. The statement, "All men who regularly take birth control pills avoid becoming pregnant," belongs to the class of nomological statements in the wider sense, for it follows from the statement, "No man ever becomes pregnant," which I take to be a law-statement.

Although the class of nomological statements in the wider sense serves as a basis for defining modalities (logical and physical), Reichenbach maintains that it is necessary to define a different class of nomological statements for the purpose of treating counterfactual conditionals and other "reasonable operations." He therefore defines the class of *admissible statements* (or *nomological statements in the narrower sense*). In order to qualify as an admissible statement, a statement must be nomological in the wider sense, and it must satisfy the additional condition of *quasi-exhaustiveness* (p. 686 above).

In order to state the condition of quasi-exhaustiveness, Reichenbach distinguishes three orders of truth and falsity. Laws of logic are true of order three, and their denials are false of the same order. Laws of nature are true of order two, and their denials are false of the same order. True contingent statements are true of order one, and false contingent statements are false of the same order. If a statement p is true of order k, and if it has any true residuals, these residuals must be true of an order lower than k if p is to qualify as quasi-exhaustive. Clearly, the statement, "All men who regularly take birth control pills avoid becoming pregnant," is nomological, so it is true of order two, but since "No men ever become pregnant" is also true of order two the truth of the residual of formula (2) is of the same order, and not a lower order. Hence, it does not qualify as admissible.[28]

Reichenbach has defined two classes of law-statements, nomological statements (in the wider sense) and admissible statements. The former class serves as the basis for the concept of physical modality, but it is too broad to furnish a useful concept of an admissible operation. Nomological statements in the wider sense do not always support counterfactuals. The class of admissible statements provides the basis for counterfactuals and other reasonable operations. When we ask whether the law-statements which appear in D–N explanations are to be taken as nomological statements in the narrower or the wider sense, it seems plausible to say that such laws should be restricted to the narrower class. The law-statements which serve as covering laws in scientific explanations, it seems fair to say, ought to support counterfactuals; at any rate, that seems to be the opinion of the main advocates of the D–N model. In that case, the "explanation" of the fact that John Jones did not become pregnant, given above, does not qualify as a D–N explanation after all, for its explanans does not contain a genuine law statement. This feature of Reichenbach's theory of nomological statements may constitute a good reason for taking another look at the theory and its rationale.

University of Arizona

NOTES

* This essay is the Foreword to Reichenbach (1976); it is reprinted here (with minor alterations) by permission of the University of California Press.

[1] A good general survey of work in modal logic can be found in G. Hughes *et al.* (1968).

[2] This point applies equally to 'inductive logic', if there is such a thing, for it cannot make a single step in this direction without the aid of concrete empirical evidence.

[3] The recent book by the Italian physicist, Aldo Bressan, *A General Interpreted Modal Calculus* (1972), is *the* striking exception. It was apparently undertaken precisely in order to deal with problems in the logical foundation of physics, and it was evidently done completely independently of the work of logicians in the main stream of modal logic. Whether Bressan's development of modal logic will shed light on the nature of physical necessity I am not prepared to say. However, this work is considered by Nuel Belnap – a leading authority in the field of modal logic – to represent a breakthrough of significant proportions. See Belnap's Foreword to this book.

[4] David Lewis (1973, pp. 5–7) makes explicit use of physical laws in characterizing physical possibility, but he provides no analysis of laws.

[5] G. H. von Wright is a distinguished contributor; see his (1968).

[6] A. N. Prior's (1957) and (1968) are well-known works in this area.

[7] See A. N. Prior (1967) for a survey.

[8] See Hempel (1965, pp. 291–295) for a 1964 Postscript to Hempel *et al.* (1948).

[9] The choice of this particular example is unfortunate. My dictionary makes the color green a defining property of emeralds, thus rendering the first of these two generalizations analytic. Synthetic examples are, however, easily devised.

[10] See the following exchange between Goodman and Carnap: Goodman (1946), Carnap (1947a), Goodman (1947a), Carnap (1947b).

[11] It is a little known fact that Bertrand Russell (1948, pp. 404–405) actually stated and offered a plausible resolution of this problem. My attempt at resolution, which I still consider fundamentally sound, is in Salmon (1963, pp. 256 ff).

[12] In the 1964 Postscript to a classic article on confirmation, Hempel (1965, pp. 47–51) acknowledges the significance of Goodman's problem as it bears upon confirmation. He might have noted its bearing on explanation as a corollary.

[13] Reichenbach cites a similar example in (1954, p. 11).

[14] For further references see Sosa (1975), a recent anthology of essays on this topic. In this

book Reichenbach's final chapter of *Elements of Symbolic Logic* rates one brief mention in a footnote, while *Nomological Statements and Admissible Operations* is totally ignored.

[15] See Hempel (1954).

[16] Reichenbach uses the term 'operation' for negation, conjunction, disjunction, implication, and equivalence. These are often called 'connectives' in the contemporary literature. I shall follow his terminology here, for as we are about to see, he introduces the term 'connective operation'. It would be awkward to alter his terminology and refer to 'connective connectives'.

[17] Reichenbach uses a bar over the top of a formula as a symbol for negation.

[18] The 'basic laws of logic' – as represented by analytic original nomological statements – will not correspond with those statements usually taken as axioms of standard logical systems.

[19] Obviously the term '*C*' which effects the restriction must not contain explicit or implicit reference to '*B*'. '*C*' must function somewhat as a 'place selection' in the sense of Richard von Mises.

[20] I have tried to clarify this approach in Salmon (1967), Chap. VII.

[21] See, for example, Definitions 15–16 and Requirement 1.5.

[22] See Definition 14. At no place in Chapter III does Reichenbach offer an example of an original nomological statement with an initial existential operator.

[23] Karl Popper would, I should think, deny even derivative nomological status to such statements, but he has given little attention to statements of mixed quantification.

[24] I mentioned above that *some* implications can qualify as original nomological even if the corresponding equivalence is true, but the present example is not one which qualifies.

[25] See Evan K. Jobe's essay below.

[26] An especially nice example is exhibited in the Richter cartoon which is reprinted as the frontispiece in Salmon (1971).

[27] In Salmon (1971) p. 49, I formulate the *requirement of maximal class of maximal specificity* and suggest that it replace Hempel's *requirement of maximal specificity*. On p. 50, I point out that it can be used in conjunction with deductive models to block just such examples as we are here discussing. Hempel (1977) rejects this alternative requirement, but he does not say how he intends to handle these troublesome examples.

[28] The situation is somewhat more complicated, since statements which satisfy the condition of quasi-exhaustiveness are fully admissible. There is, in addition, a class of semi-admissible statements (see pp. 687–688 above, defined by still another condition (supplementability). The class of admissible statements contains both the fully admissible and the semi-admissible statements. Our example is not affected by this additional complication. For details see Reichenbach (1976), chap. V.

BIBLIOGRAPHY

Anderson, A. *et al.*: 1975, *Entailment, The Logic of Relevance and Necessity*, Vol. I, Princeton University Press, Princeton.

Bressan, A.: 1972, *A General Interpreted Modal Calculus*, Yale University Press, New Haven.

Carnap, R.: 1936–1937, 'Testability and Meaning', *Philos. Sci.* **3**, 420–471; **4**, 2–40. Reprinted (with corrigenda and additional bibliography) by the Graduate Philosophy Club, Yale University, New Haven, 1950.

Carnap, R.: 1947a, 'On the Application of Inductive Logic', *Philos. Phenomenol. Res.* **8**, 133–147.

Carnap, R.: 1947b, 'Reply to Nelson Goodman', *Philos. Phenomenol. Res.* **8**, 461–462.

Chisholm, R.: 1946, 'The Contrary-to-Fact Conditional', *Mind* **55**, 289–307.

Goodman, N.: 1946, 'A Query on Confirmation', *J. Philos.* **63**, 383–385.

Goodman, N.: 1947a, 'On Infirmities of Confirmation Theory', *Philos. Phenomenol. Res.* **8**, 149–151.

Goodman, N.: 1947b, 'The Problem of Counterfactual Conditionals', *J. Philos.* **44**, 113–128.

Goodman, N.: 1965, *Fact, Fiction, and Forecast*, 2nd ed., Bobbs-Merrill Co., Indianapolis. (1st ed. Harvard University Press, Cambridge Mass., 1955.)

Hempel, C. *et al.*: 1948, 'Studies in the Logic of Explanation', *Philos. Sci.* **15**, 135–175. Reprinted, with Postscript in Hempel, 1965.

Hempel, C.: 1954, 'A Logical Appraisal of Operationism', *Scientific Monthly* **79**, 215–220. Reprinted in Hempel 1965.

Hempel, C.: 1965, *Aspects of Scientific Explanation*, The Free Press, New York.

Hempel, C.: 1966, *Philosophy of Natural Science*, Prentice-Hall, Inc., Englewood Cliffs, N.J.

Hempel, C.: 1977, 'Nachwort 1976' in *Aspekte wissenschaftlicher Erklärung*, Walter de Gruyter, Berlin, pp. 98–123.

Hughes, G. *et al.*: 1968, *An Introduction to Modal Logic*, Methuen and Co., London.

Kyburg, H.: 1965, 'Comments," *Philosophy of Science* **32**, 147–151.

Lewis, D.: 1973, *Counterfactuals*, Harvard University Press, Cambridge, Mass.

Mates, B.: 1972, *Elementary Logic*, 2nd ed., Oxford University Press, New York.

Prior, A.: 1957, *Time and Modality*, Clarendon Press, Oxford.

Prior, A.: 1967, 'Logic, Modal', in P. Edwards (ed.), *The Encyclopedia of Philosophy*, Macmillan Publishing Co., New York.

Prior, A.: 1968, *Past, Present and Future*, Clarendon Press, Oxford.

Reichenbach, H.: 1947, *Elements of Symbolic Logic*, The Macmillan Co., New York.

Reichenbach, H.: 1949, *The Theory of Probability*, University of California Press, Berkeley and Los Angeles.

Reichenbach, H.: 1954, *Nomological Statements and Admissible Operations*, North-Holland Publishing Co., Amsterdam.

Reichenbach, H.: 1958, *The Philosophy of Space and Time*, Dover Publications, New York.

Reichenbach, H.: 1976, *Laws, Modalities, and Counterfactuals*, University of California Press, Berkeley and Los Angeles. Reprint, with a new Foreword by W. Salmon, of Reichenbach (1954).

Russell, B.: 1948, *Human Knowledge, Its Scope and Limits*, Simon and Schuster, New York.

Salmon, W.: 1963, 'On Vindicating Induction', *Philos. of Sci.* **30**, 252–261. Also published in H. Kyburg *et al.* (eds.), *Induction: Some Current Issues*, Wesleyan University Press, Middletown, Conn., 1963.

Salmon, W.: 1967, *The Foundations of Scientific Inference*, Pittsburgh University Press, Pittsburgh.

Salmon, W.: 1971, *Statistical Explanation and Statistical Relevance*, University of Pittsburgh Press, Pittsburgh.

Sosa, E. (ed.): 1975, *Causation and Conditionals*, Oxford University Press, New York.

von Wright, G.: 1968, *An Essay in Deontic Logic and the General Theory of Action*, North-Holland Publishing Co., Amsterdam.

EVAN K. JOBE

REICHENBACH'S THEORY OF NOMOLOGICAL STATEMENTS

My primary aim in this paper is to discuss and evaluate Reichenbach's theory of nomological statements. The importance of this topic is evident from the fact that Reichenbach offers the concept of a synthetic nomological statement as an explication for the recalcitrant notion of a law of nature. An early version of this theory was presented in the last chapter of his *Elements of Symbolic Logic* (1947), but I shall here deal with the considerably revised version contained in his *Nomological Statements and Admissible Operations* (1954). All parenthetical page citations in the sequel (unless otherwise identified) will refer to this book.

In Reichenbach's presentation the theory of nomological statements is embedded within his analysis of what he calls 'reasonable operations', a distinct but related topic of considerable interest. Some care is required in sorting out those considerations that are actually essential to the theory of nomological statements as such. Since doing this is one of the essential goals of this paper, it will be necessary to make some reference to certain aspects of the theory of reasonable operations. I shall therefore begin by touching very briefly on the theory of reasonable operations. Then I shall move on to present a summary of the theory of nomological statements. The last part of the paper will be largely devoted to a critical discussion of selected aspects of the theory of nomological statements, and will include an evaluation of the theory as a whole.

I. REICHENBACH'S ANALYSIS OF REASONABLE OPERATIONS

It is a notorious fact that the language we speak every day is not a truth-functional one. This is especially obvious in the case of subjunctive and counter-factual conditionals. But the fact is that the ordinary connectives 'if', 'if and only if' – and their equivalents – are rarely truth-functional, while the connective 'or' also frequently fails to be so. This is of course a major reason why the usual 'translations' of ordinary sentences into logistic symbolism are so often distressingly inadequate. It

W. C. Salmon (ed.), Hans Reichenbach: Logical Empiricist, 697–720.

does not follow that there is anything 'wrong' with either the connectives of ordinary language or those of logistic. The latter, however, are gratifyingly clear, while the truth conditions for sentences involving the former are not so easy to articulate. Reichenbach's work on this topic is essentially an investigation into the conditions under which a true sentence in the language of standard predicate logic is also true when its major sentential connective is reinterpreted with the force of the corresponding connective of ordinary English. When this is the case Reichenbach says that the connective is being used in a 'reasonable' sense or is a 'reasonable operation'. The analysis of reasonable operations thus amounts to an explication of the meanings of the connectives of ordinary language.

One of the key concepts utilized in Reichenbach's theory of reasonable operations is the concept of a nomological statement. The total class of nomological statements comprises the laws of nature together with all analytic statements. Since the notion of a law of nature could hardly be taken as sufficiently clear for use in further explication, he proceeded to lay down a set of requirements to delineate the class of nomological statements. It is worth noting that his analysis of reasonable operations represents a substantial piece of work beyond the particular explication of the concept of a nomological statement. The two lines of investigation can to a large extent be evaluated separately.

Reichenbach found no simple set of conditions which are both necessary and sufficient for the reasonableness of an operation. He did, however, develop alternative sets of sufficient conditions for reasonableness. In barest outline his procedure was as follows. By laying down certain restrictions he demarcates a subclass of nomological statements which he calls 'admissible statements'. The major operations of these statements, the so-called 'admissible operations', are reasonable operations. (Actually, admissible statements may be either 'fully admissible' or only 'semi-admissible', depending on precisely what set of restrictions they satisfy. Reasonableness turns out to be a matter of degree, and semi-admissible operations are not as reasonable as fully admissible operations.) Another kind of reasonableness – one that is particularly prominent in everyday talk – is attained by conditionals and biconditionals that are not even nomological but which follow from some admissible statement taken in conjunction with certain background knowledge, assuming certain additional conditions are met. The vast majority of subjunctive and counterfactual conditionals actually occurring in everyday speech appear to correspond to statements falling in this category. Finally, there are what Reichenbach calls "counterfactuals of non-interference" – statements such as 'If John had studied more he would still have flunked'. Reichenbach analyzes these without making use of the concept of a nomological statement.

The details of Reichenbach's analysis of reasonable operations are quite complex, but perhaps not surprisingly so. What seems more surprising at any rate is the measure of success that he appears to have achieved in this difficult enterprise. This work provides impressive corroboration for his faith in the usefulness of logical tools for the clarification of distinctions inherent in our ordinary language.

II. SUMMARY OF REICHENBACH'S THEORY OF NOMOLOGICAL STATEMENTS

In his theory of nomological statements Reichenbach attempts to demarcate in a precise way the class of nomological statements, a body of statements that is to be coextensive with the class of laws of nature together with all logical truths of first-order logic with identity. This is of course to be accomplished without making use of such notions as physical possibility or necessity, and without the use of counterfactuals or other non-truth-functional locutions of ordinary language. It is part of Reichenbach's total program to explicate the physical modalities and counterfactuals in terms of the concept of nomological statement. Any essential utilization of such concepts in his analysis of nomological statements would therefore be viciously circular.

Reichenbach's procedure is first to lay down requirements defining a class of *original* nomological statements. The class of nomological statements in general is then defined as comprising all those statements that are derivable from the class of original nomological statements. Since a logically true statement is derivable from any statement whatsoever, any such statement is a nomological statement. It is obvious, therefore, that the requirements laid down for *original* nomological statements are relevant to the demarcation of nomological statements in general only insofar as they restrict the body of *synthetic* nomological statements. In substance, then Reichenbach's theory is a *theory* of synthetic nomological statements. It furnishes an explication for the concept of a law of nature. On the other hand, in the theory of reasonable operations it is important to consider analytic statements, since not all analytic statements are *admissible* statements. Since I am here concerned only with criteria for demarcating nomological statements as such, I shall in the sequel confine my consideration to synthetic statements.

Reichenbach points out that the class of nomological statements is well-defined only on the basis of a given 'rational reconstruction' of natural language, but he does not specify precisely what this should amount to. It appears, first of all, to involve a transcription of the natural language into the language of first-order predicate logic with identity, since his definitions and rules are usually stated in such a way as to be directly applicable to such a language rather than to ordinary English. It also appears to involve the formulation of a set of decisions determining which terms are to be taken as primitive, which are to be defined, and what the definition of each of the terms in the latter class is to be. Such a reconstruction could of course hardly be carried out in actual practice except in a piecemeal manner, but it may be assumed as an ideal prerequisite for the theory of nomological statements. Any such regimentation of natural language is bound to diverge more or less from ordinary usage at some points, but Reichenbach points out that we are not wholly without some criteria of adequacy for such a reconstruction. He says:

> If a language is to be rationally reconstructed, it is advisable not merely to look for definitions that coincide extensionally with usage of terms, but also to adjust definitions to the usage of such categories as 'Law of nature', 'analytic', etc. Only the total reconstruction, as a whole, can be judged as adequate. For instance, defining 'human being' as 'featherless biped' would correspond extensionally to usage; but it would contradict such statements as 'featherless bipeds are not necessarily human beings', which we also find among linguistic usage. (p. 35)

I shall now turn to the substance of Reichenbach's analysis. Reichenbach explains what may be regarded as the principal strategy behind his formulation of qualifications for nomological statements. He says:

> What we are looking for in a definition of nomological statements is not a method of verifying such statements, but a set of rules which guarantee that inductive verification is actually used for these statements, in as much as they are synthetic. The requirements laid down in the definition of nomological statements, in fact, represent a set of restrictions which exclude from such statements all synthetic forms that can be verified without

inductive extension. More than that, the restrictions single out, among inductively verified statements, a special group of all-statements associated with a very high degree of probability; and they are so constructed that they allow us to assume that these all-statements are true without exceptions. Merely factual truth, though in itself found by inductive inference, is thus distinguished from nomological truth in that it does not assert an inductive generality; and the requirements introduced for nomological statements are all governed by the very principle that factual truth must never be sufficient to verify deductively a statement of this kind. (p. 13)

As noted earlier, a nomological statement is to be any statement that is derivable from the set of *original* nomological statements. Altogether there are nine requirements that must be satisfied by any original nomological statement. The first five requirements (p. 40) are said to be invariant, since each is such that if a given statement satisfies it then any statement logically equivalent to the given statement also satisfies it. The last four requirements (p. 48) are said to be variant ones, since whether a given statement satisfies them will depend on its form. A statement *p* is an original nomological statement if and only if it meets the following requirements:

Requirement 1.1. The statement *p* must be verifiably true.
Requirement 1.2. The statement *p* must be universal.
Requirement 1.3. The statement *p* must be an all-statement.
Requirement 1.4. The statement *p* must be unrestrictedly exhaustive in elementary terms.
Requirement 1.5. If the statement *p* is synthetic, it must be general in self-contained factors.
Requirement 1.6. The statement *p* must be reduced.
Requirements 1.7–1.9. (I shall not state these rather complex requirements here, but their general nature will be discussed below.)

I shall now attempt to explain the content of each of these requirements and discuss what I take to be the rationale behind each one.

Requirement 1.1 The statement *p* must be verifiably true.

In addition to insuring that nomological statements are true statements this requirement has another crucial role that can be understood only in the light of one important result achieved by Requirements 1.2 through 1.5 taken together. The latter requirements jointly exclude statements that can be *conclusively* verified. If, then, a statement that meets these latter requirements is also true and verifiable, it is what Reichenbach calls an 'inductive generality' and is therefore capable of performing the distinctive functions of a law.

It might be thought that any true generalization would be verifiably true in the sense that it could in principle be inductively confirmed to any desired degree on the basis of a sufficient number of instances. It is now widely recognized, however, that this is not the case. As Nelson Goodman (1955, pp. 29f) says:

> . . . in the case of sentences like
>
> Everything in my pocket is silver
>
> or
>
> No twentieth-century president of the United States will be between 6 feet 1 inch and 6 feet $1\frac{1}{2}$ inches tall,
>
> not even the testing with positive results of all but a single instance is likely to lead us to accept the sentence and predict that the one remaining instance will conform to it . . .

Reichenbach is not, however, free to interpret the term 'verifiably true' in the normal sense of 'true and verifiable (highly confirmable)' but only in the peculiar sense of 'true and verified (highly confirmed) at some time or other'. He says:

> The word 'verifiable' includes a reference to possibility. Since physical possibility is a category to be defined in terms of nomological statements, it would be circular to use, in the definition of such statements, this category. For this reason, I defined the term 'verifiably true' as meaning verified at some time, in the past or in the future. It has been argued against this definition that there may be laws of nature which will never be discovered by human beings. In the present investigation I shall show that the latter statement, indeed, can be given a meaning, and that we can define a term *verifiably true in the wider sense* which covers this meaning. But in order to define this term, I shall begin with the narrower term, and proceed later to the introduction of the wider term . . . (pp. 12–13)

Using the narrower interpretation of 'verifiably true' it is clear that the term 'nomological statement' cannot be an explicans for 'law of nature' but rather for 'law of nature highly confirmed at some time'. Using this class of actually ascertained laws as a base Reichenbach attempts to define a wider sense of 'verifiably true' which, when utilized in Requirement 1.1, will result in a class of nomological statements corresponding to all existing laws of nature. This attempt to broaden the notion of 'verifiably true' without the utilization of modal concepts is perhaps the most difficult aspect of Reichenbach's analysis. I shall therefore defer further discussion of it until a later point in the paper.

Requirement 1.2. The statement *p* must be universal.

The definition of 'universal statement' is the following:

A synthetic statement is *universal* if it cannot be written in a reduced form which contains an individual-term. (p. 33)

When this definition is unpacked by reference to the definitions of 'can be written' (p. 19) and 'reduced' (p. 21), it amounts to defining a (synthetic) universal statement as a statement that does not contain an essential occurrence of an individual-term. Reichenbach's definition of 'individual-term' is as follows:

An *individual-term* is a term which is defined with reference to a certain space-time region, or which can be so defined without change of meaning. (p. 32)

Reichenbach makes it clear that the class of individual-terms is to include proper names and definite descriptions designating concrete objects or space-time regions, and also such predicates as 'terrestrial', 'solar', etc. I shall postpone a closer examination of the adequacy of this definition to a later point.

The purpose of this requirement is clearly to exclude statements that at least in part merely describe the characteristics of particular individuals. Such exclusion certainly has intuitive support in that we tend to feel that laws of nature should describe 'the way things work' in general and should not be dependent on the existence or characteristics of this or that particular object. Such exclusion is also desirable in that many statements involving individual reference can be conclusively verified – e.g., 'all the persons now in this room are blonde'. It will be recalled that it is a function of Requirements other than 1.1 that they jointly eliminate statements that can be conclusively verified. There are in fact statements which are obviously not laws of nature but which are ruled out only by this requirement. Thus, to vary one of Reichenbach's examples, if it is true that Mary is as tall as Jane, then the statement 'For every *x*, *x* is as tall as Mary if and only if *x* is as tall as Jane' meets all of the requirements except Requirement 1.2.

Requirement 1.3. The statement p must be an all-statement.

Reichenbach calls a synthetic statement an *all-statement* if and only if the quantifier set of its 'D-form' or 'C-form' contains at least one universal quantifier (p. 29). These forms are logically equivalent ways of writing a statement and require, among other things, that all of the quantifiers in them be non-redundant and in prenex position. The requirement thus rules out purely existential statements such as 'Copper exists', 'Something is blue', and 'Somebody loves somebody'. This is not only in accord with our intuitions concerning laws of nature, but it also rules out another type of statement that is subject to conclusive verification.

Requirement 1.4. The statement p must be unrestrictedly exhaustive in elementary terms.

This requirement actually involves two seemingly diverse components which Reichenbach manages to convey in a unified manner by use of a certain technical device. For purposes of exposition, however, it is probably best to consider these components separately. In order to meet this requirement a statement must first of all be exhaustive in elementary terms. This is in general quite a complex concept, but it may suffice for present purposes to illustrate the idea by means of a particularly simple example. Consider the statement 'All metals that are heated expand'. If for simplicity we ignore the temporal aspects of this statement we may symbolize it roughly as

$$(1) \qquad (x)[(Mx \,\&\, Hx) \supset Ex]$$

Assuming that 'M', 'H', and 'E' are primitive or 'elementary' terms, the D-form of (1) is given by

$$(2) \qquad \begin{aligned} &(x)[(Mx \,\&\, Hx \,\&\, Ex) \vee (Mx \,\&\, -Hx \,\&\, Ex) \\ &\vee (-Mx \,\&\, Hx \,\&\, Ex) \vee (-Mx \,\&\, -Hx \,\&\, Ex) \\ &\vee (Mx \,\&\, -Hx \,\&\, -Ex) \vee (-Mx \,\&\, Hx \,\&\, -Ex) \\ &\vee (-Mx \,\&\, -Hx \,\&\, -Ex)] \end{aligned}$$

The result of cancelling one or more of the disjuncts in (2) is called a 'disjunctive residual in elementary terms'. A statement is said to be exhaustive in elementary terms provided that it is verifiably true but none of its disjunctive residuals in elementary terms is verifiably true. Statement (1) is, under the assumed interpretation, a verifiably true statement, and so of course is (2), which is logically equivalent to (1). The question of exhaustiveness is then decided by determining whether (2) remains verifiably true upon cancellation of one or more of its disjuncts. Now, since we know that there are things that are metal and heated and expand, also things that are not metal and heated and expand, etc., it is clear that the cancellation of any of the disjuncts would result in a false statement. None of the disjunctive residuals is verifiably true. Therefore, 'All metals that are heated expand' is exhaustive in elementary terms. Suppose, however, that 'M' stands for 'magnetized', 'H' stands for 'hotter than 1000°', and 'E' stands for 'evaporates'. Since it is known that heat destroys magnetization it is verifiably true that there are no bodies that are both magnetized and hotter than 1000°. Since (1) has a verifiably false antecedent for each value of x, (1) is itself verifiably true, and so therefore is (2). In the latter, however, it is clear that cancelling the always-false first disjunct 'Mx & Hx & Ex' will also result in a true statement. Therefore, the statement 'All magnetized bodies hotter than 1000° evaporate', while true, is not exhaustive in elementary terms. This statement meets all of the requirements except Requirement 1.4.

The essential function of the requirement of exhaustiveness in elementary terms is its role in excluding statements that contain one or more true, closed, existential clauses – such clauses as 'Copper exists', 'Something is blue', etc. This exclusion is accomplished by the joint action of Requirements 1.1, 1.4, and 1.5, as Reichenbach proves (pp. 43–48). (Actually, Reichenbach's discussion is equivalent to such a proof, although he does not explicitly refer to Requirement 1.1 – he merely utilizes the fact that a statement of the type under discussion is a *true* statement.) The need for the exclusion of such clauses can be seen by considering a particular example. Consider the law 'Every magnet contains a north pole', which we might symbolize as:

$$(3) \qquad (x)[(Mx \supset (\exists y)(Ny \ \& \ Cxy)].$$

It should be stressed that the existential clause in (3), which contains a free variable and is thus not closed, is not the type of clause under consideration. But from (3) and the true statement 'Something is blue' –

in symbols '$(\exists z)Bz$' – we can derive

(4) $(x)\{Mx \supset [(\exists y)(Ny \,\&\, Cxy) \,\&\, (\exists z)Bz]\}$.

Now, (4) satisfies all of the requirements for an original nomological statement except Requirement 1.4. Intuitively (4) may seem unsatisfactory as a nomological statement, representing as it does an admixture of lawlike and merely 'accidental' components. More decisively, however, from (4) we can derive

(5) $(x)[Mx \supset (\exists z)Bz]$.

Now, if (4) were an original nomological statement, (5) would be nomological. But (5) can surely be rejected as a law of nature both on intuitive grounds and also because it follows from such accidentally true statements as 'Something is blue' or 'This pencil is blue'. It is clearly verifiable without the use of inductive extension. Such statements as (4) must therefore not be allowed to count as original nomological statements. Requirement 1.4 plays an essential role in excluding such statements from that category.

We may now turn to the consideration of the other component of Requirement 1.4. The requirement that p be *unrestrictedly* exhaustive is stronger than a requirement that p be exhaustive in elementary terms. What is additionally required is that no predicate occurring essentially in p may be such that its extension is *verifiably* confined to a certain restricted space-time region (pp. 36–38). That is, for such a predicate P it must not be the case that there is a restricted space-time region R such that the extension of P is verifiably confined to R. This provision is introduced to exclude a class of statements involving what Reichenbach regards as covert reference to particular individuals even though the statements meet the criterion of universality laid down in Requirement 1.2. To cite Reichenbach's example (p. 36), the statement 'All stars which H. v. Helmholtz saw were at least of the 11th magnitude' might be known to be true in view of the limitations of nineteenth-century telescopy. This statement is of course clearly non-universal, as is the statement 'H. v. Helmholtz was the first man to see a living human retina'. From these two non-universal statements, however, we can derive the statement 'All stars seen by any man who saw a living human retina before any other man saw one were at least of the 11th magnitude'. Even though the truth of this statement depends on certain facts about a particular individual it nevertheless seems to be universal by the criteria utilized in Requirement 1.2. The statement is, however, excluded by Requirement 1.4. The reason is that it contains the predicate 'man who saw a living human retina before any other man saw one', which is only true of H. v. Helmholtz and which therefore has its extension verifiably included within the particular space-time region occupied by Helmholtz during his lifetime.

Requirement 1.5. If the statement p is synthetic, it must be general in self contained factors.

The precise unpacking of the concept of generality in self-contained factors is fairly complex, but the requirement is that if the statement p is equivalently rewritten as a conjunction of two logically independent statements then each of these is an all-statement in the sense of 'all-statement' used in Requirement 1.3.

It has already been noted that Requirements 1.1, 1.4, and 1.5 conjointly insure the exclusion of true, closed, existential clauses. The reader may wonder why false, closed, existential clauses are not likewise undesirable. After all, such a clause preceded by a negation sign might also constitute a merely accidentally true statement and might seem to pose the sort of threat illustrated above. The crucial difference is that if an existential clause is true it can be verified without the use of inductive extension. It is then both accidentally and verifiably true. On the other hand, the falsity of an existential clause is equivalent to the truth of a universal clause. If this universal clause is not nomological it will not in general be verifiable, and so cannot affect the verifiability of a statement in which it occurs. If it is nomological it will not constitute a source of 'contamination' for the statement in which it occurs.

In the last paragraph I have spoken rather loosely of the 'exclusion' of true, closed, existential clauses. Actually, the matter is somewhat more complicated in view of two facts. First, a statement that contains an occurrence of such a clause may have a logically equivalent version in which such a clause does not occur. Clearly, the second statement is not less objectionable than the first. Second, every statement has a logically equivalent version that does contain such a clause. For example, any statement 'p' can be equivalently rewritten as 'p & $[(\exists x)Fx \vee -(\exists x)Fx]$'.

Here the occurrences of the existential clauses are redundant and therefore harmless. Roughly, the occurrence of a clause within a synthetic statement is redundant if its cancellation together with the cancellation of adjacent binary connectives results in a statement logically equivalent to the original statement – also, clauses occurring within redundant clauses must be counted as redundant. For a fuller discussion the reader may consult Reichenbach (pp. 20–21). Any statement, then, that is logically equivalent to a statement containing a non-redundant occurrence of a true, closed, existential clause is to be proscribed. One more complication threatens. A statement such as '$(x)(Ax \supset Dx)$', appears to be acceptable by the test just stated. But suppose that the particular logical reconstruction in which this statement occurs happens to have among its stock of definitions the following rather pointless but perfectly proper one: '$Dx =_{\text{df}} Bx \,\&\, (\exists y)Cy$', where '$C$' is a predicate known to have at least one instance. Clearly, then, the cited statement is objectionable after all, and a further qualification is needed. What is really meant, then, when one says that Requirements 1.1, 1.4, and 1.5 jointly insure the exclusion of true, closed, existential clauses is that these requirements jointly rule out any statement which, when expanded in primitive terms, is logically equivalent to a statement that contains a non-redundant occurrence of a true, closed, existential clause.

It should perhaps be mentioned that many occurrences of such clauses are quite innocuous. The crucial fact is that some are not innocuous, and it does not appear that any such occurrences are actually needed in formulating what we would ordinarily consider a law of nature. It therefore seems permissible to ban all of them. This is, in effect, what Reichenbach does.

Requirement 1.6. The statement p must be reduced.

The process of statement reduction outlined by Reichenbach (pp. 19–21) is essentially a method of eliminating from a statement redundant clauses, vacuous quantifiers and non-essential occurrences of predicates, so as to result, when applicable, in a more economically expressed logically equivalent version of the given statement. For reasons which will be explained immediately below in connection with Requirements 1.7–1.9, Requirement 1.6 has no essential role in the theory of nomological statements. It does play an important role in the theory of reasonable operations, however, since it expresses one of a number of necessary conditions for a statement's being a fully admissible statement.

Requirements 1.7–1.9.

I have refrained from stating these rather prolix requirements since their inclusion among the requirements for original nomological statements has no effect on the membership of the class of nomological statements as such. That is, these requirements are irrelevant to the question as to which statements qualify as laws of nature according to this explication. These requirements exclude statements harboring certain more subtle forms of redundancy than are taken account of by Requirement 1.6. Each statement that meets these requirements, however, has exactly the same set of logical consequences as its more inflated but logically equivalent counterparts that are excluded. Thus, the set of statements derivable from the set of original nomological statements – the class of nomological statements – is not affected by these requirements. Like Requirement 1.6, however, they are among the requirements for a statement's being a fully admissible statement, and so they have an important place in the theory of reasonable operations.

The class of original nomological statements as defined by Requirements 1.1 through 1.9 turns out to be a subset of the class of fully admissible statements. One might suppose this result to be of considerable importance as insuring that at least the 'fundamental' laws of nature have the form of 'reasonable' statements. It is a mistake, however, to think that original nomological statements correspond to 'fundamental' laws in any scientifically relevant sense of that term. The statement 'Alcohol is more intoxicating than water' is, when suitably transcribed, an original nomological statement, although hardly of central importance in the scheme of things. On the other hand, any really 'fundamental' law has an infinity of what would ordinarily be thought of as logically equivalent versions that fail to be original nomological statements, simply because they are stated in unnecessarily complex ways. The notion of *original* nomological statement is therefore best understood as a technical device whose importance lies in the role it plays in demarcating the class of nomological statements as such. Requirements 1.7–1.9 do not contribute anything toward the functioning of this device and are therefore superfluous within the theory of nomological statements.

It should perhaps be remarked that while Requirements 1.7–1.9 have an important role within Reichenbach's theory of reasonable operations this role is not that of helping to insure that original nomological statements as such belong to the class of fully admissible statements. Any *nomological* statement meeting Requirements 1.6–1.9 together with one additional special requirement qualifies as a fully admissible statement. The fact that *original* nomological statements are fully admissible is therefore as inconsequential for the theory of reasonable operations as it is for the theory of nomological statements.

III. COMMENTARY AND EVALUATION

In this final section I shall discuss certain aspects of the theory of nomological statements that seem to me to call for special elucidation or criticism. The critical commentary will result in an assessment of the theory as a whole.

1. *Dependency on Rational Reconstruction*

As Reichenbach points out, the question as to whether a given statement of natural language is nomological can, strictly speaking, only be answered relative to a particular rational reconstruction of the language. While this feature of Reichenbach's theory is not in itself objectionable, it does seem desirable that any two reconstructions which on other grounds are equally adequate should not result in different verdicts for a given statement of natural language. Actually, I believe such variability is probably negligible and in any case is less than Reichenbach himself supposed. Using the term 'polar bear' as an example, he points out that in one reconstruction one might define this term in such a way as to make reference to the polar regions of the Earth. In such a reconstruction 'polar bear' would be an individual-term, and so no statement containing it could be nomological. In another reconstruction, however, 'polar bear' might be defined by means of purely qualitative predicates. In such a reconstruction, Reichenbach thinks, a suitable generalization about polar bears would be nomological (p. 34). But here I think one must consider a point that has been advanced by J. J. C. Smart (1963, pp. 50–58). To relate Smart's point to the present example, suppose we are considering the statement 'All polar bears are white', and suppose we have defined 'polar bear' in terms of some likely set of purely qualitative predicates P_1, $P_2, \ldots P_n$ (not including 'white') of the sort commonly found in biological discourse. Now, could we really have any very good reason to suppose that every creature anywhere in the universe, past, present, and future, which has the properties $P_1, P_2, \ldots, P_n$ also has the property of being white? As Smart says, the universe being as large as it is and organisms being as complicated and as subject to possible variations as they are, it appears that the answer must be negative. But then the statement 'All polar bears are white' is not verifiably true and hence again not nomological. (Of course, if one included 'white' among the defining predicates then the generalization would indeed become verifiably true, but now the statement, no longer synthetic, would again not qualify as a law of nature.) The importance of this example stems from the fact that it is principally biological terms that one tends to feel could perhaps equally well be construed as either individual-terms or as purely qualitative terms. That is, a given type of organism might in actual practice be identified either by its place in the terrestrial evolutionary tree, in which case there is an implicit reference to the planet Earth, or, on the other hand, it could be identified as having certain purely qualitative characteristics. Such terms therefore generate an important class of statements for which an indeterminacy as to nomological status might appear to exist. In view of the considerations presented by Smart, it appears that this indeterminacy is largely illusory.

2. *The Concept of Individual-Term*

In order to be an original nomological statement a statement must be *universal*, i.e., it must not contain an essential occurrence of an individual-term. The notion of 'individual-term' is therefore a crucial one in Reichenbach's analysis.

Reichenbach's definition is:

> An individual-term is a term which is defined with reference to a certain space-time region, or which can be so defined without change of meaning. (p. 32)

The class of individual-terms is clearly intended to include proper names and definite descriptions designating physical objects, places, and times, and also such other terms, e.g., 'terrestrial', that would normally be defined in terms of these. Reichenbach also wishes to take account of languages (or reconstructions of language) in which a term such as 'terrestrial' might be taken as primitive. In this case he holds that the term, while not explicitly defined in terms of other individual-terms, *can* be so defined *without change of meaning*. (He notes that the phrase 'can be defined' refers to logical possibility and hence is a permissible concept in this context.)

One might quibble with the wording in the earlier part of the definition – one doesn't normally speak of proper names and definite descriptions as being *defined*. It is clear, however, from his supplementary remarks that what is intended could be expressed as follows: An *individual-term* is a singular term designating a certain space-time region, or a predicate defined with reference to such a region, or a predicate which can be so defined without change of meaning. It is the allusion to *meaning* in the last clause of the definition that deserves special scrutiny. The sameness of meaning of 'x is terrestrial' and some such phrase as 'x is of the Earth' clearly cannot be mere sameness of reference. Otherwise such a purely qualitative predicate as 'x is blue' would qualify as an individual-term by virtue of its having the same extension as the phrase 'x is color-similar to Joe's pen' – Joe's one and only pen being blue. Likewise, the putative synonymy cannot be anything so straightforward as logical equivalence, since 'x is terrestrial if and only if x is of the Earth' is clearly not logically true. The claim must rather be that 'x is terrestrial if and only if x is of the Earth' is *analytic*, whereas 'x is blue if and only if x is color-similar to Joe's pen' is not *analytic*. This claim is surely correct.

The important thing to note is that this definition introduces into the theory of nomological statements a second concept falling outside the domain of logic and formal semantics: the concept of *synonymy* in the broad and non-formal sense of that term. (The first such concept was that of inductive confirmation implicit in the statement of Requirement 1.1.) Assuming that the notion of synonymy can eventually be explicated without reliance on the concept of law or the physical modalities, there seems to be nothing intrinsically objectionable in employing this concept in this context. It is certainly interesting, however, that this major attempt to break the charmed circle of concepts such as the physical modalities, counterfactuals, and law – by defining law from outside this circle – should have to draw on that other circle inhabited by such concepts as synonymy and analyticity. Nevertheless, when properly understood, Reichenbach's attempt to characterize individual-terms can be regarded as reasonably successful. Certainly the history of the topic does not encourage us to suppose that a better definition will shortly be forthcoming.

I might point out that there is no indication elsewhere in the book that such notions as synonymy or analyticity in the sense required here will be needed. Reichenbach consistently uses the term 'analytic' to mean 'logically true', and his definition of 'equisignificance' for statements amounts essentially to the requirement that they be logically equivalent when expanded in primitive terms (p. 19). If the term 'terrestrial' is taken as primitive, then the statements 'The Atlantic Ocean is terrestrial' and 'The Atlantic Ocean is of the Earth' are not equisignificant according to Reichenbach's own definition of that term. Yet, they must 'have the same meaning' in a sense of 'meaning' related to that used in the definition of 'individual-term'. This is just a bit of terminological confusion, but the failure to note it might lead the reader to overlook a crucial feature of Reichenbach's definition of 'individual-term' and hence of the theory of nomological statements.

3. *On Simplifying Requirement* 1.4

It is part of the content of Requirement 1.4 that the statement in question must not contain any predicate whose extension is verifiably confined to a certain restricted space-time region. That is, given a restricted space-time region R, it must not be the case that the extension of the predicate is verifiably confined to R. Predicates qualifying as individual-terms, such as 'terrestrial', appear to be of this sort, but statements containing that type of term are already excluded by Requirement 1.2. Reichenbach thinks, however, that Requirement 1.4 is needed to rule out some purely qualitative predicates having a certain internal structure. The example cited was 'man who saw a living human retina before any other man saw one'. Since we might know that this predicate applies to H. v. Helmholtz we could pick a restricted space-time region within which the extension of the predicate is verifiably confined. But *do* we know that this predicate is true of Helmholtz? Here our earlier discussion of J. J. C. Smart's point concerning biological terms becomes relevant. If we are using the term 'man' – as we ordinarily would – to refer to a particular species that has evolved here on the Earth, and 'human retina' to refer to a certain bodily part of members of that species, then it may be reasonably certain that Helmholtz is the man in question. In this case, however, these are individual-terms having a covert reference to the Earth. Statements containing such terms are already disqualified by Requirement 1.2. If, on the other hand, we are really using 'man' and 'human retina' as purely qualitative terms, then it is difficult to see how we could ever really highly confirm that Helmholtz was indeed the first 'man' in the (possibly infinite) universe to see a 'living human retina'. It therefore appears that Requirement 1.4 is not really needed to rule out the predicate in question or other similar predicates. Such complex predicates are indeed satisfied by at most one individual, but if the component predicates are purely qualitative it seems that we cannot know for any given restricted space-time region that the individual is in *that* region. It thus appears that this component of Requirement 1.4 could be dropped. The requirement would then reduce to the stipulation that the statement p must be exhaustive in elementary terms.

4. *The Derivability Requirement for Nomological Statements*

In outlining his principal strategy Reichenbach makes it clear that a synthetic nomological statement is to be such as to assert an inductive generality – that no such statement is to be verifiable without the use of inductive extension. He takes this to be the basic distinction between nomological truth and merely factual truth. Synthetic nomological truth is to be distinguished by the fact that "... factual truth must never be sufficient to verify deductively a statement of this kind" (p. 13). This characteristic appears to be assured for the class of *original* nomological statements by the set of requirements laid down for that class. Nomological statements in general, however, are to be any statements derivable from the set of original nomological statements. As I shall presently show, this requirement of derivability is by no means sufficient to insure the desired result for nomological statements as such.

The basic ways in which a derived statement may fail to inherit the desired characteristic can be readily illustrated. First, suppose that '(x) $(Cx \supset Bx)$' is an original nomological statement (e.g., 'All copper sulphate is blue'). From this we can derive '$(\exists x)(Cx \supset Bx)$', which should therefore also be a nomological statement. But the latter statement follows from the purely factual statement 'Ba' (e.g., 'This pencil is blue'), and is therefore verifiable without the use of inductive extension. Second, from the same original nomological statement we can derive '$Ca \supset Ba$', which again follows from the purely factual statement 'Ba'. Finally, from the same original nomological statement we can derive the statement '$(x)(Cx \supset Bx) \vee Ba$', which again follows from 'Ba' alone. In each case we have the unacceptable consequence that a 'law of nature' is conclusively verifiable by means of a single observation statement.

The derivability requirement for nomological statements is clearly too weak. A move having much to recommend it is to replace this requirement by the stronger requirement that a statement is nomological if and only if it is logically equivalent to some original nomological statement. This not only rules out the sort of cases just considered but also has the advantage of insuring that nomological statements so defined are always universal in the sense of Requirement 1.2. This accords with our intuitive idea that laws of nature should not make reference to this or that particular spatiotemporal object or region. The proposed revision would in fact insure that any nomological statement will meet all of the requirements for original nomological statements with the possible exception of Requirements 1.6 through 1.9. These latter requirements concern the form rather than the content of the statements they refer to.

There is, however, a technical matter that makes it necessary to frame a somewhat more complex definition. It certainly seems reasonable that the conjunction (or the disjunction) of any two nomological statements should also be nomological, as is of course the case under Reichenbach's derivability requirement. Now, as Reichenbach points out, the conjunction of two original nomological statements may fail to be exhaustive in elementary terms (pp. 64–65). In this case the conjunction will not be an original nomological statement and will also fail to be logically equivalent to any original nomological statement. But then the conjunction would not qualify as a nomological statement at all by the criterion proposed above.

This complication cannot be handled merely by revising the requirement to read as follows: A statement is nomological if and only if it is logically equivalent to some original nomological statement or to the conjunction of two or more such statements. That this will not suffice can be seen from the following example. Suppose that '*p*', '*q*', and '*r*' are original nomological statements, but that the conjunction '*p* & *q*' is not. By this new criterion '*p* & *q*' is now nomological, but the statement '(p & q) $\vee$ *r*' need not be. The following revision of the requirement does appear to be satisfactory:

> A statement *p* is a nomological statement if and only if
> (1) *p* is logically equivalent to some original nomological statement
>
> or
>
> (2) *p* is logically equivalent to a statement built up by the use of disjunction and/or conjunction from statements satisfying (1).

5. *On the Wider Sense of 'Verifiably True'*

It was noted earlier that when the expression 'verifiably true' is taken in Reichenbach's narrower sense a statement does not satisfy Requirement 1.1 unless it is not only true but also actually verified (highly confirmed) at some time or other. The resulting class of nomological statements then corresponds to those laws of nature that are actually confirmed at some time or other. In order to remove this limitation while still avoiding the use of forbidden modal concepts, Reichenbach defines a wider sense of 'verifiably true' intended to be such that the class of nomological statements will correspond to all existing laws of nature (pp. 84–87).

The essentials of this definition may be summarized as follows. Call the set of nomological statements resulting from the narrower sense of 'verifiably true' 'S_0'. Call the set of all observational procedures that are ever developed 'P_0'. Call the set of all observational data that are ever actually obtained 'R_0'. Now, it is logically possible to describe particular applications of procedures from the class P_0 which are in fact never actually carried out. The results of such never-performed experiments and observations will in some cases be accurately predictable on the basis of S_0. In other cases only the general character of the result is so predictable. In still other cases this system may permit only the prediction that some result or other will ensue. In any case, within the system S_0, P_0,

R_0 it is permissible to use the *expression* 'the result of such-and-such an observation' even though that specific application of an observational procedure is never made. As Reichenbach says, the system determines a set of possible observational *questions*, but it does not determine all of the answers in a precise manner. The precise answers to all such observational questions form a set of observational data which Reichenbach calls 'R_1' and which, as he says, is determined by the nature of the physical world. The expression 'verifiably true in the wider sense' can now be given as follows: A statement is verifiably true in the wider sense if and only if it is true and it is logically possible to verify it on the observational basis R_1. When this sense of 'verifiably true' is used in Requirement 1.1 there results a class of nomological statements that Reichenbach calls 'S_1'. While Reichenbach regards the set of nomological statements S_1 as adequately corresponding to the notion of all existing laws of nature, he points out that, if desired, the step leading from S_0 to S_1 could be iterated indefinitely.

It is a striking feature of this definition that it utilizes a frankly anthropocentric base. It is tempting to suppose that one might obviate this feature and still renounce all reference to physical modalities by taking a more "abstract" approach, such as the following: A statement p is verifiably true in the wider sense if and only if it is true and there exists a set of true non-all-statements which jointly confirm p to a very high degree. This sort of approach presupposes that for every law of nature there exists (tenselessly speaking) a naturally occurring body of evidence highly confirming it. The question as to whether this assumption is reasonable probably cannot be answered until we have a better understanding of the nature of inductive confirmation. At any rate, Reichenbach needs no such assumption. His approach involves reference to the sorts of procedures by which humans actually create as well as select evidentially relevant conditions for the testing of hypotheses.

I do not think, however, that Reichenbach's attempt to extend the class of nomological statements in this way can be considered successful. Its crucial defect can be brought out by focussing on how members of the set of observational data R_1 are to be precisely characterized. Since R_1 includes the set of actual data R_0, it is only those members of R_1 which are 'merely possible data' that require closer scrutiny. Call this subset 'R_1^*'. What are the necessary and sufficient conditions for a given datum d

being a member of R_1^*? The clear answer seems to be that d is a member of R_1^* if and only if d is an observational datum that *would* be obtained if some application (which in fact is never made) of one of the observational procedures in the class P_0 *were* to be made. Now, such a counterfactual in this context would be permissible if it were clear that only laws in the class S_0 were required for its support. But this is clearly not always the case. As Reichenbach points out, the system S_0 will not always suffice to determine the answers to those possible observational questions. In such cases, then the use of the sort of counterfactual in question involves a covert reference to the set of laws S_1. In Reichenbach's own presentation the covert reference to S_1 occurs when he says that "... a merely possible datum is not determined by R_0, P_0, S_0, but depends on the nature of the physical world" (p. 86). Thus, there seems to be no way of characterizing the complete set of data R_1 independently of the set of laws S_1. But then S_1 cannot without circularity be defined in terms of R_1, as Reichenbach's approach requires.

6. *A Simplification of the Theory of Nomological Statements*

Requirement 1.4, in addition to banning certain complex predicates – a function criticized earlier – also requires that the statement p be exhaustive in elementary terms. It may be recalled that this part of the requirement functions jointly with Requirements 1.1 and 1.5 to insure that original nomological statements are free from the sort of contamination represented by true, closed, existential clauses. Other than this joint function, Requirements 1.4 and 1.5 play no essential role in the theory of nomological statements. It is true that Requirement 1.4 along with Requirements 1.6–1.9 individually help to insure that original nomological statements fall within the class of fully admissible statements. We have seen earlier, however, that this goal is superfluous.

In view of all this, it appears that a significant simplification of the requirements for original nomological statements could be achieved by dropping Requirements 1.6–1.9 and by replacing Requirements 1.4–1.5 with the single requirement that the statement p is not, when expanded in primitive terms, logically equivalent to any statement containing a non-redundant occurrence of a true, closed, existential clause. This reduces the number of requirements for original nomological statements to four.

These requirements are such that the conjunction or disjunction of two statements satisfying them also satisfies them. It therefore becomes reasonable to define a nomological statement simply as a statement logically equivalent to an original nomological statement. But here another simplification becomes automatic. The four requirements are such that if a given statement satisfies them then every statement logically equivalent to that statement also satisfies them. It follows that the distinction between nomological statements and original nomological statements vanishes. It is therefore possible to define a nomological statement simply as a verifiably true, universal all-statement that is not, when expanded in primitive terms, logically equivalent to any statement containing a non-redundant occurrence of a true, closed, existential clause.

7. *The Nomological Agenda*

With only the narrower sense of 'verifiably true' available for use in Requirement 1.1, Reichenbach's theory cannot be considered to furnish a satisfactory explication of the concept 'law of nature'. It does appear to be successful, when appropriately revised, in demarcating the class of laws that are highly confirmed at some time or other. What is still needed, however, is either a satisfactory non-modal definition for the wider sense of 'verifiably true' or a total replacement for Requirement 1.1.

We might examine these alternatives more closely. Let us start by supposing that a non-modal definition for the wider sense of 'verifiably true' were available. We would then indeed have a non-circular way of demarcating the class of laws from the class of non-laws, and this would certainly be an important achievement. Even such a theory, however, might possibly be felt as furnishing something less than a completely satisfactory explication. For, such a theory might well throw no light of its own on the following interesting question: Among those statements that meet all requirements *other* than Requirement 1.1, what is there about those statements which *also* meet Requirement 1.1 that distinguishes them from those that do not? It is possible that future progress in the area of confirmation theory would eventually enable us to answer this question. Even so, the theory of nomological statements might be criticized as throwing us into unseemly dependence on quite a different theory for this vital illumination. Also, assuming that the differentiating characteristics in question could be expressed in non-modal terms, these characteristics might well be used to supplant Requirement 1.1 altogether. There is therefore some reason to suspect that a theory based on the first alternative would eventually lead to a theory based on the second alternative. One might, of course, attempt to embrace the second alternative directly by replacing Requirement 1.1 by a requirement that involves no concepts from the area of confirmation theory. The attempt to formulate such a requirement is certainly even now one reasonable approach to the problem of explicating the concept of law.

Reichenbach's work on the nature of scientific law is truly impressive. He has surely come to grips with more of the relevant complexities of the problem than has any other philosopher, and he has done so in a painstakingly precise manner. This work, in both its unsuccessful and its successful aspects, forms a body of instructive material that no investigator into this problem area can afford to ignore.

Texas Tech University

BIBLIOGRAPHY

Goodman, N.: 1955, *Fact, Fiction, and Forecast*, Harvard University Press, Cambridge.

Reichenbach, H.: 1947, *Elements of Symbolic Logic*, The Macmillan Co., New York.

Reichenbach, H.: 1954, *Nomological Statements and Admissible Operations*, North-Holland Publishing Co., Amsterdam.

Smart, J. J. C.: 1963, *Philosophy and Scientific Realism*, Routledge and Kegan Paul, London.

CYNTHIA SCHUSTER

APPRECIATION AND CRITICISM OF REICHENBACH'S META-ETHICS: ACHILLES' HEEL OF THE SYSTEM?

Non-cognitivism in ethics is strictly entailed by Reichenbach's epistemology. Having settled for a functional conception of knowledge, with prediction as the one and only function or purpose of knowledge, Reichenbach was committed to ethical non-cognitivism, i.e. to the view that there is no normative or prescriptive knowledge, that there are no moral truths, that moral judgements are neither true nor false.

However, Reichenbach did much more than reiterate the ethical noncognitivism of the Vienna Circle and of A. J. Ayer. He replaced their emotivism, which reduced moral judgements to expressions of the emotions of the speaker, by his 'volitional ethics', which focuses on the volitional, directive, imperative nature of the decisions expressed by moral judgements. This enabled him to draw some sharp distinctions: (1) between personal directives and moral directives, the latter being volitional decisions by which individuals make demands on other individuals; and (2) between 'primary' moral decisions or goals and entailed decisions, the latter relating means to the ends adopted by the primary decisions. Through these distinctions he discusses the roles of logic and volition in our moral lives and in ethical disputes. Empirical knowledge plus logic can inform us whether a goal we have adopted by a volitional decision is attainable and is compatible with our other goals – compatible, given the entailed decisions concerning means. Thus if the relevant knowledge is available, knowledge can resolve disputes over means (entailed decisions) if the disputants agree on ends.

Although Reichenbach had adumbrated this volitional ethics in a short piece published in 1947, he elaborated it only in his relatively popular, relatively non-technical book *The Rise of Scientific Philosophy*, published in 1951.[1] That book lacks the rigor of his other writings, and, as I shall contend later, his chapter on ethics is perhaps its least rigorous part; nevertheless I think his contribution there to the ethical noncognitivism of the 1930's and 1940's was very considerable. His shift from the earlier

W. C. Salmon (ed.), Hans Reichenbach: Logical Empiricist, 721–730.

focus on the speaker's emotion of the moment to an emphasis on volitional decisions which make demands on others and which can be organized into coherent systems surely brings non-cognitivism closer to our actual moral experience. In this I think he rendered a service to the empiricist interpretation of moral judgements comparable to the service he and Carnap rendered to the philosophy of science when they revised the verifiability theory of meaning, replacing the requirement of complete verifiability by the weaker requirement of testability, and thus brought the empiricist criterion of cognitive meaning closer to the actual practice of empirical science.

Reichenbach also made fully explicit what is implicit in any non-cognitivism, viz. the double life of the empiricist as knower and as moral agent. So far as he concerns himself with ethics, the activity of the scientist or of the non-cognitivist philosopher, *qua* philosopher, must be restricted to value-neutral meta-ethics, i.e. to the search for the correct interpretation of moral judgements and to the discovery of their logical status; that same person in his social activity makes moral decisions with imperative meanings that make demands on others and engages in the 'friction' of competing volitional decisions. Reichenbach expresses this vividly in an imaginary conversation between himself and the reader of *The Rise of Scientific Philosophy*, a conversation which comes after the more familiar defenses of non-cognitivism ('is' never entails 'ought'; facts yield no moral directives; etc.). The reader has expressed fear that this volitional ethics will result in moral anarchism, that the logic of the analysis compels Reichenbach to give everyone the right to do anything. Reichenbach answers:

> Logic does not compel me to do anything. The directives I set up are not consequences of my conception of ethics, either; nor does logic tell me what imperatives I should regard as obligatory for all persons. I set up my imperatives as my volitions, and the distinction between personal and moral directives is also a matter of my volition. Directives of the latter kind, you remember, are those which I regard as necessary for the group and which I demand everybody to comply with.
>
> Now you are in complete despair. You retort: "Maybe what you say is true, logically speaking; but do you really think – you, the author of a book on scientific philosophy – that you are the man to give moral directives to the whole world? Why should we follow you?"
>
> I am sorry, friend, I did not intend to convey this impression. I was looking for the path of truth; but for this very reason I am not going to give the moral directives, which by their nature cannot be true. I have my moral directives, that is true. But I shall not write them down here. I do not wish to discuss moral issues, but to discuss the nature of morality. I even

have some fundamental moral directives, which, I think, are not so very different from yours. We are products of the same society, you and I. So we were imbued with the essence of democracy from the day of our birth. We may differ in many respects, perhaps about the question of whether the state should own the means of production, or whether the divorce laws should be made easier, or whether a world government should be set up that controls the atom bomb. But we can discuss such problems if we both agree about a democratic principle which I oppose to your anarchist principle:

Everybody is entitled to set up his own moral imperatives and to demand that everyone follow these imperatives.

This democratic principle supplies the precise formulation of my appeal to everybody to trust his own volitions, ... (pp. 294–295).

A few pages later he adds:

Whoever wants to study ethics, therefore, should not go to the philosopher; he should go where moral issues are fought out. He should live in the community of a group where life is made vivid by competing volitions, be it the group of a political party, or of a trade union, or of a professional organization, or of a ski club, or a group formed by common study in a classroom. There he will experience what it means to set his volition against that of other persons and what it means to adjust oneself to group will. If ethics is the pursuit of volitions, it is also the conditioning of volitions through a group environment. The exponent of individualism is shortsighted when he overlooks the volitional satisfaction which accrues from belonging to a group. (p. 297).

Whether one condemns this stance as the ultimate schizoid split dividing a person as knower from himself as moral agent or praises it as consummate honesty, it at least carries non-cognitivism into the arena of actual social-moral affairs.

So much for my appreciations of Reichenbach's work in meta-ethics. Now to my negative criticism. I find a strain of dogmatism in *The Rise of Scientific Philosophy*, issuing in dogmatic statements about knowledge and about ethics – a dogmatism inconsistent with the more rigorous epistemology of the earlier *Experience and Prediction* (1938).

Before I elaborate that objection, let me forestall misunderstanding by divorcing myself from a silly polemic against Reichenbach's whole system, the criticism not infrequently heard from anti-empiricists who attend to his manner of presentation but not to his arguments. This kind of polemic takes the form of sarcasm: 'He preaches probability in a certainty tone of voice.' That is worse than irrelevant; it is a failure of understanding. The arguments Reichenbach offers in support of his conclusion that scientific knowledge is a web of probabilities that hangs on no peg of certainty are arguments justifying a firmly assertive tone for that conclusion in meta-science.

My own objection to his dogmatism in meta-ethics should not be confused with the above mentioned nonsense. I shall try to show that he makes claims inconsistent with one of his own fundamental theses in epistemology. That thesis, which I regard as one of his truly great insights, is the thesis that 'theories' of cognitive meaning have the logical status of decisions. They are not themselves verifiable, are neither true nor false, and must be classified as proposals or definitions adopted because they serve our ends.

Reichenbach presented that thesis in *Experience and Prediction*. There, with careful consistency, he defended his own probability version of the verifiability theory of meaning on the grounds that it is in accord with our science and serves the purpose for which we seek knowledge, namely successful prediction. However, in *The Rise*, where he develops his thoughts on ethics, he soft-pedals this thesis of the decisional nature of the criteria of cognitive meaning (the criteria of or definition of knowledge). He does not deny it, and to be sure he even introduces it on *one* of the three hundred odd pages of the book, beclouding it with an obscure distinction between 'subjective meanings' and the meanings selected by the empiricist theory of meaning. (p. 258) That 'theory', he says, is "a rule proposed for the form of language and advisable for good reasons: it defines the kind of meaning which, if assumed for a person's words, makes his words compatible with his actions. [Those] who adopt the empiricist theory of meaning speak a language consistent with their behavior." (p. 258)

Let us not pause over the undoubtedly false notion that empiricists and only empiricists speak a language consistent with their behavior. The point I want to stress is that in his polemics against Plato and all others who claim moral knowledge, and in his own work in meta-ethics, Reichenbach assumes that meta-level statements about the cognitive or non-cognitive nature of moral judgements are cognitive (are true or false), and *that*, I submit, is inconsistent with his thesis that definitions of knowledge and criteria of cognitive meaning are decisional. (See the first quote above from pp. 294–295, particularly: "I was looking for the path of truth; but for this very reason I am not going to give you moral directives, which by their nature cannot be true.") Yet every meta-ethics is entailed by a conception of knowledge, and if that has the status of a decision (neither true nor false) whatever it entails inherits the same status, as he himself pointed out in his work on entailed decisions. Did he overlook the logical truth that even a decision 'adopted for good reasons' remains a decision?

This streak of dogmatism, inconsistent with the thesis that conceptions of knowledge are decisions, occurs throughout the book whenever he condemns the knowledge-claims of traditional philosophers on the grounds that they had an 'erroneous conception of knowledge', a phrase he uses frequently. I quote just two of the passages in which it occurs: "The search for moral directives thus becomes an extralogical motive interfering with the logical analysis of knowledge, and it must now be shown to what extent its product, the ethico-cognitive parallelism [defined earlier as the theory that ethical insight is a form of cognition] has . . . become a major source of erroneous theories of knowledge." (p. 62) "The two-thousand-year-old plan to establish ethics on a cognitive basis results from a misunderstanding of knowledge, from the erroneous conception that knowledge contains a normative part." (p. 277)

Some careful students of Reichenbach's work may object to my objection by claiming that he attributes decisional status to theories (criteria) of meaning, but not explicitly to conceptions or definitions of knowledge. I answer that he explicitly says that the kind of meaning defined by the verifiability theory of meaning is *cognitive* meaning, (p. 282), and explicitly links that theory of meaning to the 'cognitive content' of language. (p. 256) Also, he offers his 'fundamental conception of knowledge' in the following way: "In this interpretation, knowledge does not refer to another world, but portrays things of this world so as to perform a function serving a purpose, the purpose of predicting the future." (p. 255)

I think that here, in the phrases '*a* purpose' and '*the* purpose', we find the Achilles' heel of Reichenbach's (and many another) whole system of thought. He assumes that predicting the future is *the one and only* purpose served by knowledge. Since *that* purpose is best served by empirical knowledge *à la* modern science, he concludes that his conception of knowledge is the 'correct' or true conception; all other conceptions are 'erroneous'. Here, to repeat, he unwittingly contradicts his insight that the cornerstone of every epistemology (its definition of knowledge; its criteria of cognitive meaning) is a decision (neither true nor false) by claiming that *his* conception of knowledge is true.

This is not a minor inconsistency to be remedied by cleaning up the phraseology. It hits the vitals of any epistemology attempting to delve beneath the surface without resorting to dogma. I hope to show that my critique on this point is directly relevant to foolish disputes still current concerning what is to count as evidence, as explanation, as knowledge.

Now, insofar as Reichenbach was doing meta-*science*, describing the methods of modern science and prescribing means which serve *the* purpose of successful prediction, he could consistently make truth claims for his theories. However, like all logical positivists and logical empiricists, he claims that his account of scientific (predictive) knowledge is a comprehensive account of *all* human knowledge. To make good that claim, he assumes that knowledge serves only one purpose, prediction, – an obviously false assumption. The purposes for which people seek knowledge are many and various. In the kitchen and in the Department of Defense successful prediction may be the sole purpose, but some people at some times seek knowledge of their political obligations (and gain that knowledge by careful, consistent thought about the commitments they have explicitly or tacitly undertaken); some people at some times seek knowledge just for the sake of understanding; some people at some times seek knowledge so as to be able to humiliate their mothers-in-law . . . . For Plato, the purpose of elementary, specialized knowledge was to prepare the self for knowledge of the good.

Reichenbach is not unique in vitiating enquiry by a single-purpose assumption. To his claim that *the* purpose of knowledge is prediction we hear the counter-claims: 'No; *the* purpose of knowledge is explanation' or '*the* purpose of knowledge is to be able to guide action by obligation and good taste' or . . . . Whether it is a vestige of puritanical single-mindedness or merely the yen for facile unity within the human self, epistemologists too often presuppose singleness of purpose or singleness of passional motive when they try to anchor their theories in human goals. Even Humeans, committed to the dictum that reason is the slave of the passion*s*, but forgetful of Hume's two hundred pages on the multiplicity and complexity of the passions, tend to formulate questions about motives in the singular: 'What is *the* passion satisfied by explanation? by morality? by knowledge?'

The vicious role played by the singleness-of-purpose assumption is obscured by such labels as 'the verifiability theory of cognitive meaning.'

If Reichenbach had been more faithful to his insight that that 'theory' is a decision, he would have renamed it 'the empiricist decision concerning what is to count as evidence confirming knowledge-claims.' Clearly, what is at stake is whether aesthetic and moral experiences are to be admitted as evidence confirming normative conclusions, or whether 'evidence' is to be limited to the range of experience called 'observation' in Reichenbach's revised verifiability 'theory'. Now there may be good *moral* reasons for limiting knowledge-claims to those claims which are testable by observations, for observational experience is the range of experience most widely similar in human beings, and hence when sensory observation is the only kind of experience allowed to count as evidence, agreement among people's knowledge-claims is maximized. (It is on some such moral grounds that I am myself an empiricist.) Furthermore, it may be extremely difficult to find sound logical relations between ethical cognitivists' moral or aesthetic experience and their over-generalized claims to moral knowledge (though in 1975 empiricists are hardly in a position to throw stones), but to short-cut enquiry as to what is to count as evidence by assuming (legislating?) that knowledge serves only one purpose, viz. prediction of future *observational experience*, is to close the cellar door while seeking to get at the foundations.

Reichenbach and other Logical Empiricists have paid heavily for closing that door. They were so bent on closing out dogmatic and authoritarian claims to know about supernatural realities that they overlooked the fact that Hobbes' 'mortal god' can be created by acts of commitment. It was not Logical Empiricists but rebels against the single-purpose-of-knowledge dogma who undertook the investigation of the logical relations between promise-acts and objectively knowable obligations. Ethical non-cognitivists failed to examine those non-predictive uses of language which create *knowable* bonds of obligation, for they remain locked within a subjective-objective dualism as absolute as Descartes' substantival subject-object dualism, and hence could not incorporate the 'arena of competing volitions' into an account of our social-moral lives which can be at once more realistic and more idealistic.

Another way of putting my point would be to challenge Reichenbach's assertion that it is 'erroneous' to suppose that 'knowledge contains a normative part'. (1951, p. 277) To be sure, if you start with his decision concerning cognitive meaning, i.e. his decision to limit evidence to observational experience, then your body of knowledge will contain no normative conclusions. However the starting point, the decision, is as normative as the Ten Commandments. By placing that decision on the meta-level, Reichenbach can claim that it is not 'part' of knowledge, but since it determines what is and what is not to be counted as knowledge it is implicitly present in every knowledge-claim at least as an *aspect* of knowledge, and hence is part of the total epistemological system within which he claimed to offer meta-level truth about the 'nature of morality.' (See again the quote from pp. 294–295). To be cute, one could say that within any epistemological system the justification of every is-statement is derived from a basic 'ought'.

Still another way of putting my point would be to accuse Reichenbach of the vicious circle of which second-rate ethical non-cognitivists are frequently guilty, but I am not sure whether Reichenbach is really guilty of that circle. Although I am content to rest my case on the contradiction between his insight that conceptions of knowledge are decisions and his claim to know that there can be no normative knowledge, let me state the circle, leaving the reader to decide whether Reichenbach can fairly be accused of this *petitio principii*: One first assumes an absolute dichotomy of objective facts and subjective values–emotions–volitions; one then defends a definition of knowledge by showing that it includes objective facts and excludes values (as happens when predictive science is taken as the paradigm of all knowledge); finally one 'proves' the truth of the fact-value dichotomy by deriving it from the adopted conception of knowledge.

One thing makes me suspect that Reichenbach was at least close to that circle: he conceives of obligations exclusively as subjective *feelings* of obligation. So did Hume, and we know what a mess that got Hume into when he tried to understand how promises entail obligations whether the promiser *feels* an obligation or not.

The above remarks move within the nominalistic–empiricist tradition to which Reichenbach belonged. Even when I spoke of the multiple purposes of knowledge, and mentioned purposes rejected by empiricists as unattainable, I used the nominalistic idiom which assumes that the ultimate social reality is separate individuals with their individual desires and their individual volitional decisions. To go deeper would, I think, require a critique of nominalism and of its entailed social atomism, assumed in Reichenbach's meta-ethics as it is assumed in the ethical non-cognitivism of the Vienna Circle, Ayer and Stevenson.

Polemics against that non-cognitivism are epistemologically irresponsible and metaphysically superficial when they issue merely from the passional desire to know moral truth but offer no criteria of moral truth beyond personal moral-religious intuitions or invalid inferences from history, while leaving unchallenged the assumptions of nominalistic individualism. The deeper attack, which I shall adumbrate but not attempt to follow through, would start with a concept of the individual person as by nature a political animal (or a part of an organic social whole, or a species being, a mammal, or whatever way you prefer to label constitutive group membership). Social reality would then be conceived as larger-than-individual wholes, as social-moral bonds to some extent already there to be discovered, to some extent created and destroyed by individuals.

Such a critique would aim at the nominalism rather than at the ethical non-cognitivism of the Humean-positivist school. To stick with our present muttons, it would question whether *within Reichenbach's social atomism* (within his conception of moral issues as 'fought out' . . . by 'competing volitions') it makes sense for any individual to make moral demands on others. If that makes no sense, and if 'trusting one's volitions' reduces to a psychological state of all the competing wills, Reichenbach's distinction between personal directives and moral directives seems to break down. The moral life reduces to the individual's enjoyment of working with others who happen to agree with his value-decisions.

These are sketchy suggestions intended only to open the possibility of a kind of thinking about morality which might combine the holistic (anti-nominalistic) part of dialectical reasoning with the insistence on epistemological accountability of thinkers like Reichenbach.

Despite my negative criticisms, I still think Reichenbach's contributions should remain part of the basic training in honesty in moral and political philosophy. His work on the logical status of primary and entailed decisions stands. His conception of democracy fits all too well societies composed of competing ego-centered individuals – people whose self-image tends to make social atomism become true. And however inadequate his analyses may be, he showed that ethical non-cognitivism must somehow take account of the difference between personal whim-desire and socially concerned moral directives. Above all, he furthered the crusade against epistemologically irresponsible moralists who claim to know moral truths but are unable to answer the question 'How do you know what you claim to know?' (He, himself, used this term 'crusade' in this context in a conversation with me in 1950.)

University of Montana

NOTE

[1] Before his death in 1953, Reichenbach wrote some notes on the explication of moral judgments. These notes were organized by Maria Reichenbach and published in *Modern Philosophy of Science*, Humanities Press, 1959. Since I find nothing there relevant to the issues I discuss, I have not mentioned that work in the text.

BIBLIOGRAPHY

Reichenbach, H.: 1938, *Experience and Prediction*, University of Chicago Press, Chicago.
Reichenbach, H.: 1947, 'Philosophy: Speculation or Science', *Nation* **164**, 20–22.
Reichenbach, H.: 1951, *The Rise of Scientific Philosophy*, University of California Press, Berkeley & Los Angeles.

BIBLIOGRAPHY

Abbott, J. C., 1967, 'Semi-Boolean Algebra', *Matematicki Vesnik* **4**, 177–198.

Abbott, J. C., 1979, *Sets, Lattices, and Boolean Algebras*, Allyn and Bacon, Boston.

Achinstein, P., 1967, 'Hans Reichenbach', in P. Edwards (ed.), *The Encyclopedia of Philosophy*, Vol. 7, Macmillan Publishing Co., New York, 115–118.

Adams, E. N., 1960, *Physical Review* **120**, 675.

Anderson, A. *et al.*, 1975, *Entailment, The Logic of Relevance and Necessity*, Vol. I, Princeton University Press, Princeton.

Aristotle, 1928, *Posterior Analytics* in W. D. Ross (ed. & trans.), *The Works of Aristotle*, Vol. I, Clarendon Press, Oxford.

Arnauld, Antoine, 1964, *The Art of Thinking*, Bobbs-Merrill, Indianapolis.

Arzelies, H., 1966, *Relativistic Kinematics*, Pergamon, New York, ch. IX, 'The Rotating Disc', 204–243.

Ash, R. B., 1972, *Real Analysis and Probability*, Academic Press, New York.

Asped, A., 1975, *Physics Letters* **54A**, 117.

Asped, A., 1976, *Physical Review* **D14**, 1944.

Atwater, H. A., 1971, *Nature Physical Science* **230**, 197–198.

Audi, M., 1973, *The Interpretation of Quantum Mechanics*, University of Chicago Press, Chicago.

Ayer, A., 1946, *Language, Truth and Logic*, 2nd ed., Dover Publications, New York.

Bar-Hillel, Y., 1954, 'Logical Syntax and Semantics', *Language* **30**, 230–237.

Bar-Hillel, Y., 1964, *Language and Information*, Addison-Wesley, Reading, Mass.

Barker, S., 1961, 'On Simplicity in Empirical Hypotheses', *Philosophy of Science* **28**, 162–171.

Bell, J. S., 1964, 'On the Einstein-Podolsky-Rosen Paradox', *Physics* **1**, 195–200.

Bell, J. S., 1971, 'Introduction to the Hidden Variable Question', in B. d'Espagnat (ed.), *Foundations of Quantum Mechanics*, Academic Press, New York.

Belnap, N., 1970, 'Conditional Assertion and Restricted Quantification', *Nous* **4**, 1–12.

Berry, George D. W., 1949, 'Review: Hans Reichenbach, *Elements of Symbolic Logic*', *Journal of Symbolic Logic* **14**, 50–52.

Bierwisch, M., 1971. 'On Classifying Semantic Features', in D. Steinberg *et al.* (eds.), *Semantics*, The University Press, Cambridge, 1971.

Birkhoff, G. and von Neumann, J., 1936, 'The Logic of Quantum Mechanics', *Annals of Mathematics* **37**, 823–843.

Birkhoff, G., 1961, 'Lattices in Applied Mathematics', in *American Mathematical Society Proceedings of Symposia in Pure Mathematics* **2**, American Mathematical Society, Providence, R. I.

Black, M., 1954, *Problems of Analysis*, Cornell University Press, Ithaca, N. Y.

Black, M., 1962, *Models and Metaphors*, Cornell University Press, Ithaca, N. Y.

Bohm, D., 1951, *Quantum Theory*, Prentice-Hall, Englewood Cliffs, N. J.

Bohm, D., 1971, 'On the Role of Hidden Variables in the Fundamental Structure of

Physics', in T. Bastin (ed.), *Quantum Theory and Beyond*, The University Press, Cambridge, 95–116.

Bohr, N., 1935, 'Can Quantum Mechanical Description of Physical Reality be Considered Complete?' *Physical Reivew* **48**, 696–702.

Bohr, N., 1948, 'On the Notions of Causality and Complmentarity', *Dialectica* **2**, 312–319.

Bohr, N., 1963, 'Quantum Physics and Philosophy: Causality and Complementarity', in N. Bohr, *Essays 1958–1962 on Atomic Physics and Human Knowledge*, Interscience, New York.

Braithwaite, R., 1953, *Scientific Explanation*, The University Press, Cambridge.

Bressan, A., 1972, *A General Interpreted Modal Calculus*, Yale University Press, New Haven.

Bridgman, P., 1961, *The Logic of Modern Physics*, Macmillan, New York.

Bridgman, P., 1962, *A Sophisticate's Primer of Relativity*, Wesleyan University Press, Middletown, Conn.

Brillouin, L., 1956, *Science and Information Theory*, Academic Press, New York.

Broad, C. D., 1914, *Perception, Physics and Reality*, Cambridge University Press.

Bub, J., 1973, 'On the Completeness of Quantum Mechanics', in Hooker (1973).

Bub, J., 1973a, 'On the Possibility of a Phase-Space Reconstruction of Quantum Statistics: A Refutation of the Bell-Wigner Locality Argument', *Foundations of Physics* **3**, 29–44.

Bub, J., 1974. *The Interpretation of Quantum Mechanics*, D. Reidel, Dordrecht.

Bub, J. and Demopoulos, W., 1974a, 'The Interpretation of Quantum Mechanics', in R. Cohen and M. Wartofsky (eds,), *Boston Studies in the Philosophy of Science*, Vol. 13, D. Reidel, Dordrecht.

Bub, J., 1976, 'Randomness and Locality in Quantum Mechanics', in Suppes (1976), 397–420.

Carnap, R., 1922, *Der Raum*, Reuter aund Reichard, Berlin.

Carnap, R., 1926, *Physikalische Begriffsbildung*, Braun, Karlsruhe; reprint: Wissenschaftliche Buchges,. Darmstadt, 1966.

Carnap, R., 1928, *Der logische Aufbau der Welt*, Berlin-Schlachtensee; reprint: Meiner, Hamburg, 1961.

Carnap, R., 1936–37, 'Testability and Meaning', *Philosophy of Science* **3**, 420–471; **4**, 1–40.

Carnap, R., 1947, 'On the Application of Inductive Logic', *Philosophy and Phenomenological Research* **8**, 133–147.

Carnap, R., 1947a, 'Reply to Nelson Goodman', *Philosophy and Pehnomenological Research* **8**, 461–462.

Carnap, R., 1950, *The Logical Foundations of Probability*, University of Chicago Press, Chicago. (2nd ed. 1962).

Carnap, R., 1952, *The Continuum of Inductive Methods*, University of Chicago Press, Chicago.

Carnap, R., 1958, *An Introduction to Symbolic Logic and its Applications*, Dover Publications, New York.

Carnap, R., 1958a, 'Introductory Remarks to the English Edition' in Reichenbach (1958), v-vii.

Carnap, R., 1959, *The Logical Syntax of Language*, Littlefield, Adams, and Co., Paterson, N. J.
Carnap, R., 1962, 'The Aim of Inductive Logic', in E. Nagel *et al.* (eds.), *Logic, Methodology and Philosophy of Science*, Stanford University Press, Stanford, 303–318.
Carnap, R., 1963, 'Replies and Systematic Expositions', in Schilpp (1963).
Carnap, R., 1963a, 'Intellectual Autobiography', in Schilpp (1963).
Carnap, R., 1966, *Philosophical Foundationss of Physics*, Basic Books, New York. Reissued as *An Introduction to the Philosophy of Science*, Harper Torchbooks.
Carnap, R., 1971, 'A Basic System of Inductive Logic', in R. Carnap, *et al.* (eds.), *Studies in Inductive Logic and Probability*, Vol. I, University of California Press, Berkeley and Los Angeles.
Cartan, E., 1923, 'Sur un théorème fondamental de M. H. Weyl dans la théories de l'espace métrique,' *Journal des Mathématiques Pures et Appliques* **9**, 167–192.
Cartwright, N., 1974, 'Correlations without Distributions in Quantum Mechanics', *Foundations of Physics* **4**, 127–136.
Cavalleri, G., 1972, *Lett. Nuovo Cim.* **3**, 608.
Chisholm, R., 1946, 'The Contrary-to-Fact Conditional', *Mind* **55**, 289–307.
Chomsky, N., 1955. 'Logical Syntax and Semantics, their Linguistic Relevance', *Language* **31**, 36–45.
Chomsky, N., 1957, *Syntactic Structures*, Mouton, The Hague.
Chomsky, N., 1964, 'Current Issues in Linguistic Theory', in J. Fodor and J. Katz (eds.), *The Structure of Language*, Prentice-Hall. Englewood Cliffs, N. J., 50–118.
Chomsky, N., 1965, *Aspects of the Theory of Syntax*, MIT Press, Cambridge, Mass.
Church, A., 1949, 'Review of A. J. Ayer, *Language, Truth and Logic*', *Journal of Symbolic Logic* **14**, 52–53.
Clauser, J. *et al.*, 1969, *Physical Review Letters* **23**, 880.
Clauser, J. and Horne, M., 1974, 'Experimental Consequences of Objective Local Theories', *Physical Review* **D10**, 526–535.
Clauser, J., 1976, *Physical Review Letters* **36**, 1223.
Clendinnen, F. J., 1966, 'Induction and Objectivity', *Philosophy of Science* **33**, 215–229.
Clendinnen, F. J., 1970, 'A Response to Jackson', *Philosophy of Science* **37**, 444–448.
Coffa, J., 1974, 'Hempel's Ambiguity', *Synthese* **28**, 161–162.
Cohen, L., 1966, 'Can Quantum Mechanics be Formulated as a Classical Probability Theory', *Philosophy of Science* **33**, 317–322.
Cohen, R., and M. Reichenbach (eds.), 1978, *Hans Reichenbach: Selected Writings, 1909–1953*, D. Reidel, Dordrecht.
Colodny, R. (ed.), 1972, *Paradigms and Paradoxes*, University of Pittsburgh Press, Pittsburgh.
Copi, I., 1972, *Introduction to Logic*, 4th ed., Macmillan, New York.
Costa de Beauregard, O., 1952, 'Irréversibilité quantique, phénomène macroscopique' in *Louis de Broglie, Physicien et Penseur*, A. George (ed.), Albin Michel, Paris, 401.
Costa de Beauregard, O., 1958, *Cahiers de Physique* **86**, 323.
Costa de Beauregard, O., 1964, 'Irreversibility Problems' in Y. Bar Hillel (ed.), *Proceedings of the International Congress on Logic. Methodology and Philosophy of Science*, North-Holland, Amsterdam, 313.
Costa de Beauregard, O., 1965, *Dialectica* **19**, 280.

Costa de Beauregard, O., 1967, *Précise de Mécanique Quantique Relativiste,* Dunod, Paris.

Costa de Beauregard, O., 1968, *Dialectica* **22**, 187.

Costa de Beauregard, O., 1970, in P. T. Landsberg, (ed.), *Proceedings of the International Conference on Themodynamics held in Cardiff,* Butterworths, London, 539.

Costa de Beauregard, O., 1971, *Studium Generale* **24**, 10.

Costa de Beauregard, O., 1976, *Epistemological Letters,* association F. Gouseth, 14.

Costa de Beauregard, O., 1977, *Epistemological Letters,* association F. Gouseth, 15.

Curry, H., 1963, *Foundations of Mathematical Logic,* McGraw-Hill, New York.

Danto, Arthur, 1968, *What Philosophy Is,* Harper & Row, New York.

Demopoulos, W., 1976, 'The Possibility Structure of Physical Systems', in W. Harper and C. A. Hooker (1976), vol. 3, 55–80.

D'Espagnat, B., 1971, *Conceptual Foundations of Quantum Mechanics,* W. A. Benjamin, Menlo Park, California.

Destouches-Fevrier, P., 1951, *Structure des Théories Physiques,* Paris.

Dirac, P., 1958, *Principles of Quantum Mechanics,* 4th ed., Oxford University, Oxford.

Dunn, J. M., 1975, 'Axiomatizing Belnap's Conditional Assertion', *Journal of Philosophical Logic* **4**, 383–97.

Duval, P. and Montredon, E., 1968, *J. Parapsychology* **32**, 153.

Einstein, A., 1905, 'On the Electrodynamics of Moving Bodies', *The Principle of Relativity: A Collection of Original Memoirs,* Dover Publications, New York, 1952.

Einstein, A. 1916 'The Foundation of the General Theory of Relativity', *The Principle of Relativity: A Collection of Original Memoirs,* Dover Publications, New York, 1952.

Einstein, A., 1918a, 'Prinzipielles zur allgemeinen Relativitätstheorie', *Annalen der Physik* **55**, 241–244.

Einstein, A., 1918b, Review of (Weyl, 1918), *Die Naturwissenschaften,* Heft 25, 373.

Einstein, A., 1918c, 'Dialog über Einwände gegen die Relativitätstheorie', *Die Naturwissenschaften,* Heft 48, 697–702.

Einstein, A., 1928, *Rapports et Discussions du 5e Conseil Solvay,* Gauthier Villars, Paris, 253–256.

Einstein, A., 1928a, Review of (Reichenbach, 1928), *Deutsche Literaturzeitung,* Heft 1, columns 19–20.

Einstein, A., Podolsky, B., and Rosen, N., 1935, 'Can Quantum Mechanical Description of Physical Reality Be Considered Complete', *Physical Review* **47**, 770–780.

Einstein, A., 1949, 'Autobiographical Notes', in Schilpp (1949), 1–96.

Einstein, A., 1949a, 'Remarks concerning the Essays Brought Together in this Cooperative Volume', Schilpp (1949), 665–668.

Ellis, B. *at al.*, 1967, 'Conventionality in Distant Simultaneity', *Philosophy of Science* **34**, 116–136.

Farber, M., 1942, 'Logical Systems and the Principles of Logic', *Philosophy of Science* **9**, 40–54.

Feenberg, E., 1974, 'Conventionality in Distant Simultaneity', *Foundations of Physics* **4**, 121–126.

Feigl, H., 1950, 'De Principiis Non Disputandum . . . ?', in M. Black (ed.), *Philosophical Analysis,* Cornell University Press, Ithaca, 119–156.

Feigl, H. and Maxwell, G., eds., 1961, *Current Issues in the Philosophy of Science,* Holt Rinehart Winston, N. Y.

Feinberg, G., 1972, 'Philosophical Implications of Contemporary Particle Physics', in Colodny (1972), 33–46.

Fetzer, J., 1971, 'Dispositional Probabilities', in R. Buck and R. Cohen (eds.), *Boston Studies in the Philosophy of Science*, Vol. VIII, D. Reidel, Dordrecht, 473–482.

Fetzer, J., 1974, 'A Single Case Propensity Theory of Explanation', *Synthese* **28**, 171–198.

Fetzer, J., 1974a, 'Grünbaum's "Defense" of the Symmetry Thesis', *Philosophical Studies* **25**, 173–187.

Fetzer, J., 1974b, 'Statistical Probabilities: Single Case Propensities vs Long Run Frequencies', in W. Leinfellner and E. Köhler (eds.), *Developments in the Methodology of Social Science*, D. Reidel, Dordrecht, 387–397.

Fetzer, J., 1974c, 'Statistical Explanations', in K. Schaffner and R. Cohen (eds.), *Boston Studies in the Philosophy of Science*, Vol. XX, D. Reidel, Dordrecht, 337–348.

Fetzer, J., 1975, 'On the Historical Explanation of Unique Events', *Theory and Decision* **6**, no. 1, 89–91.

Fetzer, J., 1976, 'The Likeness of Lawlikeness', *Boston Studies in the Philosophy of Science*, Vol. XXXII, ed. by A. Michalos and R. Cohen, D. Reidel, Dordrecht.

Fetzer, J., 1976a, 'Elements of Induction', in R. Bogdan (ed.), *Local Induction*, D. Reidel, Dordrecht, 145–170.

Fetzer, J., 1977, 'A World of Dispositions', *Synthese* **34**, 397–422.

Feyerabend, P., 1958, 'Reichenbach's Interpretation of Quantum Mechanics', *Philosophical Studies* **9**, 47–59.

Feyerabend, P., 1970, 'Against Method' in M. Radner and S. Winokur (eds.), *Minnesota Studies in the Philosophy of Science*, Vol. IV, University of Minnesota Press, Minneapolis, 17–130.

Feynman, R. *et al.*, 1965, *The Feynman Lectures on Physics*, Addison-Wesley, Reading, Mass.

Finch, P.D., 1969, 'On the Structure of Quantum Logic', *Journal of Symbolic Logic* **34**, 275–282.

Finch, P.D., 1976, 'Quantum Mechanical Physical Quantities as Random Variables', in W. Harper and C.A. Hooker (1976), Vol. III, 81–103.

Fine, A., 1973, 'Probability and the Interpretation of Quantum Mechanics', *British Journal for the Philosophy of Science* **24**, 1–37.

Fine, A., 1974, 'On the Completeness of Quantum Mechanics', *Synthese* **29**, 257–89. Reprinted in P. Suppes (ed.), 1976, 249–281.

Fine, T., 1973, *Theories of Probability*, Academic Press, New York.

Fock, V., 1948, *Dokl. Adad. Nauk SSSR* **60**, 1157.

Foulis, D. J., 1960, 'Bear *-Semigroups', *Proceedings of the American Mathematical Society* **11**, 648–654.

Foulis, D. J., and Randall, C. H., 1972, 'Operational Statistics. I. Basic Concepts', *Journal of Mathematical Physics* **13**, 1667–1675.

Foulis, D. J., and Randall, C. H., 1974, 'Empirical Logic and Quantum Mechanics', *Synthese* **29**, 81–111.

Freudenthal, H., 1960, 'Zu den Weyl-Cartanschen Raumproblem', *Archiv der Mathematik* **11**, 107–115.

Freudenthal, H., 1965, 'Lie Groups in the Foundations of Geometry', *Advances in Mathematics* **1**, 145–190.

Friedman, S., and Clauser, J., 1972, *Physical Review Letters* **28**, 938.

Friedman, S., 1972, Ph.D. Thesis, University of California.

Fry, E. and Thomson, R., 1976, *Physical Review Letters* **37**, 405.

Galasiewicz, Z. (ed.), 1971, *Helium 4*, Pergamon Press, London.

Gardner, M., 1972, 'Two Deviant Logics for Quantum Theory: Bohr and Reichenbach', *British Journal for the Philosophy of Science* **23**, 89–109.

Gibbs, J. W., 1914, *Elementary Principles in Statistical Mechanics*, Yale University Press, New Haven, Conn.

Giere, R. N., 1973, 'Objective Single-Case Probabilities and the Foundations of Statistics', in P. Suppes, *et al.* (eds), *Logic, Methodology, and Philosophy of Science*, North-Holland, Amsterdam, 467–483.

Giere, R. N., 1975, 'The Epistemological Roots of Scientific Knowledge', in G. Maxwell and R. Anderson (eds.), *Minnesota Studies in the Philosophy of Science*, Vol. VI, University of Minnesota Press, Minneapolis, 212–261.

Gleason, A. M., 1957, 'Measures on Closed Subspaces of Hilbert Space', *Journal of Mathematics and Mechanics* **6**, 885–893.

Glymour, C., 1971, 'Theoretical Realism and Theoretical Equivalence', in R. Buck *et al.* (eds.), *PSA 1970*, D. Reidel, Dordrecht, 275–288.

Glymour, C., 1972, 'Topology, Cosmology and Convention', *Synthese* **24**, 195–218.

Glymour, C., 1977, 'Indistinguishable Space-Times and the Fundamental Group' in Earman, J. *et al.* (eds.), *Minnesota Studies in the Philosophy of Science*, Vol. VIII, University of Minnesota Press, Minneapolis, 50–60.

Glymour, C., Forthcoming, 'Physics and Evidence', in the *University of Pittsburgh Series in the Philosophy of Science*, University of Pittsburgh Press, Pittsburgh.

Gödel, K., 1964, 'What is Cantor's Continuum Problem?', in P. Benacerraf and H. Putnam (eds.), *Philosophy of Mathematics*, Prentice-Hall, Englewood Cliffs, 258–273.

Gold, T., 1962, *American Journal of Physics* **30**, 403.

Goodman, N., 1946, 'A Query on Confirmation', *Journal of Philosophy* **63**, 383–385.

Goodman, N., 1947a, 'On Infirmities of Confirmation Theory', *Philosophy and Phenomenological Research* **8**, 149–151.

Goodman, N., 1947b, 'The Problem of Counterfactual Conditionals', *Journal of Philosophy* **44**, 113–128.

Goodman, N., 1955, *Fact, Fiction, and Forecast*, Harvard University Press, Cambridge. 2nd ed., 1965, Bobbs-Merrill, Indianapolis.

Goodstein, R., 1965, *Essasys in the Philosophy of Mathematics*, Leicester University, Leicester.

Greechie, R. J., and Gudder, S. P., 1973, 'Quantum Logics', in Hooker (1973), 143–173.

Grelling, K., 1930, 'Die Philosophie der Raum-Zeit-Lehre', *Philosophischer Anzeiger* **4**, 101–128.

Grøn, Ø., 1975, 'Relativistic Description of a Rotating Disk', *American Journal of Physics* **43**, 869–876.

Grossman, N., 1972, 'Quantum Mechanics and the Interpretation of Probability', *Philosophy of Science* **39**, 451–460.

Grossman, N., 1974, 'The Ignorance Interpretation Defended', *Philosophy of Science* **41**, 333–344.

Grünbaum, A., 1960, 'The Duhemian Argument', *Philosophy of Science* **27**, 75–87.

Grünbaum, A., 1962, 'Geometry, Chronometry, and Empiricism', in H. Feigl *et al.*

(eds.), *Minnesota Studies in the Philosophy of Science*, Vol. III, University of Minnesota Press, Minneapolis, 405–526.

Grünbaum, A., 1962, *Philosophy of Science* **29**, 146.

Grünbaum, A., 1963, *Philosophical Problems of Space and Time*, A. A. Knopf, New York. 2nd ed., D. Reidel, Dordrecht, 1973.

Grünbaum, A., 1967, *Modern Science and Zeno's Paradoxes*, Wesleyan University Press, Middletown, Conn. 2nd ed., slightly revised, 1968, George Allen and Unwin, London.

Grünbaum, A., 1967, 'Theory of Relativity', in P. Edwards (ed.), *The Encyclopedia of Philosophy*, Macmillan, New York, Vol. 7, 133–140.

Grünbaum, A., 1968, *Geometry and Chronometry in Philosophical Perspective*, University of Minnesota Press, Minneapolis.

Grünbaum, A. (*et al.*), 1969, 'A Panel Discussion of Simultaneity by Slow Clock Transport in the Speical and General Theories of Relativity', *Philosophy of Science* **36**, 1–81.

Grünbaum, A., 1970, 'Space, Time, and Falsifiability', *Philosophy of Science* **37**, 469–588.

Grünbaum, A., 1971, 'The Meaning of Time', in E. Freeman *et al.* (eds.), *Basic Issues in the Philosophy of time*, Open Court, La Salle, Ill., 195–228.

Grünbaum, A., 1973, *Philosophical Problems of Space and Time* 2nd enlarged ed., D. Reidel, Dordrecht.

Grünbaum, A., 1974, 'Popper's Views on the Arrow of Time', in P. Schilpp (ed.), *The Philosophy of Karl Popper*, Open Court, La Salle, Ill., 775–797.

Grünbaum, A., 1977, 'Absolute and Relational Theories of Space and Space-Time', in J. Earman *et al.* (eds.), *Minnesota Studies in the Philosophy of Science*, Vol. VIII, University of Minnesota Press, Minneapolis, 303–373.

Gudder, S. P., 1972, 'Partial Algebraic Structures Associated with Orthomodular Posets', *Pacific Journal of Mathematics* **41**, 717–730.

Haack, S., 1974, *Deviant Logic*, Cambridge University Press, Cambridge.

Hacking, I., 1965, *Logic of Statistical Inference*, Cambridge, Cambridge University Press.

Hacking, I., 1968, 'One Problem About Induction', in I. Lakatos (ed.), *The Problem of Inductive Logic*, North-Holland, Amsterdam, 44–59.

Hamblin, C., 1967, 'Questions', in P. Edwards (ed.), *The Encyclopedia of Philosophy*, Vol. 7, The Macmillan Co., New York, 49–53.

Hardegree, G., 1974, 'The Conditional in Quantum Logic', *Synthese* **29**, 63–80. Reprinted in Suppes (1976) 55–72.

Hardegree, G., 1975, 'Compatibility and Relative Compatibility in Quantum Mechanics', unpublished paper read at the International Congress of Logic, Methodology and Philosophy of Science, London, Ontario, August, 1975.

Hardegree, G., 1975a, 'Stalnaker Conditionals and Quantum Logic', *Journal of Philosophical Logic* **4**, 399–421.

Hardegree, G., 1976, 'The Modal Interpretation of Quantum Mechanics', *Proceedings of the Fifth Biennial Meeting of the Philosophy of Science Assocation*, October, 1976, Chicago.

Hardegree, G., 1976a, 'Semantics and the Interpretation of Quantum Mechanics. Part I', unpublished ms.

Hardegree, G. M., 1977, 'Relative Compatibility in Conventional Quantum Mechanics', *Foundations of Physics* **7**, 495–510.

Hardegree, G. M., 1978, 'The Conditional in Abstract and Concrete Quantum Logic', in C. A. Hooker, (ed.), *The Logico-Algebraic Approach to Quantum Mechanics*, Vol. II, D. Reidel, Dordrecht, 49–108.

Harper, W. L., and Hooker, C.A., eds., 1976, *Foundations of Probability Theory, Statistical Inference, and Statistical Theories of Science*, D. Reidel, Dordrecht.

Harrah, D., 1963, *Communication: A Logical Model*, The MIT Press, Cambridge, Mass.

Hawking, S. and Ellis, G., 1973, *The Large Scale Structure of Space-Time*, Cambridge University Press, Cambridge.

Hawking, S. *et al.*, 1976, 'A New Topology for Curved Space-time Which Incorporates the Causal, Differential, and Conformal Structures', *Journal of Mathematics and Physics* **17**, 174–181.

Heisenberg, W., 1930, *The Physical Principles of the Quantum Theory*, University of Chicago Press, Chicago.

Helmholtz, H., 1962, *Popular Scientific Lectures*, Dover, New York.

Hempel, C., 1945, 'Review of H. Reichenbach, *Philosophic Foundations of Quantum Mechanics*', *Journal of Symbolic Logic* **10**, 97–100.

Hempel, C. *et al.*, 1948, 'Studies in the Logic of Explanation', *Philosophy of Science* **15**, 135–175. Reprinted in Hempel (1965).

Hempel, C., 1951, 'The Concept of Cognitive Significance: A Reconsideration', *Proceedings of the American Academy of Arts and Sciences* **80**, 61–77.

Hempel, C., 1954, 'A Logical Appraisal of Operationism', *Scientific Monthly* **79**, 215–220. Reprinted in Hempel (1965).

Hempel, C., 1962, 'Deductive-Nomological vs. Statistical Explanation', in H. Feigl *et al.* (eds.), *Minnesota Studies in the Philosophy of Science*, Vol. III, University of Minnesota Press, Minneapolis, 98–169.

Hempel, C., 1965, *Aspects of Scientific Explanation*, New York, The Free Press.

Hempel, C., 1966, *Philosophy of Natural Science*, Prentice-Hall, Englewood Cliffs, N. J.

Hempel, C., 1968, 'Lawlikeness and Maximal Specificity in Probabilistic Explanation', *Philosophy of Science* **35**, 116–133.

Hempel, C., 1970, 'On the "Standard Conception" of Scientific Theories', in M. Radner and S. Winokur (eds.), *Minnesota Studies in the Philosophy of Science*, Vol. IV, University of Minnesota Press, Minneapolis, 142–163.

Hempel, C., 1977, 'Nachwort 1976' in *Aspekte wissenschaftlicher Erklärung*, Walter de Gruyter, Berlin, 98–123.

Henkin, L., 1960, 'Review of H. Putnam, "Three-Valued Logic"; P. K. Feyerabend, "Reichenbach's Interpretation of Quantum Mechanics"; and I. Levi, "Putnam's Three Truth Values' ", *Journal of Symbolic Logic* **25**, 289–291.

Hilbert, D., 1902, *Foundations of Geometry*, Open Court Publishing Co., La Salle, Ill.

Hooker, C., 1972, 'The Nature of Quantum Mechanical Reality: Einstein Versus Bohr', in Colodny (1972).

Hooker, C., ed., 1973, *Contemporary Research in the Foundations and Philosophy of Quantum Mechanics*, D. Reidel Publishing Co., Dordrecht & Boston.

Hooker, C., 1974, 'Systematic Realism', *Synthese* **26**, 409.

Hooker, C., ed., 1975, *The Logico-Algebraic Approach to Quantum Mechanics*, Vol. 1: *Historical Evolution*, D. Reidel, Dordrecht.

Hooker, C., and van Fraassen, B., 1976, 'A Semantic Analysis of Niels Bohr's Philosophy of Quantum Theory', in Harper and Hooker (1976), vol. 3, 221–241.

Horne, M., 1970, Ph.D. Thesis, Boston University.

Hughes, G. *et al.*, 1968, *An Introduction to Modal Logic*, Methuen and Co., London.
Hume, D., 1888, *A Treatise of Human Nature*, L. Selby-Bigge (ed.), Clarendon Press, Oxford.
Hutten, E., 1956, *The Language of Modern Physics*, The Macmillan Co., New York.
Ivič, M., 1965, *Trends in Linguistics*, Mouton, The Hague.
Jackson, J., 1962, *Classical Electrodynamics*, John Wiley & Sons, New York.
Jammer, M., 1966, *The Conceptual Development of Quantum Mechanics*, McGraw-Hill, New York.
Jømmer, M., 1974, *The Philosophy of Quantum Mechanics*, McGraw-Hill, New York.
Jauch, J., 1968, *Foundations of Quantum Mechanics*, Addison-Wesley Publishing Co., Reading, Mass.
Jauch, J., 1973, *Are Quanta Real?*, Indiana University Press, Bloomington.
Jauch, J., 1974, 'The Qantum Probability Calculus', *Synthese* **29**, 131–154. Reprinted in Suppes (1976).
Jeffrey, R., 1965, *The Logic of Decision*, McGraw-Hill Book Co., New York.
Jeffrey, R., 1969, 'A Statistical Explanation vs Statistical Inference', in N. Rescher (ed.), *Essays in Honor of Carl G. Hempel*, D. Reidel, Dordrecht, 104–113. Reprinted in Salmon (1971).
Jøgensen, J., 1951, 'The Development of Logical Empiricism', *International Encyclopedia of Unified Science* **2**, No. 9, University of Chicago Press, Chicago-London.
Kac, M., 1959, *Statistical Independence in Probability, Analysis and Number Theory*. American Mathematical Association, Carus Mathematical Monograph # 12.
Kamlah, A., 1976, 'Erläuterungen, Bemerkungen and Verweise', in Reichenbach (1976), vol. 2.
Katz, J., 1962, *The Problem of Induction and Its Solution*, University of Chicago Press.
Katz, J., 1966, *The Philosophy of Language*, Harper and Row, New York.
Kirk, G. *et al.*, 1966, *The Presocratic Philosophers*, Cambridge University Press, Cambridge.
Klein, F., 1924, *Elementary Mathematics from an Advanced Standpoint*, Dover, New York.
Kochen, S. and Specker, E., 1967, 'The Problem of Hidden Variables in Quantum Mechanics', *Journal of Mathematical Mechanics* **17**, 59–87.
Kolmogorov, A. N., 1933, *Foundations of the Theory of Probability* (tr. Nathan Morrison), Chelsea Publishing Co., New York, 1950 (orig. 1933).
Kretzmann, N., 1967, 'Semantics, History of', in P. Edwards (ed.), *The Encyclopedia of Philosophy*, Macmillan, New York-London, Vol. 7, 358–406.
Kuhn, T., 1957, *The Copernican Revolution*, Harvard University Press, Cambridge.
Kyburg, H., and Nagel, E. (eds.), 1963, *Induction, Some Current Issues*, Wesleyan University Press, Middletown, Conn.
Kyburg, H., 1965, 'Comments', *Philosophy of Science* **32**, 147–151.
Kyburg, H., 1970, 'Discussion: More on Maximal Specificity', *Philosophy of Science* **37**, 295–300.
Kyburg, H., 1974, 'Propensities and Probabilities', *British Journal for the Philosophy of Science* **25**, 358–375.
Lacey, H., 1968, 'The Causal Theory of Time: A Critique of Grünbaum's Version', *Philosophy of Science* **35**, 332–354.

Lakatos, I., 1963, 'Proofs and Refutations', *British Journal for the Philosophy of Science* **14**, 1–25.

Lakatos, I. (ed.), 1968, *The Problem of Inductive Logic*, North-Holland Publishing Co., Amsterdam.

Lambert, K., 1969, 'Logical Truth and Microphysics', in K. Lambert (ed.), *The Logical Way of Doing Things*, Yale Press, New Haven.

Latzer, R., 1972, 'Nondirected Light Signals and the Structure of Time', *Synthese* **24**, 236–280.

Laplace, P., 1951, *A Philosophical Essay on Probabilities*, Dover Publications, New York.

Laugwitz, D., 1958, 'Ueber eine Vermutung von Hermann Weyl zum Raumproblem', *Archiv der Mathematik* **9**, 128–133.

Laugwitz, D., 1965, *Differential and Riemannian Geometry*, Academic Press.

Lenard, P., 1918, *Ueber Relativitätsprinzip, Aether, Gravitation*, Leipzig.

Levi, I., 1959, 'Putnam's Three Truth Values', *Philosophical Studies* **10**, 65–69.

Levi, I., 1969, 'Are Statistical Hypotheses Covering Laws?', *Synthese* **20**, 297–307.

Levi-Civita, T., 1917, 'Nozione di Parallelismo in una varietà qualunque e Conseguente Specificazione Geometrica della Curvatura Riemanniana', *Rendiconti del Circolo de Palermo* **42**, 173–215.

Lewis, C., 1946, *An Analysis of Knowledge and Valuation*, Open Court Publishing Co., La Salle, Ill.

Lewis, D., 1970, 'General Semantics', *Synthese* **22**, 18–67.

Lewis, D., 1973, *Counterfactuals*, Harvard University Press, Cambridge, Mass.

Lewis, G., 1930, *Science* **71**, 569.

London, F. and Bauer, E., 1939, *La Théorie de l'Observation en Mécanique Quantique*, Hermann, Paris.

Lucasiewicz, J., 1920, 'On Three-Valued Logic' (in Polish), *Ruch Filozoficzny* **5**, 169–170.

Mackey, G., 1963, *Mathematical Foundations of Quantum Mechanics*, Benjamin, New York.

Malament, D., 1977, 'Observationally Indistinguishable Space-Times', in Earman, J. *et al.* (eds.), *Minnesota Studies in the Philosophy of Science*, Vol. VIII, University of Minnesota Press, Minneapolis, 61–80.

Malament, D., 'The Class of Continuous Timelike Curves Determines the Topology of Spacetime', p. 11, Theorem 2, unpublished.

Malament, Ph.D. Thesis (Rockefeller University), chap. II, unpublished.

Margenau, H., 1963, 'Measurements and Quantum States', I and II, *Philosophy of Science* **30**, 1–16, 138–157.

Marsh, G., 1971, *Nature of Physical Science* **230**, 197.

Martin, R., 1971, *Logic, Language and Metaphysics*, New York University Press, New York.

Mates, B., 1972, *Elementary Logic*, 2nd ed., Oxford University Press, New York.

McLennan, J., 1960, *Physics Fluids* **3**, 193.

McMahon, 1976, *Hans Reichenbach's Philosophy of Grammar*, Mouton, The Hague.

Mehlberg, H., 1949/50, 'The Idealistic Interpretation of Atomic Physics', *Studia Philosophica* **4**, 171–235.

Mehlberg, H., 1961, 'Physical Laws and Time Arrow', in Feigl and Maxwell (1961).

Mellor, D., 1971, *The Matter of Chance*, The University Press, Cambridge.
Menger, K., 1952, *Calculus, A Modern Approach*, University of Chicago, Chicago.
Messiah, A., 1958, *Quantum Mechanics*, North-Holland, Amsterdam, (trans. by G.M. Temmer).
Meyerhoff, H., 1951, 'Emotive and Existentialist Theories of Ethics', *Journal of Philosophy* **48**, 769–783.
Montague, R., 1968, 'Pragmatics' in R. Klibansky (ed.), *Contemporary Philosophy: A Survey*, La Nuova Italia Editrice, Florence, 102–122. Reprinted in Montague (1974).
Montague, R., 1970, 'Pragmatics and Intensional Logic', *Synthese* **22**, 68–94.
Montague, R., 1974, *Formal Philosophy*, R. Thomason (ed.), Yale University Press, New Haven.
Møller, C., 1972, *The Theory of Relativity*, 2nd ed., Oxford University Press, Oxford.
Nagel, E., 1939, Principles of the Theory of Probability, in O. Neurath *et al.* (eds.), *International Encyclopedia of Unified Science*, Vol. 1, no. 6, University of Chicago Press, Chicago.
Nagel, E., 1945, 'Review of Hans Reichenbach's *Philosophic Foundations of Quantum Mechanics*', *Journal of Philosophy* **42**, 437–444.
Nagel, E., 1946, 'Professor Reichenbach on Quantum Mechanics: A Rejoinder', *Journal of Philosophy* **43**, 247–250.
Nagel, E., 1955, 'Review of Reichenbach, *The Theory of Probability*', *The Journal of Philosophy* **47**, 551–555.
Nagel, E., 1961, *The Structure of Science*, Harcourt, Brace & World, New York & Burlingame.
Nelson, E., 1967, *Dynamical Theories of Brownian Motion*, Princeton University Press.
Newburg, R., 1972, *Lett Nuovo Cim.* **5**, 387–388.
Niiniluoto, I., 1976, 'Inductive Explanation, Propensity, and Action', in *Essays on Explanation and Understanding*, J. Manninen and R. Tuomela (eds.), D. Reidel, Dordrecht.
Noonan, T., 1971, *Nature of Physical Science* **230**, 197.
Pap, A., 1949, *Elements of Analytic Philosophy*, Macmillan, New York.
Park, J., and Margenau, H., 1968, 'Simultaneous Measurability in Quantum Theory', *International Journal of Theoretical Physics* **1**, 211–283.
Park, J., and Margenau, H., 1971, 'Simultaneous Measurability in Quantum Theory', in W. Yourgrau and A. van der Merwe (eds.), *Perspectives in Quantum Theory: Essays in Honor of Alfred Lande*, MIT Press, Cambridge, Mass.
Park, J. and Margenau, H., 1971a, 'The Logic of Noncommutability of Quantum Mechanical Operators and Its Empirical Consequences', in W. Yourgrau and A. van der Merwe (eds.), *Perspectives in Quantum Theory*, MIT Press, Cambridge, 37–70.
Pauli, W., 1964, 'Reviewing Study of Hans Reichenbach's *Philosophic Foundations of Quantum Mechanics*', in R. Kronig and V.F. Weisskopf (eds.), *Collected Scientific Papers*, Vol. 2, Interscience Publishers, New York.
Pauling, L., 1959, *No More War*, Dodd, Mead & Co., New York.
Pearson, K., 1957, *The Grammar of Science*, 3rd ed., Meridian Books, New York.
Peirce, C., 1931, *The Collected Papers of Charles Sanders Peirce*, C. Hartshorne *et al.* (eds.), Harvard University Press, Cambridge, Mass.
Penrose, O., and Percival, I., 1962, *Proceedings of Physical Society* **79**, 493.
Piron, C., 1964, 'Axiomatique Quantique', *Helvetia Physica Acta* **37**, 439–468.

Piron, C., 1972, 'Survey of General Quantum Physics', *Foundations of Physics* **2**, 287–314.

Poincaré, H., 1908, *Science et Méthode*, Flammarion, Paris.

Poincaré, H., 1914, *Science and Method*, London.

Poincaré, H., 1952, *Science and Hypothesis*, Dover Publications, New York.

Pool, J.C.T., 1963, 'Simultaneous Observability and the Logic of Quantum Mechanics', Unpublished doctoral dissertation, State University of Iowa.

Pool, J.C.T., 1968, 'Baer *-Semigroups and the Logic of Quantum Mechanics', *Communications in Mathematical Physics* **9**, 118–141.

Popper, K., 1957, 'The Propensity Interpretation of the Calculus of Probability, and the Quantum Theory', in S. Körner (ed.), *Observation and Interpretation*, Butterworths Scientific Publications, London, 65–70.

Popper, K., 1959, 'The Propensity Interpretation of Probability', *British Journal for the Philosophy of Science* **10**, 25–42.

Popper, K., 1959a, *The Logic of Scientific Discovery*, Basic Books, New York. Revised edition, Hutchinson, London, 1968.

Popper, K., 1967, 'Quantum Mechanics without "The Observer"', in M. Bunge, ed., *Quantum Theory and Reality*, Springer-Verlag, New York, 7–44.

Popper, K., 1972, *Objective Knowledge*, Clarendon Press, Oxford.

Post, E., 1921, 'Introduction to a General Theory of Elementary Propositions', *American Journal of Mathematics* **43**, 163ff.

Prior, A., 1957, *Time and Modality*, Clarendon Press, Oxford.

Prior, A., 1967, 'Logic, Modal', in P. Edwards (ed.), *The Encyclopedia of Philosophy*, Macmillan Publishing Co., New York.

Prior, A., 1968, *Past, Present and Future*, Clarendon Press, Oxford.

Putnam, H., 1951, *The Meaning of the Concept of Probability in Application to Finite Sequences*, Ph.D. Thesis, University of California at Los Angeles.

Putnam, H., 1957, 'Three-Valued Logic', *Philosophical Studies* **8**, 73–80.

Putnam, H., 1963, 'An Examination of Grünbaum's Philosophy of Geometry', in B. Baumrin (ed.), *Philosophy of Science, The Delaware Seminar*, Vol. 2, Interscience, New York.

Putnam, H., 1970, 'Is Logic Empirical?', in R.S. Cohen and M. Wartofsky (eds.), *Boston Studies in the Philosophy of Science*, Vol. 5, D. Reidel, Dordrecht.

Putterman, S. and Rudnick, I., 1971, 'Quantum Nature of Superfluid Helium', *Physics Today* **24**, 40ff.

Quine, W., 1960, *Word and Object*, MIT Press, Cambridge, Mass.

Quine, W., 1963, 'Carnap and Logical Truth' in Schilpp (1963), 385–406.

Ramsey, F., 1931, *The Foundations of Mathematics and other Logical Essays*, Routledge and Kegan Paul, London.

Randall, C.H., and Foulis, D.J., 1973, 'Operational Statistics, II. Manuals of Operations and their Logics', *Journal of Mathematical Physics* **14**, 1472–1480.

Randall, C.H., and Foulis, D.J., 1976, 'A Mathematical Setting for Inductive Reasoning', in W. Harper and C.A. Hooker (1976) Vol. III, 169–205.

Randall, C.H., and Foulis, D.J., 1978, 'The Operational Approach to Quantum Mechanics', in C.A. Hooker (ed.), *Physics as a Logico-Operational Structure*, D. Reidel, Dordrecht, Holland, in press.

Reichenbach, H., 1912, 'Studentenschaft und Katholizismus', in W. Ostwald (ed.),

Das Monistische Jahrhundert, No. 16, 533–538. English translation in R. Cohen *et al.*, (1978).

Reichenbach, H., 1913, 'Die freistudentische Idee. Ihr Inhalt als Einheit', in *Freistudententum. Versuch einer Synthese der freistudentischen Ideen.* In verbindung mit Karl Landauer, herausgegeben von Hermann Kranold, Max Steinebach, München, 1913, 23–40. English translation to be published in R. Cohen *et al.*, (1978).

Reichenbach, H., 1916, 'Der Begriff der Wahrscheinlichkeit für die mathematische Darstellung der Wirklichkeit', *Zeitschrift für Philosophie und philosophische Kritik* **161**, 210–239; **162**, 98–112, 223–253.

Reichenbach, H., 1920, *Relativitätstheorie und Erkenntnis Apriori*, Springer, Berlin. English translation: H. Reichenbach (1965).

Reichenbach, H., 1921, 'Der gegenwärtige Stand der Relativitätsdiskussion', *Logos* **10**, 316–378; translated (with discussion of Weyl's work omitted) in Reichenbach (1959).

Reichenbach, H., 1924, *Axiomatik der Relativistischen Raum-Zeit-Lehre*, Vieweg, Braunschweig. English translation, Reichenbach (1969).

Reichenbach, H., 1924a, 'Die Bewegungslehre bei Newton, Leibniz und Huyghens', *Kantstudien* **29**, 416–438. English translation in Reichenbach (1959).

Reichenbach, H., 1925, 'Die Kausalstruktur der Welt und der Unterschied von Vergangenheit und Zukunft', *Bayerische Akademie der Wissenschaften, Sitzungsberichte*, 133–175.

Reichenbach, H., 1925a, 'Planetenuhr und Einsteinsche Gleichzeitigkeit', *Zeitschrift für Physik* **33**, 628–634.

Reichenbach, H., 1925b, 'Ueber die physikalischen konsequenzen der relativistischen Axiomatik', *Zeitschrift für Physik* **34**, 34–48.

Reichenbach, H., 1928, *Philosophie der Raum-Zeit-Lehre*, Walter de Gruyter, Berlin & Leipzig. English translation, Reichenbach (1958). 2nd ed. Reichenbach (1976) vol. 2.

Reichenbach, H., 1929, 'Ziele und Wege der physikalischen Erkenntnis', in H. Geiger and K. Scheel (eds.), *Handbuch der Physik*, Springer, Berlin, Vol. 4, 1–80.

Reichenbach, H., 1930, *Atom und Kosmos*, Deutsche Buch-Gemeinschaft, Berlin.

Reichenbach, H., 1930a, 'Kausalität und Wahrscheinlichkeit' *Erkenntnis* **1**, 158–188. English translation in Reichenbach (1959).

Reichenbach, H., 1932, *Atom and Cosmos*, George Allen & Unwin, London.

Reichenbach, H., 1932a, 'Die Kausalbehauptung und die Möglichkeit ihrer empirischen Nachprufung', *Erkenntnis* **3**, 32–64. English trans. in Reichenbach (1959).

Reichenbach, H., 1946a, 'Philosophy and Physics', in *Faculty Research Lectures – University of California*, #19, University of California Press, Berkeley and Los Angeles.

Reichenbach, H., 1946b, 'Reply to Ernest Nagel's Criticism of My Views on Quantum Mechanics', *Journal of Philosophy* **43**, 239–247.

Reichenbach, H., 1947, *Elements of Symbolic Logic*, The Macmillan Co., New York.

Reichenbach, H., 1947a, 'Philosophy: Speculation or Science', *Nation* **164**, 20–22.

Reichenbach, H., 1948, 'Rationalism and Empiricism: An Inquiry into the Roots of Philosophical Error', *Philosophical Review* **57**, 330–346. Reprinted in Reichenbach (1959).

Reichenbach, H., 1948a, 'Theory of Series and Gödel's Theorems', in Cohen *et al.* (1978).

Reichenbach, H., 1948b, 'The Principle of Anomaly in Quantum Mechanics', *Dialectica* **2**, 337–350.

Reichenbach, H., 1949, 'The Philosophical Significance of the Theory of Relativity', in Schilpp (1949), 287–311.

Reichenbach, H., 1949a, *The Theory of Probability*, 2nd ed., University of California Press, Berkeley and Los Angeles.

Reichenbach, H., 1951, *The Rise of Scientific Philosophy*, University of California Press, Berkeley-Los Angeles; German translation: Reichenbach (1965); 2nd ed., Reichenbach (1976), vol. 1.

Reichenbach, H., 1951a, 'The Verifiability Theory of Meaning', *Proceedings of the American Academy of Arts and Sciences* **80**, 46–60.

Reichenbach, H., 1951b, 'Über die erkenntnistheoretische Problemlage und den Gebrauch einer dreiwertigen Logik in der Quantenmechanik', *Z. Naturforschung* **6a**, 569–575.

Reichenbach, H., 1952, 'Are Phenomenal Reports Absolutely Certain?', *The Philosophical Review* **61**, 147–159.

Reichenbach, H., 1953, 'Les fondements logiques de la mécanique des quanta', *Extraits des Annales de l'Institut Henri Poincaré* **13**, 109–158.

Reichenbach, H., 1954, *Nomological Statements and Admissible Operations*, North-Holland Publishing Co., Amsterdam.

Reichenbach, H., 1956, *The Direction of Time*, University of California Press, Berkeley & Los Angeles.

Reichenbach, H., 1958, *The Philosophy of Space and Time*, Dover Publications, New York. English Translation of Reichenbach (1928).

Reichenbach, H., 1959, *Modern Philosophy of Science*, Routledge & Kegan Paul, London.

Reichenbach, H., 1965, *Der Aufstieg der wissenschaftlichen Philosophie* (translation of Reichenbach, 1951), Vieweg, Wiesbaden.

Reichenbach, H., 1965a, *The Theory of Relativity and A Priori Knowledge*, University of California Press, Berkeley & Los Angeles. English translation of Reichenbach (1920).

Reichenbach, H., 1969, *Axiomatization of the Theory of Relativity*, University of California Press, Berkeley & Los Angeles. Translation of Reichenbach (1924).

Reichenbach, H., 1976, *Gesammelte Werke in 9 Bänden*, Vieweg, Wiesbaden.

Reichenbach, H., 1976a, *Laws, Modalities, and Counterfactuals*, University of California Press, Berkeley & Los Angeles. Reprint of Reichenbach (1954).

Renyi, A., 1955, 'On a New Axiomatic Theory of Probability', *Acta Mathematica Hungarica* **6**, 285–333.

Rescher, N., 1966, *The Logic of Commands*, Dover Publications, New York.

Rescher, N., 1969, *Many-Valued Logic*, McGraw-Hill, New York.

Rescher, N., 1970, *Scientific Explanation*, The Free Press, New York.

Rescher, N. *et al.*, 1971, *Temporal Logic*, Springer-Verlag, New York.

Robb, A., 1914, *A Theory of Time and Space*, Cambridge.

Rorty, R. (ed.), 1967, *The Linguistic Turn: Recent Essays in Philosophical Method*, Chicago-London.

Russell, B., 1897, *An Essay on the Foundations of Geometry*, The University Press, Cambridge. Reprinted, 1956, Dover Publications, New York.

Russell, B., 1922, *Our Knowledge of the External World,* George Allen & Unwin, London.
Russell, B., 1926, *Unser Wissen von der Aubenwelt* (German translation of *Our Knowledge of the External World*), Meiner, Leipzig.
Russell, B., 1940, *An Inquiry into Meaning and Truth,* W. W. Norton & Co., New York.
Russell, B., 1948, *Human Knowledge, Its Scope and Limits,* Simon & Schuster, New York.
Salmon, W., 1953, 'The Frequency Interpretation and Antecedent Probabilities', *Philosophical Studies* **4**, 44–48.
Salmon, W., 1955, 'The Short Run', *Philosophy of Science* **22**, 214–221.
Salmon, W., 1957, 'The Predictive Inference', *Philosophy of Science* **24**, 180–190.
Salmon, W., 1957a, 'Should We Attempt to Justify Induction?' *Philosophical Studies* **8**, 33–48.
Salmon, W., 1961, 'Vindication of Induction', in Feigl and Maxwell, (1961), 245–256.
Salmon, W., 1963, 'On Vindicating Induction', *Philosophy of Science* **30**, 252–261. Also published in Kyburg and Nagel (1963).
Salmon, W., 1965, 'The Concept of Inductive Evidence', in 'Symposium on Inductive Evidence', *American Philosophical Quarterly* **2**, 1–16.
Salmon, W., 1966, 'The Foundations of Scientific Inference', in R. Colodny (ed.), *Mind and Cosmos,* University of Pittsburgh Press, Pittsburgh, 135–275.
Salmon, W., 1966a, 'Verifiability and Logic', in P. Feyerabend *et al.* (ed.), *Mind, Matter, and Method,* University of Minnesota Press, Minneapolis, 354–376.
Salmon, W., 1967, *The Foundations of Scientific Inference*, University of Pittsburgh Press, Pittsburgh.
Salmon, W., 1967a, 'Carnap's Inductive Logic', *Journal of Philosophy* **64**, 725–740.
Salmon, W., 1968, 'The Justification of Inductive Rules of Inference', in Lakatos (1968), 24–43.
Salmon, W., 1968a, 'Reply' (to discussion of Salmon 1968), in Lakatos (1968), 74–93.
Salmon, W., ed., 1970, *Zeno's Paradoxes,* Bobbs-Merrill, Indianapolis.
Salmon, W., *et al.*, 1971, *Statistical Explanation and Statistical Relevance,* University of Pittsburgh Press, Pittsburgh.
Salmon, W., 1974, 'Comments on "Hempel's Ambiguity" by J. Alberto Coffa', *Synthese* **28**, 165–169.
Salmon, W., 1974a, 'Russell on Scientific Inference or Will the Real Deductivist Please Stand Up?', in G. Nakhnikian (ed.), *Bertrand Russell's Philosophy,* Gerald Duckworth & Co., London, 183–208.
Salmon, W., 1974b, 'Memory and Perception in Human Knowledge', in G. Nakhnikian, (ed.), *Bertrand Russell's Philosophy,* Gerald Duckworth & Co., London, 139–167.
Salmon, W., 1975, 'Theoretical Explanation', in S. Körner (ed.), *Explanation,* Basil Blackwell, Oxford, 118–145.
Salmon, W., 1977, 'An "At-At" Theory of Causal Influence', *Philosophy of Science* **44**, 215–224.
Salmon, W., 1977a, 'A Third Dogma of Empiricism', in R. Butts *et al.* (eds.), *Basic Problems of Methodology and Linguistics*, D. Reidel, Dordrecht, 149–166.
Salmon, W., 1977b, 'Hans Reichenbach', in *Dictionary of American Biography, Supplement Five, 1951–1955*, Charles Scribner's Sons, New York, 562–563.
Savage, L., 1954, *Foundations of Statistics*, John Wiley & Sons, New York.

Scheibe, E., 1954, 'Ueber das Weylsche Raumproblem, '*Journal für Mathematik* **197**, 162–207.

Schilpp, P. (ed.), 1946, *The Philosophy of Bertrand Russell,* Library of Living Philosophers, Evanston.

Schilpp. P. (ed.), 1949, *Albert Einstein: Philosopher-Scientist*, Tudor, Evanston.

Schilpp, P. (ed.), 1963, *The Philosophy of Rudolf Carnap,* Open Court, La Salle.

Schleichert. H., ed., 1975, *Logischer Empirismus-Der Wiener Kreis* (Essays of the Vienna Circle) Fink, München.

Schlesinger, G., 1963, *Methods in the Physical Sciences*, Routledge and Kegan Paul, London.

Schlesinger, G., 1974, *Confirmation and Confirmability*, Clarendon Press, Oxford.

Schlick, M., 1918, *Allgemeine Erkenntnislehre*, Springer, Berlin.

Schlick, M., 1925, 2nd ed. of Schlick (1918).

Schlick, M., 1949, 'Meaning and Verification', *Philosophical Review* **44** (1936); reprinted in H. Feigl and W. Sellars (eds.), *Readings in Philosophical Analysis*, Appleton-Century-Crofts, New York, 146–170.

Schlick, M., 1963, *Space and Time in Contemporary Physics*, Dover Publications, New York.

Schlick, M., 1974, *General Theory of Knowledge*, (trans. of Schlick, 1925) Springer, Wien-New York.

Schrödinger, E., 1935, *Naturwiss* **23**, 807, 823 and 844.

Schrödinger, E., 1950, *Proceedings of the Royal Irish Academy* **53**, A189.

Schmidt, H., 1970, *Journal of Parapsychology* **34**, 255.

Schuster, C., 1975, 'Hans Reichenbach', in C. Gillispie (ed.), *Dictionary of Scientific Biography*, Vol. 11, Charles Scribner's Sons, New York, 355–359.

Scriven, M., 1975, 'Causation as Explanation', *Nous* **9**, 3–16.

Seelig, C., 1972, *Albert Einstein und die Schweiz*, Europa Verlag.

Shimony, A., 1971, in *Foundations of Quantum Mechanics*, B. d'Espaquat (ed.), Academic Press, New York.

Shimony, A., 1973, 'The Status of Hidden Variable Theories', in P. Suppes *et al.* (eds.), *Logic Methodology and Philosophy of Science* IV, North-Holland, Amsterdam.

Sklar, L., 1974, *Space, Time, and Spacetime*, University of California, Berkeley.

Sklar, L., 1977, 'Facts, Conventions and Assumptions in the Theory of Spacetime', in J. Earman *et al.* (eds.), *Minnesota Studies in the Philosophy of Science*, vol. VIII, 206–274.

Skyrms, B., 1965, 'On Failing to Vindicate Induction', *Philosophy of Science* **32**, 253–68.

Skyrms, B., 1975, *Choice and Chance,* 2nd ed., Dickenson, Encino, Calif.

Smart, J., 1963, *Philosophy and Scientific Realism,* Routledge and Kegan Paul, London.

Smith, N., ed., 1929, *Kant's Inaugural Dissertation and Early Writings on Space*, J. Handiside, Open Court, La Salle, Ill.

Sneed, J., 1970, 'Quantum Mechanics and Classical Probability Theory', *Synthese* **21**, 34–64.

Sneed, J., 1971, *The Logical Structure of Mathematical Physics*, D. Reidel, Dordrecht.

Sosa, E., ed., 1975, *Causation and Conditionals*, Oxford University Press, New York.

Stalnaker, R.C., 1968, 'A Theory of Conditionals', in N. Rescher (ed.), *Studies in Logical Theory,* Blackwell, Oxford.

Stalnaker, R.C., 1970, 'Probability and Conditionals', *Philosophy of Science* **37**, 68–80.

Stalnaker, R., and Thomason, R., 1970, 'A Semantic Analysis of Conditional Logic', *Theoria* **36**, 23–42.

Stein, H., 1967, 'Newtonian Space-Time', *Texas Quartely* **10**, 174–200.

Stein, H., 1972, 'On the Conceptual Structure of Quantum Mechanics', in Colodny (1972), 367–438.

Stevenson, C., 1944, *Ethics and Language*, Yale University Press, New Haven.

Strauss, M, 1937–38, 'Mathematics as Logical Syntax – a Method to Formalize the Language of a Physical Theory', *Erkenntnis* **7**, 147–153.

Strauss, M., 1972, *Modern Physics and Its Philosophy*, D. Reidel Publishing Co., Dordrecht and Boston.

Strauss, M., 1974, 'Rotating Frames in Special Relativity', *International Journal of Theoretical Physics* **11**, 107–123.

Strawson, P., 1952, *Introduction to Logical Theory*, Methuen & Co., London.

Suppe, F., 1967, *Meaning and Use of Models in Mathematics and the Exact Sciences*, Doctoral Dissertation, University of Michigan.

Suppe, F., 1972, 'Theories, Their Formulations, and the Operational Imperative', *Synthese* **25**, 129–164.

Suppes, P., 1965, 'Logics Appropriate to Empirical Theories', in J. Addison *et al.* (eds.), *The Theory of Models*, North-Holland Publishing Co., Amsterdam.

Suppes, P., 1966, 'The Probabilistic Argument for a Non-Classical Logic in Quantum Mechanics', *Philosophy of Science* **23**, 14–21.

Suppes, P., 1967, 'What is a Scientific Theory?', in S. Morgenbesser (ed.), *Philosophy of Science Today*, Basic Books, New York.

Suppes, P., 1967, 'Set-Theoretical Structures in Science', Mimeo'd Stanford University.

Suppes, P., 1970, *A Probabilistic Theory of Causation*, North-Holland, Amsterdam.

Suppes, P, (ed.), 1976, *Logic and Probability in Quantum Mechanics*, D. Reidel, Dordrecht.

Suzuki, M., 1971, *Nature of Physical Science* **230**, 13.

Tarski, A., 1944, 'The Semantic Conception of Truth and the Foundations of Semantics', *Philosophy and Phenomenological Research* **4**, 341–375.

Terletsky, J., 1970, *Journal of Physics* **21**, 680.

Thomason, R., 1973, 'Philosophy and Formal Semantics', in H. Leblanc (ed.), *Truth, Syntax and Modality*, North-Holland, Amsterdam.

Törnebohm, H., 1957, 'On Two Logical Systems Proposed in the Philosophy of Quantum Mechanics', *Theoria* **23**, 84–101.

Turquette, A., 1945, 'Review of Reichenbach's *Philosophic Foundations of Quantum Mechanics*', *Philosophical Review* **54**, 513–516.

Urmson, J., 1956, 'Some Questions concerning Validity', in A. Flew (ed.), *Essays in Conceptual Analysis*, Macmillan, London, 120–133.

van Fraassen, B., 1968, 'Presupposition, Implication, and Self-Reference', *Journal of Philosophy* **65**, 136–152.

van Fraassen, B., 1969, 'Presuppositions, Supervaluations, and Free Logic', in K. Lambert (ed.), *The Logical Way of Doing Things*, Yale University Press, New Haven, 67–91.

van Fraassen, B., 1970, *An Introduction to the Philosophy of Time and Space*, Random House, New York.

van Fraassen, B., 1970a, 'On the Extension of Beth's Semantics of Physical Theories', *Philosophy of Science* **37**, 325–339.

van Fraassen, B., 1971, *Formal Semantics and Logic*, Macmillan, New York.
van Fraassen, B., 1973, 'Semantic Analysis of Quantum Logic', in Hooker (1973).
van Fraassen, B., 1974, 'Hidden Variables in Conditional Logic', *Theoria* **40**, 176–190.
van Fraassen, B., 1974, 'The Labyrinth of Quantum Logics', in R. Cohen *et al.* (eds.), *Logical and Epistemological Studies in Contemporary Physics*, D. Reidel, Dordrecht, 224–254.
van Fraassen, B., 1975, 'Construction of Popper Probability Functions,' reprint.
van Fraassen, B., 1975, 'Incomplete Assertion and Belnap Connectives', in D. Hockney *et al.* (eds.), *Contemporary Research in Philosophical Logic and Linguistic Semantics*, Reidel, Dordrecht, 43–70.
van Fraassen, B., 1976, 'Probabilities of Conditionals', in Harper and Hooker (1976) Vol. 1, 261–301.
van Fraassen, B., 1977, 'The Pragmatics of Explanation', *American Philosophical Quarterly* **14**, 143–150.
Varadarajan, V., 1962, 'Probability in Physics and a Theorem on Simultaneous Observability', *Communications in Pure and Applied Mathematics* **15**, 189–217.
Varadarajan, V., 1968, *Geometry of Quantum Theory* (Vol. I), Van Nostrand, Princeton.
Venn, J., 1866, *The Logic of Chance*, Macmillan & Co., London and Cambridge.
von Mises, R., 1957, *Probability, Statistics and Truth*, The Macmillan Co., New York.
von Mises, R., 1964, *Mathematical Theory of Probability and Statistics*, Academic Press, New York.
von Neumann, J., 1955, *Mathematical Foundations of Quantum Mechanics*, Princeton University Press, Princeton.
von Waals, J., 1911, *Phys. Zeit* **12**, 547.
von Weizsacker, C., 1939, *Annals of Physics* **36**, 275.
von Weizsacker, C., 1955, 'Komplementarität und Logik', *Naturwissenschaften* **42**, 521–29 and 545–55.
von Wright, G., 1968, *An Essay in Deontic Logic and the General Theory of Action*, North-Holland, Amsterdam.
Wang, H., 1961, 'Process and Existence in Mathematics', in *Essays on the Foundations of Mathematics*, Y. Bar-Hillel *et al.* (eds.), The Magnes Press, Jerusalem, 328–351.
Watanabe, S., 1951, *Physical Review* **84**, 1008.
Watanabe, S., 1952, 'Reversibilité contre irreversibilité' in A. George (ed.), *Louis de Broglie, Physicien et Penseur*, Albin Michel, Paris.
Watanabe, S., 1975, *Physical Review* **27**, 179.
Watkins, G., 1971, *Proceedings of the Parapsychological Association* **8**, 23.
Watkins, J., 1968, 'Non-Inductive Corroboration', in Lakatos (1968) 61–66.
Weinreich, U., 1963, 'On the Semantic Structure of Language', in Joseph H. Greenberg (ed.), *Universals of Language*, MIT Press, Cambridge, Mass., 114–115.
Weyl, Hermann, 1918a, *Raum-Zeit-Materie*, lst edition, Springer, Berlin.
Weyl, Hermann, 1918b, 'Reine Infinitesimalgeometrie', *Mathematische Zeitschrift* **2**, 384–411; also in (Weyl, 1968), Vol. II, 1–28.
Weyl, Hermann, 1918c, 'Gravitation und Elektrizität', *Sitzungsberichte der Königlich Preussischen Akademie der Wissenschaften zu Berlin*, 465–480, also in (Weyl, 1968), Vol. II, 29–42.
Weyl, Hermann, 1919, 'Eine neue Erweiterung der Relativitätstheorie', *Annalen der Physik* **59**, 101–133; also in (Weyl, 1968), Vol. II, 55–87.

Weyl, Hermann, 1920a, 'Die Einsteinsche Relativitätstheorie', *Schweizerland; Schweizerische Bauzeitung*, 1921; also in (Weyl, 1968), Vol. II, 123–140.

Weyl, Hermann, 1920b, 'Elektrizität und Gravitation,' *Physikalische Zeitschrift* **21**, 649–650; also in (Weyl, 1968), Vol. II, 141–142.

Weyl, Hermann, 1921a, *Raum-Zeit-Materie*, 4th edition, Springer, Berlin.

Weyl, Hermann, 1921b, 'Feld und Materie,' *Annalen der Physik* **65**, 541–563; also in (Weyl, 1968), Vol. II, 237–259.

Weyl, Hermann, 1921c, 'Electricity and gravitation,' *Nature* **106**, 800–802; also in (Weyl, 1968), Vol. II, 260–262.

Weyl, Hermann, 1922a, *Space-Time-Matter*, Dover, New York (translation of (Weyl, 1921a)).

Weyl, Hermann, 1922b, 'Die Einzigartigkeit der Pythagoreischen Massbestimmung', *Mathematische Zeitschrift* **12**, 114–146; also in (Weyl, 1968), Vol. II, 263–295.

Weyl, Hermann, 1923a, *Raum-Zeit-Materie*, 5th edition, Springer, Berlin.

Weyl, Hermann, 1923b, Commentary to Riemann's Inaugural Dissertation in H. Weyl, *Das Kontinuum und andere Monographien*, Chelsea Publishing Co., New York (no date).

Weyl, Hermann, 1923c, *Mathematische Analyse des Raumproblems*, Springer, Berlin.

Weyl, Hermann, 1924a, Review of (Reichenbach, 1924), *Deutsche Literaturzeitung*, no. 30, columns 2122–2128.

Weyl, Hermann, 1924b, 'Massenträgheit und Kosmos. Ein Dialog,' *Die Naturwissenschaften* **12**, 197–204; also in (Weyl, 1968), Vol. II, 478–485.

Weyl, Hermann, 1927, *Philosophie der Mathematik und Naturwissenschaft*, München und Berlin.

Weyl, Hermann, 1928, 'Diskussionsbemerkungen zu dem zweiten Hilbertschen Vortrag über die Grundlagen der Mathematik,' *Abhandlungen aus dem mathematischen Seminar der Hamburgischen Universität* **6**, 86–88; also in (Weyl, 1968), Vol. III, 147–149.

Weyl, Hermann, 1929a, 'Consistency in Mathematics', *The Rice Institute Pamphlet* **16**, 245–265; also in (Weyl, 1968), Vol. III, 150–170.

Weyl, Hermann, 1929b, 'On the Foundations of Infinitesimal Geometry', *Bulletin of the American Mathematical Society* **35**, 716–725; also in (Weyl, 1968), Vol. III, 207–216.

Weyl, Hermann, 1929c, 'Gravitation and the Electron', *The Rice Institute Pamphlet* **16**, 280–295; also in (Weyl, 1968), Vol. III, 229–244.

Weyl, Hermann, 1929d, 'Elektron und Gravitation', *Zeitshcrift für Physik* **56**, 330–352; also in (Weyl, 1968), Vol. III, 245–267).

Weyl, Hermann, 1931, 'Geometrie und Physik', *Die Naturwissenschaften* **19**, 49–58; also in (Weyl, 1968), Vol. III, 336–345.

Weyl, Hermann, 1932, *The Open World*, Yale University Press, New Haven.

Weyl, Hermann, 1934, *Mind and Nature*, University of Pennsylvania Press, Philadelphia.

Weyl, Hermann, 1940, 'The Ghost of Modality', in M. Farber (ed.), *Philosophical Essays in Memory of Edmund Husserl*, Harvard Press, Cambridge.

Weyl, Hermann, 1946, 'Review: The Philosophy of Bertrand Russell', *The American Mathematical Monthly* **53**, 208–214; also in (Weyl, 1968), Vol. IV, 500–605.

Weyl, Hermann, 1949a, *Philosophy of Mathematics and Natural Science*, Princeton University Press, Princeton, N. J.

Weyl, Hermann, 1949b, 'Wissenschaft als symbolische Konstruktion des Menschen', *Eranos-Jahrbuch* 1948, 375–431; also in (Weyl, 1968), Vol. IV, 289–345.

Weyl, Hermann, 1950, '50 Jahre Relativitätstheorie', *Die Naturwissenschaften* **38**, 73–83; also in (Weyl, 1968), Vol. IV, 421–431.

Weyl, Hermann, 1952, *Symmetry*, Princeton University Press, Princeton, N.J.

Weyl, Hermann, 1953, 'Über den Symbolismus der Mathematik und mathematischen Physik', *Studium generale* **6**, 219–228; also in (Weyl, 1968), Vol. IV, 527–536.

Weyl, Hermann, 1954, 'Erkenntnis und Besinnung, (Ein Lebensrückblick)', *Studia Philosophica, Jahrbuch der Schweizerischen Philosophischen Gessellschaft, Annuaire de la Société Suisse de Philosophie*; also in (Weyl, 1968), Vol. IV, 631–649.

Weyl, H., 1963, *Philosophy of Mathematics and Natural Science*, Atheneum, New York.

Weyl, Hermann, 1968, *Gesammelte Abhandlungen*, 4 volumes, Springer, Berlin.

Wheeler, J., 1962, 'Curved Empty Space-Time as the Building Material of the Physical World', in E. Nagel *et al.* (eds.), *Logic, Methodology and Philosophy of Science*, Stanford University Press, Stanford, 361–374.

Wheeler, J., 1971, 'Forward to *The Conceptual Foundations of Contemporary Relativity Theory*, by J. Graves', The MIT Press, Cambridge, Mass.

Wheeler, J., *et al.*, 1973, *Gravitation*, W. H. Freeman & Co., San Francisco.

Whitbeck, C., 1969, 'Simultaneity and Distance', *Journal of Philosophy* **66**, 329–340.

Whitehead, A., and Russell, B., 1910–13, *Principia Mathematica*, The University Press, Cambridge.

Whitehead, A., 1929, *Process and Reality*, The Macmillan Co., New York.

Whitmire, D., 1972, *Nature of Physical Science* **235**, 175–176.

Wigner, E., 1970, 'On Hidden Variables and Quantum Mechanical Probabilities', *American Journal of Physics* **38**, 1005–1009.

Will, C., 1974, 'Gravitation Theory', in *Scientific American* **231**, 24–33.

Wilson, C., 1969, 'From Kepler's Laws, So-called, to Universal Gravitation: Empirical Factors', *Archive for the History of Exact Sciences* **6**.

Winnie, J., 1970, 'Special Relativity without One-way Velocity Assumptions', *Philosophy of Science* **37**, 81–99, 223–238.

Winnie, J., 1972, 'The Twin-Rod Thought Experiment', *American Journal of Physics* **40**, 1091–1094.

Winnie, J., 1977, 'The Causal Theory of Space-time' in J. Earman *et al.* (eds.), *Foundations of Space-Time Theories, Minnesota Studies in the Philosophy of Science*, Vol. VIII, University of Minnesota Press, Minneapolis.

Wu, T., and Rivier, D., 1961, *Helvetia Physica Acta* **34**, 661.

Yanase, N., 1957, *Annals of the Japanese Association for Philosophy of Science* **2**, 131.

Zeeman, E., 1967, 'The Topology of Minkowski Space', *Topology* **6**, 161–170.

INDEX OF NAMES

(Compiled by J/P. Thomas)

ANALYTICAL INDEX OF SUBJECTS

(Compiled by J/P. Thomas)

This index has been carefully constructed to aid the reader in tracing important themes through the various essays in which they are treated.

SYNTHESE LIBRARY

Studies in Epistemology, Logic, Methodology, and Philosophy of Science

1. J. M. Bocheński, *A Precis of Mathematical Logic.* 1959, X + 100 pp.
2. P. L. Guiraud, *Problèmes et méthodes de la statistique linguistique.* 1960, VI + 146 pp.
3. Hans Freudenthal (ed.), *The Concept and the Role of the Model in Mathematics and Natural and Social Sciences. Proceedings of a Colloquium held at Utrecht, The Netherlands, January 1960.* 1961, VI + 194 pp.
4. Evert W. Beth, *Formal Methods. An Introduction to Symbolic Logic and the Study of Effective Operations in Arithmetic and Logic.* 1962, XIV + 170 pp.
5. B. H. Kazemier and D. Vuysje (eds.), *Logic and Language. Studies Dedicated to Professor Rudolf Carnap on the Occasion of His Seventieth Birthday.* 1962, VI + 256 pp.
6. Marx W. Wartofsky (ed.), *Proceedings of the Boston Colloquium for the Philosophy of Science 1961-1962,* Boston Studies in the Philosophy of Science (ed. by Robert S. Cohen and Marx W. Wartofsky), Volume I. 1963, VIII + 212 pp.
7. A. A. Zinov'ev, *Philosophical Problems of Many-Valued Logic.* 1963, XIV + 155 pp.
8. Georges Gurvitch, *The Spectrum of Social Time.* 1964, XXVI + 152 pp.
9. Paul Lorenzen, *Formal Logic.* 1965, VIII + 123 pp.
10. Robert S. Cohen and Marx W. Wartofsky (eds.), *In Honor of Philipp Frank,* Boston Studies in the Philosophy of Science (ed. by Robert S. Cohen and Marx W. Wartofsky), Volume II. 1965, XXXIV + 475 pp.
11. Evert W. Beth, *Mathematical Thought. An Introduction to the Philosophy of Mathematics.* 1965, XII + 208 pp.
12. Evert W. Beth and Jean Piaget, *Mathematical Epistemology and Psychology.* 1966, XII + 326 pp.
13. Guido Küng, *Ontology and the Logistic Analysis of Language. An Enquiry into the Contemporary Views on Universals.* 1967, XI + 210 pp.
14. Robert S. Cohen and Marx W. Wartofsky (eds.), *Proceedings of the Boston Colloquium for the Philosophy of Science 1964-1966, in Memory of Norwood Russell Hanson,* Boston Studies in the Philosophy of Science (ed. by Robert S. Cohen and Marx W. Wartofsky), Volume III. 1967, XLIX + 489 pp.

15. C. D. Broad, *Induction, Probability, and Causation. Selected Papers.* 1968, XI + 296 pp.
16. Günther Patzig, *Aristotle's Theory of the Syllogism. A Logical-Philosophical Study of Book A of the Prior Analytics.* 1968, XVII + 215 pp.
17. Nicholas Rescher, *Topics in Philosophical Logic.* 1968, XIV + 347 pp.
18. Robert S. Cohen and Marx W. Wartofsky (eds.), *Proceedings of the Boston Colloquium for the Philosophy of Science 1966-1968,* Boston Studies in the Philosophy of Science (ed. by Robert S. Cohen and Marx W. Wartofsky), Volume IV. 1969, VIII + 537 pp.
19. Robert S. Cohen and Marx W. Wartofsky (eds.), *Proceedings of the Boston Colloquium for the Philosophy of Science 1966-1968,* Boston Studies in the Philosophy of Science (ed. by Robert S. Cohen and Marx W. Wartofsky), Volume V. 1969, VIII + 482 pp.
20. J.W. Davis, D. J. Hockney, and W. K. Wilson (eds.), *Philosophical Logic.* 1969, VIII + 277 pp.
21. D. Davidson and J. Hintikka (eds.), *Words and Objections: Essays on the Work of W.V. Quine.* 1969, VIII + 366 pp.
22. Patrick Suppes, *Studies in the Methodology and Foundations of Science. Selected Papers from 1911 to 1969.* 1969, XII + 473 pp.
23. Jaakko Hintikka, *Models for Modalities. Selected Essays.* 1969, IX + 220 pp.
24. Nicholas Rescher *et al.* (eds.), *Essays in Honor of Carl G. Hempel. A Tribute on the Occasion of His Sixty-Fifth Birthday.* 1969, VII + 272 pp.
25. P. V. Tavanec (ed.), *Problems of the Logic of Scientific Knowledge.* 1969, XII + 429 pp.
26. Marshall Swain (ed.), *Induction, Acceptance, and Rational Belief.* 1970, VII + 232 pp.
27. Robert S. Cohen and Raymond J. Seeger (eds.), *Ernst Mach: Physicist and Philosopher,* Boston Studies in the Philosophy of Science (ed. by Robert S. Cohen and Marx W. Wartofsky), Volume VI. 1970, VIII + 295 pp.
28. Jaakko Hintikka and Patrick Suppes, *Information and Inference.* 1970, X + 336 pp.
29. Karel Lambert, *Philosophical Problems in Logic. Some Recent Developments.* 1970, VII + 176 pp.
30. Rolf A. Eberle, *Nominalistic Systems.* 1970, IX + 217 pp.
31. Paul Weingartner and Gerhard Zecha (eds.), *Induction, Physics, and Ethics: Proceedings and Discussions of the 1968 Salzburg Colloquium in the Philosophy of Science.* 1970, X + 382 pp.
32. Evert W. Beth, *Aspects of Modern Logic.* 1970, XI + 176 pp.
33. Risto Hilpinen (ed.), *Deontic Logic: Introductory and Systematic Readings.* 1971, VII + 182 pp.
34. Jean-Louis Krivine, *Introduction to Axiomatic Set Theory.* 1971, VII + 98 pp.
35. Joseph D. Sneed, *The Logical Structure of Mathematical Physics.* 1971, XV + 311 pp.
36. Carl R. Kordig, *The Justification of Scientific Change.* 1971, XIV + 119 pp.
37. Milič Čapek, *Bergson and Modern Physics,* Boston Studies in the Philosophy of Science (ed. by Robert S. Cohen and Marx W. Wartofsky), Volume VII. 1971, XV + 414 pp.

38. Norwood Russell Hanson, *What I Do Not Believe, and Other Essays* (ed. by Stephen Toulmin and Harry Woolf), 1971, XII + 390 pp.
39. Roger C. Buck and Robert S. Cohen (eds.), *PSA 1970. In Memory of Rudolf Carnap,* Boston Studies in the Philosophy of Science (ed. by Robert S. Cohen and Marx W. Wartofsky), Volume VIII. 1971, LXVI + 615 pp. Also available as paperback.
40. Donald Davidson and Gilbert Harman (eds.), *Semantics of Natural Language.* 1972, X + 769 pp. Also available as paperback.
41. Yehoshua Bar-Hillel (ed.), *Pragmatics of Natural Languages.* 1971, VII + 231 pp.
42. Sören Stenlund, *Combinators, λ-Terms and Proof Theory.* 1972, 184 pp.
43. Martin Strauss, *Modern Physics and Its Philosophy. Selected Papers in the Logic, History, and Philosophy of Science.* 1972, X + 297 pp.
44. Mario Bunge, *Method, Model and Matter.* 1973, VII + 196 pp.
45. Mario Bunge, *Philosophy of Physics.* 1973, IX + 248 pp.
46. A. A. Zinov'ev, *Foundations of the Logical Theory of Scientific Knowledge (Complex Logic),* Boston Studies in the Philosophy of Science (ed. by Robert S. Cohen and Marx W. Wartofsky), Volume IX. Revised and enlarged English edition with an appendix, by G. A. Smirnov, E. A. Sidorenka, A. M. Fedina, and L. A. Bobrova. 1973, XXII + 301 pp. Also available as paperback.
47. Ladislav Tondl, *Scientific Procedures,* Boston Studies in the Philosophy of Science (ed. by Robert S. Cohen and Marx W. Wartofsky), Volume X. 1973, XII + 268 pp. Also available as paperback.
48. Norwood Russell Hanson, *Constellations and Conjectures* (ed. by Willard C. Humphreys, Jr.). 1973, X + 282 pp.
49. K. J. J. Hintikka, J. M. E. Moravcsik, and P. Suppes (eds.), *Approaches to Natural Language. Proceedings of the 1970 Stanford Workshop on Grammar and Semantics.* 1973, VIII + 526 pp. Also available as paperback.
50. Mario Bunge (ed.), *Exact Philosophy – Problems, Tools, and Goals.* 1973, X + 214 pp.
51. Radu J. Bogdan and Ilkka Niiniluoto (eds.), *Logic, Language, and Probability. A Selection of Papers Contributed to Sections IV, VI, and XI of the Fourth International Congress for Logic, Methodology, and Philosophy of Science, Bucharest, September 1971.* 1973, X + 323 pp.
52. Glenn Pearce and Patrick Maynard (eds.), *Conceptual Change.* 1973, XII + 282 pp.
53. Ilkka Niiniluoto and Raimo Tuomela, *Theoretical Concepts and Hypothetico-Inductive Inference.* 1973, VII + 264 pp.
54. Roland Fraïssé, *Course of Mathematical Logic* – Volume 1: *Relation and Logical Formula.* 1973, XVI + 186 pp. Also available as paperback.
55. Adolf Grünbaum, *Philosophical Problems of Space and Time.* Second, enlarged edition, Boston Studies in the Philosophy of Science (ed. by Robert S. Cohen and Marx W. Wartofsky), Volume XII. 1973, XXIII + 884 pp. Also available as paperback.
56. Patrick Suppes (ed.), *Space, Time, and Geometry.* 1973, XI + 424 pp.
57. Hans Kelsen, *Essays in Legal and Moral Philosophy,* selected and introduced by Ota Weinberger. 1973, XXVIII + 300 pp.
58. R. J. Seeger and Robert S. Cohen (eds.), *Philosophical Foundations of Science. Proceedings of an AAAS Program, 1969,* Boston Studies in the Philosophy of

Science (ed. by Robert S. Cohen and Marx W. Wartofsky), Volume XI. 1974, X + 545 pp. Also available as paperback.

59. Robert S. Cohen and Marx W. Wartofsky (eds.), *Logical and Epistemological Studies in Contemporary Physics*, Boston Studies in the Philosophy of Science (ed. by Robert S. Cohen and Marx W. Wartofsky), Volume XIII. 1973, VIII + 462 pp. Also available as paperback.
60. Robert S. Cohen and Marx W. Wartofsky (eds.), *Methodological and Historical Essays in the Natural and Social Sciences. Proceedings of the Boston Colloquium for the Philosophy of Science 1969-1972*, Boston Studies in the Philosophy of Science (ed. by Robert S. Cohen and Marx W. Wartofsky), Volume XIV. 1974, VIII + 405 pp. Also available as paperback.
61. Robert S. Cohen, J. J. Stachel and Marx W. Wartofsky (eds.), *For Dirk Struik. Scientific, Historical and Political Essays in Honor of Dirk J. Struik*, Boston Studies in the Philosophy of Science (ed. by Robert S. Cohen and Marx W. Wartofsky), Volume XV. 1974, XXVII + 652 pp. Also available as paperback.
62. Kazimierz Ajdukiewicz, *Pragmatic Logic*, transl. from the Polish by Olgierd Wojtasiewicz. 1974, XV + 460 pp.
63. Sören Stenlund (ed.), *Logical Theory and Semantic Analysis. Essays Dedicated to Stig Kanger on His Fiftieth Birthday*. 1974, V + 217 pp.
64. Kenneth F. Schaffner and Robert S. Cohen (eds.), *Proceedings of the 1972 Biennial Meeting, Philosophy of Science Association*, Boston Studies in the Philosophy of Science (ed. by Robert S. Cohen and Marx W. Wartofsky), Volume XX. 1974, IX + 444 pp. Also available as paperback.
65. Henry E. Kyburg, Jr., *The Logical Foundations of Statistical Inference*. 1974, IX + 421 pp.
66. Marjorie Grene, *The Understanding of Nature: Essays in the Philosophy of Biology*, Boston Studies in the Philosophy of Science (ed. by Robert S. Cohen and Marx W. Wartofsky), Volume XXIII. 1974, XII + 360 pp. Also available as paperback.
67. Jan M. Broekman, *Structuralism: Moscow, Prague, Paris*. 1974, IX + 117 pp.
68. Norman Geschwind, *Selected Papers on Language and the Brain*, Boston Studies in the Philosophy of Science (ed. by Robert S. Cohen and Marx W. Wartofsky), Volume XVI. 1974, XII + 549 pp. Also available as paperback.
69. Roland Fraïssé, *Course of Mathematical Logic* – Volume 2: *Model Theory*. 1974, XIX + 192 pp.
70. Andrzej Grzegorczyk, *An Outline of Mathematical Logic. Fundamental Results and Notions Explained with All Details*. 1974, X + 596 pp.
71. Franz von Kutschera, *Philosophy of Language*. 1975, VII + 305 pp.
72. Juha Manninen and Raimo Tuomela (eds.), *Essays on Explanation and Understanding. Studies in the Foundations of Humanities and Social Sciences*. 1976, VII + 440 pp.
73. Jaakko Hintikka (ed.), *Rudolf Carnap, Logical Empiricist. Materials and Perspectives*. 1975, LXVIII + 400 pp.
74. Milič Čapek (ed.), *The Concepts of Space and Time. Their Structure and Their Development*, Boston Studies in the Philosophy of Science (ed. by Robert S. Cohen and Marx W. Wartofsky), Volume XXII. 1976, LVI + 570 pp. Also available as paperback.

75. Jaakko Hintikka and Unto Remes, *The Method of Analysis. Its Geometrical Origin and Its General Significance,* Boston Studies in the Philosophy of Science (ed. by Robert S. Cohen and Marx W. Wartofsky), Volume XXV. 1974, XVIII + 144 pp. Also available as paperback.
76. John Emery Murdoch and Edith Dudley Sylla, *The Cultural Context of Medieval Learning. Proceedings of the First International Colloquium on Philosophy, Science, and Theology in the Middle Ages – September 1973,* Boston Studies in the Philosophy of Science (ed. by Robert S. Cohen and Marx W. Wartofsky), Volume XXVI. 1975, X + 566 pp. Also available as paperback.
77. Stefan Amsterdamski, *Between Experience and Metaphysics. Philosophical Problems of the Evolution of Science,* Boston Studies in the Philosophy of Science (ed. by Robert S. Cohen and Marx W. Wartofsky), Volume XXXV. 1975, XVIII + 193 pp. Also available as paperback.
78. Patrick Suppes (ed.), *Logic and Probability in Quantum Mechanics.* 1976, XV + 541 pp.
79. Hermann von Helmholtz: *Epistemological Writings. The Paul Hertz/Moritz Schlick Centenary Edition of 1921 with Notes and Commentary by the Editors.* (Newly translated by Malcolm F. Lowe. Edited with an Introduction and Bibliography, by Robert S. Cohen and Yehuda Elkana), Boston Studies in the Philosophy of Science (ed. by Robert S. Cohen and Marx W. Wartofsky), Volume XXXVII. 1977, XXXVIII+204 pp. Also available as paperback.
80. Joseph Agassi, *Science in Flux,* Boston Studies in the Philosophy of Science (ed. by Robert S. Cohen and Marx W. Wartofsky), Volume XXVIII. 1975, XXVI + 553 pp. Also available as paperback.
81. Sandra G. Harding (ed.), *Can Theories Be Refuted? Essays on the Duhem-Quine Thesis.* 1976, XXI + 318 pp. Also available as paperback.
82. Stefan Nowak, *Methodology of Sociological Research: General Problems.* 1977, XVIII + 504 pp.
83. Jean Piaget, Jean-Blaise Grize, Alina Szeminska, and Vinh Bang, *Epistemology and Psychology of Functions,* Studies in Genetic Epistemology, Volume XXIII. 1977, XIV+205 pp.
84. Marjorie Grene and Everett Mendelsohn (eds.), *Topics in the Philosophy of Biology,* Boston Studies in the Philosophy of Science (ed. by Robert S. Cohen and Marx W. Wartofsky), Volume XXVII. 1976, XIII + 454 pp. Also available as paperback.
85. E. Fischbein, *The Intuitive Sources of Probabilistic Thinking in Children.* 1975, XIII + 204 pp.
86. Ernest W. Adams, *The Logic of Conditionals. An Application of Probability to Deductive Logic.* 1975, XIII + 156 pp.
87. Marian Przełęcki and Ryszard Wójcicki (eds.), *Twenty-Five Years of Logical Methodology in Poland.* 1977, VIII + 803 pp.
88. J. Topolski, *The Methodology of History.* 1976, X + 673 pp.
89. A. Kasher (ed.), *Language in Focus: Foundations, Methods and Systems. Essays Dedicated to Yehoshua Bar-Hillel,* Boston Studies in the Philosophy of Science (ed. by Robert S. Cohen and Marx W. Wartofsky), Volume XLIII. 1976, XXVIII + 679 pp. Also available as paperback.
90. Jaakko Hintikka, *The Intentions of Intentionality and Other New Models for Modalities.* 1975, XVIII + 262 pp. Also available as paperback.

91. Wolfgang Stegmüller, *Collected Papers on Epistemology, Philosophy of Science and History of Philosophy*, 2 Volumes, 1977, XXVII + 525 pp.
92. Dov M. Gabbay, *Investigations in Modal and Tense Logics with Applications to Problems in Philosophy and Linguistics*. 1976, XI + 306 pp.
93. Radu J. Bogdan, *Local Induction*. 1976, XIV + 340 pp.
94. Stefan Nowak, *Understanding and Prediction: Essays in the Methodology of Social and Behavioral Theories*. 1976, XIX + 482 pp.
95. Peter Mittelstaedt, *Philosophical Problems of Modern Physics*, Boston Studies in the Philosophy of Science (ed. by Robert S. Cohen and Marx W. Wartofsky), Volume XVIII. 1976, X + 211 pp. Also available as paperback.
96. Gerald Holton and William Blanpied (eds.), *Science and Its Public: The Changing Relationship*, Boston Studies in the Philosophy of Science (ed. by Robert S. Cohen and Marx W. Wartofsky), Volume XXXIII. 1976, XXV + 289 pp. Also available as paperback.
97. Myles Brand and Douglas Walton (eds.), *Action Theory. Proceedings of the Winnipeg Conference on Human Action, Held at Winnipeg, Manitoba, Canada, 9-11 May 1975*. 1976, VI + 345 pp.
98. Risto Hilpinen, *Knowledge and Rational Belief*. 1979 (forthcoming).
99. R. S. Cohen, P. K. Feyerabend, and M. W. Wartofsky (eds.), *Essays in Memory of Imre Lakatos*, Boston Studies in the Philosophy of Science (ed. by Robert S. Cohen and Marx W. Wartofsky), Volume XXXIX. 1976, XI + 762 pp. Also available as paperback.
100. R. S. Cohen and J. J. Stachel (eds.), *Selected Papers of Léon Rosenfeld*, Boston Studies in the Philosophy of Science (ed. by Robert S. Cohen and Marx W. Wartofsky), Volume XXI. 1978, XXX + 927 pp.
101. R. S. Cohen, C. A. Hooker, A. C. Michalos, and J. W. van Evra (eds.), *PSA 1974: Proceedings of the 1974 Biennial Meeting of the Philosophy of Science Association*, Boston Studies in the Philosophy of Science (ed. by Robert S. Cohen and Marx W. Wartofsky), Volume XXXII. 1976, XIII + 734 pp. Also available as paperback.
102. Yehuda Fried and Joseph Agassi, *Paranoia: A Study in Diagnosis*, Boston Studies in the Philosophy of Science (ed. by Robert S. Cohen and Marx W. Wartofsky), Volume L. 1976, XV + 212 pp. Also available as paperback.
103. Marian Przełęcki, Klemens Szaniawski, and Ryszard Wójcicki (eds.), *Formal Methods in the Methodology of Empirical Sciences*. 1976, 455 pp.
104. John M. Vickers, *Belief and Probability*. 1976, VIII + 202 pp.
105. Kurt H. Wolff, *Surrender and Catch: Experience and Inquiry Today*, Boston Studies in the Philosophy of Science (ed. by Robert S. Cohen and Marx W. Wartofsky), Volume LI. 1976, XII + 410 pp. Also available as paperback.
106. Karel Kosík, *Dialectics of the Concrete*, Boston Studies in the Philosophy of Science (ed. by Robert S. Cohen and Marx W. Wartofsky), Volume LII. 1976, VIII + 158 pp. Also available as paperback.
107. Nelson Goodman, *The Structure of Appearance*, Boston Studies in the Philosophy of Science (ed. by Robert S. Cohen and Marx W. Wartofsky), Volume LIII. 1977, L + 285 pp.
108. Jerzy Giedymin (ed.), *Kazimierz Ajdukiewicz: The Scientific World-Perspective and Other Essays, 1931 - 1963*. 1978, LIII + 378 pp.

109. Robert L. Causey, *Unity of Science.* 1977, VIII+185 pp.
110. Richard E. Grandy, *Advanced Logic for Applications.* 1977, XIV + 168 pp.
111. Robert P. McArthur, *Tense Logic.* 1976, VII + 84 pp.
112. Lars Lindahl, *Position and Change: A Study in Law and Logic.* 1977, IX + 299 pp.
113. Raimo Tuomela, *Dispositions.* 1978, X + 450 pp.
114. Herbert A. Simon, *Models of Discovery and Other Topics in the Methods of Science,* Boston Studies in the Philosophy of Science (ed. by Robert S. Cohen and Marx W. Wartofsky), Volume LIV. 1977, XX + 456 pp. Also available as paperback.
115. Roger D. Rosenkrantz, *Inference, Method and Decision.* 1977, XVI + 262 pp. Also available as paperback.
116. Raimo Tuomela, *Human Action and Its Explanation. A Study on the Philosophical Foundations of Psychology.* 1977, XII + 426 pp.
117. Morris Lazerowitz, *The Language of Philosophy. Freud and Wittgenstein,* Boston Studies in the Philosophy of Science (ed. by Robert S. Cohen and Marx W. Wartofsky), Volume LV. 1977, XVI + 209 pp.
118. Tran Duc Thao, *Origins of Language and Consciousness,* Boston Studies in the Philosophy of Science (ed. by Robert S. Cohen and Marx. W. Wartofsky), Volume LVI. 1979 (forthcoming).
119. Jerzy Pelč, *Semiotics in Poland, 1894 - 1969.* 1977, XXVI + 504 pp.
120. Ingmar Pörn, *Action Theory and Social Science. Some Formal Models.* 1977, X + 129 pp.
121. Joseph Margolis, *Persons and Minds, The Prospects of Nonreductive Materialism,* Boston Studies in the Philosophy of Science (ed. by Robert S. Cohen and Marx W. Wartofsky), Volume LVII. 1977, XIV + 282 pp. Also available as paperback.
122. Jaakko Hintikka, Ilkka Niiniluoto, and Esa Saarinen (eds.), *Essays on Mathematical and Philosophical Logic. Proceedings of the Fourth Scandinavian Logic Symposium and of the First Soviet-Finnish Logic Conference, Jyväskylä, Finland, 1976.* 1978, VIII + 458 pp. + index.
123. Theo A. F. Kuipers, *Studies in Inductive Probability and Rational Expectation.* 1978, XII + 145 pp.
124. Esa Saarinen, Risto Hilpinen, Ilkka Niiniluoto, and Merrill Provence Hintikka (eds.), *Essays in Honour of Jaakko Hintikka on the Occasion of His Fiftieth Birthday.* 1978, IX + 378 pp. + index.
125. Gerard Radnitzky and Gunnar Andersson (eds.), *Progress and Rationality in Science,* Boston Studies in the Philosophy of Science (ed. by Robert S. Cohen and Marx W. Wartofsky), Volume LVIII. 1978, X + 400 pp. + index. Also available as paperback.
126. Peter Mittelstaedt, *Quantum Logic.* 1978, IX + 149 pp.
127. Kenneth A. Bowen, *Model Theory for Modal Logic. Kripke Models for Modal Predicate Calculi.* 1978, X + 128 pp.
128. Howard Alexander Bursen, *Dismantling the Memory Machine. A Philosophical Investigation of Machine Theories of Memory.* 1978, XIII + 157 pp.
129. Marx W. Wartofsky, *Models: Representation and Scientific Understanding,* Boston Studies in the Philosophy of Science (ed. by Robert S. Cohen and Marx W. Wartofsky), Volume XLVIII. 1979 (forthcoming). Also available as a paperback.
130. Don Ihde, *Technics and Praxis. A Philosophy of Technology,* Boston Studies in

the Philosophy of Science (ed. by Robert S. Cohen and Marx W. Wartofsky), Volume XXIV. 1979 (forthcoming). Also available as a paperback.

131. Jerzy J. Wiatr (ed.), *Polish Essays in the Methodology of the Social Sciences,* Boston Studies in the Philosophy of Science (ed. by Robert S. Cohen and Marx W. Wartofsky), Volume XXIX. 1979 (forthcoming). Also available as a paperback.
132. Wesley C. Salmon (ed.), *Hans Reichenbach: Logical Empiricist.* 1979 (forthcoming).

SYNTHESE HISTORICAL LIBRARY

Texts and Studies
in the History of Logic and Philosophy

Editors:

1. M. T. Beonio-Brocchieri Fumagalli, *The Logic of Abelard.* Translated from the Italian. 1969, IX + 101 pp.
2. Gottfried Wilhelm Leibniz, *Philosophical Papers and Letters.* A selection translated and edited, with an introduction, by Leroy E. Loemker. 1969, XII + 736 pp.
3. Ernst Mally, *Logische Schriften,* ed. by Karl Wolf and Paul Weingartner. 1971, X + 340 pp.
4. Lewis White Beck (ed.), *Proceedings of the Third International Kant Congress.* 1972, XI + 718 pp.
5. Bernard Bolzano, *Theory of Science,* ed. by Jan Berg. 1973, XV + 398 pp.
6. J. M. E. Moravcsik (ed.), *Patterns in Plato's Thought. Papers Arising Out of the 1971 West Coast Greek Philosophy Conference.* 1973, VIII + 212 pp.
7. Nabil Shehaby, *The Propositional Logic of Avicenna: A Translation from al-Shifā: al-Qiyās,* with Introduction, Commentary and Glossary. 1973, XIII + 296 pp.
8. Desmond Paul Henry, *Commentary on De Grammatico: The Historical-Logical Dimensions of a Dialogue of St. Anselm's.* 1974, IX + 345 pp.
9. John Corcoran, *Ancient Logic and Its Modern Interpretations.* 1974, X + 208 pp.
10. E. M. Barth, *The Logic of the Articles in Traditional Philosophy.* 1974, XXVII + 533 pp.
11. Jaakko Hintikka, *Knowledge and the Known. Historical Perspectives in Epistemology.* 1974, XII + 243 pp.
12. E. J. Ashworth, *Language and Logic in the Post-Medieval Period.* 1974, XIII + 304 pp.
13. Aristotle, *The Nicomachean Ethics.* Translated with Commentaries and Glossary by Hypocrates G. Apostle. 1975, XXI + 372 pp.
14. R. M. Dancy, *Sense and Contradiction: A Study in Aristotle.* 1975, XII + 184 pp.
15. Wilbur Richard Knorr, *The Evolution of the Euclidean Elements. A Study of the Theory of Incommensurable Magnitudes and Its Significance for Early Greek Geometry.* 1975, IX + 374 pp.
16. Augustine, *De Dialectica.* Translated with Introduction and Notes by B. Darrell Jackson. 1975, XI + 151 pp.

17. Arpád Szabó, *The Beginnings of Greek Mathematics*. 1979 (forthcoming).
18. Rita Guerlac, *Juan Luis Vives Against the Pseudodialecticians. A Humanist Attack on Medieval Logic*. Texts, with translation, introduction and notes. 1978, xiv + 227 pp. + index.

SYNTHESE LANGUAGE LIBRARY

Texts and Studies
in Linguistics and Philosophy

Managing Editors:

1. Henry Hiż (ed.), *Questions.* 1977, xvii + 366 pp.
2. William S. Cooper, *Foundations of Logico-Linguistics. A Unified Theory of Information, Language, and Logic.* 1978, xvi + 249 pp.
3. Avishai Margalit (ed.), *Meaning and Use. Papers Presented at the Second Jerusalem Philosophical Encounter, April 1976.* 1978 (forthcoming).
4. F. Guenthner and S. J. Schmidt (eds.), *Formal Semantics and Pragmatics for Natural Languages.* 1978, viii + 374 pp. + index.
5. Esa Saarinen (ed.), *Game-Theoretical Semantics.* 1978, xiv + 379 pp. + index.
6. F. J. Pelletier (ed.), *Mass Terms: Some Philosophical Problems.* 1978, xiv + 300 pp. + index.